Acoustics—A Textbook for Engineers and Physicists

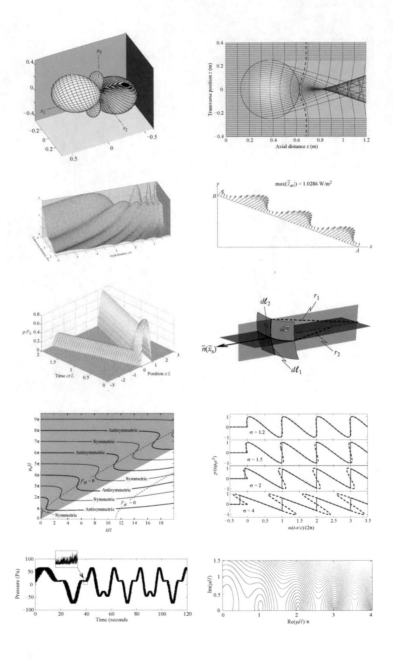

Jerry H. Ginsberg

Acoustics—A Textbook for Engineers and Physicists

Volume II: Applications

Jerry H. Ginsberg
G. W. Woodruff School of Mechanical
   Engineering
Georgia Institute of Technology
Dunwoody, GA
USA

ISBN 978-3-319-86017-6   ISBN 978-3-319-56847-8   (eBook)
DOI 10.1007/978-3-319-56847-8

Library of Congress Control Number: 2017937706

Printed on acid-free paper

This Springer imprint is published by Springer Nature
The registered company is Springer International Publishing AG
The registered company address is: Gewerbestrasse 11, 6330 Cham, Switzerland

The ASA Press

The ASA Press imprint represents a collaboration between the Acoustical Society of America and Springer dedicated to encouraging the publication of important new books in acoustics. Published titles are intended to reflect the full range of research in acoustics. ASA Press books can include all types of books published by Springer and may appear in any appropriate Springer book series.

 ASA Press

The Acoustical Society of America

On December 27, 1928, a group of scientists and engineers met at Bell Telephone Laboratories in New York City to discuss organizing a society dedicated to the field of acoustics. Plans developed rapidly and the Acoustical Society of America (ASA) held its first meeting on May 10–11, 1929, with a charter membership of about 450. Today, ASA has a worldwide membership of 7000.

The scope of this new society incorporated a broad range of technical areas that continues to be reflected in ASA's present-day endeavors. Today, ASA serves the interests of its members and the acoustics community in all branches of acoustics, both theoretical and applied. To achieve this goal, ASA has established technical committees charged with keeping abreast of the developments and needs of membership in specialized fields as well as identifying new ones as they develop.

The technical committees include acoustical oceanography, animal bioacoustics, architectural acoustics, biomedical acoustics, engineering acoustics, musical acoustics, noise, physical acoustics, psychological and physiological acoustics, signal processing in acoustics, speech communication, structural acoustics and vibration, and underwater acoustics. This diversity is one of the Society's unique and strongest assets since it so strongly fosters and encourages cross-disciplinary learning, collaboration, and interactions.

ASA publications and meetings incorporate the diversity of these technical committees. In particular, publications play a major role in the Society. *The Journal of the Acoustical Society of America* (JASA) includes contributed papers and patent reviews. *JASA Express Letters* (JASA-EL) and *Proceedings of Meetings on Acoustics* (POMA) are online, open-access publications, offering rapid publication. *Acoustics Today*, published quarterly, is a popular open-access magazine. Other key features of ASA's publishing program include books, reprints of classic acoustics texts, and videos.

ASA's biannual meetings offer opportunities for attendees to share information, with strong support throughout the career continuum, from students to retirees. Meetings incorporate many opportunities for professional and social interactions, and attendees find the personal contacts a rewarding experience. These experiences result in building a robust network of fellow scientists and engineers, many of whom became lifelong friends and colleagues.

From the Society's inception, members recognized the importance of developing acoustical standards with a focus on terminology, measurement procedures, and criteria for determining the effects of noise and vibration. The ASA Standards Program serves as the Secretariat for four American National Standards Institute Committees and provides administrative support for several international standards committees.

Throughout its history to present day, ASA's strength resides in attracting the interest and commitment of scholars devoted to promoting the knowledge and practical applications of acoustics. The unselfish activity of these individuals in the development of the Society is largely responsible for ASA's growth and present stature.

To Leah Morgan, Elizabeth Rachel,
and Abigail Rose, my grandchildren.
Each is talented, each is beautiful,
each is unique, each is amazing.
I love them.

Preface

The Basic Concept

Because you are reading this book, there is a strong likelihood that you are familiar with its companion, Volume 1. If that is so, then you are well acquainted with my core philosophy of instruction. It is evident that students should be prepared to address technical issues in the future. An essential component in doing so is the instructor, but even the most talented instructor needs support. Fulfilling that need has been the motivation for all my prior books, but even more so it is my objective for these books. To meet the present and future needs of students and instructors, treatments of various principles and concepts feature extensive explanations of the motivation and organization of the derivation, as well as thorough descriptions of the steps that are implemented. Some of the derivations and explanations I believe are unique to this book. In most cases, the derived principles are accompanied by discussions of their physical meaning.

Examples are numerous. All are my own creation. Indeed, the thought of creating a suitable set dissuaded me for a long time from beginning to write *Acoustics—A Textbook for Engineers and Physicists*, even though I believed that there was a strong need for a text like it. Most examples are more than simple applications of derived formulas. I selected many of these examples not only to illustrate the associated theory, but also to be simplified versions of issues the student might encounter in practice. Equally important to their selection was whether treating the results as the basis for small case studies would enlighten the student about the phenomena associated with that theory. I also used the examples as a vehicle to bring to the fore the fact that in many situations, alternative formulations of an analysis might be viable. I endeavored to use the examples as a way of assisting the student to recognize when these alternatives exist, and to assist them to recognize in other situations which one is best. At the same time, I endeavored to recognize the imperative that the examples be cognizant of the capabilities that can be expected of a student who is being exposed to acoustics for the first time.

Each solution explains why the example is important, why the solution proceeds as it does, how to perform unfamiliar operations, what can be learned from the results about fundamental behaviors, and why the qualitative aspects of the results are consistent with the underlying fundamental principles. Some examples analyze systems by more than one method. This serves to enhance the student's fundamental understanding of the underlying physical processes, as well as enhancing the ability to make the appropriate line of attack when confronted with a new situation. The advent and wide availability of computational software is exploited to lend greater realism to some examples. When the usage of software entails any potentially problematic aspects, especially concerning algorithms and their implementation, those issues are addressed explicitly, sometimes with program fragments. In recognition of the importance of computations, and to help students concentrate on the acoustical aspects when they solve homework exercises, the MATLAB code used to solve the examples is available for download from the Springer server.

At its inception, *Acoustics—A Textbook for Engineers and Physicists* was a single volume. The sequence of topics for the early chapters was a slightly modified version of the courses I taught at Georgia Tech. I wrote the later chapters to fill the spectrum of subjects that I consider to constitute the core concepts and techniques a student in physics or engineering is likely to encounter. Space limitations, as well as my desire that this should be a textbook, rather than a monograph, dictated the scope in the later chapters. For this reason, they were written with the notion that they should expose the student to the fundamental phenomena and provide the fundamental tools required to study and research these phenomena. No chapter attempts to bring the student up to the current state of the art in that subject.

Some instructor's might consider that some chapters delve too deeply into the subject. I do not agree with this sentiment. Some parts address questions students have asked me, and the development in other parts serves to motivate students by demonstrating interesting and enlightening phenomena. Nevertheless, if one does not desire to take advantage of the depth of treatment, the first few sections of each chapter should be adequate to proceed to the next chapter.

After I completed this book, it was evident that it would be quite large. This caused me to recall my days long ago when I was a student, and disliked carrying very large textbooks. Concurrently, when I surveyed the Table of Contents, I realized that the manuscript divided naturally into two parts. The first six chapters constitute a foundation that I consider requisite knowledge for anyone active in the physical aspects of acoustics. My examination also led me to the realization that the nature of the last seven chapters was different. Few individuals are equally familiar with all of them. This certainly was true in my case. Filling in the gaps in my knowledge, particularly in Chap. 11 on geometrical acoustics, was quite enriching and fulfilling. I wanted to provide that same sense to instructors who might not be familiar with the subject of one or more of the later chapters. I also wanted to prepare students for further study in each of those topics. I foresee using Volume 2 either in a survey course covering selected or all chapters, or else to begin specialized courses on the subject of a single chapter. Both usages would have been followed when I taught acoustics at Georgia Tech. The Acoustics II course covered

Chap. 7 on radiation from objects in free space, Chap. 8 on radiation from vibrating regions in a baffle, the beginning portion of Chap. 9 on waveguides, and the beginning portion of Chap. 10 on enclosures. There also were specialized courses on propagation on the ocean and atmosphere, nonlinear acoustics, and structural acoustics. The associated chapters would not have been adequate for those specialized courses, but I believe they would provide a strong foundation with which to initiate such courses.

Technical Content

The chapters to be found in this volume are sequenced according to what I believe is sensible in terms of the level of sophistication and analytical difficulty. The highlights of each chapter were discussed in the Preface to Volume I, but it is appropriate to summarize the scope of each chapter. Radiation from vibrating objects is the subject of Chap. 7. Configurations that are addressed are spheres and hemispheres, infinitely long cylinders, and three numerical techniques that exemplify the formulations in current use. Radiation from a piston in a baffle is encountered in a diverse set of applications, so Chap. 8 is devoted to a thorough exploration of farfield and nearfield properties. Two examples explore radiation resulting from square wave motion. The difference between these examples is whether the analysis is formulated in the frequency or time domains. The reader might find the results to be quite interesting.

The field within a waveguide is the subject of Chap. 9. First to be studied is the Webster horn equation for one-dimensional waveguides. The investigation of two-dimensional waveguides introduces the concept of modal analysis. It concludes with an investigation of coupling of the acoustic field with elastic walls that are described by plate theory. The basic theory for three-dimensional waveguides is developed. It is applied to waveguides whose cross section is rectangular and then cylindrical. Chapter 10 is devoted to the sound field within an enclosed region. It begins by using the waveguide representation to explain the alternative descriptions of the field as a set of waves that propagate in multiple directions, or as a set of cavity modes that are standing waves. Both analytical descriptions are developed, with emphasis on the situations where each is best employed. An interesting example uses an infinite series of cavity modes to describe the field within a two-dimensional rectangular enclosure due to a point source. After the analysis, an alternative using the method of images highlights the notion that the selection of an optimal analytical approach sometimes requires consideration of what the objective is. The closure of this chapter describes the Rayleigh-Ritz method and a formulation according to Dowell's approximation. The former is often included in standard texts, but the latter is relatively recent.

Chapter 11 is devoted to geometrical acoustics. Mean flow effects are excluded in order to emphasize how high-frequency rays and wavefronts may be determined. A vertically stratified fluid, which is a fundamental model for ocean acoustics, is examined first. After that, the ray tracing equations for media whose properties depend arbitrarily on location are developed. Algorithms for solving the governing equations in each case are formulated. Of particular interest is the example of the passage of a plane wave through fluid that is cylindrically heterogeneous. Demonstrations of a caustic and folding of the wavefront beyond the caustic are quite captivating. The chapter closes with a presentation of Fermat's principle. The calculus of variations is developed there.

Scattering from bodies surrounded by a fluid is the subject of Chap. 12. The Born approximation for heterogenous media is derived. Rayleigh scattering and its relation to the Born approximation are the next topic. The metrics commonly sought from a scattering study, such as target strength, are discussed. After that, Kirchhoff scattering theory and its relation to geometrical acoustics are explored. The chapter closes with the application of spherical harmonics to analyze scattering from a rigid sphere and a spherical shell. These studies shed light on the transition from Rayleigh to Kirchhoff scattering, as well as the fundamental importance of fluid loading relative to elasticity.

Chapter 13 closes the textbook with an exploration of nonlinear acoustic analyses and phenomena. The bulk of the chapter is devoted to simple plane waves described by the Riemann solution. Techniques for evaluating it are discussed and used to examine harmonic generation and depletion, followed by the propagation of shocks. The Rankine-Hugoniot relations for weak shock are derived and shown with simple mathematics to lead to the phenomena of old age and acoustic saturation. A nonlinear wave equation is derived as the basis for study of multidimensional nonlinear waves. Its first usage leads to differential equations governing the position dependence of Fourier series coefficients for a plane wave in a dissipative fluid. Perturbation analysis techniques for the nonlinear wave equation are developed for plane waves and then extended to radially symmetric spherical waves and two-dimensional waves radiated by a vibrating plate. The latter leads to demonstration of the phenomenon of self-refraction, in which the rays and wavefronts are modified by the associated pressure and particle velocity fields.

Although the overriding precept of the text is that all topics must be fully explained as they arise, two appendices are provided for further assistance. One is devoted to derivation of the coordinate transformations and vector differential operators in spherical and cylindrical coordinates. The second describes Fourier transforms and their application. Fourier transforms appear in the main body only in the treatment of radiation from an infinite cylinder and from a transducer in a baffle. Those analyses could be addressed by invoking FFT techniques, albeit without the benefit of an algebraic solution. However, Fourier transforms are a ubiquitous thread that runs through the technical literature, and the treatment of this mathematical tool in a sense ties together the dual nature of the time and frequency domains.

The chapters of Volume II of *Acoustics—A Textbook for Engineers and Physicists* are sequenced in a manner that I believe to be sensible from the viewpoint of ascending difficulty, as well as the technical sophistication. For instance, it would not make sense to study piston radiation in Chap. 8 before the Kirchoff–

Helmholtz integral theorem is developed in Chap. 7. From the opening motivational discussion of waveguides in Chap. 10 to the development of natural cavity modes, there is much reliance on the nature of propagation modes of a waveguide, which is the subject of Chap. 9. Chapter 12 on scattering relies on the KHIT and multipole expansions, which are developed in Chap. 7. Chapters 11 and 13 are exceptions. Both only require concepts developed in the first four chapters. Their placement is based on my perception of what instructors expect to find. My own research in acoustics began in the nonlinear regime, so if I followed my preference, the last chapter might have appeared earlier. In the same vein, I think that exposure of students to ray tracing for heterogeneous media would interest them greatly and thereby serve as a strong motivational tool.

Writing this volume has helped me close gaps in my knowledge of acoustics. I very much enjoyed writing it. If you are a student, I hope you learn much from it and find it useful for your future endeavors. If you are an instructor, I hope that this work captivates you and enhances your teaching efforts. If you are a reader on a self-study path, then I think you will find this volume to be exceptionally helpful and instructive, for I have gone to great effort to provide all you need to pursue your studies.

Dunwoody, GA, USA   Jerry H. Ginsberg

The original version of the book was revised: The Electronic Supplementary Materials have been included. The correction to the book is available at https://doi. org/10.1007/978-3-319-56847-8_14

Acknowledgements

I am indebted to many individuals for providing motivation to write these books. Above all, neither would exist if I had not met my good friend, Allan Pierce, when I interviewed in 1980 for a professorship at the Georgia Institute of Technology. Working with him convinced me to extend my knowledge of acoustics beyond the specialized subject of nonlinear acoustics that was part of my early career. Over the years, our discussions were quite revelatory regarding where there were gaps in my knowledge of the subject. Furthermore, I hope he is not offended by this remark, but learning from, and then teaching from, his book convinced me of the necessity that I write *Acoustics—A Textbook for Engineers and Physicists*.

In the six-year interval, during which I wrote these books what I needed most was assurance that the effort was worth pursuing. Some of my colleagues at Georgia Tech were quite supportive. Karim Sabra convinced me on several occasions that I would be filling an important need. Students solving the homework exercises should thank him because he suggested that the MATLAB code I used should be publically available. Pete Rogers provided extremely useful critical remarks for an early draft of my treatment of geometrical acoustics. My former Ph.D. students, especially J. Gregory McDaniel at Boston University and Kuangchung Wu at the NSWC Carderock Division of the Naval Sea Systems Command, were especially enthusiastic. I also am indebted to those attendees at many Acoustical Society of America meetings who I waylaid to discuss my writing efforts. They are too numerous to list, and I am sure that I have forgotten some names, but I greatly appreciate their attention. I owe Mark Hamilton of the University of Texas at Austin a special debt because he convinced me to participate in the ASA Book program under the aegis of Springer Publishing. Sara Kate Heukerott, my Editor at Springer, was quite understanding of my requests. Her expertise was a great aid as we assembled this project. Some might be surprised at the inclusion of my grand-daughter, Leah Morgan Ginsberg, in the list of folks deserving recognition. Early in the writing stage, because she was a proficient clarinetist, I sought her assistance for the discussion of music in Chap. 1. Then at the conclusion, as she approached graduation from Georgia Tech in the G. W. Woodruff School of Mechanical

Engineering, from which I had retired, she served as my sounding board and spokeswoman for students when I deliberated how best to disseminate this work.

In addition to Leah's role, my family was essential to the effort. The forbearance of my wife, Rona, while I focused on writing, ignored other responsibilities, and forgot many things that I still cannot remember, astonishes me, even now that my efforts are over. She went through this experience before when I wrote my prior books on statics, dynamics, and vibrations. However, none of those experiences could have prepared her for the intensity and duration of the present effort. My sons, Mitchell and Daniel, had similar experiences when they lived at home. Although they and their wives, Tracie and Jessica, were not as strongly impacted now, I greatly appreciate their forbearance when I was not as communicative as I should have been. My granddaughters, Leah, Beth, and Abby, inspire me by their dedication to their own activities. I hope that recognition of the pleasure their Papa derived from creating these books will inspire them.

Dunwoody, GA, USA   Jerry H. Ginsberg

Contents

Volume I: Fundamentals

1   Descriptions of Sound   1
   1.1   Harmonic Signals   4
         1.1.1   Basic Properties   4
         1.1.2   Vectorial Representation   8
         1.1.3   Complex Exponential Representation   9
         1.1.4   Operations Using Complex Exponentials   12
   1.2   Averages   17
   1.3   Metrics of Sound   27
         1.3.1   Sound Pressure Level   27
         1.3.2   Human Factors   32
         1.3.3   Frequency Bands   35
   1.4   Transfer Between Time and Frequency Domains   41
         1.4.1   Fourier Series   43
         1.4.2   Discrete Fourier Transforms   52
         1.4.3   Nyquist Sampling Criterion   55
         1.4.4   Fast Fourier Transforms   59
         1.4.5   Evaluation of Time Responses   64
   1.5   Spectral Density   72
         1.5.1   Definition   72
         1.5.2   Noise Models   78
   1.6   Closure   83
   1.7   Homework Exercises   83

2   Plane Waves: Time Domain Solutions   91
   2.1   Continuum Equations in One Dimension   92
         2.1.1   Conservation of Mass   92
         2.1.2   Momentum Equation   94
   2.2   Linearization and the One-Dimensional Wave Equation   96
   2.3   Equation of State and the Speed of Sound   102

2.4   The d'Alembert Solution 111
      2.4.1   Derivation................................... 112
      2.4.2   Interpretation.................................. 114
      2.4.3   Harmonic Waves 121
2.5   The Method of Wave Images......................... 123
      2.5.1   Initial Value Problem in an Infinite Domain......... 124
      2.5.2   Plane Waves in a Semi-infinite Domain 131
      2.5.3   Plane Waves in a Finite Waveguide 145
2.6   Analogous Vibratory Systems 170
      2.6.1   Stretched Cable................................ 170
      2.6.2   Extensional Waves in an Elastic Bar 179
2.7   Closure..................................... 183
2.8   Homework Exercises 183

3   **Plane Waves: Frequency-Domain Solutions** 191
3.1   General Solution 192
3.2   Waveguides with Boundaries........................... 199
      3.2.1   Impedance and Reflection Coefficients 200
      3.2.2   Evaluation of the Signal 207
      3.2.3   Modal Properties and Resonances.................. 212
      3.2.4   Impedance Tubes 223
3.3   Effects of Dissipation................................ 227
      3.3.1   Viscosity................................... 229
      3.3.2   Thermal Transport 231
      3.3.3   Molecular Relaxation 234
      3.3.4   Absorption in the Atmosphere and Ocean........... 238
      3.3.5   Wall Friction................................ 243
3.4   Acoustical Transmission Lines......................... 248
      3.4.1   Junction Conditions 248
      3.4.2   Time Domain 252
      3.4.3   Frequency-Domain Formulation for Long Segments... 257
3.5   Lumped Parameter Models............................ 272
      3.5.1   Approximations for Short Branches 272
      3.5.2   Helmholtz Resonator.......................... 278
3.6   Closure..................................... 285
3.7   Homework Exercises 286

4   **Principles and Equations for Multidimensional Phenomena** 295
4.1   Fundamental Equations for an Ideal Gas 296
      4.1.1   Continuity Equation 296
      4.1.2   Momentum Equation 298
4.2   Linearization 303

4.3   Plane Waves in Three Dimensions.   306
        4.3.1   Simple Plane Wave in the Time Domain   307
        4.3.2   Trace Velocity   311
        4.3.3   Simple Plane Wave in the Frequency Domain   316
4.4   Velocity Potential.   320
4.5   Energy Concepts and Principles.   324
        4.5.1   Energy and Power   324
        4.5.2   Linearization.   328
        4.5.3   Power Sources   331
4.6   Closure.   341
4.7   Homework Exercises   342

5 Interface Phenomena for Planar Waves   347
5.1   Radiation Due to Surface Waves   347
        5.1.1   Basic Analysis   348
        5.1.2   Interpretation.   354
5.2   Reflection from a Surface Having a Local Impedance   358
        5.2.1   Reflection from a Time-Domain Perspective.   358
        5.2.2   Reflection from a Frequency-Domain Perspective.   365
5.3   Transmission and Reflection at an Interface Between Fluids . . .   373
        5.3.1   Time-Domain Analysis.   373
        5.3.2   Frequency-Domain Analysis.   380
5.4   Propagation Through Layered Media.   390
        5.4.1   Basic Analysis of Three Fluids.   390
        5.4.2   Multiple Layers.   399
5.5   Solid Barriers.   408
        5.5.1   General Analysis.   409
        5.5.2   Specific Barrier Models   414
5.6   Closure.   424
5.7   Homework Exercises   424

6 Spherical Waves and Point Sources   433
6.1   Spherical Coordinates.   433
6.2   Radially Vibrating Sphere–Time-Domain Analysis   437
        6.2.1   General Solution.   438
        6.2.2   Radiation from a Uniformly Vibrating Sphere   439
        6.2.3   Acoustic Field in a Spherical Cavity.   444
6.3   Radially Vibrating Sphere–Frequency-Domain Analysis   453
        6.3.1   General Solution.   453
        6.3.2   Radiation from a Radially Vibrating Sphere.   454
        6.3.3   Standing Waves in a Spherical Cavity   462
6.4   Point Sources.   468
        6.4.1   Single Source   468
        6.4.2   Green's Function   470

        6.4.3   Point Source Arrays   483
        6.4.4   Method of Images.   493
    6.5   Dipoles, Quadrupoles, and Multipoles   505
        6.5.1   The Dipole Field.   506
        6.5.2   Radiation from a Translating Rigid Sphere   513
        6.5.3   The Quadrupole Field.   521
        6.5.4   Multipole Expansion.   532
    6.6   Doppler Effect   538
        6.6.1   Introduction   538
        6.6.2   Moving Fluid   539
        6.6.3   Subsonic Point Source   541
        6.6.4   Supersonic Point Source   548
    6.7   Closure.   554
    6.8   Homework Exercises   554

Index   569

Volume II: Applications

7   **Radiation from Vibrating Bodies.**   1
    7.1   Spherical Harmonics   2
        7.1.1   Separation of Variables.   2
        7.1.2   Description of the Pressure Field   15
        7.1.3   Arbitrary Spatial Dependence   16
    7.2   Radiation from a Spherical Body.   19
        7.2.1   Analysis   19
        7.2.2   Important Limits   21
        7.2.3   Symmetry Plane   29
        7.2.4   Interaction with an Elastic Spherical Shell   36
    7.3   Radiation from an Infinite Cylinder   53
        7.3.1   Separation of Variables.   56
        7.3.2   Transverse Dependence—Cylindrical Bessel
                 Functions   57
        7.3.3   Radiation Due to a Helical Surface Wave.   60
        7.3.4   Axially Periodic Surface Vibration   68
        7.3.5   Finite Length Effects.   81
    7.4   Kirchhoff–Helmholtz Integral Theorem   84
        7.4.1   Derivation for an Acoustic Cavity   85
        7.4.2   Acoustic Radiation into an Exterior Domain.   91
    7.5   Numerical Methods for Radiation from Arbitrary Objects   104
        7.5.1   Source Superposition   104
        7.5.2   Boundary Element Method   108
        7.5.3   Finite Element Method   115
    7.6   Homework Exercises   125

8   Radiation from a Source in a Baffle................................. 133
  8.1   The Rayleigh Integral............................... 133
  8.2   Farfield Directivity................................ 137
    8.2.1   Cartesian Coordinate Description 137
    8.2.2   Farfield of a Piston Transducer.................... 144
  8.3   Axial Dependence for a Circular Transducer 153
  8.4   An Overall Picture of the Pressure Field 158
  8.5   Radiation Impedance of a Circular Piston 166
  8.6   Time Domain Rayleigh Integral........................ 176
  8.7   Homework Exercises 180

9   Modal Analysis of Waveguides 187
  9.1   Propagation in a Horn 187
    9.1.1   The Webster Horn Equation 188
    9.1.2   Exponential Horn 193
    9.1.3   Group Velocity.............................. 202
    9.1.4   WKB Solution for an Arbitrary Horn 207
  9.2   Two-Dimensional Waveguides........................ 216
    9.2.1   General Solution............................. 217
    9.2.2   Rigid Walls................................. 218
    9.2.3   Interpretation................................ 220
    9.2.4   Flexible Walls............................... 224
    9.2.5   Orthogonality and Signal Generation 243
  9.3   Three-Dimensional Waveguides....................... 253
    9.3.1   General Analytical Procedure 253
    9.3.2   Rectangular Waveguide 257
    9.3.3   Circular Waveguide 267
  9.4   Homework Exercises 282

10   Modal Analysis of Enclosures 291
  10.1   Fundamental Issues 291
    10.1.1   Wall-Induced Signals 291
    10.1.2   Source Excitation 293
  10.2   Frequency-Domain Analysis Using Forced Cavity Modes 297
    10.2.1   Rectangular Enclosures......................... 297
    10.2.2   Spherical Cavities 311
    10.2.3   Cylindrical Enclosures 323
  10.3   Analysis Using Natural Cavity Modes.................... 336
    10.3.1   Equations Governing Cavity Modes 337
    10.3.2   Orthogonality 338
    10.3.3   Analysis of the Pressure Field.................... 340
    10.3.4   Rectangular Cavity 342
    10.3.5   Cylindrical Cavity............................ 352
    10.3.6   Spherical Cavity 358

	10.4	Approximate Methods	363
		10.4.1 The Rayleigh Ratio and Its Uses	364
		10.4.2 Dowell's Approximation	379
	10.5	Homework Exercises	395

11 Geometrical Acoustics   405
 11.1 Basic Considerations: Wavefronts and Rays   406
       11.1.1 Field Equations for an Inhomogeneous Fluid   409
       11.1.2 Reflection and Refraction of Rays   413
 11.2 Propagation in a Vertically Stratified Medium   417
       11.2.1 Snell's Law for Vertical Heterogeneity   418
       11.2.2 Intensity and Focusing Factor   428
 11.3 Arbitrary Heterogeneous Fluids   438
       11.3.1 Ray Tracing Equations   438
       11.3.2 Amplitude Dependence   443
 11.4 Fermat's Principle   463
 11.5 Homework Exercises   473

12 Scattering   479
 12.1 Background   480
 12.2 Scattering by Heterogeneity   482
       12.2.1 General Equations   483
       12.2.2 The Born Approximation   485
 12.3 Rayleigh Scattering Limit   490
       12.3.1 The Rayleigh Limit of the Born Approximation   490
       12.3.2 Mismatched Heterogeneous Region   492
       12.3.3 Scattering from a Rigid Body   494
 12.4 Measurements and Metrics   505
 12.5 High-frequency Approximation   510
 12.6 Scattering from Spheres   515
       12.6.1 Stationary Spherical Scatterer   515
       12.6.2 Scattering by an Elastic Spherical Shell   522
 12.7 Homework Exercises   532

13 Nonlinear Acoustic Waves   539
 13.1 Riemann's Solution for Plane Waves   540
       13.1.1 Analysis   540
       13.1.2 Interpretation   543
       13.1.3 Boundary and Initial Conditions   545
       13.1.4 Equations of State   549
       13.1.5 Quantitative Evaluations   553
 13.2 Effects of Nonlinearity   568
       13.2.1 Harmonic Generation   568
       13.2.2 Shock Formation   574
       13.2.3 Propagation of Weak Shocks   577

13.3   General Analytical Techniques.   597
        13.3.1   A Nonlinear Wave Equation.   598
        13.3.2   Frequency-Domain Formulation   602
        13.3.3   Regular Perturbation Series Expansion   611
        13.3.4   Method of Strained Coordinates   620
13.4   Multidimensional Systems   625
        13.4.1   Finite Amplitude Spherical Wave   625
        13.4.2   Waves in Cartesian Coordinates   637
13.5   Further Studies.   654
13.6   Homework Exercises   658

Correction to: Acoustics—A Textbook for Engineers and Physicists C1

Appendix A: Curvilinear Coordinates   665

Appendix B: Fourier Transforms   677

Index   689

About the Author

Jerry H. Ginsberg began his technical education at the Bronx High School of Science, from which he graduated in 1961. This was followed by a B.S.C.E. degree in 1965 from the Cooper Union and an E.Sc.D. degree in engineering mechanics from Columbia University in 1970, where he held Guggenheim and NASA Fellowships. From 1969 to 1973, he was an Assistant Professor in the School of Aeronautics, Astronautics, and Engineering Science at Purdue University. He then transferred to Purdue's School of Mechanical Engineering, where he was promoted to Associate Professor in 1974. In the 1975–1976 academic year, he was a Fulbright-Hayes Advanced Research Fellow at the École Nationale Supérieure d'Électricité et de Mécanique in Nancy, France. He came to Georgia Tech in 1980 as a Professor in the School of Mechanical Engineering, which awarded him the George W. Woodruff Chair in 1989. He retired in June 2008. His prior publications include five textbooks in statics, dynamics, and vibrations, most in several editions, as well as more than one hundred and twenty refereed papers covering these subjects. Dr. Ginsberg became a Fellow of the Acoustical Society of America in 1987 and a Fellow of the American Society of Mechanical Engineers in 1989. The awards and recognitions he has received include Georgia Tech Professor of the Year (1994), ASEE Archie Higdon Distinguished Educator in Mechanics (1998), ASA Trent-Crede Medal (2005),

ASME Per Bruel Gold Medal in Noise Control and Acoustics (2007), and the ASA Rossing Prize in Acoustics Education (2010). In addition to his technical activities, he is an exceptional photographer.

List of Examples

The name of the MATLAB programs used to solve the respective example are enclosed in parentheses. They are available for download from the Springer website.

Example 7.1   (Legendre_function_series.m) Evaluation of a Legendre polynomial series to fit a given function, and evaluation of the series length required to have an error below a specified level.   7

Example 7.2   (none) Determination of the spherical harmonic series for the farfield radiated by a combination of a monopole, a dipole, and a longitudinal quadrupole   17

Example 7.3   (Cap_in_sphere_radiation.m) Determination of the spherical harmonic series representation of the axisymmetric pressure field radiated by a sphere that has a cap of arbitrary size that executes a translational vibration, and is rigid elsewhere   24

Example 7.4   (Sphere_radiation_baffle_vs_free.m) Evaluation of spherical harmonic series for three cases: a hemisphere that executes a translational vibration in a rigid baffle, the same motion for a pressure-release baffle, and a rigid sphere whose forward half translates while the back half is stationary   32

Example 7.5   (Submerged_spher_shell_response.m) Evaluation of the surface vibration and pressure when a submerged spherical shell is excited by a harmonic force concentrated at the polar apex   46

Example 7.6   (Z_rad_helical.m) Determination and evaluation of the radiation impedance for a single helical wave of radial displacement on a cylinder   67

Example 7.7   (Line_source_finite.m) Determination of the pressure
                   field radiated by a uniform line source of finite length,
                   and comparison of its field properties to those of an
                   infinite line source   73
Example 7.8   (Cello_string_radiation.m) Evaluation of the pressure
                   field radiated by a cello's string based on a model
                   of the string as a cylinder whose cross sections execute
                   translational vibration   79
Example 7.9   (KHIT_spher_cavity.m) Formulation of the
                   Kirchhoff-Helmholtz integral theorem for a spherical
                   cavity by using the analytical solution to set the
                   pressure and particle velocity at the wall, with special
                   attention to the behavior as the field point approaches
                   the wall.   88
Example 7.10   (KHIT_cyl_radiation.m) Usage of the radiation
                   impedance of an infinite cylinder as the basis for an
                   approximate formulation of the Kirchhoff-Helmholtz
                   Integral Theorem for a finite cylinder, with emphasis
                   on evaluation of the integral to identify the farfield
                   directivity at several frequencies   94
Example 8.1   (Directivity_rectangular_transducer.m) Usage of the
                   Rayleigh integral to evaluate the farfield radiated by a
                   rectangular strip that executes a transverse vibration in
                   a rigid baffle   140
Example 8.2   (Example_piston_directivities.m) Comparison
                   of the farfield properties for radiation from an
                   oscillating piston or a vibrating membrane in a rigid
                   baffle.   149
Example 8.3   (Anti_piston_example.m) Determination of the axial
                   dependence of the pressure radiated by a so-called
                   anti-piston, which consists of transducer that has an
                   outer region that vibrates 180° out-of-phase from the
                   central circle   156
Example 8.4   (Rayleigh_integral_hemisphere.m) Evaluation of the
                   axial dependence and farfield radiated by a hemisphere
                   in a rigid baffle that translates as a rigid body, based on
                   phase-shifting the normal velocity to place it in the
                   plane of the baffle, followed by comparison of the
                   results to those obtained from the spherical harmonic
                   solution   163
Example 8.5   (Piston_square_wave.m) Evaluation of the time
                   dependence of the radiated power and the pressure
                   waveform at a field point on the axis of a circular
                   piston that executes a square wave oscillation.   172

Example 9.1   Usage of the general solution of the Webster horn
                equation to evaluate the two-port mobility for an
                exponential transition between waveguides having
                different cross-sectional sizes   199
Example 9.2   (Group_velocity_from_data.m) Evaluation of the
                waveform at two locations in an exponential horn that
                is driven by a biharmonic input, followed by the usage
                of those waveforms to determine the group velocity,
                and comparison of the group velocity to the value
                derived from the dispersion equation.   205
Example 9.3   (Webster_horn_numerical_vs_WKB.m)   Evaluation
                of the WKB solution for the pressure distribution in
                a plane waveguide whose cross-sectional area varies
                sinusoidally, with a small minimum, followed by
                comparison of the results to those obtained from a
                finite difference solution of the Webster horn
                equation   211
Example 9.4   (Compliant_2d_waveguide.m) Determination of the
                phase and group velocity and the transverse mode
                functions of a two-dimensional waveguide for which
                one wall is rigid and the other is purely resistive   230
Example 9.5   (Elastic_2D_waveguide.m) Determination of the
                eigenvalues of the transverse mode functions and the
                ratio of plate displacement to the pressure amplitude at
                selected frequencies for a two-dimensional waveguide
                whose walls are identical elastic plates   235
Example 9.6   (Two_dim_waveguide_excitation.m) Determination
                of the axial and transverse dependence of pressure
                amplitude in a two-dimensional waveguide whose
                floor is rigid and whose upper boundary is a free
                surface when the excitation stems from a vibrating
                ribbon transducer aligned horizontally at one end   248
Example 9.7   (Rect_waveguide_line_source.m) Determination of the
                axial and transverse dependence of pressure amplitude
                in a three-dimensional waveguide whose sides and
                floor are rigid and whose upper boundary is a free
                surface when the excitation stems from a vibrating
                ribbon transducer aligned vertically.   262
Example 9.8   (Annular_waveguide_modes.m) Determination
                of the eigenvalues and transverse mode functions
                for a waveguide that is the region bounded
                by concentric rigid cylinders   275

Example 9.9   (Piston_circular_waveguide.m) Determination
                of the axial dependence of the pressure amplitude
                in a circular waveguide when a vibrating piston of
                lesser diameter is situated concentrically at one end   280

Example 10.1   (Three_D_cavity_force_plate.m) Evaluation
                of the frequency dependence of the force required
                to move a massive plate that is part of one wall
                in a hard-walled rectangular cavity   301

Example 10.2   (Piston_transducer_rectangular_cavity.m) Evaluation
                of the frequency dependence of the pressure amplitude
                at selected locations and the force required to drive
                a vibrating circular piston at the end of a rectangular
                tank of water   305

Example 10.3   (Spherical_cavity_external_pressure.m)   Evaluation
                of the pressure field in a spherical cavity due to
                passage of an exterior plane wave.   315

Example 10.4   (Concentric_spheres.m) Evaluation of the radially
                symmetric pressure field in the region between
                concentric spheres when the inner sphere vibrates,
                with examination of the limit as the inner sphere
                shrinks to a point source   320

Example 10.5   (Cylindrical_waveguide_higher_modes_
                impedance_tube.m) Determination of the effect
                of sidewall compliance on the performance a cylinder
                that is used as an impedance tube.   326

Example 10.6   (Cylindrical_cavity_vibrating_side_wall) Evaluation
                of the volume velocity and pressure at the open end
                of a cylindrical waveguide when the cylindrical wall
                executes a specified vibration   332

Example 10.7   (Rect_cavity_modal_density.m) Evaluation of the
                natural frequencies of a rectangular tank of water with
                a free surface, with special emphasis on repeated
                natural frequencies and increasing modal density   344

Example 10.8   (Cavity_impulse_response.m) Analysis using a series
                of cavity modes of the time-dependent pressure
                radiated by an impulsive line source in a hard-walled
                rigid cavity, and discussion of an alternative
                formulation using the method of images   347

Example 10.9   (Compare_cyl_and_rect_cavity_modes.m)
                Determination of the sizes of a cylindrical tank
                and a rectangular tank such that their fundamental
                nonplanar mode has a specified natural frequency   355

Example 10.10   (none) Analysis of the modal properties of a system
                in which an inner sphere is a membrane filled
                with a fluid, and surrounded by a concentric sphere
                filled with a different fluid   360
Example 10.11   (none) Usage of the Rayleigh ratio in conjunction
                with different trial functions to estimate the
                fundamental natural frequency of a two-dimensional
                cavity   368
Example 10.12   (Rayleigh_Ritz_Ellipse.m) Evaluation of the natural
                frequencies of a two-dimensional elliptical cavity
                according to the Rayleigh-Ritz method, with emphasis
                on convergence properties and agreement with the
                separation theorem   374
Example 10.13   (Dowells_approx_2D_cavity.m) Formulation and
                solution of differential equations derived from
                Dowell's approximation for a harmonically driven
                rectangular two-dimensional cavity whose walls have
                properties that change discontinuously   388
Example 11.1   (Ray_reflection_spherical_cap.m) Evaluation
                of the rays resulting from incidence of a plane wave
                on the inner surface of a spherical cap   414
Example 11.2   (Ray_single_stratified_fluid.m) Evaluation of a ray
                path and propagation time along that ray in a water
                channel whose sound speed depends parabolically
                on depth   424
Example 11.3   (Ray_reflections_stratified_fluid.m) Computation
                of the full set of rays, including those reflected from
                the surface, for a water channel whose sound speed
                depends parabolically on depth, and usage of the result
                to evaluate a focusing factor.   434
Example 11.4   (none) Proof that the ray tracing equations for
                a channel whose sound speed varies linearly with
                depth requires that the rays be circles whose center
                is situated at the extrapolated depth where the sound
                speed vanishes   441
Example 11.5   (Rays_cylindrical_heterog.m) Solution of the ray
                tracing equations for the case of a plane wave that
                passes transversely through a cylindrically symmetric
                heterogeneous region, with special attention to the
                caustic that is created, followed by application
                of the transport equation to determine the variation
                of sound along a wavefront   451

Example 11.6   (none) Usage of geometrical acoustics to evaluate
                the eigenrays for a point source below the free
                surface in a homogeneous channel, and comparison
                of the result to that obtained from the method
                of images.   459
Example 11.7   (none) Application of Fermat's principle to determine
                the eigenrays for propagation from a point source in
                water channel of constant depth bounded above by air
                in a half-space, and below by a liquid sediment
                in a half-space   469
Example 12.1   (Born_approximation_cylinder.m) Evaluation of the
                angular dependence of the pressure scattered by a
                cylindrical region of heterogeneity according to the
                Born approximation   486
Example 12.2   (Scattering_rigid_body_Rayleigh.m) Application
                of the Rayleigh approximation of scattering to evaluate
                bistatic scattering of a rigid body   497
Example 12.3   (Scattering_movable_rigid_disk.m) Evaluation
                of low-frequency scattering at arbitrary incidence
                on a rigid circular disk that is free to move   502
Example 12.4   (Submerged_spher_shell_scattering.m)   Evaluation
                of the backscatter cross section and total scattering
                cross section for a spherical shell in water
                and in the atmosphere   527
Example 13.1   (none) Application of a graphical construction method
                to evaluate the nonlinear propagation of two plane
                wave pulses in the case where the initial waveform
                of the second is inverted from that of the first   556
Example 13.2   (Nonlin_harmonic_plane_wave.m) Computation of the
                nonlinear propagation of the nonlinear plane wave
                generated by a transient sinusoidal excitation using
                a simple algorithm   560
Example 13.3   (Nonlinear_wave_Newtons_method.m) Application of
                Newton's method to evaluate the Riemann solution for
                a nonlinear plane wave when the excitation is the sum
                of a fundamental and second harmonic, followed by
                FFT analysis to determine the frequency spectrum   565
Example 13.4   (Nonlin_harmonic_wave_shock_fitting.m)
                Implementation of the equal-area rule to fit shocks
                into the Riemann solution for an initially sinusoidal
                nonlinear plane wave   587

Example 13.5   (Frequency_domain_nonlin_ODEs.m) Solution of the
                    coupled differential equations governing the nonlinear
                    propagation of the Fourier series coefficients of an
                    initially sinusoidal plane wave in the case where the
                    attenuation coefficient increases as the square of
                    frequency.   607
Example 13.6   (none) Formulation of a regular perturbation series
                    to analyze the nonlinear wave generated by a periodic
                    sequence of beats in order to identify the growth of
                    harmonic and intermodulation distortion   607
Example 13.7   (Nonlin_spher_wave_2_harmonics.m) Evaluation
                    of the waveform of a nonlinear spherical wave
                    generated by boundary vibration that is the sum of a
                    fundamental and a third harmonic, and comparison
                    of the behavior of a nonlinear plane wave generated by
                    the same excitation   633
Example 13.8   (Nonlin_2D_waveforms.m) Application of Newton's
                    method to solve the transcendental equations resulting
                    from singular perturbation analysis of the
                    two-dimensional nonlinear plane wave radiated
                    by a periodically supported plate   651

Chapter 7
Radiation from Vibrating Bodies

The previous chapters were devoted to the analysis and interpretation of specific types of acoustic waves. In this chapter, the central theme is phenomenological. We will consider a variety of situations in which the task is to determine the acoustic field generated by a body that executes a specified vibration. These are the problems in acoustic radiation.

This chapter opens with an investigation of spherical waves that are not radially symmetric. A description of waves in terms of spherical harmonics, which may be considered to be the extension of Fourier series to a spherical geometry, will be applied to treat radiation from a sphere whose surface velocity pattern is arbitrary. At the same time, it is reasonable to ask how a velocity pattern can be imposed in reality. An object moves because forces are applied to it. The analysis of the vibration of a body excited by forces is a fundamental subject unto itself. In an acoustic system, part of the force system acting on a body is the surface pressure. It follows that to study the vibration problem, we must determine the radiated field, and that determination of the radiated field requires that we solve the vibration problem. In other words, the two problems are coupled. The joint formulation and solution of this coupled problem are the realm of *structural acoustics*. The behavior of a spherical shell immersed in a fluid will be examined in order to shed light on the analytical approach and phenomena associated with this subject.

To explore the field radiated by an infinitely long vibrating cylinder, we will develop the general solution of the Helmholtz equation in cylindrical coordinates. Spheres and infinitely long cylinders are just about the only geometrical configurations that are amenable to analysis. Their solutions shed much light on fundamental phenomena, but numerical simulation tools are required to address realistic systems. One such formulation is boundary elements, which is based on a fundamental concept called the Kirchhoff–Helmholtz integral theorem. This theorem, which we shall abbreviate to KHIT, gives the pressure at a field point in terms of an integral of the pressure and velocity on the surface of the radiating body. After the theorem is derived, its implementation as a boundary element formulation will be described. Two other numerical techniques for analyzing acoustic radiation are source superposition, which uses a set of sources to replace the vibrating body, and finite elements,

Electronic supplementary material The online version of this chapter (DOI 10.1007/978-3-319-56847-8_7) contains supplementary material, which is available to authorized users.

which requires adaptation to fit the nature of acoustic radiation. These developments close this chapter.

7.1 Spherical Harmonics

When the surface of a sphere vibrates in an irregular pattern, the radial velocity is an arbitrary function of the polar and azimuthal angles. Given the difficulties that arose for a radially symmetric wave in the time domain, it should not be surprising that we will only pursue a frequency domain analysis. Furthermore, in order to avoid excessive entanglement in the techniques of mathematical analysis, we will restrict our attention to axisymmetric situations. The consequence is that the pressure is $p = \text{Re}[P(r, \psi) \exp(i\omega t)]$. Although this will limit the generality of the situations we can address, it will not restrict the nature of the physical phenomena that arise.

7.1.1 Separation of Variables

Our objective is to determine the complex pressure amplitude $P(r, \psi)$ that is the general solution of the Helmholtz equation. The analysis will use the method of separation of variables. It begins with an ansatz that represents the dependent variable as a product of functions of each of the independent variables. In other words,

$$P = \mathcal{F}(r)\,\mathcal{G}(\psi) \tag{7.1.1}$$

It is helpful to rewrite the Helmholtz equation in axisymmetric spherical coordinates, Eq. (A.1.16), as

$$\frac{1}{r}\frac{\partial^2}{\partial r^2}(rP) + \frac{1}{r^2 \sin\psi}\frac{\partial}{\partial\psi}\left(\sin\psi\frac{\partial P}{\partial\psi}\right) + k^2 P = 0 \tag{7.1.2}$$

When we substitute the factorized representation of P into this equation, and divide by $\mathcal{F}\mathcal{G}/r^2$, the result is an equation in which each side is a function of a different position variable, specifically,

$$\frac{r}{\mathcal{F}}\frac{d^2}{dr^2}(r\mathcal{F}) + k^2 r^2 = -\frac{1}{\mathcal{G}\sin\psi}\frac{d}{d\psi}\left(\sin\psi\frac{d\mathcal{G}}{d\psi}\right) \tag{7.1.3}$$

The argument now is that the only way a function of r can equal a function of ψ is if they both equal some constant, which we denote as $\mathcal{C}$. Thus, we obtain two ordinary differential equations,

$$\boxed{\begin{aligned} \frac{d^2\mathcal{F}}{dr^2} + \frac{2}{r}\frac{d\mathcal{F}}{dr} + \left(k^2 - \frac{C}{r^2}\right)\mathcal{F} = 0 \\ \frac{1}{\sin\psi}\frac{d}{d\psi}\left(\sin\psi\frac{d\mathcal{G}}{d\psi}\right) + C\mathcal{G} = 0 \end{aligned}}$$

(7.1.4)

These equations are solved independently. We begin with $\mathcal{G}(\psi)$, because doing so will establish the separation constant C.

Actually, we will find that there are an infinite number of C values. The set of corresponding $\mathcal{G}$ functions are referred to as *spherical harmonics*. The reason for this terminology stems in part from the oscillatory nature of the functions, and also from the fact that they are the basis functions for a Fourier-like series describing an arbitrary function of the spherical angles. Because our interest is in axisymmetric situations, the spherical harmonics derived here are a subset of the general set of functions, which depend on θ, as well as ψ.

Dependence on the Polar Angle

The presence of sinusoidal terms in the coefficients of the second of Eq. (7.1.4) complicates the task of finding a solution. Fortunately, a transformation from ψ to a nondimensional variable η converts the differential equation to a form that has been widely studied. The transformation is

$$\boxed{\eta = \cos\psi \implies \sin\psi = \left(1 - \eta^2\right)^{1/2}}$$

(7.1.5)

The derivatives become

$$\frac{d\mathcal{G}}{d\psi} = -\sin\psi\frac{d\mathcal{G}}{d\eta}$$

$$\frac{1}{\sin\psi}\frac{d}{d\psi}\left(\sin\psi\frac{d\mathcal{G}}{d\psi}\right) = (\sin\psi)^2\frac{d^2\mathcal{G}}{d\eta^2} - 2\cos\psi\frac{d\mathcal{G}}{d\eta}$$

(7.1.6)

We use the transformation to eliminate ψ from the separated equation, which leads to *Legendre's equation*,

$$\left(1 - \eta^2\right)\frac{d^2\mathcal{G}}{d\eta^2} - 2\eta\frac{d\mathcal{G}}{d\eta} + C\mathcal{G} = 0$$

(7.1.7)

One of the most important aspects of this equation is that the coefficient of its highest derivative vanishes. Differential equations having this property contain a solution that is singular. Dealing with this possibility will lead to an expression for C. The fact that the coefficients in this equation are powers of η suggests that $\mathcal{G}$ may be represented as a power series. Thus, we try

$$\mathcal{G} = \sum_{j=0}^{\infty} a_j\eta^j$$

(7.1.8)

The next step is to substitute this form into Legendre's equation, and require that it be satisfied independently for each power of η. The details of these operations may be found in most textbooks on ordinary differential equations. The result is that the coefficients are obtained as a recurrence relation that gives a higher order coefficient in terms of a lower order value. The specific relation is

$$a_{j+2} = \frac{j(j+1) - \mathcal{C}}{(j+2)(j+1)} a_j, \quad j = 0, 1, 2, \ldots \tag{7.1.9}$$

The implication of this relation is that a_0 and a_1 are arbitrary and independent, from which successive coefficients are found as factors multiplying the a_0 and a_1 values.

If $\mathcal{C}$ is arbitrarily selected, it can be proven that the infinite series for $\mathcal{G}$ does not converge at $n = \pm 1$. This is how the singular nature of Eq. (7.1.7) is manifested. The locations at which $|\eta| = 1$ are the poles of the spherical coordinate system, but there is no physical aspect of the poles that would cause a singular response. The only way to avoid this behavior and thereby obtain a solution that is valid everywhere, is to select $\mathcal{C}$ such that the series becomes a finite sum. This happens if $\mathcal{C}$ makes one of the a_j coefficients to vanish, because the recurrence relation will lead to the higher coefficients being zero.

If we wish that the mth coefficient be the last one that is nonzero, then we would select

$$\mathcal{C}_m = m(m+1) \tag{7.1.10}$$

Each value of m leads to a different solution of Legendre's equation because m is the degree of the polynomial. The solutions are written as $P_m(\eta)$, which denotes that they are *Legendre polynomials of degree m*. Their definition is

$$P_m(\eta) = \sum_{j=0}^{\text{floor}(m/2)} \frac{(-1)^j (2m - 2j)!}{2^m j! (m-j)! (m-2j)!} \eta^{m} - 2j \tag{7.1.11}$$

One should exercise caution whether this formula is used for computations when m is large, because the factorial terms might be larger than the largest computable floating point number for the particular computer architecture. The lowest degree polynomials are

$$P_0(\eta) = 1, \quad P_1(\eta) = \eta, \quad P_2(\eta) = \frac{1}{2}(3\eta^2 - 1)$$
$$P_3(\eta) = \frac{1}{2}(5\eta^3 - 3\eta) \tag{7.1.12}$$

The substitution of $\eta = \cos\psi$ leads to $P_m(\cos\psi)$, which is a *Legendre function* of ψ. The first few functions are

$$P_0 \left(\cos \psi \right) = 1, \quad P_1 \left(\cos \psi \right) = \cos \psi, \quad P_2 \left(\cos \psi \right) = \frac{1}{4} \left(3 \cos 2\psi + 1 \right)$$

$$P_3 \left(\cos \psi \right) = \frac{1}{8} \left(5 \cos 3\psi + \cos \psi \right)$$

(7.1.13)

To obtain these forms directly from the Legendre polynomial with $\eta = \cos \psi$, one must apply various trigonometric identities for powers of $\cos \psi$. By virtue of their dependence on $\cos \psi$, the Legendre functions are periodic in $\Delta \psi = 2\pi$. Consequently, $P_m \left(\cos \psi \right)$ may be represented by a Fourier series. The handbook edited by Abramowitz and Stegun[1] is an excellent resource for the properties of Legendre functions.

Legendre polynomials are plotted in the upper set of graphs in Fig. 7.1. The even degrees are symmetric functions with respect to the origin, while odd degrees correspond to antisymmetric functions. As m increases, $P_m \left(\eta \right)$ oscillates over a smaller scale, but these oscillations are not harmonic. Recall that the coefficients a_0 and a_1 in Eq. (7.1.8) may be selected arbitrarily. The plotted functions have been defined such that $|P_m \left(\pm 1 \right)| = 1$. The lower set of graphs in Fig. 7.1 shows that replacing η with $\cos \left(\psi \right)$ to form a Legendre function does not fundamentally alter the appearance of the function at any degree. However, it does have the effect of compressing the region near the midpoint, which now is $\psi = \pi/2$, and expanding the region near the limits at $\eta = 0$ and π. The transformation from η to ψ leads to $(d/d\psi) P_m \left(\cos \left(\psi \right) \right)$ being identically zero at $\psi = 0$ and π. It follows that setting $\mathcal{G}$ to a Legendre function in the separation of variables form, Eq. (7.1.1), will lead to $dP/d\psi$ being zero identically

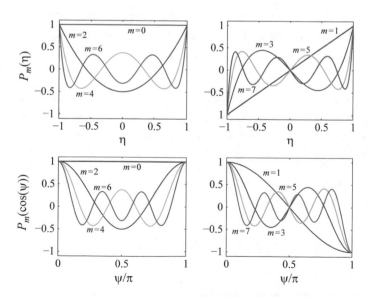

Fig. 7.1 Low-degree Legendre polynomials as a function of η, and Legendre functions as a function of ψ

[1]M.I. Abramowitz and I.A. Stegun, *Handbook of Mathematical Functions*, Dover, Chap. 8 (1965).

at the poles. The continuity equation requires that an axisymmetric field meets this condition.

The Legendre polynomials have some useful properties. The first are a set of recurrence relations for functions and derivatives at different degrees,

$$
\begin{aligned}
(m+1) P_{m+1}(\eta) &= (2m+1)\eta P_m(\eta) - m P_{m-1}(\eta) \\
\eta \frac{d}{d\eta} P_m(\eta) - \frac{d}{d\eta} P_{m-1}(\eta) &= m P_m(\eta) \\
\frac{d}{d\eta} P_{m+1}(\eta) - \frac{d}{d\eta} P_{m-1}(\eta) &= (2m+1) P_m(\eta) \\
\left(1 - \eta^2\right) \frac{d}{d\eta} P_m(\eta) &= -m\eta P_m(\eta) + m P_{m-1}(\eta)
\end{aligned}
\tag{7.1.14}
$$

A different definition of the polynomials is offered by Rodrigue's formula,

$$
P_m(\eta) = \frac{1}{2^m m!} \frac{d^m}{d\eta^m}\left[\left(\eta^2 - 1\right)^m\right]
\tag{7.1.15}
$$

An important property of Legendre polynomials is that they are mutually orthogonal. The polynomials are defined over the range $-1 \le \eta \le 1$, so orthogonality refers to an inner product defined as an integral over that interval. The orthogonality property states that if m and j are any two indices, then

$$
\int_{-1}^{1} P_m(\eta) P_j(\eta)\, d\eta = \int_{0}^{\pi} P_m(\cos\psi) P_j(\cos\psi) \sin\psi\, d\psi = \frac{2}{2m+1}\delta_{jn}
\tag{7.1.16}
$$

where δ_{jm} is the Kronecker delta, which equals one if $j = m$ and zero otherwise.

Recall that the orthogonality of harmonic functions is one of the primary reasons that a Fourier series has great utility. Orthogonality has a similar usefulness when a variable is represented by a *Legendre series*. In a Fourier series, the harmonic functions are said to be the basis functions and the series coefficients are "distances," in a linear algebra sense, in the direction of each harmonic function. A Legendre series is defined similarly, with $P_m(\eta)$ as the basis functions, specifically,

$$
F(\eta) = \sum_{m=0}^{\infty} F_m P_m(\eta) \text{ or } F(\psi) = \sum_{m=0}^{\infty} F_m P_m(\cos\psi)
\tag{7.1.17}
$$

The similarity of this representation to a Fourier series causes some individuals to say that the preceding is a *generalized Fourier series*.

Suppose we wish to determine the F_m coefficients associated with a specified F function. For an acoustics problem, ψ is the independent variable, so we shall work with the second form in Eq. (7.1.17). Both sides of the equation are multiplied by $P_j (\cos \psi) \sin \psi$, where j is a designated index. Then, both sides are integrated over $0 < \psi < \pi$. We bring the integration inside the summation, with the result that

$$\int_0^\pi F(\psi) P_j (\cos \psi) \sin \psi d\psi = \sum_{m=0}^\infty F_m \int_0^\pi P_m (\cos \psi) P_j (\cos \psi) \sin \psi d\psi$$

(7.1.18)

The orthogonality property states that all integrals vanish, except for the one where the summation index m equals the selected value of j. In that case, the integral is $2/(2j + 1)$. Thus, the orthogonality property filters from the sum all except the jth term. The result is that

$$\boxed{F_j = \frac{2j+1}{2} \int_0^\pi F(\psi) P_j (\cos \psi) \sin \psi d\psi}$$

(7.1.19)

One may attempt to evaluate the integral analytically, possibly by replacing $P_j (\cos \psi)$ with its definition as a polynomial, or else with the aid of a table of integrals. An alternative is to use numerical methods to evaluate the integral.

EXAMPLE 7.1 Legendre polynomials have been used to describe crystal vibrations.[2] That analysis used a series of Legendre polynomials to represent the dependence of displacement u on distance y from the midplane of a rectangular plate. The graph depicts $u = 0.2 \tanh^{-1} (1.98y/h)$, where h is the plate's thickness. Determine the Legendre series coefficients that represent this function. Identify at what degree the series may be truncated such that its evaluation will differ everywhere from the actual by less than 2% of the maximum value.

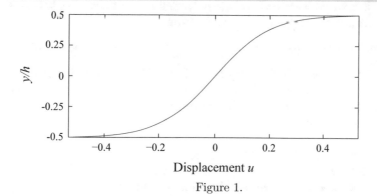

Figure 1.

[2]R.D. Mindlin, "Thickness-Shear and Flexural Vibrations of Crystal Plates,"*J. Appl. Phys.* **22**, 316–323 (1951).

Significance

In addition to showing how we may determine a Legendre series numerically, this example will give some insight to the convergence properties of this type of series.

Solution

To use Eq. (7.1.19), the range of integration must be $-1 \le \eta \le 1$, but the range for the plate is $-h/2 \le y \le h/2$. Thus, we set $2y/h = \eta$, so that $u(\eta) = 0.2 \tanh^{-1}(0.99\eta)$. Because η is the dependent variable, we use the first form in Eq. (7.1.16). This entails multiplying the series description of u by $P_j(\eta)$, then integrating over $-1 < \eta < 1$. The result is

$$F_j = \left(\frac{2j+1}{2}\right) 0.2 \int_{-1}^{1} \tanh^{-1}(0.99\eta)\, P_j(\eta)\, d\eta$$

There is no simple way to evaluate this integral analytically, so we turn to numerical methods. The details depend on the software to be used, but the basic elements are the same. A routine to evaluate the Legendre functions at a set of values of η is required, as well as a routine that performs numerical integration. If one wishes to write their own procedures, then Eq. (7.1.11) may be coded, because the degree of the Legendre polynomial will not be extraordinarily high. Numerical integration may be carried out by Simpson's rule, provided one verifies that the sampling interval is sufficiently small. The present results were obtained with MATLAB, which contains a function routine `legendre.m` that evaluates $P_m(x)$ for a vector of x values. However, it cannot be used directly because it also evaluates associated Legendre functions. They are described in Sect. 7.1.3 as functions that would arise if we were to investigate spherical waves that are not axisymmetric. The nature of `legendre.m` is that it returns an array having $m+1$ rows and n columns, where m is the degree number and n is the number of elements of x. The first row of this array holds the $P_m(x)$ values, while the other rows hold the corresponding associated functions. Thus, we evaluate the Legendre polynomials by creating a MATLAB function `leg.m` that retains only the first row, as follows:

```
function P_m = leg(x, m); full_set = legendre(m, x); P_m =full_set(1,:)
```

The integral was evaluated with one of MATLAB's numerical integration routines, `quadl.m`. Implementation of this scheme requires that we define another function to evaluate the integrand. It is a good idea to verify that everything has been implemented correctly by following the same procedure to evaluate the integral in the orthogonality condition.

The first few coefficients obtained from the numerical integration are $F_0 = 0$, $F_1 = 0.2869$, $F_2 = 0$, $F_3 = 0.0972$, $F_4 = 0$. All coefficients having even subscripts are zero because $u(\eta)$ is an odd function with respect to $\eta = 0$, and that is the symmetry of the odd-degree Legendre polynomials. To examine the convergence of the series, the odd-subscripted F_m are plotted in Fig. 2. It can be seen that the coefficients decrease rapidly as m increases.

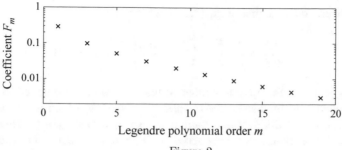

Figure 2.

It is requested that truncation of the Legendre series not lead to an error greater than 2% of the maximum value of u, which is 0.529 at $y = 0.5$. This requires that the series be synthesized at the full range of y. An estimate for the series length is obtained from the observation that the maximum magnitude of $P_m(x)$ is one, so inclusion of a term $F_N P_m(x)$ in the Legendre series can change the series sum by no more that $|F_n|$. It is specified that the error should be no more that 2% of 0.529, so we search for the smallest Legendre degree N at which $|F_N| < 0.0106$. This leads to $N = 13$. However, a certain error level in the coefficients might not be manifested by the same error when the series is synthesized.

In general, a efficient way in which a series may be evaluated at many points is to write it as a matrix product. Toward that end, we let $\{y\}$ denote a vector of y values, so that $\{\eta\} = (2/h)\{y\}$. Then, we form a rectangular array $[L]$ whose element at row j and column m is $P_m(\eta_j)$. The product of row j of $[L]$ and a vector $\{F\}$ is the same as the series evaluation of $u(y_j)$. Thus, the full set of $\{u\}$ values is found by evaluating $[L]\{F\}$. Figure 3 displays the error, defined as $|u(y_j) - u_{\text{series}}(y_j)|/u(y = h/2)$ for $N = 13$. The convergence criterion is met away from $y = \pm h/2$, but it is 4% at the limits. A larger N is required to satisfy it everywhere. Successive evaluations at increasing N revealed that $N = 19$ is the smallest degree at which truncation of the series will not lead to an error exceeding 2% at any y.

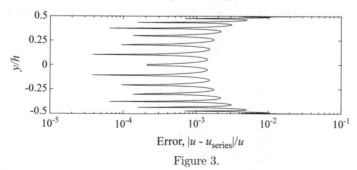

Figure 3.

Radial Dependence

The separation of variables constant C depends on the degree m of the Legendre polynomial, as given by Eq. (7.1.10). Thus, each m leads to a different radial function.

Each is governed by the first of Eq. (7.1.4), which now is

$$\frac{d^2 \mathcal{F}}{dr^2} + \frac{2}{r}\frac{d\mathcal{F}}{dr} + \left(k^2 - \frac{m(m+1)}{r^2}\right)\mathcal{F} = 0 \qquad (7.1.20)$$

At this juncture, there are a number of ways by which $\mathcal{F}$ may be determined. One approach is to implement the method of Frobenius, which is a standard tool for solving ordinary differential equations whose coefficients are not constant. The ansatz for this method is a power series multiplied by a factor that accounts for singular behavior. Another approach is suggested by the fact that far from a radiating body, spherical spreading requires that the solution decays reciprocally with r. We can anticipate this effect by changing the dependent variable according to

$$\mathcal{F} = \Phi(r)/r \qquad (7.1.21)$$

The differential equation becomes

$$\frac{d^2 \Phi}{dr^2} + \left[k^2 - \frac{m(m+1)}{r^2}\right]\Phi = 0 \qquad (7.1.22)$$

This is promising because it tells us that Φ tends to a harmonic function at larger r. Nevertheless, this equation does not appear to be easier to solve than the equation for F. Our knowledge of acoustics principles suggests how to proceed. When $m = 0$, Eq. (7.1.20) reduces to the Helmholtz equation in the case of a radially symmetric field. Thus, for $m = 0$, we know that $\mathcal{F} = Be^{\pm ikr}/r$. Furthermore, regardless of m, we know that the farfield pressure is a spherically spreading wave with an angular directivity. The observations suggest that we change the dependent variable such that

$$\mathcal{F} = \frac{1}{r}F_m(r)e^{-ikr} \qquad (7.1.23)$$

Determination of the function $F_m(r)$ is expedited by writing the differential equation as

$$\frac{1}{r}\left[\frac{d^2}{dr^2}(r\mathcal{F}) + \left(k^2 - \frac{m(m+1)}{r^2}\right)(r\mathcal{F})\right] = 0 \qquad (7.1.24)$$

When the representation of $\mathcal{F}$ is substituted into this equation, the complex exponential cancels. What remains is an ordinary differential equation for F_m,

$$\frac{d^2 F_m}{d(kr)^2} - 2i\frac{dF_m}{d(kr)} - \frac{m(m+1)}{(kr)^2}F_m = 0 \qquad (7.1.25)$$

This might not seem to be any more conducive to an analytical solution than the original equation governing $\mathcal{F}$, but it is. In the farfield, F_m must approach a constant value, which suggests a series representation in powers of $1/(kr)$, that is,

$$F_m = \sum_{j=0}^{\infty} \frac{a_j}{(kr)^j} \tag{7.1.26}$$

The coefficients a_j are determined by substituting the series into Eq. (7.1.25), followed by equating the collected coefficients of each power of $1/(kr)$ to zero. These operations lead to a recurrence relation for a_{j+1} as a proportionality to a_j, starting with $j = 0$. Examination of this relation shows that it gives $a_j = 0$ if $j > m$. In other words, although we allowed for an infinite length series, the number of terms must be one greater than the index m for the separation constant $\mathcal{C}$. The solution for $\mathcal{F}$ corresponding to each F_m is the *second kind of spherical Hankel function of order* m. A detailed analysis would lead to

$$\boxed{\mathcal{F} = h_m^{(2)}(kr) = i^{m+1}\frac{e^{-ikr}}{kr}\sum_{j=0}^{m}\frac{(m+j)!}{j!(m-j)!}\left(\frac{1}{2ikr}\right)^j} \tag{7.1.27}$$

In this form, the starting coefficient a_0 has been set to one. The lowest order functions are

$$h_0^{(2)}(kr) = \frac{ie^{-ikr}}{kr}, \quad h_1^{(2)}(kr) = -\frac{e^{-ikr}}{kr}\left(1 - \frac{i}{kr}\right)$$
$$h_2^{(2)}(kr) = -\frac{ie^{-ikr}}{kr}\left(1 - \frac{3i}{kr} - \frac{3}{(kr)^2}\right) \tag{7.1.28}$$

At this juncture, the logical question is: Why is this function called the "second kind"? We found it first, so why is it not the "first kind"? The answer lies in the fact that we have used $e^{i\omega t}$ to represent a time function. The function associated with the negative exponential temporal representation by convention is denoted as $h_m^{(1)}(kr)$, which is the *first kind of spherical Hankel function of order* m. Because the proper representation of a harmonic function adds $e^{-i\omega t}$ and $e^{i\omega t}$ terms, the coefficients of the exponential must be complex conjugates in order that the time dependence be real. Hence, it must be that

$$h_m^{(1)}(kr) = h_m^{(2)}(kr)^* \tag{7.1.29}$$

This definition of another solution could have been anticipated from the fact that the differential equation whose solution we sought, Eq. (7.1.20), is real. If a complex function satisfies a linear differential equation whose coefficients are real, then its complex conjugate must do so as well because the real and imaginary parts of the differential equation must be satisfied independently. The same reasoning leads to the recognition that the real and imaginary parts of the spherical Hankel functions must satisfy the equation independently. The real part is the *spherical Bessel function of order* m, denoted $j_m(kr)$, and the imaginary part is the *spherical Neumann function of order* m, denoted $n_m(kr)$,

$$\boxed{j_m(kr) = \mathrm{Re}\left(h_m^{(2)}(kr)\right), \quad n_m(kr) = -\mathrm{Im}\left(h_m^{(2)}(kr)\right)} \qquad (7.1.30)$$

A different way of viewing these definitions is that

$$\boxed{\begin{aligned} h_m^{(2)}(kr) &= j_m(kr) - in_m(kr), \quad h_m^{(1)}(kr) = j_m(kr) + in_m(kr) \\ j_m(kr) &= \frac{1}{2}\left[h_m^{(1)}(kr) + h_m^{(2)}(kr)\right], \quad n_m(kr) = \frac{1}{2i}\left[h_m^{(1)}(kr) - h_m^{(2)}(kr)\right] \end{aligned}}$$

$$(7.1.31)$$

The lowest order functions are

$$j_0(kr) = \frac{\sin(kr)}{kr} \qquad\qquad n_0(kr) = -\frac{\cos(kr)}{kr}$$

$$j_1(kr) = \frac{\sin(kr)}{(kr)^2} - \frac{\cos(kr)}{kr} \qquad n_1(kr) = -\frac{\cos(kr)}{(kr)^2} - \frac{\sin(kr)}{kr}$$

$$j_2(kr) = \left(\frac{3}{(kr)^3} - \frac{1}{kr}\right)\sin(kr) \quad n_2(kr) = -\left(\frac{3}{(kr)^3} - \frac{1}{kr}\right)\cos(kr)$$

$$\qquad - \frac{3}{(kr)^2}\cos(kr) \qquad\qquad\qquad - \frac{3}{(kr)^2}\sin(kr) \qquad (7.1.32)$$

When we wish to refer to these solutions generically, we will say that they are spherical Bessel functions.

Most mathematical software provides routines to evaluate Bessel functions $J_\nu(x)$ and Neumann functions $N_\nu(x)$.[3] Bessel's equation arises in problems featuring a planar circular geometry. For this reason, the functions $J_\nu(x)$ and $N_\nu(x)$ are said to be a set of cylindrical Bessel functions. The cylindrical and spherical functions are related by

$$\boxed{j_m(x) = \left(\frac{\pi}{2x}\right)^{1/2} J_{m+1/2}(x), \quad n_m(x) = \left(\frac{\pi}{2x}\right)^{1/2} N_{m+1/2}(x)} \qquad (7.1.33)$$

In the event that the available software only evaluates cylindrical Bessel function, we would evaluate the cylindrical functions for order $m + 1/2$ and argument kr and then multiply the result by $\pi^{1/2}/(2kr)^{1/2}$. The spherical Hankel functions may be found from Eq. (7.1.31).

The low-order spherical Bessel and Neumann functions are graphed in Fig. 7.2. The abscissa for the Neumann functions is split because that function is singular at

---

[3]It is standard practice to denote a Bessel function as $J_\nu(x)$, but a Neumann function often is denoted as $Y_\nu(x)$.

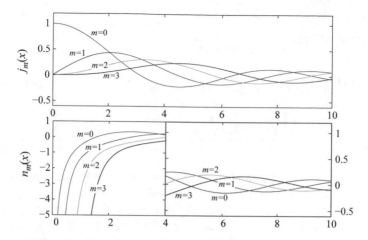

Fig. 7.2 Dependence of the spherical Bessel and Neumann functions on the value of its order m and argument x

the origin. The behavior for small x of $n_m(x)$ is quite different from that of $j_m(x)$, but both become oscillatory with decreasing amplitude as x increases.

These observations are borne out by the asymptotic approximations for small and large arguments. These behaviors for all functions may be extracted by the analysis of Eq. (7.1.27), combined with application of Eqs. (7.1.29) and (7.1.30). When the argument is very small, $x \ll 1$, the results are

$$\left.\begin{array}{l} h_m^{(2)}(x) = h_m^{(1)}(x)^* \to i\dfrac{(2m)!}{m!2^m}\left(\dfrac{1}{x}\right)^{m+1} \\[2mm] j_m(x) \to \dfrac{m!\,(2x)^m}{(2m+1)!} \\[2mm] n_m(x) \to -\dfrac{(2m)!}{m!2^m}\left(\dfrac{1}{x}\right)^{m+1} \end{array}\right\} \quad \text{as } x \to 0 \qquad (7.1.34)$$

The trends for large arguments are

$$\left.\begin{array}{l} h_m^{(2)}(x) = h_m^{(1)}(x)^* \to \dfrac{1}{x}e^{-i(x-(m+1)\pi/2)} \\[2mm] j_m(x) \to \dfrac{1}{x}\cos\left(x - \dfrac{m+1}{2}\pi\right) \\[2mm] n_m(x) \to \dfrac{1}{x}\sin\left(x - \dfrac{m+1}{2}\pi\right) \end{array}\right\} \quad \text{as } x \to \infty \qquad (7.1.35)$$

An important aspect of the preceding is that they require that the argument x be larger than the order m. Another asymptotic representation covers the situation where the order is large,

$$h_m^{(2)}(x) = h_m^{(1)}(x)^* \to i \left(\frac{2}{(2m+1)x} \right)^{1/2} \left(\frac{2m+1}{ex} \right)^{(m+1/2)}$$

$$j_m(x) \to \left(\frac{\pi}{2x} \right)^{1/2} \frac{1}{(\pi(2m+1))^{1/2}} \left(\frac{ex}{2m+1} \right)^{(m+1/2)}$$

as $m \to \infty$
with x fixed

$$n_m(x) \to -\left(\frac{2}{(2m+1)x} \right)^{1/2} \left(\frac{2m+1}{ex} \right)^{(m+1/2)}$$

$$(7.1.36)$$

The qualitative description of these relations is that only the spherical Bessel function $j_m(x)$ is finite at zero argument, with all except the zero order approaching zero. Each of the other functions is singular at $x = 0$, with an increasing growth rate as the order is increased. Also, replacing x with kr shows that the farfield behavior of $h_m^{(2)}(kr)$ matches that of a wave that propagates outward, whereas $h_m^{(1)}(kr)$ is associated with inward propagation. Furthermore, $j_m(kr)$ and $n_m(kr)$ represent standing wave patterns. The behavior at large orders is important because it affects the convergence of series that sums over all orders.

There are many formulas that relate the various spherical Bessel functions and their derivatives. A group that are particularly useful are the recurrence relations. The following uses $f_m(x)$ to denote $j_m(x)$, $n_m(x)$, $h_m^{(1)}(x)$, or $h_m^{(2)}(x)$,

$$f_{m-1}(x) + f_{m-1}(x) = \frac{2m+1}{x} f_m(x), \quad m > 0$$

$$\frac{d}{dx} f_m(x) = \frac{m}{x} f_m(x) - f_{m+1}(x)$$

$$= f_{m-1}(x) - \frac{m+1}{x} f_m(x), \quad m > 0$$

$$= \frac{1}{2m+1} \left[m f_{m-1}(x) - (m+1) f_{m+1}(x) \right], \quad m > 0 \quad (7.1.37)$$

$$\left(\frac{1}{x} \frac{d}{dx} \right)^n \left[x^{m+1} f_m(x) \right] = x^{m-n+1} f_{m-n}(x)$$

$$\left(\frac{1}{x} \frac{d}{dx} \right)^n \left[x^{-m} f_m(x) \right] = (-1)^n x^{-m-n} f_{m+n}(x)$$

The first equation may be used to evaluate higher order functions in terms of lower order values. We will have frequent need for derivatives of Bessel functions, so it is good idea to create a subprogram that implements the first equation for $df_m(x)/dx$. The last two relations seldom come into play in the analysis of a specific system. As is true for Legendre functions, the handbook compiled by Abramowitz and Stegun[4] provides a compact resource for many other formulas. The book by Sneddon[5] is a reference for those seeking detailed analyses of Bessel functions.

---

[4]M.I. Abramowitz and I.A. Stegun, *Handbook of Mathematical Functions*, Dover, Chaps. 9–11 (1965).

[5]I.A. Sneddon, *Special Functions of Mathematical Physics and Chemistry* Oliver & Boyd (1961).

7.1.2 Description of the Pressure Field

The separation of variables analysis began by seeking the conditions for which $P = \mathcal{F}(r)\,\mathcal{G}(\psi)$ is a solution of the Helmholtz equation. This led to the recognition that $\mathcal{G}(\psi)$ must be a Legendre function of integer order m and $\mathcal{F}(r)$ must be one of the types of spherical Bessel functions. Because the Helmholtz equation is homogeneous, the trial solution may be multiplied by a constant. Furthermore, the equation is linear, so a sum of solutions at various orders also is a solution.

Which type of spherical Bessel function we should use depends on the extent of the fluid's domain. We began with the stated interest of describing radiation into an infinite domain. The Sommerfeld radiation condition requires an outgoing wave, which means that we should use whichever function behaves as e^{-ikr}/r when r is large. Reference to Eq. (7.1.35) shows that this function is $h^{(2)}(kr)$. Thus, the axisymmetric pressure exterior to a sphere is described by

$$P = \sum_{m=0}^{\infty} B_m h_m^{(2)}(kr)\, P_m(\cos\psi), \quad r > a \tag{7.1.38}$$

To specialize this expression to the farfield, we use the asymptotic representation of $h_m^{(2)}(kr)$ contained in Eq. (7.1.35), which leads to

$$P_{\mathrm{ff}} = \sum_{m=0}^{\infty} B_m P_m(\cos\psi)\, \frac{e^{-ikr}}{kr} e^{i(m+1)\pi/2} \tag{7.1.39}$$

The pressure in a spherical cavity is a topic in Chap. 10. As a preliminary to that study, we recall that the analysis of the radially symmetric field in Sect. 6.3.3 began with a solution that contains inward and outwardly propagating waves. The corresponding form here would use both kinds of spherical Hankel function. The finiteness condition in the previous analysis led to the recognition that these waves must have complex amplitudes that sum to zero. A similar analysis would lead to the same conclusion here. According to Eq. (7.1.31), the sum of $h_m^{(2)}(kr)$ and $h_m^{(1)}(kr)$ gives a spherical Bessel function, while the difference gives a spherical Neumann function. The latter is singular at $r = 0$, so the general solution for the axisymmetric pressure within a spherical cavity is

$$P = \sum_{m=0}^{\infty} B_m j_m(kr)\, P_{\mathrm{ff}}(\cos\psi), \quad 0 \le r < a \tag{7.1.40}$$

7.1.3 Arbitrary Spatial Dependence

The most general situation is a pressure field that is neither radially nor axially symmetric. For example, a single dipole and a longitudinal quadrupole generate fields that are axisymmetric, but a lateral quadrupole is not. An investigation of the most general spherical wave would pursue a separation of variables analysis of the Helmholtz equation by starting with

$$P = \mathcal{F}(r)\, \mathcal{G}(\psi)\, \Theta(\theta) \tag{7.1.41}$$

Such an analysis would lead to the recognition that the azimuthal function $\Theta(\theta)$ must be harmonic. This fact is apparent on a physical basis, because circling the polar axis at a fixed radial distance and polar angle must return us to the point where we began. Thus, it must be that $P(r, \psi, \theta) = P(r, \psi, \theta + 2\pi)$. This is the condition for representing the azimuthal dependence in a Fourier series. At our option, we can take Θ to be a complex exponential, $e^{in\theta}$, where n is any positive or negative integer, or we may consider it to be a real harmonic by letting it be $\sin(n\theta)$ or $\cos(n\theta)$.

Regardless of which representation is selected, the fact that $\Theta(\theta)$ is a harmonic function whose argument is $n\theta$ reduces the task to a determination of the product $\mathcal{F}(r)\,\mathcal{G}(\psi)$. Application of the separation of variables technique to the Helmholtz equation leads to the recognition that the azimuthal harmonic number n affects the equation for $\mathcal{G}$, but not the one for $\mathcal{F}$. The solution for $\mathcal{G}$ is found to be an *associated Legendre function* $P_m^n(\cos\psi)$, which is obtained from n differentiations of an mth *degree* Legendre polynomial, followed by the substitution of $\eta = \cos\psi$. The specific definition is

$$P_m^n(\cos\psi) = \left[-\left(1 - \eta^2\right)^{1/2}\right]^m \frac{d^n}{d\eta^n} P_m(\eta)\bigg|_{\eta = \cos\psi} \tag{7.1.42}$$

Because $P_m(\eta)$ is an mth degree polynomial, any derivative above the mth degree will be identically zero, from which it follows that

$$P_m^n(\cos\psi) = 0 \text{ if } n > m \tag{7.1.43}$$

The index n is said to be the *order* of the associated Legendre function and m is its degree.

Because the separation equation for $\mathcal{F}(r)$ does not depend on the azimuthal number n, it is the same alternative set of Bessel functions as in the axisymmetric case. Thus, for each polar number m, there are azimuthal harmonics from 0 to m. The pressure for each m, n pair consists of a product of a spherical Bessel or Hankel function of order n for the radial dependence, a $P_m^n(\cos\psi)$ function, and a harmonic function of $n\theta$. The range of m is infinite, so the general solution for the pressure is a double sum. For example, in the case of radiation, with a complex Fourier series used to represent the azimuthal dependence, the solution would have the form

$$P = \frac{1}{2} \sum_{m=0}^{\infty} \sum_{n=-m}^{m} B_{m,n} \, h_m \, (kr) \, P_m^n \, (\cos \psi) \, e^{in\theta} \tag{7.1.44}$$

The factors $P_m^n \, (\cos \psi) \, e^{in\theta}$ describe an arbitrary angular dependence; they constitute the full set of spherical harmonics.

Like the axisymmetric set, the general set of spherical harmonics are orthogonal when integrated over a sphere. Indeed, procedures that are used to analyze an axisymmetric field in terms of spherical harmonics are readily modified to handle arbitrary fields. Several books[6] treat the mathematical aspects of a general spherical harmonic series expansion and its application to acoustical systems. If we were to study situations where there is no symmetry, we would not encounter much that is new from a phenomenological viewpoint. The primary reason for this is the aforementioned nature of the radial dependence, which is independent of the azimuthal harmonic number.

EXAMPLE 7.2 Consider an axisymmetric field consisting of a monopole, a dipole, and a longitudinal quadrupole. This combination of sources is collocated, and the axes of the dipole and quadrupole are aligned. What is the spherical harmonic representation of the farfield?

Significance

Experience in working with spherical harmonics is the primary emphasis. The result will highlight the different pictures of the farfield that results from a spherical harmonic series, rather than a multipole expansion.

Solution

The multipole representation of the farfield pressure is provided by Eq. (6.5.142). To use it, we will let $z \equiv x_3$ be the axis of symmetry, and set the dipole such that its moment vector is aligned along z. A longitudinal quadrupole that is aligned in this direction has $Q_{3,3}$ as the only nonzero component. The spherical coordinate transformation gives $z = r \cos \psi$, so the given pressure is

$$P = \left[A + ik D_3 \cos \psi - k^2 Q_{3,3} \, (\cos \psi)^2 \right] \frac{e^{-ikr}}{r} \tag{1}$$

where the strength factors are unspecified.

The spherical harmonic description of an axisymmetric field is Eq. (7.1.38), and Eq. (7.1.39) is the corresponding farfield approximation. The B_m coefficients are found by applying the orthogonality property. It is somewhat easier to formulate the

---

[6]J.O. Hirschfelder, C.F. Curtiss, and R.B. Bird, *Molecular Theory of Gases and Liquids*, Wiley-Interscience (1964).

analysis in terms of $\eta = \cos \psi$. The actual and series descriptions of P are equated, and both sides are multiplied by $P_j (\eta)$. Integration over $-1 \leq \eta \leq 1$ leads to

$$\sum_{m=0}^{\infty} \int_{-1}^{1} \left[B_m P_m (\eta) \frac{e^{-ikr}}{kr} e^{i(m+1)\pi/2} \right] P_j (\eta)\, d\eta$$

$$= \int_{-1}^{1} \left[A + ik D_3 \eta - k^2 Q_{3,3} \eta^2 \right] \frac{e^{-ikr}}{r} P_j (\eta)\, d\eta$$

All terms in the summation integrate to zero, except for the term for which $m = j$. The result is

$$B_j = \left(\frac{2j + 1}{2} \right) k e^{-i(j+1)\pi/2} \int_{-1}^{1} \left[A + ik D_3 \eta - k^2 Q_{3,3} \eta^2 \right] P_j (\eta)\, d\eta \qquad (2)$$

We could evaluate the integral numerically as we did in the previous example. However, the facts that $P_j (\eta)$ is a polynomial in η, and that it is multiplied by powers of η, suggest we try analytical integration. One approach would be to work with the series representation of Legendre functions, Eq. (7.1.11), but a simpler analysis is available. Equation (7.1.12) indicates that $P_0 (\eta) = 1$, $P_1 (\eta) = \eta$, and $[2P_2 (\eta) + P_0 (\eta)] /3 = \eta^2$, so Eq. (2) may be written as

$$B_j = \left(\frac{2j + 1}{2} \right) k e^{-i(j+1)\pi/2} \int_{-1}^{1} \left[A P_0(\eta) + ik D_3 P_1 (\eta) \right.$$

$$\left. - k^2 Q_{3,3} \frac{2P_2 (\eta) + P_0 (\eta)}{3} \right] P_j(\eta) d\eta$$

The orthogonality property describes the integral of each term. Specifically, $P_n (\eta)$ $P_j (\eta)$ integrates to zero unless $j = n$, in which case the integral is $2/ (2n + 1)$. Thus, the result is that

$$B_0 = -ikA + ik^3 Q_{33}, \quad B_1 = -ik^2 D_3, \quad B_2 = - (2/3)\, i Q_{3,3}$$

The most notable attribute is that the spherical harmonics are not in one-to-one correspondence to the point source distributions. As the source distribution goes up in order, more harmonics are required to represent it. In general, the field of an axisymmetric n-pole will be proportional to $\cos \psi$ raised to powers up to n. The resulting B_j coefficients for $j \leq n$ will contain a contribution from that pole if j and n have the same parity.

Given that several spherical harmonics are required to describe the higher multipoles, the utility of spherical harmonics might seem to be questionable. This is a limited view. The multipole construction requires that the sources be a compact set, whereas spherical harmonics may be used to describe any spherical field. Furthermore, the orthogonality property enables a complicated field to be mapped into a simpler representation as a series.

7.2 Radiation from a Spherical Body

7.2.1 Analysis

Determination of the field radiated by an axisymmetrically vibrating sphere is our first application of a spherical harmonic series. The axis of symmetry is designated as the polar axis z. The radial vibration of the sphere is an arbitrary function of the polar angle, $V(\psi)$. The sphere, whose radius is a, is surrounded by an ideal fluid. Cases in which the sphere executes a breathing mode vibration, or oscillates as a rigid body, which we studied in Chap. 6, are special cases. Generalization to an arbitrary vibration pattern will lead to insights about what aspects of the vibration pattern are important to the pressure field and radiated power.

In addition to the spherical harmonic series for the pressure in a radiation problem, Eq. (7.1.38), we will require the associated radial velocity. It is obtained by applying Euler's equation to the pressure field. These expressions are

$$P = \sum_{m=0}^{\infty} B_m h_m(kr) P_m(\cos \psi)$$
$$V_r = \frac{i}{\rho_0 c} \sum_{m=0}^{\infty} B_m h'_m(kr) P_m(\cos \psi) \tag{7.2.1}$$

Explicit designation of the Hankel function as being the second kind has been abandoned because that is the only kind that is relevant to the analysis. (If we had adopted the $\exp(-i\omega t)$ convention for a harmonic function, it would use the first kind of Hankel function.) Also, the prime designates differentiation of the function with respect to its argument, that is,

$$h'_m(kr) \equiv \frac{d}{d\xi} h_m^{(2)}(\xi) \Big|_{\xi=kr} \tag{7.2.2}$$

The preceding is descriptive of any axisymmetric field. To fit it to the present radiation problem, we match the radial particle velocity at the sphere's surface to that of the sphere. Toward that end, the surface velocity is represented as a spherical harmonic series,

$$V(\psi) = v_0 \sum_{m=0}^{\infty} V_m P_m(\cos \psi) \tag{7.2.3}$$

The parameter v_0 is a measure of the magnitude of the surface velocity, such as its maximum. It has been factored out of the series in order that the V_m coefficients be dimensionless. An expression for these coefficients is found by invoking the orthogonality property of the spherical harmonics, which gives

$$V_m = \left(m + \frac{1}{2}\right) \int_0^\pi \frac{V(\psi)}{v_0} P_m(\cos\psi) \sin\psi d\psi$$

$$= \left(m + \frac{1}{2}\right) \int_{-1}^1 \frac{V(\cos^{-1}(\eta))}{v_0} P_m(\eta) d\eta \qquad (7.2.4)$$

Which of these forms is more conducive to the evaluation of the integral depends on the nature of the $V(\psi)$ function.

The continuity condition requires that $V_r = V(\psi)$ at $r = a$. We perform this matching by using Eq. (7.2.3) to represent the sphere's motion and Eq. (7.2.1) to represent the response of the fluid. Thus, we have

$$\frac{i}{\rho_0 c} \sum_{m=0}^\infty B_m h_m'(ka) P_m(\cos\psi) = v_0 \sum_{m=0}^\infty V_m P_m(\cos\psi) \qquad (7.2.5)$$

The Legendre functions constitute an orthogonal set, so the preceding relation must be satisfied separately by each term. This leads to

$$B_m = \rho_0 c v_0 \frac{V_m}{i h_m'(ka)} \qquad (7.2.6)$$

The resulting expressions for particle velocity and pressure are

$$\boxed{\begin{aligned} P &= \rho_0 c v_0 \sum_{m=0}^\infty V_m \frac{h_m(kr)}{i h_m'(ka)} P_m(\cos\psi) \\ V_r &= v_0 \sum_{m=0}^\infty V_m \frac{h_m'(kr)}{h_m'(ka)} P_m(\cos\psi) \end{aligned}} \qquad (7.2.7)$$

Although these expressions call for infinite summations, in most cases, adequate accuracy will be obtained when the summations are truncated at a relatively small number of terms.

It is useful to verify that these expressions are consistent with prior developments. In the case of a sphere that executes a breathing mode vibration, the surface velocity $V(\psi) = v_0$. Because $P_0(\cos\psi) = 1$, the only nonzero velocity coefficient is $V_0 = 1$. From Eq. (7.1.28), we have

$$h_0(x) = \frac{i e^{-ix}}{x} \implies h_0'(x) = \frac{x - i}{x^2} e^{-ix} \qquad (7.2.8)$$

These expressions, evaluated at $x = kr$ for the former and $x = ka$ for the latter, are substituted into the $m = 0$ term in the summation, which yields

$$P = \rho_0 c v_0 \frac{ie^{-ikr}/r}{i\left[\dfrac{ka - i}{(ka)^2}\right]e^{-ika}} \tag{7.2.9}$$

When this expression is simplified, the result is the same as the first of Eq. (6.3.7).

Proof that the spherical harmonic description reduces to the previous result for a translating sphere is a little more complicated. In this case, $V(\psi) = v_0 \cos\psi$, so $V_1 = 1$ is the sole nonzero coefficient. According to Eq. (7.1.28), the first-order Hankel function and its derivative are

$$h_1(x) = \left(\frac{-x + i}{x^2}\right)e^{-ix} \implies h_1'(x) = \left(\frac{ix^2 + 2x - 2i}{x^3}\right)e^{-ix} \tag{7.2.10}$$

Substitution of these terms into the $m = 1$ term in the summation gives

$$P = \rho_0 c v_0 \frac{\left[\dfrac{-kr + i}{(kr)^2}\right]e^{-ikr}}{i\left[\dfrac{i(ka)^2 + 2ka - 2i}{(ka)^3}\right]} \tag{7.2.11}$$

Simplification of this expression leads to Eq. (6.5.102).

7.2.2 Important Limits

Not much about the nature of the field is apparent in the general solution, so we turn to asymptotic trends, beginning with the farfield. When kr is large, Eq. (7.1.35) indicates that

$$h_m(kr) \to \frac{e^{-ikr}}{kr}e^{i(m+1)\pi/2}, \quad h_m'(kr) \to -ih_m(kr) \text{ as } kr \to \infty \tag{7.2.12}$$

Application of these trends to Eq. (7.2.7) leads to

$$\boxed{P_{\text{ff}} = \rho_0 c \, (V_r)_{\text{ff}} = \rho_0 c v_0 \frac{e^{-ikr}}{kr}\sum_{m=0}^{\infty} V_m \frac{e^{im\pi/2}}{h_m'(ka)} P_m(\cos\psi)} \tag{7.2.13}$$

Typically, the farfield property of interest is the directivity. Previously, we defined $\mathcal{D}(\psi)$ such that its maximum value is one. It is more convenient here to use as the reference pressure $\rho_0 c v_0 (a/r)$. Only the pressure magnitude is considered. Thus, we have

$$\mathcal{D}(\psi) \equiv \left| \frac{P_{\text{ff}}}{\rho_0 c v_0} \right| \left(\frac{r}{a} \right) = \frac{1}{ka} \left| \sum_m \frac{V_m e^{im\pi/2}}{h'_m(ka)} P_m(\cos \psi) \right| \tag{7.2.14}$$

Another use of the farfield pressure is to evaluate the radiated power. Toward that end, we form the time-averaged intensity at large kr, which is $(I_r)_{\text{av}} = P_{\text{ff}} P_{\text{ff}}^* / (2\rho_0 c)$. To evaluate the product, we change the summation index for one factor, which leads to

$$(I_r)_{\text{av}} = \frac{1}{2} \rho_0 c v_0^2 \left(\frac{1}{kr} \right)^2 \sum_{j=0}^{\infty} \sum_{m=0}^{\infty} V_j V_m^* \frac{e^{i(j-m)\pi/2}}{h'_j(ka) \left(h'_m(ka) \right)^*} P_j(\cos \psi) P_m(\cos \psi)$$

$$\tag{7.2.15}$$

The radiated power is found by integrating this expression over the surface of a very large sphere whose radius is r. Because of the axisymmetry of the field, a suitable differential surface element is $dS = 2\pi r \sin \psi (r d\psi)$. When we bring the surface integral inside the double sum, we find that

$$\mathcal{P} = \frac{\pi \rho_0 c}{k^2} \sum_{j=0}^{\infty} \sum_{m=0}^{\infty} V_j V_m^* \frac{e^{i(j-m)\pi/2}}{h'_j(ka) \left(h'_m(ka) \right)^*} \int_0^{\pi} P_j(\cos \psi) P_m(\cos \psi) \sin \psi d\psi$$

$$\tag{7.2.16}$$

The integral is merely the orthogonality condition for Legendre functions, which means that all terms in the double sum for which $j \neq m$ vanish. This simplifies the result to

$$\mathcal{P} = \pi \rho_0 c v_0^2 a^2 \frac{1}{(ka)^2} \sum_{m=0}^{\infty} \frac{2}{2m+1} \frac{|V_m|^2}{|h'_m(ka)|^2} \tag{7.2.17}$$

This expression exhibits an important property: *The spherical harmonics of the surface velocity are uncoupled in the radiated power.* In other words, each harmonic gives a fixed contribution to the radiated power, independently of the magnitude of the other harmonics.

Another important feature is related to convergence properties. When the harmonic number m becomes increasingly large, $h'_m(ka)$ will increase rapidly, see Eq. (7.1.36). Consequently, the high-order terms in the series for the radiated power have much less importance than they do for the surface velocity. Recall that increasing spherical harmonic number represents more "wiggles" in the spatial fluctuation. It follows that this behavior is another manifestation of the general notion that waves on a surface that have a short wavelength tend to not radiate to the farfield.

Another limit of interest is a small sphere. The expressions for pressure and radial velocity depend on the radius nondimensionally as ka, so the small radius limit also is the low-frequency limit. The asymptotic behavior of $h'_m(ka)$ may be found by differentiating $h_m^{(2)}(x)$ in Eq. (7.1.34), which gives

$$h'_m (ka) \rightarrow -4i \frac{(m+1)(2m)!}{(2ka)^{m+2}} \text{ as } ka \rightarrow 0 \qquad (7.2.18)$$

With these, the pressure in Eq. (7.2.7) becomes

$$P = \rho_0 c v_0 \sum_{m=0}^{\infty} V_m \frac{(2ka)^{m+2}}{4(m+1)(2m)!} h_m (kr) P_m (\cos \psi), \quad ka \ll 1 \qquad (7.2.19)$$

Before we interpret this expression, let us consider the special case of a radially symmetric vibration. The only nonzero velocity coefficient is $V_0 = V(\psi)/v_0 = 1$. The definitions are $h_0 (kr) \equiv i e^{-ikr} / (kr)$ and $P_0 (\cos \psi) = 1$, and we also have

$$h'_0 (ka) \equiv \frac{d}{d\xi} \left(\frac{i e^{-i\xi}}{\xi} \right) \Big|_{\xi=ka} \equiv \left(\frac{ka - i}{k^2 a^2} \right) e^{-ika}, \qquad (7.2.20)$$

When V_0 is the only nonzero coefficient, Eq. (7.2.9) for arbitrary values of kr and ka gives

$$P = \rho_0 c v_0 \frac{h_0 (kr)}{i h'_0 (ka)} = \rho_0 c v_0 \left(\frac{ka}{ka - i} \right) \frac{a}{r} e^{-ik(r-a)} \qquad (7.2.21)$$

Now consider Eq. (7.2.19) when $ka \ll 1$. Each term in the summation raises ka to a higher power, so it follows that the dominant term corresponds to the lowest harmonic m for which V_m is nonzero. Usually, this corresponds to $m = 0$, in which case, this description reduces to

$$P = \rho_0 c v_0 (ka)^2 h_0 (kr) P_0 (\cos \psi) \equiv \rho_0 c v_0 (ika) \frac{a}{r} e^{-ikr} \qquad (7.2.22)$$

Because $e^{ika} \approx 1$ when $ka \ll 1$, Eq. (7.2.21) reduces to this expression. Hence, we conclude that in the small sphere/low-frequency limit, the only aspect of the surface motion that is important is the volume velocity, $\hat{Q} = 4\pi a^2 v_0$. Only if the volume velocity is zero does the pattern of the surface vibration affect the acoustic field.

The high-frequency/large sphere limit also leads to an interesting observation. Let M denote the lowest spherical harmonic at which the summation in Eq. (7.2.3) may be truncated without reducing the accuracy below the desired value. The definition of high frequency is that $ka \gg M$. In that case, an expression for $h'_m (ka)$ at all significant orders m may be obtained by differentiation of the large argument approximation of $h_m^{(2)} (ka)$, Eq. (7.1.35), which gives

$$h'_m (ka) \rightarrow \frac{-i}{ka} e^{-i(ka-(m+1)\pi/2)} \text{ as } ka \rightarrow \infty \qquad (7.2.23)$$

This approximation converts Eq. (7.2.13) to

$$P_{\text{ff}} \rightarrow \rho_0 c v_0 \frac{a}{r} e^{-ik(r-a)} \sum_{m=0}^{M} V_m P_m \left(\cos \psi\right) \qquad (7.2.24)$$

The sum is the truncated representation of $V\left(\psi\right)/v_0$, so the corresponding directivity is $\mathcal{D} = |V\left(\psi\right)|/v_0$. In other words, the directivity replicates the pattern of the sphere's radial velocity pattern.

This limiting behavior is readily explained. Because $P_m\left(\eta\right)$ is an mth degree polynomial, it follows that $P_m\left(\cos\psi\right)$ may be represented as a Fourier series in ψ whose highest harmonic is $\cos\left(m\psi\right)$. The wavelength of a surface vibration on the sphere at this harmonic is the circumference divided by m, $\lambda_m = 2\pi a/m$. The shortest wavelength that is significant to the vibration is λ_M. The acoustic wavelength is $2\pi/k$. Thus, defining high frequency to be $ka \gg M$ is equivalent to stating that the high-frequency regime is such that all significant λ_m are much greater than $2\pi/k$. In other words, spherical harmonics whose surface wavelength is much larger than the acoustic wavelength are responsible for most of the radiation to the farfield. This is reminiscent of the behavior that was observed for surface waves on an infinite planar surface, see Sect. 5.1.

EXAMPLE 7.3 A spherical cap is the region subtending an angle ψ_0 relative to a radial line that is designated as the z-axis. This cap executes a translational vibration Re $\left(v_0 e^{i\omega t}\right)$ parallel to z-axis, whereas the remainder of the sphere is rigid and stationary. The corresponding complex amplitude of the radial velocity on the surface $r = a$ is $V_r = v_{\text{cap}} \bar{e}_z \cdot \bar{e}_r = v_0 \cos \psi$ if $0 \le \psi \le \psi_0$, $V_r = 0$ if $\psi_0 < \psi \le \pi$. Derive expressions for the coefficients V_m of the spherical harmonic series. Then, evaluate these coefficients for the case where $\psi_0 = 30°$, and identify the harmonic number N for which a reconstruction of V_r deviates by no more than $0.2v_0$ from the actual V_r. Use this value of N to evaluate the pressure along the z-axis, $\psi = 0$, for $ka = 12$. Use that result to estimate the value of kr at which the farfield begins. Then, use the same value of N to evaluate the farfield directivity, and compare that result to the high-frequency asymptotic trend.

Significance

A simplified version of this problem has appeared in many books. The simplification entails considering the velocity of the spherical cap to be radial, rather than being proportional to $\cos\psi$. The availability of mathematical software, coupled with a little ingenuity regarding the application of the recurrence relations for Legendre polynomials, makes it possible to carry out a consistent analysis. The outcome will show the utility of some of the asymptotic properties of the pressure field.

Solution

We begin by evaluating the V_m coefficients, which are given by Eq. (7.2.4). The function of interest here is zero beyond $\psi = \psi_0$, so we have

$$V_m = \left(m + \frac{1}{2}\right) \int_0^{\psi_0} \cos\psi\, P_m(\cos\psi)\sin\psi\, d\psi$$

$$= \left(m + \frac{1}{2}\right) \int_{\cos\psi_0}^1 \eta P_m(\eta)\, d\eta \tag{1}$$

The question now is how to evaluate the integral? We could use a numerical integration algorithm. If we were to follow that route, we would encounter difficulty because a rather large number of terms are required to meet the imposed convergence criterion, and $P_m(\eta)$ oscillates rapidly when m is large. An alternative is to replace $P_m(\eta)$ with its series representation. Here too, large m would lead to difficulty because the factorials in the numerator and denominator of the series for $P_m(\eta)$ would be very large. Our approach relies on the recurrence relations. We use the first of Eq. (7.1.14) to replace $\eta P_m(\eta)$ with term that are purely Legendre polynomials. To that result, we will apply the third of Eq. (7.1.14) in order to replace those polynomials with derivatives of Legendre polynomials. These derivatives may be integrated directly. The operations are

$$V_m = \left(m + \frac{1}{2}\right) \int_{\cos\psi_0}^1 \frac{(m+1)P_{m+1}(\eta) + mP_{m-1}(\eta)}{2m+1}\, d\eta$$

$$= \frac{1}{2}\int_{\cos\psi_0}^1 \left\{ \frac{m+1}{2m+3}\left[\frac{d}{d\eta}P_{m+2}(\eta) - \frac{d}{d\eta}P_m(\eta)\right] \right.$$

$$\left. + \frac{m}{2m-1}\left[\frac{d}{d\eta}P_m(\eta) - \frac{d}{d\eta}P_{m-2}(\eta)\right] \right\} d\eta$$

All Legendre polynomials are unity at $\eta = 1$, so we find that

$$V_m = -\frac{1}{2}\left[\frac{m+1}{2m+3}P_{m+2}(\cos\psi_0) + \frac{(2m+1)}{(2m+3)(2m-1)}P_m(\cos\psi_0) \right.$$

$$\left. - \frac{m}{2m-1}P_{m-2}(\cos\psi_0) \right] \tag{2}$$

These relations are valid for $m \geq 2$. The first two coefficients may be found directly by substituting $P_0(\eta) \equiv 1$ and $P_1(\eta) \equiv \eta$. The result is

$$V_0 = \frac{1}{4}\left[1 - (\cos\psi_0)^2\right], \quad V_1 = \frac{1}{2}\left[1 - (\cos\psi_0)^3\right] \tag{3}$$

For comparison, the simplified version, in which $V_r = v_0$, would not have the η factor in the integrand. The third of the recurrence relations in Eq. (7.1.14) would suffice in that case, which would lead to

$$(V_m)_{\text{simp}} = \left(m + \frac{1}{2}\right) \int_{\cos\psi_0}^{1} P_m(\eta)\, d\eta$$

$$= -\frac{1}{2}\left[P_{m+1}(\cos\psi_0) - P_{m-1}(\cos\psi_0)\right], \quad m \geq 1 \qquad (4)$$

$$(V_0)_{\text{simp}} = \frac{1}{2}\left[1 - (\cos\psi_0)\right]$$

The consistent and simplified V_m coefficients are described by Fig. 1. The two sets of values are quite close. This is a consequence of the fact that for $0 \leq \psi \leq 30°$, the value of $\cos\psi$ is close to unity. The discrepancy would grow if ψ_0 were larger.

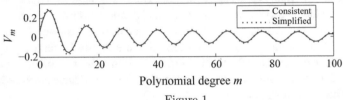

Figure 1.

The V_m values decrease with increasing m, but slowly and in an oscillatory manner. As we add more terms to the series for $V(\psi)$, sometimes the additional terms will add, and sometimes they will subtract. Consequently, examination of the magnitude of the coefficients gives an uncertain picture of the significance of each term in the series. We must assess convergence by constructing $V(\psi)$ according to Eq. (7.2.3). Let $V(\psi, N)$ denote the value obtained when the series is truncated at $m = N$. The error metric of interest is $|V(\psi, N) - \cos(\psi)[h(\psi) - h(\psi - \psi_0)]|$. The following table lists the maximum error for $0 \leq \psi \leq \pi$ for several N.

N	20	40	60	80	100
Error (%)	37	32	25	21	16

The error at $N = 80$ is very close to the requested value $0.2v_0$, so that is what we will use for the remainder of this investigation.

Figure 2 depicts the reconstruction of $V(\psi)/v_0$ obtained for two values of N. The synthesis of Eq. (7.2.3) can be done with the matrix algorithm described in Example 7.1. Let $\{V\}$ be the column vector of the $N + 1$ velocity coefficients V_m, and let $\{\psi\}$ be a set of J polar angles at which it is desired to evaluate the series, so that $\{\eta\} = \cos(\{\psi\})$. A rectangular $J \times (N + 1)$ array $[L]$ is defined such that column m is $P_m(\{\eta\})$ for $m = 0, 1, ..., N$. Then, the radial velocity on the surface is given by $\{V_{\text{surf}}\} = [L]\{V\}$. The result for $N = 20$ exhibits a good resemblance of the actual function, while $N = 80$ is very close, with a minor Gibb's phenomenon at $\psi = \psi_0 = 30°$.

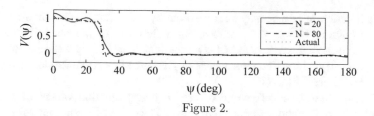

Figure 2.

The implication thus far is that a good representation of the surface velocity is necessary to describe the pressure field adequately, but this is not so. Let us rewrite the solution for pressure in Eq. (7.2.7) as

$$P_{\text{ff}} = \rho_0 c v_0 \sum_{m=0}^{\infty} \Gamma_m h_m \left(kr \right) P_m \left(\cos \psi \right), \quad \Gamma_m = \frac{V_m}{i h'_m \left(ka \right)}$$

Correspondingly, Eq. (7.2.14) for the directivity and Eq. (7.2.17) for the radiated power become

$$\mathcal{D} \left(\psi \right) = \frac{1}{ka} \left| \sum_{m=0}^{\infty} \Gamma_m e^{im\pi/2} P_m \left(\cos \psi \right) \right|$$

$$\mathcal{P} = \pi \rho_0 a^2 c v_0^2 \sum_{m=0}^{\infty} \frac{2}{2m+1} \frac{|\Gamma_m|^2}{(ka)^2}$$

The Γ_m values indicate which spherical harmonics are important to the pressure field. Figure 3 displays the Γ_m values and the magnitude of each term in the series for radiated power. These graphs would be unaltered if we truncated the series for $V (\psi)$ at $N = 20$, so it is evident that the small scale details of the surface motion are unimportant for the radiated field.

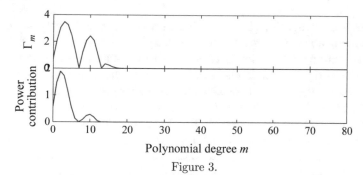

Figure 3.

The next feature to examine is the farfield and where it begins. The primary aspect of the farfield in any case is that the pressure decays as $1/r$ along any radial line. Thus, a plot of $kr \left| P/(\rho_0 c v_0) \right|$ as a function of kr will approach a constant value as the kr reaches into the farfield. The only radial line we were requested to consider is $\psi = 0$, but one should carry out such computations along several lines in order to assure that what is observed on a specific line is not anomalous. An evaluation of

$|P|$ as a function of kr also may be carried out by the matrix algorithm. Specifically, $\{kr\}$ is a column vector of kr_j values, and column m of $[h]$ holds the values of $h_m(\{kr\})$ for $m = 0, ..., N$, and $[L]$ is a diagonal $N + 1$ array whose elements are the $P_m(\cos\{\psi\})$ values. Then, the values of pressure at locations kr_j along the radial line at fixed ψ are $[h][L]\{V\}$.

The center line is $\psi = 0$, so all of the Legendre function values are one. The result of this computation is graphed in Fig. 4. The onset of a constant value of $kr\,P$ is gradual, but it seems that a good estimate would be that the product approaches a constant value in the vicinity of $kr = 100$. The value of ka is 12, so the observation that the farfield begins at $kr = 100$ suggests that the farfield begins at a kr value whose order of magnitude is $(ka)^2$. This turns out to be a good empirical rule in general. Specifically, if a is a measure of the size of an object that radiates an acoustic signal, then the farfield will begin at a nondimensional distance $kr_{\mathrm{ff}} = \beta(ka)^2$, where β is a value having unit order of magnitude. The identified value of β for the present system is approximately 0.75.

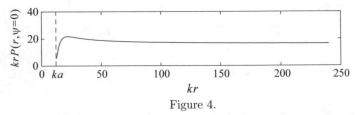

Figure 4.

The last property to consider is the farfield directivity. From Eq. (7.2.14), we have

$$\mathcal{D}(\psi) = \frac{1}{ka}\left|\sum_{m=0}^{\infty}\Gamma_m e^{i(m+1)\pi/2}P_m(\cos\psi)\right|$$

The matrix-based algorithm for evaluating a series may also be implemented here. The directivity obtained from the consistent and simplified analyses appear in Fig. 5. To avoid obscuring details, the data has been displayed as a rectangular plot, rather than the standard polar plot. It is not surprising, given the closeness of the two sets of coefficients, that the directivity obtained from the simplified representation of $V(\psi)$ is quite good. We can see a hint of the high-frequency limit, in which the directivity is proportional to the surface velocity. However, the value at $\psi = 0$ is 40% greater than the limiting value, and $\mathcal{D}$ does not drop abruptly from $\cos\psi_0$ to zero at ψ_0.

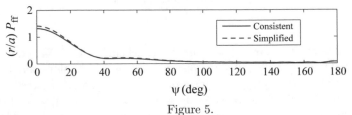

Figure 5.

7.2.3 Symmetry Plane

Anyone who is interested in sound reproduction systems is aware that speakers usually are mounted on a baffle. A baffle is a boundary, usually flat, whose purpose is to direct the radiated sound to one side. A system that features a sphere in a baffle is a hemispherical dome tweeter, which is commonly used in speaker systems to create and radiate the higher frequencies in the audible spectrum. In this arrangement, a circular cap is mounted on the baffle, but is free to vibrate. In practice, the cap might not be a full hemisphere, but we will employ that idealization because it has an analytical solution. The example will compare the radiated field obtained with and without the baffle.

A hemispherical dome tweeter is not the only configuration in which a baffle guides spherical waves. A great simplification of a surface ship represents it as a sphere. Suppose that the sphere's buoyancy is such that its equator coincides with the free surface. The characteristic impedance of water is much greater than that of air. To a wave in the water, the surface appears to be pressure-release, whereas it appears to be rigid to an acoustic wave in the air. Thus, the model for acoustic radiation from the (idealized) ship into the air would consider a hemisphere in an infinite, planar, rigid baffle. In contrast, the model for acoustic radiation from the ship into the water would consider a hemisphere in an infinite, planar, pressure-release baffle.

Figure 7.3 describes the geometry of a hemisphere in an infinite baffle. The portion of the boundary that is the hemisphere is defined by $r = a$, $0 \leq \psi < \pi/2$, and the baffle is defined by $r > a$, $\psi = \pi/2$. Velocity continuity requires that the particle velocity matches the hemisphere's radial velocity. In addition, if the baffle is rigid, the particle velocity normal to the baffle must vanish. Because $\bar{e}_r$ is parallel to the baffle and $\bar{e}_\psi$ is perpendicular to it, this requirement is that $\bar{V} \cdot \bar{e}_\psi = 0$ at $\psi = \pi/2$ for $r > a$. The case of a pressure-release baffle requires that $P = 0$ at the same locations. Thus, the problem to solve is

Fig. 7.3 A hemisphere mounted on an infinite planar baffle

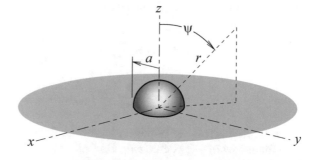

$$
\left\{
\begin{array}{l}
\nabla^2 P + k^2 P = 0, \quad r > a, \, 0 \le \psi < \pi/2 \\[4pt]
\dfrac{\partial P}{\partial r} = -i\omega\rho_0 V_h, \quad r = a, \, 0 \le \psi < \pi/2 \\[6pt]
\left.
\begin{array}{l}
\text{Rigid baffle: } \dfrac{\partial P}{\partial \psi} = 0 \\[6pt]
\text{Pressure-release baffle: } P = 0
\end{array}
\right\}, \quad r > a, \, \psi = \pi/2
\end{array}
\right.
\tag{7.2.25}
$$

where V_h is the complex amplitude of the radial velocity of the hemisphere.

The response with either type of baffle may be obtained as a special case of a general axisymmetric wave. To see how to do so, we recall the method of images. If the boundary of an infinite half-space is rigid, the field may be obtained by placing a mirror image(s) behind the boundary of the actual source(s), with all images vibrating in-phase with their matching actual source. Here, the image of a hemisphere is also hemisphere, so the source and its image form a full sphere. If V_h (ψ) is the radial velocity at location ψ on the hemisphere, then the mirror image point is at $\pi - \psi$. Hence, the radial velocity distribution for the image sphere is

$$
\text{Rigid baffle: } V\,(\psi) =
\begin{cases}
V_h\,(\psi), & 0 \le \psi < \pi/2 \\
V_h\,(\pi - \psi), & \pi/2 \le \psi < \pi
\end{cases}
\tag{7.2.26}
$$

In the case where the boundary of a half-space is pressure-release, the mirror image hemisphere vibrates 180° out-of-phase relative to the source hemisphere. Thus, the radial velocity of the full sphere in this case is

$$
\text{Pressure-release baffle: } V\,(\psi) =
\begin{cases}
V_h\,(\psi), & 0 \le \psi < \pi/2 \\
-V_h\,(\pi - \psi), & \pi/2 \le \psi < \pi
\end{cases}
\tag{7.2.27}
$$

The acoustic field that is radiated must share the same symmetry properties as the excitation. Thus, if the baffle is rigid, the pressure must be an even function of ψ relative to $\psi = \pi/2$, in other words, $P\,(r, \pi - \psi) = P\,(r, \psi)$. This requires that the only nonzero spherical harmonics in the general solution, Eq. (7.2.1), be those whose degree is even, P_0 (cos ψ), P_2 (cos ψ), ... Thus, the method of images tells us that the field radiated by a hemisphere mounted on an infinite rigid baffle must be describable as

$$
\begin{aligned}
P &= \rho_0 c v_0 \sum_{m=0}^{\infty} V_{2m} \frac{h_{2m}\,(kr)}{i h'_{2m}\,(ka)} P_{2m}\,(\cos \psi) \\
V_r &= v_0 \sum_{m=0}^{\infty} V_{2m} \frac{h'_{2m}\,(kr)}{h'_{2m}\,(ka)} P_{2m}\,(\cos \psi)
\end{aligned}
\tag{7.2.28}
$$

To verify the correctness of this ansatz, we evaluate V_ψ at $\psi = \pi/2$. This should be zero for any set of V_{2m} coefficients. By Euler's equation, this condition is obtained if $\partial P/\partial \psi = 0$ at this surface. The only quantity that depends on ψ in the preceding is P_{2m} (cos ψ). By virtue of being an even function relative to $\psi = \pi/2$, it follows that $(\partial/\partial \psi)\, P_{2m}$ (cos ψ) $\equiv 0$ at $\psi = \pi/2$, which verifies that the boundary condition is satisfied identically.

Similar reasoning applies in the case where the hemisphere is mounted on a pressure-release baffle. The pressure in this case is an odd function of ψ relative to $\psi = \pi/2$, so that $P(r, \pi - \psi) = -P(r, \psi)$. The Legendre functions that fit this symmetry specification are those whose degree is odd. Thus, the field radiated by a hemisphere in an infinite pressure-release baffle is given by

$$
\begin{aligned}
P &= \rho_0 c^2 \sum_{m=0}^{\infty} V_{2m+1} \frac{h_{2m+1}(kr)}{ih'_{2m+1}(ka)} P_{2m+1}(\cos \psi) \\
V_r &= c \sum_{m=0}^{\infty} V_{2m+1} \frac{h'_{2m+1}(kr)}{h'_{2m+1}(ka)} P_{2m+1}(\cos \psi)
\end{aligned}
\tag{7.2.29}
$$

The odd-degree Legendre functions vanish at the midpoint of their range, $P_{2m+1}(0) \equiv 0$, so the above representation identically satisfies the boundary condition that $P = 0$ on the baffle.

The velocity coefficients in Eqs. (7.2.28) and (7.2.29) describe the composite $V(\psi)$ function in Eq. (7.2.26) or (7.2.27). An alternative to integrating over this sphere is to use orthogonality over the hemisphere. The even-degree Legendre polynomials are a mutually orthogonal set over $0 \le \psi \le \pi/2$, as are the odd-degree polynomials. (The even and odd sets are not orthogonal with respect to each other in this range.) Thus, we have

$$
\begin{aligned}
\int_0^1 P_{2m}(\eta) P_{2j}(\eta) \, d\eta &= \int_0^{\pi/2} P_{2m}(\cos \psi) P_{2j+1}(\cos \psi) \sin \psi \, d\psi \\
&= \frac{1}{2(2m)+1} \delta_{jn} \\
\int_0^1 P_{2m+1}(\eta) P_{2j+1}(\eta) \, d\eta &= \int_0^{\pi/2} P_{2m+1}(\cos \psi) P_{2j+1}(\cos \psi) \sin \psi \, d\psi \\
&= \frac{1}{2(2m+1)+1} \delta_{jn}
\end{aligned}
\tag{7.2.30}
$$

When we apply these orthogonality properties to the respective series for V_r, we find that the velocity coefficients are given by

$$
\text{Rigid baffle: } V_{2m} = (4m+1) \int_0^{\pi/2} \frac{V(\psi)}{c} P_{2m}(\cos \psi) \sin \psi \, d\psi
$$
$$
\text{Pressure-release baffle: } V_{2m+1} = (4m+3) \int_0^{\pi/2} \frac{V(\psi)}{c} P_{2m+1}(\cos \psi) \sin \psi \, d\psi
$$
$$
\tag{7.2.31}
$$

The representations in Eqs. (7.2.28) and (7.2.29) are like those for a sphere in an unbounded space, except that only the even or odd spherical harmonics are relevant. The fact that the domain is a half-space affects the radiated power. The power is obtained by integrating the time-averaged intensity over a large hemisphere. Because $(I_r)_{\text{av}}$ is formed by multiplying P and V_r^* product, and both are either even or odd functions relative to $\psi = \pi/2$, $(I_r)_{\mathrm{av}}$ is an even function. Thus, the integrals over the hemisphere are half the value of what they would be over the full sphere. Correspondingly, the time-averaged radiated power is obtained by halving Eq. (7.2.17) and using only the appropriate set of V_j coefficient. In other words

$$\mathcal{P} = \begin{cases} \pi \rho_0 a^2 c v_0^2 \dfrac{1}{(ka)^2} \displaystyle\sum_{m=0}^{\infty} \dfrac{1}{4m+1} \dfrac{|V_{2m}|^2}{\left|h'_{2m}(ka)\right|^2} : \text{rigid} \\[4mm] \pi \rho_0 a^2 c v_0^2 \dfrac{1}{(ka)^2} \displaystyle\sum_{m=0}^{\infty} \dfrac{1}{4m+3} \dfrac{|V_{2m+1}|^2}{\left|h'_{2m+1}(ka)\right|^2} : \text{pressure-release} \end{cases} \qquad (7.2.32)$$

Somewhere in the midst of this discussion, you might have wondered about the relevance of the concept of an infinite baffle, given that nothing has infinite extent. An important general aspect is that the pressure on a baffle surrounding a radiator tends to fall off rapidly with increasing distance from the center. Thus, the baffle only needs to extend sufficiently far to attain this decrease. In most cases, especially at high frequencies, this distance has the order of magnitude of an acoustic wavelength, which corresponds to an outer radius of the baffle being $R_{\mathrm{baffle}} = a\,(1 + 2\pi/ka)$. In some designs, the apparatus used to mount the transducer is sufficiently large that it is a good approximation of an infinite baffle.

EXAMPLE 7.4 A tweeter consists of a hemisphere that translates as a rigid body at $v_0 \sin(\omega t)$ in the direction of the axis of symmetry. It is desired to compare the field that this transducer radiates when it is mounted on an infinite rigid or pressure-release baffle to a reference configuration. The latter places the sphere in an unbounded fluid with translational motion only over the forward half, $0 \le \psi \le \pi/2$, with the back half, $\pi/2 < \psi \le \pi$, stationary. Compare the farfield distribution $(r/a)\,|p|\,/\,(\rho_0 c v_0)$ for both types of baffles to that obtained from the reference sphere. Also, evaluate the radiated power and the pressure distribution acting on the surface of the hemisphere for each system. Frequencies of interest are $ka = 1$, 4 and 20.

Significance

In addition to gaining experience working with spherical harmonics, this example will exhibit some important general features of acoustic radiation and provide insight to the behavior of a common device.

Solution

In each configuration, the complex amplitude of the radial velocity for $0 \le \psi \le \pi/2$ is $V_r = v_0 \cos\psi$. Both baffle cases give rise to an image of the hemisphere that covers $\pi/2 < \psi \le \pi$, thereby creating a virtual complete sphere. For a rigid baffle, the radial velocity must be the same on the image, so that $V_r = -v_0 \cos\psi$. The union of these for the virtual sphere is $V_r = v_0\,|\cos\psi|$. For a pressure-release baffle, the radial velocity for $\psi = \pi/2 + \Delta > \pi/2$ must be equal in magnitude but in the opposite sense from the value at $\psi = \pi/2 - \Delta$. Thus, the radial velocity on the virtual sphere is $V_r = v_0 \cos \psi$. The radial velocity on the reference full sphere is $V_r = v_0 \cos \psi$ for $0 \le \psi \le \pi/2$ and $V_r = 0$ for $\pi/2 < \psi \le \pi$.

The radial velocity on the virtual sphere for the rigid baffle case is an even function with respect to $\psi = \pi/2$. This means that only the only nonzero velocity coefficients are V_{2m}, $m = 0, 1, 2, ...$ According to the first of Eq. (7.2.31), these coefficients are

$$V_{2m} = (4m + 1) \int_0^{\pi/2} (\cos \psi)\, P_{2m} (\cos \psi) \sin \psi d\psi$$

This is like Eq. (1) in Example 7.3 with $\psi_0 = \pi/2$, except that $2m$ replaces m and the integral is doubled. When we make these alterations, the result is

$$(V_{2m})_{\text{rigid}} = -\left[\frac{2m + 1}{4m + 3} P_{2m+2} (0) + \frac{(4m + 1)}{(4m + 3)(4m - 1)} P_{2m} (0)\right.$$
$$\left. - \frac{2m}{4m - 1} P_{2m-2} (0)\right], \quad m > 0$$

$$(V_0)_{\text{rigid}} = \frac{1}{2}$$

In the case of a pressure-release baffle, the radial velocity function for the equivalent full sphere is an odd function relative to $\psi = \pi/2$. In other words, it is $\cos \psi$. Thus, the surface velocity of the image sphere is that of a sphere that translates as a rigid body. Because $P_1 (\cos \psi) \equiv \cos \psi$, orthogonality of the Legendre functions informs us that $V_1 = 1$ is the only nonzero coefficient.

The radial velocity on the reference full sphere has no special symmetry properties with respect to $\psi = \pi/2$. The velocity coefficients are given directly by Eq. (7.2.4) for $m = 0, 1, 2, ...$ Setting $V_r = 0$ for $\psi > \pi/2$ leads to

$$V_m = \left(m + \frac{1}{2}\right) \int_0^1 \eta P_m (\eta) \, d\eta$$

This equation is the same as Eq. (1) in Example 7.3 with $\psi_0 = \pi/2$, so we quote the results directly as

$$(V_m)_{\text{ref}} = -\frac{1}{2} \frac{v_0}{c} \left[\frac{m + 1}{2m + 3} P_{m+2} (0) + \frac{(2m + 1)}{(2m + 3)(2m - 1)} P_m (0)\right.$$
$$\left. - \frac{m}{2m - 1} P_{m-2} (0)\right], \quad m > 1$$

$$(V_0)_{\text{ref}} = \frac{1}{4} \frac{v_0}{c}, \quad V_1 = \frac{1}{2} \frac{v_0}{c}$$

Despite the similarity of the surface velocity in the three configurations, the V_j coefficients are significantly different, as is shown in Fig. 1. This figure also shows that truncation of the spherical harmonic series at $N = 12$ should be adequate for the rigid and reference cases.

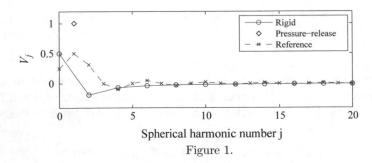

Figure 1.

The computation of the directivity proceeds as though we had a full sphere, except that ψ only extends to $\pi/2$ when there is a baffle, and the velocity coefficients are those of the relevant configuration. Equation (7.2.14) for the three cases gives

$$\mathcal{D}_{\text{rigid}}(\psi) = \frac{1}{ka}\left|\sum_{m=0}^{\infty}\frac{(V_{2m})_{\text{rigid}}}{ih'_{2m}(ka)}e^{i(2m+1)\pi/2}P_{2m}(\cos\psi)\right|$$

$$\mathcal{D}_{\text{p.r.}}(\psi) = \frac{1}{ka}\left|\frac{(V_1)_{\text{p.r.}}}{ih'_1(ka)}\right|\cos\psi = \frac{v_0}{c}\frac{(ka)^2}{\left|2+2ika-(ka)^2\right|}\cos\psi$$

$$\mathcal{D}_{\text{ref}}(\psi) = \frac{1}{ka}\left|\sum_{m=0}^{\infty}\frac{(V_m)_{\text{ref}}}{ih'_m(ka)}e^{i(m+1)\pi/2}P_m(\cos\psi)\right|$$

The directivities for each specified frequency are displayed in Fig. 2.

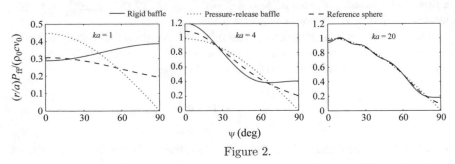

Figure 2.

Because the pressure-release baffle corresponds to an image sphere that oscillates as a rigid body in the z direction, the directivity is always proportional to $\cos\psi$, but its maximum value is frequency dependent. At the lowest frequency, $ka = 1$, the rigid baffle and the reference sphere radiate approximately uniform. This is consistent with the general observation that a body that executes a low-frequency vibration acts like a monopole whose strength is proportional to the body's volume velocity. In contrast, at the highest frequency, $ka = 20$, each directivity fits the asymptotic trend that it will be proportional to the surface velocity distribution.

The radiated power in each case may be found from Eq. (7.2.17) using the appropriate set of V_j coefficients. For the configurations where there is a baffle, the formula is halved because a baffle reduces the domain to a half-space. The results for $\mathcal{P}/\left(\pi\rho_0 a^2 c v_0^2\right)$ are tabulated below.

	$\mathcal{P}$ (nondim.)		
ka	1	4	20
Rigid baffle	0.12588	0.32242	0.33331
Pressure-release baffle	0.066667	0.32821	0.33333
Reference sphere	0.096272	0.32531	0.33332

At low frequencies, the rigid baffle configuration gives twice as much power radiation as the pressure-release case, while the reference sphere is midway between the two baffle configurations. In contrast, at high frequencies, all configurations give the same radiated power. This is to be expected because the directivity of all configurations at large ka is essentially the same for $\psi < 90°$, and it is close to zero for $\psi > 90°$ in the case of the reference sphere.

The evaluation of surface pressure follows a direct implementation of Eq. (7.2.7), with the V_j coefficients for the rigid baffle case being those for even j, while the only nonzero coefficient for the pressure-release baffle is V_1. Of course, only $0 \le \psi \le \pi/2$ is relevant. Here too, the matrix algorithm for the evaluation of a series representation may be applied.

The radial velocity on the surface is real and positive. Thus, Re (P) is the resistive part, and Im (P) is the reactive part, with Im $(P) > 0$ for an inertance. The patterns in Fig. 3 indicate that when the frequency is low, the surface pressure for both types of baffles has a large inertance. The resistive part for the rigid baffle at low frequency is approximately uniform, whereas Re (P) is proportional to $\cos \psi$ for the pressure-release baffle. High frequencies lead to a surface pressure that is $\rho_0 c v_0 \cos \psi$ for both configurations. This occurs because the surface, $r = a$, effectively is in the farfield when ka is very large.

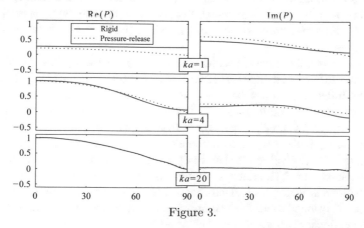

Figure 3.

In systems for sound reproduction, it usually is desirable to emit the sound omni-directionally. This is the role of a hemispherical dome tweeter. Because a is relatively small (typically less than 40 mm), the value of ka will be small even at the upper limits of the audible spectrum. Thus, their use leads to radiation of high frequencies over a broad angular range. The rigid baffle enhances the radiated power at low frequencies. A designer would be interested in the surface pressure for design purposes. For example, the fact there is a large reactive part means that the electronic amplifier that drives the tweeter must be able to output an instantaneous power that is substantially greater than the time-averaged power.

7.2.4   Interaction with an Elastic Spherical Shell

A situation we have not yet addressed arises when the surface vibration is produced by a force applied to an elastic body. The vibration in that case must be determined by solving laws of structural dynamics. Such an analysis must account for the pressure applied to the surface by the acoustic signal that is generated. But that pressure depends on the surface vibration. This is not circular logic. Rather, it requires a simultaneous approach that concurrently formulates the governing acoustic and structural equations. Problems such as this fall into the category of *fluid–structure interaction*. If the shape of the body and the nature of the structure are complicated, as in the case of a submarine, then accurate determination of the radiated field at all frequencies would require the most powerful computational tools available, and even such resources might not be adequate. The decoupling properties of spherical harmonics make the analysis of radiation from elastic spheres quite tractable.

We previously encountered an elastic plate, which refers to a planar sheet whose thickness is much less than the dimensions of the sheet. A *shell* consists of a thin sheet whose surface is curved. Doing so alters the nature of the internal stresses in comparison to a plate, which thereby complicates the equations of motion. We will begin with a *membrane shell model*, in which the internal stresses come from stretching the sheet.

To some extent, the stresses resemble how a balloon carries a load, but there is a fundamental difference. In the case of a balloon, the sheet is inflated before it is subject to loading. Inflation stretches the sheet, thereby inducing a prestress. Displacement of the sheet produces additional stress. The work that is done in this displacement, which is stored as potential energy, comes from displacing the system in opposition to the total stress. In the case of a membrane shell, there is no prestress, so the internal work stems from displacing the shell in opposition to the stretching stress induced by that displacement. A simpler system that has the same alternatives is a cable. The inflated balloon is analogous to a tensioned cable, whereas a membrane shell is like a taut, but untensioned cable. Interestingly, the taut cable has no load carrying ability according to linear elasticity theory. This is not true for the membrane shell because of its curvature.

The membrane shell theory we will employ becomes progressively less valid as the thickness/curvature ratio increases. It also is not valid if the frequency is so high that the wavelength of the surface displacement is comparable to the thickness. Improved shell theories that address both situations are available. (The author once joked to a colleague that there might be more improved shell theories than the number of people who have been involved in the development of those theories.) We shall begin with membrane shell theory because it is the simplest model. However, a breakdown of this theory will cause an anomaly of the series, which we will address by introducing a correction that accounts for flexural effects.

Equations of Motion

Figure 7.4 shows the midsurface of a spherical shell, which is the surface that is midway between the outer and inner surfaces. We will limit our consideration to situations in which the shell response is axisymmetric. This means that the displacement of a point at (a, ψ) on the midsurface will consist of a radial component w and meridional displacement u tangent to the surface. A positive displacement component is one that moves the midsurface in the sense of $\bar{e}_r$ or $\bar{e}_\psi$. The restriction to axisymmetric motions limits u and w to be functions of ψ and t only. The shell's thickness h is much less than a. Thus, we may also use a as the radius of the outer surface when we formulate the equations governing the acoustic field exterior to the sphere.

Fig. 7.4 Definition of the radial displacement w and polar displacement u of a spherical shell

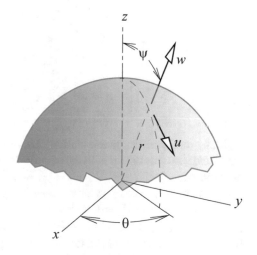

Junger and Feit[7] employed equations of motion that incorporate the effects of flexural rigidity. Omitting those effects leads to the following simplified set of equations of membrane shell theory,

[7]M.C. Junger and D. Feit, *Sound, Structures, and Their Interaction*, Acoustical Society; Second edition, pp. 229–230 (1993 reprint).

$$c_{\mathrm{e}}^2 \left[\frac{\partial^2 u}{\partial \psi^2} + \cot \psi \frac{\partial u}{\partial \psi} - \left(\nu + (\cot \psi)^2 \right) u + (1 + \nu) \frac{\partial w}{\partial \psi} \right] - a^2 \frac{\partial^2 u}{\partial t^2} = 0$$

$$-c_{\mathrm{e}}^2 (1 + \nu) \left[\frac{\partial u}{\partial \psi} + (\cot \psi) u + 2w \right] - a^2 \frac{\partial^2 w}{\partial t^2} = \frac{a^2}{\rho_{\mathrm{e}} h} \left(p_{\mathrm{acoustic}}|_{r=a} - q_{\mathrm{applied}} \right)$$

$$(7.2.33)$$

In this expression, q_{applied} represents the outward radial force per unit surface area that is the excitation, ρ_{e} is the density of the shell's material, ν is Poisson's ratio, and c_{e} is the phase speed of an extensional wave of plane strain in a plate of the same thickness,

$$c_{\mathrm{e}} = \left[\frac{E}{\rho_{\mathrm{e}} \left(1 - \nu^2 \right)} \right]^{1/2} \tag{7.2.34}$$

We know that the pressure exterior to a sphere may be represented by a spherical harmonic series. Also, the acceleration term in the second of Eq. (7.2.33) is proportional to w in the case of a harmonic response. Because the w term in that equation is not differentiated with respect to ψ, it seems reasonable to try a Legendre function series for w. In both equations of motion, most of the u terms differ from the w terms by one differentiation with respect to ψ. This feature suggests that we should try a series representation of u, with the derivatives of the Legendre functions as the basis functions. Thus, the ansatz we begin with is

$$p = \mathrm{Re} \sum_{m=0}^{\infty} B_m h_m (kr) P_m (\cos \psi) e^{i\omega t}$$

$$w = \mathrm{Re} \sum_{m=0}^{\infty} W_m P_m (\cos \psi) e^{i\omega t} \tag{7.2.35}$$

$$u = \mathrm{Re} \sum_{m=0}^{\infty} U_m \frac{d}{d\psi} P_m (\cos \psi) e^{i\omega t}$$

Many mathematical manipulations are required to prove that the basis functions for w and u are consistent with the equations of motion, and to identify the equations governing the coefficients W_m and U_m. Both Baker[8] and Junger and Feit[9] apply the transformation $\eta = \cos \psi$ prior to identifying the basis functions. In both analyses, the series for w uses $P_m (\eta)$, which is equivalent to the above. The u basis function used by Junger and Feit is $\left(1 - \eta^2 \right)^{1/2} P_m' (\eta)$, whereas the basis function for u in Baker's analysis is the associated Legendre polynomial $P_m^1 (\eta)$. Equation (7.1.42) states this function is $P_m^1 (\eta) = (d/d\psi) P_m (\cos \psi) = - \left(1 - \eta^2 \right)^{1/2} P_m' (\eta)$. Thus, Baker's series matches the one used here, whereas Junger and Feit's work corresponds to U_m coefficients that are the negative of the present definition.

---

[8]W.E. Baker, "Axisymmetric Modes of Vibration of Thin Spherical Shells," *J. Acoust. Soc. Am.* **33**, 1749–1758 (1961).

[9]M.C. Junger and D. Feit, ibid.

The equations that result from substitution of Eq. (7.2.35) into the equations of motion, followed by application of the derivative identities in Eq. (7.1.14), are

$$\sum_{m=0}^{\infty} \left\{ \left[-K_{UU} + (k_e a)^2 \right] U_m + K_{UW} W_m \right\} P_m^1 (\eta) = 0$$

$$\sum_{m=0}^{\infty} \left\{ K_{WU} U_m + \left[-K_{WW} + (k_e a)^2 \right] W_m \right\} P_m (\eta)$$

$$= \frac{a^2}{\rho_s h c_e^2} \left[\sum_{m=0}^{\infty} B_m h_m (ka) P_m (\eta) - \hat{q}_{applied} \right]$$

(7.2.36)

where the K coefficients are functions of m given by

$$K_{UU} = m (m + 1) - (1 - \nu), \quad K_{UW} = (1 + \nu)$$
$$K_{WU} = (1 + \nu) m (m + 1), \quad K_{WW} = 2 (1 + \nu)$$

(7.2.37)

The wavenumber k_e is defined relative to c_e, rather than c, in order to avoid confusion when we consider the shell in a vacuum. Thus,

$$k_e = \frac{\omega}{c_e}, \quad k = \frac{c_e}{c} k_e$$

(7.2.38)

The first of Eq. (7.2.36) yields uncoupled equations because the $P_m^1 (\eta)$ are a linearly independent set. To uncouple the second equation, we employ the orthogonality property of Legendre functions, Eq. (7.1.16). The result of multiplying the equation by a specific $P_n (\eta)$, then integrating over $-1 \leq \eta \leq 1$, is

$$\left[K_{UU} - (k_e a)^2 \right] U_m - K_{UW} W_m = 0$$
$$-K_{WU} U_m + \left[K_{WW} - (k_e a)^2 \right] W_m = -\frac{a^2}{\rho_e h c_e^2} B_m h_m (ka) + \frac{a^2}{\rho_s h c_e^2} F_m$$

(7.2.39)

The F_m coefficients are those of a Legendre series for the applied excitation,

$$F_m = \frac{2m + 1}{2} \int_{-1}^{1} \hat{q}_{applied} P_m (\eta) \, d\eta$$

(7.2.40)

These values are considered to be known, so there are three unknowns in Eq. (7.2.39).

The additional equation comes from enforcing continuity at the sphere's surface. The relation between the spherical harmonic coefficients of pressure and radial velocity on a sphere's surface is provided by Eq. (7.2.6). The radial velocity of the surface is $\dot{w}$, so the velocity coefficients are $i\omega W_m$, from which it follows that

$$B_m = \rho_0 c \omega \frac{1}{h'_m (ka)} W_m \qquad (7.2.41)$$

We now have a solvable triad of equations. The most direct solution uses the preceding to eliminate B_m and the first of Eq. (7.2.36) to eliminate U_m. This leads to

$$U_m = \frac{K_{UW}}{K_{UU} - (k_e a)^2} W_m$$

$$\left[K_{WW} - (k_e a)^2 - \frac{K_{UW} K_{WU}}{K_{UU} - (k_e a)^2} + \frac{\rho_0 a}{\rho_s h} \left(\frac{c}{c_e}\right)^2 \frac{kah_m (ka)}{h'_m (ka)} \right] \frac{W_m}{a} = \frac{a}{h} \frac{F_m}{\rho_e c_e^2}$$

$$(7.2.42)$$

Terms have been grouped to emphasize the manner in which the system parameters affect the response. The pressure excitation is referenced to $\rho_e c_e^2$, which is a bulk modulus for the shell, and displacement is referenced to the radius. This nondimensional displacement depends parametrically on the density and sound speed ratios, Poisson's ratio ν, and the nondimensional frequency.

Evaluation of the response at any frequency entails solving the second of the above equations for the radial displacement coefficient W_m for a range of harmonic numbers m, then using those coefficients to evaluate the corresponding pressure coefficients B_m according to Eq. (7.2.41). The meridional displacement typically is not of interest, but if it were, we would use the first equation above to evaluate the U_m values. The displacement and pressure fields would be synthesized according to Eq. (7.2.35). The farfield description of the pressure is obtained by replacing $h_m (kr)$ by its asymptotic behavior for large kr, which yields

$$p = \mathrm{Re} \left\{ \frac{e^{-ikr}}{kr} \sum_{m=0}^{\infty} B_m e^{i(m+1)\pi/2} P_m (\cos \psi) e^{i\omega t} \right\} \qquad (7.2.43)$$

In-vacuo Vibration

An important property of the shell is its natural frequencies, at which the system may vibrate without application of an external excitation. No free system can execute a steady-state harmonic response if energy is removed from it. As we will soon see, that is what the fluid always does. Thus, the natural frequencies correspond to the harmonic motion of the in-vacuo shell when there is no excitation. The effect of fluid loading is described by the last term in the bracket in the second of Eq. (7.2.42). Dropping this term and setting $F_m = 0$ leads to the characteristic equation, which is a quadratic equation whose roots are the natural frequencies Ω_m of the mth spherical harmonic,

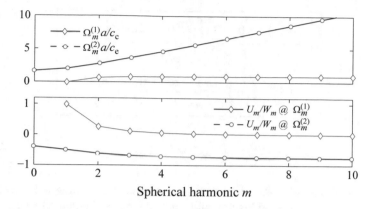

Fig. 7.5 Natural frequencies and modal displacement ratios of a spherical shell according to membrane shell theory

$$\chi\left(\frac{\Omega_m a}{c_e}\right) = \left[K_{WW} - \left(\frac{\Omega_m a}{c_e}\right)^2\right]\left[K_{UU} - \left(\frac{\Omega_m a}{c_e}\right)^2\right] - K_{UW}K_{WU} = 0$$

(7.2.44)

Thus, there are two natural frequencies $\Omega_m^{(1)} < \Omega_m^{(2)}$ at a specific m. At either natural frequency, Eq. (7.2.39) with $B_m = F_m = 0$ ceases to be linearly independent. Consequently, we may only determine the ratio U_m/W_m from either of those equations. When the natural frequencies are plotted as functions of m, they form two curves called *branches*. They are depicted in Fig. 7.5 along with the displacement ratio U_m/W_m for each branch. The common parlance is to say that the displacement pattern in a free vibration at a system's natural frequency is a *mode function* or *mode shape*. The radial displacement in a mode is the product of W_m and $P_m(\cos\psi)$, while the meridional displacement is the product of U_m and $(d/d\psi)P_m(\cos\psi)$.

The lower branch has no natural frequency at $m = 0$. (The mathematical solution is not physical, because it corresponds to an imaginary natural frequency.) At $m = 1$, the lower natural frequency is zero, which means that it corresponds to a rigid body mode in which the shell moves without deformation. This is verified by the modal displacement ratio $U_1^{(1)}/W_1^{(1)} = 1$, which corresponds to $w = \text{Re}\left(W_1^{(1)}\cos\psi e^{i\Omega^{(1)}t}\right)$, $u = \text{Re}\left(-W_1^{(1)}\sin\psi e^{i\Omega^{(1)}t}\right)$. This pattern is translational oscillation in the axial direction.

If the natural frequencies are known, the second of Eq. (7.2.42) in the case of forced response may be written as

$$\left[\frac{\left[\left(\kappa_m^{(1)} \right)^2 - (k_e a)^2 \right] \left[\left(\kappa_m^{(2)} \right)^2 - (k_e a)^2 \right]}{\left(\kappa_m^{(0)} \right)^2 - (k_e a)^2} \right.$$
$$\left. + \frac{\rho_0 a}{\rho_e h} \left(\frac{c}{c_e} \right)^2 \frac{k a h_m (ka)}{h_m' (ka)} \right] \frac{W_m}{a} = \frac{a}{h} \frac{F_m}{\rho_e c_e^2} \qquad (7.2.45)$$

where

$$\kappa_m^{(1)} \equiv \Omega_m^{(1)} \frac{a}{c}, \quad \kappa_m^{(2)} \equiv \Omega_m^{(2)} \frac{a}{c}, \quad \kappa_m^{(0)} = (K_{UU})^{1/2} \qquad (7.2.46)$$

The special case of a shell in a vacuum is obtained by setting $\rho_0/\rho_e = 0$. It is evident that removal of fluid loading will cause W_m to be singular when $k_e a$ equals $\kappa_m^{(1)}$ or $\kappa_m^{(2)}$. This condition marks a resonance, which we will examine more closely in the following section. In contrast, this equation indicates that W_m will be zero when $k_e a = \kappa_m^{(0)}$. In control system theory, $\omega = \Omega_m^{(1)}$ or $\omega = \Omega_m^{(2)}$ are "poles" of the in-vacuo system, and $\omega = (c_e/a) (K_{UU})^{1/2}$ is a "zero".

A troubling aspect of the displacement in any case becomes evident if we consider the value of W_m as m is increased with $k_e a$ fixed. According to Fig. 7.5, $\kappa_m^{(1)}$ tends to a constant value as m increases. Also, examination of Eq. (7.2.37) indicates that K_{UU} approaches m^2, so $\kappa_m^{(0)}$ approaches m. Consequently, for very large m at fixed ka, the first term in Eq. (7.2.45) approaches a constant value. The asymptotic expansion of $h_m (ka)$ and its derivative for large m, see Eq. (7.1.36), indicate that the second term also tends to a constant value. If the F_m coefficients do not decrease with increasing m, the W_m values will not decrease. In that event, none of the spherical harmonic series will converge.

The source of this difficulty is the usage of membrane shell theory. In particular, the fault lies in its prediction that the natural frequencies on the lower branch $\kappa_m^{(1)}$ approach a constant value as m is increased. Any shell theory that incorporates flexural deformation effects will yield natural frequencies that increase with increasing harmonic number. For this reason, we shall henceforth use the full theory presented by Junger and Feit. The parameter that scales flexural effects relative to membrane effects is

$$\beta^2 = \frac{h^2}{12 a^2} \qquad (7.2.47)$$

Fortunately, adoption of this improved shell theory only entails altering the definitions of the K coefficients, which now are

$$\begin{aligned} K_{UU} &= \left(1 + \beta^2 \right) [m (m+1) - (1 - \nu)] \\ K_{UW} &= (1 + \nu) + \beta^2 [m (m+1) - (1 - \nu)] \\ K_{WU} &= m (m+1) \left[(1 + \nu) + \beta^2 m (m+1) - \beta^2 (1 - \nu) \right] \\ K_{WW} &= 2 (1 + \nu) + \beta^2 m^2 (m+1)^2 - \beta^2 (1 - \nu) m (m+1) \end{aligned} \qquad (7.2.48)$$

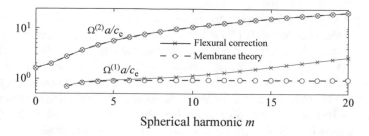

Fig. 7.6 Comparison of natural frequencies of a spherical shell according to membrane theory to those obtained when flexural effects are included, $a/h = 50$

The natural frequencies that result from inclusion of flexural effects are depicted in Fig. 7.6. There is no perceptible effect on the natural frequencies of the upper branch. The frequencies of the lower branch increase as expected, and the rate of increase is more rapid than a simple proportionality to m. This increase is sufficient to assure that the W_m values obtained from Eq. (7.2.45) decrease for sufficiently large m, regardless of the nature of the excitation, q_{applied}. Decreasing a/h increases the values of $\Omega_m^{(1)}$, but has little effect on the values of $\Omega_m^{(2)}$. The modal displacement ratios $U_m^{(j)}/W_m^{(j)}$ are not displayed because the values associated with either branch are indistinguishable from those in Fig. 7.5.

Even with the inclusion of flexural effects, an important limitation must be recognized. Simple shell theory is not valid if the wavelength along the surface is comparable to the thickness, because other deformation effects must be included. A qualitative guideline for validity may be identified from the property that $P_m (\cos \psi)$ has m zeros in the range $0 < \psi < \pi$. If we ignore the fact that these zeros are not equidistant, we arrive at a surface half-wavelength of $2\pi a/m$. A reasonable criterion is that this distance should be no less than $10h$. This leads to a guideline that the shell theory employed here can be expected to lose accuracy if $m > (\pi/5)\, a/h$. For example, if $a/h = 50$, it is desirable that the spherical harmonic series converge prior to $m = 32$. If they do not, we might obtain a mathematically convergent solution that does not accurately describe the actual physical response.

Interpretation

In the present context, the in-vacuo vibration properties provide a lens for interpreting the fluid-loaded response. To see why, let us compare Eq. (7.2.42) to the expression for the displacement of a one-degree-of-freedom oscillator subjected to a harmonically varying force. Let K, M, and D, respectively, denote stiffness, mass, and dashpot constants, nondimensionalized based on the usage of ka to represent frequency. In terms of a nondimensional velocity $V = ika\,(W/a)$, the steady-state amplitude of the oscillator is governed by

$$ZV = F, \quad Z = i\left(kaM - \frac{K}{ka}\right) + D \tag{7.2.49}$$

The imaginary part of the impedance is the reactance. It is negative if ka is below the natural frequency, which is $(K/M)^{1/2}$, and it is positive if ka exceeds the natural frequency. The dashpot term is a positive real value at any frequency. If the dashpot was not present, a frequency sweep would reveal a true resonance at the natural frequency, in which the value of V is infinite.[10] The dashpot removes energy at any frequency, so the response never shows a true resonance if the dashpot is present.

To apply this perspective to the spherical shell, we convert Eq. (7.2.45) to an impedance form by factoring $ik_e a$ out of the coefficient. The coefficients in the first line of this expression represent the effects of the structure, whereas the term in the second line is due to the fluid. Let us denote these contributions as the structural impedance $(Z_e)_m$ and the fluid surface impedance $(Z_f)_m$, respectively. Thus, we rewrite the equation as

$$
\left[(Z_e)_m + (Z_f)_m \right] \left(ik_e a \frac{W_m}{a} \right) = \frac{a}{h} \frac{F_m}{\rho_e c_e^2}
$$

$$
(Z_e)_m = \frac{k_e a}{i} \frac{\left[\left(\kappa_m^{(1)} \right)^2 / (k_e a)^2 - 1 \right] \left[\left(\kappa_m^{(2)} \right)^2 / (k_e a)^2 - 1 \right]}{\left(\kappa_m^{(0)} \right)^2 / (k_e a)^2 - 1}
$$

$$
(Z_f)_m = -i \frac{\rho_0 a}{\rho_e h} \left(\frac{c}{c_e} \right) \frac{h_m (ka)}{h_m' (ka)}
$$

(7.2.50)

The frequency dependence of $(Z_e)_m + (Z_f)_m$ does not match that of a simple oscillator, but we can see that $(Z_e)_m$ is imaginary, so it is a reactance. If $k_e a < \kappa_m^{(1)}$, then it is negative imaginary, corresponding to a compliance. Also, $(Z_e)_m$ is infinite when $k_e a = \kappa_m^{(0)}$, and it grows without bound to a large positive imaginary value (inertance) with increasing $k_e a$ beyond $\kappa_m^{(2)}$. A large impedance leads to small velocity and displacement amplitudes.

The fluid impedance, $(Z_f)_m$, is complex and nonsingular. The frequency dependence of both impedances for the lowest spherical harmonics is described by Fig. 7.7 for the case of a steel spherical shell submerged in water. Several features are noteworthy. The most important are that Im $(Z_f) > 0$ and Re $(Z_f) > 0$ for any frequency. In other words the fluid loading adds an inertance and a resistance.

We have seen that if the shell vibrates in a vacuum, it will resonate when $k_e a = \kappa_m^{(1)}$ or $k_e a = \kappa_m^{(2)}$. Both conditions correspond to $(Z_e)_m = 0$. (In an actual system, internal dissipation would limit W_m to a finite value.) Now consider what happens when the impedance of the fluid is added to that of the shell. Because Im (Z_f) is always greater than zero, a plot of Im $((Z_e)_m + (Z_f)_m)$ as a function of $k_e a$ would yield a curve at each m that is above the corresponding plot of Im$(Z_e)_m$ in Fig. 7.7. Consequently, the total inertance crosses the zero axis at a lower frequency. It is convenient to refer

---

[10] An infinite value of W_m is an artifact. If any linear system is excited at a natural frequency, the response grows in time without bound. This behavior was observed in Sect. 2.5.3. The mathematical explanation may be found in *Mechanical and Structural Vibration*, J.H. Ginsberg, John Wiley and Sons, Inc., Sect. 3.3 (2001).

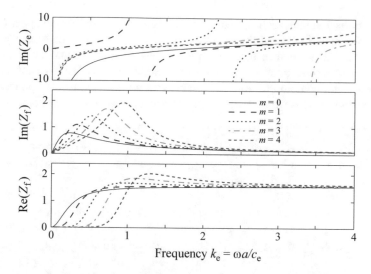

Fig. 7.7 Structural and fluid impedances for several spherical harmonics as a function of frequency. The media are steel and water, and $a/h = 50$

to any condition in which the total reactance vanishes as a *fluid-loaded resonance*. These resonances occur at a lower frequency than the corresponding (in-vacuo) natural frequencies. Despite the terminology, it is essential that one not forget that fluid resistance is always present. This resistance, being in-phase with the surface velocity, plays the same role as a dashpot. The difference is that a dashpot dissipates energy internally, whereas the power required to overcome the fluid resistance is transferred into the fluid as the radiated pressure field. The name given to this effect is *radiation damping*.

Another important observation stems from the fact that the magnitude of Z_f is proportional $(\rho_0 c)/(\rho_e c_e)$, that is, the ratio of characteristic impedances. For water and steel, this ratio is approximately 0.03, whereas it is approximately $0.9 \left(10^{-5}\right)$ in the case of air and steel. The usual terminology is to say that the former constitutes *heavy fluid loading*, whereas the latter is *light fluid loading*. In a lightly fluid-loaded system, the fluid impedance will be much smaller than the structural impedance, except at the in-vacuo natural frequencies, where $\mathrm{Im}\,(Z_e) = 0$.

This attribute leads to a simplification that often is invoked in the cases of light fluid loading. In it, the structural response is determined with the in-vacuo model. The effect of fluid resistance when the frequency is close to in-vacuo natural frequencies is estimated, typically on the basis of prior experience with similar systems. The vibratory displacement is then used to predict the pressure field as a standard radiation problem. The important aspect of this simplified approach is that increasing $\rho_0 c/\rho_e c_e$ will lead to a decrease in the peak amplitudes at resonances, but it will not account for the downshift of the frequency at which these resonances occur.

The discussion thus far has focused on the behavior of individual spherical harmonics. We have examined the factors influencing the displacement and pressure coefficients, and have seen that the structure has an infinite number of in-vacuo modes that have some relevance to resonant-type phenomena. We have not seen how these behaviors combine to affect the displacement and acoustic responses to an actual excitation. This we shall do by investigating a prototypical system.

EXAMPLE 7.5 A spherical shell $(a/h = 40)$ composed of aluminum is submerged in water. It is excited by a time-harmonic point force that acts in the radial direction. (a) Determine the amplitude of the displacement coefficients W_m as a function of frequency for $m \leq 4$ and $0 < k_e a < 8$. (b) Determine the amplitude of the radial displacement at the point where the force is applied and at the diametrically opposite point for $0 < k_e a < 8$. (c) Determine the farfield pressure amplitude $r |P|$ for $0 < k_e a < 8$ in the direction $\psi = 0$ and $\psi = 180°$, where ψ is measured relative to the radial line through the point where the force is applied. (d) Determine the directivity of the farfield pressure amplitude. Carry out the analysis for three frequencies: the frequency in the range $k_e a < 1$ for which $r |P|$ at $\psi = 0$ is maximized, $k_e a = 2.5$, and the frequency in the range $k_e a < 5$ for which $r |P|$ at $\psi = 0$ is maximized.

Significance

An important part of any analysis is interpretation and explanation of the results. This example will explore how to do so when elastic and acoustic effects interact. Along the way, issues regarding efficient computational approaches and series convergence will come to the fore.

Solution

The only deviation of this system from radial symmetry is the point force. If its location is defined as the $\psi = 0$ pole for a spherical coordinate system, the system is axisymmetric with respect to the z-axis. The value of W_m as a function of frequency is given by Eq. (7.2.42). Solution of those equations requires specification of the F_m coefficients, which are described by Eq. (7.2.40). That expression is based on a distributed force, that is, a force per unit area, whereas the present excitation is a concentrated force $\text{Re} \left(\hat{F} e^{i\omega t} \right)$ applied at $\eta = 0$. We could use a Dirac delta to describe this force, but a less mathematical approach is available. Let us consider the force to be distributed over a small spherical cap defined by $0 \leq \psi \leq \Delta$, where $\Delta \ll 1$. The point force coefficients will emerge by recognizing the smallness of Δ.

The distributed force is $q_{\text{applied}} = \hat{F}/\mathcal{A}$ for $\psi \leq \Delta$ and $q_{\text{applied}} = 0$ for $\psi > \Delta$, where $\mathcal{A}$ is the area of the cap,

$$\mathcal{A} = \int_0^\Delta (2\pi a \sin \psi) \, (a d\psi) = 2\pi a^2 \, (1 - \cos \Delta) \approx \pi a^2 \Delta^2$$

It is convenient to use $\eta = \cos\psi$ to change the integration variable in Eq. (7.2.40), so that

$$F_m = \frac{2m+1}{2} \int_0^{\Delta} \frac{\hat{F}}{\pi a^2 \Delta^2} P_m\,(\cos\psi)\,\sin\psi d\psi$$

Because Δ is extremely small, the integration variable is always small. Therefore, the Legendre functions are well approximated as $P_m\,(\cos\psi) \approx 1$, and we may set $\sin\psi \approx \psi$. The result is

$$F_m = \left(\frac{2m+1}{4}\right)\frac{\hat{F}}{\pi a^2} \tag{1}$$

Before we proceed to any computations, it is useful to review the equations to be solved. First, the nondimensional natural frequencies $\kappa_m^{(1)}$ and $\kappa_m^{(2)}$, which are independent of ka, will be evaluated by finding the roots of Eq. (7.2.44). With those quantities known for each azimuthal harmonic m, we may solve Eq. (7.2.45) for all W_m/a. The right side of that equation is $(a/h)F_m/\left(\rho_e c_e^2\right)$. In view of Eq. (1), we shall nondimensionalize this term by defining a variable χ_m that replaces W_m/a, such that

$$\frac{W_m}{a} = \frac{\hat{F}}{\rho_e c_e^2 ah}\chi_m \tag{2}$$

The result of changing variables in this manner is to convert Eq. (7.2.45) to

$$\left[\frac{\left[\left(\kappa_m^{(1)}\right)^2 - (k_e a)^2\right]\left[\left(\kappa_m^{(2)}\right)^2 - (k_e a)^2\right]}{\left(\kappa_m^{(0)}\right)^2 - (k_e a)^2} \right. $$
$$\left. + \frac{\rho_0 a}{\rho_e h}\left(\frac{c}{c_e}\right)^2\frac{kah_m\,(ka)}{h'_m\,(ka)}\right]\chi_m = \frac{2m+1}{4\pi} \tag{3}$$

The values of χ_m may be used to evaluate the coefficients of the pressure series. Introducing Eq. (2) into (7.2.41) leads to

$$B_m = \rho_0 c^2 \frac{ka}{h'_m\,(ka)}\frac{W_m}{a} = \left(\frac{\hat{F}}{ah}\right)\frac{\rho_0 c^2}{\rho_e c_e^2}\frac{ka}{h'_m\,(ka)}\chi_m \tag{4}$$

Thus, the χ_m values found from Eq. (3) are the (dimensional) displacement coefficients ratioed to $\hat{F}/\left(\rho_e c_e^2 h\right)$. Using those values to evaluate, Eq. (4) yields the pressure coefficients ratioed to $\hat{F}/(ah)$. The displacement and pressure at any location are found by synthesizing the series in Eq. (7.2.35).

Part (a) requested the W_m coefficients, which now are χ_m, for $m \le 4$. However, evaluation of the displacement and pressure series requires computations for increasing m until the series converge. Thus, our strategy is to solve Eq. (3) at the full range of frequencies, which we will designate $(k_e a)_j$. The result will be a matrix $\left[\tilde{\chi}\right]$ whose element at row $m+1$ and column j is $\tilde{\chi}_{m,j} = \chi_m$ at $(k_e a)_j$. It is logical to follow the evaluation of the χ_m coefficients with an evaluation of the corresponding B_m coefficients. Doing so will yield a matrix $[B]$ that matches $[\chi]$.

Standard properties for aluminum are $\rho_e = 2100$ kg/m^3, $c_e = 6420$ m/s, and $\nu = 0.33$. The in-vacuo natural frequencies are tabulated below.

	Spherical harmonic m				
	0	1	2	3	4
$\Omega_m^{(1)}$	–	0	0.7014	0.8327	0.8895
$\Omega_m^{(2)}$	1.6125	1.9749	2.7221	3.6351	4.5967

Figure 1a shows the frequency dependence of $|\chi_m|$ for $m = 0$ to 4. As expected, each amplitude has a zero at $k_e a = \kappa_m^{(0)}$, and it falls off with increasing frequency. Although the theory indicates that each spherical harmonic should show two fluid-loaded resonances because there are two natural frequencies at each m other than $m = 0$, only one resonance appears in the figure. Figure 1b provides a zoomed view of this range. It indicates that $|W_0|$ does not feature a strong resonance. The fluid-loaded resonance of W_1 occurs at $k_e a = 0$. Increasing m beyond one shifts the peak of $|W_m|$ to a higher frequency, with a narrower width and greater height. (In the study of mechanical and electrical oscillations, this feature would be said to be that the Q factor of the resonance increases.) These fluid-loaded resonances occur at substantially lower frequencies than the corresponding in-vacuo natural frequency.

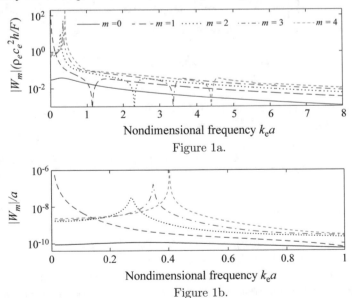

Figure 1a.

Figure 1b.

Reasonable questions to ask at this juncture are: Why don't any of the curves show two peaks if there are two natural frequencies for each $m > 0$? Why do peaks occur at

frequencies that are much less than either natural frequency for that m? Why doesn't
the $m = 0$ curve have a sharp peak? Why does the $m = 1$ curve rise without bound
as the frequency approaches zero? The answer to each may be found in the general
analysis, but it is helpful to examine them from a slightly different perspective. Rather
than examining impedances that relate the applied force to velocity coefficients, let
us write Eq. (3) for the displacement coefficients as

$$(K_e + K_f)\, \chi_m = \frac{2m+1}{4\pi}$$

The stiffness of an elastic spring is the ratio of the internal force to the elongation.
The preceding describes a similar relation, so K_e and K_f are, respectively, called
the structural and fluid *dynamic stiffnesses*. A comparison of the above definition to
Eq. (7.2.50) shows that these quantities are related to the respective impedances by

$$K_e = ik_e a Z_e, \quad K_f = ik_e a Z_f$$

We begin with an examination of the structural reactance. The above definitions show
that the real part of a dynamic stiffness is $-k_e a$ times the reactance. The imaginary
part of a dynamic stiffness is $k_e a$ times the resistance. Dissipation in the shell material
has been neglected, so the sole contributor to $\mathrm{Im}\,(K_e + K_f)$ is the fluid resistance,
which means that $\mathrm{Im}\,(K_e + K_f) > 0$ at any frequency.

Let us examine how the behavior of the dynamic stiffnesses affects each χ_m,
beginning with $m = 2$. Fig. 2 depicts the nonzero parts of K_e and K_f. If fluid loading
was not present, a zero of $\mathrm{Re}\,(K_e)$ would lead to a singular value of χ_m. Close
inspection shows that this curve does indeed cross the zero axis at $k_e a = \Omega_2^{(1)}$ and
$\Omega_2^{(2)}$. Now suppose that fluid loading was such that $\mathrm{Im}\,(K_f)$ were zero. Then, a singular
value of W_m would occur at any frequency at which $\mathrm{Re}\,(K_e + K_f) = 0$. The plotted
data indicates that this condition occurs at $k_e a = 0.276$, which is much less than
$\Omega_2^{(1)}$, and at $k_e a = 2.668$, which is slightly less than $\Omega_2^{(2)}$. Of course, $\mathrm{Im}(K_f)$ is not
zero. However, in the vicinity of $k_e a = 0.276$, the value of $\mathrm{Im}(K_f)$ is comparable to
the small value of $\mathrm{Re}\,(K_e + K_f)$. Consequently, division of the F_2 force by $K_e + K_f$
yields a large value of $|\chi_2|$. In fact, the maximum is $|\chi_2|/a = 10.704$ at $k_e a = 0.274$.

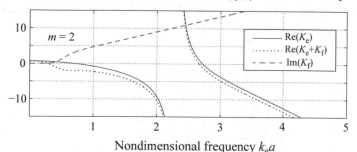

Nondimensional frequency $k_e a$

Figure 2.

Now consider the situation for χ_2 in the vicinity of $k_e a = 2.668$, at which $\text{Re}(K_e + K_f) = 0$. The value of $\text{Im}(K_f)$ in this range is much greater than $\text{Re}(K_e + K_f)$. This means that radiation damping is high, so there is no frequency in this range at which $|K_e + K_f|$ is noticeably diminished. It is for this reason that $|\chi_2|$ does not feature a peak in this range.

The situation for $m > 2$ is similar to what we have seen for $m = 2$. The behaviors for $m = 0$ and $m = 1$ are different because of the special nature of the natural frequencies for each case. Figure 3 displays the dynamic stiffnesses for $m = 0$. There is only one natural frequency, so there is only one frequency at which $\text{Re}(K_e) = 0$. The addition of fluid loading leads to $\text{Re}(K_e + K_f)$ being zero at a slightly lower value of $k_e a$, but $\text{Im}(K_f)$ is quite large in this range. This means that there is no frequency at which $|K_e + K_f|$ is greatly reduced, so is there no frequency at which $|\chi_0|$ shows a sharp peak.

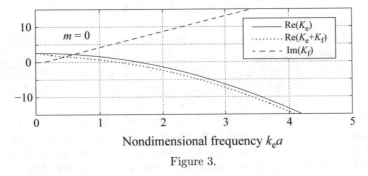

Figure 3.

Figure 4 explains the $m = 1$ behavior. The lower natural frequency is $\Omega_1^{(1)} = 0$, and $K_f = 0$ at that frequency because low frequencies correspond to low velocities. Hence, $K_e + K_f = 0$ at $k_e a = 0$, which leads to a singular value of $|W_1|/a$. The behavior in the vicinity of the higher natural frequency is the same as it was for $m = 2$. Adding K_f to K_e adds a large imaginary part. Thus, even though there is a frequency at which $\text{Re}(K_e + K_f)$ is zero, nowhere in that frequency range is $|K_e + K_f|$ small. The consequence is that large values of $|W_1|/a$ only occur at low frequencies.

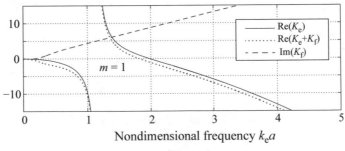

Figure 4.

For Part (b), we observe that the displacement at polar angles $\psi_1 = 0$ and $\psi_2 = \pi$ when the frequency is a specific value $(ka)_n$ may be computed by the matrix algorithm. Let $[L]$ be a $2 \times (N+1)$ array whose elements are $L_{j,m} = P_m(\cos\psi_j)$, that is, row j has the Legendre functions at all orders for ψ_j. Also, define a rectangular array $[\chi]$ whose columns hold the χ_m values at each $(ka)_n$. Then, the displacement series in Eq. (7.2.35) reduces to

$$\frac{\rho_e c_e^2 h}{\hat{F}} [w] = [L][\chi]$$

It is at this juncture that the question of series convergence arises. Given that the F_m coefficients in Eq. (2) increase with increasing m, it is reasonable to anticipate that the spherical harmonic M at which the summation is halted will be large. If we perform the first calculation with a large M, then results for smaller M may be found by deleting columns from the right of $[L]$, and deleting the same number of rows from the bottom of $[\chi]$. Results for $M = 100$, 67, and 33 are described by Fig. 5. There is no perceptible difference between the results for the two longer series. Thus, it is reasonable to conclude that $M = 67$ is adequate for the range $k_e a < 8$. The series for $M = 33$ agrees well up to $k_e a = 4$, then diverges drastically for $k_e a > 6$. There is a simple explanation for this behavior. Each χ_m, other than χ_0, has a narrow peak in the vicinity of a single frequency. Because the displacement w is formed from a weighted series of the χ_m coefficients, w/a may be expected to have peaks at each of these frequencies. The highest peak for $M = 33$ occurs at $k_e a = 5.908$, which is where the series begins to diverge from the correct solution. The conclusion is that the series length must be sufficiently long to capture all fluid–loaded "resonances" that occur within the frequency range of interest. Unfortunately, this criterion is not much use a priori, because we find these frequencies at part of the analysis.

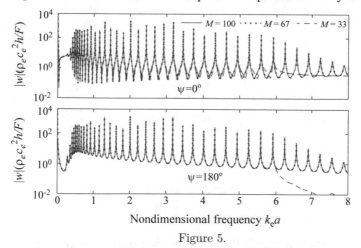

Figure 5.

A different perspective for convergence results from considering the criterion for accuracy of the shell theory. It was stated as requiring that $m < (\pi/5)\,a/h$. For the present system, this requires $m \leq 25$. A graph like Fig. 5 would show that a truncation of the series at $m = 25$ would yield a result that is convergent only for $k_e a < 3.3$. Given that, we should not assume that our results for higher $k_e a$ are what would be observed in an experiment.

The farfield pressure is given by Eq. (7.2.43), which also may be computed as a matrix product. A rectangular array $[C]$ is computed by multiplying each row $[B_m]$ by a phase delay for that m, specifically

$$[C_m] = [B_m] \, e^{i(m+1)\pi/2}$$

The farfield locations at which the pressure is to be evaluated for Part (c) also are $\psi = 0$ and $\psi = \pi$, so we may also use $[L]$ for this evaluation. Therefore, we evaluate

$$\left[r P_{\text{ff,axis}} \right] = [L][C]$$

Because of the manner in which the χ_m values are defined, the quantities that are evaluated in this manner are $(r/a) P_{\text{ff}} (r, \psi = 0 \text{ or } \pi) \, e^{ikr} (ah/\hat{F})$.

Figure 6 displays the frequency dependence of the farfield pressures on the axis of symmetry. A zoomed view gives a clearer picture of the many low-frequency resonances. The frequencies at which peaks occur are those that maximize one of the χ_m coefficients. An interesting attribute is that the radiation in the forward direction, $\psi = 0$, is much greater than the backward direction, $\psi = 180°$, for nonresonant $k_e a$ values between 0.5 and 5.

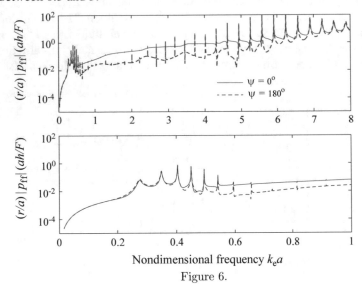

Figure 6.

The directivity requested in Part (d) can be found similarly to the farfield pressure at a specific angle. Let $(k_e a)_j$ denote the frequency of interest, and let $\{C_j\}$ denote the jth column of $[C]$. The farfield pressure at many angles ψ_j may be found by defining a matrix $[L']$ that differs from $[L]$ only by the fact that its has many rows, $L_{j,m} = P_m (\cos \psi_j)$.

A scan of the data shows that the largest value of $r\,|P|$ for $k_e a < 1$ occurs at $k_e a = 0.404$, and the largest value in the interval $k_e a < 5$ occurs at $k_e a = 4.935$. Polar plots describing $r\,|P|$ at these frequencies and $k_e a = 2.50$ constitute Fig. 7. The response at the lowest and highest frequencies corresponds to peaks of χ_n for $m = 4$ and $m = 30$, respectively. Consequently, the farfield pressure at these frequencies is proportional to the Legendre functions $P_4\,(\cos\psi)$ and $P_{30}\,(\cos\psi)$. In contrast, the middle frequency, $k_e a = 2.50$, is not close to any resonance. Hence, all spherical harmonics contribute significantly to the pressure. The consequence is that the directivity plot at $k_e a = 2.50$ does not resemble that of a single Legendre function. It will be noted that the overall amplitude at $k_e a = 2.5$ is comparable to that at the lower frequency, even though $|\chi_4|$ at $k_e a = 0.404$ is two orders of magnitude greater than any $|\chi_m|$ value at $k_e a = 2.5$. This result is a consequence of the fact that the $h'_m\,(ka)$ appears in the denominator of Eq. (7.2.43). These values are much larger at the lower frequency. This is another manifestation of the inefficiency of radiation at low frequencies.

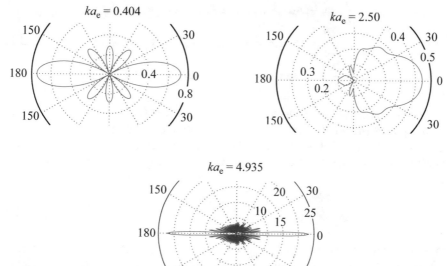

Figure 7.

7.3 Radiation from an Infinite Cylinder

It is known that the separation of variables approach may be used to solve the Helmholtz equation in eleven orthogonal coordinate systems, such as the spherical coordinates of the previous section.[11] However, it does not follow that we will always be able to use the technique to analyze radiation from a body whose shape

[11]Morse and Feshbach, *Methods of Theoretical Physics*, vol. 1 (1953) pp. 494–523, 655–666.

Fig. 7.8 An otherwise rigid
box undergoing a uniform
vibration at velocity
amplitude v_0 along the edge
$x = a/2$

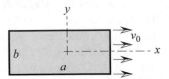

matches one of these coordinate systems. The apparent paradox results from the
nature of boundary conditions.

Consider a rectangular box, which is simplified to a two-dimensional system.
Suppose the right edge translates, and the other edges are stationary. This is the
situation in Fig. 7.8.

The origin of xy has been placed at the center. Thus, the boundary conditions are

$$\frac{\partial P}{\partial x} = -i\rho_0\omega V_0 \text{ on } x = a/2, \quad -b/2 < y < b/2$$

$$\frac{\partial P}{\partial x} = 0 \text{ on } x = -a/2, \quad -b/2 < y < b/2 \qquad (7.3.1)$$

$$\frac{\partial P}{\partial y} = 0 \text{ on } y = \pm b/2, \quad -a/2 < x < a/2$$

The separation of variables ansatz is that $P = F(x)\,G(y)$. It is possible to find
ordinary differential equations for $F(x)$ and $G(y)$ whose satisfaction assures that P
is a solution of the Helmholtz equation. The difficulty is with the boundary conditions.
Substitution of the ansatz into the second of the preceding set of boundary conditions
gives

$$\left(\frac{dF}{dx}\bigg|_{x=-a/2}\right) G(y) = 0, \quad -b/2 < y < b/2 \qquad (7.3.2)$$

In cannot be that $G(y) = 0$, so it must be that dF/dx is zero at $x = -a/2$. However,
if it is zero along this edge, then $\partial P/\partial x$ is zero at $x = -a/2$ for all y, not just the
edge of the box. In other words, the resulting solution would describe the box plus
a rigid sheet at $x = -a/2$ whose extent is infinite. Similar situations occur for the
rigid edges at $y = \pm b/2$. Thus, the system that is implied by making the separation
of variables solution satisfying the rigid boundary conditions is the one depicted in
Fig. 7.9. Clearly, this system is not at all like the one we set out to analyze, and it has
no physical significance.

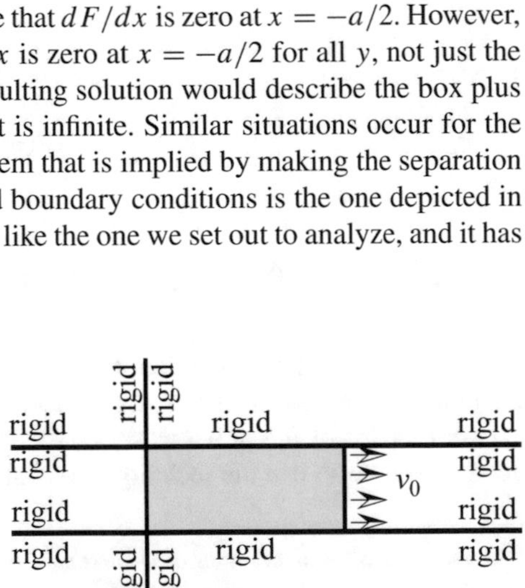

Fig. 7.9 System implied by
making a separation of
variables ansatz satisfying
the boundary conditions for
the vibrating box in Fig. 7.8.
The rigid lines represent
rigid thin planes that extend
to infinity

rigid	rigid ⋮ rigid	rigid	rigid
rigid		rigid	v_0 rigid
rigid		rigid	rigid
rigid	rigid ⋮ rigid	rigid	rigid

Given this observation, it is reasonable to wonder what body shapes can be analyzed for their radiated field by the separation of variables method? If we add the requirement that the analysis not be excessively complicated, and that the system is reasonably common, the answer is: Very few! Spheres were covered in the previous section. Variants featuring ellipses, such as spheroids, have been analyzed, but those studies require great effort because of the nature of the associated separation functions.

Radiation from cylindrical bodies is of great interest because it is an issue that arises in several contexts, including HVAC and automotive exhaust system, and piping and pressure vessel applications. Also, in underwater acoustics, the shape of the pressure hull of a modern submarine usually is a cylinder closed by rounded caps. Unfortunately, the separation of variables solution for a cylindrical geometry does not describe any of these systems. This is so because it is limited to situations where the cylindrical boundary has infinite length. Otherwise, we encounter the same difficulty as that in Fig. 7.9, with a cylindrical boundary extended beyond the ends of the cylinder. Despite this limitation, we will pursue the analysis of infinite cylinders. In part, the justification for doing so is that the investigation will provide results that are good approximations in some regions of the actual field. In addition, analyses like those pursued here often are used to develop simpler models that capture basic phenomena encountered in finite length cylinders.

The system we will explore is described in Fig. 7.10. The z-axis is defined to coincide with the axis of a cylinder whose radius is a when the cylinder is at rest. The fluid domain extends outward from the cylinder's surface without limit. The surface of the cylinder executes a small amplitude vibration such that its velocity component normal to the surface is $\text{Re}\left(V_s e^{i\omega t}\right)$, where V_s is an arbitrary specified function of the circumferential angle θ and axial position z. The unit vector $\bar{e}_R$ at an arbitrary field point is said to be oriented in the transverse direction relative to the z-axis. The circumferential direction is $\bar{e}_\theta$, and the axial direction is $\bar{e}_z$. Both $\bar{e}_R$ and $\bar{e}_\theta$ depend on θ, but are independent of R and z, whereas $\bar{e}_z$ has a fixed orientation.

Fig. 7.10 Coordinate systems and unit vectors for describing a surface vibration of a cylinder

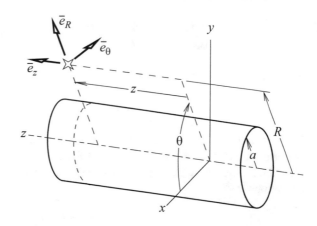

7.3.1   Separation of Variables

The pressure field radiated by a vibrating cylinder is a function of R, θ, and z. Because $\bar{e}_R$ is perpendicular to the surface, the continuity condition requires that the transverse velocity of the fluid equals V_s. The frequency domain equations governing the field are the Helmholtz equation subject to a boundary condition derived from Euler's equation. The gradient and Laplacian in terms of cylindrical coordinates are derived in Appendix A. The governing equations are

$$\frac{\partial^2 P}{\partial R^2} + \frac{1}{R}\frac{\partial P}{\partial R} + \frac{1}{R^2}\frac{\partial^2 P}{\partial \theta^2} + \frac{\partial^2 P}{\partial z^2} + k^2 P = 0, \quad \begin{cases} R > a \\ -\pi \le \theta < \pi \\ -\infty < z < \infty \end{cases} \quad (7.3.3)$$

$$\left.\frac{\partial P}{\partial R}\right|_{r=a} = -i\rho_0 \omega V_s, \quad \begin{cases} -\pi \le \theta < \pi \\ -\infty < z < \infty \end{cases}$$

We could immediately proceed to apply the method of separation of variables to the Helmholtz equation. However, the circular geometry allows us to simplify identification of the dependence on θ.

If we were to start from any point and follow a circle defined by constant values of R and z, an increment of θ by 2π would bring us back to the starting point. It follows that P must be θ-periodic in 2π, that is, $P(R, \theta, z) = P(R, \theta + 2\pi, z)$, so we introduce a Fourier series. The coefficients of this series may be functions of R and z. It is convenient to begin by using a complex series, so we know that

$$P = \frac{1}{2}\sum_{n=-\infty}^{\infty} F_n(R, z)\, e^{in\theta} \quad (7.3.4)$$

Because of the linearity of the Helmholtz equation, each term of the Fourier series must constitute a solution. If we substitute a generic term, the harmonic function of θ appears throughout and may be eliminated. What remains is

$$\frac{\partial^2 F_n}{\partial R^2} + \frac{1}{R}\frac{\partial F_n}{\partial R} + \frac{\partial^2 F_n}{\partial z^2} + \left(k^2 - \frac{n^2}{R^2}\right) F_n = 0 \quad (7.3.5)$$

At this juncture, we introduce the separation of variable ansatz,

$$F_n = f(R)\, g(z) \quad (7.3.6)$$

We substitute this form into the differential equation, divide by fg, then group terms that are functions of R and z on either side of the equality. The k^2 term could be placed on either side, but it preferable to keep it with the $1/R^2$ term. Thus, these operations lead to

$$\frac{1}{f}\left(\frac{d^2 f}{dR^2} + \frac{1}{R}\frac{df}{dR}\right) + \left(k^2 - \frac{n^2}{R^2}\right) = -\frac{1}{g}\frac{d^2 g}{dz^2} \tag{7.3.7}$$

The terms on the left are a function of R, while the one on the right depends only on z. This can only be true if both sides equal the same constant. The value of this constant is not known a priori, but we do know that $g(z)$ cannot grow as $|z|$ increases. This requires that the separation constant be positive or zero, because a negative value would lead to solution for g that is a growing exponential. Thus, we shall denote the separation constant as μ^2. The corresponding separated differential equations are

$$\frac{d^2 f}{dR^2} + \frac{1}{R}\frac{df}{dR} + \left(\kappa^2 - \frac{n^2}{R^2}\right)f = 0$$
$$\frac{d^2 g}{dz^2} + \mu^2 g = 0 \tag{7.3.8}$$
$$\kappa^2 = k^2 - \mu^2$$

The axial function $g(z)$ is harmonic. In keeping with the usage of a complex Fourier series for the radial distance, this function is written in a similar form as

$$g = Ae^{-i\mu z} + Be^{+i\mu z} \tag{7.3.9}$$

The values of the *axial wavenumber* μ and the coefficients A and B are arbitrary in regard to satisfaction of the Helmholtz equation.

7.3.2 Transverse Dependence—Cylindrical Bessel Functions

The value of μ affects $f(R)$ by altering the constant κ. To begin, we will consider μ to be less than k, so that κ is a real number, which may be taken to be positive. The case where $\mu > k$ is important and will be studied after the primary development.

The differential equation for $f(R)$ is reminiscent of the equation governing the radial function for spherical waves. However, the absence of a two factor in the second term, $(1/R)\,df/dR$, is important. It is possible to derive a transformation that converts the present equation to the spherical form. An alternative would derive the solution by following the method of Frobenius, which modifies a power series in R with a term that is an unspecified fractional power. Some readers might have encountered a differential equation like this in other subjects. In any event, the present endeavor is well served by merely observing that the first of Eq. (7.3.8) is *Bessel's equation*. Its solutions are Bessel and Neumann functions. We encountered them in the previous section, where it was noted there that standard algorithms compute these functions as intermediate steps in the evaluation of spherical Bessel functions.

A rigorous terminology would refer to these functions as cylindrical Bessel functions, but their occurrence is so common that the adjective "cylindrical" usually is dropped. To distinguish them from the spherical functions, capital letters are used for the cylindrical functions: J for the Bessel function, N for the Neumann (although Y also is often used), and H for the Hankel functions that we soon will define. The value of n, which is the *order* of the function, appears in the subscript. Furthermore, if we were to divide the first of Eq. (7.3.8) by κ, we would see that the independent variable for this form of Bessel's equation is κR, so that is the argument of the functions. Among the many ways to evaluate Bessel functions, the least useful is from their definitions. Nevertheless, they are listed here for the sake of completeness,

$$J_n(x) = \sum_{k=0}^{\infty} \frac{(-1)^k}{k!(n+k)!}\left(\frac{z}{2}\right)^{2k+n}$$

$$N_n(x) = \frac{2}{\pi}\ln\left(\frac{z}{2}\right)J_n(x) - \frac{1}{\pi}\sum_{k=0}^{n-1}\frac{(n-k-1)!}{k!}\left(\frac{z}{2}\right)^{2k-n} \qquad (7.3.10)$$

$$-\frac{1}{\pi}\sum_{k=0}^{\infty}\frac{(-1)^k}{k!(n+k)!}[\Psi(k+1) + \Psi(n+k+1)]\left(\frac{z}{2}\right)^{2k+n}$$

In the second expression, $\Psi(\ )$ is the psi or digamma function, which is defined in standard references on special functions.[12]

Both $J_n(x)$ and $N_n(x)$ are real. Therefore, neither by itself can represent a wave propagating in the transverse direction. For spherical waves, the spherical Hankel functions fit this specification. The same is true here. Hankel functions of the first and second kind and order n are defined to be

$$H_n^{(2)}(x) = H_n^{(1)}(x)^* = J_n(x) - iN_n(x) \qquad (7.3.11)$$

As was true for spherical waves, the second Hankel function represents an outgoing wave at large kR corresponding to our representation of harmonic response as $\mathrm{Re}\left(Pe^{i\omega t}\right)$. This fact is seen when we consider the asymptotic behavior of each function as its argument κR increases,

$$\left.\begin{array}{l} H_n^{(2)}(\kappa R) = H_n^{(1)}(\kappa R)^* \rightarrow \left(\dfrac{2}{\pi\kappa R}\right)^{1/2} e^{-i(\kappa R - (2n+1)\pi/4)} \\[12pt] J_n(\kappa R) \rightarrow \left(\dfrac{2}{\pi\kappa R}\right)^{1/2}\cos\left(\kappa R - \dfrac{2n+1}{4}\pi\right) \\[12pt] N_n(\kappa R) \rightarrow \left(\dfrac{2}{\pi\kappa R}\right)^{1/2}\sin\left(\kappa R - \dfrac{2n+1}{4}\pi\right) \end{array}\right\} \text{ as } \kappa R \rightarrow \infty$$

$$(7.3.12)$$

The Bessel and Neumann functions will arise when we explore cylindrical cavities and waveguides, but we will have no need for the first kind Hankel function. Thus,

---

[12]M.I. Abramowitz and I.A. Stegun, *Handbook of Mathematical Functions*, 9th ed., Dover, Sect. 6.3. (1965).

we will drop the superscript "two" notation for acoustical analyses, as we did for spherical waves, so that $H_n(kR)$ will always refer to the second Hankel functions.

The behavior of the cylindrical Bessel functions at small arguments also is important. The Bessel function is finite at $x = 0$, but the Neumann and Hankel functions are singular. The trends are

$$
\begin{rcases}
H_n^{(2)}(x) = H_n^{(1)}(x)^* \to
\begin{cases}
-\dfrac{2i}{\pi} \ln(x) & \text{if } n = 0 \\[2mm]
\dfrac{i}{\pi}(n-1)! \left(\dfrac{2}{x}\right)^n & \text{if } n \geq 1
\end{cases} \\[10mm]
J_n(x) \to \dfrac{1}{n!}\left(\dfrac{x}{2}\right)^n \\[8mm]
N_n(x) \to
\begin{cases}
\dfrac{2}{\pi} \ln(x) & \text{if } n = 0 \\[2mm]
-\dfrac{1}{\pi}(n-1)! \left(\dfrac{2}{x}\right)^n & \text{if } n \geq 1
\end{cases}
\end{rcases}
\text{as } x \to 0 \quad (7.3.13)
$$

The limiting behaviors for small and large arguments are evident in Fig. 7.11. The Bessel functions are finite at $x = 0$, with all except the zero order being zero there. In contrast, all Neumann functions are negatively infinite at the origin, with a growth rate that increases with increasing order. For large arguments, a Neumann function resembles the Bessel function at that order with a lag of $\pi/2$. Both functions decay with increasing values of their argument at fixed order. However, the decrease is inversely proportional to the square root of their argument, which is a slower decay than the decay rate for the spherical Bessel functions.

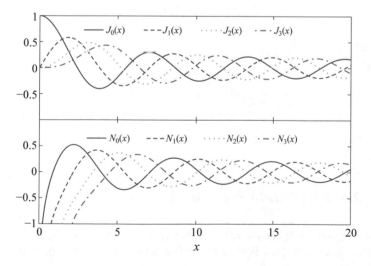

Fig. 7.11 Low-order Bessel and Neumann functions as functions of κR

The order n affects the magnitude of the argument x at which it is appropriate to employ Eq. (7.3.12). Specifically, the large argument trends will be observed if $x > n + 2\pi$. The opposite trend is the behavior when the order is much greater than the argument. In this case, we have

$$\left. \begin{array}{l} J_n(\kappa R) \to \left(\dfrac{1}{2\pi n}\right)^{1/2} \left(\dfrac{e\kappa R}{2n}\right)^n \\[3mm] N_n(\kappa R) \to -\left(\dfrac{2}{\pi n}\right)^{1/2} \left(\dfrac{e\kappa R}{2n}\right)^{-n} \end{array} \right\} \quad \text{if } n \gg \kappa R \qquad (7.3.14)$$

Most standard mathematics software contains routines for evaluating the Bessel and Neumann functions. We also will need to evaluate their derivatives. As always, a prime denotes differentiation of a function with respect to its argument. The following recurrence relations resemble those for spherical Bessel functions. Those for the derivatives will be useful when it is necessary to satisfy velocity boundary conditions, whereas the relations featuring functions at different orders are a core component of most numerical algorithms by which the functions are evaluated. In the following, $F_n(x)$ represents any of the cylindrical functions.

$$\begin{array}{c} \dfrac{2n}{x} F_n(x) = F_{n-1}(x) + F_{n+1}(x) \\[3mm] F_n'(x) = \dfrac{1}{2}\left[F_{n-1}(x) - F_{n+1}(x)\right] \\[3mm] = F_{n-1}(x) - \dfrac{n}{x} F_n(x) \\[3mm] = -F_{n+1}(x) + \dfrac{n}{x} F_n(x) \\[3mm] F_0'(x) = -F_1(x) \\[3mm] F_{-n}(x) = (-1)^n F_n(x) \\[3mm] \dfrac{d^j}{dx^j} F_n(x) = \dfrac{1}{2^j}\left[F_{n-j}(x) - \dfrac{j!}{1!(j-1)!} F_{n-j+1}(x) \right. \\[3mm] \left. + \dfrac{j!}{2!(j-2)!} F_{n-j+2}(x) - \cdots + (-1)^j F_{n+j}(x) \right] \end{array} \qquad (7.3.15)$$

Many other relations featuring the various Bessel functions are available.[13]

7.3.3 Radiation Due to a Helical Surface Wave

The first case we shall consider is that in which the surface motion consists of a single term in the complex Fourier series for the circumferential dependence, and

[13] M.I. Abramowitz and I.A. Stegun, *Handbook of Mathematical Functions*, Dover, (1965) Chaps. 9–11.

Fig. 7.12 Depiction of a helical wave vector $\bar{k}_s$. **a** Unwrapped cylindrical surface. **b** Tangent to the helix

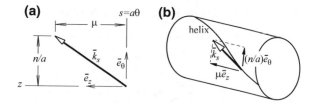

the z dependence is a sinusoidal wave that propagates in the positive direction. In other words, the radial velocity on the surface is taken to be

$$v_s = \text{Re}\left(V_0 e^{i(\omega t - \mu z - n\theta)}\right) \tag{7.3.16}$$

Clearly, this represents a wave, but in what direction is it propagating? To answer this, note that the arclength along the cylinder in the circumferential direction is $s = a\theta$. Let us unwrap the surface of the cylinder and lay it on a plane, as is shown in Fig. 7.12a. Position along this flat surface is $\bar{x} = z\bar{e}_z + s\bar{e}_\theta$, where the unit vectors are those in Fig. 7.10. This suggests that we rewrite the phase variable of v_s as $\omega t - \bar{k}_s \cdot \bar{x}$, where the surface wavenumber vector is

$$\bar{k}_s = \mu\bar{e}_z + \frac{n}{a}\bar{e}_\theta \tag{7.3.17}$$

This vector forms a constant angle with $\bar{e}_z$, which means that on this plane the wave follows a straight line. When the plane is wrapped back onto the cylinder in Fig. 7.12b, this straight line becomes a helix. Correspondingly, Eq. (7.3.16) is said to be a *helical surface wave*. In the case where $n = 0$, the surface wave is axisymmetric and the wavenumber vector is parallel to the axis of the cylinder.

In addition to the boundary condition at the cylinder's surface, the acoustic response must be an outgoing wave having suitably decaying amplitude as R increases. The only radial dependence fitting this specification with our use of the $e^{i\omega t}$ convention is the Hankel function of the second kind. Furthermore, because $\bar{V} \cdot \bar{e}_R$ is obtained from a derivative of P with respect to R, the manner in which pressure depends on z and θ must match the dependence of V_s. Thus, we shall try

$$P = B H_n(\kappa R) e^{-i(\mu z + n\theta)}, \quad \kappa = \left(k^2 - \mu^2\right)^{1/2} \tag{7.3.18}$$

The boundary condition is

$$\nabla P \cdot \bar{e}_R|_{R=a} \equiv \frac{\partial P}{\partial R}\bigg|_{R=a} = -i\omega\rho_0 V_s = -i\omega\rho_0 V_0 e^{-i(\mu z + n\theta)} \tag{7.3.19}$$

Our ansatz will satisfy this condition if

$$B\kappa H_n'(\kappa a) = -i\omega\rho_0 V_0 \equiv -ik\left(\rho_0 c V_0\right) \tag{7.3.20}$$

The corresponding pressure and particle velocity expressions are

$$
\begin{aligned}
P &= -i\rho_0 c V_0 \frac{k H_n (\kappa R)}{\kappa H_n' (\kappa a)} e^{-i(\mu z + n\theta)} \\
\bar{V} &= -\frac{1}{i\omega\rho_0} \left(\frac{\partial P}{\partial R} \bar{e}_R + \frac{1}{R} \frac{\partial P}{\partial \theta} \bar{e}_\theta + \frac{\partial P}{\partial z} \bar{e}_z \right) \\
&= V_0 \left[\frac{H_n' (\kappa R)}{H_n' (\kappa a)} \bar{e}_R - i \left(\frac{n}{R} \bar{e}_\theta + \mu \bar{e}_z \right) \frac{H_n (\kappa R)}{\kappa H_n' (\kappa a)} \right] e^{-i(\mu z + n\theta)}
\end{aligned}
\tag{7.3.21}
$$

An important aspect of this expression is that n may be positive, negative, or zero. Because of the way the Hankel, Bessel, and Neumann functions are defined, functions of negative order are related to the positive order functions by an alternating sign, specifically

$$
H_{-n} (x) = (-1)^n H_n (x) , \quad J_{-n} (x) = (-1)^n J_n (x) , \quad N_{-n} (x) = (-1)^n N_n (x)
\tag{7.3.22}
$$

The sign change has no effect on Eq. (7.3.21) because the terms are ratios of a Hankel function and its derivative at the same order.

The farfield decay of a cylindrical wave is different from the spherical spreading of the field that radiates from a finite body. To see why, suppose we select a field point at a very large value of R. The line from the field point to points in the cylinder cross section at the same z is perpendicular to the cylinder's axis. In contrast, lines from the field point to the distant ends of the cylinder are essentially parallel to the z-axis. There is no radial distance at which all lines from the field point to the surface are parallel, so the prior theorems regarding farfield behavior do not apply.

To identify the farfield limit, we employ the first of Eq. (7.3.12) to write

$$
P_{\text{ff}} = -i\rho_0 V_0 \frac{k e^{i(2n+1)\pi/4}}{\kappa H_n' (\kappa a)} \left(\frac{2}{\pi \kappa R} \right)^{1/2} e^{-i(\kappa R + n\theta + \mu z)}
\tag{7.3.23}
$$

The particle velocity in the farfield is obtained by substituting P_{ff} into Euler's equation. The fact that R is large renders the circumferential velocity negligible compared to $\bar{V} \cdot \bar{e}_R$ and $\bar{V} \cdot \bar{e}_z$. The result is

$$
\bar{V}_{\text{ff}} = \frac{P_{\text{ff}}}{\rho_0 c} \left(\frac{\kappa}{k} \bar{e}_R + \frac{\mu}{k} \bar{e}_z \right)
\tag{7.3.24}
$$

The relation of pressure and particle velocity in a plane wave is $\bar{V} = (P/\rho_0 c) \bar{k}/k$. The above expression suggests a similar form. The phase variable of the farfield pressure, Eq. (7.3.23), is $\kappa R + n\theta + \mu z$. It may be expressed in terms of a wavenumber vector $\bar{k}'$ if we let $\bar{x}$ be the position off the cylinder after it is unwrapped, that is, $\bar{x} = R\bar{e}_r + s\bar{e}_\theta + z\bar{e}_z$. Then, setting $\bar{k}' \cdot \bar{x} = \kappa R + n\theta + \mu z$ leads to

$$
\bar{k}' = \kappa \bar{e}_r + \frac{n}{R} \bar{e}_\theta + \mu \bar{e}_z
\tag{7.3.25}
$$

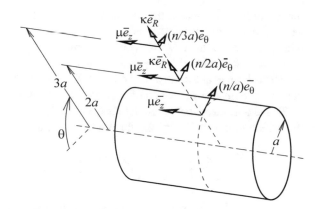

Fig. 7.13 Wavenumber vectors for the acoustic wave generated by a helical wave on the surface of a cylinder

Figure 7.13 shows $\bar{k}'$ at two locations along a transverse line, as well as $\bar{k}_s$ for the surface wave. (In a strict sense, we should not depict $\bar{k}'$ close to the surface unless $\kappa a \gg 1$, because Eq. (7.3.23) is only valid if $\kappa R \gg 1$.)

At the surface, the portion of $\bar{k}'$ that is parallel to the surface, that is the $\bar{e}_\theta$ and $\bar{e}_z$ components, matches the surface wavenumber $\bar{k}_s$. In other words, the trace wavenumbers match, and the surface wave is supersonic. At very large R, the $\bar{e}_\theta$ component of $\bar{k}'$ is negligible. Thus, Eq. (7.3.24) states that

$$\bar{V}_{\mathrm{ff}} = \frac{P_{\mathrm{ff}}}{\rho_0 c} \lim_{R \to \infty} \left(\frac{\bar{k}'}{k} \right) \tag{7.3.26}$$

Furthermore, it follows from the definition of κ that $|\bar{k}'| \to k$ as $R \to \infty$. In other words, in the farfield, the wave propagates in the plane formed by the field point and the cylinder's axis. Over short propagation distances, it seems to be locally planar, but cylindrical spreading is observable for large changes in R.

Whether a surface wave results in radiation depends only on the scale of the axial dependence. The condition that $\mu < k$ for a wave to radiate to the farfield is equivalent to saying that the axial wavelength $2\pi/\mu$ is greater than the acoustic wavelength $2\pi/k$, or equivalently, that the axial propagation speed c_s exceeds the speed of sound, $c \equiv \omega/k$. Each is a manifestation of trace matching. It is evident that the circumferential component of $\bar{k}_s$ is not relevant to the matching of trace velocities. In essence, as waves propagate along the cylinder, the axial features do not change, whereas the surface falls away in the circumferential direction. Thus, the circumferential effect is local, whereas the axial effect is global. (This is the reason why the phenomenon of matching trace velocity is not evident in radiation from spheres.)

Subsonic Surface Waves—Modified Bessel Functions

Recognition that the axial trace velocity of the acoustic wave must equal the axial speed of the surface wave leads to the same question as that for a surface wave on a plane, specifically, what if the surface wave is subsonic? The axial component of the surface wave's phase velocity is ω/μ. The supersonic case is obtained if $\mu < k$, in which case, the value of κ^2 is positive. Conversely, a subsonic wave corresponds to $\mu > k$, in which case, κ^2 is negative. When this situation occurs on a planar boundary, the result is an evanescent acoustic wave. The same is true here, but the fact that the dependence on R is described by a Hankel function is a slight complication.

To make it clear that κ^2 is negative, we replace it with $-\beta^2$, with $\beta > 0$. Then, the differential equation resulting from the separation of variables procedure is

$$\frac{d^2 f}{dR^2} + \frac{1}{R} - \left(\beta^2 + \frac{n^2}{R^2}\right) f = 0, \quad \beta^2 = \mu^2 - k^2 \qquad (7.3.27)$$

This is a modified Bessel's equation, and its solutions are the real functions $K_n(\beta R)$ and $I_n(\beta R)$, which, respectively, are the modified Bessel functions of the first and second kind. The $K_n(\beta R)$ function is singular at the origin and decays to zero as $\beta R \to \infty$. In contrast, $I_n(\beta R)$ is finite at $\beta R = 0$ and grows without bound as $\beta R \to \infty$. The dependence of the lowest order functions is depicted in Fig. 7.14.

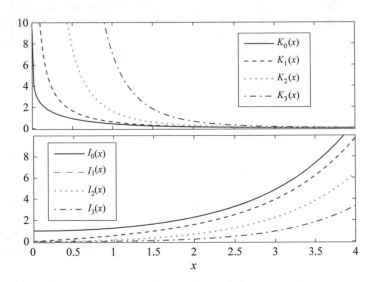

Fig. 7.14 Modified Bessel functions for orders zero to four

Asymptotic approximations of the modified functions are

$$\left.\begin{array}{l} K_0(x) \to -\ln(x), \quad K_n(x) \to \dfrac{1}{2}(n+1)!\left(\dfrac{2}{x}\right)^n \\[4mm] I_n(x) \to \dfrac{1}{(n+2)!}\left(\dfrac{x}{2}\right)^n \end{array}\right\} \text{ as } x \to 0 \qquad (7.3.28)$$

$$K_n(x) \to \left(\frac{\pi}{2x}\right)^{1/2} e^{-x}, \quad I_n(x) \to \left(\frac{1}{2\pi x}\right)^{1/2} e^x \text{ as } x \to \infty$$

The recurrence relations for these functions are slightly different from those for the regular functions. In the following, $F_m(x)$ is either $I_n(x)$ or $(-1)^n K_n(x)$,

$$\frac{2n}{x} F_n(x) = F_{n-1}(x) - F_{n+1}(x)$$

$$F_n'(x) = \frac{1}{2}\left[F_{n-1}(x) + F_{n+1}(x)\right]$$

$$= F_{n-1}(x) - \frac{n}{x} F_n(x) \qquad (7.3.29)$$

$$= F_{n+1}(x) + \frac{n}{x} F_n(x)$$

$$K_0'(x) = -K_1(x), \quad I_0'(x) = I_1(x)$$

Although both $I_n(\beta R)$ and $K_n(\beta R)$ satisfy Eq. (7.3.27), the former grows exponentially with increasing R. This behavior is counter to the Sommerfeld radiation condition, so it is discarded. The modified function of the second kind decays exponentially as R increases, so it represents an evanescent wave. Thus, the pressure generated from a high wavenumber on the surface is

$$P_{ev} = B K_n(\beta R) e^{-i(\mu z + n\theta)}, \quad \mu > k \qquad (7.3.30)$$

The corresponding particle velocity is

$$\bar{V} = -\frac{B}{i\omega\rho_0}\left[\beta K_n'(\beta R)\bar{e}_R - i K_n(\beta R)\left(\frac{n}{R}\bar{e}_\theta + \mu\bar{e}_z\right)\right]e^{-i(\mu z + n\theta)} \qquad (7.3.31)$$

The transverse component must match the surface velocity in Eq. (7.3.16), which sets B. When that expression is substituted into the above general solution, the result is

$$\boxed{P_{ev} = -i\rho_0 c V_0 \frac{k K_n(\beta R)}{\beta K_n'(\beta a)} e^{-i(\mu z + n\theta)}} \qquad (7.3.32)$$

The fact that subsonic surface waves do not radiate to the farfield is the same property as subsonic waves on the planar boundary of a half-space. Another aspect becomes evident when we recognize that the modified Bessel functions are real. This means that the transverse component of velocity, $\bar{V} \cdot \bar{e}_R$, is everywhere 90° out-of-phase from the pressure. One consequence is that power in not radiated when $\mu > k$. It also has implications for fluid–structure interaction, which is touched on in the next example.

In some applications, such as computer programs, it is inconvenient to monitor whether $\mu > k$, and then select the appropriate description of P. In the case of waves on a planar surface, we found that we could use the equations for supersonic waves to describe the subsonic case. With the objective of doing so here, we note that the modified functions are related to the regular Bessel functions, which should not be surprising in view of the fact that the equations they satisfy only differ by a sign. The relations are

$$I_n(x) = (-i)^n J_n(ix)$$
$$K_n(x) = \frac{\pi}{2}(-i)^{n+1} H_n(-ix)$$

(7.3.33)

In the present context, the values of x of interest are βa and βR. By definition, $\kappa = \pm i\beta$. If we use the positive sign, then $H_n(\kappa R)$ will become $H_n(i\beta R)$, in which the above relation indicates to be proportional $K_n(-\beta R)$. We wish that $H_n(\kappa R)$ convert to $K_n(+\beta R)$, so we set $\kappa = -i\beta$ for subsonic surface waves. Thus, we may employ the supersonic solution, Eq. (7.3.21), to treat any helical surface wave, provided that the transverse wavenumber is set according to

$$\kappa = \begin{cases} \left(k^2 - \mu^2\right)^{1/2} & \text{if } \mu < k \\ -i\left(\mu^2 - k^2\right)^{1/2} & \text{if } \mu > k \end{cases}$$

(7.3.34)

Of course, this scheme is only useful if the computational routine that evaluates Hankel functions allows the argument to be imaginary.

The alteration from a regular to a modified Bessel function has an important influence on the radiation impedance, which is a measure of how effectively a surface motion generates a pressure wave. The impedance of a locally reacting surface is the ratio of the surface pressure to the inward normal surface velocity. In contrast, the *radiation impedance* is the proportionality of the surface acoustic pressure resulting from a known outward surface velocity. The relation is $Z_{\text{rad}} = P / \bar{V}_s \cdot \bar{n}$. Despite the apparent similarity in their definitions, the two types of impedances are fundamentally different. A local impedance is a property of the surface, such as its viscoelastic properties. The radiation impedance is an acoustical property that depends on the shape of the radiating body and the nature of the surface motion. If we know it as a function of the fundamental parameters of the system, such as the wavenumbers and frequencies of the surface motion, it may be used to describe the behavior in a variety of situations.

The radiation impedance of a helical wave is obtained by evaluating the pressure at $R = a$, then dividing it by the surface velocity. The velocity is given by Eq. (7.3.16). The surface pressure in the case of a supersonic propagation in the axial direction is found by setting $R = a$ in Eq. (7.3.21), whereas setting $R = a$ in Eq. (7.3.32) gives the surface pressure in the subsonic case. The radiation impedance that results is given by

$$\frac{Z_{\text{rad}}}{\rho_0 c} = \begin{cases} -i\dfrac{ka}{\kappa a}\dfrac{H_n(\kappa a)}{H_n'(\kappa a)} & \text{if } \mu a < ka \\ -i\dfrac{ka}{\beta a}\dfrac{K_n(\beta a)}{K_n'(\beta a)} & \text{if } \mu a > ka \end{cases}$$

(7.3.35)

The nondimensional form of this expression tends to facilitate numerical evaluations, which are conducted in the next example.

A fundamental difference between the supersonic and subsonic cases is evident in this expression for Z_{rad}. The regular Hankel function is complex, which means that Z_{rad} has both a resistive and a reactive component. The resistive part is associated with transfer of power from the surface into the fluid. In contrast, the modified Bessel functions are real, so Z_{rad} is imaginary for subsonic waves. In other words, it is purely a reactance. Because the pressure is 90° out-of-phase from the velocity, the time-averaged power flow into the fluid will be zero.

EXAMPLE 7.6 The task here is to determine the radiation impedance associated with a helical surface wave. Evaluate and graph the result as a function of μa, with $ka = 2$ and 10, and $n = 0$, 2, and 4.

Significance

If the radiation impedance is known at the outset of the analysis, it can be used to describe the pressure effect on the structure without performing a parallel acoustical analysis. The parameter dependencies that will be disclosed in the course of this example have much significance.

Solution

How we proceed depends on whether our computational software can evaluate Hankel function for an imaginary argument. If so, then the field for any set of wavenumbers may be computed by employing Eq. (7.3.34) in conjunction with the first of Eq. (7.3.35) to evaluate Z_{rad} for any value of μ. If not, then both parts of Eq. (7.3.35) must be constructed explicitly. A vector $\{\mu a\}$ that covers the full range of axial wavenumbers m consists of two partitions. The upper vector $\{\kappa a\}$ corresponds to values of m for which $\mu_m a > \kappa a$, so its elements are defined according to Eq. (7.3.18). The lower vector $\{\beta a\}$ corresponds to $\mu_m a > \kappa a$, so Eq. (7.3.27) defines its elements. A vector $\{Z_{\mathrm{rad}}\}$ then is obtained by stacking the values obtained from $\{\beta a\}$ below those obtained from $\{\kappa a\}$.

The Hankel function is complex, so Z_{rad} for supersonic waves is complex. In contrast, the modified Bessel functions are real and positive, and their derivative is real and negative. Therefore, Z_{rad} for subsonic waves is always positive imaginary, that is, it is an inertance. Figure 1 shows the low-frequency behavior. There is a singularity at $\mu = k$ for $n = 0$, but not for $n > 0$. This difference stems from the behavior of the Bessel functions as their argument approaches zero. For $n = 0$, $H_0 (\kappa a)$ has a logarithmic singularity as $\kappa a \to 0$. Equation (1) indicates that Z_{rad} in that case is proportional to $\ln(\kappa a)$. The behavior for $\beta a \to 0$ is similar. In contrast, for $n > 0$, the trend is that $H_n (\kappa a)$ is proportional to $(\kappa a)^{-n}$ as $\kappa a \to 0$. The result is that Z_{rad} is finite in the limit. Another interesting aspect is that $\mathrm{Im}\,(Z_{\mathrm{rad}}) > 0$ for all values of μ, which means that its reactive part always is an inertance. We also see that the impedance at any value of μa decreases as n increases.

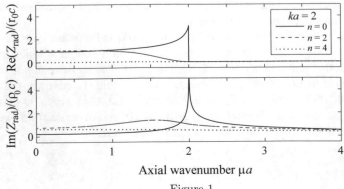

Figure 1.

Increasing the frequency changes some of these trends. The singular behavior in Fig. 2 around $\mu = k$ is like Fig. 1, but both parts of Z_{rad} for $n > 0$ are maximized slightly below $\mu = k$. Also, the intertances above $\mu = k$ are greater than they were at the lower frequency.

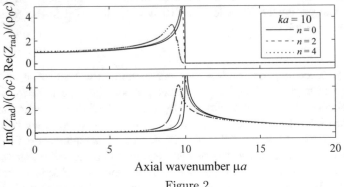

Figure 2.

From the viewpoint of the structure, a large value of $|Z_{\mathrm{rad}}|$ means that any movement will be strongly opposed by the fluid. Larger values of $|Z_{\mathrm{rad}}|$ in the vicinity of $\mu = k$ mean that it would be more difficult to coerce the structure to move in the transition from supersonic to subsonic surface waves.

7.3.4 Axially Periodic Surface Vibration

A helical wave is a building block from which more complicated patterns of surface vibration may be synthesized. How that synthesis is implemented depends on the extent of the vibration along the axis of the cylinder. The simpler situation is that of an axially periodic vibration. In this case, the surface velocity may be described by combining a Fourier series for the z dependence with a Fourier series in θ. This pattern is replicated by the pressure. The more difficult situation to analyze is that in which the surface vibration covers a finite length of the cylinder, with the remainder of the surface being quiescent. That case is taken up in the next section.

Equation (7.3.4) described the surface velocity as a Fourier series composed of circumferential harmonics. Our interest here is the situation where the surface velocity also is periodic over an axial distance L. Such a condition requires that the coefficient of each circumferential harmonic has that period in the axial direction. Thus, the surface velocity may be described by a double Fourier series, with coefficients denoted as $V_{m,n}$ to indicate the associated axial harmonic m and circumferential harmonic n. Hence, the surface velocity in the transverse direction is represented as

$$V_s = \frac{1}{4} \sum_{n=-\infty}^{\infty} \sum_{m=-\infty}^{\infty} V_{m,n} e^{-i(2m\pi x/L + n\theta)} \tag{7.3.36}$$

This velocity distribution is a superposition of helical surface waves whose wavenumbers are $\mu = 2m\pi/L$ in the z direction and n in the θ direction. The corresponding pressure is obtained by adding the contribution associated with each helical wave, which is given by the first of Eq. (7.3.21) for supersonic waves, and Eq. (7.3.32) for subsonic surface waves. To distinguish the two in the summation, we define a cutoff axial harmonic that marks the highest axial harmonic for which $\mu < k$. This value is

$$M = \text{floor}\left(\frac{kL}{2\pi}\right) \tag{7.3.37}$$

The axial harmonics for $m \le M$ are supersonic, so they form the radiated pressure P_{rad}, while the higher harmonics form the evanescent field P_{ev},

$$
\boxed{
\begin{aligned}
P &= P_{\text{rad}} + P_{\text{ev}} \\
P_{\text{rad}} &= -\frac{1}{4} i \rho_0 c \sum_{n=-\infty}^{\infty} \sum_{m=-M}^{M} V_{m,n} \frac{k H_n(\kappa_m R)}{\kappa_m H_n'(\kappa_m a)} e^{-i2m\pi z/L} e^{-in\theta} \\
P_{\text{ev}} &= -\frac{1}{4} i \rho_0 c \sum_{n=-\infty}^{\infty} \sum_{m=M+1}^{\infty} \left[V_{m,n} e^{-i2m\pi z/L} + V_{(-m),n} e^{i2m\pi z/L} \right] \frac{k K_n(\beta_m R)}{\beta K_n'(\beta_m a)} e^{-in\theta}
\end{aligned}
}
$$

$$\tag{7.3.38}$$

A subscript m has been assigned to the wavenumbers of the Bessel functions to indicate that they depend on the axial harmonic number. Specifically,

$$\kappa_m = \left(k^2 - \frac{4m^2\pi^2}{L^2}\right)^{1/2}, \quad \beta_m = \left(\frac{4m^2\pi^2}{L^2} - k^2\right)^{1/2} \tag{7.3.39}$$

The signal generated by a subsonic surface wave does not reach the farfield. The farfield representation is obtained by dropping P_{ev} and applying Eq. (7.3.12) to approximate $H_n(\kappa_n R)$. This leads to

$$P_{\text{ff}} = -\frac{1}{4} i \rho_0 c \sum_{n=-\infty}^{\infty} \sum_{m=-M}^{M} \frac{k V_{m,n}}{\kappa_m H_n'(\kappa_m a)} \left(\frac{2}{\pi \kappa_m a}\right)^{1/2} \left(\frac{a}{R}\right)^{1/2} e^{i(2n+1)\pi/4}$$
$$\times e^{-i(\kappa_m R + n\theta + 2m\pi z/L)}$$

(7.3.40)

Equation (7.3.36) describes V_s as a superposition of helical waves that propagate in the positive z direction ($m > 0$) and negative z direction ($m < 0$), with each set consisting of waves that twist in the direction of increasing θ ($n > 0$) and decreasing θ ($n < 0$). Some individuals prefer to describe the dependencies on θ and z as a superposition of standing waves, rather than propagating helical waves. Such a form may be obtained by applying Euler's identity to the complex exponentials. The result would be four terms that are either $\cos(n\theta)$ or $\sin(n\theta)$ multiplied by either $\cos(2m\pi x/L)$ or $\sin(2m\pi x/L)$.

It is sufficient to determine the time-averaged power radiated from a segment of the cylinder whose length is L because the spatial field is replicated over this distance. An evaluation of radiated power begins by surrounding the radiating cylinder with an infinitely long concentric cylindrical surface. The radius of the surrounding cylinder is taken to be sufficiently large that the farfield approximation applies. The outward normal to the surrounding cylinder is $\bar{e}_R$, so we must construct $(I_R)_{\text{av}}$. The farfield pressure is described by Eq. (7.3.40). When the terms in that series are denoted as $(P_{\text{ff}})_{m,n}$, the particle velocity for each term is described by Eq. (7.3.24). The transverse velocity component is associated with radiation, so we have

$$(I_R)_{\text{av}} = \frac{1}{2\rho_0 c} \sum_{n=-\infty}^{\infty} \sum_{m=-M}^{M} \sum_{j=-\infty}^{\infty} \sum_{\ell=-M}^{M} \text{Re}\left[\frac{\kappa_m}{k} (P_{\text{ff}})_{m,n} (P_{\text{ff}})_{\ell,j}^*\right] \quad (7.3.41)$$

The time-averaged radiated power crossing the surrounding cylinder over an axial distance L is given by

$$\mathcal{P} = \int_{-L/2}^{L/2} \int_{-\pi}^{\pi} (I_R)_{\text{av}} (R d\theta)\, dx \quad (7.3.42)$$

Because the θ and z exponentials in a $(P_{\text{ff}})_{m,n}$ term constitute an orthogonal set, the only terms that give a nonzero contribution are those for which $\ell = m$ and $n = j$. This reduces the description to a double sum, which is i.e. P = 1/8, etc.

$$\mathcal{P} = \frac{1}{8} \rho_0 c L \sum_{n=-\infty}^{\infty} \sum_{m=-M}^{M} \frac{(k/\kappa_m)^2}{H_n'(\kappa_m a)^2} |V_{m,n}|^2 \quad (7.3.43)$$

There is no cross-coupling between individual terms having different axial or circumferential wavenumbers. In other words, each helical surface wave contributes independently to the radiated power. If we had chosen to describe the circumferential variation of the surface velocity in terms of a real Fourier series, we would have found that the individual $\sin(n\theta)$ and $\cos(n\theta)$ terms do not couple with the radiated power.

The expression for $\mathcal{P}$ is consistent with the earlier theorem regarding radiated power. The region contained between the radiating and surrounding cylinders is not dissipative. Hence, the time-averaged power that flows into the fluid at the surface must flow across the surrounding cylinder. The pressure and radial velocity spread as $1/R^{1/2}$, so the radial intensity is proportional to $1/R$. The area of the surrounding cylinder over a spatial period L is $2\pi RL$, so the power that flows across the cylinder over a period does not depend on R.

Axisymmetric Cylindrical Waves

The $n = 0$ circumferential harmonic is descriptive of the vibration of a tube that contains a flowing fluid. If the tube is filled with a fluid, and a pressure wave propagates through that fluid, the walls will be pushed outward, thereby inducing an axisymmetric motion of the walls. All points in the cylinder's surface at a specific axial position z share the same radial velocity, so the pressure field is axisymmetric. The following study of the $n = 0$ harmonic shows the close relation of the result to the field of an infinite line source.

The surface normal velocity of an axisymmetric wave propagating in the z direction is $V_s = V_0 e^{-i2\pi Z/L}$. The corresponding pressure signal may be obtained from Eq. (7.3.38) by letting $m = 1$ and setting $V_{0,1} = 4V_0$ as the sole nonzero coefficient. (In the case, where the wave propagates in the negative z direction, the $m = -1$ velocity coefficient $V_{0,-1}$ would be the nonzero term.) The farfield pressure obtained by dropping all terms other than $n = 0$ in Eq. (7.3.38), then setting $V_{0,1} = 4V_0$, is

$$P = -i\rho_0 c V_0 \left(\frac{ka}{\kappa_1 a}\right) \frac{H_0(\kappa_1 R)}{H_0'(\kappa_1 a)} e^{-i2\pi z/L} \tag{7.3.44}$$

It is instructive to examine the farfield behavior as a function of frequency. There are two independent parameters to consider: ka and L/a. Fig. 7.15 is the result of a calculation for four values of L/a.

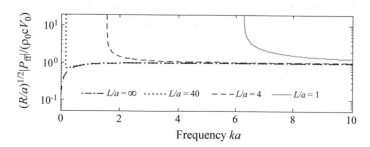

Fig. 7.15 Normalized farfield pressure radiated by a supersonic axisymmetric surface wave propagating axially along an infinite cylinder

The quantity that is plotted is $(R/a)^{1/2} |P_{\rm ff}| / (\rho_0 c V_0)$, which is analogous to the manner in which the farfield of a spherical wave is described, except for the difference in the spreading factor. The minimum plotted frequency for each L/a value is $ka = 2\pi a / L$, because no signal radiates to the farfield below that cutoff. At high frequencies, the plotted quantity approaches one. The singularity at $ka = 2\pi a / L$ stems from Eq. (7.3.40) having $(\kappa_1 a)^{3/2} H_0' (\kappa_1 a)$ in the denominator. If $ka \to 2\pi a / L$, then $\kappa_1 a \to 0$. Differentiation of the first of Eq. (7.3.13) shows that $H_0' (\kappa_1 a) \to -2i/(\pi \kappa_1 a)$ for $\kappa_1 a \ll 1$. This means that the denominator approaches zero as the frequency is reduced to the cutoff value. This singularity means that if we were to induce a vibration amplitude that does not vary with frequency, then a very large pressure would be obtained. The opposite view is that a large radiation impedance indicates that a very large force would be required to sustain the vibration amplitude at a constant value.

The result for infinite L at low frequencies is not singular. The behavior for $L/a = 40$ appears to be a transition between the extremes of infinite and small axial wavelengths. An infinite value of L corresponds to a field that is the same at all z. Because this field only depends on R and θ, it represents a two-dimensional model. Such models often are used to test concepts and examine physical phenomena. Thus, it is useful to examine this case. We set $2\pi/L = 0$ and $\kappa_1 = k$ in Eq. (7.3.38), which gives

$$P = -i \rho_0 c V_0 \frac{H_0 (kR)}{H_0' (ka)} \tag{7.3.45}$$

There is no axial wave in this case, so there is no minimum frequency for radiation to the farfield. When $H_0 (kR)$ is replaced in the above with its approximation for large kR, and $H_0' (ka)$ is approximated as $2i/(\pi ka)$ for small ka, we find that the farfield pressure is proportional to $(ka)^{1/2}$, rather than the $1/(\kappa_1 a)^{1/2}$ proportionality when L is finite and $ka \to 2\pi/L$. Thus, in the two-dimensional case, $(R/a)^{1/2} |P| / (\rho_0 c V_0)$ approaches zero as $ka \to 0$, whereas the finite L case shows a singularity at the lowest ka.

We obtained a point source by shrinking a radially vibrating sphere to zero radius. Similarly, we obtain an infinite *line source* by shrinking the two-dimensional vibrating cylinder to zero radius with the volume velocity per unit length held fixed. The complex amplitude of the radial velocity on the surface is V_0. If we consider a unit length of the cylinder, the surface area is $2\pi a$, so the volume velocity per unit length is $\hat{Q}_s = 2\pi a V_0$. We use this definition to replace V_0, and also invoke the small ka approximation of $H_0' (ka)$, see Eq. (7.3.13). Doing so converts the two-dimensional solution to the line source pressure,

$$P = \frac{1}{4} \rho_0 \omega \hat{Q}_s H_0 (kR) \tag{7.3.46}$$

Recall that the free-space Green's function was defined in Eq. (6.4.29) to be the solution of the Helmholtz equation with a Dirac delta inhomogeneous term, with the additional specification that it has unit mass acceleration. The mass acceleration per unit length of the line source is $i\omega\rho_0 \hat{Q}_s$, so the two-dimensional free-space Green's function is

$$G\left(\bar{x}, \bar{x}_s\right) = \frac{1}{4i} H_0\left(kR\right)$$
$$R = |\bar{x} - \bar{x}_s| = \left[(x - x_s)^2 + (y - y_s)^2\right]^{1/2}$$

(7.3.47)

It often is more convenient to use the monopole amplitude of a line source, rather than the volume velocity. The monopole amplitude A for a point source was defined to be the coefficient of the position-dependent terms. We adopt the same definition here, so that the line source pressure may be written in either of two forms,

$$P = i\omega\rho_0 \hat{Q}_s G\left(\bar{x}, \bar{x}_s\right) = A H_0\left(kR\right)$$
$$\hat{Q}_s = 2\pi a V_0, \quad A = \frac{1}{4}\omega\rho_0 \hat{Q}_s$$

(7.3.48)

This development is important because concepts and procedures that feature Green's functions for three-dimensional situations are equally applicable for a two-dimensional model. For example, we may employ the method of images to analyze the field radiated by a line source above a rigid or pressure-release plane.

EXAMPLE 7.7 The surface of a cylinder whose length is L undergoes a uniform radial vibration, such that $v_R = \mathrm{Re}\left(v_0 e^{i\omega t}\right)$ everywhere. The cylinder has finite length L, and its radius is very small relative to the acoustic wavelength $2\pi/k$. The origin of the xyz coordinate system is situated at the center of the cylinder, with the z-axis coincident with the cylinder's centerline. Derive an expression for the pressure at an arbitrary field point. Use this expression to evaluate $|P|$ as a function of distance from the centerline along the x-axis. Consider $kL = 0.5$, 5, and 50, and compare each result to the field of an infinite line source.

Significance

This is one of the few problems featuring a finite length cylinder that may be solved without recourse to advanced numerical techniques. In addition to providing a quantitative picture that suggests when it might be acceptable to take a cylinder's length to be infinite, it is a useful reminder of the technique for superposing point source fields.

Solution

The cylinder reduces to a line source when its radius is much less than a wavelength. The case of an infinite length is described by Eq. (7.3.48), which may be applied directly. To evaluate the finite length cases, we return to the concept of a source distribution. The cylinder at zero radius may be considered to be a continuous distribution of point sources that are differential segments of the line. A sketch of the system shows the radial distance r from the cylinder segment to the field point, whereas R is the perpendicular distance from the field point to the cylinder. The distance from the origin to a differential segment is denoted as s in order to distinguish this distance from the coordinate z of the field point. Correspondingly, ds is the length of the segment. The field is axisymmetric, so the field point may be situated in the xz plane, as it is shown in Fig. 1.

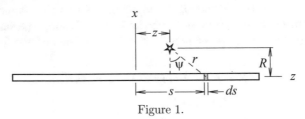

Figure 1.

The radial velocity of a differential segment is $V_R = v_0$ and its surface area is $2\pi a\,ds$, so the complex amplitude of its volume velocity is $d\hat{Q} = \hat{Q}_s ds = v_0 (2\pi a\,ds)$. We retain $\hat{Q}_s$ rather than av_0 because v_0 is not meaningful in the limit as $a \to 0$. The corresponding mass acceleration of the element is $i\omega\rho_0 \hat{Q}_s ds$. The radial distance from the differential element to the field point is $r = [R^2 + (s-z)^2]^{1/2}$, and the (three-dimensional) Green's function is $G(\bar{x}, \bar{x}_s) = \exp(-ikr)/4\pi r$. The pressure at the field point is the sum of the signal received from each differential segment. In other words, it is an integral,

$$P(R, z) = \int G(\bar{x}, \bar{x}_s) \left(i\omega\rho_0 \hat{Q}_s ds \right)$$

$$= i\omega\rho_0 \hat{Q}_s \int_{-L/2}^{L/2} \frac{e^{-ik[R^2+(s-z)^2]^{1/2}}}{4\pi \left[R^2 + (s-z)^2 \right]^{1/2}} (i\omega\rho_0)(2\pi a v_0)\, ds \qquad (1)$$

The requested evaluation places the field point on the x-axis. Setting $z = 0$ in the general expression gives

$$P(R, 0) = i\omega\rho_0 \frac{\hat{Q}_s}{2\pi} \int_0^{L/2} \frac{e^{-ik(R^2+s^2)^{1/2}}}{(R^2 + s^2)^{1/2}}\, ds \qquad (2)$$

where the fact that the integrand only depends on s^2 allows us to double the integral over positive s.

The integral in Eq. (2) is not one that can be found in a standard tabulation. Nevertheless, let us consider an analytical integration. The square root is a complicating feature, so it is reasonable to consider changing the integration variable. We can transform from s to the angle ψ in the sketch. The trigonometric relations are $s = R \tan \psi$, $(R^2 + s^2)^{1/2} = R \sec \psi$, and $ds = R d\psi (\sec \psi)^2$. This does not make the integrand more amenable to analysis. Another transformation that eliminates the square root is $s = R \sinh \zeta$, for which $(R^2 + s^2)^{1/2} = R \cosh \zeta$ and $ds = R (\cosh \zeta) d\zeta$. This is promising because it converts the integral to

$$P(R, 0) = i\omega\rho_0 \frac{\hat{Q}_s}{2\pi} \int_0^{\sinh^{-1}\left(\frac{L}{2R}\right)} e^{-ikR\cosh\zeta} d\zeta \qquad (3)$$

The complex exponential may be decomposed into its real and imaginary parts. A search through integral tables yields results for the case where L is infinite, but not finite. It is worth the effort to consider that limit, because it should agree with Eq. (7.3.48), and thereby confirm our analysis. The tabulated formulas[14] are

$$J_0(kR) = \frac{2}{\pi} \int_0^\infty \sin(kR\cosh\zeta) d\zeta$$

$$N_0(kR) = -\frac{2}{\pi} \int_0^\infty \cos(kR\cosh\zeta) d\zeta$$

Thus, if L were infinite, Eq. (3) would be

$$P(R, 0) = i\omega\rho_0 \frac{Q_0}{2\pi} \int_0^\infty [\cos(kR\cosh\zeta) - i\sin(kR\cosh\zeta)] d\zeta$$

$$= i\omega\rho_0 \frac{Q_0}{2\pi} \left(\frac{\pi}{2}\right) [-N_0(kR) - iJ_0(kr)]$$

The term in the bracket is $-iH_0(kR)$, so the pressure for the case of infinite L reduces to

$$P_{\text{inf}}(R, 0) = i\omega\rho_0 \hat{Q}_s \left(\frac{1}{4i} H_0(kR)\right) \qquad (4)$$

This is the same as (7.3.48). In addition to confirming the analysis, this result tells us that evaluating Eq. (3) for increasing values of L/R should yield results that are close to those obtained for an infinite length.

The only recourse for a finite value of L is a numerical evaluation of the integral. The numerical integration scheme we employ could be one of our own construction using a technique like Simpson's rule. Instead, we shall use a routine provided by MATLAB, which is `quadl`. Although the integrand in Eq. (2) does not have a singularity, there is a potential difficulty in evaluating it for large values of kR. The cause of this difficulty is that kR is like the frequency of harmonic terms. A large value of kR leads to an integrand that is a rapidly oscillating function of ζ. Some say that the integrand has a singularity at infinite ζ. This property must be acknowledged in the numerical integration, either by using an adaptive routine, or by explicitly increasing the number of integration points as the value of kR increases.

The maximum value of kR has not been specified. We know that if $R \gg L$, then the three-dimensional farfield approximation should apply. Because $R/L \equiv$

---

[14]Equation (9.1.23) in M.I. Abramowitz and I.A. Stegun, *Handbook of Mathematical Functions*, Dover, p. 360 (1965).

$kR/(kL)$, this necessitates adjusting the range of kR commensurate with kL. Thus, the computation first sets the kL value. A program loop evaluates Eq. (2) at each kR value from zero to $60kL$, with an increment of $kL/2$.

Figures 2–4 display the results for each kL. The data indicates that the finite and infinite line sources for $kL \geq 4$ agree well in the region from $R = 0$ out to the distance at which $kR = O(kL)$, in other words, $0 \leq R \leq O(L)$. Beyond that range, the infinite line source model overpredicts the pressure. For the shortest length, $kL = 0.5$, the pressure for an infinite line source everywhere exceeds the pressure radiated by the finite cylinder.

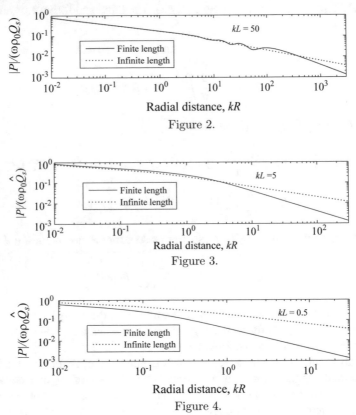

Figure 2.

Figure 3.

Figure 4.

These trends have a simple explanation. If the length is comparable to, or larger than, one wavelength, a field point that is not too distant from the cylinder will most strongly be affected by the signals that radiate from the closest source points ($z \approx 0$). As the field point is moved farther outward, the contribution of the sources near the ends becomes relatively more significant because all points on the cylinder approach being equidistant from the field point. In the case where L is much smaller than a wavelength, the cylinder is acoustically compact. The best simple model in that case is a point source, rather than an infinite line source.

Numerical simulations and experimental measurements of a variety of systems have found that these trends are general. This knowledge allows us to use infinite length models to investigate the nearfield of a finite length body, provided that the length of that body is not much less than a wavelength. A useful corollary is that the radiation impedance of a finite length cylinder may be expected to be close to that of an infinite cylinder if the length is comparable to, or larger than, $2\pi/k$. We will exploit this attribute in Exercise 7.10.

Field Generated by a Vibrating Cable or Beam

When a flexural wave propagates along the z-axis of a cable or beam, all points on a specific cross section displace perpendicularly to the axis by the same amount. In other words, each cross section translates without alteration of its shape. We shall define the x-axis to be such that the displacement $\bar{u}$ lies in the xz plane, so that

$$\bar{u} = \mathrm{Re}\left(\frac{V_x}{i\omega}e^{-i2\pi z/L}e^{i\omega t}\right)\bar{e}_x \tag{7.3.49}$$

The component of this displacement normal to a circular cross section is $\bar{u} \cdot \bar{e}_R$. We measure the circumferential angle θ relative to the x-axis, so $\bar{e}_x \cdot \bar{e}_R = \cos\theta$. Correspondingly, the surface velocity is

$$V_s = V_x e^{-i2\pi z/L}\cos\theta \equiv \frac{1}{2}V_x e^{-i2\pi z/L}\left(e^{-i\theta}+e^{i\theta}\right) \tag{7.3.50}$$

This representation tells us that the surface wave consists of $n = 1$ and $n = -1$ circumferential harmonics. A comparison with Eq. (7.3.36) shows that the only nonzero coefficients in this motion are $V_{1,(-1)} = V_{1,1} = 2V_x$. Thus, the pressure generated by a supersonic wave propagating along a cable or beam is

$$P = -i\rho_0 c V_x \frac{k H_1(\kappa_1 R)}{\kappa_1 H_1'(\kappa_1 a)}e^{-i2\pi z/L}\cos\theta \tag{7.3.51}$$

Figure 7.16 depicts the frequency dependence of the farfield pressure amplitude normalized by $\rho_0 c V_0 (R/a)^{1/2}$. The minimum frequency for radiation is $ka = 2\pi a/L$. The high-frequency limit is seen to be a value of one, as it was for $n = 0$. Rather than being singular, the transition from a subsonic to supersonic surface wave is marked by a zero value for the pressure. To see why this is so, we observe that $H_1(\kappa_1 R)$ is proportional to $1/(\kappa_1 R)^{1/2}$ at large $\kappa_1 R$ (farfield), while $H_1'(\kappa_1 a) \to -(2i/\pi)/(\kappa_1 a)^2$ as $\kappa_1 a \to 0$ (small radius). The pressure that results from combining these limiting forms is

$$P = \rho_0 c V_x \left(\frac{\pi}{2}\right)^{1/2} ka\,(\kappa_1 a)^{1/2}\left(\frac{a}{R}\right)^{1/2}e^{-i(\kappa_1 R - 3\pi/4 - 2\pi z/L)}\cos\theta \tag{7.3.52}$$

Thus rather than P being singular, it approaches a proportionality to $(\kappa_1 a)^{1/2}$ as $k \to 2\pi/L$. However, if L/a is not too large, the figure indicates that values of ka slightly larger than $2\pi a/L$ give rise to large pressures.

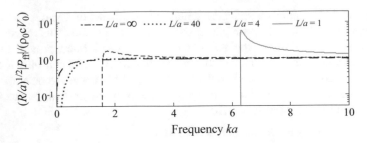

Fig. 7.16 Normalized farfield pressure radiated by a supersonic flexural wave propagating axially along an infinite cylinder

A two-dimensional model of a translating beam is obtained by letting $L \to \infty$. When we studied point sources, we found that a translating sphere generates a dipole field, and that a dipole field may be obtained from a Taylor series expansion of the field generated by two oppositely phased monopoles. Let us see whether the same is true in the two-dimensional case. We consider two line sources at $x = d/2$ and $x = -d/2$, where their placement in the xz plane gives symmetry relative to that plane, which is the property of $\cos\theta$. The mass accelerations are $\pm i\omega\rho_0 \hat{Q}_s$, so that

$$P_{\text{dipole}} = i\omega\rho_0 \hat{Q}_s G\left(\bar{x}, \bar{x}_0 + \frac{d}{2}\bar{e}_x\right) - i\omega\rho_0 \hat{Q}_s G\left(\bar{x}, \bar{x}_0 - \frac{d}{2}\bar{e}_x\right) \quad (7.3.53)$$

where the Green's function is as given in Eq. (7.3.47) and $\bar{x}_0$ is situated on the z-axis. Furthermore, for a two-dimensional analysis, we take $z = 0$ for both the field and source points. We apply a Taylor series to both terms, and switch the gradient at $\bar{x}_0$ for the gradient at $\bar{x}$, which changes the sign associated with each gradient. Thus, we have

$$P = i\omega\rho_0 \hat{Q}_s \left[-d\bar{e}_x \cdot \nabla G\left(\bar{x}, \bar{x}_0\right)\right] \quad (7.3.54)$$

Now we observe that $G(\bar{x}, \bar{x}_0)$ depends only on the distance R between the points in the xy plane, and that the gradient of R at the field point is a unit vector pointing in the direction of increasing R, which is $\bar{e}_R$. This leads to

$$\begin{aligned}
P &= -i\omega\rho_0 \hat{Q}_s d\,(\bar{e}_x \cdot \bar{e}_R)\, \frac{\partial}{\partial R} G\left(\bar{x}, \bar{x}_0\right) \\
&= -\frac{1}{4}\omega\rho_0 \hat{Q}_s d\,(\cos\theta)\, H_1\,(kR)
\end{aligned} \quad (7.3.55)$$

The dipole moment D is the product of the monopole amplitude and the separation distance. Introduction of A from Eq. (7.3.48) into the preceding leads to

$$P = D H_1\,(kR)\cos\theta, \quad D = Ad = -\frac{1}{4}\omega\rho_0 \hat{Q}_s d \quad (7.3.56)$$

A comparison of this result with Eq. (7.3.51) for $a/L = 0$ shows that the dipole moment associated with a translating rigid cylinder is

$$D_{\text{cyl}} = -\frac{i\rho_0 c V_x}{H_1'(\kappa_1 a)} \qquad (7.3.57)$$

In principle, we could mirror every development using three-dimensional point sources with an analogous one for the two-dimensional field for a set of infinite line source. For example, we could develop a multipole expansion in two dimensions. In practice, this is seldom done because of the artificiality of the infinite length model.

EXAMPLE 7.8 When the string of a musical instrument is plucked, the free vibration that ensues is a standing wave that consists of a series of mode functions. The displacement of mode number m is proportional to $\sin(m\pi z/\ell)\cos\theta$ and its frequency is $m\omega_1$, where ω_1 is the fundamental frequency and ℓ is the string's length. In the situation of interest, the displacement amplitude U_m of modes 1 to 10 is known to be inversely proportional to the square of the mode number, and the contributions of higher modes are negligible. Thus, $U_m = U_1/m^2$ if $m = 1, ..., 10$, $U_m = 0$ if $m > 10$, where U_1 is the amplitude of the fundamental. Consider a model of a cello string that ignores the presence of the cello body and takes the string to be an infinitely cylinder that undergoes an axially periodic vibration. The string's length is 670 mm and its diameter is 1.20 mm. For the case where the sound pressure level at 2 m from the string is measured to be 75 dB and the fundamental frequency is 440 Hz, estimate the RMS displacement amplitude of the string.

Significance

Inverse radiation problems, in which a system's vibration is deduced from measured properties of the pressure field, help reinforce understanding of an analytical solution. They also mimic, in a small way, the nature of some experimental investigations.

Solution

We know the pressure at a field point and must deduce properties of the vibration that generated it. Our strategy is to reverse the problem. The only quantity not specified for the displacement is the amplitude U_1 of the $m = 1$ mode. It will be determined by computing $(p^2)_{\text{av}}$ in terms of U_1, then matching that result to the value corresponding to 75 dB//20 μPa, which is 0.01265 Pa^2.

The specified displacement of the cable, which we take to be in the x direction, is

$$\bar{u} = \text{Re} \sum_{m=1}^{10} \frac{U_1}{m^2} \sin\left(\frac{m\pi z}{\ell}\right) e^{im\omega_1 t} \bar{e}_x \qquad (1)$$

We differentiate this expression with respect to time, then take the radial component in order to determine the normal velocity on the surface. Because $\bar{e}_x \cdot \bar{e}_r = \cos\theta$, the resulting expression is

$$v_s = \mathrm{Re} \sum_{m=1}^{10} \frac{i\omega_1 U_1}{m} \sin\left(\frac{m\pi z}{\ell}\right) (\cos\theta)\, e^{im\omega_1 t}$$

We will use Eq. (7.3.51) to describe the pressure, so we employ Euler's identity to decompose v_s into waves traveling in opposite directions, which gives

$$v_s = \mathrm{Re} \sum_{m=1}^{10} \frac{U_1 \omega_1}{4m} \left(e^{im\pi z/\ell} - e^{-im\pi z/\ell}\right) \left(e^{i\theta} + e^{-i\theta}\right) e^{im\omega_1 t} \tag{2}$$

This expression tells us that each time harmonic consists of a pair of helical waves at $n = \pm 1$ that spiral about the z-axis in opposite senses. The velocity coefficients are $V_{m,1} = V_{m,-1} = -V_{-m,1} = -V_{-m,-1} = \omega U_1/(2m)$.

The nature of the pressure generated by each wave depends on whether it is supersonic. The frequency of a term in Eq. (2) is $m\omega_1$ and its axial wavenumber is $m\pi/\ell$, so that

$$\kappa_m = \left[\left(\frac{m\omega_1}{c}\right)^2 - \left(\frac{m\pi}{\ell}\right)^2\right]^{1/2} = m\kappa_1 = 6.780m \text{ m}^{-1} \tag{3}$$

All values of κ_m are real, which means that all of the terms in Eq. (1) are supersonic helical waves.

Each term in Eq. (2) occurs at a different frequency, so the time factor must be retained when Eq. (7.3.51) is used to evaluate the pressure. In each term of that expression, we replace k with $m\omega_1/c$, κ_1 with $m\kappa_1$, and $2/L$ with m/ℓ. The result is

$$\begin{aligned}
p &= \mathrm{Re}\left[-i\rho_0 c \sum_{m=1}^{10} \frac{U_1 \omega_1}{2m} \left(e^{im\pi z/\ell} - e^{-im\pi z/\ell}\right) \left(\frac{\omega_1}{c\kappa_1}\right) \frac{H_1(m\kappa_1 R)}{H_1'(m\kappa_1 a)} (\cos\theta)\, e^{im\omega_1 t}\right] \\
&= \mathrm{Re}\left[-i\rho_0 \frac{\omega_1^2}{\kappa_1} U_1 \sum_{m=1}^{10} \frac{H_1(m\kappa_1 R)}{m H_1'(m\kappa_1 a)} \sin\left(\frac{m\pi z}{\ell}\right) (\cos\theta)\, e^{im\omega_1 t}\right]
\end{aligned} \tag{4}$$

Each term in Eq. (4) represents an oscillation at a distinct frequency, so the mean-squared pressure is a sum of squares of the individual amplitudes. Thus, the predicted mean-square pressure is

$$\left(p^2\right)_{av} = \frac{1}{2} \sum_{m=1}^{\infty} P_m P_m^* = \frac{1}{2}\left(\frac{\rho_0 \omega_1^2}{\kappa_1}|U_1|\right)^2 \sum_{m=1}^{10} \left|\frac{H_1(m\kappa_1 R)}{m H_1'(m\kappa_1 a)}\right|^2 \left[\sin\left(\frac{m\pi z}{\ell}\right)\right]^2 (\cos\theta)^2 \tag{5}$$

The transverse distance for the pressure measurement is specified to be $R = 2$ m, but the values of z and θ at which the pressure was measured are not given. It is reasonable to assume that the value of θ is that which leads to a maximum, which is

$\theta = 0$ or π, both of which give $(\cos\theta)^2 = 1$. In contrast, no value of z is evident as leading to a maximum. For example, if we select the midpoint, $z = \ell/2$, the terms for even m will be zero. Thus, let use the value of $(p^2)_{\text{av}}$ averaged over the length of the string, which is

$$(p^2)_{\text{av,av}} = \frac{1}{\ell}\int_0^\ell (p^2)_{\text{av,av}}\Big|_{\theta=0}\, dz = \frac{1}{4}\left(\frac{\rho_0\omega_1^2}{\kappa_1}|U_1|\right)^2 \sum_{m=1}^{10}\left|\frac{H_1(m\kappa_1 R)}{mH_1'(m\kappa_1 a)}\right|^2 \quad (6)$$

The result of substituting all quantities into Eq. (6) is $(p^2)_{\text{av,av}} = 784.3\,|U_1|^2$. Matching this to the measured mean-squared pressure of 0.01265 Pa^2 leads to $|U_1| = 4.016$ mm. With this, we have fully characterized the displacement. Its mean-squared value is obtained by averaging $\bar{u} \cdot \bar{u}$ in Eq. (1) over $0 < t < \pi/\omega_1$ and $0 < x < \ell$. The result is

$$\left(u_x^2\right)_{\text{av}} = \frac{|U_1|^2}{4}\sum_{m=1}^{10}\frac{1}{m^4} = 4.363\left(10^{-6}\right)\text{ m}^2 \implies (u_x)_{\text{rms}} = 2.08\text{ mm} \quad (7)$$

The actual displacement amplitude of the strings of a cello is barely perceptible to the eye. The value we have determined is too large. The discrepancy between our analysis and reality is only partially the consequence of taking the string's vibration to be spatially periodic and using a model of an infinitely long string. The primary fault is that the strings are an extremely minor contributor to the sound emitted by a stringed musical instrument. The role of the strings is to induce vibration of the instrument's body. This vibration radiates sound directly, and it also excites the cavity enclosed by the body. The sound created in the cavity is emitted through the open ports of the body.

7.3.5   Finite Length Effects

Two aspects of the vibrating cylinder model we have created limit its practical utility. The most obvious one is taking the cylinder to be infinitely long. Unfortunately, there are no analytical solutions for radiation of finite length cylinders. If one requires an accurate solution of such problems, all available tools use numerical techniques. We will examine a few later in this chapter.

Even within the context of a model that takes a cylinder to have infinite length, there is another unrealistic aspect. We have considered helical waves and axially periodic vibrations, which leads to the question of how would one generate such a disturbance? A more realistic model allows the surface velocity to depend in an arbitrary manner on the axial distance, subject to the condition that it be negligible at large distances in either direction. In order to not obscure the analysis, we will restrict attention to the case of a single circumferential harmonic. The solutions may be superposed to create the Fourier series for an arbitrary circumferential excitation.

The surface velocity for this investigation is

$$v_s = \mathrm{Re}\left(f\left(z\right) e^{-in\theta} e^{i\omega t}\right), \quad \lim_{|z|\to\infty} |f\left(z\right)| \to 0 \tag{7.3.58}$$

This distribution is not periodic in the axial direction, which is equivalent to saying that its period is infinite. The conditions imposed on $f\left(z\right)$ fit the requirements for its Fourier transform to exist. The transform, which is denoted as $\tilde{F}\left(\mu\right)$, is

$$\tilde{F}\left(\mu\right) = \int_{-\infty}^{\infty} f\left(z\right) e^{+i\mu z} dz \tag{7.3.59}$$

As explained in Appendix B, the Fourier transform of $f\left(z\right)$ essentially describes a Fourier series whose wavenumber is a continuous spectrum μ, rather than a set of discrete values $n\mu_1$. The series coefficient at each μ is $\tilde{F}\left(\mu\right) d\mu$. This perspective is especially evident when the inverse Fourier transform is used to synthesize $f\left(z\right)$. It tells us that

$$V_s = \frac{1}{2\pi} \int_{-\infty}^{\infty} \tilde{F}\left(\mu\right) e^{-i(\mu z + n\theta)} d\mu \tag{7.3.60}$$

The elementary view of an integral is that it is an infinite sum of terms described by its integrand. From that viewpoint, the above description states that the surface velocity is an infinite sum of helical waves whose axial wavenumber is μ, whose circumferential wavenumber is n, and whose complex amplitude is $\tilde{V}\left(\mu\right) d\mu$. The fact that the number of terms in the sum is infinite does not affect the validity of superposing the pressure field for each helical wave. Equation (7.3.21) gives the pressure for a single helical wave, so the superposition gives

$$P\left(R, \theta, z\right) = -i\rho_0 c \int_{-\infty}^{\infty} \tilde{F}\left(\mu\right) \frac{k H_n\left(\kappa\left(\mu\right) R\right)}{\kappa\left(\mu\right) H_n'\left(\kappa\left(\mu\right) a\right)} e^{-i(\mu z + n\theta)} d\mu \tag{7.3.61}$$

where

$$\kappa\left(\mu\right) = \left(k^2 - \mu^2\right)^{1/2} \tag{7.3.62}$$

The transverse wavenumber has been written as $\kappa\left(\mu\right)$ as a reminder that its dependence must be recognized when the integral is evaluated. The portion of the integrand in Eq. (7.3.61) that is a factor of $e^{-i\mu z}$ is a Fourier transform. Thus, we have obtained a representation of P as an inverse Fourier transform.

The preceding is a heuristic derivation. A more rigorous analysis provides understanding as to how one may apply the Fourier transform in other situations. We begin with the argument that the dependence of P on the axial position also may be represented by a Fourier transform. In addition to depending on the axial wavenumber μ, this transform may depend on the transverse distance R, whereas $\exp\left(-in\theta\right)$ must describe the circumferential dependence. If we knew the transformed function, which is denoted as $\tilde{P}\left(R, \mu\right)$, then the inverse Fourier transform would enable us to recreate $P\left(R, \theta.z\right)$ according to

$$P(R, \theta, z) = \frac{1}{2\pi} \int_{-\infty}^{\infty} \tilde{P}(R, \mu) e^{-i\mu z} e^{-in\theta} d\mu \qquad (7.3.63)$$

This ansatz must satisfy the Helmholtz equation in cylindrical coordinates. Spatial derivatives of the above are

$$\frac{\partial}{\partial R} P(R, \theta, z) = \frac{1}{2\pi} \int_{-\infty}^{\infty} \frac{\partial}{\partial R} \tilde{P}(R, \mu) e^{-i\mu z} e^{-in\theta} d\mu$$

$$\frac{\partial}{\partial \theta} P(R, \theta, z) = \frac{1}{2\pi} \int_{-\infty}^{\infty} (-in) \tilde{P}(R, \mu) e^{-i\mu z} e^{-in\theta} d\mu \qquad (7.3.64)$$

$$\frac{\partial}{\partial z} P(R, \theta, z) = \frac{1}{2\pi} \int_{-\infty}^{\infty} (-i\mu) \tilde{P}(R, \mu) e^{-i\mu z} e^{-in\theta} d\mu$$

Second-order derivatives follow directly from these forms, so substitution of the assumed solution into the Helmholtz equation leads to

$$\int_{-\infty}^{\infty} \left[\frac{d^2 \tilde{P}}{d R^2} + \frac{1}{R} \frac{d\tilde{P}}{dR} + \left(\kappa(\mu)^2 - \frac{n^2}{R^2} \right) \tilde{P} \right] e^{-i\mu z} e^{-in\theta} d\mu = 0 \qquad (7.3.65)$$

The preceding must be satisfied at all z, which will only be true if the bracketed term in the integrand vanishes. Thus, $\tilde{P}(R, \mu)$ must be a solution of Bessel's equation. As previous, an outgoing wave is obtained by taking it to be a Hankel function of the second kind. It may be multiplied by a coefficient that depends on μ and n, so we now have

$$P(R, \theta, z) = \int_{-\infty}^{\infty} B(\mu, n) H_n(\kappa(\mu) R) e^{-i\mu z} e^{-in\theta} d\mu \qquad (7.3.66)$$

The coefficient $B(\mu, n)$ is set by the boundary condition, which matches V_s to the complex particle velocity in the radial direction at $R = a$. A derivative with respect to R for Euler's equation operates on the Hankel function, so using Eq. (7.3.60) to represent the surface velocity leads to

$$\int_{-\infty}^{\infty} \kappa(\mu) B(\mu, n) H_n'(\kappa(\mu) a) e^{-i\mu z} e^{-in\theta} d\mu = -i\omega\rho_0 \int_{-\infty}^{\infty} \tilde{F}(\mu) e^{-i(\mu z + n\theta)} d\mu$$

$$(7.3.67)$$

Here too, the equality must hold for all z, so the integrands must be equal. From this, we solve for $B(\mu, n)$. Backsubstituting that results into the integral representation of P leads to Eq. (7.3.61), which verifies the heuristic approach.

The spectrum of surface waves consists of supersonic waves for $|\mu| < k$ and subsonic waves for $|\mu| > k$. For subsonic waves, let us set $\kappa(\mu) = -i\beta(\mu)$, where $\beta = (\mu^2 - k^2)^{1/2}$. It is helpful for evaluations to rewrite Eq. (7.3.61) in a form that splits the two in order to explicitly display the contributions of the supersonic and subsonic spectra. For this, we note that neither κ nor β depends on the sign of μ, so we may reduce the integral to extend only over $\mu \geq 0$. These operations lead to

$$P (R, \theta, z) = -i\rho_0 c \int_0^k \left[\tilde{F} (\mu) + \tilde{F} (-\mu) \right] \frac{k H_n (\kappa (\mu) R)}{\kappa (\mu) H_n' (\kappa (\mu) a)} e^{-i(\mu z + n\theta)} d\mu$$
$$+ \rho_0 c \int_k^\infty \left[\tilde{F} (\mu) + \tilde{F} (-\mu) \right] \frac{k K_n (\beta (\mu) R)}{\beta (\mu) K_n' (\beta (\mu) a)} e^{-(\mu z + n\theta)} d\mu$$

$$(7.3.68)$$

An important aspect of this representation is that the contribution from $\mu > k$ decays rapidly with increasing R and z. Therefore, the second integral may be ignored if we seek a farfield description.

Equation (7.3.68) is an inverse Fourier transform. It is an implementation of a general analytical approach known as an angular spectrum decomposition. We will encounter in the next chapter another system that has been analyzed by this technique. Many research papers have derived solutions in the form of inverse Fourier transforms, and then used sophisticated tools of mathematical analysis to identify important properties. Such analysis exceeds the scope of this book. An alternative is to employ numerical methods. The last part of Appendix B describes how to use an FFT numerical procedure to evaluate the Fourier transform of a spatial function, and how to convert a Fourier transform back to a spatial function by application of an inverse FFT procedure. Thus, it might seem viable to employ FFT techniques to evaluate Eq. (7.3.61). Despite the apparent simplicity of this approach, there are several issues regarding aliasing and wraparound error, which were discussed in Chap. 1, that must be addressed. An additional complication arises as either R or z is increased, because the integrands become rapidly oscillating functions of μ. This raises the sampling requirements, and sometimes leads to numerical instabilities.

Ultimately, evaluation of the pressure field would have limited use. As was shown in Example 7.7, the fact that the cylinder is not infinite becomes important at radial distances that exceed the length. Furthermore, even if we are only interested in the field close to a cylinder, end effects cannot be ignored at locations that are close to an end.

7.4 Kirchhoff–Helmholtz Integral Theorem

It should be apparent at this juncture that we need tools other than separation of variables solutions of the Helmholtz equation. The first step toward that objective is application of the definition of a Green's function to derive the Kirchhoff–Helmholtz integral theorem, which we will refer to as the KHIT. It is the basis for some common numerical modeling technique, but it is not a solution of any problem. Rather, it will tell us how to evaluate the pressure at a field point when we know the pressure and normal velocity on the surface from which this signal emanates. The laws of mechanics, as well as our experience thus far, tell us that both quantities cannot be specified at the same location. (This fact is merely an extension of Newton's second law for a particle, which tells us that we can determine the velocity as a function of time if we know the force history, or we can determine the force required to attain a specified velocity history, but we cannot specify both histories.) The KHIT version we will derive pertains to the response in the frequency domain.

7.4.1 Derivation for an Acoustic Cavity

The development begins with the simultaneous consideration of two acoustic fields: the frequency domain pressure radiated by vibration of the boundary and a Green's function. Usually, the free-space Green's function is the one we would use, but all steps will be equally valid of the Green's function satisfies boundary conditions for the system at hand. We let $\mathcal{V}$ denote the domain of interest, and $\mathcal{S}$ is its boundary. For our initial effort, we consider $\mathcal{V}$ to be enclosed by $\mathcal{S}$. In other words, it is a cavity whose shape is arbitrary. The configuration is depicted in Fig. 7.17.

Fig. 7.17 Configuration of an acoustic cavity

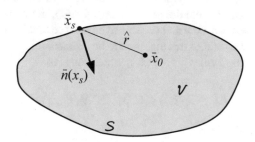

The pressure at a field point $\bar{x}$ satisfies the Helmholtz equation. The Green's function for this point due to a source at $\bar{x}_0$, which is $G(\bar{x}_0, \bar{x})$, satisfies an inhomogeneous version of that equation, specifically

$$\nabla^2 P + k^2 P = 0, \quad \nabla^2 G + k^2 G = -\delta(\bar{x} - \bar{x}_0) \tag{7.4.1}$$

We multiply the equation for P by $G(\bar{x}_0, \bar{x})$ and the equation for G by $P(\bar{x})$. Taking the difference yields

$$G(\bar{x}_0, \bar{x}) \nabla^2 P(\bar{x}) - P(\bar{x}) \nabla^2 G(\bar{x}_0, \bar{x}) = P(\bar{x}) \delta(\bar{x} - \bar{x}_0) \tag{7.4.2}$$

This expression is multiplied by a differential element $d\mathcal{V}$ at $\bar{x}$ and integrated over the entire domain. Gradients of products are covered by the same rule as the scalar derivative of a product, and $\nabla^2 \equiv \nabla \cdot \nabla$, so we have

$$\iiint\limits_{\mathcal{V}} \left[G\left(\bar{x}_0, \bar{x}\right) \nabla^2 P\left(\bar{x}\right) - P\left(\bar{x}\right) \nabla^2 G\left(\bar{x}_0, \bar{x}\right) \right] d\mathcal{V}$$

$$\equiv \iiint\limits_{\mathcal{V}} \nabla \cdot \left[G\left(\bar{x}_0, \bar{x}\right) \nabla P\left(\bar{x}\right) - P\left(\bar{x}\right) \nabla G\left(\bar{x}_0, \bar{x}\right) \right] d\mathcal{V}$$

$$= \iiint\limits_{\mathcal{V}} P\left(\bar{x}\right) \delta\left(\bar{x} - \bar{x}_0\right) d\mathcal{V} \qquad (7.4.3)$$

The filtering property of a Dirac delta function simplifies the right side. If $\bar{x}_0$ is inside $\mathcal{V}$, then the integral will give P at $\bar{x}_0$. This is the case that usually interests us, but there also is reason to consider $\bar{x}_0$ outside $\mathcal{V}$. Let us denote $\chi\left(\bar{x}_0\right)$ as a coefficient that captures both alternatives,

$$\chi = \begin{cases} 1 \text{ if } \bar{x}_0 \in \mathcal{V} \\ 0 \text{ if } \bar{x}_0 \notin \mathcal{V} \end{cases} \qquad (7.4.4)$$

The left side of Eq. (7.4.3) is described by the divergence theorem. The differential element is situated at $\bar{x}$ within $\mathcal{V}$. To assure that we recognize that the resulting surface integral applies to field points that are on $\mathcal{S}$, we designate those points as $\bar{x}_s$ and denote the gradient as ∇_s. Also, the usual statement of the divergence theorem uses the normal vector on the surface that is oriented outward from $\mathcal{V}$, but our studies have taken this direction to be oriented into the fluid. Thus, we shall use the divergence theorem with $-\bar{n}\left(\bar{x}_s\right)$, which leads to

$$\iint\limits_{\mathcal{S}} \left(-\bar{n}\left(\bar{x}_s\right)\right) \cdot \left[G\left(\bar{x}_0, \bar{x}_s\right) \nabla_s P\left(\bar{x}_s\right) - P\left(\bar{x}_s\right) \nabla_s G\left(\bar{x}_0, \bar{x}_s\right) \right] d\mathcal{S} = \chi\left(\bar{x}_0\right) P\left(\bar{x}_0\right)$$

$$(7.4.5)$$

The gradient of $P\left(\bar{x}_s\right)$ is related to the complex particle velocity amplitude by Euler's equation, whose application leads to the final form of the Kirchhoff–Helmholtz integral theorem (KHIT),

$$\chi\left(\bar{x}_0\right) P\left(\bar{x}_0\right) = \iint\limits_{\mathcal{S}} \left[i\omega\rho_0 \bar{n}\left(\bar{x}_s\right) \cdot \bar{V}\left(\bar{x}_s\right) G\left(\bar{x}_0, \bar{x}_s\right) \right. \\ \left. + P\left(\bar{x}_s\right) \bar{n}\left(\bar{x}_s\right) \cdot \nabla_s G\left(\bar{x}_0, \bar{x}_s\right) \right] d\mathcal{S} \qquad (7.4.6)$$

This relation has a simple interpretation in terms of point sources. For the first term in the integrand, we observe that $i\omega\rho_0 \bar{n}\left(\bar{x}_s\right) \cdot V\left(\bar{x}_s\right) d\mathcal{S}$ is a complex amplitude of the mass acceleration across a differential patch of the boundary. The pressure radiated by a point source having this mass acceleration is $i\omega\rho_0 \bar{n}\left(\bar{x}_s\right) \cdot V\left(\bar{x}_s\right) d\mathcal{S} G\left(\bar{x}_0, \bar{x}_s\right)$. The interpretation of the second term follows from the observation that a similar term occurred in Eq. (6.5.90) for a dipole. A comparison with that term shows that the $-P\left(\bar{x}_s\right) \bar{n}\left(\bar{x}_s\right) d\mathcal{S}/4\pi$ is a differential dipole moment that is oriented perpendicularly to the surface. Thus, we may regard the KHIT as a statement that a vibrating surface acts as though it was a continuous sheet of monopoles and dipoles.

The KHIT that is found in a text or a research paper might not exactly replicate Eq. (7.4.6). One difference might arise from the use of Re $(\exp(-i\omega t))$ to represent a harmonic variation. Here, like everywhere else in this book, this alternate convention leads to a complex conjugate representation. Another difference might lie in the definition of the Green's function, which might insert a minus sign and/or omit the 4π factor. Differences in the definitions of the Green's function cease to be an issue if the definition is explicitly substituted into the theorem. The free-space Green's function we have defined is

$$G(\bar{x}_0, \bar{x}_s) = \frac{1}{4\pi\hat{r}}e^{-ik\hat{r}}, \quad \hat{r} = |\bar{x}_s - \bar{x}_0| \tag{7.4.7}$$

(The use of $\hat{r}$ to denote the source-field point distance is intended to avoid confusion with r as the radial coordinate.) The meaning of $\nabla_s G(\bar{x}_0, \bar{x}_s)$ in the integrand is that it is the gradient at $\bar{x}_s$ with $\bar{x}_0$ held fixed. This gradient was evaluated in Sect. 6.5.1 when we used a Taylor series to derive the dipole field. The direction in which $\hat{r}$ increases most rapidly when $\bar{x}_0$ is fixed is the direction of increasing $\hat{r}$. Therefore, the procedure we followed earlier yields

$$\bar{n}(\bar{x}_s) \cdot \nabla_s G(\bar{x}_0, \bar{x}) = \bar{n}(\bar{x}_s) \cdot \nabla_s \hat{r} \frac{\partial}{\partial \hat{r}}\left(\frac{1}{4\pi\hat{r}}e^{-ik\hat{r}}\right)$$
$$= -\bar{n}(\bar{x}_s) \cdot \left(\frac{\bar{x}_s - \bar{x}_0}{\hat{r}}\right)\left(\frac{ik\hat{r}+1}{4\pi\hat{r}^2}\right)e^{-ik\hat{r}} \tag{7.4.8}$$

Thus, the KHIT version that explicitly displays the free-space Green's function is

$$\boxed{\begin{aligned}\chi(\bar{x}_0)P(\bar{x}_0) = \frac{1}{4\pi}\iint_{\mathcal{S}}&\left[i\omega\rho_0\bar{n}(\bar{x}_s) \cdot \bar{V}(\bar{x}_s)\right.\\&\left. - \bar{n}(\bar{x}_s) \cdot \left(\frac{\bar{x}_s - \bar{x}_0}{\hat{r}}\right)\left(ik + \frac{1}{\hat{r}}\right)P(\bar{x}_s)\right]\frac{e^{-ik\hat{r}}}{\hat{r}}d\mathcal{S}\end{aligned}}$$

$$\tag{7.4.9}$$

The KHIT has not been used frequently to study two-dimensional systems. If one wishes to do so, it would be incorrect to employ the preceding form. The proper formulation would use Eq. (7.4.6) with the two-dimensional free-space Green's function in Eq. (7.3.47), which is $G(\bar{x}, \bar{x}_s) = H_0(kR)/(4i)$.

As was noted before we began the derivation, the pressure and particle velocity cannot both be specified on $\mathcal{S}$. Thus, it is reasonable to question whether the KHIT has practical use. Indeed, it does! It may be used to verify a solution that was obtained by another method, but it seldom is. A common application uses the KHIT to evaluate a farfield pressure field when the surface response has been obtained by a numerical simulation technique, or by experimental measurements. An important version of the

KHIT lies in the case not covered by the definition of $\chi(\bar{x}_0)$, Eq. (7.4.4), specifically $\bar{x}_0$ being situated on $\mathcal{S}$. In that case, the KHIT gives an equation for the surface pressure at a specific field point on the surface in terms of the surface pressure and velocity distributions. Just as a differential equation relates a variable to its derivatives, this version of the KHIT is an integral equation for the pressure. The solution of this equation by numerical methods is known as a boundary element formulation. This concept is explored in Sect. 7.5.2.

EXAMPLE 7.9 Consider a spherical cavity in the case where the container executes a radially symmetric oscillation. In that case, the complex radial velocity on the boundary, $r = a$, is $V_r = V_0$. This field was analyzed in Sect. 6.3.3. It is desired to use KHIT to verify the analytical solution for the radially symmetric pressure in the interior. To do so, use the analytical solution to determine the pressure at $r = a$, then use that distribution to formulate the KHIT. Evaluate the pressure as a function of kr according to KHIT, and compare it to the analytical solution for the pressure field. The frequency for the evaluation is $ka = \pi$.

Significance

The formulation of KHIT for a system whose geometry is not too complicated allows us to focus on the meaning of each term. The computed results will suggest an attribute that is crucial to the application of the KHIT as the foundation of a boundary element formulation.

Solution

Equation (6.3.27) states that the radially symmetric pressure and particle velocity inside a vibrating sphere are

$$
\begin{aligned}
P &= -i\rho_0 c V_0 \left(\frac{a}{r}\right) \frac{ka\sin(kr)}{ka\cos(ka) - \sin(ka)} \\
V_r &= V_0 \left(\frac{a}{r}\right)^2 \frac{kr\cos(kr) - \sin(kr)}{ka\cos(ka) - \sin(ka)}
\end{aligned}
\tag{1}
$$

At $r = a$, these expressions give

$$
P = -i\rho_0 c V_0 \frac{ka\sin(ka)}{ka\cos(ka) - \sin(ka)}, \quad V_r = V_0
\tag{2}
$$

The task is to use Eq. (2) to form the integrand of the KHIT, then evaluate the pressure within the cavity, and compare it to the pressure in Eq. (1). The field is radially symmetric, so there is no difference in the dependence of P along any radial line. We shall designate this line as the polar axis z because doing so simplifies the formulation. Figure 1 shows a generic surface point $\bar{x}_s$ at polar angle ψ, and field point $\bar{x}_0$ at distance z_0 on the z-axis. (This picture is descriptive of all points at the same value of ψ for any azimuthal angle θ.)

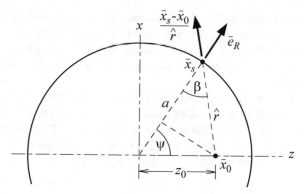

Figure 1.

We could obtain a description of the terms in the KHIT integrand from a trigonometric analysis. However, one objective here is to get an idea of how to formulate the theorem in a general situation where the geometrical configuration is less amenable to a pictorial representation. Thus, all vectorial quantities will be represented in terms of components relative to xyz. The normal direction points into the spherical cavity, so the terms appearing in Eqs. (7.4.7) and (7.4.8) are

$$\bar{x}_s = a \left(\sin \psi \bar{e}_x + \cos \psi \bar{e}_z \right), \quad \bar{x}_0 = z_0 \bar{e}_z$$

$$\hat{r} = |\bar{x}_s - \bar{x}_0| = \left(a^2 + z_0^2 - 2az_0 \cos \psi \right)^{1/2}$$

$$\bar{n}\,(\bar{x}_s) = -\bar{e}_r = -\sin \psi \bar{e}_x - \cos \psi \bar{e}_z \tag{1}$$

$$-\bar{n}\,(\bar{x}_s) \cdot \frac{\bar{x}_s - \bar{x}_0}{\hat{r}} = \cos \beta = \frac{a - z_0 \cos \psi}{\hat{r}}$$

An axisymmetric differential area element is $2\pi \,(a \sin \psi) \,(ad\psi)$, so we have characterized the geometric variables required to form the KHIT. For $P\,(\bar{x}_s)$, we use the first of Eq. (2). Because the normal direction is defined to point into the fluid domain, the surface velocity is $\bar{n}\,(\bar{x}_s) \cdot \bar{V}\,(\bar{x}_s) = -V_r = -V_0$. All parameters are specified in nondimensional form, so we shall replace a, $\hat{r}$, and z_0 with ka, $k\hat{r}$, and kz_0 everywhere.

The descriptions are substituted into the KHIT, with the result that we must evaluate

$$\frac{\chi P\,(\bar{x})}{\rho_0 c v_0} = -\frac{i}{2}\,(ka)^2 \left[\mathcal{I}_1 + \frac{ka \sin (ka)}{ka \cos (ka) - \sin (ka)} \mathcal{I}_2 \right] \tag{2}$$

where

$$\mathcal{I}_1 = \int_0^\pi \frac{e^{-ik\hat{r}}}{k\hat{r}} \sin \psi d\psi$$

$$\mathcal{I}_2 = \int_0^\pi \frac{e^{-ik\hat{r}}}{k\hat{r}} \,(\cos \beta)\,(\sin \psi)\, d\psi \tag{3}$$

Equation (1) give $k\hat{r}$ and $\cos\beta$ as functions of ψ for specified values of ka and kz_0, so we proceed to the evaluation. If we were to search through tables of integrals, we might find expressions for $\mathcal{I}_1$ and $\mathcal{I}_2$. However, the idea here is to use numerical methods because we would have no expectation of an analytical result for more complicated shapes.

The data in Fig. 2 was obtained with MATLAB. For each value of kz_0, the anonymous function capability was used to define `kr_hat` as a function of `psi`, which then was used to define `cos_beta` as a function of `psi`. Both were then used to define the integrands of $\mathcal{I}_1$ and $\mathcal{I}_2$, and the integrals were found by invoking the `quadl` routine. The error tolerance was decreased to 10^{-8}.

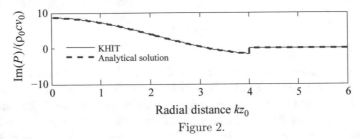

Figure 2.

According to Eq. (2), $P/(\rho_0 c V_a)$ within the cavity is purely imaginary. The value obtained from the KHIT nowhere differs from the analytical value by more than $1.4\left(10^{-7}\right)$, and the real part is no greater than $\pm 1.8(10^{-10})$. The discontinuity at $kz_0 = 4$ results because the pressure obtained from KHIT outside the domain should be zero. For $kz_0 > ka$, the computed value is less than $6\left(10^{-8}\right)$ out to $kz_0 = 1.5ka$.

The evaluation skipped $kz_0 = ka$ because both integrands in that case are singular at $\psi = 0$. Nevertheless, it is interesting to wonder what would happen if kz_0 were very close to ka. To test this, a sequence of evaluations with $kz_0 = ka \mp 10^{-n}$ were carried out. The results are tabulated below.

$kz_0 - ka$	$p/(\rho_0 c v_0)$ KHIT	Analytical
$-\left(10^{-5}\right)$	$-5.1281\left(10^{-9}\right) - 1.6295i$	$-1.6295i$
$-\left(10^{-6}\right)$	$-5.1282\left(10^{-9}\right) - 0.8146i$	$-1.6295i$
$-\left(10^{-7}\right)$	$-5.1282\left(10^{-9}\right) - 0.8145i$	$-1.6295i$
$-\left(10^{-8}\right)$	Singular	$-1.6295i$
$\left(10^{-8}\right)$	Singular	0
$\left(10^{-7}\right)$	$-5.1282\left(10^{-9}\right) - 0.8144i$	0
$\left(10^{-6}\right)$	$-5.1282\left(10^{-9}\right) - 0.8143i$	0
$\left(10^{-5}\right)$	$-5.1283\left(10^{-9}\right) - 0.8127i$	0
$\left(10^{-4}\right)$	$-5.1288\left(10^{-9}\right) + 6.5779(10^{-7})i$	0

The table indicates that as the field point approaches the surface from the inside, the pressure drops from the analytical value to approximately half that value, but it is singular at the closest distance. The trend for points outside the sphere begins with a singularity at the closest distance, then approximately half the value on the surface, then zero. This behavior will be examined in our development of boundary elements, where it will be proven that $\chi(\bar{x}_0) = 0.5$ if $\bar{x}_0$ actually is situated on a smooth surface.

7.4.2 Acoustic Radiation into an Exterior Domain

This situation we consider here is one where a vibrating surface S radiates into a surrounding fluid domain V, and the extent of V is infinite. We cannot apply KHIT directly because its derivation required that the vibrating boundary encloses the domain. However, the derivation of KHIT did not require that the boundary be continuous. Consequently, it is valid if we consider the boundary of the fluid domain to be the composite of the radiating surface S and a very large sphere S_O having radius r_O that encloses S. The fluid domain V is the region between S and S_O. Figure 7.18 depicts the decomposition of the region. After we modify KHIT to describe this configuration, the limit as $r_O \to \infty$ will provide the theorem we seek.

Fig. 7.18 Acoustic region for derivation of the Kirchhoff–Helmholtz integral theorem describing radiation from vibrating body S. The surrounding sphere is virtual

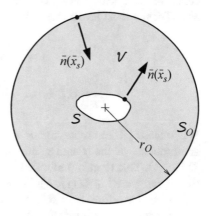

The integral in the KHIT now becomes the sum of integrals over the vibrating surface S and the virtual S_O. The first has the same form as the left side of Eq. (7.4.6). The only difference stems from the convention that $\bar{n}(\bar{x}_s)$ always is oriented into the fluid. Hence, for the radiation problem, it is outward from S, whereas it is oriented inward when S is the outer boundary of a cavity. The other part of the surface integral is the contribution from S_O. The normal direction for this portion of the boundary is $-\bar{e}_r$, so that $\bar{n}(\bar{x}_s) \cdot \nabla = -\partial/\partial r$ and $\bar{n}(\bar{x}_s) \cdot \bar{V}(\bar{x}_s) = -V_r$. Also, in terms of spherical coordinates, we have $dS_O = r_O^2 \sin\psi d\theta d\psi$. Thus, the contribution of the surrounding sphere is

$$\iint\limits_{S_O} [i\omega\rho_0\bar{n}\,(\bar{x}_s) \cdot V\,(\bar{x}_s)\,G\,(\bar{x}_0,\bar{x}_s) + P\,(\bar{x}_s)\,\bar{n}\,(\bar{x}_s) \cdot \nabla_s G\,(\bar{x}_0,\bar{x}_s)]\,dS$$

$$= \int_0^\pi \int_{-\pi}^\pi \left[-i\omega\rho_0 V_r\,(\bar{x}_s)\,G\,(\bar{x}_0,\bar{x}_s) - P\,(\bar{x}_s)\,\frac{\partial}{\partial r}G\,(\bar{x}_0,\bar{x}_s) \right] r_O^2 \sin\psi d\theta d\psi$$

$$(7.4.10)$$

Our study of radiated power in Sect. 4.5.3, which led to the Sommerfeld radiation condition for a three-dimensional domain, proved that if r_O is very large, then the pressure should exhibit spherical spreading. The specific description was

$$P = \frac{1}{r_O}f\,(\psi,\theta)\,e^{-ikr_O} + O\left(\frac{1}{r_O^2}\right) \tag{7.4.11}$$

Application of Euler's equation shows that the radial velocity is

$$V_r = \frac{1}{\rho_0 c r_O}f\,(\psi,\theta)\,e^{-ikr_O} + O\left(\frac{1}{r_O^2}\right) \tag{7.4.12}$$

where the two variables have been written separately, rather than as $P = \rho_0 c V_r$ to emphasize that this property is only satisfied to the leading order in r_O. The Green's function $G\,(\bar{x}_s,\bar{x}_0)$ also will satisfy the farfield approximation when $\bar{x}_s$ is situated on S_O. This is so because the source location $\bar{x}_0$ is at a fixed finite distance from the center of the sphere, and r_O can be as large as necessary to make this statement be true. Thus, it must be that for any $\bar{x}_s \in S_O$,

$$G\,(\bar{x}_0,\bar{x}_s) = \frac{1}{r_O}g\,(\psi,\theta)\,e^{-ikr_O} + O\left(\frac{1}{r_O^2}\right)$$

$$\frac{\partial}{\partial r}G\,(\bar{x}_0,\bar{x}_s) = -\frac{ik}{r_O}g\,(\psi,\theta)\,e^{-ikr_O} + O\left(\frac{1}{r_O^2}\right) \tag{7.4.13}$$

When these farfield representations are substituted into Eq. (7.4.10), the terms that are products of the f and g functions cancel. What remains in the integrand is $O\left(1/r_O^3\right)$. This term is multiplied by r_O^2 in the differential area dS, so the quantity that is integrated is $O\left(1/r_O\right)$. Consequently, the integral vanishes in the limit as $r_O \to \infty$. Therefore, the KHIT only requires integration over S.

The consequence is that Eq. (7.4.6) *also applies when V is the infinite fluid regions surrounding a vibrating surface S*. It is helpful to restate the theorem. The general form leaves the Green's function unspecified, so it is valid for two- and three-dimensional problems. This description is

$$\boxed{\begin{aligned} \chi\,(\bar{x}_0)\,P\,(\bar{x}_0) = \iint\limits_{S} &[i\omega\rho_0\bar{n}\,(\bar{x}_s) \cdot \bar{V}\,(\bar{x}_s)\,G\,(\bar{x}_0,\bar{x}_s) \\ &+ P\,(\bar{x}_s)\,\bar{n}\,(\bar{x}_s) \cdot \nabla_s G\,(\bar{x}_0,\bar{x}_s)]\,dS \end{aligned}} \tag{7.4.14}$$

The usual application is to three-dimensional systems. Explicitly, representing the free-space Green's function in the preceding leads to

$$
\chi(\bar{x}_0) P(\bar{x}_0) = -\frac{1}{4\pi} \iint_S \left[\bar{n}(\bar{x}_s) \cdot \left(\frac{\bar{x}_s - \bar{x}_0}{\hat{r}} \right) \left(ik + \frac{1}{\hat{r}} \right) P(\bar{x}_s) \right.
$$
$$
\left. - i\omega\rho_0 \bar{n}(\bar{x}_s) \cdot \bar{V}(\bar{x}_s) \right] \frac{e^{-ik\hat{r}}}{\hat{r}} dS, \quad \hat{r} = |\bar{x}_s - \bar{x}_0|
$$

(7.4.15)

This is the form in which KHIT typically appears.

The fact that KHIT has the same appearance for the interior and exterior problem can cause confusion. Compare the following situations: (1) Surface S is the exterior boundary of a cavity, (2) the same surface S is the vibrating surface of a transducer surrounded by fluid. The normal velocity distribution in both systems is the same. So what is the difference in KHIT for these systems? The answer lies in the definition of the normal direction on the surface, $\bar{n}(\bar{x}_s)$ as the normal that points into the fluid domain. Thus, if S is the surface that encloses a cavity, $n(\bar{x})$ is oriented onto the region bounded by S. In contrast, if S is the surface of a body that is surrounded by fluid, then $n(\bar{x})$ points outward from the region bounded by S. In other words, $\bar{n}(\bar{x}_s)$ is reversed between the two cases. The only other change is that $\chi(\bar{x}_0)$ is one if $\bar{x}_0$ is situated within the cavity, or if it is exterior to the transducer, and $\chi(\bar{x}_0)$ is zero if $\bar{x}_0$ is outside the cavity or inside the transducer.

As was the case for the interior cavity, the KHIT is not a solution to a problem because we cannot specify both the pressure and normal velocity at a location. A common use is to determine the pressure, especially in the farfield, when the surface response has been determined by another method. Example 7.10, which follows, describes such an application in conjunction with an approximation of the surface pressure.

If only the farfield pressure is of interest, the KHIT may be written in a form that is somewhat simpler. This is done by introducing the same approximations as those employed to describe the farfield of a collection of point sources. The origin is inside S. As part of the restriction to the farfield, the distance to $\bar{x}_0$ from the origin must be much greater than the maximum distance from the origin to a point on the surface, that is, $|\bar{x}_0| \gg \max(|\bar{x}_s|)$ In that case, the line from $\bar{x}_s$ to $\bar{x}_0$, which is $\bar{x}_0 - \bar{x}_s$, will approach parallelism to $\bar{x}_0$. Thus, the distance $\hat{r}$ from the surface point to the field point may be approximated as the radial distance $r = |\bar{x}_0|$ from the origin less the projection of $\bar{x}_s$ onto this radial line. The unit vector for the radial line is $\bar{x}_0/r$, so we have

$$
r = |\bar{x}_0|, \quad \hat{r} \equiv |\bar{x}_s - \bar{x}_0| \approx r - \bar{x}_s \cdot \frac{\bar{x}_0}{r}
$$

(7.4.16)

where the approximation becomes increasingly accurate with increasing $|\bar{x}_0|$. We use this approximation to factorize the phase of the free-space Green's function and its normal derivative in Eq. (7.4.8). The result is that

$$G\left(\bar{x}_s, \bar{x}_0\right) \approx \left(\frac{e^{-ikr}}{4\pi r}\right) e^{ik\bar{x}_s \cdot \bar{x}_0/r}$$

$$\bar{n}\left(\bar{x}_s\right) \cdot \nabla_s G\left(\bar{x}_s, \bar{x}\right) \approx \bar{n}\left(\bar{x}_s\right) \cdot \frac{\bar{x}_0}{r} \left(\frac{ike^{-ikr}}{4\pi r}\right) e^{ik\bar{x}_s \cdot \bar{x}_0/r}$$

(7.4.17)

The corresponding form of the KHIT is

$$P_{\text{ff}}\left(\bar{x}_0\right) = ik\frac{e^{-ikr}}{4\pi r} \iint_S \left[\rho_0 c \bar{V}\left(\bar{x}_s\right) \cdot \bar{n}\left(\bar{x}_s\right) + P\left(\bar{x}_s\right)\bar{n}\left(\bar{x}_s\right) \cdot \frac{\bar{x}_0}{r}\right] e^{ik\bar{x}_s \cdot \bar{x}_0/r} dS$$

(7.4.18)

The sole dependence of the integrand on the location of the field point is the value of r and the unit vector $\bar{x}_0/r$. When we use the polar angle ψ_0 and azimuthal angle θ_0 to locate that point, the farfield version of KHIT reduces to

$$P_{\text{ff}}\left(\bar{x}_0\right) = ik\frac{e^{-kr}}{4\pi r} f\left(\psi_0, \theta_0\right)$$

(7.4.19)

This is the same form as that in Eq. (4.5.43). The difference is that the KHIT tells us how to evaluate the directivity factor, $f\left(\psi_0, \theta_0\right)$. Of course, this assumes that we know the pressure and normal velocity on the radiating body's surface.

As was true for the interior problem, $G\left(\bar{x}_0, \bar{x}_s\right)$ need not be the free-space Green's function, but it usually is because any other is difficult to obtain. However, one notable exception is the Green's function used to describe radiation from an infinite planar boundary. Chapter 8 is devoted to this topic. The foundation for that study is the Rayleigh integral, which essentially is the KHIT with a Green's function that is derived from application of the method of images.

EXAMPLE 7.10 A cylindrical tank of length L is submerged. Pulsations in the feed line result in internal pressure fluctuations that induce an axisymmetric vibration of the cylinder. The velocity of the cylinder's surface in the normal direction is $v_R = \text{Re}(v_0 \cos(\pi z/L) e^{i\omega t})$ for $-L/2 \le z \le L/2$, which is a single lobe of a sine function that extends over the length of the cylinder. The ends of the cylinder are immobile. It is desired to determine the farfield pressure distribution. The radiation impedance of an infinite cylinder provides an approximation of the surface pressure generated by the specified velocity. Use this approximation to formulate the farfield pressure according to the KHIT. Evaluate the result for the directivity $(r/a)\left|P_{\text{ff}}\left(\bar{x}_0\right)\right|/\left(\rho_0 c v_0\right)$ as function of the spherical angle locating field point $\bar{x}_0$. Parameters for this evaluation are $ka = 1$ and 6, and $L/a = 4$ and 20.

Significance

In addition to showing how the KHIT may be formulated for three-dimensional bodies, this example has the purpose of showing that in some occasions, one may use knowledge of fundamentals to construct approximate solutions to problems that otherwise would require sophisticated simulation software.

Solution

The analysis is based on the suggestion in Example 7.7 that the field close to a cylinder might be reasonably well predicted by a model in which the cylinder is infinite. It is an approximation to ignore the fact that the helical surface waves do not propagate to infinity. In addition, deformation of the cylinder would result in movement of the ends, even if they are made very rigid. Furthermore, even if the ends did not move, the pressure on the ends would not be zero. Nevertheless, the analysis we will pursue is useful because it enables us to understand some phenomena and it provides an order of magnitude check for more exact analyses. Some might refer to this as a "back of the envelope" analysis, but that description is only appropriate in comparison with the effort required for a faithful simulation.

The decomposition of the given radial velocity into its constituent helical surface waves is

$$v_R = \mathrm{Re}\left[\frac{v_0}{2}\left(e^{i\mu z} + e^{-i\mu z}\right)e^{i\omega t}\right], \quad \mu = \frac{\pi}{L} \tag{1}$$

Both are axisymmetric, $n = 0$. In comparison with Eq. (7.3.16), this motion is a superposition of two helical waves whose velocity coefficients are $V_{1,0} = V_{-1,0} = v_0/2$. The radial wavenumber for both is

$$\kappa a = \left((ka)^2 - \pi^2 \frac{a^2}{L^2}\right)^{1/2} \tag{2}$$

All combinations of the specified values of ka and L/a lead to κa being real, so the helical waves are supersonic. The radiation impedance in Eq. (7.3.35) does not depend on the sign of the axial wavenumber, so it is the same for both helical waves. Adding the contribution of each leads to the complex amplitude of surface pressure being

$$P = \rho_0 c Z_{\mathrm{rad}} v_0 \cos(\mu z), \quad Z_{\mathrm{rad}} = -i\frac{\kappa a \, H_0(\kappa a)}{\kappa a \, H_0'(\kappa a)} \tag{3}$$

The next step is to describe the geometric variables in the farfield representation of the KHIT, Eq. (7.4.18). We do so by using a set of xyz axes to describe vector components. The variables to be represented appear in Fig. 1.

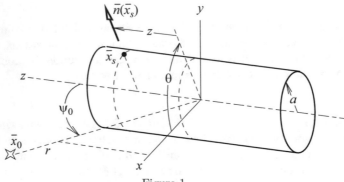

Figure 1.

The field point $\bar{x}_0$ has been placed on the xz plane at polar angle ψ_0. This placement will yield a general result because an axisymmetric surface vibration generates a field that also is axisymmetric. The surface point has cylindrical coordinates (a, θ, z), and the transverse direction $\bar{e}_R$ is normal to the surface. It is stated that the ends of the cylinder are immobile. We shall assume that the pressure on the ends also is zero. (This assumption often is based on the conjecture that radiation from the ends is unimportant if L/a is not small.) The quantities of interest are

$$\bar{x}_0 = r \sin \psi_0 \bar{e}_x + r \cos \psi_0 \bar{e}_z$$
$$\bar{x}_s = a \cos \theta \bar{e}_x + a \sin \theta \bar{e}_y + z \bar{e}_z$$
$$\bar{n}(\bar{x}_s) = \bar{e}_R = \cos \theta \bar{e}_x + \sin \theta \bar{e}_y$$

These variables depend on θ, so we use $(a d\theta)\, dz$ as the differential area element and integrate over $-\pi < \theta \le \pi$, $-L/2 \le z \le L/2$. Substitution of the preceding terms and $P = Z_{\text{rad}} V_R$ into Eq. (7.4.18) leads to

$$\frac{P_{\text{ff}}(\bar{x}_0)}{\rho_0 c v_0} = \frac{ika}{4\pi r} e^{-ikr} \int_{-L/2}^{L/2} \int_{-\pi}^{\pi} \cos(\mu z)\, (1 \tag{4}$$
$$+ Z_{\text{rad}} \cos \theta \sin \psi_0)\, e^{ik(a \cos \theta \sin \psi_0 + z \cos \psi_0)} d\theta dz$$

The exponential may be split into factors that depend solely on θ and z, and Z_{rad} is independent of θ and z. Furthermore, the θ integral depends only on $\cos \theta$, so it is the same as twice the integral over $0 < \theta \le \pi$. Recognition of these features reduces Eq. (4) to

$$\frac{P_{\text{ff}}(\bar{x}_0)}{\rho_0 c v_0} = \frac{ika}{4\pi r} e^{-ikr}\, (\mathcal{I}_1 + \mathcal{I}_2 Z_{\text{rad}} \sin \psi_0)\, \mathcal{I}_z \tag{5}$$

Integration over z is described by $\mathcal{I}_z$. It is

$$
\begin{aligned}
\mathcal{I}_z &= \int_{-L/2}^{L/2} \cos(\mu z)\, e^{ikz \cos \psi_0} dz \\
&= \frac{\sin\left[(kL \cos \psi_0 + \mu L)/2\right]}{(k \cos \psi_0 + \mu)} + \frac{\sin\left[(kL \cos \psi_0 - \mu L)/2\right]}{(k \cos \psi_0 - \mu)}
\end{aligned}
$$

The θ integration is described by $\mathcal{I}_1$ and $\mathcal{I}_2$. They appear in Eq. (9.1.21) in Abramowitz and Stegun as

$$
\int_0^{\pi} e^{ika \cos \theta \sin \psi_o} d\theta = \pi J_0 (ka \sin \psi_0)
$$

$$
\int_0^{\pi} e^{ika \cos \theta \sin \psi_o} \cos \theta d\theta = i\pi J_1 (ka \sin \psi_0)
$$

Substitution of these integrals into the Eq. (4) leads to

$$
\begin{aligned}
\frac{P_{ff}(\bar{x}_0)}{\rho_0 c v_0} &= \frac{ikL}{2} \left(\frac{a}{r}\right) e^{-ikr} \left[J_0 (ka \sin \psi_0) + i Z_{rad} \sin \psi_0 J_1 (ka \sin \psi_0) \right] \\
&\quad \times \left\{ \frac{\sin\left[(kL \cos \psi_0 + \mu L)/2\right]}{(kL \cos \psi_0 + \mu L)} + \frac{\sin\left[(kL \cos \psi_0 - \mu L)/2\right]}{(kL \cos \psi_0 - \mu L)} \right\}
\end{aligned}
\tag{6}
$$

It should be noted that the farfield pressure decays reciprocally to the radial distance. This must be so because no matter how long the cylinder is, the farfield specification places the field point sufficiently far away that the cylinder seems to occupy a single point.

For each combination of ka and L/a, we compute $(r/a)\,|P_{ff}|\,/\,(\rho_0 c v_0)$ as a function of ψ_0. (The parameter kL is written as $ka\,(L/a)$ for this computation.) Figure 2 displays polar plots for the four parameter combinations. Proper interpretation of these plots requires that one recognize that because of the axisymmetry of the system, these plots describe the view in any plane that contains the cylinder's axis. The overall picture is that only the shorter cylinder at the lower frequency radiates strongly in all directions. This is not surprising because our earlier studies have indicated that at very low frequencies, bodies radiate as a monopole whose strength depends on the complex amplitude of the volume velocity. For the longer cylinder in both cases, as well as the shorter cylinder at high frequency, the radiation is confined to a small range of angles around $\psi_0 = 90°$, which is said to be "broadside" or "beam aspect," both of which are nautical in their origin. Another trend displayed by the graphs is that increasing the frequency increases the maximum farfield pressure.

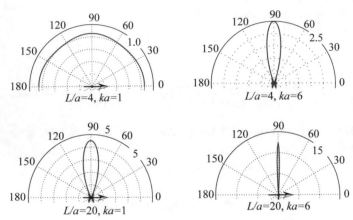

Figure 2.

A fault of polar plots is that they obscure data close to the origin. For this reason, the same data is plotted in Fig. 3 in abscissa–ordinate form. These graphs vividly show how narrowly the radiation is aimed in all cases except the short cylinder at the low frequency. In the intermediate cases, there are many side lobes, but they are much smaller than the main lobe. The side lobes for the long cylinder at the high frequency are negligibly small.

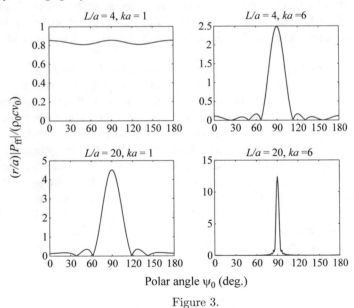

Figure 3.

Multipole Expansion of the KHIT

The initial discussion noted that the KHIT essentially describes the pressure field as the superposition of monopoles and dipoles on the vibrating surface. However, the description of the directivity derived in the previous section shows no evidence of a set of sources. The first effort in the following will resolve this apparent conflict. By itself, the development will merely provide a different perspective. However, further analysis based on a restriction to acoustically compact radiators (size substantially less than a wavelength) will lead to a recipe for actually evaluating the pressure field.

General Description We begin with a Taylor series expansion of the Green's function terms. A central point C inside the surface $\mathcal{S}$, such as the centroid of the vibrating body, is defined to be a reference location. Then, the location of any point on $\mathcal{S}$ is described by the relative position $\bar{\xi}$, which is

$$\bar{\xi} = \bar{x}_s - \bar{x}_C \tag{7.4.20}$$

Because $\bar{x}_C$ is a specified fixed point, we may consider the surface response to be functions of $\bar{\xi}$. An explicit description of the Green's function as a function of $\bar{\xi}$ results from application of Eq. (6.5.82). Doing so gives

$$G\left(\bar{x}_0, \bar{x}_s\right) = G\left(\bar{x}_0, \bar{x}_C\right) + \bar{\xi} \cdot \nabla_C G\left(\bar{x}_0, \bar{x}_C\right) + \frac{1}{2}\left(\bar{\xi} \cdot \nabla_C\right)^2 G\left(\bar{x}_0, \bar{x}_C\right) + \cdots \tag{7.4.21}$$

where the notation ∇_C serves to emphasize that the gradient is evaluated at $\bar{x}_C$. The normal derivative of $G\left(\bar{x}_0, \bar{x}_s\right)$ in the KHIT is taken at $\bar{x}_s$. To put it on the same basis as the above representation of $G\left(\bar{x}_0, \bar{x}_s\right)$, we invoke Eq. (6.5.86), which states that $\nabla_s G\left(\bar{x}_0, \bar{x}_s\right) = -\nabla_0 G\left(\bar{x}_0, \bar{x}_C\right)$. For the present purpose, it is adequate to truncate Taylor series at second derivatives, so we use the preceding to represent the Green's function. Doing so yields

$$\begin{aligned}
\bar{n}\left(\bar{\xi}\right) \cdot \nabla_s G\left(\bar{x}_0, \bar{x}_s\right) &= \left[-\bar{n}\left(\bar{\xi}\right) \cdot \nabla_0\right]\left[G\left(\bar{x}_0, \bar{x}_C\right) + \bar{\xi} \cdot \nabla_C G\left(\bar{x}_0, \bar{x}_C\right) + \cdots\right] \\
&= -\bar{n}\left(\bar{\xi}\right) \cdot \nabla_0 G\left(\bar{x}_0, \bar{x}_C\right) \\
&\quad + \left(\bar{n}\left(\bar{\xi}\right) \cdot \nabla_0\right)\left(\bar{\xi} \cdot \nabla_0\right) G\left(\bar{x}_0, \bar{x}_C\right) + \cdots
\end{aligned} \tag{7.4.22}$$

The dipole and quadrupole strengths were described in Chap. 6 in terms of derivatives at the field point, with the source point defined to be the origin of a Cartesian coordinate system. Therefore, we convert the preceding expression to scalar form by explicitly describing the gradients. In the following, x_1, x_2, x_3 are the Cartesian coordinates of $\bar{x}_0$. (In most situations, we use the free-space Green's function and designate $\bar{x}_C$ as the origin.) Substitution of the series expansion of $G\left(\bar{x}_0, \bar{x}_s\right)$ and its gradient into the KHIT, followed by collection of the coefficients of each order of derivative, yields

$$P(\bar{x}_0) = 4\pi \left[AG(\bar{x}_0, \bar{x}_C) + \sum_{j=1}^{3} D_j \frac{\partial}{\partial x_j} G(\bar{x}_0, \bar{x}_C) \right.$$
$$\left. + \sum_{j=1}^{3} \sum_{m=1}^{3} Q_{j,m} \frac{\partial^2}{\partial x_j \partial x_m} G(\bar{x}_0, \bar{x}_C) + \cdots \right]$$

(7.4.23)

The derivatives of the Green's function are described by Eq. (6.5.123). The source strength, dipole moments, and quadrupole strengths appearing in this expression are

$$A = \frac{1}{4\pi} \iint_S ik\rho_0 c \bar{V}(\bar{\xi}) \cdot \bar{n}(\bar{\xi}) d\mathcal{S}$$

$$D_j = -\frac{1}{4\pi} \iint_S \left[ik\rho_0 c \bar{V}(\bar{\xi}) \cdot \bar{n}(\bar{\xi}) \xi_j + P(\bar{\xi}) n_j(\bar{\xi}) \right] d\mathcal{S}$$

$$\hat{Q}_{j,m} = \frac{1}{8\pi} \iint_S \left\{ ik\rho_0 c \bar{V}(\bar{\xi}) \cdot \bar{n}(\bar{\xi}) \xi_j \xi_m + P(\bar{\xi}) \left[n_j(\bar{\xi}) \xi_m + n_m(\bar{\xi}) \xi_j \right] \right\} d\mathcal{S}$$

(7.4.24)

The expression for $\hat{Q}_{j,m}$ has a symmetric form because a mixed derivative does not depend on the order of differentiation. The total strength of the lateral quadrupoles is $Q_{j,m} \equiv 2\hat{Q}_{m,j}$.

By itself, this representation generally is not useful because convergence will require many higher order poles. The exception is low frequencies, in which case, the body is compact. That is, its largest dimension a is much smaller than the acoustic wavelength, so $ka \ll 1$. In such situations, the preceding should be adequate. Indeed, the monopole contribution, whose strength is proportional to the volume velocity of the source, often is sufficient.

A Vibrating Rigid Body The multipole expansion provides an interesting perspective when the object from which the signal emanates is a rigid body. We restrict our attention to acoustically compact bodies, or equivalently, to very low frequencies. The theorems of kinematics state that the movement of such an object is the combination of a translation that follows an arbitrary point in the body, and a rotation in which the arbitrary point does not move. We shall use the centroid C of the body as the arbitrary point, and define this point to be the origin of the coordinate system.

The translational velocity $\bar{v}_C$ and angular velocity $\bar{\Omega}$ are taken to be harmonic at the same frequency. The complex velocity amplitude of any point in the body is

$$\bar{V} = \bar{V}_C + \bar{\Omega} \times \bar{x}$$

(7.4.25)

The normal velocity on the surface is found by evaluating this expression at surface point $\bar{x}_s$, then taking the dot product with the surface normal $\bar{n}(\bar{x}_s)$.

The result is used to evaluate the source strengths in Eq. (7.4.24). For the monopole amplitude, we have

$$A = \frac{1}{4\pi} ik\rho_0 c \iint\limits_{S} \bar{n}\left(\bar{\xi}\right) \cdot \left[\bar{V}_C + \left(\bar{\Omega} \times \bar{\xi}\right)\right] dS \qquad (7.4.26)$$

We apply the divergence theorem to this expression, and recognize that $\bar{V}_C$ is independent of position, so that $\nabla \cdot \bar{V}_C = 0$. In addition, $\nabla \cdot \left(\bar{\Omega} \times \bar{x}\right)$ is identically zero. Consequently, $A = 0$. The monopole amplitude is zero because the nature of rigid body motion is that outward movement of some regions of the surface is balanced by inward motion of other regions, so the net volume velocity is zero.

The dipole moment obtained from Eq. (7.4.24) is

$$D_j = -\frac{1}{4\pi} ik\rho_0 c \iint\limits_{S} \left[\bar{V}_C \xi_j \cdot \bar{n}\left(\bar{\xi}\right) + \left(\bar{\Omega} \times \bar{\xi}\right)\xi_j \cdot \bar{n}\left(\bar{\xi}\right)\right] dS$$
$$- \frac{1}{4\pi} \iint\limits_{S} P\left(\bar{\xi}\right) n_j\left(\bar{\xi}\right) dS, \quad j = 1, 2, 3 \qquad (7.4.27)$$

We use the divergence theorem to convert the first integral to a volume integral,

$$D_j = -\frac{1}{4\pi} ik\rho_0 c \iiint\limits_{\mathcal{V}} \left[\nabla \cdot \left(\bar{V}_C \xi_j\right) + \nabla \cdot \left(\left(\bar{\Omega} \times \bar{\xi}\right)\xi_j\right)\right] d\mathcal{V}$$
$$- \frac{1}{4\pi} \iint\limits_{S} P\left(\bar{\xi}\right) n_j\left(\bar{\xi}\right) dS, \quad j = 1, 2, 3 \qquad (7.4.28)$$

Both terms in the first integrand have the form $\nabla \cdot \left(\bar{U}\xi_j\right)$. The gradient is taken at the surface point. Thus, $\nabla = \bar{e}_1\left(\partial/\partial\xi_1\right) + \bar{e}_2\left(\partial/\partial\xi_2\right) + \bar{e}_3\left(\partial/\partial\xi_3\right)$. Expanding $\bar{U}$ into its components gives

$$\nabla \cdot \left(\bar{U}\xi_j\right) = \frac{\partial}{\partial\xi_1}\left(U_1\xi_j\right) + \frac{\partial}{\partial\xi_2}\left(U_2\xi_j\right) + \frac{\partial}{\partial\xi_3}\left(U_3\xi_j\right) = U_j + \xi_j\nabla \cdot \bar{U} \quad (7.4.29)$$

In the first term in Eq. (7.4.28), $\bar{U} = \bar{V}_C$. This term is independent of $\bar{\xi}$, so $\nabla \cdot \bar{V}_C = 0$. For the second term, $\bar{U} = \bar{\Omega} \times \bar{\xi}$. The same vector identity that led to vanishing of the monopole amplitude states that $\nabla \cdot \left(\bar{\Omega} \times \bar{\xi}\right)$ is zero. Thus, we have reduced the dipole moment components to

$$D_j = -\frac{1}{4\pi} ik\rho_0 c\mathcal{V}\left(\bar{V}_C \cdot \bar{e}_j\right) - \frac{1}{4\pi} ik\rho_0 c \iiint\limits_{\mathcal{V}} \left(\bar{\Omega} \times \bar{\xi}\right) \cdot \bar{e}_j d\mathcal{V}$$
$$- \frac{1}{4\pi} \iint\limits_{S} P\left(\bar{\xi}\right) n_j\left(\bar{\xi}\right) dS \qquad (7.4.30)$$

The fact that we have designated point C to be the centroid of the body means that the first moment of position is zero, that is,

$$\iiint\limits_{\mathcal{V}} \xi_j d\mathcal{V} = 0 \tag{7.4.31}$$

Thus, the middle integral vanishes. The dipole moment vector that results from adding the individual components of the remaining terms is

$$\bar{D} = -\frac{1}{4\pi} \left[\rho_0 \mathcal{V} \left(i\omega \bar{V}_C \right) + \iint\limits_{S} P\left(\bar{\xi} \right) \bar{n}\left(\bar{\xi} \right) dS \right] \tag{7.4.32}$$

To interpret the dipole moment, we observe that $\rho_0 \mathcal{V}$ is the mass m_{disp} of the body of fluid displaced by the vibrating body, and the complex amplitude of the acceleration of point C is $i\omega \bar{V}_C$. Hence, the first term in the bracket is the force that would be required to impart to the displaced fluid the acceleration of the vibrating body. The second term also has a familiar explanation. Recall that $\bar{n}\left(\bar{\xi} \right)$ is the normal to the vibrating body pointing into the fluid. The normal force exerted by the fluid on a surface patch dS of the vibrating body is $-\bar{n}\left(\bar{\zeta} \right) P\left(\bar{\xi} \right) dS$. The resultant pressure force $\bar{F}_p$ exerted by the fluid is the surface integral of this term,

$$\bar{F}_P = -\iint\limits_{S} P\left(\bar{\xi} \right) \bar{n}\left(\bar{\xi} \right) dS \tag{7.4.33}$$

It follows that the dipole moment in Eq. (7.4.32) may be written as

$$\bar{D} = -\frac{1}{4\pi} \left(m_{\text{disp}} i\omega \bar{V}_C - \bar{F}_P \right) \tag{7.4.34}$$

The force exerted *on the fluid by the body* is the reaction, that is, it is $-\bar{F}_p$. Thus, the dipole moment is $-1/4\pi$ times the sum of the force required to accelerate the displaced fluid mass and the resultant force exerted on the fluid by the vibrating body.

To make this relation useful, it is necessary to quantify $\bar{F}_P$. Doing so requires an analysis specific to the vibrating body's shape. Nevertheless, certain aspects can be anticipated. For low frequencies, the pressure should be proportional to the body's acceleration, but opposed to it. Therefore, $\bar{F}_p$ must be proportional to $-i\omega \bar{V}_C$. Given that the first term in the dipole moment is proportional to the displaced mass, it is reasonable to anticipate that $\bar{F}_P$ is proportional to the density of the fluid. An aspect that might not be apparent is that $\bar{F}_P$ is not necessarily parallel to the acceleration. A vibrating thin disk exemplifies this behavior because the resultant of the pressure distribution must be perpendicular to the flat surface of the disk regardless of how the disk is oriented.

The corollary of the possible misalignment of $\bar{F}_P$ and $i\omega \bar{V}_C$ is that the proportionality is tensorial in nature. To describe it, we shall switch to a matrix description.

Let $\{F_P\}$ and $\{V_C\}$ be column vectors that are populated by the components of the respective quantities. The preceding general observations lead to definition of the factor of proportionality as a square matrix $[W]$, such that

$$\{F_P\} = -i\omega\rho_0 [W] \{V_C\} \tag{7.4.35}$$

The meaning of relation becomes somewhat clearer if we represent $[W]$ in partitioned form as a column vector, that is

$$\begin{Bmatrix} F_x \\ F_y \\ F_z \end{Bmatrix} = -i\omega\rho_0 \begin{bmatrix} \{W^{(x)}\}^{\mathrm{T}} \\ \{W^{(y)}\}^{\mathrm{T}} \\ \{W^{(z)}\}^{\mathrm{T}} \end{bmatrix} \{V_C\} \tag{7.4.36}$$

This representation shows that the components of $\bar{F}_P$ in the x, y, and z directions are, respectively, the projection of $\bar{V}_C$ onto vectors $\bar{W}^{(x)}$, $\bar{W}^{(y)}$, and $\bar{W}^{(z)}$. Only if $[W]$ is proportional to the identity matrix will $\bar{F}_P$ be parallel to $\bar{V}_C$. Dimensional consistency requires that the elements of $[W]$ be volume quantities. Newton's second law states that $\bar{F}_e + \bar{F}_P = M\left(i\omega\bar{V}\right)$, where $\bar{F}_e$ is the external force causing the body to move and M is the body's mass. Substitution for $\bar{F}_p$ gives

$$\{F_e\} = i\omega\left[M[I] + \rho_0[W]\right]\{V_C\} \tag{7.4.37}$$

This expression leads to $\rho_0[W]$ being referred to as the *virtual mass*, or the *added mass*.

Although determination of $[W]$ requires further analysis specific to the body's shape, certain properties are general. The same *vector* $\bar{F}_p$ must result from a specified vector $\bar{V}_C$, regardless of the orientation of the coordinate system associated with the respective components. This condition requires that $[W]$ be symmetric. A special property applies if the body is axisymmetric. Let this axis be z. The resultant force then must lie in the plane formed by $\bar{V}_c$ and $\bar{e}_z$ because the differential surface pressure resultant $-P\left(\bar{\xi}\right)\bar{n}\left(\bar{\xi}\right)dS$ on points on either side of this plane balance. In addition, the force component transverse to z must be invariant as the body is rotated about that axis with $\bar{V}_C$ fixed. These conditions require that $[W]$ be diagonal, with $W_{1,1} = W_{2,2}$. Two shapes of interest are a sphere and a thin disk,[15] for which the nonzero elements of the respective matrices are

$$\left(W_{\text{sphere}}\right)_{j,j} = \frac{2}{3}\pi a^3, \quad j = 1, 2, 3; \quad \left(W_{\text{disk}}\right)_{3,3} = \frac{8}{3}a^3 \tag{7.4.38}$$

Determination of $[W]$ allows us to describe the dipole moment in matrix form. The relation is

$$\{D\} = -\frac{i\omega\rho_0}{4\pi}[V[I] + [W]]\{V_c\} \tag{7.4.39}$$

---

[15]Pierce, *ibid*, p. 427.

Knowledge that the monopole part of the radiated field is zero, combined with the ability to evaluate the dipole contribution, usually is adequate to determine the pressure radiated by a rigid body vibrating at low frequencies. This is so because the dipole component will dominate the higher order terms. However, in some cases, it will be found that $\bar{D} = \bar{0}$. This situation will occur if the body oscillates in a pure rotation about the centroid C, so that $\bar{V}_C$ is zero. The pressure resultant is zero in such a motion because some pairs of points on the body's surface move in opposite sense as a result of the rotation. Consequently, the surface pressure distribution shows a 180° change of phase, with the net result that different regions cancel. In other words, the small ka field radiated by a body that executes a pure rotation will be dominated by various quadrupoles.

7.5   Numerical Methods for Radiation from Arbitrary Objects

The techniques we have developed are limited to specific configurations. Determination of the field radiated by a body that has an arbitrary shape requires a numerical simulation. We will examine three approaches having fundamentally different foundations. The first is the *source superposition method,* which is based on the notion that the radiating body may be replaced by a set of point sources in free space. Another approach converts the Kirchhoff–Helmholtz integral theorem to an integral equation for the surface pressure. A numerical analysis of this equation leads to the *boundary element method.* Both of these approaches are founded on acoustical analyses we have already carried out. The third is the *finite element method,* which is a general tool for solving field equations associated with the various media. We will see that each method has positive aspects and liabilities. The objective is to familiarize the reader with the available alternatives. Hence, our investigations will focus on the fundamental issues, rather than the details of their implementation. Each approach has several variants, but we will not delve into them.

7.5.1   Source Superposition

The concept here is quite simple—replace the radiating body with a collection of point sources. The idea is quite natural, but it leads to questions of how many sources are needed, where should they be placed, and how strong should each source be? Answering these questions is the contribution of the work by Koopman, Song, and Fahnline.[16] They began by considering a continuous sheet of sources distributed over a surface interior to the radiating body. This was done to prove that the method is equivalent to the Kirchhoff–Helmholtz integral theorem. That is the junction at which we begin.

---

[16]Koopman, Song, and Fahnline, "A method for computing acoustic fields based on the principle of wave superposition," J. Acoust. Soc. Am., 86 (1989), 2433–2438.

Fig. 7.19 Sources interior to
a body whose normal
component of surface
velocity is $V_s(\bar{x}_s)$

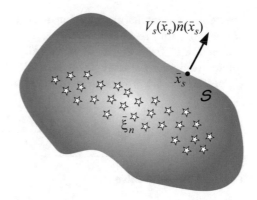

Figure 7.19 depicts a body surrounded by an ideal fluid. Its surface S executes a
specified vibration in which the velocity normal to the surface, positive into the fluid,
is Re $\left(V_s(\bar{x}_s) e^{i\omega t} \right)$. Distributed throughout the region surrounded by S is a set of N
point sources whose strengths will be determined. The objective is to determine the
pressure at field point $\bar{x}_0$.

The source superposition method may be used to analyze two- and three-
dimensional situations. For the two-dimensional case, $G(\bar{x}, \bar{x}_n)$ is described by
Eq. (7.3.47), and Eq. (6.4.31) describes the three-dimensional version. The pressure
is the superposition of the contribution of each source. The strength of that contribution is the complex amplitude of the mass acceleration, which is $i\omega\rho_0 \hat{Q}_n$, with $\hat{Q}_n$
being the volume velocity amplitude. Thus, the pressure at any point $\bar{x}$ is given by

$$P(\bar{x}) = \sum_{n=1}^{N} i\omega\rho_0 \hat{Q}_n G(\bar{x}, \bar{x}_n) \qquad (7.5.1)$$

An expression for the particle velocity is obtained by applying Euler's equation
to the preceding. The gradient is evaluated at $\bar{x}$, and the free-space Green's function
only depends on the distance between $\bar{x}_n$ and $\bar{x}$, so that

$$\bar{V}(\bar{x}) = -\sum_{n=1}^{N} \hat{Q}_n \left(\frac{\bar{x} - \bar{x}_n}{r_n} \right) \frac{d}{dr_n} G(\bar{x}, \bar{x}_n), \quad r_n = |\bar{x} - \bar{x}_n| \qquad (7.5.2)$$

To determine the unknown volume velocities, we require that the particle velocity
matches the known normal velocity distribution $V_s(\bar{x}_s)$. We cannot do so at every
point on the surface, so we select a set of surface points, designated $\bar{\xi}_m$, at which this
condition will be imposed. Let M be the number of surface points selected for this
purpose. Then, we have

$$\sum_{n=1}^{N} \hat{Q}_n \bar{n}(\bar{\xi}_m) \cdot \left(\frac{\bar{\xi}_m - \bar{x}_n}{r_n} \right) \frac{d}{dr_n} G(\bar{\xi}_m, \bar{x}_n) = -V_s(\bar{\xi}_m), \quad m = 1, 2, ..., M \quad (7.5.3)$$

The only quantities that are not set are the volume velocities. Thus, we have developed M equations for these parameters.

Solution of these equations is expedited by writing them in matrix form. The volume velocities are arranged sequentially in a column vector. The coefficient of $\hat{Q}_n$ in equation #m is an element of the $M \times N$ rectangular array $[U]$,

$$U_{m,n} = \bar{n}\left(\bar{\xi}_m\right) \cdot \left(\frac{\bar{\xi}_m - \bar{x}_n}{r_n}\right) \frac{d}{dr_n} G\left(\bar{\xi}_m, \bar{x}_n\right) \qquad (7.5.4)$$

Thus, the task is to solve

$$[U]\left\{\hat{Q}\right\} = -\{V_s\} \qquad (7.5.5)$$

At this juncture, selection of the $\bar{x}_n$ locations for the sources and of the $\bar{\xi}_m$ locations for velocity matching enters the picture. If the number M of the latter is less than N, then there are more sources than the number of available equations. Thus, it is necessary that M at least equals N. Koopman, Song, and Fahnline set $M = N$, but this choice is problematic. If we were to select a different set of points to perform the matching, then it is likely that the $\hat{Q}_n$ values would be different. However, based on the proof that the source superposition and KHIT formalisms are equivalent, we expect the volume velocities to be unique. Thus, setting $M = N$ and increasing that number should eventually lead to a convergent result.

An alternative that has been followed matches the surface velocity at more points than the number of sources, $M > N$. Suppose we have identified a tentative solution for $\{\hat{Q}\}$ using $M > N$. Because there are more equations than the number of unknowns, it is unlikely that $\{\hat{Q}\}$ will satisfy all equations. A velocity matching error may be formed by substituting the solution into Eq. (7.5.5), then comparing the two sides of the equation. This gives an error residual $\{E\}$ whose definition is

$$\{E\} = [U]\left\{\hat{Q}\right\} + \{V_s\} \qquad (7.5.6)$$

A single error metric E^2 is the summed magnitude squared of all elements of $\{E\}$. This quantity is useful because it cannot be negative, so minimizing it assures that all elements of $\{E\}$ are a minimum. Another term for this metric is that it is the square of the Euclidean norm of $\{E\}$, which is obtained as a Hermitian dot product. The Hermitian of a matrix is the complex conjugate of its transpose, so the product is

$$E^2 \equiv \sum_{m=1}^{M} |E_m|^2 \equiv \{E\}^{\mathrm{H}}\{E\} = \left\{\hat{Q}\right\}^{\mathrm{H}}[U]^{\mathrm{H}}[U]\left\{\hat{Q}\right\} + \left\{\hat{Q}\right\}^{\mathrm{H}}[U]^{\mathrm{H}}\{V_s\}$$
$$+ \{V_s\}^{\mathrm{H}}[U]\left\{\hat{Q}\right\} + \{V_s\}^{\mathrm{H}}\{V_s\} \qquad (7.5.7)$$

We wish to find the set of source strengths that minimize E^2, which means that the value of E^2 must be stationary with respect to each $\hat{Q}_j$. References usually only discuss the case where all quantities are real. The case where these quantities are complex is slightly more complicated, but the result is only different by replacing transpose operations by Hermitians, that is,

$$\boxed{[U]^{\mathrm{H}}[U]\left\{\hat{Q}\right\} = -[U]^{\mathrm{H}}\{V_s\}} \qquad (7.5.8)$$

This formulation is said to be the *method of linear least squares*. The coefficient matrix $[U]$ has M rows (the number of velocity equations) and N columns (the number of sources). Thus, $[U]^{\mathrm{H}}[U]$ is $N \times N$. The inverse of this matrix is called the Morse–Penrose inverse, but any method may be used to solve for $\hat{Q}$. After those values have been obtained, the pressure at any field point may be determined from Eq. (7.5.1).

The formulation seems to be straightforward, but its implementation requires that several issues be addressed. Most significant is the possibility that $[U]^{\mathrm{H}}[U]$ is ill-conditioned. As explained by Ochmann,[17] this situation might occur if the radiating body's shape is drastically different from spherical, as would be the case with a long cylinder. Another reason might be that the distribution of sources is inconsistent with the symmetry of the body. For example, a rectangular box should have sources placed equally on both sides of each of its midplanes. Ill-conditioned situations typically are addressed by application of singular value decomposition,[18] but doing so requires greater computational effort.

The question still remains as to where the sources should be located. Koopman et al. provide some guidelines, as does Ochmann.[19] Because the equations are quite easy to formulate, practitioners typically use a very large number of sources placed consistently with the symmetry of the system. For example, consider radiation from a cylinder whose surface vibration is axisymmetric. Many sources should be distributed along the cylinder's axis in order that the field has no variation around any circumferential circle. When this tactic is used, then the number of surface points $\bar{\xi}_m$ for matching the surface velocity would be chosen to be much greater than N. Some users distribute the source and surface locations uniformly, and some use a random number generator for that purpose.

Regardless of how one selects these locations, the result should be verified. One way of doing so is to redo the analysis with another set of source and surface points. Another way to verify the solution for $\hat{Q}_n$ values is to use Eq. (7.5.2) to evaluate $\bar{n}(\bar{x}_s) \cdot \bar{V}(\bar{x}_s)$ at several points on the surface that were not used to formulate Eq. (7.5.8). The degree to which this set of normal velocities differ from the known values of $V_s(\bar{x}_s)$ provides a measure of the error in the method.

---

[17]M. Ochmann, "The full-field equations for acoustic radiation and scattering," J. Acoust. Soc. Am. **105** (1999) 2574–2584.

[18]W.H. Press, S.A. Teukolsky, W.T. Vetterling, & B.P. Flannery, *Numerical Recipes*, 3rd Ed., Chap. 12, Cambridge University Press (2007).

[19]M. Ochmann, "The source simulation technique for acoustic radiation problems," *Acustica* **81**, 512–527 (1995).

7.5.2 Boundary Element Method

A very different approach is founded on the KHIT. A preview of this method arose
in Example 7.9, where the KHIT was used to evaluate the pressure at field points
that are very close to the surface. We begin with the radiation version of KHIT, in
which the field point $\bar{x}_0$ is exterior to the vibrating body. The fundamental equation
we seek has the field point on the surface. This is not a trivial matter because the
integrand of the KHIT is singular when the surface point $\bar{x}_s$ at which the integrand
is evaluated coincides with $\bar{x}_0$.

The Surface Helmholtz Integral Equation

The singularity of the KHIT integrand is handled by bringing the field point to the
surface in a limiting process. In Fig. 7.20, the field point is at distance ε from the
surface measured in the direction of the surface normal that intersects this field point.
The surface point for this normal is designated $\bar{x}_0'$, so that the field point's position
is $\bar{x}_0 = \bar{x}_0' + \varepsilon \bar{n}\left(\bar{x}_0'\right)$.

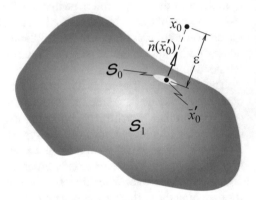

Fig. 7.20 Placement of a
field point close to a
vibrating surface

The idea is to decompose the surface into two parts: S_0 is a small region that
surrounds $\bar{x}_0'$ and S_1 is the remainder of S. Because $\bar{x}_0$ is situated in the fluid domain,
we set $\chi = 1$ in Eq. (7.4.6), so the KHIT may be written as

$$P\left(\bar{x}_0\right) = \mathcal{I}_0 + \mathcal{I}_1$$
$$\mathcal{I}_j = \iint\limits_{S_j} \left[i\omega\rho_0\bar{n}\left(\bar{x}_s\right) \cdot V\left(\bar{x}_s\right) G\left(\bar{x}_0, \bar{x}_s\right) + P\left(\bar{x}_s\right) \bar{n}\left(\bar{x}_s\right) \cdot \nabla G\left(\bar{x}_0, \bar{x}_s\right)\right] dS$$

$$(7.5.9)$$

In order for $\bar{n}\left(\bar{x}_0'\right)$ to be a unique normal vector, it must be that S is smooth at that
location, that is, there must be a unique tangent plane there. (This excludes sharp
corners, which we will address after the basic derivation.) When the geometry fits
this specification, we may consider S_0 to be a small circle of radius Δ lying in the
tangent plane and centered on $\bar{x}_0'$. This is the configuration depicted in Fig. 7.21.

The procedure we will follow is to consider the limiting behavior of the integral
over S_0 as $\varepsilon \to 0$ with Δ held at a small but finite value. After that limit is established,

Fig. 7.21 Definition of a
spherical coordinate system
for evaluating the principal
part of the KHIT when the
field point is situated on the
vibrating surface

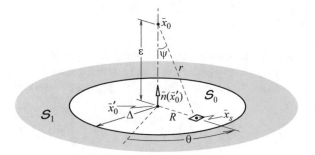

we will take the limit as $\Delta \to 0$. The resulting expression for the integral over $\mathcal{S}_0$ is called the *Cauchy principal part* of the integral over $\mathcal{S}$. We will derive the result for a three-dimensional system, so the free-space Green's function will be $G(\bar{x}_0, \bar{x}_s) = e^{-ikr}/(4\pi r)$.

The geometric properties of the terms in the integral over $\mathcal{S}_0$ are described in term of a spherical coordinate system centered on $\bar{x}_0$. The radius Δ is small, and it eventually will be reduced to zero, which means that the surface pressure and normal velocity may be taken to be constant over $\mathcal{S}_0$ at their values at the center point. Thus, the portion of KHIT associated with this region is

$$\mathcal{I}_0 \approx i\rho_0 \omega V_s\left(\bar{x}_0'\right) \iint\limits_{\mathcal{S}_0} G\left(\bar{x}_0, \bar{x}_s\right) d\mathcal{S} + P\left(\bar{x}_0'\right) \iint\limits_{\mathcal{S}_0} \bar{n}\left(\bar{x}_s\right) \cdot \nabla G\left(\bar{x}_0, \bar{x}_s\right) d\mathcal{S}$$

$$= \frac{1}{4\pi} i\rho_0 \omega V_s\left(\bar{x}_0'\right) \iint\limits_{\mathcal{S}_0} \frac{e^{-ikr}}{r} d\mathcal{S} + \frac{1}{4\pi} P\left(\bar{x}_0'\right) \iint\limits_{\mathcal{S}_0} \bar{e}_z \cdot \left(\frac{\bar{x}_s - \bar{x}_0}{r}\right) \frac{d}{dr}\left(\frac{e^{-ikr}}{r}\right) d\mathcal{S}$$

$$(7.5.10)$$

Each of the terms in the integrand is readily described in terms of the spherical coordinate system in Fig. 7.21. The axis of this coordinate system is normal to the circle, so the geometry is asymmetric. Consequently, we may use a ring element to describe the differential area, so that $d\mathcal{S} = 2\pi R dR$. The value of R is related to the radial distance by $R^2 = r^2 - \varepsilon^2$. The differential of this relation is $R dR = r dr$ because ε is the constant distance to the surface. The range of radial distances covered by $\mathcal{S}_0$ is $\varepsilon \le r \le \left(\varepsilon^2 + \Delta^2\right)^{1/2}$. This change of variables converts the integral to

$$\mathcal{I}_0 = \frac{1}{2} i\rho_0 \omega V_s\left(\bar{x}_0'\right) \int_\varepsilon^{\left(\varepsilon^2 + \Delta^2\right)^{1/2}} e^{-ikr} dr$$

$$+ \frac{1}{2} P\left(\bar{x}_0'\right) \int_\varepsilon^{\left(\varepsilon^2 + \Delta^2\right)^{1/2}} \bar{e}_z \cdot \left(\frac{\bar{x}_s - \bar{x}_0}{r}\right)\left(-ik - \frac{1}{r}\right) e^{-ikr} dr \qquad (7.5.11)$$

The dot product in the second integral is expressible in terms of ε and the radial distance, specifically,

$$\bar{e}_z \cdot \frac{(\bar{x}_s - \bar{x}_0)}{r} = -\cos\psi = -\frac{\varepsilon}{r} \qquad (7.5.12)$$

In addition, the smallness of ε and Δ leads to simplification of both integrals, because we may set $\exp(-ikr) = 1$ and approximate the last factor in the second integrand as $-1/r$. The result is that the integral over $\mathcal{S}_0$ reduces to

$$\mathcal{I}_0 = \frac{1}{2} i\rho_0\omega V_s(\bar{x}_0') \int_\varepsilon^{(\varepsilon^2+\Delta^2)^{1/2}} dr + \frac{1}{2} P(\bar{x}_0') \int_\varepsilon^{(\varepsilon^2+\Delta^2)^{1/2}} \frac{\varepsilon}{r^2} dr$$

$$= \frac{1}{2} i\rho_0\omega V_s(\bar{x}_0') \left[(\varepsilon^2+\Delta^2)^{1/2} - \varepsilon\right] + \frac{1}{2} P(\bar{x}_0') \varepsilon \left[\frac{1}{\varepsilon} - \frac{1}{(\varepsilon^2+\Delta^2)^{1/2}}\right]$$

$$(7.5.13)$$

The limit of $\mathcal{I}_0$ as $\varepsilon \to 0$ is

$$\mathcal{I}_0 = \frac{1}{2} i\rho_0\omega V_s(\bar{x}_0') \Delta + \frac{1}{2} P(\bar{x}_0') \qquad (7.5.14)$$

When we take the limit as $\Delta \to 0$, the first term vanishes but the second is unaltered. In regard to the integral over $\mathcal{S}_1$, we recognize that it approaches the original integral over the entire surface $\mathcal{S}$, except that the singular point, $\bar{x}_s = \bar{x}_0'$, is excluded from the domain. (How this is done in a computational scheme will be discussed.) Thus, to evaluate Eq. (7.5.9), we carry out a regular integration over the entire surface S with $\bar{x}_0$ excluded, then add $P(\bar{x}_0')/2$. When we bring the latter term to the left side, we find that $P(\bar{x}_0')/2$ equals the regular integral.

Let us step back to take an overview of the KHIT before we write the final form. The left side is $\chi P(\bar{x}_0)$. Previously, we found that $\chi = 1$ if $\bar{x}_0$ is outside the radiating body, and $\chi = 0$ if $\bar{x}_0$ is inside the radiating body. Here, we brought $\bar{x}_0$ to the surface. Doing so led to $\chi = 1/2$. This result was derived by assuming that the surface is smooth. If $\bar{x}_0$ is situated at a corner, then the smoothness condition is not met. That situation is addressed by the concept of a solid angle. Figure 7.22 shows the four cases for $\bar{x}_0$: (a) exterior, (b) interior, (c) smooth surface, and (d) corner. For each, a small sphere of radius δ is centered on $\bar{x}_0$.

The portion of the surface area of the sphere that lies in the fluid domain is defined to be $\Gamma\delta^2$, where Γ is the *solid angle*. Thus, $\Gamma = 4\pi$ for case (a), $\Gamma = 0$ for case (b),

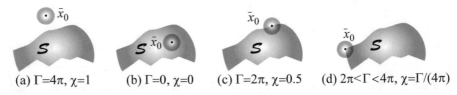

(a) $\Gamma=4\pi$, $\chi=1$ (b) $\Gamma=0$, $\chi=0$ (c) $\Gamma=2\pi$, $\chi=0.5$ (d) $2\pi<\Gamma<4\pi$, $\chi=\Gamma/(4\pi)$

Fig. 7.22 Parameter χ for the KHIT is obtained by surrounding the field point by a small sphere of radius δ. The solid angle $\delta^2\Gamma$ is the portion of the sphere's surface area that is situated in the fluid

$\Gamma = 2\pi$ for case (c), and Γ is between 2π and 4π for case (d) because more than half the sphere is in the fluid at that corner. The value of Γ at a corner depends on whether the slope is discontinuous in one direction or two orthogonal directions. For example, if $\bar{x}_0$ is at the edge of cylinder with flat ends, then $\Gamma = 3\pi$, whereas $\Gamma = (7/8)\,4\pi$ if $\bar{x}_0$ is at the corner of a rectangular box. The parameter $\chi\,(\bar{x}_0)$ describes the fraction of a full sphere that lies in the fluid. In general, it is

$$\chi\,(\bar{x}_0) = \frac{\Gamma}{4\pi} \qquad (7.5.15)$$

An extremely useful feature stems from the fact that KHIT applies to acoustic cavities, as well as radiation problems. The sole modifications required to treat either situation is that $\bar{n}$ must point into the fluid and the surface normal velocity must be defined to be positive if it is in the sense of $\bar{n}$. It follows that a similar modification of the SHIE will make it applicable to cavities. In that case, the solid angle would be defined by the extent of the interior fluid surrounding $\bar{x}_0$.

The KHIT for all cases is described by

$$\begin{aligned}
\chi\,(\bar{x}_0)\,P\,(\bar{x}_0) = \frac{1}{4\pi}\iint\limits_{S} &\left[i\omega\rho_0 V_s\,(\bar{x}_s) - P\,(\bar{x}_s)\,\bar{n}\,(\bar{x}_s)\cdot\left(\frac{\bar{x}_s - \bar{x}_0}{|\bar{x}_n - \bar{x}_s|}\right)\right. \\
&\left. \times\left(\frac{ik\,|\bar{x}_n - \bar{x}_s| + 1}{|\bar{x}_n - \bar{x}_s|}\right)\right]\frac{e^{-ik|\bar{x}_n - \bar{x}_s|}}{|\bar{x}_n - \bar{x}_s|}dS
\end{aligned} \qquad (7.5.16)$$

It is implicit to this expression that $\bar{x}_s = \bar{x}_0$ is excluded from the integration domain if $\bar{x}_0$ is on the surface. In that case, the equation is called the *surface Helmholtz integral equation*. We will use the abbreviation SHIE to refer to it.

Discretized Implementation

In a radiation problem, we seek the pressure field generated by a specified velocity distribution on the vibrating surface. The SHIE states that the pressure on the surface depends on the properties of the pressure on the surface. Just as an integral is sometimes called an antiderivative, such a relation could be considered to be an "antidifferential equation." (Its proper description is that is Fredholm integral equation of the second kind.) It often is necessary to use numerical methods to solve a differential equation, and that is the situation for the SHIE. This process begins by dividing the surface into a set of small patches, which we will denote as S_j. Each patch is one element situated on the surface, from which the description as a *boundary element* formulation follows.

The integral over a patch depends on the geometrical properties of the surface and the spatial distribution of $P\,(\bar{x}_s)$ and $V_s\,(\bar{x}_s)$. Some formulations have used interpolating functions to describe these properties. We will adopt a simpler approach in which both fields are considered to be constant over a patch. This is the formulation that was implemented to develop CHIEF,[20] which was one of the first effective numerical

---

[20]H.A. Schenck, "Improved Integral Formulation for Acoustic Radiation Problems," *J. Acoust. Soc. Am.* 44 (1968) pp. 41–58.

codes given wide distribution. The meaning of the abbreviation will emerge in the course of the development.

Let $\bar{x}_j$ denote the location of the center of element S_j, and define $P_j \equiv P(\bar{x}_j)$ and $V_j \equiv V_s(\bar{x}_j)$. Because $\bar{x}_0$ in the SHIE is situated on the surface, $\bar{x}_0$ is taken to be one of these center points. The pressure and velocity are factored out of the integral over a patch because the requisite smallness of all S_j makes negligible their variation over that region, Thus, Eq. (7.5.16) becomes

$$
4\pi\chi_n P_n = \sum_{j=1}^{N} i\omega\rho_0 V_j \iint_{S_j} \frac{e^{-ik|\bar{x}_n - \bar{x}_s|}}{|\bar{x}_n - \bar{x}_s|} dS
$$
$$
- \sum_{j=1}^{N} P_j \iint_{S_j} \bar{n}(\bar{x}_s) \cdot (\bar{x}_s - \bar{x}_n) \left(\frac{ik|\bar{x}_n - \bar{x}_s| + 1}{|\bar{x}_n - \bar{x}_s|^3} \right) e^{-ik|\bar{x}_n - \bar{x}_s|} dS
$$

(7.5.17)

Both integrands depend only on the geometrical properties of S and the selection of the points on S. Thus, the integrals constitute a set of coefficients that are independent of the pressure and surface velocity,

$$
B_{n,j} = \iint_{S_j} \bar{n}(\bar{x}_s) \cdot (\bar{x}_s - \bar{x}_n) \left(\frac{ik|\bar{x}_n - \bar{x}_s| + 1}{|\bar{x}_n - \bar{x}_s|^3} \right) e^{-ik|\bar{x}_n - \bar{x}_s|} dS
$$
$$
C_{n,j} = \iint_{S_j} \frac{e^{-ik|\bar{x}_n - \bar{x}_s|}}{|\bar{x}_n - \bar{x}_s|} dS
$$

(7.5.18)

These coefficients are evaluated by two different schemes. No singularity arises in the off-diagonal terms, $j \neq n$, because $\bar{x}_n$ is not situated in S_j. These terms are evaluated by a two-dimensional Gaussian integration scheme, which uses the value of the integrand at points interior to S_j. A low-order scheme using four points is illustrated in Fig. 7.23, but some investigations[21] have suggested that more points are required for accurate results.

$$
\iint_{S_j} f(x,y)\,dx\,dy = 4ab \sum_{m=1}^{4} 0.25 f(x_m, y_m)
$$

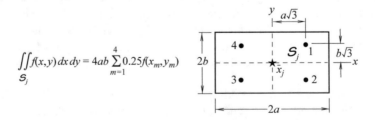

Fig. 7.23 A four-point Gaussian integration rule for integrating over surface patch S_j

[21] A.L. van Buren, "A Test of the Capabilities of CHIEF in the Numerical Evaluation of Acoustic Radiation from Arbitrary Surfaces," NRL Report 7160 (1970).

Evaluation of the diagonal coefficients $B_{n,n}$ and $C_{n,n}$ recognizes that the Cauchy principal part gives the contribution of the first integral at $\bar{x}_s = \bar{x}_n$. The numerical integration scheme excludes this point by further dividing the patch in Fig. 7.23 into four quadrants indicated there by dashed lines. A sixteen-point Gaussian integration rule is used for each subpatch. Because the Gaussian points are internal to each subregion, evaluation of the integrands at $\bar{x}_n$ is excluded.

The discretized form of the SHIE, Eq. (7.5.17), may be written in matrix form as

$$\left[4\pi\,[\chi] + [B]\right]\{P\} = [C]\{V\} \tag{7.5.19}$$

where $[\chi]$ is a diagonal array of the value of χ at each center point $\bar{x}_n$; so $\chi_{n,n} = 1/2$ except for points at corners and edges. Both $[B]$ and $[C]$ are $N \times N$ complex arrays, which is the number of unknown $P(x_n)$. It might seem that solving the equations for the $\{P\}$ given $\{V\}$ is a straightforward task. However, this is only true if the domain is an interior cavity. For the exterior radiation problem, there are certain frequencies at which difficulties arise.

All cavities are like closed one-dimensional waveguides and spherical cavities, in the sense that an infinite number of modes exist if the boundary is rigid. A mode is a pressure field that can exists without excitation; the frequency of that field is the natural frequency. Why is this important to the radiation problem? The KHIT for interior and exterior domains have similar appearance. It was proven by Schenck[22] that if the frequency matches a natural frequency of the domain inside the vibrating surface, then Eq. (7.5.19) has no unique solution. Some individuals refer to this condition as a "forbidden internal cavity resonance." If the frequency for a specific computation was exactly one of these natural frequencies, the matrix $[4\pi\,[\chi] + [B]]$ would be rank-deficient, so its inverse would not exist. The more common occurrence is that the overall computation entails a frequency sweep, and one or more of the frequencies in the swept range is close to a natural frequency for the enclosed region. Then, the matrix will be ill-conditioned. If one proceeds to solve for $\{P\}$ corresponding to this matrix, numerical errors will be greatly magnified, making the solution unusable.

How to proceed in this situation depends on what one knows about the modes of the domain contained inside S. In some special cases, like a sphere or a flat-ended circular cylinder, the modal solution can be obtained analytically. In that case, one could simply omit any results obtained from Eq. (7.5.19) for frequencies close to natural frequencies. However, for arbitrary shapes, we seldom know what the interior natural frequencies are.

Loss of uniqueness at the forbidden frequencies has been addressed in several ways, each resulting in a different boundary element computer code. CHIEF apparently was the first general procedure. It is based on a very simple idea. The approximation of the SHIE in Eq. (7.5.19) has reduced rank at a forbidden frequency, which means that the number of independent equations is less than the size of $\{P\}$. Extra equations are obtained by using the original Helmholtz integral equation, Eq. (7.4.6),

---

[22]H.A. Schenck, *ibid.*

for the case where the fluid is exterior to the vibrating surface and the field point $\bar{x}_0$ is interior to it. The field point is not within the fluids, which means that $\chi = 0$, so the integral should evaluate to zero. As is done for the surface points, the values of P and V_s are factored out of the integral over each S_j, and Gaussian integration is used to evaluate the contribution of the geometric terms in each integrand. Because $\bar{x}_0$ is situated inside S, the integrand is never singular, so the Gaussian integration scheme used to evaluate the off-diagonal terms of $[B]$ and $[C]$ is used here also. Each interior field point to which this analysis we applied leads to an extra equation to supplement what has been lost. The computer code name "CHIEF", which is an acronym for Combined Helmholtz Integral Equation Formulation, was the first to implement this concept.

It is known that the rank of $[B]$ is less than N at a forbidden frequency, but the actual rank is not known. Without that knowledge, it is not known how many equations for interior points are required to make the system of equations solvable. Furthermore, as was noted, the interior field is a mode at a forbidden frequency. Such modes have nodal surfaces, along which the pressure is zero. The Helmholtz integral equation will give zero identically if $\bar{x}_0$ is situated on a nodal surface, so it will not be an additional equation to supplement the equations for points on the surface. The strategy that addresses both issues uses far more interior points than the number that is required. Let $\bar{\xi}_n$ be the positions of these interior points, whose location may be selected in any convenient manner, including randomly. Let M be the number of such points. Then, because $\chi = 0$ for interior points, the numerical approximation of the Helmholtz integral equation at such points gives

$$[B']\{P\} = [C']\{V\} \tag{7.5.20}$$

where the coefficient matrices are

$$B'_{n,j} = \iint\limits_{S_j} \bar{n}\,(\bar{x}_s) \cdot (\bar{x}_s - \bar{\xi}_n) \left(\frac{ik\,|\bar{\xi}_n - \bar{x}_s| + 1}{|\bar{\xi}_n - \bar{x}_s|^3} \right) e^{-ik|\bar{\xi}_n - \bar{x}_s|} dS$$

$$C_{n,j} = \iint\limits_{S_j} \frac{e^{-ik|\bar{\xi}_n - \bar{x}_s|}}{|\bar{\xi}_n - \bar{x}_s|} dS \tag{7.5.21}$$

The equations for the interior points are stacked below Eq. (7.5.19) for the surface points. There are M of the former and N of the latter, so the equations to be solved are

$$\begin{bmatrix} 4\pi\,[\chi] + [B] \\ [B'] \end{bmatrix} \{P\} = [C]\{V\} \tag{7.5.22}$$

This represents a total of $N + M$ equations for the M values contained in $\{P\}$. The equations are overdetermined, as they were for the source superposition method, Eq. (7.5.5). This condition is handled in the same way, specifically by invoking the method of linear least squares. The result is that the coefficient matrix is defined to be an array $[U]$ whose size is $(N + M) \times N$, and the equation to be solved is multiplied by $[U]^{\mathrm{T}}$. This leads to

$$
[U] = \begin{bmatrix} [4\pi [\chi] + [B]] \\ [B'] \end{bmatrix}
$$
$$
[U]^{\mathrm{T}} [U] \{P\} = [U]^{\mathrm{T}} [C] \{V\}
$$

(7.5.23)

After the solution for $\{P\}$ has been obtained, the pressure at any field point exterior to S is found by using the same formulation as that used for the interior points. If the field point is in the farfield, this evaluation would be based on the farfield form in Eq. (7.4.18).

CHIEF is just one of many boundary element codes. They differ in the way that the variation of $P(\bar{x}_s)$ and $V_s(\bar{x}_s)$ over a patch is described, as well as how the integral over the surface patches is obtained. In addition, some use a different integral equation to handle the forbidden frequency issue. One that has had some popularity is obtained by differentiating the Helmholtz integral equation to obtain an integral equation for V_s on the surface. The combination of that equation and SHIE at the forbidden frequencies is called the Burton–Miller formulation.[23]

Boundary element formulations have some common attributes. One is that they tend to have a large number of variables. This is a consequence of a general guideline that the surface patches should be approximately square, with each side being no bigger than one-sixth of the acoustic wavelength at the frequency for the evaluation. The number of equations is increased by the supplemental equations required to handle the forbidden frequency issue. The requisite computation resources increase substantially as the frequency is increased in a sweep. This is so because one has the choice of setting the surface mesh to fit the one-sixth wavelength requirement at the highest frequency, or else rezoning the mesh as the frequency increases. Furthermore, all coefficient matrices are functions of frequency, so they must be recomputed at each step in the sweep. Another attribute that causes a boundary element model to require substantial computational resources is the fact that $[B]$ and $[B']$ are full matrices, so only standard solution algorithms may be employed. This is contrasted by the finite element method, whose matrices are diagonally dominant. Many algorithms that greatly reduce core memory requirements and the number of operations have been developed to solve such equations.

7.5.3 Finite Element Method

A variety of finite element formulations have been developed. We will consider a simple one in order to expose the broad concepts, and then discuss ways in which

---

[23] A.J. Burton and G.F. Miller, "The application of integral equation methods to the numerical solution of some exterior boundary-value problems", Proc. R. Soc. A **323** (1971) 201–210.

it may be used to study radiation into an infinite domain. As with the preceding developments, we will only consider the frequency domain.

One can find finite element formulations for acoustics that employ displacement as a primary variable, either individually, or in combination with pressure. We will examine a simpler approach that only uses pressure. The derivation will apply the Galerkin method to the Helmholtz equation. The first step is to decompose the domain $\mathcal{V}$ of the fluid into small pieces, which are the finite elements. We shall denote a generic element as $\mathcal{V}_n$. Each element may have any shape; common ones are a box or a tetrahedron. In general, a set of mesh points is defined in each $\mathcal{V}_n$. Some of these points may be interior to the element. However, placing some on the surface of the element will assure continuity of pressure because those points will also lie on one or more adjacent elements. Interpolation functions, typically polynomials, are used to represent the pressure within an element.

To understand the basic aspects of the formulation, a one-dimensional waveguide is described in Fig. 7.24. The waveguide is segmented into elements that are numbered sequentially by index $n = 1, 2, ... N$. The length of element n is L_n, which need not be the same for all elements. A mesh point is located at the end of each element, and no interior points are defined. A local coordinate x_n is used to measure the axial position within element n.

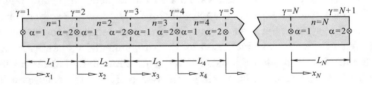

Fig. 7.24 Numbering scheme for a finite element model of a one-dimensional waveguide

The mesh points within an element are numbered with the index $\alpha = 1$ for the left node and $\alpha = 2$ for the right node. A superscript will be used to indicate which element is under consideration if there is ambiguity. Therefore, the pressure at a node is denoted as $P_\alpha^{(n)}$. The lowest order interpolating function is linear. The corresponding description of the pressure distribution within an arbitrary element is

$$P^{(n)}(x_n) = \begin{cases} P_1^{(n)}\left(1 - \dfrac{x_n}{L_n}\right) + P_2^{(n)}\dfrac{x_n}{L_n}, & 0 \le x_n \le L_n \\ 0 \text{ otherwise} \end{cases} \qquad (7.5.24)$$

The interpolating functions are denoted as $\mathcal{N}_\alpha^{(n)}(x_n)$, where n is the element number and α is the mesh point index. Thus, the pressure within an element may be written as

$$P^{(n)}(x_n) = \sum_{\alpha=1}^{2} \mathcal{N}_\alpha^{(n)}(x_n) P_\alpha^{(n)} \qquad (7.5.25)$$

If one were to use higher order interpolating functions, the sole alteration in this expression would be adjustment of the summation range to match the commensurate increase in the number of mesh points per element. Because the interpolating functions for an element are defined to be zero outside their element, the pressure at any position may be considered to be a sum over all elements of the preceding expression, that is,

$$P = \sum_{n=1}^{N} \sum_{\alpha=1}^{2} \mathcal{N}_{\alpha}^{(n)}(x_n) P_{\alpha}^{(n)} \tag{7.5.26}$$

This representation does not account for the fact that some mesh points are shared by elements, that is, $P_2^{(1)} = P_1^{(2)}$, $P_2^{(2)} = P_1^{(3)}$, ..., $P_2^{(N-1)} = P_1^{(N)}$. This overlap is described by assigning a unique index γ to each mesh point, and defining $\hat{P}_{\gamma}$ to be the pressure at that point. These values are the *global pressures*, whereas the $P_{\alpha}^{(n)}$ are the *local* values. The latter may be obtained by using a filtering matrix $\left[S^{(n)}\right]$ to pick out the appropriate elements of the global set $\left\{\hat{P}\right\}$,

$$\left\{ \begin{matrix} P_1^{(n)} \\ P_2^{(n)} \end{matrix} \right\} = \left[S^{(n)}\right]\left\{\hat{P}\right\} \tag{7.5.27}$$

For example, the filtering matrices for the first two elements in Fig. 7.24 are

$$\left[S^{(1)}\right] = \begin{bmatrix} 1\ 0\ 0\ 0 \cdots \\ 0\ 1\ 0\ 0 \cdots \end{bmatrix}, \quad \left[S^{(2)}\right] = \begin{bmatrix} 0\ 1\ 0\ 0 \cdots \\ 0\ 0\ 1\ 0 \cdots \end{bmatrix} \tag{7.5.28}$$

The $[S_n]$ matrices are more commonly referred to as the *connectivity matrix*.

Although these relations have been derived for planar waves in a waveguide, they are applicable with a few modifications to a three-dimensional field. Position within an element is described by a local coordinate system $x_n y_n z_n$. Mesh points are distributed over an element, and more than two points per element are required. Other than the higher dimensionality, the basic relations are as above. The pressure at any position $\bar{x}_n$ within an element is a sum of terms consisting of an interpolating function multiplied by the associated mesh pressure. The matrix representation of this ansatz is

$$\boxed{P(\bar{x}_n) = \left[\mathcal{N}^{(n)}(\bar{x}_n)\right]\left\{P^{(n)}\right\}} \tag{7.5.29}$$

This is the fundamental description of the pressure field. Only the mesh pressures are unknown. As before, the number of columns for the interpolating matrix equals the number of mesh points in an element, and it has one row.

The connectivity matrix $\left[S^{(n)}\right]$ still filters the pressures $\left\{P^{(n)}\right\}$ at an element's mesh points from the global set of pressures. The matrix representation of these relations is

$$\boxed{\left\{P^{(n)}\right\} = \left[S^{(n)}\right]\left\{\hat{P}\right\}} \tag{7.5.30}$$

Of course, the number of rows for $\left[S^{(n)}\right]$ will be the number of mesh points in an element, which is greater than two. Equation (7.5.29) gives the pressure in a specific element, and the interpolating functions are defined to give zero if the position is outside that element. Thus, substitution of Eq. (7.5.30) into (7.5.29) gives the contribution of the mesh points within an element to the entire field. Adding these relations for all elements gives a representation of the pressure throughout the fluid,

$$P\left(\bar{x}\right) = \sum_{n=1}^{N} \left[\mathcal{N}^{(n)}\left(x_n, y_n, z_n\right)\right] \left[S^{(n)}\right] \left\{\hat{P}\right\} \tag{7.5.31}$$

Various rules for finite elements correspond to using different interpolating functions, and/or different shapes for the elements. The preceding will accommodate any rule.

The next task is to derive the equations governing the global pressures. Equation (7.5.31) will not satisfy the Helmholtz equation. The remainder after its substitution represents an error whose value will depend on the location within $\mathcal{V}$. The Galerkin method requires that this error be orthogonal to an arbitrary test function Ψ that lies in the same functional space as the series for P. This means that the test function is like Eq. (7.5.31), except that its coefficients are *an arbitrary set of values* η_n. In other words

$$\Psi = \sum_{n=1}^{N} \left[\mathcal{N}^{(n)}\left(x_n, y_n, z_n\right)\right] \left[S^{(n)}\right] \{\eta\} \tag{7.5.32}$$

The orthogonality condition is defined relative to an inner product of functions over the entire domain $\mathcal{V}$. Thus, the condition required in the Galerkin method is

$$\iiint_{\mathcal{V}} \left(\nabla^2 P + k^2 P\right) \Psi \, d\mathcal{V} = 0 \tag{7.5.33}$$

We introduce the identity that $\Psi \nabla^2 P \equiv \nabla \cdot (\Psi \nabla P) - \nabla \Psi \cdot \nabla P$. Green's theorem converts the first term to a surface integral, so we now have

$$\iiint_{\mathcal{V}} \left(\nabla P \cdot \nabla \Psi - k^2 P \Psi\right) d\mathcal{V} = -\iint_{\mathcal{S}} \left[\left(\bar{n} \cdot \nabla P\right) \Psi\right] d\mathcal{S} \tag{7.5.34}$$

Note that the signs in this expression correspond to definition of $\bar{n}$ as the surface normal that points into the fluid domain.

The series in Eqs. (7.5.31) and (7.5.32) are used to describe P and Ψ. To use the matrix form, the components of the gradient operator are placed in a column vector $\{\nabla\}$,

$$\{\nabla\} \equiv \left[\frac{\partial}{\partial x_n} \quad \frac{\partial}{\partial y_n} \quad \frac{\partial}{\partial z_n}\right]^{\mathrm{T}} \tag{7.5.35}$$

It will be noted that the gradient is expressed in terms of the local $x_n y_n z_n$ coordinate system, because that is the coordinate system used for the interpolating functions.

The pressure gradient is obtained by operating on Eq. (7.5.31), which leads to

$$\{\nabla\} P = \begin{Bmatrix} \partial P/\partial x_n \\ \partial P/\partial y_n \\ \partial P/\partial z_n \end{Bmatrix} = \sum_{n=1}^{N} \{\nabla\}[\mathcal{N}^{(n)}(x_n, y_n, z_n)][S^{(n)}]\{\hat{P}\} \tag{7.5.36}$$

In this expression, only the interpolating functions depend on position. If J is the total number of mesh points, then $\{\nabla\}[\mathcal{N}^{(n)}(x_n, y_n, z_n)]$ is a $3 \times J$ position-dependent array given by

$$\{\nabla\}[\mathcal{N}^{(n)}(x_n, y_n, z_n)] = \begin{bmatrix} \partial \mathcal{N}_1^{(n)}/\partial x_n & \partial \mathcal{N}_2^{(n)}/\partial x_n & \partial \mathcal{N}_3^{(n)}/\partial x_n & \cdots \\ \partial \mathcal{N}_1^{(n)}/\partial y_n & \partial \mathcal{N}_2^{(n)}/\partial y_n & \partial \mathcal{N}_3^{(n)}/\partial y_n & \cdots \\ \partial \mathcal{N}_1^{(n)}/\partial z_n & \partial \mathcal{N}_2^{(n)}/\partial z_n & \partial \mathcal{N}_3^{(n)}/\partial z_n & \cdots \end{bmatrix} \tag{7.5.37}$$

Changing P to Ψ in the preceding gives the matrix representation of $\nabla\Psi$. A dot product is obtained as a product of row and column vectors, so the first term in Eq. (7.5.34) may be written as

$$\begin{aligned} \nabla P \cdot \nabla \Psi &= (\{\nabla\}\Psi)^{\mathrm{T}} (\{\nabla\}P) \\ &= \sum_{m=1}^{N}\sum_{n=1}^{N} \{\eta\}^{\mathrm{T}} [S^{(m)}]^{\mathrm{T}} (\{\nabla\}[\mathcal{N}^{(m)}(x_n, y_n, z_n)])^{\mathrm{T}} \\ &\quad \times (\{\nabla\}[\mathcal{N}^{(n)}(x_n, y_n, z_n)])[S^{(n)}]\{\hat{P}\} \end{aligned} \tag{7.5.38}$$

Both $\{P\}$ and $\{\eta\}$ are global parameters, so they may be brought outside the sum. Furthermore, the interpolating functions are zero outside their element. This means that all terms in the double sum for which the elements are different are identically zero. This reduces the double sum to a single sum over all elements, so that

$$\begin{aligned} \nabla P \cdot \nabla \Psi = \{\eta\}^{\mathrm{T}} \Bigg[\sum_{n=1}^{N} [S^{(n)}]^{\mathrm{T}} (\{\nabla\}[\mathcal{N}^{(n)}(x_n, y_n, z_n)])^{\mathrm{T}} \\ \times (\{\nabla\}[\mathcal{N}^{(n)}(x_n, y_n, z_n)])[S^{(n)}]\Bigg]\{\hat{P}\} \end{aligned} \tag{7.5.39}$$

The scalar product $P\Psi$ is

$$P\Psi = \{\eta\}^{\mathrm{T}} \Bigg[\sum_{n=1}^{N} [S^{(n)}]^{\mathrm{T}} [\mathcal{N}^{(n)}(x_n, y_n, z_n)]^{\mathrm{T}} [\mathcal{N}^{(n)}(x_n, y_n, z_n)][S^{(n)}]\Bigg]\{\hat{P}\} \tag{7.5.40}$$

Now let us consider the right side of Eq. (7.5.34). This term is evaluated on the boundary of the acoustic domain, so that $\bar{n} \cdot \nabla P = -i\rho_0 \omega V_s$. Thus this integrand may be written as

$$(\bar{n} \cdot \nabla P)\Psi = -i\omega\rho_0 \{\eta\}^{\mathrm{T}} \sum_{n=1}^{N} [S^{(n)}]^{\mathrm{T}} [\mathcal{N}^{(n)}(x_n, y_n, z_n)]^{\mathrm{T}} V_s(x_n, y_n, z_n) \tag{7.5.41}$$

The result of substituting these expressions into the Galerkin principle, Eq. (7.5.34), is

$$\{\eta\}^T [A] \{\hat{P}\} - k^2 \{\eta\}^T [B] \{\hat{P}\} = i\omega\rho_0 \{\eta\}^T \{\hat{V}\} \qquad (7.5.42)$$

The matrices $[A]$ and $[B]$ are square and symmetric. Their definitions are

$$
\begin{aligned}
[A] &= \sum_{n=1}^{N} \iiint_{\mathcal{V}_n} \left[S^{(n)} \right]^T \left(\{\nabla\} [\mathcal{N}^{(n)} (x_n, y_n, z_n)] \right)^T \left(\{\nabla\} [\mathcal{N}^{(n)} (x_n, y_n, z_n)] \right) \\
&\qquad \times \left[S^{(n)} \right] d\mathcal{V} \\
[B] &= \sum_{n=1}^{N} \iiint_{\mathcal{V}_n} \left[S^{(n)} \right]^T \left[\mathcal{N}^{(n)} (x_n, y_n, z_n) \right]^T \left[\mathcal{N}^{(n)} (x_n, y_n, z_n) \right] \left[S^{(n)} \right] d\mathcal{V}
\end{aligned}
$$

$$(7.5.43)$$

The column vector $\{\hat{V}\}$ contains the effect of the surface velocity that generates the pressure field. It is given by

$$\left\{\hat{V}\right\} = \sum_{n=1}^{N} \iint_{S} \left[S^{(n)} \right]^T \left[\mathcal{N}^{(n)} (x_n, y_n, z_n) \right]^T V_s (x_n, y_n, z_n) \, dS \qquad (7.5.44)$$

The evaluation of $\{\hat{V}\}$ entails an integration over the surface, but most finite elements are on the interior. Those elements may be omitted from the sum. Also, it seldom is convenient to evaluate the integral for the actual V_s function, especially in the context of a finite element computer code. Rather, the usual practice is to use a set of interpolating functions to represent the dependence of V_s on the surface position in terms its value at those mesh points that lie on the surface.

Now we come to the crucial step. The test function Ψ is arbitrary, which means that the η_j coefficients are arbitrary. It is necessary that Eq. (7.5.42) be satisfied for any $\{\eta\}$, which will only be true if the coefficients of $\{\eta\}^T$ in both sides of the equality are equal. The result is the standard form of the finite element equations of motion

$$\left[[A] - k^2 [B] \right] \{\hat{P}\} = i\omega\rho_0 \{\hat{V}\} \qquad (7.5.45)$$

Because k^2 is proportional to ω^2, an individual who is conversant with the principles of structural dynamics might be tempted to call $[B]$ the inertia matrix, and $[A]$ the stiffness matrix. In fact, this is the opposite of the effect each represents. The kinetic energy density is $(\rho_0/2)\, \bar{v} \cdot \bar{v}$, and $\bar{v} = -\,\mathrm{Re}\,((1/i\rho_0\omega)\,\nabla P)$. The integrand for $[A]$ contains the dot product of gradients, so it is the inertial effect. Similarly, the acoustic potential energy per unit volume is $p^2/(2\rho_0 c)$, and the integrand for $[B]$ contains a product of pressure functions. Hence, it is the stiffness term. The difference from the structural dynamics equations stems from the fact that we have used

Fig. 7.25 Finite element
domain used to analyze
radiation from the vibrating
body at the center

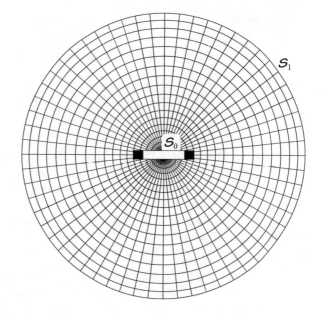

pressure, which is a force-like quantity, as the basic variable, whereas displacement
is the basic variable in structural dynamics.

The formulation leading to Eq. (7.5.45) is quite general, but it does not address
the unique nature of a radiation problem in which energy is transported into an
unbounded space. The problem is that it is not possible to divide an infinite domain
into a finite number of elements. Individuals who have enormous computing power
at their disposal have represented that domain as a very large sphere. This is the finite
element model depicted in Fig. 7.25, where the radiating body is a hydrophone that
consists of a cylinder with vibrating pistons at both ends. The mesh depicted there
adheres strictly to a spherical coordinate grid, but it would likely be better to distort
the mesh such that the radial lines in the vicinity of the hydrophone are close to being
perpendicular to the surface.

Why is a spherical domain used? The answer lies in the farfield behavior of the field
radiated by any finite-sized object. We know that in terms of spherical coordinates
centered inside a radiating body, the pressure at a very large r is a spherical wave with
directivity. The complex pressure and particle velocity in that region are described by

$$P = \frac{f(\psi, \theta)}{r} e^{-ikr}, \quad V_r = \frac{P}{\rho_0 c} \qquad (7.5.46)$$

The surface S is a composite of the exterior of the hydrophone S_0 and the outer
spherical surface S_1. Correspondingly, Equation (7.5.44) for $\{\hat{V}\}$ is split into contributions from each surface. The surface velocity on the hydrophone is known. On
S_1, we use the farfield approximation. Recall that V_s was defined to be positive if it is
into the fluid, so $V_s = -V_r$ on the spherical boundary. Thus, the velocity coefficients

on $\mathcal{S}_1$ become

$$\{\hat{V}\}_1 = \sum_{n=1}^{N} \iint\limits_{\mathcal{S}_1} \left[S^{(n)}\right]^{\mathrm{T}} \left[\mathcal{N}^{(n)}(x_n, y_n, z_n)\right]^{\mathrm{T}} V_s(x_n, y_n, z_n) \, d\mathcal{S}$$

$$= \frac{1}{\rho_0 c} \sum_{n=1}^{N} \iint\limits_{\mathcal{S}_1} \left[S^{(n)}\right]^{\mathrm{T}} \left[\mathcal{N}^{(n)}(x_n, y_n, z_n)\right]^{\mathrm{T}} P(x_n, y_n, z_n) \, d\mathcal{S}$$

(7.5.47)

The pressure in the integrand is described by evaluating Eq. (7.5.31) on the spherical boundary. Doing so reduces the integrand to a product of interpolating functions, with $\{\hat{P}\}$ brought outside the integral. The result is an equation for $\{\hat{V}\}_1$ that depends on $\{\hat{P}\}_1$. Bringing these terms to the left side of Eq. (7.5.45) yields a set of equations for the mesh pressures in which the velocity coefficients on the radiating body are the inhomogeneous terms,

$$\boxed{\left[[A] + k^2 [B] + ik [C]\right] \left\{\hat{P}\right\} = i\omega\rho_0 \left\{\hat{V}\right\}_0}$$

(7.5.48)

where the additional contribution to the coefficient matrix is

$$[C] = \sum_{n=1}^{N} \iint\limits_{\mathcal{S}_1} \left[S^{(n)}\right]^{\mathrm{T}} \left[\mathcal{N}^{(n)}(x_n, y_n, z_n)\right]^{\mathrm{T}} \left[\mathcal{N}^{(n)}(x_n, y_n, z_n)\right] \left[S^{(n)}\right] d\mathcal{S} \quad (7.5.49)$$

An important feature is that the term containing $[C]$ is 90° out-of-phase from the contributions of $[A]$ and $[B]$. As noted previously, $[A]$ and $[B]$ are associated, respectively, with the kinetic and potential energy of the fluid contained in the finite element model. Both effects are conservative. In contrast, $[C]$ represents a loss of energy, specifically, radiation damping resulting from power flowing out of the domain across the spherical boundary.

Although creating a model based on surrounding the radiating body with a very large sphere seems to be a reasonable procedure, it has a serious flaw. We do not know precisely where the farfield begins, but we do know that the sphere's radius r_1 must be sufficiently large that adjacent lines from a point on the outer surface to the vibrating body are essentially parallel. This means that the sphere's radius r_1 must be much greater than the *largest dimension* of the radiating body. The farfield approximation also requires that $kr_1 \gg 1$, that is, r_1 must be much greater than an acoustic wavelength. Furthermore, the general requirement that the element size's be one-sixth of an acoustic wavelength, which was identified for boundary elements, also applies to finite elements. Meeting these criteria would require that the model contains many mesh points. That is why this approach typically requires substantial computational resources to analyze realistic radiation problems.

What should the rest of us do? Many concepts have been developed, all with the idea of shrinking the size of the sphere, and possibly, adjusting the shape of the outer boundary to conform more closely to the shape of the vibrating body. The first modifies the spherical plane wave approximation to obtain a modified boundary condition. If the radius of the sphere is sufficiently large, setting $V_r = P/(\rho_0 c)$ on the outer body will result in the reflection coefficient being zero. If a boundary had a zero reflection coefficient for any type of wave, we would say that is *perfectly transmitting*. The reflection coefficient for a plane wave obliquely incident on a locally reacting plane is

$$R = \frac{Z - Z_1}{Z + Z_1}, \quad Z_1 = \frac{\rho_0 c}{\cos \psi_I} \tag{7.5.50}$$

One could think of selecting Z to minimize $|R|$ over a range of ψ_I, and then using that Z as the boundary condition for the exterior surface S of the finite element mesh. That surface need not be spherical, so the approximate transmitting boundary condition would be

$$P = -Z\bar{n} \cdot \bar{V} \tag{7.5.51}$$

where the negative sign results from defining $\bar{n}$ to point into the domain represented by the mesh.

This approximation does not work too well, so the next idea is to introduce dissipation, in which case the result will be an *absorbing boundary condition*. This can be done by allowing Z to be complex, or by introducing a dissipation operator into the plane wave approximation. A simple absorbing boundary condition would be

$$\rho_0 c\, \bar{n} \cdot \bar{V} = -(P + \alpha \bar{n} \cdot \nabla P) \tag{7.5.52}$$

where the value of α would be selected to minimize the signal that is reflected back into the acoustic domain. The absorbing boundary conditions that are used are much more complicated than these. An archival work is the paper by Bayliss and Turkel[24]; a more accessible description is the paper by Assaad et al.[25]

The latest development is the concept of an *infinite element*. Such an element uses the basic form of a radially diverging wave as the basis for its shape functions. These elements are placed outside the fluid domain that is described by regular finite elements. Most derivations have considered this domain to be spherical, which is the configuration depicted in Fig. 7.26. The exception is the work of Burnett[26] in which the outer boundary of the fluid domain is a prolate spheroid, which can enclose a long slender object without extending far in the broadside direction.

As shown in Fig. 7.26, some mesh points lie on the outer boundary S_1 of the finite element mesh. The field within the element is taken to be an outgoing spherical wave with directivity. This dependence is expressed in terms of spherical coordinates centered on the sphere, which leads to

---

[24] A. Bayliss and E. Turkel, "Radiation BoundaryConditions for Wave-Like Equations," *Comm. Pure Appl. Math.*, **33** (1980) pp.707–725.

[25] J. Assaad, J.-N. Decarpigny, C. Bruneel, R. Bossut, and B. Hamonic, "Two-dimensional radiation problems," J. Acoust. Soc. Am., **94** (1993) 562–573.

[26] D.S. Burnett, "A three-dimensional acoustic infinite element based on a prolate spheroidal multipole expansion," *J. Acoust. Soc. Am.*, **96**, (1994) 2798–2816.

Fig. 7.26 Configuration of
an infinite element used to
approximate the free field in
a finite element model

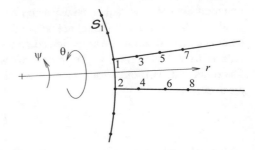

$$P = \frac{e^{-ikr}}{r} \sum_{m=0}^{M} \frac{F_m\,(\psi,\theta)}{r^m} \tag{7.5.53}$$

Several methods have been used to set the $F_m\,(\psi,\theta)$ functions. One, described by Burnett,[27] selects the $m=0$ function to match the regular finite element functions in the two transverse directions, and then finds the others by fitting the expression to the angular terms in the Laplacian operator. A different derivation based on a mapping of the cylindrical coordinates into a hyperbolic grid was used by Astley et al.[28] In any event matching, the standard interpolation form in Eq. (7.5.31) to the diverging spherical wave form leads to the interpolation functions for the infinite element. At that juncture, the contribution of the infinite elements to $[A]$ and $[B]$ is folded into those of the regular finite elements.

The advantages of the finite element formulation might not be apparent at this juncture, especially because the one-sixth wavelength requirement also applies to a finite element model. A boundary element model only entails a mesh on the surface, whereas finite elements discretize the three-dimensional domain of the fluid. Hence, a finite element model will have more mesh points, and therefore more equations to solve, than a comparable boundary element model. The advantageous aspects of the finite element model follow from its basic features. First, the coefficient matrices $[A]$, $[B]$, and $[C]$ are constants independent of frequency. Thus, in the likely event that it is necessary to perform a frequency sweep, the basic terms for the equations of motion are only evaluated once. In contrast, the coefficient matrices for a boundary element model depend on frequency. Another difference is that the integrals for the evaluation of coefficient matrices can be derived as standard formulas for a specified set of interpolation functions. In contrast, the integrals for boundary element coefficients contain singularities and therefore constitute a more formidable task to program and evaluate. Some individuals find the absence of the "forbidden frequency" issue to be a major asset of finite elements.

---

[27] *ibid.*

[28] R.J. Astley, G.J. Macaulay, J.-P. Coyette, and L. Cremers, "Three-dimensional wave-envelope elements of variable order for acoustic radiation and scattering. Part I. Formulation in the frequency domain," *J. Acoust. Soc. Am.* **103** (1998), 49–63.

Perhaps the most significant difference between finite elements and boundary elements may be found in the fundamental mathematical properties of their equations. As was noted earlier, the coefficient matrix for boundary elements is full. In contrast, the mesh points for a finite element model are shared by only a few elements, so only a few elements are directly coupled. This means that the coefficient matrices for a finite element model will be banded around the diagonal. Many highly efficient methods have been developed to solve equations whose system matrices are banded. Thus, although the finite element model has more mesh points, it tends to require less computational resources.

An interesting feature of the finite element technique is that all developments, including absorbing boundary conditions and infinite elements, have analogous formulations in the time domain. Whether this is useful depends on what properties of the response one wishes to obtain. Often, a complex frequency response is the desired result. Extracting time domain data from a frequency domain simulation would require FFT processing, thereby adding to the computational effort. On the other hand, solving time domain equations of motion must be done with a numerical differential equation solver, which require a great deal of computational resources if it is necessary to a large number of coupled equations. In any event, the time domain formulation lies more in the research arena because some features are still being explored.

It is worth noting that the finite element formulation is equally valid for the analysis of a cavity. The same is true for the boundary element formulation, if the direction of the surface normal is reversed. Several commercial codes of both types are available for such applications. Both are simplified by the fact that a cavity is a finite domain. Hence, there is no need for absorbing boundary conditions or infinite elements, and the forbidden frequency issue is irrelevant.

7.6 Homework Exercises

Exercise 7.1 The farfield radiation of a vibrating sphere is observed to match the pattern of a longitudinal quadrupole. The fluid is air and the radius of the sphere is 100 mm. When the frequency is 1.2 kHz, the maximum pressure at 5 m from the center of the sphere is 0.3 Pa. What vibration of the sphere's surface would produce a pressure field having these attributes?

Exercise 7.2 The surface velocity on a sphere is given by $v_r = \text{Re}\,[V_4 \cos (4\phi)\,\exp (-i\omega t)]$. Derive an expression for the complex pressure amplitude in the radiated field. Evaluate $|P/\rho_0 c V_4|$ along the polar axis, $\phi = 0$. Plot the result as a function of kr when $ka = 4$ and $ka = 24$. Based on these graphs, identify the value of kr at which the farfield approximation is valid for each ka.

Exercise 7.3 A sphere vibrates over a band surrounding its equator, specifically, $v_r = \text{Re}\,[V_0 \exp (i\omega t)]$ if $\pi/3 < \psi < 2\pi/3$, $v_r = 0$ otherwise. Derive an expression for the acoustic pressure in the farfield. From this, evaluate the directivity at $ka = 1$, 10, and 50. Interpret these results relative to what one would expect qualitatively.

Exercise 7.4 Scanning the farfield of a vibrating sphere has led to identification of the coefficients of a spherical harmonic series describing the polar angle dependence of the pressure at a specific radial distance $r_1 = ka^2$. Specifically, it is known that only spherical harmonics above $m = 8$ are insignificant, with the contribution of the lower harmonics being

$$P(r_1, \psi) = P = \rho_0 c^2 \sum_{m=0}^{8} C_m P_m (\cos \psi), \quad C_m = \frac{1}{(m + 0.5)^2}$$

Determine the radial velocity on the surface of the sphere for $ka = 2$ and $ka = 20$. Compare the polar pattern of the surface velocity to that of the measured pressure field.

Exercise 7.5 Due to measurement error, the coefficients C_m in Exercise 7.4 have a $\pm 20\%$ error in amplitude and a $\pm 40°$ error in phase relative to the nominal value given there. The errors for each coefficient have a uniform probability distribution. Create a new set of coefficients by adding the random error to each of the stated values. Determine the radial velocity for $ka = 0.5$, 4, and 20 at the surface corresponding contaminated coefficients. Compare that result to the surface velocity when the coefficients have their nominal values. What conclusions can be drawn regarding this inverse identification process?

Exercise 7.6 A vibrating sphere of radius a is surrounded by a polymeric material that has been molded into a sphere of radius b that is concentric with the inner sphere. Outside these spheres, the fluid is water. The shear strength of the polymeric material is sufficiently small that the material may be considered to be a liquid whose density is ρ_1 and whose sound speed is c_1. The vibration of the inner sphere is radially symmetric, so that $v_r = v_a \sin(\omega t)$ at $r = a$. Derive an expression for the pressure in the water. *Hint:* The pressure and radial particle velocity must be continuous at $r = b$.

Exercise 7.7 A hemispherical balloon is fastened on its edges to a rigid baffle. A jet of air internally induces a vibration that has a bulbous distribution given by $v_r = v_0 \cos(2\psi)\sin(\omega t)$ for $r = a$ and $0 \le \psi \le \pi/2$. (a) Derive a spherical harmonic series describing the radiated pressure field. (b) Evaluate the farfield directivity for $ka = 0.5$, 5, and 20.

Exercise 7.8 A hemispherical balloon is fastened on its edges to an infinite pressure-release baffle. A jet of air internally induces a vibration that has a bulbous distribution given by $v_r = v_0 \cos(2\psi)\sin(\omega t)$ for $r = a$ and $0 \le \psi \le \pi/2$. (a) Derive a spherical harmonic series describing the radiated pressure field. (b) Evaluate the farfield directivity for $ka = 0.5$, 5, and 20.

Exercise 7.9 A hemispherical steel shell is fastened at its rim, $\psi = 90°$, to a rigid baffle. The attachment is such that displacement parallel to the baffle is possible, but displacement perpendicular to the baffle is not. In terms of the shell displacements, these conditions require that $u = 0$ at $\psi = \pi/2$. The internal pressure is a uniform distribution that fluctuates harmonically, so that $p_s(\psi, t) = P_0 \cos(\omega t)$. The shell's diameter is a factor of 120 greater than its thickness, its density is $5\rho_0$, and the sound speed of the shell is $c_e = 3c$. (a) Derive spherical harmonic series describing the radiated pressure field and the shell displacement. (b) Evaluate the farfield directivity for $ka = 0.5$, 5, and 20.

Exercise 7.10 A hemispherical steel shell is fastened at its rim, $\psi = 90°$, to a pressure-release baffle. The attachment is such that displacement perpendicular to the baffle is possible, but radial displacement parallel to the baffle is not. In terms of the shell displacements, these conditions require that $w = 0$ at $\psi = \pi/2$. The internal pressure is a uniform distribution that fluctuates harmonically, so that $p_s(\psi, t) = P_0 \cos(\omega t)$. The shell's diameter is a factor of 120 greater than its thickness, its density is $5\rho_0$, and the sound speed of the shell is $c_e = 3c$. (a) Derive spherical harmonic series describing the radiated pressure field and the shell displacement. (b) Evaluate the farfield directivity for $ka = 0.5$, 5, and 20.

Exercise 7.11 It is suspected that a long cylindrical section of an HVAC duct is vibrating excessively, thereby radiating an unacceptable sound level. The diameter of this section is 240 mm. Measurements at a distance of 2.5 m from the centerline indicate that the sound pressure level is 95 dB//20 μPa, and that the wavenumber is $\bar{k}' = 20\bar{e}_R + 1.6\bar{e}_\theta + 15\bar{e}_z$ m^{-1}. What properties of the duct's vibration can be deduced from these measurements?

Exercise 7.12 An extremely long pipeline is submerged in a deep region of the ocean. The diameter of the pipe is 700 mm. A disturbance at a distant end generates a flexural wave in the pipe at wavenumber. A reference location for the radiated pressure is $R - 600$ mm, $z = 0$. The pressure measured at this location is $940 \cos(3770t + 0.1160)$. Then, the position of the hydrophone is shifted gradually. Measurements at various circumferential angles lead to the conclusion that field is axisymmetric. The radial distance R is increased gradually. It is found that $R_2 = 2.48895$ m is the smallest distance greater than $R_1 = 0.6$ m at which the pressure at the two locations is 180° out-of-phase. After this measurement is made, the hydrophone is shifted axially from the reference location, with R held constant. This new measurement indicates that the pressure at $R_3 = 0.6$, $z_3 = 1.6017$ also is 180° out-of-phase from the reference signal. (a) Determine the axial and radial wavenumbers. (b) Determine the complex amplitude of the surface vibration.

Exercise 7.13 A pressure field is the superposition of a plane wave that propagates in the axial direction of a cylindrical pipe and an axisymmetric cylindrical standing wave. Specifically, the pressure at an arbitrary field point is $p = B H_0(\kappa R) \cos(\pi z/L) \cos(\omega t) + C \cos(\omega t - kz)$. Derive an expression for the time-averaged power per unit axial length radiated by the cylinder.

Exercise 7.14 The surface normal velocity on a very long cylinder consists of a bulging motion on one side of the circumference, with a sinusoidal variation in the axial direction. The specific form is $\bar{v} \cdot \bar{e}_R = V_0 \cos(\theta) \cos(\pi z/L)$ if $|\theta| \le \pi/2$, $\bar{v} \cdot \bar{e}_R = 0$ if $\pi/2 < |\theta| \le \pi$. The wavelength is twice the cross section's radius, $L = 2a$, and the frequency is $ka = 0.75\pi$. (a) Derive an expression for the radiated pressure as a function of R, z, and θ. (b) Evaluate the directivity function $(R/a)^{1/2} |P_{\mathrm{ff}}| / (\rho_0 c V_0)$ as a function of θ in the plane $z = 0$. (c) Evaluate $(R/a)^{1/2} |P| / (\rho_0 c V_0)$ as a function of R along the line defined by $\theta = 0$, $z = 0$. From that result, estimate where the farfield begins.

Exercise 7.15 A cylindrical pipeline is supported at intervals spaced at distance L. It transports a liquid whose flow velocity is high, with the consequence that the cross sections undergo a translational displacement in a fixed transverse direction, which is defined to be x. In addition, pressure fluctuations within the flowing liquid induce a uniform expansion and contraction of a cross section. The result is that the normal displacement on the cylinder's surface is $w = [W_0 + W_1 \cos(\pi z/L) \cos\theta] \sin(\omega t)$, where z is the axial distance from a reference point. The cylinder's diameter is $2a$ and the half-wavelength is $L = a$. Derive an expression for the dependence of the pressure on the radial distance R, axial distance z, and circumferential angle θ measured from the x-axis. It may be assumed for this analysis that end effects are negligible. (b) Post-processing of data acquired in an experiment indicates that the pressure amplitude when W_0 and W_1 are nonzero is the twice the amplitude when $W_0 = 0$. The location for this observation is situated on the transverse line for which $\theta = 0$ and $z = 0$. Determine the ratio W_0/W_1 for each of the following combinations of frequency and location: (1) $ka = 1$ and $R = a$, (2) $ka = 1$ and $R = 10a$, (3) $ka = 10$ and $R = a$, (4) $ka = 10$ and $R = 10a$.

Exercise 7.16 The surface of a very long cylinder whose radius is a consists of a sequence of circular bands of length L. These bands are piezoceramic elements, each of which vibrates axisymmetrically at frequency ω with no axial variation. Each band's velocity is 180° out-of-phase from those to which it is adjacent. The result is that the surface velocity has a square wave pattern in the axial direction, given by $v_R = \mathrm{Re}\,(V(z) \exp(i\omega t))$, where the complex amplitude is $V(z) = V_0$ if $0 < z < L$, $V(z) = -V_0$ if $L < z < 2L$, $V(z) = V(z \pm 2L)$. Derive an expression for the farfield pressure. Plot $(R/a)^{1/2} |p/(\rho_0 c V_0)|$ as a function of z for the case where $ka = 1$ and $ka = 10$. Does either result resemble the square wave pattern on the surface?

Exercise 7.17 The field radiated by a sphere that executes an oscillatory rigid body translation has a known solution as a dipole, see Sect. 6.5.2. Use the results derived there to evaluate the surface pressure when a sphere's surface velocity is $\mathrm{Re}\,(V_1 \exp(i\omega t)) \cos\psi$. Then, use this surface response to formulate the KHIT for the pressure at an arbitrary field point. In the case where $ka = 5$, evaluate the result of that formulation along the polar axis, $\psi = 0$, for $0 < r < 4a$. Compare the result to the analytical description, Eq. (6.5.102).

Exercise 7.18 A cylindrical rod whose length is L and diameter is $2a$ vibrates like a simply supported beam. This means that all points on a cross section displace by the same amount in a fixed transverse direction. In terms of cylindrical coordinates whose origin is at the midpoint, the resulting velocity of a surface point is $\bar{v} = \text{Re}\,(V_1 \exp\,(i\omega t))\sin\,(\pi z/L)\,(\bar{e}_r \cos\theta + \bar{e}_\theta \sin\theta)$. It is desired to ascertain the farfield directivity, so it is not acceptable to approximate the rod as an infinite length cylinder. Use the formulation in Example 7.10, in which the radiation impedance for an infinite cylinder is used to estimate the surface pressure corresponding to the stated surface velocity. An integral expression for the far field pressure can be obtained by substituting the surface velocity and approximate surface pressure into the KHIT. (a) Derive this integral expression. (b) Use analytical or numerical methods to evaluate and plot $|R|\,P_{\text{ff}}/\,(\rho_0 c V_1)$ in the plane $\theta = 0$ as a function of the polar angle ψ. Parameters are $ka = 1$ and 10, $L/a = 6$.

Exercise 7.19 The rigid disk in the sketch executes an oscillatory transverse vibration as a rigid body. The normal velocity on the front surface, whose normal is $\bar{e}_z$, is $\bar{v}_{\text{front}} \cdot \bar{n} = \text{Re}\,(V_1 \exp\,(i\omega t))$, whereas on the back surface, it is $\bar{v}_{\text{back}} \cdot \bar{n} = \text{Re}\,(-V_1 \exp\,(i\omega t))$. It is very thin, so the surface response on the edge has negligible effect. Because the front and back motions are $180°$ out-of-phase, but otherwise alike, the surface pressures on opposite sides also must be $180°$ out-of-phase. Furthermore, the pressure distributions must be axisymmetric relative to the z-axis, so it must be that $p_{\text{front}} = -p_{\text{back}} = \text{Re}\,(P\,(R)\exp\,(i\omega t))$. Another consideration comes from the theory of diffraction. According to it, the zero thickness approximation leads to a discontinuity at $R = a$. This requires that the pressure at the edge be zero with an infinite slope, $dp_{\text{front}}/dR = 0$. A trial function fitting these requirements is $P_{\text{front}} = P_C\,(1 - R^2/a^2)^{1/2}$. (a) Show that using this guess to formulate the farfield version of the KHIT, Eq. (7.4.18), leads to

$$P_{\text{ff}}\,(\bar{x}_0) = ik\frac{e^{-ikr}}{r}\,(\cos\psi)\int_0^a P_C\left(1 - \frac{R^2}{a^2}\right)^{1/2} J_0(kR\sin\psi)R\,dR$$

(b) In the limit as $ka \to 0$, this expression should yield the same result as Eqs. (7.4.38) and (7.4.39), which is the (low-frequency) multipole expansion for a transversely oscillating disk. Use this fact to derive an expression for P_C.

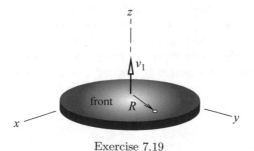

Exercise 7.19

Exercise 7.20 The sketch shows a rigid box mounted on a vertical shaft. Its rotation rate is $\Omega = \Omega_0 \sin(\omega t)$. Identify the restrictions that must be imposed on Ω_0 and ω in order that a linearized analysis be valid. Given that these restrictions are met, describe the multipole expansion that would fit the field radiated by the box. In what directions can the pressure amplitude be expected to have a maximum value? In what directions can the pressure be expected to be zero? Estimate the largest pressure amplitude that would be observed at a fixed distance r from the origin.

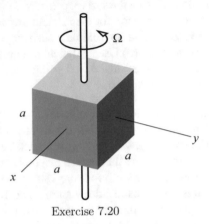

Exercise 7.20

Exercise 7.21 The surface Helmholtz integral equation was derived in Sect. 7.5.2 for the case of radiation by letting $\bar{x}_0$ be an exterior field point that approaches the surface S. Show that the same result is obtained if $\bar{x}_0$ is a field point inside the radiating body.

Exercise 7.22 The field radiated by a sphere that executes an oscillatory rigid body translation has a known solution as a dipole, see Sect. 6.5.2. Thus, it offers the opportunity to test an implementation of the source superposition formulation. The arrangement of sources and field points is described in the sketch. There are four sources that are situated along the z-axis, which is the direction of the translational velocity $\bar{v} = V_1 \sin(\omega t) \bar{e}_z$. These points are spaced at $0.4a$ centered on the sphere's center. There are nine surface points spaced equally along a meridian. (a) Explain why it is preferable that all of sources be situated on the z-axis. (b) Explain why it would be incorrect to use source or surface points at the same polar angle along more than one meridian. (c) Without solving the superposition equations, anticipate how the source strengths left of the center should be related to those to the right of the center. (d) Formulate and solve the source superposition equations for the case where $ka = 8$. Compare the pressure on the surface at $\psi = 30°$ to the dipole solution.

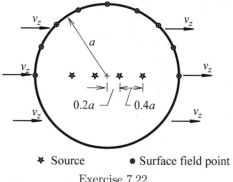

✽ Source   ● Surface field point

Exercise 7.22

Exercise 7.23 The sketch shows a model for an underwater projector. It is a cylinder whose surfaces and one end are stationary. The end that is visible in the sketch executes a translational oscillation, such that all points on that surface vibrate in the axial direction according to $v_z = V \cos(\omega t)$. The task is to use the source superposition method to evaluate the radiation. The eventual objective would be to evaluate the radiated pressure field, but here it is sufficient to evaluate the source strengths. The arrangement of sources and surface points is such that both are spaced at one-fifth of the cylinder's length centered on the midpoint in the axial direction, with an additional six field points, three on each end. (a) Explain why it is preferable that all sources be situated on the z-axis. (b) Explain why it would be incorrect to use source points at the axial location along other circumferential angles. (c) Explain why it would be unwise to locate a surface point at the junction of the cylinder and its end. (d) For the case where $L = 3a$ and $ka = 4$, determine the source strengths as nondimensional quantities, $\hat{Q}_n / (a^2 V)$.

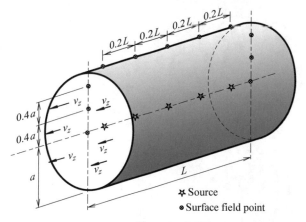

✽ Source
● Surface field point

Exercise 7.23

Chapter 8
Radiation from a Source in a Baffle

Section 3.5.1 introduced a correction for the pressure-release condition at the open end of a one-dimensional waveguide. The basis for that development is the radiation properties of a circular piston flush mounted in a wall whose extent is large. The wall is referred to as a *baffle*. A piston serves well as a model of real transducers that might be piezoceramics in an underwater projector. It may be employed as an approximate representation of the movable cone of a loudspeaker. The term "piston" is intended to convey the notion that all points in the surface have the same velocity. Our study will concentrate on such situations, but allowance will be made in the development for more general velocity distributions.

Because the concern here is with the signal that is generated by a type of surface vibration, it could logically be included in the previous chapter. However, the phenomena that arise are extremely relevant to a number of every day systems. Consequently, many aspects of radiation from a transducer in a planar baffle have been thoroughly analyzed. The subject appears in its own chapter as a way of emphasizing its importance.

The KHIT is the fundamental principle that is the foundation for the developments that follow. Rayleigh identified a Green's function that removes the need to know the surface pressure, with the result that the pressure at a field point is described as a surface integral that requires specification of the normal velocity. Given that the KHIT we derived governs the frequency domain, that is the description with which we begin, but we will also examine a time-domain solution.

8.1 The Rayleigh Integral

Figure 8.1 depicts the system to be analyzed. We wish to determine the pressure $\mathrm{Re}\left(P\left(\bar{x}_0\right) e^{i\omega t}\right)$ at an arbitrary field point $\bar{x}_0$ that results from a specified normal velocity distribution $\mathrm{Re}\left(\bar{n}\left(\bar{x}_s\right) \cdot \bar{V}\left(\bar{x}_s\right) e^{i\omega t}\right)$ on $\mathcal{S}$. This plane is the boundary of a

Electronic supplementary material The online version of this chapter (DOI 10.1007/978-3-319-56847-8_8) contains supplementary material, which is available to authorized users.

Fig. 8.1 An aribitrary
velocity distribution over
points $\bar{x}_s$ on a flat plane that
is the boundary S of a
half-space

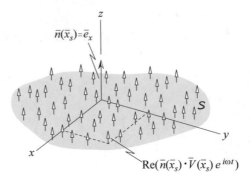

half space. The origin of the xyz coordinate system is situated on S, with the z-axis
pointing into the fluid domain. The only requirement imposed on $\bar{V}(\bar{x}_s)$ is that it
approach zero with increasing distance in all directions from the origin.

The KHIT entails integration over the surface that bounds the domain V in which
the signal occurs. In the present situation that domain is a half space. The plane is
only a part of the boundary, with the other part extending to infinity. The concept
of an enclosing virtual surface allows us to apply the KHIT to this configuration.
In Sect. 7.4.2, a sphere that surrounds the vibrating body was used to extend the
KHIT to radiation problems. Here, we use a hemisphere S_0 centered on the origin,
whose radius r_0 is allowed to grow without limit. The reasoning that previously led
to recognition that the contribution of S_0 to the KHIT is negligible if r_0 is very large
also applies here. The consequence is that the KHIT only requires integration over
the planar surface S.

The discussion in Sect. 6.4.2 showed that in addition to the free-space Green's
function, other functions that satisfy boundary conditions can be defined. If we can
find a Green's function for which $\bar{n}(\bar{x}_s) \cdot \nabla G(\bar{x}_0, \bar{x}_s)$ is identically zero, then the
integrand in the KHIT will not contain the surface pressure. Such an arrangement
would make it possible to determine the pressure field from a direct evaluation of
the integral.

The first step in identifying the requisite Green's function is to remove any ambiguity in what we seek. The position of the field point $\bar{x}_0$ is represented by its coordinates
x_0, y_0, and z_0 and a source point $\bar{x}$, not necessarily on the surface, is situated at coordinates x, y, and z. If we reduce z to zero, the source point will be the surface point
$\bar{x}_s$. In other words, $\bar{x} = \bar{x}_s + z\bar{e}_z$. The gradient in the KHIT is evaluated at the source
point, and we want the source point to be on the surface. Thus, we seek a Green's
function that satisfies

$$\bar{n}(\bar{x}_s) \cdot \nabla_s G(\bar{x}_0, \bar{x}_s) \equiv \lim_{z \to 0} \frac{\partial}{\partial z} G(\bar{x}_0, \bar{x}) = 0 \qquad (8.1.1)$$

Our experience with a point source in a half space suggests where to search for the desired function. According to the method of images, the field associated with a point source situated above a rigid plane is obtained by adding an equal point source at the same distance below the plane. Figure 8.2 depicts a point source at distance z above the surface and its image at distance z below the surface. The projection of $\bar{x}'$ onto the surface is also $\bar{x}_s$.

Fig. 8.2 Green's function for a half space bounded by a rigid planar boundary. The source is at $\bar{x}$ and the image is at $\bar{x}'$

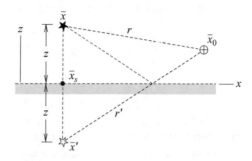

The mass acceleration of the point source is defined to be one. The image location is $\bar{x}' = \bar{x}_s - z\bar{e}_z$. The sum of this source and its image is a Green's function for the half space. This is so because the field radiated by the pair satisfies the inhomogeneous Helmholtz equation, $\nabla^2 G + k^2 G = -\delta(\bar{x}_0 - \bar{x})$ for any $\bar{x}$ and $\bar{x}_0$ within the half space, as well as the radiation condition, and the boundary condition of zero normal velocity on the surface. To distinguish the half-space function G from the free-space Green's function let us use a subscript "fs" to identify the latter. Thus, the function we have constructed is

$$G(\bar{x}, \bar{x}_0) = G_{\text{fs}}(\bar{x}_s + z\bar{e}_z, \bar{x}_0) + G_{\text{fs}}(\bar{x}_s - z\bar{e}_z, \bar{x}_0) \tag{8.1.2}$$

We wish to reduce z to zero. Before we do so, let us evaluate $\partial G/\partial z$ when z is not zero. The radial distances to the field point from the source and image are

$$r = |\bar{x}_s + z\bar{e}_z - \bar{x}_0| = \left[(x - x_0)^2 + (y - y_0)^2 + (z - z_0)^2\right]^{1/2}$$
$$r' = |\bar{x}_s - z\bar{e}_z - \bar{x}_0| = \left[(x - x_0)^2 + (y - y_0)^2 + (-z - z_0)^2\right]^{1/2} \tag{8.1.3}$$

Because the free-space Green's function depends only on the radial distance between the pair of points, the z derivative may be evaluated according to

$$\frac{\partial}{\partial z} G(\bar{x}_0, \bar{x}) = \frac{\partial r}{\partial z}\frac{d}{dr}G_{\text{fs}}(\bar{x}_s + z\bar{e}_z, \bar{x}_0) + \frac{\partial r'}{\partial z}\frac{d}{dr'}G_{\text{fs}}(\bar{x}_s - z\bar{e}_z, \bar{x}_0)$$
$$= \frac{z - z_0}{r}\frac{d}{dr}G_{\text{fs}}(\bar{x}_s + z\bar{e}_z, \bar{x}_0) + \frac{z + z_0}{r'}\frac{d}{dr'}G_{\text{fs}}(\bar{x}_s - z\bar{e}_z, \bar{x}_0) \tag{8.1.4}$$

When we reduce z to zero, both $\bar{x}$ and $\bar{x}'$ become $\bar{x}_s$. Consequently, $G_{\text{fs}}(\bar{x}_s + z\bar{e}_z, \bar{x}_0)$ and $G_{\text{fs}}(\bar{x}_s - z\bar{e}_z, \bar{x}_0)$ both become $G_{\text{fs}}(\bar{x}_s, \bar{x}_0)$, so that dG_{fs}/dr'

approaches dG_{fs}/dr. Furthermore, in the limit, the fractions preceding each G_{fs} term in the above expression become $-z_0/r$ and z_0/r. Thus, we conclude that the Green's function defined in Eq. (8.1.2) with $z = 0$ satisfies Eq. (8.1.1). In other words, the Green's function we seek is

$$\boxed{G\left(\bar{x}_0, \bar{x}_s\right) = 2G_{fs}\left(\bar{x}_0, \bar{x}_s\right)} \tag{8.1.5}$$

The result of using this Green's function to form KHIT is

$$\boxed{P\left(\bar{x}_0\right) = 2i\omega\rho_0 \iint\limits_{S} \bar{n}\left(\bar{x}_s\right) \cdot V\left(\bar{x}_s\right) G_{fs}\left(\bar{x}_0, \bar{x}_s\right) dS} \tag{8.1.6}$$

In retrospect, Eq. (8.1.5) could have been identified heuristically. A free-space Green's function represents the limit as its radius shrinks to zero of a radially symmetric spherical source, whose mass acceleration is unity. When this sphere is placed on a rigid surface at $\bar{x}_s$, its mass flow is confined to the hemisphere above the boundary. If the mass flow is held at a unit value, the particle velocity on the hemisphere is double that for the full sphere in free space. This results in doubling the radiated pressure relative to what would have been obtained if the surface were not present. In regard to the gradient property, consider a spherical source in free space at location $\bar{x}_s$. Changing the z coordinate of the source will change the distance to a field point, and therefore the pressure at that point. In contrast, if a hemispherical source is situated on the boundary, moving it to a small distance above the boundary moves its image below the boundary by the same distance. The change in the distance from the image to the field point is the negative of the change in the distance from the source to the field point. The pressure changes for each cancel.

Equation (8.1.6) is valid for two-dimensional, as well as three-dimensional models. This is so because the only place where dimensionality was considered was in the representation of r_1 and r_2, and setting y_s and y_0 to zero for a two-dimensional geometry would not alter any result. However, our primary interest is in three-dimensional configurations. When $G_{fs}\left(\bar{x}_s, \bar{x}_0\right)$ is specified as the three-dimensional version, the form of Eq. (8.1.6) that results is known as the *Rayleigh integral*.[1] It is

$$\boxed{P\left(\bar{x}_0\right) = \frac{i\omega\rho_0}{2\pi} \iint\limits_{S} V_s\left(\bar{x}_s\right) \frac{e^{-ik|\bar{x}_0-\bar{x}_s|}}{|\bar{x}_0 - \bar{x}_s|} dS} \tag{8.1.7}$$

where $V_s\left(\bar{x}_s\right)$ is the velocity normal to a surface at location $\bar{x}_s$,

$$\boxed{V_s\left(\bar{x}_s\right) = \bar{n}\left(\bar{x}_s\right) \cdot \bar{V}\left(\bar{x}_s\right)} \tag{8.1.8}$$

---

[1] J.W. Strutt Lord Rayleigh, *Theory of Sound*, vol. 2, 2nd ed., Dover (1945 reprint) Sect. 78.

The Rayleigh integral describes situations where S is the planar boundary of an infinite half space. We will see in Sect. 8.4 that it may be used for a finite baffle, provided that the baffle extends at least one acoustic wavelength beyond the portion of S that moves. However, it has been used as an approximation when S is not planar. Doing so requires a thorough verification through computational modeling or experiment because it is not justified by the derivation. This issue will be examined in Example 8.4.

A detailed mapping of the pressure field requires that the Rayleigh integral be evaluated numerically. How to perform such an evaluation is addressed in Sect. 8.4. However, our initial explorations will examine analytical solutions that are descriptive of the field in specific regions.

8.2 Farfield Directivity

Suppose the surface vibration is confined to a finite portion of S, which will be designated as S_v. If the distance from the origin to the field point $\bar{x}_0$ is much greater than the maximum distance between two points in S_v, then we may invoke the farfield approximation of the radial distance from a surface point to $\bar{x}_0$.

8.2.1 Cartesian Coordinate Description

We consider first the case where the outline of the vibrating region is best described in terms of the Cartesian coordinates x and y that lie on the surface. A point in S_v is $\bar{x}_s$, and $\bar{x}_0$ is a field point. The farfield approximation requires that $|\bar{x}_0|$ be much greater than the largest $|\bar{x}_s|$. When this condition applies, the radial line from $\bar{x}_0$ to any $\bar{x}_s$ is essentially parallel to the radial line from $\bar{x}_0$ to the origin of xyz, so that the length of radial lines differs by the projection of $\bar{x}_s$ onto $\bar{x}_0$. The magnitude of $\bar{x}_0$ is the radial distance r, so the approximation is

$$|\bar{x}_s - \bar{x}_0| \approx |\bar{x}_0| - \bar{x}_s \cdot \frac{\bar{x}_0}{|\bar{x}_0|} = r - \bar{x}_s \cdot \bar{e}_r \tag{8.2.9}$$

The location of the field point is described in Fig. 8.3 in terms of spherical coordinates. The radial unit vector is

$$\boxed{\bar{e}_r = \sin \psi \cos \theta \bar{e}_x + \sin \psi \sin \theta \bar{e}_y + \cos \psi \bar{e}_z} \tag{8.2.10}$$

The location of the surface point is $\bar{x}_s = x_s \bar{e}_x + y_s \bar{e}_y$, so we have

$$|\bar{x}_s - \bar{x}_0| \approx r - x_s \sin \psi \cos \theta - y_s \sin \psi \sin \theta \tag{8.2.11}$$

Fig. 8.3 Locations of a surface point $\bar{x}_s$ and a field point $\bar{x}_0$ for implementation of the farfield approximation of the Rayleigh integral

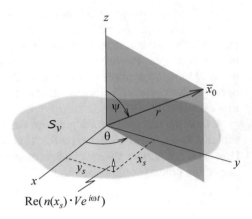

The condition that $\bar{x}_0$ be a farfield point requires that $r \gg x_s$ and y_s for any point in the surface. In that case $|\bar{x}_s - \bar{x}_0|$ in the denominator is well approximated as $1/r$. The normal velocity is $V_s(x_s, y_s)$, and an area element is $dx_s dy_s$. The result is that Eq. (8.1.7) becomes

$$P_{\text{ff}}(\bar{x}_0) = \frac{i\omega\rho_0}{2\pi r} e^{-ikr} \iint\limits_{S_v} V_s(x_s, y_s)\, e^{ik(x_s \sin\psi \cos\theta + y_s \sin\psi \sin\theta)} dx_s dy_s \qquad (8.2.12)$$

Suppose that S_v is a rectangle defined by $-a \leq x_s \leq a$, $-b \leq y_s \leq b$, and that the velocity distribution has the special form of a product of functions of x_s and y_s, that is, $V_s(\bar{x}_s) = X(x_s) Y(y_s)$. Because $dS = dx_s\, dy_s$ and the integration limits are constants, the integrals over x_s and y_s may be evaluated independently. In this case, the farfield pressure is given by

$$P_{\text{ff}}(\bar{x}_0) = \frac{i\omega\rho_0}{2\pi r} e^{-ikr} \int_{-a}^{a} X(x_s)\, e^{ikx_s \sin\psi \cos\theta} dx_s \int_{-b}^{b} Y(y_s)\, e^{iky_s \sin\psi \sin\theta} dy_s \quad (8.2.13)$$

Each integral fits the definition of a Fourier transform, which is defined in Eq. (B.1.5). The factors appearing in the exponent in each integral are the wavenumbers $\mu_x = k \sin\psi \cos\theta$ and $\mu_y = k \sin\psi \sin\theta$. The *angular spectrum* $\tilde{F}(\mu_x, \mu_y)$ is defined to be the product of the individual transforms, according to

$$\tilde{F}(\mu_x, \mu_y) = \int_{-a}^{a} X(x_s)\, e^{i\mu_x x_s} dx_s \int_{-b}^{b} Y(y_s)\, e^{i\mu_y y_s} dy_s \qquad (8.2.14)$$

In terms of this function, the farfield pressure is

$$\boxed{P_{\text{ff}}(\bar{x}_0) = i\omega\rho_0 \frac{e^{-ikr}}{2\pi r} \tilde{F}(k \sin\psi \cos\theta, k \sin\psi \sin\theta)} \qquad (8.2.15)$$

If $\mathcal{S}_v$ is not rectangular, or if V_s is not a product of x_s and y_s functions, the area integral does not factorize. Nevertheless, integration over the full range of x_s and y_s leaves only the wavenumbers as variables. The angular spectrum in this case is a surface integral that constitutes a *two-dimensional Fourier transform*. Thus, Eq. (8.2.15) is descriptive of the farfield pressure in any situation, with the angular spectrum defined to be

$$\tilde{F}\left(\mu_x, \mu_y\right) = \int_{-\infty}^{\infty} \int_{-\infty}^{\infty} V_s\left(x_s, y_s\right) e^{i\left(\mu_x x_s + \mu_y y_s\right)} dx_s dy_s \qquad (8.2.16)$$

In the case where the angular spectrum factorizes as in Eq. (8.2.14), tabulations of functions and their Fourier transforms are available, see Tables B.1 and B.2 in Appendix B. FFT technology may be employed to handle truly complicated velocity functions. However, the normal velocity typically is characterized by an elementary function, which makes it feasible to evaluate the integral analytically. Situations in which the vibrating portion of $\mathcal{S}_v$ is circular may be analyzed by using polar coordinates. Such a formulation is the topic of the next section.

Identification that the farfield pressure is directly related to the two-dimensional Fourier transform of the surface velocity leads to an interesting interpretation. Suppose we wish to determine the pressure at a specific field point in the farfield. The radial unit vector to that point is $\bar{e}_r$, which is defined in Eq. (8.2.10). The wavenumber of the farfield signal propagating in this direction is $k\bar{e}_r$. The trace of this vector onto the xy plane is its projection, $\bar{k}_s = k \sin \psi \cos \theta \bar{e}_x + k \sin \psi \sin \theta \bar{e}_y$ in Fig. 8.4. These components are the wavenumbers μ_x and μ_y for the angular spectrum in Eq. (8.2.15).

This construction makes it evident that $\left|\bar{k}_s\right| \leq k$, which leads to an interpretation of the farfield radiation. We construct a surface $\tilde{F}\left(\mu_x, \mu_y\right)$ in a Cartesian wavenumber

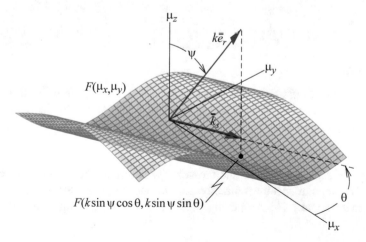

Fig. 8.4 Trace wavenumber vector $\bar{k}_s$ for the farfield spherical wave in a designated direction $\bar{e}_r$

space such that the height of the surface is the value of $\tilde{F}$, as is done in Fig. 8.4. In the $\mu_x\mu_y$ plane, we construct a circle of radius k. Any point within this circle corresponds to a signal that propagates to the farfield at polar angle ψ and azimuthal angle θ. These angles are related to the wavenumbers μ_x and μ_y of the selected point by $\cos\theta = \mu_x/(\mu_x^2 + \mu_y^2)^{1/2}$, $\sin\theta = \mu_y/(\mu_x^2 + \mu_y^2)^{1/2}$, and $\cos\psi = (\mu_x^2 + \mu_y^2)^{1/2}/k$. Selection of a point in the $\mu_x\mu_y$ plane for which $|\bar{k}_s| > k$ corresponds to subsonic surface waves that do not radiate to the farfield.

This interpretation of the relationship between the surface wave spectrum and the signal that propagates in a specified direction can be used as a design tool. Suppose we wish that the radiated pressure in a certain direction designated by spherical angle ψ_0 and θ_0 be the maximum. For a specified frequency, we can form the wavenumber vector in this direction. The components of this vector in the x and y directions are the corresponding wavenumbers, $\mu_x = k\sin\psi_0\cos\theta_0$, $\mu_y = k\sin\psi_0\cos\theta_0$. The maximum condition is attained if the two-dimensional Fourier transform of $\bar{n}(\bar{x}_s) \cdot V(\bar{x}_s)$ has a maximum magnitude at this pair of wavenumbers. Finding a surface vibration pattern whose Fourier transform has this attribute would not be a trivial task, but tabulations of Fourier transform pairs like those in Appendix B would be useful.

EXAMPLE 8.1 An electrostatic speaker essentially is a pair of oppositely charged plates, one of which is fixed. The free plate vibrates in response to oscillations of the voltage difference. This vibrating plate may be taken to be rigid, so its normal velocity is $v_0\cos(\omega t)$ everywhere on its surface $\mathcal{S}_v$. The plate is surrounded on all sides by a large rigid baffle. Two alternative designs are contemplated. Both have equal surface area, but one has a square face whose sides are 200 mm, and the other is 100 mm by 400 mm. The frequency is 2 kHz. Determine the farfield pressure as a function of the polar angles ψ and θ for each design, and graph $\left(r/\sqrt{S_v}\right)|P_{\text{ff}}|/(\rho_0 c v_0)$ as a function of the spherical angles.

Significance

The Fourier transform of the vibration pattern will be found with little effort, which will allow us to focus on the correlation between the properties of the vibration and those of the pressure field.

Solution

The shape of $\mathcal{S}_v$ is rectangular and the velocity function is a constant, so we shall evaluate the angular spectrum directly. Let $2a$ and $2b$ be the lengths of the edges parallel to the x and y axes, respectively. Then the complex vibration amplitude is

$$V_s\left(x_s, y_s\right) = \begin{cases} v_0 & \text{if } |x_s| \le a \text{ and } |y_s| \le b \\ 0 & \text{otherwise} \end{cases}$$

The two-dimensional Fourier transform of this function is

$$\tilde{F}\left(\mu_x, \mu_y\right) = \int_{-a}^{a}\int_{-b}^{b} v_0 e^{i\left(\mu_x x + \mu_y y\right)}\, dy_s dx_s$$

$$= 4v_0\left(\frac{\sin\left(\mu_x a\right)}{\mu_x}\right)\left(\frac{\sin\left(\mu_y b\right)}{\mu_y}\right)$$

A compact way to write this expression is to use the sinc function, which is defined as $\mathrm{sinc}(\xi) \equiv \sin\left(\pi \xi\right) / \left(\pi \xi\right)$. Figure 1 is a plot of the sinc function.

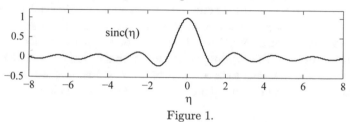

Figure 1.

At $\mu_x = 0$ or $\mu_y = 0$, the value of $\tilde{F}\left(\mu_x, \mu_y\right)$ is the finite value obtained as a limit, which gives $\mathrm{sinc}(0) = 1$. The transform of the surface velocity is thereby found to be

$$\tilde{F}\left(\mu_x, \mu_y\right) = 4v_0 ab\, \mathrm{sinc}\left(\frac{\mu_x a}{\pi}\right)\mathrm{sinc}\left(\frac{\mu_y b}{\pi}\right)$$

To evaluate the farfield radiation pattern, we define a hemispherical mesh formed from meridional semicircles through the z-axis along which θ is a constant value between $-\pi$ and π, and circles centered on the z-axis along which ψ is a constant between zero and $\pi/2$. The radial direction $\bar{e}_r$ at each mesh point is described by Eq. (8.2.10). The trace of $k\bar{e}_r$ on the xy plane is $k\sin\psi\left(\cos\theta\bar{e}_x + \sin\theta\bar{e}_y\right)$, so the surface wavenumbers are $\mu_x = k\sin\psi\cos\theta$ and $\mu_y = k\sin\psi\sin\theta$. With an eye toward computations, we collect variables as nondimensional groups by multiplying and dividing $\tilde{F}\left(\mu_x, \mu_y\right)$ by ab. Correspondingly, Eq. (8.2.15) gives

$$P_{\mathrm{ff}}\left(\bar{x}_0\right) = i\rho_0 c v_0 \frac{2kab}{\pi}\frac{e^{-ikr}}{r}\mathrm{sinc}\left(\frac{ka\sin\psi\cos\theta}{\pi}\right)\mathrm{sinc}\left(\frac{kb\sin\psi\sin\theta}{\pi}\right)$$

The value of ab for the alternative designs is the same, and the maximum value of the sinc function is one, so the alternative design gives similar pressure amplitudes. When $b > a$, the oscillation of $\tilde{F}\left(\mu_x, \mu_y\right)$ in the μ_y direction will be more rapid than that in the μ_x direction. Maxima of $\tilde{F}\left(\mu_x, \mu_y\right)$ correspond to the directions in which the farfield pressure is a maximum, and zeros of $\tilde{F}\left(\mu_x, \mu_y\right)$ correspond to zero farfield pressure in the corresponding $\bar{e}_r$ direction. It follows that the slender rectangle configuration will exhibit more side lobes along $\theta = 90°$ and $270°$ than it does along $\theta = 0$ and $180°$. To see the details, we need to evaluate the pattern. The area on which the vibration occurs is $4ab$, so the specified quantity to plot is given by

$$\left(\frac{r}{2\,(ab)^{1/2}}\right) \frac{|P_{\text{ff}}\,(\bar{x}_0)|}{\rho_0 c v_0} = \frac{k\,(ab)^{1/2}}{\pi} \left| \text{sinc}\left(\frac{ka \sin\psi \cos\theta}{\pi}\right) \text{sinc}\left(\frac{kb \sin\psi \sin\theta}{\pi}\right) \right|$$

Figure 2 describes the square configuration at 2 kHz. The data is displayed as a spherical plot in which the radial distance is the quantity defined above. This is accompanied by polar plots at $\theta = 0$, which is the xz plane, and $\theta = 90°$, which is the yz plane. (The range of ψ was set as $-\pi/2 < \psi < \pi/2$ to depict the data on both sides of the z-axis.) Because $a = b$, the polar plots are identical. The spherical plot is close to being axisymmetric relative to the z-axis, and the pressure falls off rapidly as ψ increases from zero. This type of pattern, in which the signal is essentially aimed in a certain direction, is said to be a *sound beam*. There are several side lobes, but the pressure amplitude in them is quite low relative to the main lobe.

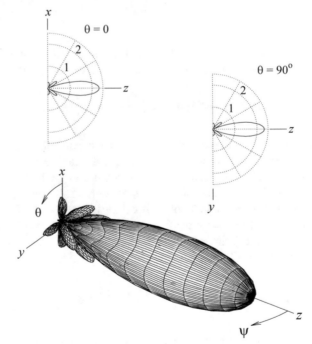

Figure 2. $\left(r/\left(ab^{1/2}\right)\right)|P|/\left(\rho_0 c v_0\right)$ at 2 kHz,
$a = b = 200$ mm.

Figure 3 describes the case where the plate is a narrow rectangle at 2 kHz, as in the previous case. This field also is a sound beam. The polar plots indicate that the signal in the yz plane is much more tightly confined to the z-axis than it is in the xz plane. This might seem to be counterintuitive because the width of the plate in the x direction is large. To explain this feature, consider the limit in which the plate's dimensions a and b become infinite. In that case, the radiated field would consist of a simple plane wave that radiates in the z direction. Such a wave would be zero in all directions other than $\psi = 0$.

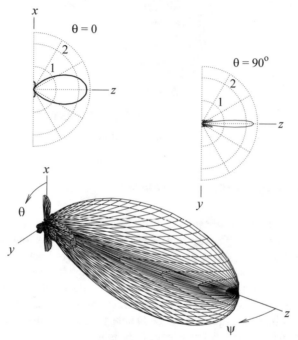

Figure 3. $\left(r/\left(ab^{1/2}\right)\right) |P|/\left(\rho_0 c v_0\right)$ at 2 kHz,
$a = 100$ mm, $b = 400$ mm.

We have encountered a number of situations in which decreasing the size of the vibrating object (or the frequency) produces a field that varies more gradually in space. Figure 4 exhibits this feature. The values of a and b are like those of Fig. 2, but the frequency now is 200 Hz. Even though the rectangular plate is long and narrow (we say that its aspect ratio b/a is high), the field is almost axisymmetric, and there is little variation from $\psi = 0$ to $\psi = 90°$. The nondimensional parameters setting the behavior are ka and kb, which are 0.37 and 1.48, respectively, in this last case. Thus, both sides are much smaller than the acoustic wavelength, $2\pi/k$. As a general rule, a fairly uniform farfield will be obtained if the size of the radiating body is substantially less than the acoustic wavelength.

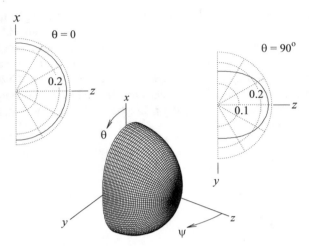

Figure 4. $\left(r/\left(ab^{1/2}\right)\right) |P| / \left(\rho_0 c v_0\right)$ at 200 Hz,
$a = 100$ mm, $b = 400$ mm.

8.2.2   Farfield of a Piston Transducer

In the previous section, the surface vibration was considered to be best described in terms of a Cartesian coordinate system. Here, we consider the situation in which the outline of the vibrating region is circular. A set of polar coordinates (R_s, θ_s) centered on this region are most suitable to formulate the Rayleigh integral. The development will begin by allowing the velocity distribution to depend arbitrarily on R_s and θ_s. However, in most cases of interest the vibration pattern is axisymmetric, which means that the vibration amplitude depends solely on R_s. A special case is a piston transducer, for which the vibration amplitude is uniform across the circular area. This model often is used to represent a common loudspeaker. Some very interesting phenomena are encountered in this case, so it will receive the most attention.

Figure 8.5 depicts a vibrating circular region S_v surrounded by a rigid baffle. The z-axis for a cylindrical coordinate system is normal to the surface. The field point $\bar{x}_0$ is located by its spherical coordinates (r, ψ, θ), and a generic point on S_v has polar coordinates (R_s, θ_s), so its position is $\bar{x}_s = R_s \cos\theta_s \bar{e}_x + R_s \sin\theta_s \bar{e}_y$.

The farfield approximation of the distance between a surface point and a field point is $|\bar{x} - \bar{x}_s| = r - \bar{x}_s \cdot \bar{e}_r$. In view of Eqs. (8.2.9) and (8.2.10), this reduces to

$$|\bar{x} - \bar{x}_s| = r - R_s \sin\psi \cos\left(\theta_s - \theta\right) \tag{8.2.17}$$

In combination with a polar coordinate description of dS, this expression converts the farfield approximation of the Rayleigh integral, Eq. (8.2.12), to

$$\boxed{P_{\text{ff}}\left(\bar{x}_0\right) = \frac{i\omega\rho_0}{2\pi r} e^{-ikr} \int_0^a \int_{-\pi}^{\pi} V_s\left(R_s, \theta_s\right) e^{ikR_s \sin\psi \cos(\theta_s - \theta)} R_s d\theta_s dR_s} \tag{8.2.18}$$

Fig. 8.5 Coordinate
descriptions of a surface
point and a field point
suitable for the analysis of a
circular transducer in an
infinite baffle

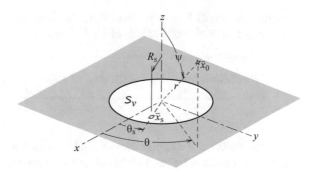

where a is the radius of the area over which the vibration occurs.

Usually the velocity distribution is axisymmetric, so V_z is independent of θ_s. This greatly simplifies the analysis because it allows the integration over θ_s to be factored out. An integration over θ_s with R_s held fixed represents the contribution to the farfield pressure of a circular ring whose radius is R_s and whose radial width is dR_s. The normal velocity along this ring is $V_s(R_s)$, so that

$$dP_{\text{ring}} = \frac{i\omega\rho_0}{2\pi r}e^{-ikr}V_s(R_s)R_sdR_s\int_{-\pi}^{\pi}e^{ikR_s\sin\psi\cos(\theta_s-\theta)}d\theta_s \qquad (8.2.19)$$

The value of θ is fixed by selection of the field point. This allows us to change the integration variable to $\theta' = \theta_s - \theta$. Furthermore, the integrand is periodic in any 2π interval, and it is an even function of θ'. This allows us to integrate over $0 \le \theta' \le \pi$, then double the result. Hence, the pressure radiated by a ring whose radius is R_s and width is dR_s is given

$$dP_{\text{ring}} = \frac{i\omega\rho_0}{\pi r}e^{-ikr}V_s(R_s)R_sdR_s\int_{0}^{\pi}e^{ikR_s\sin\psi\cos\theta_s}d\theta_s \qquad (8.2.20)$$

The integral may be found in most references on Bessel functions,[2] which leads to

$$\boxed{dP_{\text{ring}} = i\rho_0cV_s(R_s)kR_sdR_s\frac{e^{-ikr}}{r}J_0(kR_s\sin\psi)} \qquad (8.2.21)$$

There is no dependence on the azimuthal angle θ for the field point because any point-symmetric vibration pattern will generate an axisymmetric pressure field.

Equation (8.2.21) may be used for an actual ring by replacing the differential width dR_s with the actual width ΔR_s and taking R_s to be the mean radius. However, doing so requires that $\Delta R_s/R_s$ be sufficiently small to that any fluctuation of the V_z value from the inner to outer radius is negligible. The more important use is to

---

[2] M.I. Abramowitz and I.A. Stegun, *Handbook of Mathematical Functions*, Dover (1965) p. 360, Eq. (9.1.21).

construct the signal for a full circular region. The Rayleigh integral in that case is essentially a sum of contributions of all differential elements, so that

$$P_{\mathrm{ff}}(\bar{x}_0) = i\rho_0 ck \frac{e^{-ikr}}{r} \tilde{f}(k\sin\psi), \quad \tilde{f}(\mu) = \int_0^a V_s(R_s)J_0(\mu R_s) R_s dR_s$$

$$(8.2.22)$$

This expression has the appearance of Eq. (8.2.15). The function $\tilde{f}(\mu)$ is the angular spectrum for a decomposition of the surface velocity into waves that propagate along the surface radially outward from the center. From a mathematical perspective, it is the Hankel transform of the axisymmetric $V_s(R_s)$. The Hankel transform may be derived by transforming the two-dimensional Fourier transform to polar coordinates.[3] Depending on the nature of $V_s(R_s)$, it might be possible to evaluate $\tilde{f}(\mu)$ analytically. If not, it can be evaluated numerically.

A particularly important case is that of a piston or translating rigid disk, for which $V_s = V_0$ for $R_s \le a$. Introduction of $u = kR_s \sin\psi$ into the integrand of the angular spectrum leads to a standard integral,[4] with the eventual result that

$$P_{\mathrm{ff}}(\bar{x}_0) = i\rho_0 c V_0 \frac{J_1(ka\sin\psi)}{\sin\psi} \frac{a}{r} e^{-ikr} \tag{8.2.23}$$

The *Rayleigh distance* R_0 is a reference length often used to describe acoustic radiation. We will see in the next section that it indicates the radial distance at which the signal is well represented by its farfield approximation. The definition of R_0 is that it is the ratio of the area over which the vibration occurs to the wavelength. For a circular piston, it is

$$R_0 = \frac{S_v}{\lambda} = \frac{\pi a^2}{2\pi/k} = \frac{1}{2}ka^2 \tag{8.2.24}$$

A simple expression for the maximum pressure at a specified radial distance, which occurs at $\psi = 0$, results from using the Rayleigh distance in Eq. (8.2.23). As $\psi \to 0$, the first order Bessel function approaches $(1/2)\,ka\sin\psi$, so that

$$\max(|P_{\mathrm{ff}}(\bar{x}_0)|) = \rho_0 c V_0 \frac{R_0}{r} \quad @ \ \psi = 0 \tag{8.2.25}$$

The corresponding farfield pressure is

$$P_{\mathrm{ff}}(\bar{x}_0) = i\rho_0 c V_0 \left(\frac{2J_1(ka\sin\psi)}{ka\sin\psi}\right)\left(\frac{R_0}{r}\right) e^{-ikr} \tag{8.2.26}$$

---

[3] I.A. Sneddon, *Fourier Transforms*, Dover (1995 reprint) Chap. 2.
[4] Abramowitz and Stegun, *ibid.*, #11.3.20.

The directivity is the ratio of the pressure magnitude as a function of ψ to the value on-axis, with r fixed in the farfield. Thus,

$$\mathcal{D}(\psi) = \frac{|P_{\text{ff}}|}{\max(|P_{\text{ff}}|)}\Bigg|_{r>R_0} = \frac{2\,|J_1\,(ka\sin\psi)|}{ka\sin\psi} \tag{8.2.27}$$

An interesting comparison is with the pressure radiated by a ring having the same radius a and vibrational velocity V_0. We find from Eq. (8.2.21) that the ratio $(P_{\text{ff}})_{\text{ring}} / (P_{\text{ff}})_{\text{piston}} = 2\Delta R_s/a$ on the z-axis, where $\psi = 0$. The expression for a ring is based on the width ΔR_s being much less than the radius, from which it follows that the ring radiates much more weakly as a consequence of the smaller face area. The directivity factor for a ring is $J_0\,(ka\sin\psi)$. The nulls of the directivity in each case correspond to parameters for which $ka\sin\psi$ is a zero of the respective Bessel functions.

The number of nulls that occurs is limited by the requirement that $\sin\psi < 1$, from which it follows that this number increases in a stepwise manner approximately in proportion to the value of ka. Thus, below $ka = 3.832$ for a piston and $ka = 2.405$ for a ring, the directivity exhibits no nulls. Above these frequencies, the interval from the axis to the first null is the main lobe, and the other intervals are the side lobes. The maxima in the side lobes decrease rapidly with increasing angle from the z-axis. Figure 8.6 displays some typical patterns. The field in each case is axisymmetric, so a polar plot over $-\pi/2 \le \psi \le \pi/2$ is sufficient.

The rapid decrease in the magnitude of the side lobes obscures them in a polar plot. A rectangular plot, such as that in Fig. 8.7, conveys a clearer quantitative picture.

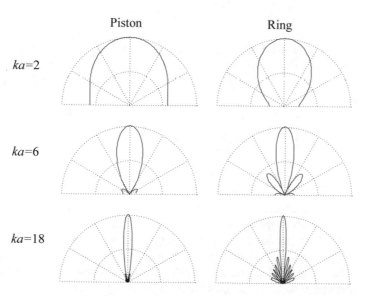

Fig. 8.6 Directivity of a piston and a ring having radius a embedded in an infinite baffle

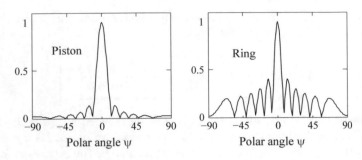

Fig. 8.7 Cartesian plot of the directivity of a piston and a ring embedded in an infinite baffle, $ka = 18$

We can discern some general trends from Fig. 8.7 and Table 8.1. For a given value of ka, the main lobe is narrower with a ring than it is with a piston. The ring has more side lobes, and the side lobes are stronger. These trends, in combination with the stronger signal radiated by the piston, indicate that if one's objective is to aim a signal in a certain direction at a specified frequency, they should use the piston configuration. The piston's radius should be as large as feasible, because a large ka will yield a small value of ψ for the first null.

Table 8.1 Polar angles at which the directivity of a piston and a ring have a null

	$ka \sin \psi$			
Null #	1	2	3	4
Piston	3.832	7.016	10.173	13.324
Ring	2.405	5.520	8.654	11.792

A quantity of interest is the *beamwidth* Δ_{bw}, which is the cone angle at which the intensity is half the maximum on-axis value. In the farfield, the radial particle velocity is $p/(\rho_0 c)$ and the other velocity components are negligible, so the time-averaged intensity is

$$(I_r)_{\mathrm{av}} = \frac{|P^2|}{2\rho_0 c} = \frac{\rho_0 c \, |V_0|^2}{2} \left(\frac{R_0}{r}\right) \mathcal{D}(\psi)^2 \qquad (8.2.28)$$

The half-power point corresponds to $\mathcal{D}(\psi) = 1/\sqrt{2}$. For the piston directivity in Eq. (8.2.27), this condition occurs at $ka \sin \psi = 1.62$, whereas Table 8.1 indicates that the first null occurs at $ka \sin \psi = 3.82$. Hence, if large ka is large, the beamwidth approximately subtends the center half of the main lobe.

A question that sometimes is posed is why is the piston model used to describe the radiation properties of transducers that do not look like a piston? One such device is a speaker cone embedded in a planar baffle, which is a common driver for sound reproduction systems. This arrangement is shown in Fig. 8.8.

The field is symmetric about the z-axis. The depth of the transducer below the xy plane of the baffle is some function $Z(R)$. A well-designed speaker cone translates in the z direction, so that all points on the cone have the same velocity

Fig. 8.8 Cross-sectional view of a conical speaker embedded in a rigid baffle

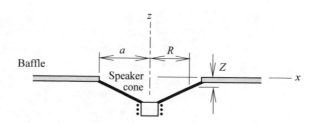

$\text{Re}\left(V_z e^{i\omega t}\right)$. Because the angle of the cone is close to $90°$, the normal to the cone's surface is nearly parallel to the z-axis, so we may say that the (time domain) surface velocity is

$$n\left(\bar{x}_s\right) \cdot \bar{v}\left(\bar{x}_s\right) = \begin{cases} \text{Re}\left(V_z e^{i\omega t}\right) @ z = -Z\left(R\right) \text{ if } |R| < a \\ 0 \text{ if } |R| > a \end{cases} \tag{8.2.29}$$

An approximation of the field close to the surface is that each point emits a plane wave that propagates in the axial direction. (This is a crude representation of a complicated process, but it is adequate for the present purpose.) In this approximation, a signal that is emitted from a point on the transducer's face at radial distance R requires an interval of $Z\left(R\right)/c$ to arrive at the xy plane. This is a lag, so the phase delay for a harmonic signal at frequency ω is $kZ\left(R\right)$. The complex surface velocity on the xy plane in this approximation is

$$V_s = \begin{cases} V_z e^{-ikZ(R)} \text{ if } |R| < a \\ 0 \text{ if } |R| > a \end{cases} \tag{8.2.30}$$

In most cases, the intent is to radiate sound over a broad range of polar angles, in which case ka will be small. The preceding description of V_s states that there is a position-dependent phase lag, which may be written as $ka\left(Z\left(R\right)/a\right)$. If ka is not much larger than one and the cone angle is close to $90°$, so that $Z\left(R\right)/a$ is small everywhere, then it is reasonable to ignore this phase lag. A different way of viewing this is to write the phase lag as $2\pi Z\left(R\right)/\lambda$, where λ is the wavelength of a plane wave at frequency ω. In this view, a nonplanar piston can be considered to behave like a piston if its maximum depth is much less than the plane wavelength. Conversely, in many cases, especially ultrasonic applications like acoustical microscopy, a transducer is driven at a very large ka in order to obtain a narrow main lobe. In such cases, it would be wrong to use a piston model to analyze the radiated field. Equation (8.2.30) will be used in Example 8.4 to develop an approximate nonpiston model for the pressure radiated by a nonplanar transducer at high ka.

EXAMPLE 8.2 Two designs for a circular transducer embedded in a rigid baffle are under consideration. One is a conventional piston, whose surface velocity is $v_0 \cos\left(\omega t\right)$ everywhere on its face. The face of the other is a membrane that is stretched across the circumference. Its surface velocity is well approximated be a parabolic distribution, that is, $V_s = v_1 \left(1 - R^2/a^2\right)$. The medium in which these devices will operate is water. It is desired that the pressure amplitude at a distance of 100 m for either transducer be 0.8 atm, and that the first null of the farfield directivity be at $\psi = 2°$ in either case. The frequency at which the transducers will operate is 30 kHz in order to test the hearing of sea mammals. For each design, determine the required radius and the maximum velocity on its face. Then, determine the polar angle of its second null and the maximum pressure in the side lobe between the first and second nulls.

Significance

Design problems like this improve our understanding of the relationship between system parameters and performance.

Solution

The first task is to determine the directivity associated with the parabolic velocity profile. The farfield description in Eq. (8.2.22) for the given velocity distribution is simplified slightly by changing the integration variable to $\xi = R_s/a$, which gives

$$P_{\text{ff}}(\bar{x}_0) = i\rho_0 c v_1 ka^2 \frac{e^{-ikr}}{r} \int_0^1 \left(1 - \xi^2\right) J_0\left(ka\xi \sin \psi\right) \xi d\xi \tag{1}$$

We could evaluate the integral numerically for each value of $ka\xi \sin \psi$, but an analytical evaluation is possible. A tabulated integral[5] states that

$$\int_0^{\pi/2} J_\mu\left(z \sin t\right) \left(\sin t\right)^{\mu+1} \left(\cos t\right)^{2\nu+1} dt \tag{2}$$
$$= \frac{2^\nu \Gamma\left(\nu+1\right)}{z^{\nu+1}} J_{\mu+\nu+1}(z), \quad \text{Re}\left(\mu\right) > -1 \text{ and } \text{Re}\left(\nu\right) > -1$$

where Γ denotes the gamma function. Our interest is with integer values, in which case $\Gamma\left(\nu+1\right) \equiv \nu!$ If this formula is to apply to Eq. (1), then the order of the Bessel functions and the argument should match. This suggests that we should set $\mu = 0$, $\sin t = \xi$, and $z = ka \sin \psi$. Furthermore, setting $\sin t = \xi$ leads to matching integration limits. This change of variables gives $(\cos t)\, dt = d\xi$ and $(\cos t)^{2\nu} = \left(1 - \xi^2\right)^\nu$. Thus, setting $\xi = \sin t$, $ka \sin \psi = z$, $\mu = 0$, and $\nu = 1$ transforms Eq. (1) to the integral in Eq. (2). The result is that the farfield for the membrane transducer is

$$P_{\text{ff}}(\bar{x}_0) = i\rho_0 c v_1 ka^2 \left[\frac{2J_2\left(ka \sin \psi\right)}{\left(ka \sin \psi\right)^2}\right] \frac{e^{-ikr}}{r} \tag{3}$$

---

[5] Abramowitz and Stegun #11.4.10.

As we did for the piston in Eq. (8.2.27), we define the directivity relative to the maximum farfield pressure. At $\psi = 0$, the denominator is singular, but the asymptotic expansion of the Bessel function gives $J_2\,(ka \sin \psi) \approx (ka \sin \psi)^2\,/8$, so we find that

$$\max\,(|P_{\mathrm{ff}}\,(\bar{x}_0)|) = \rho_0 c v_1 \frac{a}{r}\left(\frac{ka}{4}\right) \equiv \frac{1}{2}\rho_0 c v_1 \frac{R_0}{r} \; @ \; \psi = 0$$

$$D\,(\psi) = \frac{|P_{\mathrm{ff}}\,(\bar{x}_0)|}{\max\,(|P_{\mathrm{ff}}\,(\bar{x}_0)|)} = \frac{8\,|J_2\,(ka \sin \psi)|}{(ka \sin \psi)^2}$$

(4)

Before we address the design criteria, it is worth noting that if v_1 and ka for the membrane equal v_0 and ka for the piston, then the maximum farfield pressure in Eq. (4) is half the value for a piston in Eq. (8.2.26). Separately, we know that at very low frequencies the radiated pressure is omnidirectional, with an amplitude that is proportional to the complex volume velocity amplitude. This value for each design is

$$\left(\hat{Q}_s\right)_{\mathrm{piston}} = \iint_S V_s dS = v_0\,(\pi a^2)$$

$$\left(\hat{Q}_s\right)_{\mathrm{parabolic}} = \iint_S v_1\left(1 - \frac{R_s^2}{a^2}\right) 2\pi R_s dR_s = v_1\left(\frac{1}{2}\pi a^2\right)$$

Thus, the ratio of maximum farfield pressures for the two designs is the same as the ratio of volume velocities.

One design criterion is the value of ψ for the first null, which corresponds to the first positive zero of the respective Bessel function. For $J_1\,(ka \sin \psi)$, we have Table 8.1. We may find the zeros of $J_2\,(ka \sin \psi)$ by inspection of a graph of $J_2\,(x)$, and a high precision value may be obtained by using numerical methods to find the roots of $J_2\,(x) = 0$. The first few values are

$$J_2\,(x) = 0 \; @ \; x = 5.136,\; 8.417,\; 11.620,\; 14.796$$

The frequency is $\omega = 60,\,000\pi$ rad/s, and we take $c = 1480$ m/s. For a first null at $\psi = 2°$, matching $ka \sin \psi$ to the first zero of the respective functions gives

$$\text{Piston: } ka = \frac{3.832}{\sin\,(2°)} = 109.80 \implies a = 0.862 \text{ m}$$

$$\text{Membrane: } ka = \frac{5.136}{\sin\,(2°)} = 147.15 \implies a = 1.155 \text{ m}$$

Both are enormous designs, but such transducers can be found on some ocean-going research vessels.

The other design criterion sets the maximum pressure at 100 m to 0.8 atm $= 81.06$ kPa. The Rayleigh distance for the membrane transducer, which is the larger value, is $R_0 = 85$ m, so $r = 100$ for the specified pressure is in the farfield. Matching the maximum farfield pressure to the specified value in each case gives

$$\text{Piston: } \rho_0 c v_1 \frac{a}{r} \left(\frac{ka}{4} \right) = 1000\,(1480)\, v_0 \left(\frac{0.862}{100} \right) \left(\frac{24.70}{2} \right) = 81 \left(10^3 \right) \implies$$

$$v_0 = 0.1156 \text{ m/s}$$

$$\text{Membrane: } \rho_0 c v_1 \frac{a}{r} \left(\frac{ka}{4} \right) = 1000\,(1480)\, v_1 \left(\frac{1.155}{100} \right) \left(\frac{33.11}{4} \right) = 81 \left(10^3 \right) \implies$$

$$v_1 = 0.1287 \text{ m/s} \tag{8.2.31}$$

The required maximum surface velocities for the transducers are close, because the membrane transducer has a larger radius, which compensates for its smaller volume velocity per unit surface area.

The last factor to consider pertains to the first side lobe, which is bounded by the first and second nulls. The second null is located at $ka \sin \psi = 7.016$ for the piston, and $ka \sin \psi = 8.417$ for the membrane, which corresponds to $\psi = 3.66°$ and $\psi = 3.28°$, respectively. We could perform a mathematical search for the maximum in each side lobe, but it is easier to scan the data for a plot of each directivity. The angular dependence for both configurations are depicted in Fig. 1. The first side lobe maxima are $\mathcal{D} = 0.1323$ at $\psi = 2.69°$ for the piston and $\mathcal{D} = 0.0586$ at $\psi = 2.49°$ for the membrane transducer. The graphs show that although the main lobes are closely matched, the side lobes of the membrane transducer are narrower and much lower in amplitude. These differences are consequences of the fact that v_s is a continuous function of R_s for the membrane transducer, whereas it is discontinuous at the baffle for the piston. Discontinuities generally lead to the occurrence of diffraction effects.

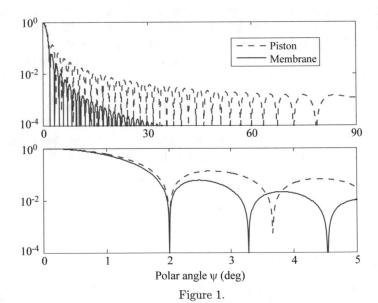

Figure 1.

8.3 Axial Dependence for a Circular Transducer

The farfield approximation yields valuable information, but it does not provide any insight regarding the range at which it is applicable. The analysis in this section evaluates the Rayleigh integral for any distance from the surface. However, the analysis that follows is not general. It requires that the vibrating region of the surface be circular, that the pattern of the surface vibration be axisymmetric, and that the field point be situated on the axis of symmetry. One of its benefits is that the result will shed light on the nearfield properties, and how those properties transition to the farfield.

The description of the location of a surface point and a field point in Fig. 8.5 are suitable here. The field point is on the z-axis, so $\psi = 0$. Then, the distance from a surface point to a field point is $|\bar{x}_0 - \bar{x}_s| = \left(z^2 + R_s^2\right)^{1/2}$. As a consequence of taking the surface vibration to be axisymmetric, the surface velocity is $V_s\left(R_s\right)$. Furthermore, because $|\bar{x} - \bar{x}_s|$ does not depend on the polar angle, we may use a ring of radius R_s and width dR_s as the differential element of area, so $dS = 2\pi R_s s R_s$. According to Eq. (8.1.7), the contribution of this ring to the pressure at the field point is

$$dP_{\text{axial}}\left(z\right) = i\omega\rho_0 V_s\left(R_s\right) \frac{e^{-ik\left(z^2+R_s^2\right)^{1/2}}}{\left(z^2 + R_s^2\right)^{1/2}} R_s dR_s \tag{8.3.1}$$

This differential contribution is integrated over the surface area, where the vibration occurs. Thus, we find that

$$P_{\text{axial}}\left(z\right) = i\omega\rho_0 \int_0^a V_s\left(R_s\right) \frac{e^{-ik\left(z^2+R_s^2\right)^{1/2}}}{\left(z^2 + R_s^2\right)^{1/2}} R_s dR_s \tag{8.3.2}$$

In most situations, evaluation of this integral will require numerical methods, but an analytical integration is possible in the special case of a translating piston, for which $V_s\left(R_s\right) - V_0$. Toward that end, we change the integration variable from R_s to the radial distance r from a point on a ring to the field point. Because $r^2 = R_s^2 + z^2$ and z is constant in the integral, differentiation of this relation gives $rdr = R_s dR_s$. Thus, the change of variables leads to

$$P_{\text{axial}}\left(z\right) = i\omega\rho_0 V_0 \int_z^{\left(z^2+a^2\right)^{1/2}} e^{-ikr} dr \tag{8.3.3}$$

This change of variables gives the integrand the appearance of a plane wave, in that there is no $1/r$ term to account for spreading loss. This perception seems to be borne out by the integrated form, which is

$$\boxed{P_{\text{axial}}\left(z\right) = \rho_0 c V_0 \left(e^{-ikz} - e^{-ik\left(z^2+a^2\right)^{1/2}}\right)} \tag{8.3.4}$$

The first term is the *center wave* because its phase is that of a wave that propagates distance z from the center of the piston to the field point. The distance from the edge of the piston to the field point is $(z^2 + a^2)^{1/2}$, so the second term is the *edge wave*. In this view, Eq. (8.3.3) represents a superposition of waves from rings that begin at the center and end at the edge. The integrated form says that the contributions of all the ring waves between the center and edge annihilate each other, but there is nothing coming from beyond $R_s = a$, which leaves the edge wave as the cancelling effect.

A clearer picture is obtained by using the identities that

$$z = \frac{1}{2}\left[z + (z^2 + a^2)^{1/2}\right] + \frac{1}{2}\left[z - (z^2 + a^2)^{1/2}\right]$$
$$(z^2 + a^2)^{1/2} = \frac{1}{2}\left[z + (z^2 + a^2)^{1/2}\right] - \frac{1}{2}\left[z - (z^2 + a^2)^{1/2}\right]$$

(8.3.5)

Substitution of these expressions into Eq. (8.3.4) leads to

$$P_{\text{axial}}(z) = 2i\rho_0 c V_0 e^{-i(k/2)\left[z + (z^2 + a^2)^{1/2}\right]} \sin\left[\frac{k}{2}\left((z^2 + a^2)^{1/2} - z\right)\right]$$

(8.3.6)

The complex exponential factor gives the phase lag relative to the surface velocity. This lag does not increase linearly with increasing z. The sine term is more important because it tells us that the magnitude of the pressure on the z-axis is

$$\boxed{|P_{\text{axial}}(z)| = 2\rho_0 c \, |V_0| \left|\sin\left(\frac{k(a^2 + z^2)^{1/2} - kz}{2}\right)\right|}$$

(8.3.7)

The maximum pressure that is observed on the axis of symmetry is $2\rho_0 c V_0$, which is twice the plane wave value. Maxima occur at discrete locations called *antinodes* that are separated by *nodes* at which the pressure is zero. Although the function describing this oscillation is sinusoidal, the argument of the function is not proportional to z, so the nodes and antinodes are not spaced evenly along the z-axis. Nodes occur at locations where the argument is $n\pi$, with $n > 0$ because $z \geq 0$. This condition corresponds to

$$\text{Nodes: } kz = \frac{(ka)^2 - (2n\pi)^2}{4n\pi}, \quad n = 1, 2, \ldots, \text{floor}\left(\frac{ka}{2\pi}\right)$$

(8.3.8)

For antinodes, the argument of the sine function must be $(2n - 1)\pi/2$, also with $n > 0$. This leads to

$$\text{Antinodes: } kz = \frac{(ka)^2 - (2n - 1)^2 \pi^2}{2(2n - 1)\pi}, \quad n = 1, 2, \ldots, \text{floor}\left(\frac{ka + \pi}{2\pi}\right)$$

(8.3.9)

The corollary of these relations is that nodes do not exist if $ka < 2\pi$, and antinodes do not exist if $ka < \pi$. Below $ka = \pi$, the pressure decreases monotonically with increasing z. The pressure amplitude at $z = 0$, which marks the center of the piston, is found from Eq. (8.3.7) to be $2\rho_0 c V_0 \sin(ka/2)$. These observations lead us to recognize that decreasing the frequency below $ka = \pi$ will have the effect of decreasing the pressure everywhere along the axis of symmetry. Above $ka = 2\pi$, the farthest antinode ($n = 1$) is always more distant than the farthest node.

An important aspect is the behavior at large z. A binomial expansion of Eq. (8.3.6) based on a/z being small gives

$$P_{\text{axial}}(z) = 2i\rho_0 c V_0 e^{-ikz} \sin\left[\frac{kz}{2}\left(1 + \frac{a^2}{z^2}\right)^{1/2} - \frac{kz}{2}\right] \approx 2\rho_0 c V_0 e^{-ikz} \sin\left(\frac{R_0}{2z}\right)$$

(8.3.10)

where R_0 is the Rayleigh distance defined in Eq. (8.2.24). This expression is valid whenever $z \gg a$. If z is sufficiently greater than R_0, then the sine function may be approximated by its argument. As it must to be consistent with the general property of spherical spreading in the farfield, the pressure amplitude tends to a $1/z$ dependence,

$$P_{\text{axial}}(z)_{\text{ff}} = \rho_0 c V_0 \frac{R_0}{z} e^{-ikz}$$

(8.3.11)

When we set $\psi = 0$ in Eq. (8.2.27), so that $r = z$, the result is identical to this expression. This equivalence suggests that $r > R_0$ marks the beginning of the farfield. Although this suggestion is based on the axial behavior, it is confirmed by mappings of the field obtained from numerical evaluations of the Rayleigh integral, as we will see in the next section.

Figure 8.9, for $ka = 30$, is a typical axial pattern. There are four nodes and five antinodes. The spacing between these locations increases with increasing axial distance, with the increment to the last antinode being much greater than the others. The antinodes occur sharply, but peaks also occur over distances that are a small fraction of the piston radius. This is an important feature if one wishes to measure the nearfield at high frequencies. It means that the microphone/hydrophone must have a very small diameter in order to probe these small regions, and the position must be precisely controlled. Figure 8.9 also shows that the farfield approximation is quite good from the Rayleigh distance outward.

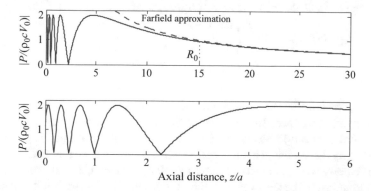

Fig. 8.9 Dependence on distance along the axis of symmetry of the pressure radiated by a piston in an infinite baffle, $ka = 30$

EXAMPLE 8.3 The chief engineer of Weird Acoustical Devices, Inc. has the idea to make an "anti-piston" transducer to be embedded in a baffle. The idea is to use piezoceramic elements that form an inner circle that covers $0 \leq R_s < a/2$, and a ring covering $a/2 \leq R_s < a$. Each set of elements will be driven by a separate circuit in order that the surface velocity on the outer ring will have the same magnitude but $180°$ out-of-phase relative to the inner ring's surface velocity. Derive an expression for the axial pressure radiated by this device. Compare it to the axial field of a conventional piston transducer at $ka = 10$ and 40. In particular, compare the pattern of nodes and antinodes and the distance at which the farfield approximation becomes accurate.

Significance

This example will greatly enhance our understanding of the use of the Rayleigh integral to evaluate an axial field, and the results will highlight the significance of center and edge waves.

Solution

The given surface velocity is

$$V_s = \begin{cases} V_0 \text{ if } 0 \leq R_s < a/2 \\ -V_0 \text{ if } a/2 \leq R_s < a \end{cases}$$

The on-axis version of the Rayleigh integral given by Eq. (8.3.1) is valid for any axisymmetric surface velocity. The discontinuous velocity distribution is handled by decomposing the integral into two parts,

$$P_{\text{axial}}(z) = i\omega\rho_0 V_0 \left[\int_0^{a/2} \frac{e^{-ik\left(z^2+R_s^2\right)^{1/2}}}{\left(z^2+R_s^2\right)^{1/2}} R_s dR_s - \int_{a/2}^a \frac{e^{-ik\left(z^2+R_s^2\right)^{1/2}}}{\left(z^2+R_s^2\right)^{1/2}} R_s dR_s \right]$$

The change of variables $r^2 = R_s^2 + z^2$ is appropriate here, so we find that

$$P_{\text{axial}}(z) = i\omega\rho_0 V_0 \left(\int_z^{\left(z^2+a^2/4\right)^{1/2}} e^{-ikr} dr - \int_{\left(z^2+a^2/4\right)^{1/2}}^{\left(z^2+a^2\right)^{1/2}} e^{-ikr} dr \right)$$

$$= \rho_0 c V_0 \left(e^{-ikz} - 2e^{-ik\left(z^2+a^2/4\right)^{1/2}} + e^{-ik\left(z^2+a^2\right)^{1/2}} \right)$$

This expression resembles Eq. (8.3.6), but here there are three edge waves. The first term is a signal that departs from the center, the second arrives from the middle radius, and the third comes from the perimeter. The strength of each signal is proportional to the velocity discontinuity at the location where it departs. That is, at the center the surface velocity is $+V_0$, then it decreases by $-2V_0$ at $R_s = a/2$, and then it increases by $+V_s$ to return to the zero value on the baffle. This expression is not amenable to the type of manipulation that led to an explicit expression for the magnitude of the pressure, so we will merely evaluate it as a complex function.

Each complex exponential in the preceding has a different argument. It follows that the maxima and zeros of each term occur independently. Thus, at some locations they might all have maxima, leading to a pressure amplitude, that is, close to $4\rho_0 c \,|V_0|$, which is twice the maximum for a piston. It might also happen that at some locations the first and last term are maxima, while the second is a minimum, or vice versa, in which case the pressure would be close to zero. Figure 1, for $ka = 10$, confirms this expectation. The peak pressure occurs close to the node for the circular piston, and no antinode is evident. However, the farfield pressure is approximately half that for a piston.

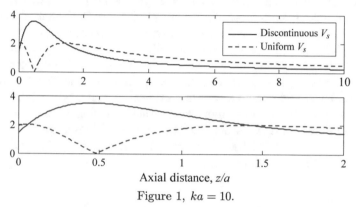

Figure 1, $ka = 10$.

High values of ka lead to the pressure being very small at some z, but not zero, as shown in Fig. 2 for $ka = 40$. The pressure amplitude is at or close to $4\rho_0 c V_0$ at several locations, but the number of such locations is fewer than for a piston. Only one minimum is close to zero, and the pressure at another minimum is close to the maximum for the piston. Overall, the "antipiston" generates a larger pressure in the nearfield in comparison to the piston, but the pressure is lower in the farfield.

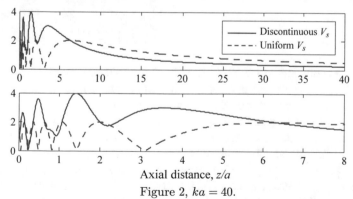

Figure 2, $ka = 40$.

Whether this is a useful conceptual design depends on the application. If the intent is to generate a higher pressure on-axis in the nearfield that varies less and is larger overall than a piston, then the antipiston might be useful. In contrast, if one requires a large pressure in the farfield, then the antipiston is not a good idea.

8.4 An Overall Picture of the Pressure Field

At this juncture, we can evaluate the pressure at any point in the farfield. In contrast, the only aspect of the nearfield that we can evaluate readily is the pressure for a field point on-axis in the case where the surface vibration occurs over a circular region. An accurate determination of the pressure at an arbitrary location in the nearfield requires numerical methods. We shall do so here for the case of a circular piston, but the modifications to treat other systems are straightforward.

To integrate over S_v as required for the Rayleigh integral, Eq. (8.1.7), we will approximate the region as a set of patches. We will use the radial distance from the center of a patch to the field point to describe the contribution of that patch to the Rayleigh integral. There are more sophisticated numerical algorithms that use multiple points on a patch, analogously to Simpson's and higher order rules for integration over a single variable. The scheme we shall implement is not too difficult to program, but it will require a very fine grid. The same procedure was used by Zemanek[6] to perform a detailed study that considered a wide range of parameters.

The procedure begins by dividing the circular area S_v into N_R rings and a center circle, as shown in Fig. 8.10. The circle's radius is set to half the width ΔR of a ring, which leads to $\Delta R = 2a/(2N_R + 1)$.

Fig. 8.10 Decomposition of a circular region into patches

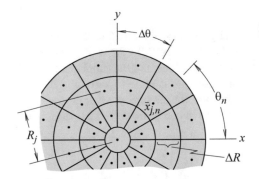

Each ring is divided into N_θ sectors, which leads to the sectorial increment being $\Delta\theta = 2\pi/N_\theta$. Each patch is associated with a pair of indices j and n, such that its central point has polar coordinates (R_j, θ_n), where $R_j = n\Delta R$ and $\theta_n = (n - 1/2)\Delta\theta$. The center circle is assigned to $j = n = 0$, while $1 \leq j \leq N_R$ and $1 \leq n \leq N_\theta$ for the other patches. The inner and outer radii of ring j are $r_{\text{in}} = (n - 1/2)\Delta R$ and $r_{\text{out}} = (n + 1/2)\Delta R$, and the angle subtended by a patch is $\Delta\theta$. Thus, the area of a patch is $\Delta S_{j,n} = (r_{\text{out}}^2 - r_{\text{in}}^2)\Delta\theta/2 = n\Delta\theta(\Delta R)^2$, with $\Delta S_{0,0} = \pi(\Delta R)^2/4$.

A map of the pressure in the nearfield is best obtained by depicting it in a cylindrical coordinate grid. The distance from a field point $\bar{x}_0$, whose cylindrical coordinates are (R, θ, z), to the central point $\bar{x}_{j,n}$ on a surface patch is

[6]J. Zemanek, "Beam behavior within the nearfield of a vibrating piston," J. Acoust. Soc. Am. **49** (1971) pp. 181–191.

$$\left|\bar{x}_0 - \bar{x}_{j,n}\right| = \left[\left(R\cos\theta - R_j\cos(\theta_n)\right)^2 + \left(R\sin\theta - R_j\sin(\theta_n)\right)^2 + z^2\right]^{1/2}$$

$$\equiv \left[R^2 + R_j^2 - 2RR_j\cos(\theta - \theta_n) + z^2\right]^{1/2} \tag{8.4.1}$$

The integral in Eq. (8.1.7) is approximated as a sum of the contribution of each surface patch, with that contribution based on using $\left|\bar{x} - \bar{x}_{j,n}\right|$ for the entire patch. The resulting discretized form of the Rayleigh integral is

$$
P(\bar{x}_0) = \frac{i\omega\rho_0}{2\pi} \left\{ V_s(0,0) \frac{e^{-ik|\bar{x}_0|}}{|\bar{x}_0|} \frac{\pi}{4}\left(\Delta R^2\right) + \right.
$$
$$
\left. + \sum_{n=1}^{N_\theta}\sum_{j=1}^{N_R} V_s(R_j,\theta_n) \frac{e^{-ik|\bar{x}_0 - \bar{x}_{j,n}|}}{|\bar{x}_0 - \bar{x}_{j,n}|} n\Delta\theta\,(\Delta R)^2 \right\} \tag{8.4.2}
$$

Equation (8.4.2) may be used to evaluate the field resulting from any surface velocity distribution. In the special case of a translating piston, we set $V_s(R_j,\theta_n) = V_0$. The field in that case is axisymmetric, so it is sufficient to compute P only on the xz plane, where $\theta = 0$. Another saving is that the xz plane divides the surface into two parts whose contributions are equal, so only noncentral surface patches covering $0 \le \theta \le \pi$ need to be evaluated, with the result doubled. The field is computed on a grid of R and z values, and depicted as a three-dimensional surface plot, in which pressure is the third coordinate. The results of such computations are shown in Figs. 8.11 and 8.12. These computations used $N_R = 400$ and $N_\theta = 180$, that is, the sectors subtended $2°$. These values were selected by progressively increasing each number until the plot of $|P|$ as a function of z along $R = 0$ matched the on-axis solution, Eq. (8.3.6). Equally important, the grid of field points must be sufficiently fine to capture the spatial fluctuations. The plotted results depict a grid whose points are separated by $\Delta z = a/50$ in both directions. Another consideration is that placement of the field point on the vibrating surface will lead to a singularity in the integrand if the field point should happen to fall on one of the central points for a patch. This singularity can be addressed mathematically, as was done to derive the Surface Helmholtz Integral Equation. Here, the issue was avoided by setting $\min(z) = \Delta z$. All of these considerations mean that the second plot, for $ka = 24$, for which the maximum value of z is $R_0/2$, constitutes a grid of $6\left(10^4\right)$ field points, and the evaluation of the pressure at each field point required evaluation of a sum of $3.6\left(10^4\right)$ terms. Remarkably, this computation required less than four minutes on a desktop computer with dual quad 2.7 GHz Intel processors using MATLAB as the programming language.

The first case to be considered is $ka = 6$, which is chosen because it is slightly lower than the frequency at which the on-axis pressure has a null. The maximum axial distance in Fig. 8.11 is twice the Rayleigh distance, and the width of the plotted region is twice the piston radius. The most evident feature is the smoothness of the spatial variation away from the piston's face. The farfield directivity, which is shown in Fig. 8.6, is discernible here too. For example, the pressure amplitude along $R = 2a$ is close to zero at $z \approx 2.4a$, which corresponds to a polar angle of

$\tan^{-1}(2/2.4) = 39.8°$, whereas the zero of $\mathcal{D}(\psi)$ at this frequency is $39.7°$. Another feature worth noting is that the pressure very close to the baffle falls off quickly for transverse distances that are greater than a.

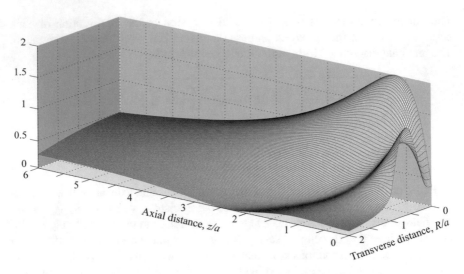

Fig. 8.11 Pressure amplitude in the nearfield of a piston in an infinite baffle, $ka = 6$

The nearfield at high frequencies exhibits significant small-scale fluctuations, as may be seen in Fig. 8.12. This attribute is a consequence of cancelation between the signals that arrive at a field point from each location on the piston. As the frequency increases, the wavelength of these signals decreases, thereby increasing the

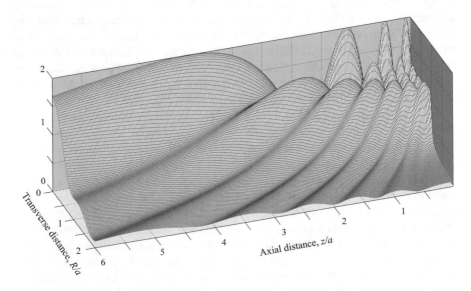

Fig. 8.12 Pressure amplitude in the nearfield of a piston in an infinite baffle, $ka = 24$

possibility of constructive and destructive interference. On-axis, such interference is perfect at the nodes and antinodes and its transitions to the farfield directivity with increasing radial distance.

The spatial profiles in Fig. 8.13 provide a supplementary view of the data. The pressure out to $z = a$ fluctuates in the transverse direction, but it is reminiscent of what happens if a square wave is synthesized from a Fourier series that has an insufficient number of terms. Furthermore, the phase along these lines changes little for $R < a$, which is the characteristic of a simple plane wave. As we saw in the case when $ka = 6$, the pressure close to the baffle falls off beyond $R > a$, but the decay is much more rapid at $ka = 24$. At the larger distances, the farfield behavior can be seen to take over beyond the farthest maximum of the on-axis pressure. This distance is approximately half the Rayleigh distance.

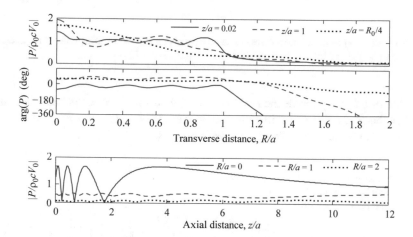

Fig. 8.13 Spatial profiles of the pressure radiated by a piston in a baffle. The plotted data describe variations parallel and transverse to the axis of symmetry, $ka = 24$

These trends are magnified in the case where $ka = 72$, whose profiles are depicted in Fig. 8.14. The pressure within $R < a$ along the transverse lines close to the piston and at $z = a$ varies little and has a nearly constant phase. The general features are like those for $ka = 24$.

The fall-off of pressure on the baffle beyond the piston radius explains why we may use analytical results for an infinite baffle model to describe a finite baffle. When a is used to define dimensionless distances, an acoustic wavelength is $2\pi/(ka)$. Maps such as Figs. 8.11 and 8.12 indicate that the pressure field close to the baffle, $z/a \ll 1$, is very small for $R/a > 1 + \pi/(ka)$. Furthermore, in this region, the pressure changes quite gradually with increasing R and z. As a consequence, both particle velocity components are small in this region. Because the surface velocity is inherently small at sufficiently large R/a, there is no need for an infinite baffle to attain that condition; surrounding the piston with a rigid circular ring whose width equals the acoustic wavelength should be adequate to mimic the infinite baffle condition. The qualitative picture of a *sound beam* emerges from these computations. It is especially

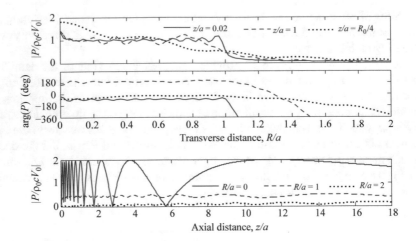

Fig. 8.14 Spatial profiles of the pressure radiated by a piston in a baffle. The plotted data describes variations parallel and transverse to the axis of symmetry, $ka = 72$

suitable as a description of the radiated sound field for a nonacoustician, although the fact that it is only correct in a macroscopic sense often is forgotten by practicing individuals. Figure 8.15 conveys the salient aspects.

Fig. 8.15 Conceptual model of the sound beam radiated by a circular piston in a baffle

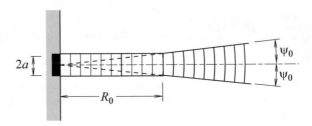

The nearfield is considered to exist in the cylindrical region extending out from the piston's face to the Rayleigh distance. Within this region, the field is represented as a planar wave. The signal beyond the Rayleigh distance is considered to be a spherical wave within a cone whose semi-vertex angle is ψ_0, with the origin of this wave being the center of the piston. The angle is found from Fig. 8.15 to be $\tan^{-1}(a/R_0)$. The model is intended for situations, where $ka \gg 1$, in which case $a/R_0 = 2/(ka)$ is small. This reduces the arctangent to its argument, which leads to

$$\psi_0 \approx \frac{a}{R_0} = \frac{2}{ka} \tag{8.4.3}$$

For comparison, the maximum polar angle at which the intensity is at least half the on-axis maximum is half of the beamwidth, which was discussed in in Sect. 8.2.2.

For large ka, this angle was found to be $\Delta_{bw}/2 \approx 1.62/ka$. The conceptual angle ψ_0 for the radiated field is slightly greater than this angle.

EXAMPLE 8.4 Equation (8.2.30) gives a velocity distribution on the plane of a rigid baffle that supposedly describes the behavior of a transducer whose moving face translates but is not flush with the baffle. The function $Z(R)$ is the depth of the transducer's face at transverse distance R, with positive $Z(R)$ corresponding to the point being below the xy plane. The discussion of Eq. (8.2.30) suggested that the phase delay associated with propagation from the transducer's face to the xy plane may be ignored if ka is sufficiently small. On the other hand, if ka is very large, then the nearfield should be reasonably close to a plane wave, so the phase-shifted velocity distribution should be accurate. If so, then using it as the input to the Rayleigh integral should yield a reasonably accurate description of the pressure field. This hypothesis may be tested by considering the field of a hemispherical transducer mounted on a rigid baffle that executes a vibration as a rigid body. The radiated field was analyzed in Example 7.4 by formulating a spherical harmonic solution. The hemisphere's shape is $Z = -\left(a^2 - R^2\right)^{1/2}$, where the minus sign applies because the hemisphere is above the xy plane. Correspondingly, the approximate surface velocity on the baffle is

$$V_s = \begin{cases} V_z e^{ik(a^2 - R^2)^{1/2}} & \text{if } R < a \\ 0 \text{ if } R > a \end{cases}$$

This function could be used to map the entire field numerically in conjunction with Eq. (8.4.2). It is much simpler, although less definitive, if only the farfield and on-axis fields are evaluated according to the Rayleigh integral. Thus, the task is to evaluate the farfield and on-axis pressure by using the preceding expression for V_s as the input to the Rayleigh integral, then to compare the results to those obtained from the spherical harmonic solution. Cases to consider are $ka = 3$ and $ka = 24$.

Significance

Many applications use a transducer whose vibrating face does not lie in the plane of the baffle. Accurate evaluations of their pressure field requires numerical modeling techniques like those in Sect. 7.5. A simpler description like the one examined here would be especially useful for design studies.

Solution

The farfield approximation of the Rayleigh integral for an arbitrary circular transducer is Eq. (8.2.22). The angular spectrum corresponding to the phase-shifted surface velocity is

$$\tilde{f}(k \sin \psi) = \int_0^a e^{ik(a^2 - R_s^2)^{1/2}} J_0(kR_s \sin \psi) R_s dR_s$$

This expression cannot be integrated analytically, so we will perform a numerical integration based on the nondimensional distance $\xi = R_s/a$, which leads to

$$\tilde{f}(k \sin \psi) = a^2 \int_0^1 e^{ika(1-\xi^2)^{1/2}} J_0(ka\xi \sin \psi) \xi d\xi$$

How to evaluate the integral depends on the software package to be used. The procedure implemented in MATLAB began by setting ka and ψ. Then an anonymous function that evaluates the integrand was used as the input to another function that performs the integration. The anonymous function specification is F=@(xi) exp(1i*ka*sqrt(1-xi.^2)).*besselj(0,ka*xi*sin(psi)).*xi;. (The period before the exponent and the multiplication signs causes the operations to apply element by element to data vectors.) Then the integral is found according to quadl(F,0,1). Both steps must be repeated for each value of ψ from zero to $\pi/2$ in order to construct a directivity pattern.

The on-axis version of the Rayleigh integral is Eq. (8.3.1). For the present V_s function, this becomes

$$P_{\text{axial}}(z) = i\omega\rho_0 V_z \int_0^a e^{ik(a^2 - R_s^2)^{1/2}} \frac{e^{-ik(z^2 + R_s^2)^{1/2}}}{(z^2 + R_s^2)^{1/2}} R_s dR_s$$

Changing the integration variable to the radial distance for a field point slightly simplifies the integral. Thus, we define $a\xi = (z^2 + R_s^2)^{1/2}$. This expression is solved for R_s^2 in order to transform the integral to

$$P_{\text{axial}}(z) = i\omega\rho_0 V_z \int_{z/a}^{(z^2/a^2 + 1)^{1/2}} e^{-ika\left[\xi - (z^2/a^2 + 1 - \xi^2)^{1/2}\right]} d\xi$$

This too is not integrable analytically. A MATLAB anonymous function of ξ that evaluates the integrand would be G=@(xi) exp(-1i*ka*(xi-sqrt(z_a.^2+1-xi.^2));. The integral correspondingly is evaluated according to quadl (G,z_a,sqrt(z_a.^2+1)), where z_a is the value of z/a at the field point.

The spherical harmonic analysis of the pressure field radiated by a translating hemisphere in a rigid baffle was described in Example 7.4. The first operation evaluated the velocity coefficients $(V_{2m})_{\text{rigid}}$. Division of these values by the respective values of $H'_{2m}(ka)$ gave a set of pressure coefficients,

$$B_m = \frac{(V_{2m})_{\text{rigid}}}{iH'_{2m}(ka)}$$

The resulting farfield directivity was

$$\mathcal{D}_{\text{rigid}}(\psi) = \left|\frac{P}{\rho_0 c V_z}\right|\left(\frac{r}{a}\right) = \frac{1}{ka}\left|\sum_{m=0}^{M/2}B_m e^{i(2m+1)\pi/2}P_{2m}(\cos\psi)\right|$$

where M is the minimum spherical harmonic index required for convergence to the desired precision. This series must be recomputed for each value of ψ.

The spherical harmonic series for the on-axis pressure is obtained by evaluating the general solution in Eq. (7.2.28) at $\psi = 0$. The Legendre polynomials appearing in the series at this angle are $P_{2m}(0) \equiv 1$, and $r = z$ on the axis of symmetry. Thus, the spherical harmonic series reduces to

$$P_{\text{axial}}(z) = \rho_0 c v_z \sum_{m=0}^{M/s}B_m h_{2m}\left(ka\left(\frac{z}{a}\right)\right)$$

The results of this computation for a range of z values beginning at the hemisphere, $z/a = 1$, are plotted as $(z/a)|P/(\rho_0 c V_z)|$. Doing so compensates for spherical spreading, which makes it easier to compare data over a large range of z values.

Figure 1 compares the farfield and axial dependencies obtained from the approximate Rayleigh integral and the spherical harmonic series for $ka = 3$. The directivity patterns are similar for all ψ, with the approximate on-axis value being less than 2 dB higher than the correct one. (This error increases rapidly as ka is decreased below three.) The on-axis data tends to a constant value at $z/a \approx 3$, which suggests that this is the distance at which the farfield begins. For comparison, the Rayleigh distance is $R_0 = ka/2 = 1.5$.

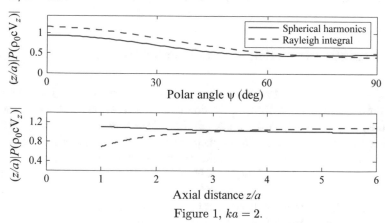

Figure 1, $ka = 2$.

Setting $ka = 24$ leads to very close agreement in the farfield directivity out to $\psi = 60°$, as can be seen in Fig. 2. The Rayleigh distance for a piston at this frequency is $R_0 a = ka/2 = 12$, which is consistent with the spherical harmonic prediction of the value of z/a at which $r|P|$ on axis is a constant value. This is contrasted by the approximate model, which indicates that a constant value of $r|P|$ is attained at $z/a \approx 20$.

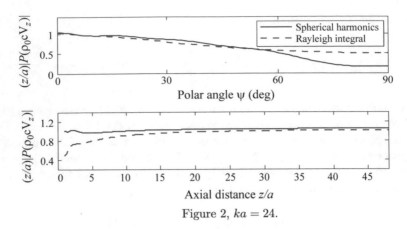

Figure 2, $ka = 24$.

The data for the axial pressure suggests that the approximate analysis is quite good for the farfield, but exhibits significant errors in the nearfield. It is reasonable to generalize the preceding observations. If one has a transducer whose vibrating face is not flush to the plane of the baffle, it might be acceptable to use the Rayleigh integral with a phase-shifted velocity distribution to evaluate the farfield at high frequencies. Furthermore, although it might not be very accurate for the farfield at low frequencies and for the nearfield at any frequency, predictions obtained from this model give a qualitative picture that resembles the actual field. Such simulations might represent a useful alternative to numerical modeling techniques, and they can also be used to verify results obtained by such techniques.

8.5   Radiation Impedance of a Circular Piston

An important aspect of a transducer is the power required to drive it at a specified vibration amplitude. Aside from dissipation within the transducer, this is the same as the radiated power. Up to now, we have evaluated the time-averaged radiated power by using the farfield properties. We could perform a similar analysis here. Instead, we will determine the properties by analyzing the power that flows out of the piston. The virtue of such analysis is that it will provide a description of the instantaneous power output of the piston, which is important for such tasks as designing the electronics to drive the piston. We begin with a determination of the resultant force exerted by the fluid on the piston. This will lead to the radiation impedance, which we previously used in Sect. 3.5.1 to form the end correction at the opening of a waveguide. We will now see how that correction is derived.

The radiation impedance Z_{rad} for a vibrating piston is defined as the complex amplitude of the average pressure on the piston's face divided by the complex amplitude of the velocity, V_0. The average pressure is the total force divided by the area, and the total force is obtained by integrating $P d\mathcal{S}$ over the face. Thus,

$$Z_{\text{rad}} = \frac{F_s}{\pi a^2 V_0}, \quad F_s = \iint_{S_v} P(\bar{x}_s)\, dS \tag{8.5.1}$$

A slightly different interpretation of this definition, which is useful as an extension to all transducers, is that the radiation impedance is the resultant force divided by the volume velocity. If we use the Rayleigh integral to evaluate P at a point $\bar{x}$ on the surface, then that evaluation itself is an integral over the surface,

$$P(\bar{x}_s) = \frac{i\omega\rho_0}{2\pi} V_0 \iint_{S_v} \frac{e^{-ik|\bar{x}_s - \bar{x}_s'|}}{|\bar{x}_s - \bar{x}_s'|}\, dS' \tag{8.5.2}$$

Because $\bar{x}_s$ is situated somewhere on S_v, the integrand contains a singularity at $\bar{x}_s = \bar{x}_s'$.

Rayleigh[7] figured out an ingenious way to circumvent this singularity. Consider a pair of surface points $\bar{x}_s$ and $\bar{x}_s'$. Let us define $\tau(\bar{x}_s, \bar{x}_s')$ as the contribution to pressure at $\bar{x}_s$ due to a unit surface velocity for a differential patch at $\bar{x}_s'$, that is,

$$\tau(\bar{x}_s, \bar{x}_s') = \left(\frac{i\omega\rho_0}{2\pi}\right) \frac{e^{-ik|\bar{x}_s - \bar{x}_s'|}}{|\bar{x}_s - \bar{x}_s'|} \tag{8.5.3}$$

In terms of this quantity, $p(\bar{x}_s)$ is an integral $\tau(\bar{x}_s, \bar{x}_s')\, V_0 dS'$ over S_v, and the resultant force is an integral of $p(\bar{x}_s)\, dS$, also over S_v. Thus, F_s may be found by evaluating

$$F_s = \iint_{S_v} \iint_{S_v} \tau(\bar{x}_s, \bar{x}_s')\, V_0 dS' dS \tag{8.5.4}$$

Rayleigh's idea was to exploit the reciprocity property that $\tau(\bar{x}_s, \bar{x}_s') = \tau(\bar{x}_s, \bar{x}_s')$, which is obvious from its definition, as well as the principle of reciprocity. Rather than performing the double integral in a single operation, he decomposed it into two parts. To assist the discussion, we shall refer to x_s' as the source location and $\bar{x}_s$ as the pressure location. The first integral only considers sources that are closer to the center than the pressure location, that is, $|\bar{x}_s'| < |\bar{x}_s|$. This is the attribute of the lightly shaded circle S_{in} in Fig. 8.16.

The second integral accounts for sources that are more distant than the pressure location. For every pair of points in the first integral, the same pair of points arise in the second integral. To recognize this, let $\bar{x}_a$ and $\bar{x}_b$ denote a specific pair of points, with $\bar{x}_b$ being the point that is closer to the center. Then, the first integral contains $\tau(\bar{x}_a, \bar{x}_b)\, V_0 dS_b dS_a$, while the second integral contains $\tau(\bar{x}_b, \bar{x}_a)\, V_0 dS_a dS_b$. Reciprocity of $\tau(\bar{x}_b, \bar{x}_a)$ tells us that these contributions are equal. A similar equality applies for every pair of points. Therefore, the evaluation of F_s becomes one of the integrating over the circle S_{in} whose radius is the radial distance to an arbitrary point

[7]J.W. Strutt Lord Rayleigh, *Theory of Sound*, vol. 2, 2nd ed., Dover (1945 reprint) Sect. 78.

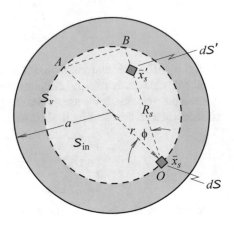

Fig. 8.16 Decomposition of the face area of a vibrating piston for the purpose of evaluating the resultant force

$\bar{x}_s$ on its perimeter. The location of this perimeter point is held fixed in the integral over $\mathcal{S}_{\text{in}}$. It follows that $\bar{x}'_s$ is the only position variable on which this integral depends. An integral of this functional dependence over $\mathcal{S}_{\text{in}}$ accounts for all sources that are closer than the pressure location. Doubling that result accounts for all source locations outside the pressure location, so the evaluation becomes

$$F_s = 2 \iint\limits_{\mathcal{S}_v} \iint\limits_{\mathcal{S}_{\text{in}}} \tau \left(\bar{x}_s, \bar{x}'_s\right) dS' dS \qquad (8.5.5)$$

The second part of Rayleigh's contribution is the way in which the integral over $\mathcal{S}_{\text{in}}$ is carried out. The definition of $\tau \left(\bar{x}, \bar{x}'_s\right)$ depends on the distance between the points. Recall that $\bar{x}_s$ is held fixed in the integral over $\mathcal{S}_{\text{in}}$. Designating it as the origin for a set of polar coordinates (R_s, ϕ) is very useful because $R_s = \left|\bar{x}_s - \bar{x}'_s\right|$, as depicted in Fig. 8.16. The integration is performed first over R_s, then over ϕ. The length of diameter OA is $2r \equiv 2|\bar{x}_s|$. This diameter is the hypotenuse of right triangle OAB, so the range of radial distances is $0 \le R_s \le 2a \cos \phi$, and the range of polar angles is $-\pi/2 \le \phi \le \pi/2$. In this polar coordinate system, $dS' = R_s dR_s d\phi$, which cancels the occurrence of R_s in the denominator of $\tau \left(\bar{x}, \bar{x}'_s\right)$, thereby removing the singularity. Thus, we find that

$$\iint_{\mathcal{S}_{\text{in}}} \tau \left(\bar{x}_s, \bar{x}'_s\right) dS' = \frac{i\omega\rho_0}{2\pi} V_0 \int_{-\pi/2}^{\pi/2} \int_0^{2r\cos\phi} e^{-ikR_s} dR_s d\phi$$

$$= \frac{1}{2\pi} \rho_0 c V_0 \int_{-\pi/2}^{\pi/2} \left[1 - \cos\left(2kr\cos\phi\right) + i\sin\left(2kr\cos\phi\right)\right] d\phi$$

$$(8.5.6)$$

Because $\cos \phi$ is an even function relative to $\phi = 0$, the integral is twice the integral over $0 \le \phi \le \pi/2$, so that

$$\iint_{S_{\text{in}}} \tau \left(\bar{x}_s, \bar{x}_s'\right) dS' = \frac{1}{\pi} \rho_0 c V_0 \int_0^{\pi/2} [1 - \cos(2kr\cos\phi) + i\sin(2kr\cos\phi)]\, d\phi$$

$$= \frac{1}{\pi} \rho_0 c V_0 \left[\frac{\pi}{2} - \mathcal{I}_c + i\mathcal{I}_s\right]$$

$$\mathcal{I}_c = \int_0^{\pi/2} \cos(2kr\cos\phi)\, d\phi, \quad \mathcal{I}_s = \int_0^{\pi/2} \sin(2kr\cos\phi)\, d\phi \qquad (8.5.7)$$

An integral like $\mathcal{I}_c$ arose in the derivation of the farfield directivity of a ring. The tabulated integral[8] used there extended over $0 \le \phi \le \pi$. The integrand of $\mathcal{I}_c$ is an even function with respect to $\phi = \pi/2$, so $\mathcal{I}_c$ is half the value of the tabulated entry, that is,

$$\mathcal{I}_c = \frac{\pi}{2} J_0(2kr) \qquad (8.5.8)$$

The integration range for $\mathcal{I}_s$ does not match any entry for Bessel functions in a basic resource for special functions. However, it does appear in a chapter on *Struve functions*, which are denoted as $\mathbf{H}_\nu(x)$, where ν is the order. Alternative definitions of this function describe it as a power series or as an integral,[9]

$$\mathbf{H}_\nu(x) = \left(\frac{x}{2}\right)^{\nu+1} \sum_{k=0}^{\infty} \frac{\left(-x^2/4\right)^k}{\Gamma(k+3/2)(k+\nu+3/2)}$$

$$= \frac{2(x/2)^\nu}{\pi^{1/2}\Gamma(\nu+1/2)} \int_0^{\pi/2} \sin(x\cos\theta)(\sin\theta)^{2\nu}\, d\theta \qquad (8.5.9)$$

These alternative definitions feature the *Gamma function*, denoted as $\Gamma(z)$. It is another special function described in our basic reference.[10] It follows the recurrence relation $\Gamma(z+1) = z\Gamma(z)$. Thus, for integer arguments, it is related to a factorial according to $\Gamma(n+1) = n!$ For a noninteger positive argument z, it may be computed by evaluating the function at the fractional part of the argument, $z_f = z - \text{floor}(z)$, followed by application of the recurrence relation. The fractional part for the preceding formulas is $z_f = 1/2$ and $\Gamma(1/2) - \pi^{1/2}$. If the chosen mathematical software does not provide a routine that evaluates the Struve function, it may be evaluated numerically according to either representation in Eq. (8.5.9). However, both forms will encounter difficulty with round-off error if the argument is large, e.g. $x > 40$. An asymptotic approximation is quite effective in such situations.

A comparison of the integral representation in Eq. (8.5.9) and the definition of $\mathcal{I}_s$ show that

$$\mathcal{I}_s = \frac{\pi^{1/2}\Gamma(1/2)}{2} \mathbf{H}_0(2kr) == \frac{\pi}{2} \mathbf{H}_0(2kr) \qquad (8.5.10)$$

[8] M.I. Abramowitz and I.A. Stegun, *Handbook of Mathematical Functions*, 9th ed., Dover (1965), Eq. (9.1.21).

[9] M.I. Abramowitz and I.A. Stegun, *ibid*, Eqs. (12.1.3) and (12.1.7).

[10] M.I. Abramowitz and I.A. Stegun, *ibid*, Chap. 5.

Correspondingly, we find that

$$\iint_{\mathcal{S}_{\text{in}}} \tau\left(\bar{x}_s, \bar{x}'_s\right) V_0 d\mathcal{S}' = \frac{1}{2}\rho_0 c V_0 \left[1 - J_0\left(2kr\right) + i\mathbf{H}_0\left(2kr\right)\right] \qquad (8.5.11)$$

The total resultant force is found by integrating this term over $\mathcal{S}_v$. Because r is the distance from the center to $\bar{x}_s$, the integration is done by changing to a polar coordinate system centered on $\mathcal{S}_v$. Thus, we must evaluate

$$F_s = \iint_{\mathcal{S}_v} 2 \iint_{\mathcal{S}_{\text{in}}} \tau\left(\bar{x}_s, \bar{x}'_s\right) d\mathcal{S}' d\mathcal{S} = \int_0^a \int_{-\pi}^{\pi} \rho_0 c V_0 [1 - J_0(2kr) + i\mathbf{H}_0(2kr)] r d\theta dr \qquad (8.5.12)$$

The θ integration gives a 2π factor. To integrate the Bessel and Struve function terms, we invoke a recurrence relation that has a similar form for both, specifically,

$$\frac{d}{dy}\left(y^n J_n\left(y\right)\right) = y^n J_{n-1}\left(y\right), \quad \frac{d}{dy}\left(y^n \mathbf{H}_n\left(y\right)\right) = y^n \mathbf{H}_{n-1}\left(y\right) \qquad (8.5.13)$$

The result is

$$\boxed{F_s = \pi a^2 V_0 Z_{\text{rad}}, \quad Z_{\text{rad}} = \rho_0 c\, \chi\left(ka\right)} \qquad (8.5.14)$$

where the specific impedance is

$$\boxed{\chi\left(ka\right) = \left(1 - \frac{J_1\left(2ka\right)}{ka}\right) + i\left(\frac{\mathbf{H}_1\left(2ka\right)}{ka}\right)} \qquad (8.5.15)$$

Figure 8.17 repeats from Chap. 3 the graph of χ as a function of ka.

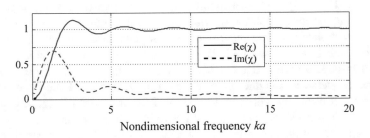

Fig. 8.17 Plot of the frequency dependence of the specific acoustic impedance χ for a vibrating circular piston of radius a

For small ka, the first few terms of a Taylor series expansion are adequate,

$$\chi\left(ka\right) = \left[\frac{1}{2}\left(ka\right)^2 - \frac{1}{12}\left(ka\right)^4 + \cdots\right] + i\frac{4}{\pi}\left[\frac{2}{3}\left(ka\right) - \frac{8}{45}\left(ka\right)^3 + \cdots\right] \qquad (8.5.16)$$

The specific impedance tends to $\chi(ka) = 1$ as $ka \to \infty$, which is merely the plane wave limit associated with high frequencies.

One reason that the specific impedance is important is that it is needed to predict the voltage and current required to drive the piston transducer at a specified velocity. Knowledge of the specific impedance provides a simple way to evaluate the power required to drive a piston. Aside from losses within the transducer, this is the same as the power radiated by the piston. Let us consider this quantity as a function of time. All points on the piston's face translate at the same velocity, so the instantaneous power is the product of the resultant force and that velocity. The instantaneous power input to the fluid at the piston face, therefore, is

$$
\begin{aligned}
\mathcal{P}(t) &= \frac{1}{4} \operatorname{Re}\left[\left(F_s e^{i\omega t} + \text{c.c.}\right)\left(V_0 e^{i\omega t} + \text{c.c.}\right)\right] \\
&= \frac{1}{2} \operatorname{Re}\left[F_s\left(V_0\right)^* + F_s V_0 e^{2i\omega t}\right] \\
&= \frac{\pi a^2}{2}\left[\left(\operatorname{Re} Z_{\text{rad}}\right)|V_0|^2 + \operatorname{Re}\left(\left(V_0\right)^2 Z_{\text{rad}} e^{2i\omega t}\right)\right]
\end{aligned}
\tag{8.5.17}
$$

The invariant part of this expression is the time-averaged power,

$$
\mathcal{P}_{\text{av}} = \frac{\pi a^2}{2} \operatorname{Re}\left(Z_{\text{rad}}\right)|V_0|^2
\tag{8.5.18}
$$

The fluctuating part is the power that flows into and out of the fluid across the piston's face. It oscillates at a frequency 2ω with an amplitude that is $\left(\pi a^2/2\right)|Z_{\text{rad}}||V_0|^2$, so

$$
|\mathcal{P}_{\text{fluct}}| = \mathcal{P}_{\text{av}} \frac{|Z_{\text{rad}}|}{\operatorname{Re}\left(Z_{\text{rad}}\right)}
\tag{8.5.19}
$$

It follows that the average power flow across the piston face never exceeds the fluctuating part, so there is a time interval during which the power flows from the fluid into the piston. In this condition, the pressure resultant *acting on the piston* is in the same sense as the piston's velocity.

It is enlightening to compare the power flow across the piston face to the power that flows across a very large surrounding hemisphere. We know that the average portions should be the same, but what about the fluctuating parts? To answer this, we us a large hemisphere whose radius is sufficiently large that we may use the farfield approximation that $P/V_r = \rho_0 c$ at any polar angle. The time-domain version of Eq. (8.2.23) is

$$
p_{\text{ff}}\left(\bar{x}_0, t\right) = \rho_0 c \frac{J_1\left(ka\sin\psi\right)}{\sin\psi} \frac{a}{r}\left(\frac{1}{2}\right)\left(V_0 e^{-ikr} e^{i\omega t} + \text{c.c.}\right)
\tag{8.5.20}
$$

The instantaneous radial intensity is $p_{\text{ff}}^2/\left(\rho_0 c\right)$, and the field is axisymmetric, so $dS_0 = 2\pi r^2 d\psi$. Hence, the instantaneous power flowing across a hemisphere of radius r is given by

$$\mathcal{P}_{ff}(t) = \int_0^{\pi} \frac{1}{\rho_0 c} \left[p_{ff}(\bar{x}_0) \right]^2 2\pi r^2 d\psi$$

$$= \pi \rho_0 c a^2 \left[\int_0^{\pi} \left(\frac{J_1(ka \sin \psi)}{\sin \psi} \right)^2 d\psi \right] \left[|V_0|^2 + \mathrm{Re}\left((V_0)^2 e^{2i(kr - \omega t)} \right) \right]$$

$$(8.5.21)$$

There is no need to evaluate the integral because the time-averaged power is the constant part of this expression. The net power flowing into any region of an ideal fluid must have a zero average value, so this portion of $\mathcal{P}_{ff}(t)$ must equal $\mathcal{P}_{av}$ in Eq. (8.5.18). Hence, writing V_0 in polar form leads to

$$\mathcal{P}_{ff}(t) = \frac{\pi a^2}{2} \mathrm{Re}\left(Z_{rad}\right) |V_0|^2 \left[1 + e^{2i(kr - \omega t + \arg(V_0))} \right] \qquad (8.5.22)$$

This description of $\mathcal{P}_{ff}(t)$ highlights the fact that the fluctuating part has an amplitude that equals the mean value. This is different from the behavior at the transducer face, where the magnitude of the fluctuating part of the power flow exceeds the average. At any instant, the difference between the fluctuating parts of the power radiated by the piston and the power that flows into the farfield represents the rate of change of the fluid's mechanical energy in the region close to the piston.

EXAMPLE 8.5 A circular 450 mm diameter piston is mounted in a baffle on the floor in an anechoic chamber. In one test, the electrical signal creates a square wave alternating between $+V_0$ and $-V_0$ with a fundamental frequency of 1 kHz. The sound pressure on-axis 4 meters from the piston is measured to be 125 dB/20 μPa. This measurement is cutoff at 18 kHz in order to account for audible tones only. Determine the power output as a function of time. Also determine the waveform measured at the 4 m distance on-axis. Graph both time-dependent quantities.

Significance

Beyond the obvious objective of performing an evaluation of radiated power, we will see that power requirements are significantly affected by the harmonic content. In addition, the pressure waveform provides a preview of an interesting phenomenon that arises in a time-domain analysis.

Solution

The value of V_0 is not specified. We will determine it by evaluating the pressure at $z = 4$ m on-axis in terms of V_0, which will yield a value of the mean-squared pressure at that location. The square wave is specified to be a periodic signal defined by

$$v_s(t) = V_0 \left[h(t) - 2h(t - T/2) \right] \text{ if } 0 \le t \le T, \quad v_s(t \pm T) = v_s(t) \qquad (1)$$

The value of ω_1 is 2000π rad/s. The Fourier series representation of this signal is

$$v_s = \text{Re} \sum_{n=1}^{\infty} V_n e^{i\omega_n t}, \quad \omega_n = n\omega_1, \quad \omega_1 = \frac{2\pi}{T}$$

$$V_n = \frac{2}{i\pi n} \left[1 - (-1)^n\right] V_0 \tag{2}$$

Only the V_n values for odd n are not zero. Each harmonic contained in the surface vibration is associated with a different wavenumber,

$$k_n = n\omega_1/c \tag{3}$$

Each value of k_n leads to a contribution to the on-axis pressure according to Eq. (8.3.6). The resulting Fourier time series is

$$p(z, t) = \text{Re} \sum_{\substack{n=1 \\ n \text{ odd}}}^{\infty} P_n(z) e^{in\omega_1 t}$$

$$P_n(z) = 2i\rho_0 c \left(\frac{4V_0}{i\pi n}\right) e^{-i(k_n/2)\left[z + (z^2 + a^2)^{1/2}\right]} \sin\left[\frac{k_n}{2}\left((z^2 + a^2)^{1/2} - z\right)\right] \tag{4}$$

We take $c = 340$ m/s and $\rho_0 = 1.2$ kg/m^3 as representative values for air. The pressure coefficients at $z = 4$ m, $R = 0$ of the odd harmonics are tabulated below. They are cutoff at 18 kHz, which for a 1 kHz fundamental corresponds to $n = 18$.

n	1	3	5	7	9
$\text{Re}\left(\dfrac{P_n}{V_0}\Big\|_{z=4}\right)$	9.1155	−26.4029	40.9712	−51.3675	56.6292
$\text{Im}\left(\dfrac{P_n}{V_0}\Big\|_{z=4}\right)$	59.9788	−54.3143	43.6173	−29.0723	12.2696

n	11	13	15	17
$\text{Re}\left(\dfrac{P_n}{V_0}\Big\|_{z=4}\right)$	−56.3883	50.9007	−40.9956	27.9556
$\text{Im}\left(\dfrac{P_n}{V_0}\Big\|_{z=4}\right)$	4.9911	−20.9190	33.9445	−42.8971

The given sound pressure level of 125 dB at $z = 4$ m corresponds to a mean-squared pressure of $1.2649 \left(10^3\right)$ Pa^2. At the same time, the mean-squared pressure is the incoherent sum of the mean-squared pressure of each harmonic in the tabulation, that is,

$$\left(p^2\right)_{av} = \frac{1}{2} |V_0|^2 \sum_{n \text{ odd}}^{18} \left|\frac{P_n}{V_0}\Big|_{z=4}\right|^2 = 14719 \, |V_0|^2 = 1.2649 \left(10^3\right)$$

From this, we find that the piston's velocity amplitude is

$$V_0 = 0.2932 \text{ m/s}$$

With this value determined, the actual V_n and P_n coefficients can be computed. The primary factor deciding the time increment for synthesis of the Fourier series is how well we wish to capture the rise and fall of the square wave. Let us select $\Delta t = T/100$ for this purpose. Presumably, the full harmonic content of the pressure is measured and then filtered for the content below 18 kHz. It is possible that despite such filtering, the existence of higher harmonics is important, especially for considerations related to the power output of the transducer. To ascertain whether such speculation is correct, all series shall be synthesized for a series length of $N = 100$, as well for $N = 18$. The first evaluation tests the adequacy of either truncation by evaluating the input velocity waveform according to Eq. (2).

We could use an inverse FFT routine to create the waveforms, but we shall implement the matrix algorithm in Eq. (1.4.39). Thus, v_s at instant t_j is indicated by Eq. (1) to be

$$v_s(t_j) = \text{Re} \sum_{\substack{n=1 \\ n \text{ odd}}}^{N} V_n e^{i\omega_n t_j} \implies \{v_s\} = [E]^{\text{H}} \{V\} \tag{5}$$

where

$$\{V\} = [V_1 \ V_3 \ \cdots]^{\text{T}}, \quad [E] = \begin{bmatrix} e^{-i\omega_1 t_1} & e^{-i\omega_1 t_2} & \cdots \\ e^{-i\omega_2 t_1} & e^{-i\omega_2 t_2} & \cdots \\ \vdots & \vdots & \ddots \end{bmatrix}$$

The superscript in Eq. (5) specifies the Hermitian of $[E]$, which is the complex conjugate transpose of the array.

Figure 1 shows that either truncation gives an adequate reproduction of the square wave, although the ringing for the smaller N is noticeable. Note that this ringing stems from the fact that $N = 18$ does not meet the Nyquist criterion.

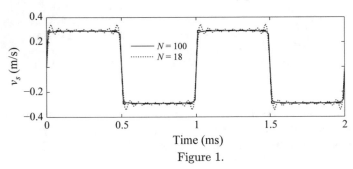

Figure 1.

Figure 2 depicts the pressure waveform at $z = 4$ m. It was computed by adapting the matrix algorithm in Eq. (5) to Eq. (4). The result is somewhat surprising because it indicates that the waveform is a sequence of spikes. One clue why this is so is found in the Rayleigh distance $(R_0)_n = k_n a^2/2$ for each harmonic. This distance for the first harmonic is 0.468 m, and it is not until $n = 9$ that $z = 4$ m is less than the Rayleigh distance. Thus, for the most significant harmonics, the field point is in the farfield. According to Eq. (8.3.11), the contribution of each harmonic at fixed z in the farfield is proportional to $(R_0)_n V_n$. The V_n values are inversely proportional to n, and the Rayleigh distance of each harmonic is proportional to ω_n, which means that the lower $P_n (z)$ values should have nearly equal magnitude. This expectation is borne out by the tabulation of $P_n (z) / V_0$. A Fourier series whose coefficients have equal magnitude, with suitable relative phase delays, represents a periodic train of impulses. This is manifested as the pressure being a series of alternating spikes. But why does this happen? That explanation will be found in the next section, which explores the time-domain Rayleigh integral.

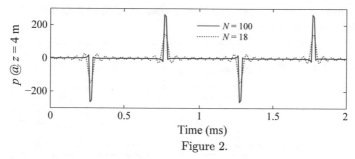

Figure 2.

It is not necessary to synthesize waveforms to evaluate the time-averaged power because the intensity is the incoherent sum of contributions from each harmonic. Therefore, $\mathcal{P}_{av}$ is the sum of terms described by Eq. (8.5.18), with Z_{rad} for each n evaluated at the respective k_n value. That is,

$$\mathcal{P}_{av} = \frac{\pi a^2 \rho_0 c}{2} \operatorname{Re} \sum_{\substack{n \\ n \text{ odd}}}^{N} \chi (k_n a) |V_n|^2 \tag{5}$$

The result for $N = 100$ is $\mathcal{P}_{av} = 5.27$ W. Cutting off the series at $N = 18$ yields $\mathcal{P}_{av} = 5.17$ W, so the shorter series length seems to be quite adequate.

Unlike the evaluation of the average power, interference between harmonics is an important feature of the instantaneous power. We begin this evaluation by describing the force resultant as a Fourier series. The complex amplitude for a harmonic is described by Eq. (8.5.14), which we adapt by evaluating Z_{rad} at the respective harmonic frequency and multiplying it by the velocity coefficient V_n in Eq. (2). The series is

$$F_s = \pi a^2 \rho_0 c \operatorname{Re} \sum_{\substack{n \\ n \text{ odd}}}^{N} \chi (k_n a) V_n e^{in\omega_1 t} \tag{6}$$

The matrix algorithm in Eq. (5) for this calculation is

$$\{F_s\} = [E]^H \{\{\chi\} \cdot \{V\}\} \tag{8.5.23}$$

where $\{\{\chi\} \cdot \{V\}\}$ denotes an element by element multiplication.

The instantaneous power may be computed by multiplying the F_s value at any t by the v_s value plotted in Fig. 1. The result of both computations is shown in Fig. 3. The force waveform resembles the piston's velocity, except that the plateaus are not flat. The radiated power is positive if v_s and F_s have the same sign, so the plot of $\mathcal{P}_{rad}$ is a periodic sequence of positive plateaus, with $T/2$ as the period. The instants at which $\mathcal{P}_{rad}$ is zero are the instants at which $v_s = 0$. There is no interval in which P_{rad} is negative. This is so because all $k_n a > 3$, so all $\chi (k_n a)$ are close to unity. The maximum radiated power, which occurs soon after the transitions of v_s, is 6.18 W. The power falls off after the peak, then drops off rapidly at the velocity transitions. The consequence is that the average power does not differ much from the peak value. An interesting aspect is that the inadequacies of using the smaller N are evident for the force, and especially the power. The shorter series length leads to max $(\mathcal{P}_{rad}) = 8.84$ W, which is more than 40% high.

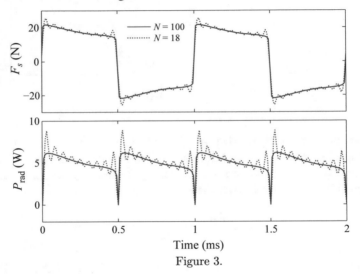

Figure 3.

8.6   Time Domain Rayleigh Integral

A time-domain description of the signal generated by a vibrating piston has advantages in some circumstances. One approach describes the field as a superposition of plane and evanescent waves of the type studied in Sect. 5.1. A two-dimensional Fourier transform of the pressure leads to a representation that matches the angular spectrum of the surface velocity. The requirement that the pressure at each wavenumber matches the normal surface velocity yields a representation of the pressure as an inverse Fourier transform. From a philosophical viewpoint, this formulation is attractive because it provides a similar perspective to that provided by the prior investigations of radiation from spheres to cylinders. The formulation is primarily intended for numerical evaluations, for which FFT technology substantially reduces the computational effort. The paper by Wu, Kazys, and Stepinski[11] explains the details. We shall not delve into this line of investigation because it does not lead to new insights as to the nature of the pressure field.

In contrast, the time-domain version of the Rayleigh integral provides a different perspective. For example, it provides a simple explanation of the impulse-like waveform observed in the previous example. However, evaluation of the pressure field at an arbitrary field point is more difficult than it is in the frequency-domain approach.

We have regarded the Rayleigh integral as a prescription for evaluating the frequency-domain response when the surface vibration is a set of point sources associated with Re $(V_s (\bar{x}_s) \exp (i\omega t))$. A different perspective views the Rayleigh integral as the Fourier transform of a time-domain representation of the radiated field. Our task is to determine what that time representation is.

We shall not call on the inverse Fourier transform to carry out this determination. Rather, we will use the notion of function/transform pairs for operations in the time domain, such as those in Table B1 of Appendix B. The first step is to recognize that the factor $i\omega V_s (\bar{x}_s)$ in the Rayleigh integral is the Fourier transform of the rate of change of $v_s (\bar{x}_s, t)$, that is, $\mathcal{F} ((\partial v_s/\partial t), \omega) = i\omega V_s (\bar{x}_s)$. Hence, the Rayleigh integral may be written as

$$P (\bar{x}_0) = \frac{\rho_0}{2\pi} \iint\limits_{S} \mathcal{F} \left(\frac{\partial}{\partial t} v_s (\bar{x}_s, t), \omega \right) \frac{e^{-ik|\bar{x}_0-\bar{x}_s|}}{|\bar{x}_0 - \bar{x}_s|} dS \qquad (8.6.1)$$

The other occurrence of ω is in the complex exponential. The coefficients of a Fourier series that are delayed by time τ relative to another function are described in Eq. (1.4.11). It is shown there that if $p (t)$ and $q (t)$ are two functions that have the same period T, with $q (t) = p (t - \tau)$, then their Fourier series coefficients are related by $Q_j = P_j e^{-i\omega_j \tau}$, where $\omega_j = j (2\pi/T)$. A Fourier transform corresponds to the coefficients of the Fourier series of a function whose period is infinite, which replaces the discrete frequencies ω_j with a continuous variable ω. Thus, if $g (t) = g (t - \tau)$, their Fourier transforms are related by $\mathcal{F} (g (t), \omega) = \mathcal{F} (f (t), \omega) e^{-i\omega\tau}$. (This property also may be identified from the coordinate shifting property in Table B1 of Appendix B, except that the sign of the exponent must be reversed because the tabulation is based on our definition of a spatial Fourier transform.) Because $k = \omega/c$, the exponential in the Rayleigh integral corresponds to delaying a function by $|\bar{x}_0 - \bar{x}_s| /c$. Correspondingly, the preceding form of the Rayleigh integral may be written as

$$P (\bar{x}_0) = \frac{\rho_0}{2\pi} \iint\limits_{S} \mathcal{F} \left(\frac{\partial}{\partial t} v_s \left(\bar{x}_s, t - \frac{|\bar{x}_0 - \bar{x}_s|}{c} \right), \omega \right) \frac{1}{|\bar{x}_0 - \bar{x}_s|} dS \qquad (8.6.2)$$

---

[11] Ping Wu, Rymantas Kazys, and Tadeusz Stepinski, "Analysis of the numerically implemented angular spectrum approach based on the evaluation of two-dimensional acoustic fields. Part I. Errors due to the discrete Fourier transform and discretization," J. Acoust. Soc. Am. vol. 99, Issue 3, pp. 1339–1348 (1996); (10 p).

By definition, the inverse Fourier transform of a Fourier transform is the original time-domain function. Because the source-field point distance is independent of ω, it is constant in the transformation. We thereby obtain the *time domain Rayleigh integral*,

$$p\left(\bar{x}_0\right) = \frac{\rho_0}{2\pi} \iint_S \frac{1}{|\bar{x}_0 - \bar{x}_s|} \frac{\partial}{\partial t} v_s \left(\bar{x}_s, t - \frac{|\bar{x}_0 - \bar{x}_s|}{c}\right) dS \qquad (8.6.3)$$

This form is readily understood in terms of source distributions. Consider a differential element of the surface. The mass acceleration of this element is $\rho_0 dS\, dv_s/dt$, and the distance from it to the field point is $r = |\bar{x}_0 - \bar{x}_s|$. The time-domain pressure radiated by this element is double the value in Eq. (6.4.3) because the sound radiates into a half space, rather than a free space. Thus, it is

$$dp\left(\bar{x}_0, t\right) = \frac{\rho_0}{2\pi r} \frac{\partial}{\partial t} dQ_s \left(t - \frac{r}{c}\right) \qquad (8.6.4)$$

The time-domain Rayleigh integral is the superposition of the signals radiated by a sheet of these sources distributed over the surface. The signals from different sources that arrive contemporaneously at a field point departed from the surface at different instants. Specifically, the signal from a differential source at $\bar{x}_s$ that arrives at the field point at time t must have left the surface at time $t - |\bar{x}_0 - \bar{x}_s|/c$. Hence, the signal observed at a field point at time t consists of contributions that departed from the surface in a time interval beginning at $t - \max\left(|\bar{x}_0 - \bar{x}_s|\right)/c$ and ending at $t - \min\left(|\bar{x}_0 - \bar{x}_s|\right)/c$.

An evaluation of the Rayleigh integral for an arbitrary location and surface velocity requires numerical methods. The technique described by Fig. 8.10 also is suitable for this application. The radial distance from the central point on a surface patch to the selected field point is used to evaluate $t - r/c$ for that patch. Then, Eq. (8.6.4) is added for each patch.

If we limit our consideration to the on-axis pressure radiated by a circular piston, we may evaluate the integral analytically. The analysis follows similar steps to that for the frequency domain. Thus, a point on the surface is located by its polar coordinates (R_s, θ_s). The distance from such a point to the field point at distance z from the surface is $r = \left(R_s^2 + z^2\right)^{1/2}$. The value of r is independent of θ_s, so an integration over this variable leads to a 2π factor. In addition, v_s for a piston is the same for all R_s. Thus, Eq. (8.6.3) becomes

$$p\left(\bar{x}_0, t\right) = \rho_0 \int_0^a \frac{1}{\left(R_s^2 + z^2\right)^{1/2}} \frac{\partial}{\partial t} v_s \left(t - \frac{\left(R_s^2 + z^2\right)^{1/2}}{c}\right) R_s dR_s \qquad (8.6.5)$$

At this stage, we change variables from R_s to the radial distance r. Because z is constant, differentiation of $r^2 = R_s^2 + z^2$ leads to $r\,dr = R_s dR_s$. Furthermore, the limits

$R_s = 0$ and $R_s = a$ correspond to $r = z$ and $r = \left(a^2 = z^2\right)^{1/2}$, which leads to

$$p\left(\bar{x}_0, t\right) = \rho_0 \int_z^{\left(a^2+z^2\right)^{1/2}} \frac{\partial}{\partial t} v_s \left(t - \frac{r}{c}\right) dr \tag{8.6.6}$$

The last step is to recognize the meaning of the derivative of v_s. Let Θ denote the argument, $\Theta = t - r/c$. Then, the chain rule for differentiation leads to

$$\frac{\partial}{\partial t} v_s \left(t - \frac{r}{c}\right) \equiv \frac{\partial \Theta}{\partial t} \frac{d}{d\Theta} v_s \left(\Theta\right) = \frac{dv_s}{d\Theta}$$

$$\frac{\partial}{\partial r} v_s \left(t - \frac{r}{c}\right) \equiv \frac{\partial \Theta}{\partial r} \frac{d}{d\Theta} v_s \left(\Theta\right) = -\frac{1}{c} \frac{dv_s}{d\Theta} = -\frac{1}{c} \frac{\partial}{\partial t} v_s \left(t - \frac{r}{c}\right) \tag{8.6.7}$$

Usage of this property to replace the time derivative with a derivative with respect to r leads to

$$p\left(z, t\right) = -\rho_0 c \int_z^{\left(a^2+z^2\right)^{1/2}} \frac{\partial}{\partial r} v_s \left(t - \frac{r}{c}\right) dr \tag{8.6.8}$$

The integrand is a perfect differential, so we find that

$$\boxed{p\left(z, t\right) = \rho_0 c \left[v_s \left(t - \frac{z}{c}\right) - v_s \left(t - \frac{\left(a^2 + z^2\right)^{1/2}}{c}\right) \right]} \tag{8.6.9}$$

This is the time-domain equivalent of Eq. (8.3.4). The first term is the center wave, which leaves the center of the piston at time z/c prior to its arrival at the field point. The second term is the edge wave, whose propagation time prior to arrival at the field point is $\left(a^2 + z^2\right)^{1/2}/c$.

Consider the case where the surface vibration is a pulse, such that $v_s\left(t\right) = 0$ for $t < 0$ or $t > T$. Then, the signal received at an axial point will be the same shape pulse in the edge wave, followed by the negative of the first pulse emanating from the edge of piston. The interval between the arrival times is $\left[\left(a^2 + z^2\right)^{1/2} - z\right]/c$. If this interval is greater than T, the pulses will appear as separate events. This inequality may be solved for the range of z in which the pulses occur individually. It is $z < \left(a^2 - c^2 T^2\right) / \left(2cT\right)$.

Some insight into the farfield behavior may be gained by considering axial locations at which $z/a \gg 1$. Then, $\left(a^2 + z^2\right)^{1/2} \approx z + a^2 / \left(2z\right)$. The second term is much less than z, so application of a Taylor series to Eq. (8.6.9) leads to

$$p\left(z, t\right) \rightarrow \frac{\rho_0 a^2}{2z} \dot{v}_s \left(t - \frac{z}{c}\right) \tag{8.6.10}$$

Thus, the pressure signal received at an axial point far from the piston will mirror the surface acceleration.

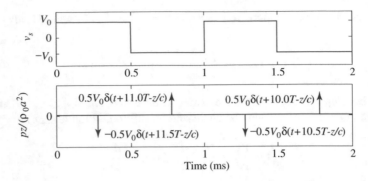

Fig. 8.18 The pressure signal received at a point on the axis of a circular piston whose velocity v_s is a square wave

An interesting consequence of this property is what is heard when a piston moves as a square wave, which was the situation in Example 8.5. No pressure is generated when the velocity is constant, so the only sound that reaches the farfield is a series of spikes at the discontinuities. This is illustrated in Fig. 8.18, which depicts the time-domain construction of v_s and the pressure waveform at $z = 4$ m, as opposed to the representation of these signals as truncated Fourier series in the aforementioned example. The value of $z/c = 11.76$ ms, so the first impulse to be observed for $t > 0$ is a negative one corresponding to the discontinuous decrease of v_s at $t = -11.5T$. Of course, it is not possible to change the velocity discontinuously, because doing so would require an infinite force. Thus, the graph is an approximation of a rapid transition, just as an impulsive force is an approximation of a large force that acts over a very short time interval.

Additional analyses have been performed for various off-axis locations. For example, Blackstock[12] describes an approximate time-domain farfield analysis for a circular piston. Pierce[13] shows how to determine the pressure radiated by a circular piston at an arbitrary field point as a single integral, rather than an integral over an area. These analyses are somewhat complicated, so we shall not explore them.

8.7 Homework Exercises

Exercise 8.1 A rectangular plate is flush mounted in an infinite rigid baffle that lies in a vertical plane xy. The plate is very rigid, so it undergoes a uniform translational vibration v_z in the normal z direction. The frequency of this vibration might be any value up to 1.2 kHz. Within this interval, the farfield mean-squared pressure in the (horizontal) xz plane at 30° from the z-axis should be no more than 6 dB less than the on-axis value. Also, the farfield mean-squared pressure at 5° from the z-axis in the

[12] D.T. Blackstock, *Fundamentals of Physical Acoustics*, John Wiley & Sons (2000) pp. 461–463.

[13] A.D. Pierce, *Acoustics*, McGraw-Hill (1981), reprinted by ASA Press (1989) pp. 227–231.

vertical yz plane should be no more than 6 dB less than the on-axis value. Determine the horizontal and vertical edge lengths that meet these specifications.

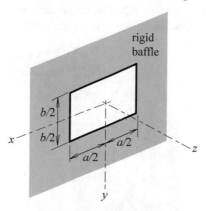

Exercise 8.1

Exercise 8.2 The sketch depicts a square plate that is embedded in a very large rigid baffle. The plate is flexible, with motion actuated 180° out-of-phase from one corner to the next, with the result that the surface normal velocity is $v_z = V_0(xy/b^2)\cos(\omega t)$, where b is the length of each edge. Derive an expression for the farfield directivity. For the case where $kb = 10$, evaluate and graph this function for points in the xz plane and yz plane, as well as a plane that contains the z-axis at 45° from the x- and y-axes.

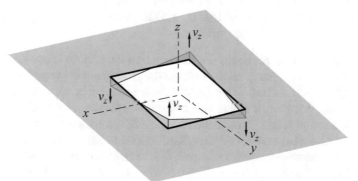

Exercise 8.2

Exercise 8.3 The sketch depicts a submerged square transducer mounted on a large rigid baffle that lies in the vertical plane. The edges of the transducer are horizontal and vertical, with the upper edge at depth h below the free surface. This depth is sufficiently large that the baffle may be considered to extend infinitely in all directions for the purpose of formulating the Rayleigh integral. The transducer executes a uniform vibration $v_0\sin(\omega t)$ in the z direction, which is perpendicular to the plane of the baffle. It is permissible to approximate the free surface as being pressure-release for waves in the water. (a) Derive an expression for the complex amplitude of the pressure in the farfield as a function of spherical coordinates for which z is the polar axis and r measures the distance from the xyz origin. (b) Consider the case where $a = 0.4$ m and the frequency is 5 kHz. Determine and plot the farfield radiation $(r/a)\,|p_{\mathrm{ff}}|\,/\,(\rho 0 c v_0)$ as a function of angle from the z-axis for points in the yz plane.

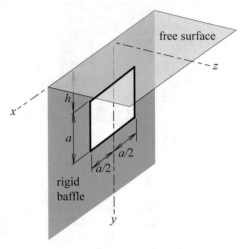

Exercise 8.3

Exercise 8.4 It is desired to build a baffled circular piston projector that generates a signal in air whose beamwidth is less than $1.5°$ for frequencies greater than 16 kHz. What is the smallest radius a of the piston that fits this specification? For this value of a, what is the beamwidth at 16 kHz, 1.6 kHz, and 160 Hz? Graph the farfield directivity at each of these frequencies.

Exercise 8.5 A Gaussian velocity distribution on a flat surface is one for which the velocity distribution depends on the polar distance R_s from the center according to $v_z = V_0 \exp\left(-\beta R_s^2\right) \sin\left(\omega t\right)$. This distribution is desirable because it limits diffraction effects that cause nulls due to a discontinuous surface velocity distribution. In reality, this concept is an approximation because R_s cannot exceed the active radius a. Thus, the Gaussian surface velocity distribution applies for $R_s < a$, and $v_z = 0$ for $R_s > a$. Use numerical methods to evaluate the farfield directivity when $ka = 12$. Perform this evaluation for $\beta = 0$, $\beta = 0.5$, $\beta = 2$, and $\beta = 8$. What conclusions can be drawn from these evaluations?

Exercise 8.6 Derive an expression for the farfield directivity of the transducer in Example 8.3. Evaluate the result for $ka = 5$ and $ka = 20$. Compare each pattern to the directivity of a baffled piston.

Exercise 8.7 A circular plate flush mounted in an infinite baffle executes a harmonic motion in which it rotates as a rigid body about a diametral line. The consequence is that the surface velocity is $v_1\left(y/a\right)\sin\left(\omega t\right)$. Derive an expression for the farfield pressure. Evaluate and graph this function for points in the xz plane and yz plane for the case where $ka = 10$.

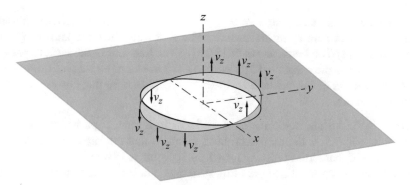

Exercise 8.7

Exercise 8.8 Drawing I shows a projector embedded in the surface of a rigid infinite baffle. The projector face is a semicircular piston, with the velocity being constant along the face. If R_s, θ_s are polar coordinates for a point on the face, with $x = R_s \cos(\theta_s)$, $y = R_s \sin(\theta_s)$, then a harmonic oscillation of the piston leads to a surface velocity described by

$$v_z = \begin{cases} v_0 \cos(\omega t), & 0 < R_s < a, \quad -\pi/2 < \theta_s < \pi/2 \\ 0 \text{ otherwise} \end{cases}$$

(a) Consider a field point $\bar{x}_0$ on the z-axis ($\psi = 0$). Starting with the general Rayleigh integral, derive an expression for the complex amplitude of the pressure at such a location. How does that result compare to the result for the case of a full circular piston? (b) Starting with the general Rayleigh integral, derive an expression for the complex amplitude of the pressure at location $\bar{x}_0$ in the farfield, $r \gg a$. The final result may be left in integral form. (c) In Drawing II, an infinite baffle is situated in the vertical plane. A vibrating piston is mounted on the baffle such that its center is at the free surface of the water. For the purpose of the analysis, the free surface of the water may be taken to be pressure-release for waves in the water and rigid for waves in the air. Let $P_b(r, \phi, \theta)$ denote the pressure derived in Part (b). Describe in terms of $P_b(r, \phi, \theta)$ the pressure in the water and in the air.

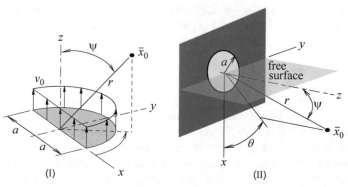

Exercise 8.8

Exercise 8.9 A circular piston in a rigid baffle executes a transverse vibration $v_0 \sin(\omega t)$ everywhere along its surface. The following facts are known: (1) The fluid is water. (2) The frequency is 50 kHz. (3) The first null in the farfield pressure occurs at $3°$ from the z-axis. (4) The pressure at 3 m along the z-axis from the plate is 14 kPa. Determine the radius a and velocity amplitude v_0 that correspond to the stated conditions.

Exercise 8.10 An ultrasonic piston transducer in a large rigid baffle transmits a signal into water. A scan of the axial field indicates that the largest pressure is 195 dB//1 μPa, and that farthest distance at which it occurs is 28 mm from the face of the piston. It also is observed that the farthest antinode occurs at 12 mm from the face. Determine the radius a of the piston, the frequency, and the amplitude of the piston's velocity. Also estimate the minimum radial distance at which farfield properties can be expected to be observed.

Exercise 8.11 A transducer is composed of multiple piezoeceramic elements that are driven in-phase at frequency ω, with the individual elements approximately filling a circle of radius a. This circle is surrounded by a rigid baffle, that is, sufficiently large to consider it to be infinite. Because this transducer is not exactly a circular piston the effective radius a_{eff} for acoustical properties might not be exactly the geometrical radius a. The value of a_{eff}, as well as the local speed of sound, can be determined from the axial distances of the farthest two minima of the on-axis pressure amplitude. Let these distances be z_A and $z_B < z_A$. (a) Derive formulas for a_{eff} and c corresponding to a known frequency ω and measured values of z_A and z_B. (b) Evaluate the results in Part (a) for the case of a 10 kHz signal with $z_A = 0.395$ m, and $z_B = 0.105$ m. (c) How many pressure antinodes will be observed on the symmetry axis when the parameters are those in Part (b)? (d) For the parameters in Part (b), what distance is a good estimate for the onset of farfield behavior? (e) What is the beamwidth corresponding to the parameters in Part (b)?

Exercise 8.12 An underwater baffled piston projector must have a maximum sound pressure level of 240 dB//1 μPa on-axis when the frequency is 8 kHz. The farthest distance at which this maximum occurs should be 3 meters. (a) What is the required radius of the piston? (b) At what distance is it reasonable to consider farfield spreading to begin? (c) How many nulls are there on-axis? (d) What is the sound pressure level at 30 m from the projector on-axis? (e) What is the beamwidth of this projector? (f) How many nulls occur in the directivity?

Exercise 8.13 An infinite plane is the boundary of a half space filled with an ideal fluid. The entire plane outside a circle of radius a executes an in-phase harmonic vibration normal to the plane. With z being the axial direction for a set of cylindrical coordinates, and $z = 0$ designated as the plane, the surface velocity is $v_z = v_0 \sin(\omega t)$ if $R > a$ and $v_z = 0$ if $R < a$. Derive expressions for the pressure along the z-axis and the pressure at an arbitrary location in the farfield. For the case where $ka = 20$, graph the axial distribution $|P|/(\rho_0 c v_0)$ as a function of z/a. Also, graph the farfield pressure $|P_{ff}|/(\rho_0 c v_0)$ as a function of distance R/a from the symmetry axis. The axial distance for this evaluation should be $z = R_0$.

Exercise 8.14 A projector in an infinite baffle consists of a ring whose inner radius is $a/2$ and whose outer radius is a. The complex normal velocity amplitude at frequency ω is v_0 over the surface of the ring, while the region inside the inner radius is motionless. Derive expressions for the farfield pressure and the pressure on-axis. Given that $ka = 25$, evaluate $(r/a) |p_{ff}/(\rho_0 c v_0)|$ as a function of the polar angle, and the axial pressure distribution $|p(\psi = 0)/(\rho_0 c v_0)|$ as a function of distance from the piston. Do nulls occur in either pattern?

Exercise 8.15 Example 8.2 analyzed the farfield radiation of a membrane transducer. The membrane's surface velocity is well approximated be a parabolic distribution, $V_s = v_1 \left(1 - R^2/a^2\right) \sin(\omega t)$ if $R < a$, $V_s = 0$ if $R \geq a$. Evaluate $|p/(\rho_0 c v_0)|$ as a function of distance along the axial center line from the plane of the baffle. Perform this evaluation for $ka = 3$ and 20.

Exercise 8.16 The sketch depicts a transducer that radiates sound into water. It consists of a piston that is flush mounted in an infinite baffle. The piston, whose mass is $M = 0.6\,\text{kg}$, is restrained by spring K. The displacement at the other end of the spring is controlled by an actuator that can exert whatever force is required to impose a specified harmonic motion $v_{\text{act}} \cos(\omega t)$. Design specifications are as follows: (1) The piston radius is 150 mm. (2) The operational frequency is 2 kHz. (3) The natural frequency of the mass-spring system in a vacuum equals the operational frequency. (4) The maximum pressure on-axis at the operational frequency is 200 kPa. Given these requirements, determine v_{act} and the force exerted by the fluid on the piston.

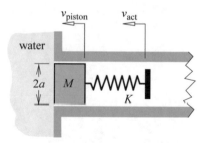

Exercise 8.16 and 8.17.

Exercise 8.17 The sketch depicts a piston transducer that is flush mounted in an infinite baffle. The piston, whose mass is M, is restrained by spring K. The displacement at the other end of the spring is controlled by an actuator that can exert whatever force is required to impose a specified harmonic motion $\text{Re}\left(V_{\text{act}} \exp(i\omega t)\right)$. The task is to determine radiation properties as a function of the frequency. Such an analysis is initiated by deriving a differential equation of motion for v_{piston} that accounts for the fluid loading acting on the piston. Solution of this equation for the specified actuator motion will yield $\left|V_{\text{piston}}/V_{\text{act}}\right|$ as a function of ω. From that result evaluate the time-averaged radiated power $P_{\text{av}}/\left(\rho_0 c |V_{\text{act}}|^2\right)$. Graph both quantities as functions of ω in the range from zero to 20% greater than the natural frequency of the mass-spring system in a vacuum. Carry out the analysis for the case where the fluid is water, then repeat the computation for the case where the fluid is air. The transducer's properties are $a = 150\,\text{mm}$, $M = 0.2\,\text{kg}$, and $K = 45\,(10^6)$ N/m.

Exercise 8.18 A piston, whose radius is 250 mm, is embedded in an infinite rigid baffle. The fluid is air. The piston was at rest until $t = 0$, at which instant it begins a parabolic pulse, in which the surface velocity is $v_s = 0.3\,(t/T - t^2/T^2)$ m/s if $0 < t < T$. Evaluate the waveform at the on-axis positions $z = 0$, $4a$, and $32a$ in the case where $T = 0.25a/c$. Compare each to the limiting form in Eq. (8.6.10). For each location, determine the time required for the signal to attain a steady state.

Exercise 8.19 The waveform plotted below is the normal velocity of a circular piston in an infinite baffle. It consists of a single cycle of a sawtooth. Evaluate and plot $p/(\rho_0 c v_0)$ as a function of the nondimensional retarded time $(c/a)(t - z/c)$ for on-axis locations. Perform the calculation for $z = a$ and $z = 4a$, with the pulse interval set at $T = 0.3a/c$ and $T = 3a/c$.

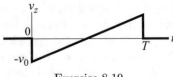

Exercise 8.19

Chapter 9
Modal Analysis of Waveguides

We return to situations where boundaries guide the direction in which sound may propagate. The difference from the one-dimensional waveguides considered in Chaps. 2 and 3 is that some feature will vary transversely to the propagation direction. It might be that the excitation varies in the transverse direction, which is analogous to the effect of a spherical source that is not radially symmetric. This situation commonly is encountered in turbine engines and ocean acoustics. Another cause for transverse variation of a propagating signal is a nonconstant spacing between the confining boundaries. Examples of this situation are musical horns and some speaker systems. Both features often are encountered in HVAC systems. Fully realistic models that account for all effects are difficult to analyze, so they are typically addressed with computer methods. However, if we make some idealizations, there are a number of configurations that are amenable to analysis. The studies that follow provide much understanding of basic phenomena, and all are descriptive of situations that one might encounter.

9.1 Propagation in a Horn

Unlike the one-dimensional waveguides considered previously, a *horn* has a cross section that is not constant along the axis of propagation. For example, conical horns, such as those commonly used by the coxswain in shell rowboat racing, have a cross-sectional radius that is proportional to the axial distance x. A configuration of particular interest is an exponential horn, in which the cross-sectional area depends exponentially on x. In principle, the shape of the cross section could change, but such configurations are rare. In any event, the actual shape of the cross section will be irrelevant to the analysis.

The idea of attaching a horn to a small source is to attain some features of an unadorned large source. One such feature is directivity. We know that a source whose

Electronic supplementary material The online version of this chapter (DOI 10.1007/978-3-319-56847-8_9) contains supplementary material, which is available to authorized users.

size is small relative to a wavelength will radiate almost omnidirectionally. The large area of a horn at its open end has the effect of increasing the size of the radiator. This aspect may be recognized by considering a conical horn. In Fig. 9.1a, a spherical cap in a baffle executes a purely radial vibration. A rigid cone is mounted on the perimeter of the cap, with the apex of the cone coincident with the center of the sphere. In Fig. 9.1b, the cone is positioned on a freely standing sphere of the same radius that executes the same radially symmetric vibration everywhere on its surface. The particle velocity in the signal radiated by the sphere in case (b) is purely radial. This is compatible with the rigidity of the cone, which requires that the particle velocity normal to its surface be zero. It follows that the field inside the cone is the same as what it would be if the cone were not present. (This assertion assumes that there is no reflection from the cone's open end.) Furthermore, what occurs outside the cone is irrelevant to the field inside the cone. This leads to recognition that the field in case (a) is a radially symmetric spherical wave within the cone. Thus, the presence of the conical horn in case (a) changes the effective size of the radiator from the diameter $2a_0$ of the cap to the diameter $2a_1$ of the opening. It also makes it unnecessary to mount the cap in the baffle.

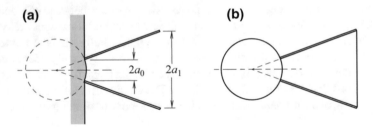

Fig. 9.1 A conical horn mounted on a spherical transducer. In case **a**, the transducer is a cap in a baffle, whereas in case **b**, it is in free space

Another reason to mount a horn on a source pertains to applications in which it is desired to have a constant output over a broad frequency range. Without the horn, attainment of this objective would require either a small transducer whose velocity amplitude increases inversely to the frequency, or else a very large transducer. Neither feature is desirable for reasons that will be disclosed when we see why a horn avoids these issues.

9.1.1   The Webster Horn Equation

The model to be constructed is based on a fundamental assumption: The pressure perturbation depends only on the axial position x and t. A corollary that follows from Euler's equation is that the particle velocity is solely in the axial direction, with v_x also being a function of x and t. Clearly, the latter cannot be exactly correct, because if it were there would be no signal in the region into which the horn expands.

Intuitively, it is reasonable to expect that this assumption is valid if the cross-sectional area is a function $\mathcal{A}(x)$ that varies slowly relative to the scale of a wavelength. The correctness of this assumption is an item we will explore.

The field equation is obtained by enforcing conservation of mass and the momentum-impulse principle. These considerations parallel those that led to the one-dimensional wave equation in Chap. 2. However, rather than using finite-sized control volumes and finite increments in time, here we shall take a less rigorous, but more expedient, approach that considers the rates of change in the state of differential control volumes. We begin with conservation of mass for the stationary control volume in Fig. 9.2.

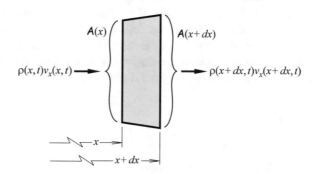

Fig. 9.2 A differential stationary control volume used to describe conservation of mass in a horn

The wall of the horn forms the sides of this region. It is impermeable, so there is no mass flow across the sides of the control volume. Thus, the net rate at which mass flows into the control volume over the two cross sections must equal the rate at which the mass of this control volume increases. For the cross section at x, positive v_x transports mass into the control volume, and ρv_x evaluated at x is the mass flow rate per unit area. Positive v_x on the face at $x + dx$ transports mass out of the control volume, and ρv_x evaluated at $x + dx$ is the rate per unit area there. According to the central limit theorem, $\rho(x, t)\mathcal{A}(x)dx$ is the mass of the control volume at time t. Thus, it must be that

$$\rho(x, t)v_x(x, t)\mathcal{A}(x) - \rho(x + dx, t)v_x(x + dx, t)\mathcal{A}(x + dx) = \frac{\partial}{\partial t}[\rho(x, t)\mathcal{A}(x)dx] \tag{9.1.1}$$

Division by dx converts the left side of the preceding to a partial derivative of $\rho v_x \mathcal{A}$, so the expression becomes

$$\mathcal{A}\frac{\partial \rho}{\partial t} + \frac{\partial}{\partial x}(\rho v_x \mathcal{A}) = 0 \tag{9.1.2}$$

The second term is nonlinear because it is the product of two state variables. A linearized analysis restricts the particle velocity to being a small quantity (relative to the speed of sound), and the density is restricted to being little changed from the ambient value ρ_0. Correspondingly, we linearize the continuity equation by replacing

ρv_x with $\rho_0 v_x$. For a homogeneous medium, ρ_0 is the same at all locations, so the linearized continuity equation for a horn is

$$A \frac{\partial \rho}{\partial t} + \rho_0 \frac{\partial}{\partial x} (A v_x) = 0 \qquad (9.1.3)$$

If the cross section is constant, this relation reduces to the continuity equation for truly one-dimensional plane waves.

The momentum equation is derived by considering a control volume that moves with the particles. Because it tracks a specific set of particles, its selection simplifies application of the momentum-impulse principle for a system of particles. The control volume in Fig. 9.3a consists of the fluid contained between cross sections at x and $x + dx$ at time t. The forces depicted there consist of the pressure resultant acting on each cross section and the distributed force exerted by the wall of the horn in reaction to the pressure the fluid exerts in the wall. It would be awkward to account for the axial component of the wall force.

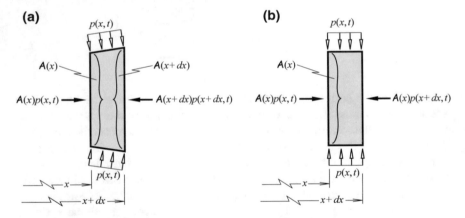

Fig. 9.3 Control volume used to form the derivative form of the momentum-impulse principle. The domain moves in unison with the particles contained between two adjacent cross sections. **a** The domain consists of all particles contained between the adjacent cross sections. **b** The domain is a cylinder whose cross section is constant

The control volume in Fig. 9.3b is more amenable to the analysis. It is a general cylinder whose cross section is $A(x)$. This shape exploits the fact that shear stresses are negligible for an ideal fluid. Thus, the force system exerted on the sides of this cylindrical domain is the stress resultant of the pressure exerted by the adjacent fluid. This resultant acts perpendicularly to the cylindrical surface, so it has no component in the axial direction. (The transverse resultant actually must be zero for the assumed motion, because there is no acceleration in that direction.)

There is no evidence of a variable cross section in the second force diagram, as in Fig. 9.3b. Thus, the time-domain Euler equation, which is obtained from conservation of momentum, may be applied to the motion in the x direction

$$\rho \frac{d}{dx} v(x, t) = -\frac{\partial p}{\partial x} \tag{9.1.4}$$

A linearized equation of motion results from dropping any terms that are products of small variables. Thus, the acceleration is approximated as a partial time derivative, and the current density is approximated by its ambient value. Doing so reduces the equation of motion for the control volume to

$$\boxed{\rho_0 \frac{\partial v_x}{\partial t} = -\frac{\partial p}{\partial x}} \tag{9.1.5}$$

This is the same as the x component of Euler's equation.

Equations (9.1.3) and (9.1.5) constitute two equations for three unknowns: p, ρ, and v_x. The third relation is the equation of state, whose linearized form is $p = c^2 (\rho - \rho_0)$. We use this equation to eliminate ρ from the continuity equation. To eliminate v_x, we multiply Eq. (9.1.5) by $\mathcal{A}$ and then differentiate it with respect to x. Because $\mathcal{A}$ is independent of t, the result may be written as

$$\rho_0 \frac{\partial^2}{\partial x \partial t} (\mathcal{A} v_x) = -\frac{\partial}{\partial x} \left(\mathcal{A} \frac{\partial p}{\partial x} \right) \tag{9.1.6}$$

Differentiation of Eq. (9.1.3) with respect to t yields the same v_x term,

$$\frac{\mathcal{A}}{c^2} \frac{\partial^2 p}{\partial t^2} + \rho_0 \frac{\partial^2}{\partial x \partial t} (\mathcal{A} v_x) = 0 \tag{9.1.7}$$

Elimination of the mixed derivative term in these two equations yields the *Webster horn equation*

$$\boxed{\frac{1}{\mathcal{A}} \frac{\partial}{\partial x} \left(\mathcal{A} \frac{\partial p}{\partial x} \right) - \frac{1}{c^2} \frac{\partial^2 p}{\partial t^2} = 0} \tag{9.1.8}$$

This partial differential equation has variable coefficients except in the case of a uniform bar. In the frequency domain, it becomes an ordinary differential equation,

$$\boxed{p = \mathrm{Re}\left(P(x) e^{i\omega t} \right) \implies \frac{d^2 P}{dx^2} + \frac{1}{\mathcal{A}} \left(\frac{d\mathcal{A}}{dx} \right) \left(\frac{dP}{dx} \right) + k^2 P = 0} \tag{9.1.9}$$

This too is referred to as the Webster horn equation. There are horn shapes $\mathcal{A}(x)$ for which it is possible to obtain an analytical solution in the frequency domain. We also will develop an approximate solution for arbitrary $\mathcal{A}(x)$. However, before we pursue these tasks, let us return to the conical horn because doing so will allow us to identify when the Webster horn equation may be used.

Figure 9.4 depicts a conical waveguide whose semi-vertex angle is β. An arbitrary point is located by its cylindrical coordinates (x, R) and its spherical coordinates

Fig. 9.4 Cylindrical and
spherical coordinates of a
field point within a conical
horn

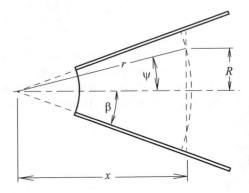

(r, ψ), with both coordinate systems defined relative to the apex of the cone. The
spherical cap at the left executes a uniform radial vibration at frequency ω.

From the discussion of Fig. 9.1, we know that the field is a radially symmetric
wave, so an outgoing wave is

$$p_{\text{true}} = \text{Re}\left(\frac{D}{r}e^{-ikr}e^{i\omega\tau}\right) \tag{9.1.10}$$

An analytical solution of the Webster horn equation for this case is available. Let us
compare it to the spherical wave solution in order to grasp the effect of the approximations embedded in that equation. The cross-sectional radius of the waveguide is
$x \tan \beta$, which leads to $A(x) = \pi (\tan \beta)^2 x^2$. The Webster horn equation in this case
is

$$\frac{\partial^2 p}{\partial x^2} + \frac{2}{x}\frac{\partial p}{\partial x} - \frac{\partial^2 p}{\partial t^2} = 0 \tag{9.1.11}$$

If, instead of x, the independent variable were r, this would be the wave equation for
a radially symmetric spherical wave. Thus, if we ignore reflections from the far end
of the horn, the solution of the Webster horn equation for an outgoing wave is

$$p_{\text{horn}} = \text{Re}\left(\frac{D'}{x}e^{-ikx}e^{i\omega t}\right) \tag{9.1.12}$$

The radial distance to a point on a specific cross section is $r = x/\cos\psi$. Setting
$D' = D$ makes this solution and the spherical coordinate solution, Eq. (9.1.10), the
same on the axis, $\psi = 0$.

With increasing distance from the axis, the solutions diverge. The relative error in
the amplitude at a specific ψ is obtained by replacing r in Eq. (9.1.10) with $x/\cos\psi$,
which leads to

$$\varepsilon_{\text{amplitude}} \equiv \frac{|P_{\text{horn}}| - |P_{\text{true}}|}{|P_{\text{true}}|} = \cos\psi - 1 \tag{9.1.13}$$

Another aspect is the phase error, which must be small if the waveform is to be positioned correctly. This quantity is

$$\varepsilon_{\text{phase}} = kx - kr = kx \left(1 - \frac{1}{\cos \psi} \right) \tag{9.1.14}$$

Both errors are underestimates that increase monotonically with increasing ψ. The largest errors at fixed x occur at the wall, where $\psi = \beta$. A trigonometric identity is $1 - \cos \beta \equiv (\sin \beta) \tan(\beta/2)$. The resulting error estimates are

$$\varepsilon_{\text{amplitude}} = -\sin \beta \tan \frac{\beta}{2}, \quad \varepsilon_{\text{phase}} = -kx \tan \beta \tan \frac{\beta}{2} \tag{9.1.15}$$

The amplitude error will be small if $\beta \ll 1$, that is, if the cross section expands slowly. However, even if the amplitude error is acceptable, the fact that $\varepsilon_{\text{phase}} = (kx/\cos \beta) \varepsilon_{\text{amplitude}}$ means that the amplitude and phase error will have comparable magnitude only if kx is not large compared to unity. In other words, β should be much less than 1 rad and x should not be much larger than a wavelength. Let us assess these criteria for the audible range of a horn in air. Consider $\beta = 10°$. The amplitude error at $\psi = \beta$ is 1.5%. At the low end of the audible spectrum, 20 Hz, the phase error will be less than 10% if $x < 17.5$ m. In contrast, at 10 kHz, the phase error will be less than 10% if $x < 35$ mm. The low-frequency case suggests that an English horn or euphonium might be well described by the Webster horn equation, but this model might not be adequate for a flute or piccolo.

These criteria may be generalized to any shape horn. Let $R_{\text{wall}}(x)$ be the transverse distance to the wall at a specified axial position. For a cone, we have $dR_{\text{wall}}/dx = \tan \beta$, so small β corresponds to small values of dR_{wall}/dx and $\tan(\beta/2) \approx (1/2) dR_{\text{wall}}/dx$. Thus, the amplitude and phase errors will not be excessive if $dR_{\text{wall}}/dx \ll 1$ and kx is not much larger than a wavelength.

9.1.2 Exponential Horn

Let us examine whether it is possible for a simple plane wave to exist within a horn. Let κ be the constant wavenumber. Then, substitution of

$$P = Be^{-i\kappa x} \tag{9.1.16}$$

into the frequency-domain Webster horn equation requires that

$$\left(k^2 - \kappa^2 \right) \mathcal{A} - i\kappa \frac{d\mathcal{A}}{dx} = 0 \tag{9.1.17}$$

The wavenumber of a simple plane wave is a real quantity. Because $\mathcal{A}$ is real, it is evident that in no circumstance, other than constant $\mathcal{A}$, is it possible for a simple wave to exist in a horn.

In the special case where $d\mathcal{A}/dx$ is proportional to $\mathcal{A}$, that is, $d\mathcal{A}/dx = b\mathcal{A}$, the area $\mathcal{A}$ factors out of Eq. (9.1.17). What remains is a complex quadratic equation for κ. The cross-sectional variation fitting this description constitutes an *exponential horn*

$$\boxed{\mathcal{A}(x) = A_0 e^{bx}} \tag{9.1.18}$$

The growth factor b may be negative, corresponding to a cross section whose radius decreases with increasing x. In most cases, the cross section is either rectangular or circular, with the shape held constant at all x, but these details are irrelevant to the analysis.

When the cross-sectional area varies exponentially, Eq. (9.1.17) leads to a quadratic equation for the wavenumber, specifically

$$\kappa^2 + ib\kappa - k^2 = 0 \tag{9.1.19}$$

If $k > |b|/2$, there are two roots complex roots. The one whose real part is positive corresponds to propagation in the positive x direction. It is

$$\boxed{\kappa = \left(k^2 - \frac{b^2}{4}\right)^{1/2} - i\frac{b}{2} \text{ if } k > \frac{|b|}{2}} \tag{9.1.20}$$

The other root corresponds to waves that propagate in the direction of decreasing x. It is

$$\kappa' = -\left(k^2 - \frac{b^2}{4}\right)^{1/2} - i\frac{b}{2} \text{ if } k > \frac{|b|}{2} \tag{9.1.21}$$

Both waves will exist if there are reflections at one end. The imaginary part of both κ and κ' correspond to waves that decrease with increasing x if b is positive. As was noted above, a horn whose cross section decreases with increasing x may be represented by a negative b. The alternative is to reverse the x-axis, so that it is oriented in the direction of increasing cross section. In either view, the amplitude decreases for a wave that propagates in the sense of increasing area, and it increases for a wave that propagates in the sense of decreasing area. The general solution when $k > |b|/2$ is

$$\boxed{P = e^{-bx/2}\left(B_1 e^{-i\left(k^2 - b^2/4\right)x} + B_2 e^{i\left(k^2 - b^2/4\right)x}\right) \text{ if } k > |b|/2} \tag{9.1.22}$$

It is possible that k is smaller than $b/2$, in which case the square root yields an imaginary value. Both wave numbers are negative imaginary values

$$\kappa = -i\left[\pm\left(\frac{b^2}{4} - k^2\right)^{1/2} + \frac{b}{2}\right] \text{ if } k < \frac{|b|}{2} \tag{9.1.23}$$

In this situation the waves that travel back and forth are evanescent waves. Neither solution corresponds to a wave that propagates. Rather, both are standing waves in which $|P|$ decreases on the sense of increasing values of bx. The evanescent field may be written as

$$P = e^{-bx/2}\left(B_1 e^{+(b^2/4-k^2)x} + B_2 e^{-(b^2/4-k^2 x)}\right) \tag{9.1.24}$$

The case where $k = |b|/2$ marks a transition from a propagating wave to evanescence. This condition marks the *cutoff frequency*,

$$\boxed{\omega_{\text{cutoff}} = bc/2} \tag{9.1.25}$$

Below the cutoff frequency, a waveguide is seldom of interest because the time-averaged power transported through the horn is essentially zero. (Power flow will be examined shortly.) The parameter κ is the *complex wavenumber*. Its real part gives the wavelength. It also defines the phase speed, according to

$$c_{\text{phase}} \equiv \frac{\omega}{|\text{Re}(\kappa)|} = \frac{c}{\left(1 - b^2/4k^2\right)^{1/2}} \tag{9.1.26}$$

Because the phase speed is a function of frequency, a signal consisting of more than one frequency will spread out, or *disperse*, as it propagates. For this reason, a relation like Eq. (9.1.26) is said to be a *dispersion equation*. However, the nature of other systems is such that it might not be possible to obtain a closed form solution for Re (κ) as a function of frequency. Thus, any equation, such as Eq. (9.1.20), that relates the complex wavenumber to the frequency will be said to be a dispersion equation.

In terms of dimensionless quantities, c_{phase}/c for an exponential horn depends only on k/b, which is the relation depicted in Fig. 9.5. At high frequencies, the phase speed is close to the plane wave speed. Decreasing the frequency causes c_{phase} to increase, until it becomes infinite at the cutoff frequency. Below $k = b/2$, the phase speed is undefined because κ is imaginary.

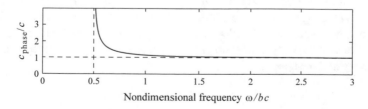

Fig. 9.5 Dispersion curve for an exponential horn

The cutoff effect results from the fact that $|\kappa| = k$ for any $|b| < 2k$. Consider a variety of waveguides at a specific frequency. As b increases from a small value with k fixed, it is necessary that Im (κ) increase in order to match b. This requires that Re (κ) decrease. At the cutoff condition, $b = 2k$, the imaginary part equals k, so Re $(\kappa) = 0$.

The simpler alternative when the larger end is excited is to let b be negative, rather than reversing the sense of x. Doing so allows us to set $x = 0$ as the driven end in any case. In the absence of reflections at the undriven end, setting $b < 0$ leads to $|P(x = L)/P(x = 0)| = e^{-bL} > 1$. The signal at the exit would be substantially enhanced if $-bL \gg 1$. Such a configuration was used for antique hearing aids, as well as in the apparatus that recorded sound on a phonograph platter. A horn amplified the signal that actuated the encoding stylus.

The usual situation is that the particle velocity at the end $x = 0$ is specified. Let us assume that the far end is terminated in a manner that prevents reflection. Accordingly, we set $B_2 = 0$ in Eq. (9.1.22). A general expression for the velocity may be obtained by substituting the pressure ansatz, Eq. (9.1.16), into Euler's equation. Doing so gives

$$V_x = \frac{\kappa}{\omega \rho_0} B e^{-i\kappa x} \equiv \frac{\kappa}{k} \frac{P}{\rho_0 c} \qquad (9.1.27)$$

The expressions for $P(x)$ and $V(x)$ are general, but the behavior is fundamentally different above and below the cutoff frequency. Above cutoff, κ is complex. This means that there is a frequency-dependent phase difference between the velocity and the pressure. Below the cutoff frequency, κ is imaginary. This corresponds to frequency-independent 90° phase difference between the particle velocity and pressure.

Satisfying the boundary condition will set the coefficient B. We equate V_x at $x = 0$ from the preceding to V_0. The value of B is set by matching the particle velocity according to this relation to the complex amplitude V_0 of the velocity at that end. The result is

$$P(x) = \rho_0 c V_0 \frac{k}{\kappa} e^{-i\kappa x} \qquad (9.1.28)$$

The development began with two reasons horns are used presently. If the horn is sufficiently long, the open aperture will be much larger than the size of the opening where the horn is excited, thereby enhancing directivity of the signal that radiates from that end. However, the signal at the aperture is much weaker due to spreading loss, so the advantages of a horn regarding radiated power are not evident. An analysis of power flow begins with the observation that there is no variation over a cross section. Consequently, the power flowing across any cross section is the product of the intensity and the area

$$\mathcal{P} = \frac{1}{4} A \left(P e^{i\omega t} + P^* e^{-i\omega t} \right) \left(V_x e^{i\omega t} + \text{c.c.} \right)$$

$$= \frac{1}{2} A \, \text{Re} \left(P V_x e^{2i\omega t} + P^* V_x \right) \tag{9.1.29}$$

This expression is applicable to any system that fits the assumptions embedded in the Webster horn equation. In the special case of an exponential horn, substitution of $P(x)$ and $A(x)$ into this expression gives

$$\mathcal{P} = \frac{1}{2} A_0 \rho_0 c \, \text{Re} \left[(V_0)^2 \, \frac{k}{\kappa} e^{2i\omega t} e^{(-2i\kappa + b)x} + |V_0|^2 \, \frac{k}{\kappa} e^{-i(\kappa \mid \kappa^* - b)x} \right] \tag{9.1.30}$$

The complex wavenumber is given in Eq. (9.1.20). If $\omega < \omega_{\text{cut}}$, then κ is negative imaginary, which leads to

$$\mathcal{P} = \frac{1}{2} A_0 \rho_0 c \, \text{Re} \left[(V_0)^2 \, \frac{k}{\kappa} e^{2i\omega t} \right] e^{-(b^2 - 4k^2)^{1/2} x} \quad \text{if } k < b/2 \tag{9.1.31}$$

Thus, for frequencies below cutoff, the time-averaged power flow is zero, and the oscillatory part of the power flow decays exponentially with increasing distance. Above the cutoff frequency, κ is complex. A few manipulations show that the power flow in this case is

$$\mathcal{P} = \frac{1}{2} A_0 \rho_0 c \, \text{Re} \left[(V_0)^2 \, \frac{k}{\kappa} e^{2i\omega(t - x/c_{\text{phase}})} + |V_0|^2 \, \frac{k}{\kappa} \right] \quad \text{if } k > b/2 \tag{9.1.32}$$

This relation indicates that above the cutoff frequency, the time-averaged power flow is independent of the cross section's location. This property is predicted by the general theorem regarding power, which tells us that the average power that flows through the waveguide must be the same at all cross sections because energy does not flow out of the walls. The fluctuating part of the power flow propagates downstream as a second harmonic at the phase speed of the pressure wave.

The invariance of the time-averaged power means that we only need know the amount that is input at $x = 0$. This is true regardless of the shape of the horn. For this reason, a quantity of interest for any type of horn is the ratio of P at $x = 0$ to the particle velocity V_0 at that end. This ratio is the *throat impedance*,

$$\boxed{Z_{\text{throat}} \equiv \frac{P|_{x=0}}{V_0}} \tag{9.1.33}$$

If we know this quantity, we may evaluate the time-averaged power input to the horn by the transducer according Eq. (9.1.29), which gives

$$\mathcal{P}|_{z=0} = \frac{1}{2} A_0 \, \text{Re} \left(Z_{\text{rad}} (V_0)^2 e^{2i\omega t} \right) + \frac{1}{2} A_0 \, \text{Re} \left(Z_{\text{rad}} \right) |Z|^2 \tag{9.1.34}$$

For the case of an exponential horn, Eq. (9.1.27) gives

$$
Z_{\text{throat}} = \rho_0 c \frac{k}{\kappa} =
\begin{cases}
\rho_0 c \left[\left(1^2 - \dfrac{b^2}{4k^2} \right)^{1/2} + i\dfrac{b}{2} \right], & \omega > \omega_{\text{cut}} \\[4mm]
\rho_0 c i \left[\left(\dfrac{b^2}{4} - k^2 \right)^{1/2} + \dfrac{b}{2} \right], & \omega < \omega_{\text{cut}}
\end{cases}
\tag{9.1.35}
$$

To gain insight to the design issues pertaining to a horn, let us compare an exponential horn to a hemispherical source in an infinite rigid baffle in the case of a radially symmetric vibration. To make the comparison, we assign the same active surface area to both devices, so $\mathcal{A}_0 = 2\pi a^2$. The time-averaged power radiated by the hemisphere is half the value in Eq. (6.3.16),

$$
(\mathcal{P}_{\text{av}})_{\text{hemi}} = \pi \rho_0 c a^2 \left| V_0' \right|^2 \frac{(ka)^2}{(ka)^2 + 1}
\tag{9.1.36}
$$

Both devices should have a relatively flat response. That is, we wish that $\mathcal{P}_{\text{av}}$ be nearly independent of frequency over as wide a band as possible. The power radiated by an exponential horn is described by Eq. (9.1.32). According to it, if k/κ is close to one, then $\mathcal{P}_{\text{av}}$ is independent of k. The value of κ approaches k as k increases beyond cutoff. Thus, we require that the frequency be substantially greater than the cutoff value, which leads to

$$
\omega \gg \omega_{\text{cutoff}} \implies (\mathcal{P}_{\text{av}})_{\text{horn}} = 2\pi \rho_0 c a^2 \left| V_0 \right|^2
\tag{9.1.37}
$$

Reference to Fig. 9.5 shows that this criterion is met if ω is as little as twice the cutoff value.

It is not immediately apparent whether the equivalent hemisphere should be large or small relative to a wavelength, so we shall consider both possibilities. Equation (9.1.36) gives $\mathcal{P}_{\text{av}} \approx \pi \rho_0 c a^2 \left| V_0' \right|^2 (ka)^2$ if $ka \ll 1$. Thus, producing a constant power output over a frequency band with a small hemisphere requires that $\left| V_0' \right| ka$ be constant. The amplitude of the surface displacement is $\left| V_0' \right| /\omega$, so sustaining a constant power output with a small diameter requires that the displacement be inversely proportional to the square of the frequency. It is quite challenging to create a small transducer that is capable of increasingly large displacement as the frequency decreases. Hence, let us consider the alternative in which the hemisphere is large relative to a wavelength. If $ka \gg 1$ at the lowest frequency in the band, then the power output would be $\mathcal{P}_{\text{av}} \approx \pi \rho_0 c a^2 \left| V_0' \right|^2$. This is a similar dependence to that of a horn, so an invariant $\left| V_0 \right|$ will yield a constant power output. The difficulty with this design is that it entails a large transducer. Suppose the minimum frequency of interest is 100 Hz. The criterion $ka \gg 1$ in air leads to $a \gg 3.4$ m! Even if we raise the low end tenfold to 1 kHz, the transducer would be enormous. Large transducers are expensive to manufacture, but there is another fundamental difficulty. The surface acceleration is proportional to $\omega |V_0|$, so the upper end of the frequency range will be required to undergo relatively high accelerations. Imparting a large acceleration to a large device requires very strong mechanical drivers to induce the vibration, and it also introduces large stresses. In contrast, the cutoff frequency of an exponential horn only sets the exponential constant b. We are free to select the horn's length and throat area to fit other specifications, such as a desired sound pressure level at the opening.

The analysis of an exponential horn has been based on the assumption that a backward propagating wave is not present. This is only true if the reflection coefficient at that end is zero. In most cases, the undriven end of the horn is open. The end correction for the open end of a constant cross-sectional waveguide, which approximated the impedance at an open end by the radiation impedance of a piston, may be used for a horn. The impedance at that end is given by Eq. (3.5.53) as $\rho_0 c \chi (ka)$. This value is very close to the fluid's characteristic impedance, $\rho_0 c$, if $ka > 3$, which corresponds to the radius of the opening exceeding half the acoustic wavelength. In this condition, the reflection coefficient is close to zero. The alternative case in which $ka < 3$ requires inclusion of backward waves resulting from reflection at the open end. (This observation may be applied to waveguides whose cross section is not circular by taking the mean radius to be $[A(L)/\pi]^{1/2}$.) The next example describes how a nonzero reflection coefficient at the opening affects the field within the horn.

EXAMPLE 9.1 A horn may be used to obtain a gradual transition between constant cross-sectional waveguides having different sizes. This transitional segment typically is conical to join circular cross sections or a truncated pyramid for square ones. However, we will consider an exponential section. It is desired to determine the relationship between the pressure and particle velocity at the ends $x = 0$ and $x = L$. This property is a key part of the algorithm developed in Sect. 3.4.3 to analyze waveguide networks. The requisite quantity is the two-port mobility matrix $[D]$, which was defined such that

$$\begin{Bmatrix} V_x (x = 0) \\ V_x (x = L) \end{Bmatrix} = [D] \begin{Bmatrix} P (x = 0) \\ P (x = L) \end{Bmatrix}$$

Derive an expression for $[D]$ in terms of the values of k, L, and b.

Significance

The analysis will explore how reflection phenomena are affected by a nonuniform cross section. It has practical application for waveguide networks.

Solution

We cannot assume that waves propagate in one direction, because that property only occurs if the open end is large and exposed to an open space. Thus, we must retain both waves in Eq. (9.1.22). Manipulations are simpler if we use symbols to represent the characteristic exponents, so we set

$$\kappa_1 = \left(k^2 - \frac{b^2}{4}\right)^{1/2} - \frac{b}{2}i, \quad \kappa_2 = \left(k^2 - \frac{b^2}{4}\right)^{1/2} + \frac{b}{2}i$$

The general solution reduces to

$$P = B_1 e^{-i\kappa_1 x} + B_2 e^{i\kappa_2 x}$$

Note that this form is valid for all k values. The particle velocity is obtained from Euler's equation

$$\rho_0 c V_x = \frac{\kappa_1}{k} B_1 e^{-i\kappa_1 x} - \frac{\kappa_2}{k} B_2 e^{i\kappa_2 x}$$

The pressure and velocity at the ends resulting from these expressions are

$$B_1 + B_2 = P_0$$
$$\frac{\kappa_1}{k} B_1 - \frac{\kappa_2}{k} B_2 = \rho_0 c V_0$$
$$B_1 e^{-i\kappa_1 L} + B_2 e^{i\kappa_2 L} = P_L$$
$$\frac{\kappa_1}{k} B_1 e^{-i\kappa_1 L} - \frac{\kappa_2}{k} B_2 e^{i\kappa_2 L} = \rho_0 c V_L$$

The process of eliminating B_1 and B_2 is facilitated by writing these equations in matrix form

$$\left\{ \begin{matrix} P_0 \\ P_L \end{matrix} \right\} = [T_p] \left\{ \begin{matrix} B_1 \\ B_2 \end{matrix} \right\}, \quad \left\{ \begin{matrix} \rho_0 c V_0 \\ \rho_0 c V_L \end{matrix} \right\} = [T_v] \left\{ \begin{matrix} B_1 \\ B_2 \end{matrix} \right\}$$

where the transfer matrices are

$$[T_p] = \begin{bmatrix} 1 & 1 \\ e^{-i\kappa_1 L} & e^{i\kappa_2 L} \end{bmatrix}$$
$$[T_v] = \begin{bmatrix} \kappa_1/k & -\kappa_2/k \\ (\kappa_1/k) e^{-i\kappa_1 L} & -(\kappa_2/k) e^{i\kappa_2 L} \end{bmatrix}$$

The pressure equations are solved for the wave amplitudes, and the result is substituted into the velocity equations. These operations yield

$$\left\{ \begin{matrix} B_1 \\ B_2 \end{matrix} \right\} = [T_p]^{-1} \left\{ \begin{matrix} P_0 \\ P_L \end{matrix} \right\}$$
$$\left\{ \begin{matrix} \rho_0 c V_0 \\ \rho_0 c V_L \end{matrix} \right\} = [T_v][T_p]^{-1} \left\{ \begin{matrix} P_0 \\ P_L \end{matrix} \right\}$$

Hence, the two-port impedance matrix is

$$[D] = \frac{1}{\rho_0 c} [T_v][T_p]^{-1}$$

The network algorithm is computational, so it would be adequate in most situations to use numerical values of $[T_p]$ and $[T_v]$ corresponding to specified values of k, b, and L. However, an algebraic evaluation will enable us to examine the effect of the backward-traveling wave. The inverse of $[T_p]$ is

$$[T_p]^{-1} = \frac{1}{e^{i\kappa_2 L} - e^{-i\kappa_1 L}} \begin{bmatrix} e^{i\kappa_2 L} & -1 \\ -e^{-i\kappa_1 L} & 1 \end{bmatrix}$$

In turn, this leads to

$$[D] = \frac{1}{\rho_0 ck \left(e^{i\kappa_2 L} - e^{-i\kappa_1 L} \right)} \begin{bmatrix} \left(\kappa_1 e^{i\kappa_2 L} + \kappa_2 e^{-i\kappa_1 L} \right) & -(\kappa_1 + \kappa_2) \\ (\kappa_1 + \kappa_2) e^{-i\kappa_1 L} e^{i\kappa_2 L} & -\left(\kappa_1 e^{-i\kappa_1 L} + \kappa_2 e^{i\kappa_2 L} \right) \end{bmatrix}$$

This expression is general. To specialize it to the case where $k > b/2$, we define

$$\kappa_1 = K - \frac{b}{2}i, \quad \kappa_2 = K + \frac{b}{2}i, \quad K = \left(k^2 - \frac{b^2}{4} \right)^{1/2}$$

Application of Euler's identity then reduces the transfer matrix to

$$[D] = \frac{1}{\rho_0 ck \, i \sin (KL)}$$
$$\begin{bmatrix} [K \cos (KL) + (b/2) \sin (KL)] & -2K e^{bL/2} \\ 2K e^{-bL/2} & [-K \cos (KL) + (b/2) \sin (KL)] \end{bmatrix}$$

A constant cross section corresponds to $b = 0$, which gives $K = k$. In that case, the above expression is identical to Eq. (3.4.14), which is the two-port mobility for a waveguide whose cross section is uniform. An interesting possibility is that it $[D]$ is singular if KL is a multiple of π. This corresponds to a resonance, like that encountered in an impedance tube.

The acoustic mobility matrix has other uses in addition to the stated purpose of analyzing waveguide networks. It describes how the state variables at the ends must be related for any set of end conditions. In the case of an isolated exponential horn, either the complex pressure amplitude or the axial velocity amplitude will be known at the end where the signal is generated. At the far end, a rigid termination requires that V be zero. Another possibility is that the far end is open. In that case, we might set $P = 0$, or it might be appropriate to use the open end impedance described by Eq. (3.5.53). Another alternative is that the far end is terminated by a material whose impedance is Z_L. In any case, one state variable is unknown at each end. The equation $\{V\} = [D]\{P\}$ gives two equations for two pressures and two particle velocities. We may solve those equations for the unknown variables in terms of the excitation amplitude, P_0 or V_0 at $x = 0$ and $P(x = L)$. Evaluation of such solutions over a frequency band gives a complex frequency response. Depending on the impedance at the far end, there might be frequencies at which the pressure is greatly enhanced. These are cavity resonances, which will be studied from a general framework in the next chapter.

9.1.3 Group Velocity

The term "dispersive" refers to waves whose waveform changes shape with increasing propagation distance. In such a wave, the phase speed of a sinusoidal wave will depend on the frequency. The waves at the higher phase speed advance farther than the slower waves, so the arrival times at a specific location differ. The fact that a wave is dispersive has another important effect, which is that the speed at which it transports energy also depends on frequency. The phase speed is merely the speed at which the fact that there is a disturbance is passed to downstream locations. For example, a specific value of the pressure at a certain phase passes from point to point at the phase speed. A simple plane wave is nondispersive, so all features of such a wave, including its energy content, propagate at the speed of sound. The situation is different if a wave is dispersive. The *group velocity* c_g is the speed at which energy is transported. (The group velocity is a scalar, so it would be more appropriate to say that it is the group speed, but that term is not used.)

The group velocity is important for a variety of waveguide systems, so we shall begin by considering it in general terms. The analysis will lead to a general formula by which the group velocity may be extracted from a dispersion equation. The analysis will close by applying that formula to an exponential horn. As implied by the adjective "group", identification of c_g requires that we consider waves at different frequencies. These frequencies will be taken to be very close, so the associated wave numbers also will be very close. Close frequencies lead to beating. The packet of waves in one beat contains a specific amount of energy that can be tracked.

Two waves of equal amplitude at frequencies ω_1 and ω_2 propagate in the x direction. We assume that the dispersion equation is known; the function describing the frequency dependence of the wavenumber is $\kappa(\omega)$. Our only interest is how the waves propagate, so it is irrelevant if the field properties are such that the amplitude depends on position transverse to x. Hence, the signal we shall consider is

$$p(x.t) = \text{Re}\left\{ B\left[e^{-i\kappa(\omega_1)x + i\omega_1 t} + e^{-i\kappa(\omega_2)x + i\omega_2 t} \right] \right\} \qquad (9.1.38)$$

The location at which the waves originated also is irrelevant to this investigation, so B may be considered to be real.

Because there is little difference between the frequencies of the individual harmonics, a waveform at any location exhibits a beating pattern as a function of t at any instant. Furthermore, because $\kappa(\omega)$ is taken to be an analytical function, a small difference in frequency leads to wavenumbers that are close. Therefore, spatial profiles also show a beating pattern. The upper graph in Fig. 9.6 shows the spatial profiles at two adjacent instants t_1 and $t_1 + \Delta t$, while the lower graph are waveforms at two adjacent positions, x_1 and x_1 and $x_1 + \Delta x$. The dotted lines are the envelope function for each beat pattern. In a time interval Δt, features of a profile, such as maxima, minima, and zeros, advance at the phase speed. Thus, these features in the upper graph shift to the right by $c_{\text{phase}} \Delta t$. The group velocity is the rate at which the envelope of the beat moves, so the envelope in the upper graph shifts to the right

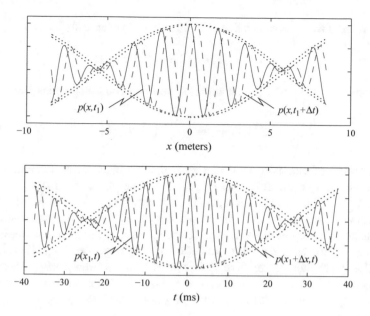

$p(x, t_1)$ $p(x, t_1 + \Delta t)$

x (meters)

$p(x_1, t)$ $p(x_1 + \Delta x, t)$

t (ms)

Fig. 9.6 Addition of two harmonic waves at slightly different frequencies creates a beating pattern in the waveform and spatial profile

by $c_g \Delta t$. If $c_{phase} > c_g$, the extrema will seem to run past the envelope's leading zero, to be replaced by extrema that enter the packet at its trailing zero. In the lower graph, features of the waveform are delayed by the time required for the phase to travel distance Δx, so this delay is $\Delta x / c_{phase}$. The delay of the envelope is $\Delta x / c_g$. In either view, the total energy density (kinetic and potential) contained between the envelope's zeros is transported at the group velocity.

The first step in identifying the group velocity entails writing the pressure in a form that makes the beating pattern explicit. To that end, the center and difference frequencies are defined to be

$$\omega_0 = \frac{1}{2} (\omega_2 + \omega_1), \quad \Delta\omega = \omega_2 - \omega_1 \tag{9.1.39}$$

This corresponds to $\omega_1 = \omega_0 - \Delta\omega/2$ and $\omega_2 = \omega_0 + \Delta\omega/2$. It is assumed that $\Delta\omega \ll \omega_{av}$, so we may use a two-term Taylor series to approximate the wavenumber at each frequency

$$\kappa(\omega_1) \approx \kappa_0 - \kappa_0' \frac{\Delta\omega}{2}, \quad \kappa(\omega_2) \approx \kappa_0 + \kappa_0' \frac{\Delta\omega}{2} \tag{9.1.40}$$

where

$$\kappa_0 \equiv \kappa(\omega_0), \quad \kappa_0' = \frac{d\kappa}{d\omega}\bigg|_{\omega=\omega_0} \tag{9.1.41}$$

Substitution of these representations into Eq. (9.1.38) eventually leads to

$$p = E(x, t) \, \text{Re} \left\{ 2B \left[e^{-i\kappa_0 x + i\omega_0 t} \right] \right\} \tag{9.1.42}$$

The function $E(x, t)$, which is an amplitude modulation function, is given by

$$E(x, t) = \cos \left[\frac{\Delta\omega}{2} \left(t - \kappa_0' x \right) \right] \tag{9.1.43}$$

The complex factor in Eq. (9.1.42) represents a nominal constant amplitude wave at the center frequency. This term propagates at the phase speed $c_{\text{phase}} = \omega_0 / \kappa_0$. The envelope of the wave is $\pm E(x, t)$. The frequency of $E(x, t)$ is $\Delta\omega/2$, which is much slower than ω_0, and its wavenumber is $\kappa_0' \Delta\omega/2$, which is much smaller than κ_0. Equation (9.1.43) indicates that all features of the envelope, including its zeros and maxima, propagate in the direction of increasing x at speed $1/\kappa_0'$. Therefore, the interval between adjacent zeros of $E(x, t)$ constitutes a *wave packet* that contains a fixed amount of energy. Because κ_0' is the derivative of the $\kappa(\omega)$ dispersion curve at the average frequency, it follows that the phase speed and group velocity at any frequency are

$$c_{\text{phase}} = \frac{\omega}{\kappa}, \quad c_g = \frac{1}{d\kappa/d\omega} \tag{9.1.44}$$

In principle, c_g could be negative, which would indicate that energy propagates oppositely to the direction of propagation. Such an occurrence is exceptional.

The analysis has taken $\kappa(\omega)$ to be real. However, the function described by Eq. (9.1.20) is complex. This often is the case, as is evidenced by the wavenumbers for plane waves in a dissipative fluid, Sect. 3.3. The real part of the wavenumber is associated with the propagation, and the group velocity also is extracted from $\text{Re}(\kappa)$. Thus, the general relations are

$$\boxed{c_{\text{phase}} = \text{Re} \left(\frac{\omega}{\kappa} \right), \quad c_g = \text{Re} \left(\frac{1}{d\kappa/d\omega} \right) \equiv \text{Re} \left(\frac{d\omega}{d\kappa} \right)} \tag{9.1.45}$$

The second form in the equation for c_g is useful if it is easier to solve the dispersion equation for the frequency corresponding to a specified wavenumber.

The preceding is generally applicable. In the case of an exponential horn, the dispersion equation is Eq. (9.1.20). It describes κ as a function of the plane wave number k, whereas Eq. (9.1.45) takes κ to be a function of ω. The chain rule for differentiation leads to

$$c_g = c \, \text{Re} \left(\frac{1}{d\kappa/dk} \right) = \begin{cases} c \left(1 - \dfrac{b^2}{4k^2} \right)^{1/2} & \text{if } k > \dfrac{b}{2} \\ 0 \text{ if } k < \dfrac{b}{2} \end{cases} \tag{9.1.46}$$

This expression tells us that the group velocity in an exponential horn decreases with decreasing frequency up to cutoff. Below the cutoff frequency, the group velocity is zero. The latter aspect could have been anticipated by the earlier analysis, which showed that the average power flow is zero below cutoff.

A comparison of the preceding relation for group velocity and Eq. (9.1.26) for the phase speed shows that

$$c_g c_{phase} = c^2 \tag{9.1.47}$$

It is not unusual in acoustics to encounter this relation, but it is not a general property. The relation in a sense helps us to resolve any uneasiness regarding the phase speed increasing without bound as ω decreases the cutoff value. We now see that increasing c_{phase} is accompanied by decreasing c_g, until energy ceases to propagate at ω_{cutoff}.

EXAMPLE 9.2 The velocity of a source at the throat of an exponential horn in air ($\rho_0 = 1.2$ kg/m^3 and $c = 340$ m/s) is $v_0 = 0.004 [\sin(500t) + \cos(501t)]$ m/s. The area growth constant is $b = 0.6$ m^{-1}, and reflections at the open end are negligible. Determine the pressure waveforms at $x = 0.60$ m and 1.20 m. Use them to determine the group velocity, and compare the result to the value obtained from the dispersion relation for an exponential horn.

Significance

The notion here is that seeing how features of a propagating wave are correlated to the group velocity will enhance intuitive understanding of these concepts.

Solution

The specified analysis calls for extraction of the group velocity directly from computed waveforms, followed by a comparison of the result to the one obtained from the dispersion equation. Thus, our first task is to construct the waveforms. The pressure in an exponential horn is described by Eq. (9.1.22) for a single harmonic. To combine two harmonics, we write the sum in time-domain form. Before we do so, we determine the amplitude coefficient for each harmonic. The pressure in each harmonic is related to the source's velocity by Eq. (9.1.35) for the throat impedance. The values for frequencies of 500 and 501 rad/s are

$$Z_{throat}(\omega_1 = 500) = 399.42 + 83.23i, \quad Z_{throat}(\omega_2 = 501) = 399.45 + 83.07i \text{ Rayl}$$

The corresponding pressure amplitudes are $P_n(x = 0) = Z_{throat} V_n$, where $V_1 = 0.004/i$ and $V_2 = 0.004$ m/s. The exact occurrence of beats requires that $|P_2| = |P_1|$. The amplitudes obtained here are sufficiently close to treat the data as a true beat.

The waveform must be sampled at a very high rate in order to determine accurately the instants at which $p = 0$. The data set we shall use corresponds to a sampling interval of $0.001 (2\pi/\omega_{av})$. The data set that is generated is

$$p = \text{Re}\left[Z_{\text{throat}}(\omega_1) \frac{0.004}{i} e^{i\omega_1 t_n} + Z_{\text{throat}}(\omega_1)(0.004) e^{i\omega_1 t_n} \right], \quad t_n = \frac{0.002\pi n}{\omega_{\text{av}}}$$

Figure 1 shows the pressure waveform at $x_1 = 0.6$ m. The existence of a beat is evident, but the oscillation at the average frequency is too rapid to see the actual waveform.

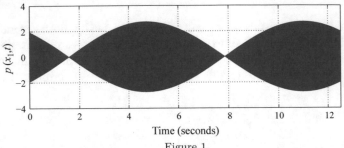

Figure 1.

Figure 2 zooms the waveforms at $x_1 = 0.6$ m and $x_2 = 1.2$ m in the interval surrounding the first zero of the envelope. We can evaluate the group velocity by dividing the distance $x_2 - x_1$ by the time required for a zero to travel between these locations. Hence, the feature we seek is a zero of the pressure, but which one? The answer emerges when we consider the first zero after $t = 1.57$ s at which p transitions from positive to negative. Both waveforms are such that the smallest peak is to the left and the smallest valley is to the right. This is what we would expect at the location where the amplitude modulation function vanishes. The other zeros correspond to oscillations at the mean frequency, so the amplitude of the oscillation will be larger than it is in the interval where the modulation function is small.

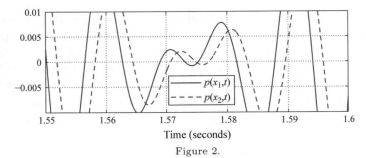

Figure 2.

A less intuitive way in which to identify the instant at which the envelope function is zero is to examine the zeros of the signal. In the interval depicted in Fig. 2, we see that $p(x_1, t) = 0$ at $t_n = 1.5501, 1.5564, 1.5627, 1.5690, 1.5730, 1.5753, 1.5815, 1.5878, 1.5941$, and 1.6004 seconds. The intervals between adjacent zeros are $t_{n+1} - t_n = 0.0063, 0.0063, 0.0063, 0.0040, 0.0023, 0.0062, 0.0063, 0.0063$, and 0.0063. A regular interval, which is 0.0063 s for the present data, corresponds to the zeros of the high-frequency oscillation at the average frequency. An extra zero due to the envelope function zero breaks up such regularity. This condition is manifested here by the subintervals of 0.0040 s followed by 0.0023 s. The instant they mark is $t_1 = 1.5730$ s. For greater precision, we interpolate between the instants at which the pressure changes sign. The value extracted in this manner is $t_1 = 1.573015$ s. A similar analysis of the x_2 data in the zoomed interval leads to identification of $t_2 = 1.574817$ s as the instant at which the envelope is zero. The group velocity is thereby estimated as

$$\left(c_g\right)_{\text{data}} \approx \frac{x_2 - x_1}{t_2 - t_1} = 330.0 \text{ m/s}$$

The analytical expression for the group velocity is Eq. (9.1.46). The value of k is set by the average frequency, so we have

$$c_g = c \left(1 - \frac{b^2}{4k^2}\right)^{1/2}\Bigg|_{k=\omega_{\text{av}}/c} = 332.9 \text{ m/s}$$

To further calibrate the accuracy of this procedure, let us evaluate the phase velocity by a similar calculation. The envelope modulates the amplitude of a wave at the center frequency that propagates at the phase speed. The best feature to track is a zero near an instant when the envelope is largest because its location can be identified with little ambiguity. The one we shall use occurs slightly later than 4.5 s. The values identified by scanning the data and interpolating across the zero crossing are $t_1 = 4.500292$ s and $t_2 = 4.502019$ s. This gives $c_{\text{phase}} = (x_2 - x_1)/(t_2 - t_1) = 347.4$ m/s, whereas Eq. (9.1.20) gives $c_{\text{phase}} = 347.3$ m/s.

Our identification scheme performed well. It should be evident from this exercise that experiments to extract phase and group velocity from waveform data measured over a small spatial range require highly accurate and precise equipment. Accuracy would be improved by using measurements at points that are farther apart. However, if that interval exceeds a wavelength at the average frequency, we will not be able to correlate zeros at two locations because of the almost periodic nature of the waveform.

9.1.4 WKB Solution for an Arbitrary Horn

There are several reasons to use a horn whose profile is neither conical or exponential. For example, to smoothly join two cylindrical tubes, it would be necessary that the transition segment be circular with a radius function $R_{\text{wall}}(x)$ for which $dR_{\text{wall}}/dx = 0$ at the ends. Except for a few cases, $\mathcal{A}(x)$ will be such that the Webster horn equation has no simple solution. A technique that has been developed is the *WKB method*.[1] It is an approximate analysis in the frequency domain based on reasonable assumptions.

---

[1] The initials recognize G. Wentzel, H. Kramers, and L.N. Brillouin, who employed the method to analyze the Shroedinger wave equation. A detailed development is provided by C.M. Bender and S.A. Orszag, *Advanced Mathematical Methods for Scientists and Engineers* (1978) Chap. 10.

Our starting point is consideration of the propagation of a plane wave in the direction of increasing x. In an ideal unbounded fluid, propagation introduces a phase lag $-kx$ to the pressure. Therefore, it is reasonable to anticipate that the phase lag of the arbitrary plane wave will be negative and increase with increasing x. In view of these considerations, it must be that the general time-domain representation of the wave may be written as

$$p(x,t) = |P(x)| \operatorname{Re}\left(e^{i(\omega t - \Phi(x))}\right) \qquad (9.1.48)$$

The phase lag for this representation is the real variable $\Phi(x)$.

The phase speed, c_{phase}, is obtained by equating the phase of p at (x, t) and $(x + dx, t + dx/c_{\text{phase}})$. Therefore,

$$\omega t - \Phi(x) = \omega\left(t + \frac{dx}{c_{\text{phase}}}\right) - \Phi(x + dx)$$

$$= \omega t + \frac{\omega dx}{c_{\text{phase}}} - \left[\Phi(x) + \frac{d}{dx}\Phi(x)\,dx\right] \qquad (9.1.49)$$

Solution of this relation for the phase speed yields

$$c_{\text{phase}} = \frac{\omega}{\dfrac{d}{dx}[\Phi(x)]} \equiv \frac{\omega}{\dfrac{d}{dx}\left[\arg(P(x))\right]} \qquad (9.1.50)$$

The local wavenumber therefore is

$$k_{\text{local}} = \frac{\omega}{c_{\text{phase}}} = \frac{d}{dx}\left[\arg(P(x))\right] \qquad (9.1.51)$$

These expressions are suitable for any plane wave propagating in a fixed direction. One of the outcomes of the WKB method will be a description of the phase speed and local wavenumber in terms of the frequency and the properties of the waveguide.

The WKB analysis of the Webster horn equation assumes that there is no reflection at the far end, so only waves that propagate in the direction of increasing x are present. The starting point is an ansatz suggested by Eq. (9.1.48), but lacking the distinction between the phase and the magnitude of P. Specifically, we begin with

$$p(x,t) = \operatorname{Re}\left(P(x)\,e^{i\omega t}\right), \quad P(x) = Be^{-i\Theta(x)} \qquad (9.1.52)$$

where B is constant. The function $\Theta(x)$ is allowed to be complex. If we write it as $\operatorname{Re}(\Theta) + i\operatorname{Im}(\Theta)$, then we have

$$|P(x)| = |B|\,e^{\operatorname{Im}(\Theta)}, \quad \Phi(x) = \operatorname{Re}(\Theta) - \arg(B) \qquad (9.1.53)$$

Substitution of Eq. (9.1.52) into the Webster horn equation gives

$$\left[-i\frac{d^2\Theta}{dx^2} - \left(\frac{d\Theta}{dx}\right)^2 - i\Gamma\frac{d\Theta}{dx} + k^2 \right] B e^{-i\Theta} = 0 \qquad (9.1.54)$$

where

$$\Gamma(x) = \frac{1}{\mathcal{A}}\frac{d\mathcal{A}}{dx} \qquad (9.1.55)$$

The exponential factor is not identically zero, so a nontrivial solution corresponds to vanishing of the bracketed term

$$i\frac{d^2\Theta}{dx^2} + \left(\frac{d\Theta}{dx}\right)^2 + i\Gamma(x)\frac{d\Theta}{dx} - k^2 = 0 \qquad (9.1.56)$$

This is a nonlinear second-order ordinary differential equation. The frequency-domain Webster horn equation is linear, so it might seem that we have made the analysis more difficult. However, such thinking ignores the difference in the behavior of the complex pressure P and phase variable Θ. In the case of an exponential waveguide, P varies rapidly over the scale of a wavelength, whereas $\Theta = \kappa x$, which is a much more gradual dependence on x. In the case of arbitrary $\mathcal{A}(x)$, we can find an exponential area function $B e^{bx}$ that closely fits the actual $\mathcal{A}(x)$ over any small range of x values. Within this range, the exponential horn solution for this approximating b should be close to the solution for the true $\mathcal{A}(x)$. This observation suggests that in the short x interval where the area functions match, it is reasonable to consider $\Theta = \kappa x$. A different exponential factor b will fit $\mathcal{A}(x)$ in an adjacent x interval. Thus, the value of κ for this interval will be different. However, unless the cross-sectional area undulates greatly over a very small scale, the adjacent value of κ should not differ greatly from the previous value; that is, $d\kappa/dx$ may be expected to be small. In other words, we can anticipate that in general $\Theta \approx \kappa(x)x$, with $\kappa(x)$ being a slowly varying function. Because $d^2\Theta/dx^2 = (d^2\kappa/dx^2)x + 2d\kappa/dx$, and $d\kappa/dx$ is taken to be small in some sense, perhaps it is reasonable to consider $d^2\Theta/dx^2$ to be even smaller. This is the WKB approximation.

The approximation is implemented as an iterative procedure. The first iteration results from setting $d^2\Theta/dx^2$ to zero in Eq. (9.1.56). We solve that equation for $d\Theta/dx$ and integrate to find the first approximation $\Theta^{(1)}(x)$. A second approximation, $\Theta^{(2)}(x)$, may be obtained by using the first approximation to estimate $d^2\Theta/dx^2$. The specific equations to be solved are

$$\begin{aligned}
\left(\frac{d\Theta^{(1)}}{dx}\right)^2 + i\Gamma(x)\frac{d\Theta^{(1)}}{dx} - k^2 &= 0 \\
\left(\frac{d\Theta^{(2)}}{dx}\right)^2 + i\Gamma(x)\frac{d\Theta^{(2)}}{dx} - k^2 &= -i\frac{d^2\Theta^{(1)}}{dx^2}
\end{aligned} \qquad (9.1.57)$$

It is not difficult to prove that the equation governing the nth iterant resembles the one for $\Theta^{(2)}$. It is

$$\left(\frac{d\Theta^{(n)}}{dx}\right)^2 + i\Gamma\,(x)\,\frac{d\Theta^{(n)}}{dx} - \left(k^2 - i\frac{d^2\Theta^{(n-1)}}{dx^2}\right) = 0; \quad n = 1, 2, \ldots \quad (9.1.58)$$

The result of solving this quadratic equation for $d\Theta^{(n)}/dx$ is a recurrence relation

$$\Theta^{(0)} = 0; \quad \frac{d\Theta^{(n)}}{dx} = -\frac{i}{2}\Gamma\,(x) \pm \left[k^2 - \frac{1}{4}\Gamma\,(x)^2 - \frac{d^2\Theta^{(n-1)}}{dx^2}\right]^{1/2}, \quad n = 1, 2, \ldots$$
$$(9.1.59)$$

We seek the wave that propagates in the positive x direction, which corresponds to Re $(\Theta) > 0$. Thus, the root for the positive sign is selected. A direct integration, with $\Theta^{(n)}\,(0) = 0$, leads to

$$\Theta^{(n)} = -\frac{i}{2}\int_0^x \Gamma\,(\eta)\,d\eta + \int_0^x \left[k^2 - \frac{1}{4}\Gamma\,(\eta)^2 - \frac{d^2\Theta^{(n-1)}}{d\eta^2}\right]^{1/2} d\eta \quad (9.1.60)$$

The definition in Eq. (9.1.55) may be written as $\Gamma = (d/dx)\ln(\mathcal{A})$, which leads to

$$\Theta^{(n)} = -i\ln\left(\frac{\mathcal{A}\,(x)^{1/2}}{\mathcal{A}\,(0)^{1/2}}\right) + \int_0^x \left[k^2 - \frac{1}{4}\Gamma\,(\eta)^2 - \frac{d^2\Theta^{(n-1)}}{d\eta^2}\right]^{1/2} d\eta \quad (9.1.61)$$

Let N be the last iteration step. Because $e^{-\ln(u)} \equiv 1/u$, the complex pressure amplitude resulting from the nth iteration is given by

$$P\,(x) = B\frac{\mathcal{A}\,(0)^{1/2}}{\mathcal{A}\,(x)^{1/2}} \exp\left(-i\int_0^x \left[k^2 - \frac{1}{4}\Gamma\,(\eta)^2 - \frac{d^2\Theta^{(N-1)}}{d\eta^2}\right]^{1/2} d\eta\right) \quad (9.1.62)$$

An important aspect of the preceding expression is that the integrand generally will be complex for $N > 1$, which is evidenced by Eq. (9.1.61). Consequently, $|P|$ and $\Phi\,(x)$ in Eq. (9.1.53) can only be extracted after the integral has been evaluated. The exception is $N = 1$. In addition, it is evident that any iteration beyond the first will be quite complicated. It is for this reason that a WKB analysis seldom goes beyond the first iteration. For $N = 1$, the integrand in Eq. (9.1.62) is real. The corresponding description of the pressure amplitude and phase angle is

$$\boxed{|P| = |B|\frac{\mathcal{A}\,(0)^{1/2}}{\mathcal{A}\,(x)^{1/2}}, \quad \Phi\,(x) = \int_0^x \left[k^2 - \frac{1}{4}\Gamma\,(\eta)^2\right]^{1/2} d\eta - \arg\,(B)} \quad (9.1.63)$$

The fact that the pressure depends inversely on the square root of the area is consistent with the requirement that the time-averaged power flowing through a cross section be independent of x. Substitution of $\Phi\,(x)$ into Eqs. (9.1.50) and (9.1.51) yields

$$k_{\text{local}} = \left[k^2 - \frac{1}{4} \Gamma (x)^2 \right]^{1/2}$$

$$c_{\text{phase}} = \frac{\omega}{\left[k^2 - \frac{1}{4} \Gamma (x)^2 \right]^{1/2}} \tag{9.1.64}$$

In practice, the integral in Eq. (9.1.62) will be quite challenging to evaluate analytically at any iteration, including the first. One exception is the exponential horn, for which $\Gamma (x) = b$. The first iteration gives

$$\Theta^{(0)} = 0 \implies \Theta^{(1)} = -\frac{i}{2} bx + \left(k^2 - \frac{b^2}{4} \right)^{1/2} x \tag{9.1.65}$$

This gives $d^2 \Theta^{(1)} / dx^2 = 0$, which means that $\Theta^{(2)}$ will be the same as $\Theta^{(1)}$, and so on. From this, we conclude that $\Theta^{(1)} = \Theta$, but the preceding expression states that $\Theta^{(1)} = \kappa x$, where κ is the complex wavenumber in Eq. (9.1.20). Hence, the WKB solution is identically correct for an exponential horn. This is as it should be, because the fact that $\Theta = \kappa x$ for an exponential horn means that $d^2 \Theta / dx^2 = 0$ is not an approximation.

When the $\mathcal{A}$ function is such that it is not possible to carry out the integral analytically, it could be evaluated numerically. However, it is reasonable to ask why we should employ the WKB method in that case, because an alternative is to use numerical methods to solve the Webster horn equation. Doing so does not entail assuming that $d^2 \Theta / dx^2$ is very small. The next example will examine the various facets of both numerical approaches.

EXAMPLE 9.3 In order to explore the effect of a constriction in a waveguide, consider the case where the cross-sectional area is $\mathcal{A} (x) = \mathcal{A}_0 \left[\varepsilon + \cos (\pi x / L)^2 \right]$, which varies periodically over length $L/2$ between a minimum of $\varepsilon \mathcal{A}_0$ and a maximum of $(1 + \varepsilon) \mathcal{A}_0$. At $x = 0$, the particle velocity is $v_x = \text{Re} \left(V_0 e^{i\omega t} \right)$, and the far end, which is $x = 2L$, is terminated with an absorptive material that has a zero reflection coefficient. The derivation of the Webster horn equation requires that the cross section changes gradually. This condition is satisfied if L is much greater than the nominal wavelength $2\pi / k$, so set $kL = 20\pi$. Use the WKB method to determine the complex pressure amplitude $P (x)$, and the corresponding local wavenumber and phase speed for the case where $\varepsilon = 0.1$. Restrict the analysis to the first iteration. Then, assess the quality of the approximate solution by using numerical methods to solve the Webster horn equation.

Significance

The WKB method is interesting for the insights derived from its solution. However, it is very difficult to perform higher iterations, so its accuracy is uncertain. A direct

numerical solution of the Webster does not entail fundamental approximations, so
we will have the opportunity to decide whether the WKB method is worth pursuing.

Solution

The first iteration of the WKB solution is indicated by Eq. (9.1.62) to be

$$P(x) = B \frac{\mathcal{A}(0)^{1/2}}{\mathcal{A}(x)^{1/2}} \exp\left(-i \int_0^x \left[k^2 - \frac{1}{4}\Gamma(\eta)^2\right]^{1/2} d\eta\right) \tag{1}$$

The value of B is set by matching V_x at $x = 0$ to V_0. This requires application
of Euler's equation. Differentiation of Eq. (1) and application of Leibnitz' rule for
differentiation of an integral leads to

$$\left.\frac{dP}{dx}\right|_{x=0} = -\frac{1}{2} B \mathcal{A}(0)^{1/2} \left(\frac{1}{\mathcal{A}(x)^{3/2}} \frac{d\mathcal{A}}{dx}\right)\Bigg|_{x=0}$$

$$- i B \frac{\mathcal{A}(0)^{1/2}}{\mathcal{A}(x)^{1/2}} \left[k^2 - \frac{1}{4}\Gamma(0)^2\right] = -i\omega \rho_0 V_0$$

By definition, $\Gamma(0) = (d\mathcal{A}/dx)/\mathcal{A}$ evaluated at $x = 0$, so the preceding yields

$$B = \frac{\omega \rho_0 V_0}{\left[k^2 - \frac{1}{4}\Gamma(0)^2\right]^{1/2} - \frac{i}{2}\Gamma(0)}$$

The function $\Gamma(x)$ for the given area function is

$$\Gamma(x) = \frac{1}{\mathcal{A}} \frac{d\mathcal{A}}{dx} = \frac{-2(\pi/L)\varepsilon \sin(2\pi x/L)}{\varepsilon + \cos(\pi x/L)^2} \tag{2}$$

At $x = 0$, we have $\Gamma(0) = 0$, so we have established that

$$B = \rho_0 c V_0 \tag{3}$$

Substitution of Eq. (2) into (1) leads to an integral that is in the class of elliptic integrals, which are tabulated functions. However, that recognition still requires
considerable sophistication to manipulate the integral to a standard form. Since the
primary purpose is to determine the accuracy of the WKB method, we shall abandon
the effort to evaluate the integral analytically. Instead, we will evaluate the integral
numerically.

It is best to use nondimensional variables for numerical studies, so let $x = \tilde{x}L$
and $P = \rho_0 c V_0 \tilde{P}$. Then, the first WKB iteration, Eq. (1), becomes

$$\tilde{P}_{\text{wkb}} = \frac{\tilde{\mathcal{A}}(0)^{1/2}}{\tilde{\mathcal{A}}(\tilde{x})^{1/2}} e^{-i\tilde{\Phi}(\tilde{x})} \tag{4}$$

where the phase angle is

$$\tilde{\Phi}\left(\tilde{x}\right) = \int_0^{\tilde{x}} \left[(kL)^2 - \frac{1}{4}\tilde{\Gamma}\left(\eta\right)^2 \right]^{1/2} d\eta \tag{5}$$

The nondimensional area properties are

$$\tilde{A}\left(\tilde{x}\right)^{1/2} \equiv \frac{A\left(\tilde{x}L\right)}{A_0} = \varepsilon + \cos\left(\pi\tilde{x}\right)^2$$

$$\tilde{\Gamma}\left(\tilde{x}\right) = L\Gamma\left(\tilde{x}L\right) = \frac{-\pi\varepsilon\sin\left(2\pi\tilde{x}\right)}{1 + \cos\left(\pi\tilde{x}\right)^2}$$

Evaluation of Eq. (5) is straightforward for a numerical routine such as MATLAB's `quadl` because the integrand has no singularities. Evaluation at a large number of locations $\tilde{x}_j$ may be done efficiently by exploiting the incremental nature of an integral, which gives

$$\tilde{\Phi}\left(\tilde{x}_j\right) = \tilde{\Phi}\left(\tilde{x}_{j-1}\right) + \int_{\tilde{x}_{j-1}}^{\tilde{x}_j} \left[(kL)^2 - \frac{1}{4}\tilde{\Gamma}\left(\tilde{\eta}\right)^2 \right]^{1/2} d\tilde{\eta}$$

Now let us turn to the direct application of numerical methods. In terms of the nondimensional variables defined above, the Webster horn equation in the frequency domain is

$$\frac{d^2\tilde{P}}{d\tilde{x}^2} + \tilde{\Gamma}\left(\tilde{x}\right)\frac{d\tilde{P}}{d\tilde{x}} + (kL)^2\tilde{P} = 0$$

Standard numerical software offers routines for solving first-order differential equations. To convert this second-order differential equation to first-order form, we define two variables Y_1 and Y_2, such that

$$Y_1 = \tilde{P} = \frac{P}{\rho_0 c v_0}, \quad Y_2 = \frac{d\tilde{P}}{d\tilde{x}} = \frac{L}{\rho_0 c V_0}\frac{dP}{dx} \tag{6}$$

The derivative identity is $dY_1/d\tilde{x} = Y_2$, and the derivative of Y_2 is the second derivative of $\tilde{P}$, whose value is given by the differential equation. Thus, the first-order equations to solve are

$$\frac{d}{d\tilde{x}}\begin{Bmatrix} Y_1 \\ Y_2 \end{Bmatrix} = \begin{Bmatrix} Y_2 \\ -(kL)^2 Y_1 - \tilde{\Gamma}\left(\tilde{x}\right) Y_2 \end{Bmatrix} \tag{7}$$

Solution of this pair of order differential equations requires values of Y_1 and Y_2 at $\tilde{x} = 0$. Euler's equation gives $dP/dx = -i\omega\rho_0 V_0$ at $x = 0$. It follows from Eq. (6) that

$$Y_2|_{x=0} = -ikL \tag{8}$$

What should we use for the boundary value of Y_1? If we do not set it correctly, we will not initiate a wave that propagates solely in the direction of increasing x. To resolve this dilemma, we recall the earlier observation that we may fit an exponential shape to $\mathcal{A}(x)$ in a small interval. Let us do so for the region around $x = 0$. Then we can use the equivalent b factor to set the relation between P and V at $x = 0$ as though the horn were exponential. For an exponential horn, $\Gamma = b$ everywhere, whereas $\Gamma(0) = 0$ in the present system. Therefore, the exponential constant should be $b_{eq} = 0$. According to Eq. (9.1.27) for an exponential horn, the relation between P and V at $x = 0$ is

$$V_0 = \frac{\kappa_{eq}}{k} \frac{P|_{x=0}}{\rho_0 c} = \left\{ \left[1 - \left(\frac{b_{eq}L}{kL} \right)^2 \right]^{1/2} - \frac{1}{2} \left(\frac{b_{eq}L}{kL} \right) \right\} V_0 \tilde{P}|_{\tilde{x}=0} \qquad (9)$$

Setting $b_{eq} = 0$ leads to

$$Y_1|_{\tilde{x}=0} = 1$$

Many techniques have been developed to solve a set of first-order differential equations. A fourth-to-fifth-order Runge-Kutta routine, which is implemented as `ode45` in MATLAB, works well for the present configuration. It requires definition of a function that evaluates the right side of the first-order equation. A program loop increments x and calls `ode45` with the function name as the first argument. The value of Y_1 at each output location x_n is saved.

Errors sometimes accumulate in numerical analyses, so it is useful to have ways of checking our work. One way of doing so is to replace the given $\tilde{\Gamma}(\tilde{x})$ with a constant value for an exponential horn. Comparison of the computed pressure to the analytical solution will serve to verify that the program steps are correct. It also is good practice to verify that reducing the x increment in the program loop and decreasing the error tolerances do not significantly alter the computed result. Another check is to compute the time-averaged power flowing through any cross section. From Eq. (6), we have $P = \rho_0 c V_0 Y_1$ and $dP/dx = (\rho_0 c V_0/L) Y_2$. Then, Euler's equation gives

$$V_x = -\frac{1}{i\omega\rho_0} \frac{dP}{dx} = i \frac{V_0}{kL} Y_2$$

Correspondingly, the power flowing across any cross section is

$$\mathcal{P}_{av} = \frac{1}{2} \mathcal{A}(\tilde{x}) \operatorname{Re}(P V_x^*) = \frac{1}{2} \mathcal{A}(\tilde{x}) \operatorname{Re}\left[(\rho_0 c V_0 Y_1) \left(V_0 \frac{i}{kL} Y_2 \right)^* \right] \qquad (10)$$

$$= \frac{1}{2} \frac{\rho_0 c |V_0|^2}{kL} \mathcal{A}(\tilde{x}) \operatorname{Re}\left(\frac{Y_1 Y_2^*}{i} \right)$$

At all locations, the numerical solution gave $\mathcal{P}_{av}/\left(\rho_0 c |V_0|^2 \mathcal{A}_0\right) = 0.5500 \pm 4\left(10^{-13}\right)$.

Figure 1 displays the numerical results for Y_1 below the $\mathcal{A}$ and Γ functions. The pressure rises where $\mathcal{A}$ is relatively small, but its features otherwise are unremarkable.

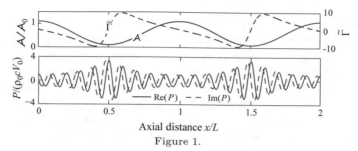

Axial distance x/L

Figure 1.

To compare the WKB and direct numerical solutions, we examine $|P|$ and $\tilde{\Phi}$. Equation (4) gives

$$\left|\tilde{P}_{\text{WKB}}\right| = \frac{\mathcal{A}(0)^{1/2}}{\mathcal{A}(x)^{1/2}}$$

and Eq. (5) gives $\Phi_{\text{WKB}}(x)$. These quantities are obtained in the direct numerical solution by post-processing of the Y_1 data. By definition $Y_1 = \tilde{P}(x)$. Conversion of the Y_1 data to polar form places Φ in a 2π range. The consequence is that Φ can change discontinuously. This data may be converted to a continuous variable by invoking an unwrap routine that is available in many software packages. Alternatively, jumps of Φ by 2π may be canceled manually by comparing the phase angle at each x location to its predecessor at an adjacent location.

Figure 2 compares the amplitude and phase for the alternative methods. The latter is plotted as ϕ/x, because that quantity would be the constant wavenumber of a simple plane wave. Both analyses are in extraordinarily good agreement, except for the phase near the source.

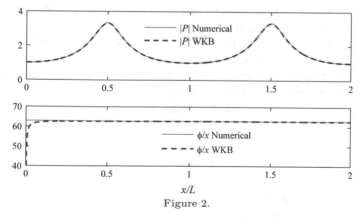

x/L

Figure 2.

It also is useful to compare the local wavenumber and phase speed. These quantities for the first iteration of the WKB method are given by Eq. (9.1.64). To find k_{local} for the differential equation solution, we use the general relation, Eq. (9.1.51), which requires values of $(d/dx)\arg(P(x)) \equiv d\Phi/dx$. We shall use a central finite difference to approximate Eq. (9.1.51), so that

$$k_{\text{local}}(x_n) = -\left.\frac{d\Phi}{dx}\right|_{x_n} \approx \left(\frac{\Phi(x_{n-1}) - \Phi(x_{n+1})}{(x_{n+1} - x_{n-1})}\right)$$

The phase speed in either case is $c_{\text{phase}}(x_n) = \omega/k_{\text{local}}(x_n)$.

Comparisons of the wavenumber and phase speed are provided by Fig. 3. Here, the agreement is less good. In particular, the differential equation solution indicates that the highest phase speed occurs where the area is smallest. The first WKB iteration misses this feature. However, the numerical solution also has a troubling aspect, specifically, the small-scale fluctuation in k_{local} and c_{phase}. It is not clear whether this is an artifact or a true property. Examination of the data in the interval $0.5 < \tilde{x} < 1.5$ indicates that there are 20 ridges. The value of kL is 20π, which means that there are 20 half-wavelengths in this interval. Furthermore, the curves were found to be unaltered when the error tolerance and time step for the Runge-Kutta solver were decreased, nor were they altered when a backward difference was used to evaluate k_{local}.

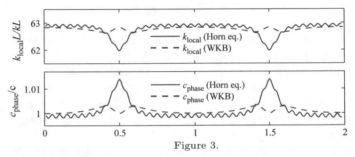

Figure 3.

Which method is preferable is discretionary in most cases. Once one has programmed either solution, adaptation of that program to analyze other configurations merely requires altering the definition of $\Gamma(x)$. The WKB first iteration is easier to program and requires very little CPU time. However, the differential equation solution also is quite fast because it only requires solving two coupled equations. The main drawback of the WKB method is that its numerical implementation is limited to the first iteration unless we wish to evaluate the second derivative of an iteration's results. Thus, if extreme accuracy is required, especially for the phase speed, then numerical solution of the Webster horn equation is preferable. However, in most situations, we are only interested in the amplitude and phase angle of P, and we can accept a fractional dB error. In such situations, the WKB method might be quite acceptable.

9.2 Two-Dimensional Waveguides

We now turn our attention to situations where the pressure within a waveguide varies transversely to the propagation. Our starting point is the simplest system, in which the fluid is situated in the region bounded by two infinite planes. It is assumed that the pressure is invariant in one direction parallel to these planes. Thus, the signal propagates parallel to the boundary, and it varies perpendicularly to the boundaries. Despite the apparent simplicity, the various configurations we shall examine feature most of the phenomena encountered in realistic systems.

9.2.1 General Solution

The model system is the fluid contained between two parallel infinite planes. There is no variation in one of the directions parallel to these planes. This is the two-dimensional situation depicted in Fig. 9.7, where x measures distance in the direction of propagation and y measures distance from one wall. A harmonic excitation applied along the plane at $x = 0$ induces a signal that propagates in the direction of increasing x. Our first task is to ascertain the general nature of its dependence on both x and y.

Fig. 9.7 A two-dimensional waveguide consisting of the region between parallel infinite planes

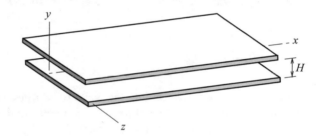

We may shortcut the separation of variables formalism by using a complex exponential $e^{-i\kappa x}$ to represent the propagation, where the constant axial wavenumber κ must be determined. Hence, we begin with

$$\boxed{P = \Phi(y)\, e^{-i\kappa x}} \tag{9.2.1}$$

Note that the above expression is equally valid if κ is replaced to $-\kappa$, which would correspond to waves traveling in the negative x direction. This ansatz for P will satisfy the Helmholtz equation if

$$\frac{d^2\Phi}{dy^2} + \mu^2\Phi = 0, \quad \mu^2 = k^2 - \kappa^2 \tag{9.2.2}$$

The y-dependence is generally described by

$$\Phi = B e^{-i\mu y} + C e^{i\mu y} \tag{9.2.3}$$

This form is useful because it will lead us to a fundamental interpretation. However, most individuals prefer trigonometric functions for analysis, so the form we shall use

$$\boxed{\Phi = B \cos (\mu y) + C \sin (\mu y)} \tag{9.2.4}$$

The coefficients B and C are not the same as they were in Eq. (9.2.3), but this is of no concern because there will be no occasion where both forms are used simultaneously.

Equation (9.2.4) constitutes the general solution. The unknown parameters in it are B, C, and μ. These will be set by satisfaction of boundary conditions at $y = 0$ and $y = H$. Determination of μ will then lead to the axial wavenumber, which is given by

$$\boxed{\kappa = \pm \left(k^2 - \mu^2\right)^{1/2}} \tag{9.2.5}$$

The plus/minus sign allows us to select the proper branch cut in situations where $\mu > k$ or μ is complex. In any case the sign of the square root must be taken to satisfy the Sommerfeld radiation condition for propagation in the direction of increasing x, which requires that $\mathrm{Re}\,(\kappa) \geq 0$ and $\mathrm{Im}\,(\kappa) \leq 0$.

9.2.2 Rigid Walls

For our initial study, we consider the walls to be rigid. Acoustic parlance is to say that this is a *hard-walled waveguide*. The general solution, Eq. (9.2.4), will satisfy the rigid wall boundary condition at $y = 0$ if

$$\left. \frac{\partial P}{\partial y} \right|_{y=0} = \mu C = 0 \tag{9.2.6}$$

It is possible that $\mu = 0$, but in that case, P is independent of y. This is the case of a plane wave. Thus, we set $C = 0$. Then, the rigid condition at $y = H$ requires that

$$\left. \frac{\partial P}{\partial y} \right|_{y=H} = -\mu B \sin (\mu H) = 0 \tag{9.2.7}$$

It cannot be that $B = 0$, because that would set P to zero. Thus, it must be that

$$\sin (\mu H) = 0 \implies \mu_n = \frac{n\pi}{H}, \quad n = 0, 1, 2, \ldots \tag{9.2.8}$$

The description of P in Eq. (9.2.1) at any n is a *waveguide mode,* and n is the *mode number.* The modes that arise in structural and mechanical vibration typically are standing waves in all directions, whereas the modes of a waveguide propagate in one direction and form a stationary pattern transversely to the propagation direction. The values of μ_n are the *eigenvalues,* and the equation whose roots give the eigenvalues is the *characteristic equation*. Somewhat confusingly, Φ, which describes the transverse variation of P corresponding to μ_n, is called a *transverse mode function*, or more concisely, a *transverse mode*.

The only aspect that we have not found is the coefficient B. The solution for P satisfies the Helmholtz equation and rigid wall conditions independently of the value of B. Furthermore, any value assigned to it would be compensated when we determine the actual signal generated by a specified excitation. We will soon define a standard definition of B, but a common practice for graphing a mode is to set B such that the maximum value of Φ_n is one. Accordingly, we shall temporarily leave this coefficient in algebraic form, so the transverse modes of a hard-walled waveguide are

$$\Phi_n = B_n \cos\left(\frac{n\pi y}{H}\right) \tag{9.2.9}$$

The axial wavenumber in the nth mode is

$$\kappa_n = \left[k^2 - \left(\frac{n\pi}{H}\right)^2\right]^{1/2} \tag{9.2.10}$$

The pressure and particle velocity corresponding to a specific n are

$$
\begin{aligned}
P &= B_n \cos\left(\frac{n\pi y}{H}\right) e^{-i\kappa_n x}, \quad n = 0, 1, 2, \dots \\
\bar{V} &= \frac{B_n}{\rho_0 c}\left[\frac{\kappa_n}{k}\cos\left(\frac{n\pi y}{H}\right)\bar{e}_x - i\frac{n\pi}{kH}\sin\left(\frac{n\pi y}{H}\right)\bar{e}_y\right]e^{-i\kappa_n x}
\end{aligned}
\tag{9.2.11}
$$

An important feature is the $n = 0$ mode, $\Phi_0 = B_0$, for which $\kappa_0 = k$. Both P and $\bar{V}$ are independent of y in this mode, and $\bar{V}$ is in the x direction; it is a one-dimensional plane wave.

As the mode number increases, the value of κ_n decreases, until n is sufficiently large that the value of κ_n becomes imaginary. The largest n for which κ_n is real is the *cutoff mode number*. For the case of rigid walls, it is

$$N_{\text{cut}} = \text{floor}\left(\frac{kH}{\pi}\right) \tag{9.2.12}$$

Although Eq. (9.2.10) for $n > N_{\text{cut}}$ is satisfied if κ_n is either a positive or negative imaginary value, only the negative value leads to decay. Specifically, we set

$$\kappa_n = -i\left(\mu_n^2 - k^2\right)^{1/2} \quad \text{if } n > N_{\text{cut}} \tag{9.2.13}$$

In this case, the x dependence of P and $\bar{V}$ is

$$e^{-i\kappa_n x} = e^{-\left(\mu_n^2 - k^2\right)^{1/2} x} \tag{9.2.14}$$

These modes evanesce. The decay constant $|\kappa_n|$ for $n > N_{\text{cut}}$ increases with increasing n when k is fixed. This means that the cutoff modes become less significant at a specific x as the number of "wiggles" of the mode in the y direction increases. The terminology we will use is to say that a *propagating mode* has an axial wavenumber that is positive real (for propagation in the sense of increasing x). An *evanescent mode* has an axial wavenumber that is negative imaginary. (These definitions will be refined when we examine situations where the axial wavenumber is complex.) The value of κ_n may be computed in one operation regardless of the frequency according to $\kappa_n = \text{conj}\big((k^2 - \mu_n^2)^{1/2}\big)$ or $\kappa_n = -\text{conj}\big((k^2 - \mu_n^2)^{1/2}\big)$. Which alternative should be selected depends on whether the software gives $\text{Im}\,(\kappa) < 0$ if $\kappa^2 < 0$.

If we consider a specific mode, that is, hold n fixed, $|\kappa_n|$ decreases as k decreases from a large value. The *cutoff frequency* of a mode is the value below which that mode becomes evanescent,

$$\omega_{\text{cut}} = \frac{n\pi c}{H} \tag{9.2.15}$$

Rather than viewing the behavior of a specific mode as the frequency decreases, we may consider $N_{\text{cut}} = \text{floor}(\omega H/(\pi c))$ as the highest mode number at which a mode at a specific frequency will propagate. The $n = 0$ (plane wave) mode never is cutoff.

The phase speed of a propagating mode is

$$c_{\text{phase}} = \frac{\omega}{\kappa_n} = \frac{c}{\left[1 - \left(\dfrac{n\pi}{kH}\right)^2\right]^{1/2}} \tag{9.2.16}$$

The group velocity obtained from Eq. (9.2.10) is

$$(c_{\text{g}})_n = c\,\frac{1}{d\kappa_n/dk} = c\left[1 - \left(\frac{n\pi}{kH}\right)^2\right]^{1/2} \tag{9.2.17}$$

For frequencies below the cutoff, the axial wavenumber is imaginary, so the phase and group velocities are zero. An interesting aspect is the similarity of these properties to those of an exponential horn. In both systems

$$c_{\text{g}}c_{\text{phase}} = c^2 \tag{9.2.18}$$

The next section will explain this relation in terms of the trace velocity.

9.2.3 Interpretation

The description of a mode in Eq. (9.2.11) indicates that a constant value of P, whose amplitude is proportional to $\cos\,(n\pi y/H)$, propagates in the x direction. This amplitude is sustained as the signal moves downstream. Thus, it is appropriate to say that a waveguide mode is a *nonuniform plane wave*. This means that the rays of constant phase are parallel and that the amplitude is not constant along the wavefronts, which are perpendicular to the rays.

An alternate picture explains several aspects, including the transition to evanescent behavior. To obtain it, we employ Euler's identity to convert the transverse mode functions to complex exponentials

$$\Phi_n = \frac{B_n}{2} \left[e^{(-in\pi y/H)} + e^{(in\pi y/H)} \right] \tag{9.2.19}$$

This is the incarnation of the general solution, Eq. (9.2.3), for the rigid wall case. When we use this representation to form the nth waveguide mode, we find that

$$P_n = \frac{B_n}{2} \left[e^{-i\bar{k}_1 \cdot \bar{x}} + e^{-i\bar{k}_2 \cdot \bar{x}} \right] \tag{9.2.20}$$

where the position is $\bar{x} = x\bar{e}_x + y\bar{e}_y$. The wavenumber vectors are

$$\begin{aligned} \bar{k}_1 &= \kappa_n \bar{e}_x + \frac{n\pi}{H} \bar{e}_y \\ \bar{k}_2 &= \kappa_n \bar{e}_x - \frac{n\pi}{H} \bar{e}_y \end{aligned} \tag{9.2.21}$$

These vectors are depicted in Fig. 9.8. The vectors are directed at angle ψ above and below the x-axis. In view of Eq. (9.2.10), $|\bar{k}_1| = |\bar{k}_2| = k$. Consequently, each wavenumber vector is the product of k and a unit vector, which is the way each is depicted in the figure.

The construction in Fig. 9.8 leads to an alternate description of the wavenumber vectors as

Fig. 9.8 Wavenumber vectors for the plane waves that are equivalent to mode n in a two-dimensional hard-walled waveguide

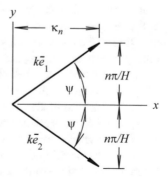

$$\bar{k}_1 = k\bar{e}_1, \quad \bar{k}_2 = k\bar{e}_2$$

$$\bar{e}_1 = (\cos\psi_n)\,\bar{e}_x + (\sin\psi_n)\,\bar{e}_y, \quad \bar{e}_2 = (\cos\psi_n)\,\bar{e}_x - (\sin\psi_n)\,\bar{e}_y$$

$$\psi_n = \tan^{-1}\left(\frac{n\pi}{\kappa_n H}\right) = \cos^{-1}\left(\frac{\kappa_n}{k}\right) = \sin^{-1}\left(\frac{n\pi}{kH}\right) \tag{9.2.22}$$

Substitution of this representation of the wave numbers into Eq. (9.2.20) leads to

$$P_n = P_n^{(1)} + P_n^{(2)}, \quad P_n^{(j)} = \frac{B_n}{2}e^{-ik\bar{e}_j\cdot\bar{x}} \tag{9.2.23}$$

This representation describes a mode $P_n(x, y)$ as a superposition of simple plane waves whose phase speed is c that propagate obliquely in opposite sense relative to the x-axis. The particle velocity is the superposition of the contribution of the individual waves.

$$\rho_0 c\bar{V} = \frac{B_n}{2}\left(\bar{e}_1 e^{-i\bar{k}_1\cdot\bar{x}} + \bar{e}_2 e^{-i\bar{k}_2\cdot\bar{x}}\right) \tag{9.2.24}$$

The picture of the signal as superposed plane waves explains many features. The transition to an evanescent mode is a consequence of the monotonic increase of ψ_n as n increases at fixed k. Ultimately, if $n > N_{\text{cut}}$, the last of Eq. (9.2.22) states that ψ_n is an angle whose sine is greater than one. This leads to a complex ψ_n, with $\cos\psi_n$ being a negative imaginary value. The corresponding x components of $\bar{k}_1$ and $\bar{k}_2$ become negative imaginary values, which converts the mode from one that propagates to one that evanesces.

Decomposition of a propagating mode into simple plane waves gives a new perspective for the phase speed in Eq. (9.2.16) and the group velocity in Eq. (9.2.17). For a propagating mode, $\cos\psi_n = \kappa_n/k = [1 - (n\pi)^2/(kH)^2]^{1/2}$. Thus, these propagation speeds may be written as

$$c_{\text{phase}} = \frac{c}{\cos\psi_n}, \quad c_g = c\cos\psi_n \tag{9.2.25}$$

The phase speed is recognizable as the trace in the x direction of the phase velocity of the oblique plane waves. In contrast, the group velocity is the component in the x direction of the phase velocity of each oblique plane wave. Increasing ψ_n, either by raising n or decreasing k, increases the trace velocity along the x-axis. Ultimately, at the cutoff frequency, the phase speed is infinite because the wavefronts of the simple waves are parallel to the x-axis, so all points on the x-axis simultaneously see a specific wavefront. The explanation of the group velocity lies in the fact that the oblique plane waves are nondispersive, so both their group and phase velocities are c. Thus, a packet of oblique waves formed from two slightly different frequencies would propagate obliquely at speed c. The projection onto the x-axis of the packet's location would move in the x direction at speed $c\cos\psi_n$.

Representation of a mode as two plane waves makes is possible to view the propagation in terms of rays. Figure 9.9a follows the ray of each plane wave that passes through an arbitrary point C. The $\bar{e}_1$ ray is incident at the upper wall, where it

Fig. 9.9 Multiple reflections
in a hard-walled
two-dimensional waveguide
observed by decomposing a
mode into its constituent
plane waves

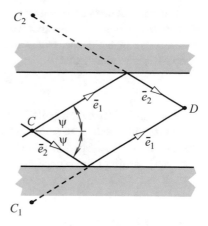

is reflected, thereby generating an $\bar{e}_2$ ray. Because the boundary is rigid, the reflection coefficient is $R = 1$, so the pressure in the reflected $\bar{e}_2$ ray matches the pressure in the incident ray. In other words, the reflected wave seems to have originated from the image C_2 of point C. A similar process applies to the $\bar{e}_2$ ray that intersects point C, so the pressure on the $\bar{e}_1$ ray at point D seems to have originated from image point C_1. In other words, the signal within a two-dimensional waveguide with rigid walls is a process of repeated reflections.

The standing wave pattern in the transverse y direction results from the requirement that the field be consistent for reflection from either wall. From the viewpoint of the lower wall $\bar{e}_2$ is the propagation direction for the incident wave, and $\bar{e}_1$ is the propagation direction for the reflected wave. The viewpoint of the upper wall is opposite, with the incident direction being $\bar{e}_1$ and the reflected direction being $\bar{e}_2$. The reflection coefficient at each wall is $R = 1$ because the walls are rigid. Suppose the angle of incidence (and reflection) is an arbitrary value χ. Then, Eq. (5.2.16) states that the pressure field due to reflection at the lower wall is

$$p(x, y, t) = \text{Re}\left[\frac{B_n}{2}e^{i(\omega t - kx \sin \psi_1)}\left(e^{iky \cos \chi} + e^{-iky \cos \chi}\right)\right]$$
$$= \text{Re}\left[B_n e^{i(\omega t - kx \sin \chi)} \cos(ky \cos \chi)\right] \qquad (9.2.26)$$

Reflection from the upper wall may be described similarly if we replace y with $y' = H - y$, so that $y' = 0$ at the upper wall and $y' = H$ at the lower wall. Then, the pressure field from this viewpoint is

$$p(x, y, t) = \text{Re}\left[B_n e^{i(\omega t - kx \sin \chi)} \cos(k(H - y) \cos \chi)\right]$$
$$\equiv \text{Re}\left\{B_n e^{i(\omega t - kx \sin \chi)}\left[\cos(kH \cos \chi) \cos(ky \cos \chi)\right.\right.$$
$$\left.\left. + \sin(kH \cos \chi) \sin(ky \cos \chi)\right]\right\} \qquad (9.2.27)$$

These alternate descriptions must represent the same pressure field, so it must be that $\sin(kH \cos \chi) = 0$. In other words, $kH \cos \chi = n\pi$. The angle χ was defined to be the angle of incidence, so the angle ψ for the decomposition into oblique waves is $\pi/2 - \chi$. It follows that

$$\sin \psi = \frac{n\pi}{kH} \tag{9.2.28}$$

This is identical to the third relation for ψ_n in Eq. (9.2.22). Thus, we have shown that the waveguide mode is the superposition of a pair of waves that propagate obliquely to the x-axis. The propagation angle for each oblique wave is such that the interference patterns resulting from reflection at the lower and upper walls are compatible.

9.2.4 Flexible Walls

The model of a waveguide with rigid walls is a useful idealization. However, the walls in some applications may be quite flexible, and they might be composed of an absorptive material, as is the case for turbine engine exhaust systems. Some coastal regions of the ocean may be considered to be a constant-depth waveguide with a pressure-release upper wall (the surface) and a compliant lower wall (the ocean bottom). In the situation to be addressed here, the planar boundary at $y = 0$ has local impedance $\rho_0 c \zeta_0$ and the boundary at $y = H$ has local impedance $\rho_0 c \zeta_H$; both parameters are considered to be the same at all locations on the respective surfaces. The normal directions into the fluid are $\bar{n} = \bar{e}_y$ and $\bar{n} = -\bar{e}_y$ at $y = 0$ and $y = H$, respectively. The local impedance condition is $P = \rho_0 c \zeta (-\bar{n} \cdot \bar{V})$. We use Euler's equation to describe $\bar{V}$, which leads to the boundary conditions at the walls being

$$P = -\rho_0 c \zeta_0 V_y = \frac{\zeta_0}{ik} \frac{\partial P}{\partial y} \text{ at } y = 0$$

$$P = \rho_0 c \zeta_H V_y = -\frac{\zeta_H}{ik} \frac{\partial P}{\partial y} \text{ at } y = H \tag{9.2.29}$$

The general solution for the pressure in any two-dimensional waveguide is given by Eq. (9.2.4). We substitute this representation into the boundary conditions. The requirement that the result be satisfied for all x allows cancelation of the $e^{-i\kappa x}$ factor. The result is a pair of linear algebraic equations for B and C,

$$\zeta_0 \mu C - ikB = 0$$
$$\zeta_H \mu [B \sin(\mu H) - C \cos(\mu H)] - ik[B \cos(\mu H) + C \sin(\mu H)] = 0 \tag{9.2.30}$$

The matrix form of these equations is

$$[D][B \ C]^{\mathrm{T}} = [0 \ 0]^{\mathrm{T}}$$

$$[D] = \begin{bmatrix} -ik & \zeta_0\mu \\ [\zeta_H\mu \sin(\mu H) - ik\cos(\mu H)] & -[\zeta_H\mu\cos(\mu H) + ik\sin(\mu H)] \end{bmatrix}$$
(9.2.31)

The coefficient equations are homogeneous, so their solution would give $B = C = 0$. We seek a nontrivial solution, which can only exist of the equations are rank-deficient. This condition is marked by the coefficient matrix having a zero determinant. The condition that $|[D]| = 0$ constitutes the characteristic equation

$$|[D]| = -\left(k^2 + \zeta_0\zeta_H\mu^2\right)\sin(\mu H) + i\left(\zeta_0 + \zeta_H\right)k\mu\cos(\mu H) = 0 \quad (9.2.32)$$

Multiplication of this equation by H^2 reveals that it contains only nondimensional parameters: $\mu H, kH, \zeta_0$, and ζ_H. The frequency is specified, so the only unknown in the characteristic equation is μH. The presence of sinusoidal terms whose argument is μH tells us that there are an infinite number of nondimensional roots, whose values we designate as η_n; the corresponding eigenvalues are $\mu_n = \eta_n/H$ and n is the mode number.

The preceding characteristic equation has a root at $\mu = 0$, but that root is extraneous and must be disregarded if either ζ_0 or ζ_H is finite. This is so because $\mu = 0$ corresponds to a simple plane wave. The particle velocity in a plane wave is solely in the x direction. The pressure in such a wave would cause the wall to move in the y direction, which is incompatible with the velocity field. The only exception is the case where both walls are rigid. It is noteworthy that the plane wave mode does not have a cutoff condition. The consequence is that only if both walls are rigid can sound propagate at any frequency, including values that are nearly zero.

The B and C coefficients corresponding to an eigenvalue must satisfy Eq. (9.2.31). Because this pair of equations are rank-deficient when μ equals any μ_n, only one of the scalar equations is independent. This means that we only can solve those equations for one coefficient as a proportionality to the other. We take B_n to be the arbitrary coefficient, where "n" subscript allows for a value that varies from mode to mode. When we replace μ with η_n/H in the first of Eq. (9.2.30) and solve for C/B, we find that the waveguide mode is

$$\boxed{\begin{array}{c} P = \Phi_n(y)\,e^{-i\kappa_n x} \\[4pt] \Phi_n = B_n\left[kH\sin\left(\eta_n\dfrac{y}{H}\right) - i\zeta_0\eta_n\cos\left(\eta_n\dfrac{y}{H}\right)\right] \\[4pt] -\left(k^2H^2 + \zeta_0\zeta_H\eta_n^2\right)\sin(\eta_n) + i\left(\zeta_0 + \zeta_H\right)kH\eta_n\cos(\eta_n) \\[4pt] \kappa_n = \pm\left(k^2 - \eta_n^2/H^2\right)^{1/2} \end{array}}$$
(9.2.33)

As was true for the case where both walls are rigid, the alternative sign for κ_n corresponds to waves that propagate in the x direction in either sense. If either wall's impedance has a resistive part, the root η_n will be complex. In such case, care must be taken to select the alternative sign. Stated differently, we must select the appropriate branch cut for the square root. Let us focus on the wave that propagates in the sense of increasing x, which means that we only are interested in η_n values that lead to Re $(\kappa_n) \geq 0$. Furthermore, the radiation condition requires that the magnitude of the pressure not grow in the direction of propagation, so it must be that Im $(\kappa_n) \leq 0$.

To identify the restriction this requirement imposes on the axial wavenumber, let us write κ_n in polar form as $\kappa_n = |\kappa_n| e^{i\phi_n}$. Thus, we seek the conditions for which a root μ_n will lead to $-\pi/2 \leq \phi_n \leq 0$. From the definition of μ_n, we have $\mu_n^2 = k^2 - |\kappa_n|^2 \exp(2i\phi_n) \equiv k^2 - |\kappa_n|^2 \cos(2\phi_n) - i |\kappa_n|^2 \sin(2\phi_n)$. If ϕ_n is in the proper range, then $-1 \leq \sin 2\phi_n \leq 0$, so Im $\left(\mu_n^2\right) \geq 0$. There is no limitation to $|\kappa_n|$, so the preceding tells us that μ_n^2 is in either the first or second quadrant, that is, $\mu_n^2 = u \exp(i\theta)$ with $u > 0$ and $0 \leq \theta \leq \pi$. DeMoivre's theorem gives two values: $\mu_n = u^{1/2} \exp(i\theta/2)$ and $\mu_n = -u^{1/2} \exp(i\theta/2)$. The second root, being the negative of the first, is accounted for as the second complex exponential in Eq. (9.2.3). Therefore, only the first value is retained. Because θ must lie in the first or second quadrant of the complex plane, it follows that the roots μ_n of the characteristic equation that are consistent with the radiation condition lie in the first quadrant of the complex plane. Roots that lie in other quadrants are extraneous.

A complex wavenumber κ_n leads to an oscillating function that decays as it propagates in the x direction. To understand the nature of the transverse dependence when κ_n is complex, we recall identities that convert sinusoidal functions having a complex argument to hyperbolic functions whose argument is real. They give

$$\begin{aligned} \sin(\alpha + \iota\beta) &\equiv \sin(\alpha)\cosh(\beta) + i\cos(\alpha)\sinh\beta \\ \cos(\alpha + \iota\beta) &\equiv \cos(\alpha)\cosh(\beta) - i\sin(\alpha)\sinh\beta \end{aligned} \tag{9.2.34}$$

If neither α nor β is negative, then both hyperbolic functions increase with increasing β. Thus, setting $\alpha + i\beta = \mu_n y$ shows that Φ_n varies in an oscillatory manner with a growing amplitude.

Cases where the eigenvalues are real are exceptional. One consists of the combination of a rigid and a pressure-release wall, which we take respectively to be $y = 0$ and $y = H$. The general characteristic equation, Eq. (9.2.32), is divided by ζ_0 to handle the infinite value of that impedance. Because $\zeta_H = 0$, what remains is $k\mu \cos(\mu H) = 0$. The cosine function must vanish because $\mu = 0$ is extraneous. Thus, the roots η_n are odd multiples of $\pi/2$. The modal properties are

$$\mu_n = \left(\frac{2n-1}{2}\right)\frac{\pi}{H}, \quad n = 1, 2, ..., \quad \kappa_n = \left[k^2 - n^2\left(\frac{\pi}{H}\right)^2\right]^{1/2}$$

$$\Phi_n = B_n \cos\left(\frac{(2n-1)}{2}\frac{\pi y}{H}\right) \tag{9.2.35}$$

$$N_{\text{cut}} = \text{floor}\left(\frac{kH}{\pi} + \frac{1}{2}\right), \quad \omega_{\text{cut}} = \left(\frac{2n-1}{2}\right)\frac{\pi c}{H}$$

As was true in the case where both walls are rigid, the variation in the y direction is a real harmonic function, which means that the pressure at all points situated at the same x oscillates in-phase.

Whereas a water channel might be approximated as having a lower rigid wall and an upper pressure-release wall, a waveguide with two pressure-release walls has no common incarnation. Such a system could be approximated by using lightly tensioned thin mylar sheets for the walls. However, this model is primarily useful as a limiting behavior. If $\zeta_0 = \zeta_H = 0$, then the general characteristic equation reduces to $\sin(\mu H) = 0$. This is the same equation as that for a rigid-rigid waveguide, except that μ_n cannot be zero here, so

$$\mu_n = n\frac{\pi}{H}, \quad \kappa_n = \left[k^2 - n^2\left(\frac{\pi}{H}\right)^2\right]^{1/2}, \quad n = 1, 2, \ldots$$

$$\Phi_n = B_n \sin\left(\frac{n\pi y}{H}\right) \tag{9.2.36}$$

$$N_{\text{cut}} = \text{floor}\left(\frac{kH}{\pi}\right), \quad \omega_{\text{cut}} = n\frac{\pi c}{H}$$

The case where one or both walls are purely reactive is representative of some common systems. For simplicity, let us take the wall at $y = 0$ to be rigid. The specific impedance of the wall at $y = H$ is taken to be $\zeta_H = i\sigma$, where σ is a real value that may be positive or negative. The characteristic equation in this case is found by dividing Eq. (9.2.32) by ζ_0 and then substituting for ζ_H. It is

$$-\sigma\mu H \sin(\mu H) + kH \cos(\mu H) = 0 \tag{9.2.37}$$

It is more difficult to extract the roots $\mu_n H = \eta_n$ of this equation, but once we do the other features follow. The transverse mode function, axial wavenumber, and cutoff parameters are described by

$$\Phi_n = B_b \cos\left(\frac{\eta_n y}{H}\right)$$

$$\kappa_n = \left(k^2 - \frac{\eta_n^2}{H^2}\right)^{1/2} \tag{9.2.38}$$

$$\eta_n < kH \text{ if } n \leq N_{\text{cut}}, \quad \omega_{\text{cut}} = \eta_n\frac{c}{H}$$

Due to the transcendental nature of the characteristic equation, numerical methods are required to find the eigenvalues. Those methods work best when they are initiated with a good estimate. It also is useful to have a general understanding of the nature of the root relative to the stiffness and the frequency. Both objectives are met by rewriting the characteristic equation as

$$\tan(\mu H) = \left(\frac{kH}{\sigma}\right)\frac{1}{\mu H} \tag{9.2.39}$$

The sole nondimensional parameter affecting the roots is the ratio of kH to σ. The case where $\sigma \ll kH$ corresponds to either a very small σ, meaning a wall that is very flexible, or else a very high frequency. The smaller roots $\mu H = \eta_n$ will be such that $\mu H \ll kH/\sigma$, so the characteristic equation in this case is well approximated as $\cot(\mu H) = 0$. The roots are the same as those of $\cos(\mu H) = 0$ for a pressure-release wall at $y = H$. The opposite case is that in which $\sigma \gg kH$, which requires that the wall be quite stiff, as well as that the frequency not be very high. In that case, the characteristic equation is approximately $\tan(\mu H) = 0$, and the approximation improves with increasing mode number because of the presence of μH in the denominator. The roots in this case are the same as those of $\sin(\mu H) = 0$ for a hard-walled waveguide.

These qualitative trends may be identified by a graphical construction of the functions of μH on either side of Eq. (9.2.39). Figure 9.10 shows the plots for a positive and a negative value of σ. The value of $kH/|\sigma|$ is 2.5π, which is not very large. Nevertheless, the first root is close to the first singularity of $\tan(\mu H)$, that is, $\cos(\mu_1 H) \approx 0$, which is the first root for a rigid-pressure-release configuration. Increasing the value of μH brings $(kH/\sigma)/(\mu H)$ closer to the zero axis, so the roots progressively become closer to the zeros of $\tan(\mu H)$. Thus, they approach the zeros of $\sin(\mu H)$, which is the characteristic equation for a hard-walled waveguide. Decreasing σ moves both hyperbolic curves farther from the zero axis. Nevertheless, the higher roots always tend to approach the zeros of $\tan(\mu H)$. In other words, regardless of how flexible a wall is, the higher eigenvalues always tend to the values for a hard-walled waveguide.

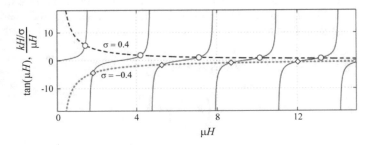

Fig. 9.10 Identification of the real eigenvalues as intersections of the functions forming the characteristic equation for a two-dimensional waveguide bounded by a rigid wall and a locally reactive wall, $kH = \pi$

An unexpected possibility is that an eigenvalue is purely imaginary, because an imaginary ζ_H can lead to a real characteristic equation. To explore this possibility, we set $\mu = i\beta$. Identities give $\cos(\mu H) \equiv \cosh(\beta H)$ and $\sin(\mu H) \equiv i \sinh(\beta H)$. A rearrangement of the terms in the characteristic equation gives

$$\tanh{(\beta H)} = -\left(\frac{kH}{\sigma}\right)\frac{1}{\beta H} \qquad (9.2.40)$$

Only positive values of β are meaningful because of the earlier analysis, which showed that μ must be in the first quadrant of the complex plane. Consequently, $\tanh{(\beta H)} > 0$, which means that a real root for βH can exist only if the function on the right side of the above equation is positive. In turn, this requires that $\sigma < 0$, which is the case of a wall that is a pure compliance. The plot of both sides of this equation in Fig. 9.11 confirms this interpretation. It also shows that if $\sigma < 0$. there is only one root. We will designate this root as $\beta H = \eta_{im}$, so that $\mu_{im} = i\eta_{im}/H$. In general, we will use a subscript "im" to denote this type of mode. As a generality, this type of mode is not possible if a wall is an inertance.

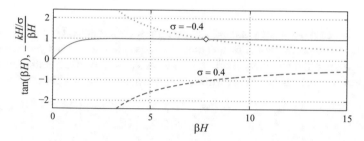

Fig. 9.11 Identification of the imaginary eigenvalue for a two-dimensional waveguide bounded by a rigid wall and a locally reactive wall, $kH = \pi$

The transverse mode function corresponding to a real eigenvalue is given Eq. (9.2.38). Substitution of $\eta_n = i\beta_{im}$ yields

$$\Phi_{im} = B \cosh{\left(\eta_{im}\frac{y}{H}\right)} \qquad (9.2.41)$$

This function, like those for the real eigenvalues, is real. This means the pressure in any mode oscillates either in-phase or 180° out-of-phase along any cross section at fixed x. Fig. 9.12 shows a few transverse mode functions. Figure 9.11 indicates that the value of the root η_{im} will be very large, unless kH/σ is very small. The consequence of a large value of η_{im} is that Φ_{im} increases greatly in the vicinity of the compliant wall. Another interesting aspect is that the axial wavenumber for this mode is greater than k because $\kappa_{im} = \left(k^2 - \mu_{im}^2\right)^{1/2} = \left(k^2 + \eta_{im}^2/H^2\right)^{1/2}$. Thus, this type of mode propagates much more slowly than the speed of sound.

Let us compare the results for cases where the wall impedance is infinite (rigid), zero (pressure-release), or imaginary (reactive), to the general situation when the real part of the impedance at either wall is nonzero. The former are idealized cases in which dissipation is absent, whereas energy is absorbed in the wall when $\mathrm{Re}\,(\zeta) > 0$. The presence of dissipation complicates the analysis by requiring the solution of a complex characteristic equation. The eigenvalues μ_n are complex, which means that

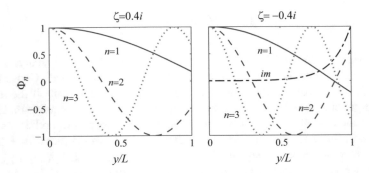

Fig. 9.12 Mode functions for a two-dimensional waveguide bounded by a rigid wall and a compliant wall, $kH = \pi$

the pressure as a function of y at fixed x shows phase differences. The axial wavenumber κ_n also is complex, which means that a wave is attenuated as it propagates. We would recognize the attenuation effect as a consequence of dissipation, even if we did not perform an analysis. Phase differences are a more subtle effect of dissipation.

EXAMPLE 9.4 The lower wall of a two-dimensional waveguide is rigid, and the upper wall is purely resistive with $Z = 0.1\rho_0 c$. Determine the phase and group velocities for the first three modes in the frequency range $0 < kH < 10$. Also, plot the transverse mode functions at $kH = 0.1$, 1.0, and 10.

Significance

A highlight is determination of the eigenvalues when the characteristic equation is complex. Another new feature is a general approach for extracting group velocities from data. The results will display an interesting phenomenon exhibited by the group velocity.

Solution

Before we can solve the characteristic equation we must consider what is meant by the specification "first three modes". If the eigenvalues were real, we would identify the mode number by sequencing μ_n in increasing order. A logical extension for complex eigenvalues is to sequence them in ascending order of their real part, $\text{Re}\,(\mu_1) < \text{Re}\,(\mu_2) < \ldots$. The characteristic equation we shall solve is the limiting form that results from multiplying Eq. (9.2.32) by H/ζ_0 and then taking the limit as $\zeta_0 \to \infty$. The result is

$$f(\mu H) = -\zeta_H(\mu H)\sin(\mu H) + i(kH)\cos(\mu H) = 0$$

where $\zeta_H = 0.1$ in the current case.

The standard methods for solving a single nonlinear equation $f(x) = 0$ when x and f are scalars also are valid for complex variables. Many software packages contain the necessary modifications. MATLAB (as of the 2009) does not. The results presented here were obtained by programming a standard Newton-Raphson procedure, which states that the jth estimate for a root is given by

$$\mu_n^{(j)} H = \mu_n^{(j-1)} H - \left. \frac{f(\mu H)}{\dfrac{d}{d\mu} f(\mu H)} \right|_{\mu H = \mu^{(j-1)} H}$$

where

$$\frac{d}{d(\mu H)} f(\mu H) = -(\zeta_H + ikh) \sin(\mu H) - \zeta_h(\mu H) \cos(\mu H)$$

Nonlinear equation solvers generally require initial estimates of the roots. A robust approach that is suitable if it is only necessary to find the eigenvalues at a few values of kH entails inspection of data contours. The characteristic equation is a complex-valued function of the complex variable μH. When it and μH are decomposed into real and imaginary parts, we may graph $\text{Re}(f(\mu H))$ and $\text{Im}(f(\mu H))$ as surfaces above the complex plane μH whose orthogonal axes are $\text{Re}(\mu H)$ and $\text{Im}(\mu H)$. If both parts of f are zero at the same point in the complex μH-plane, then that point marks a root of the characteristic equation. The difficulty is that depiction of these surfaces and their identification of the zeros would be challenging.

An alternative approach recognizes that if both parts of $f(\mu H)$ are zero at a root, then it must be that $|f(\mu H)| = 0$ at such locations. A plot of $|f(\mu H)|$ constitutes a single surface above the μH complex plane. Its value may be depicted by a contour plot. Because $|f(\mu H)|$ cannot be negative, a root must occur at a location where $|f(\mu H)|$ is a minimum. Minima in a contour plot appear as a point interior to a contour that forms a closed curve. However, closed curves also surround maxima, and a minimum value might not be zero. Thus a reasonable procedure is to draw the contours, evaluate $|f(\mu H)|$ at a point inside a closed loop, and test if that value is close to zero. If so, that value of μH is a candidate as a starting value for the root search. Recall that only the contours lying in the first quadrant should be examined because the meaningful roots are those that give $\text{Re}(\mu_n) \geq 0$ and $\text{Im}(\mu_n) \geq 0$.

This procedure is illustrated in Fig. 1, for which $kH = \pi$ and $\zeta_H = 0.1$. Because only the first quadrant is depicted there, a loop that actually is closed might appear to be open if it is situated near the real or imaginary axis. In view of this possibility, Fig. 1 suggests four roots, $\mu_n H \approx (2n-1)\pi/2 + 0.1ni$. The value of $|f(\mu H)|$ at each of these points is less than 0.1, whereas $|f(\pi + i)| = 6.4$, which suggests that the minima might be zeros. The actual roots were found to be $\mu_1 H/\pi = 0.49949 + 0.0159i$, $\mu_2 H = 1.4984 + 0.0481i$, $\mu_3 H = 2.4972 + 0.0812i$.

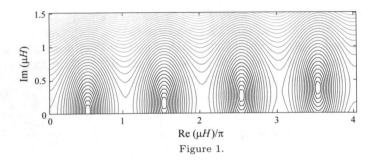

Figure 1.

A different approach followed by some individuals finds estimates by considering an ideal system, that is, one that is purely reactive. For example, in the present situation the value of ζ_H is much less than one. This suggests that the μ_n values are close to the values for $\zeta_H = 0$, which is the pressure-release condition described by Eq. (9.2.35). The real part of the estimates identified from the contour plot are these values. The potential difficulty with this approach is that it does not address the case where there is no comparable ideal system, which would be the case if Re (ζ_H) is not small.

The present task calls for evaluation of the eigenvalues for a range of frequencies. This suggests yet another approach for generating the initial guesses. The eigenvalues are continuous analytical functions of kH. Thus, if we increment kH by a small amount, we can use the values identified for the previous frequency as starting estimates for the new frequency. A benefit of following this approach is that it only requires a contour map to identify the eigenvalues at the lowest kH. We begin with a contour plot of $|f(\mu H)|$ for $kH = 0.01\pi$, which is Fig. 2. It suggests that the smallest eigenvalue is close to the origin, while the higher eigenvalues are multiples of π plus a small imaginary part. Thus, the starting estimates used to initialize the equation solver at $kH = 0.01\pi$ are $\mu_1 H \approx 0.1\pi + 0.2i$, $\mu_2 H \approx \pi + 0.1i$, $\mu_3 H = 2\pi + 0.05i$.

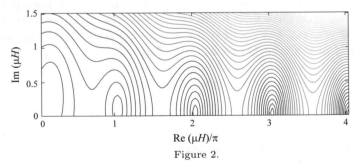

Figure 2.

The procedure that was followed set a loop in which kH was increased gradually. At each frequency, a loop over the mode number n initialized the $\mu_n H$ value with the value found at the prior frequency. This value was input to the routine that solves the characteristic equation. The output was the value of $\mu_n H$ at the current frequency. After the eigenvalues were determined, the $\kappa_n H$ were found from the definition, Eq. (9.2.33), after which the frequency was incremented and the procedure repeated. The result is described in Fig. 3. It will be noted that $\mu_n H$ is in the first quadrant, and $k_n H$ is in the fourth quadrant, as required. The fact that all plots are continuous curves is an indication that the procedure was carried out correctly.

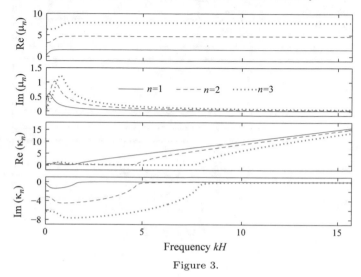

Figure 3.

Each mode shows a transition in its κ_n value as the frequency increases. Below a certain frequency, Re (κ_n) is small but nonzero, and Im (κ_n) is relatively large negatively. Above that frequency, Re (κ_n) begins to grow and Im (κ_n) is small negatively. This behavior is analogous to cutoff for situations in which the wall does not dissipate energy.

Division of a $\kappa_n H$ value by the corresponding value of kH gives c_{phase}/c. Evaluation of the group velocity requires a little more effort. The nondimensional version of Eq. (9.1.45) is

$$\frac{c_g}{c} = \text{Re}\left(\frac{d\,(kH)}{d\,(\kappa H)}\right)$$

Although we do not have a functional dispersion equation, we can use finite differences to differentiate the data. A central difference formula based on a uniform frequency increment $\Delta = (kH)_j - (kH)_{j-1}$ is

$$\left(\frac{c_{\mathrm{g}}}{c}\right)_j \approx \frac{\Delta}{\mathrm{Re}\,(\kappa H)_{j+1} - \mathrm{Re}\,(\kappa H)_{j-1}}$$

The results of these evaluations appear in Fig. 4.

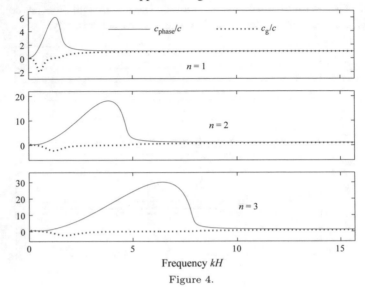

Figure 4.

As the frequency increases, c_{phase} approaches c from above, and c_{g} approaches c from below. The large values of c_{phase} occur in the low-frequency range where κ_n is essentially a negative imaginary value. This means that the waveguide mode is evanescent. The consequence is that neither the high phase speed nor the negative group velocity is meaningful, because the wave decays over a very short distance.

The transverse mode functions are computed by substituting into Eq. (9.2.33) the values of $\mu_1 H$, $\mu_2 H$, and $\mu_3 H$ at $kH = 0.01$, 1, and 10. The amplitude factor B_n for each is arbitrary in regard to a pictorial representation. Because the eigenvalues are complex, setting all $B_n = 1$ would lead to mode functions that have substantially different magnitudes. To facilitate comparisons, the transverse mode functions in Fig. 5 have been scaled by evaluating each as a function of y/H for $B_n = 1$ and then dividing each by its maximum magnitude. (The ordinate axis in Fig. 5 is y/H, because the transverse direction runs vertically in the configuration depicted in Fig. 9.7.)

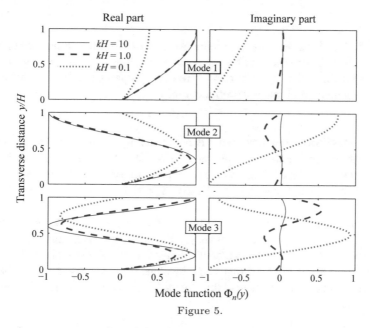

Figure 5.

A few features are worth noting. The highest frequency, $kH = 10$, exceeds the cutoff value for each of the three modes. The consequence is that the imaginary part of each mode function is much smaller than its real part. Indeed, the real parts are nearly proportional to $\cos(\pi x/(2H))$, $\cos(3\pi x/(2H))$, and $\cos(5\pi x/(2H))$, which are the first three modes for a rigid/pressure-release configuration. This behavior indicates that the impedance is sufficiently low that the surface effectively is soft at high frequencies. If kH is below the cutoff frequency for a mode, the imaginary part of the mode is comparable to the real part, and it is oscillatory in a pattern that resembles the real part

If we were to compare the present results to those for $\zeta_H = 0$, we would see that the overall effect of dissipation associated with power transfer across the locally reacting wall is quite small. This is especially true for modes that are not cutoff. For those modes, the wave is gradually attenuated as it propagates. The modes whose cutoff frequency exceeds the oscillation rate seem to be much affected by dissipation, as suggested by a negative group velocity. However, these modes are rapidly attenuated, so it would be difficult to observe such features.

EXAMPLE 9.5 The walls of a two-dimensional waveguide are identical elastic plates whose properties are Young's modulus E, density ρ_p and thickness h. The fluid is water, and the material properties of the plate are $E = 100\rho_0 c^2$, $\rho_p/\rho_0 = 5$, and $h/H = 0.02$. Frequencies of interest are $kH = 0.1$, 1, and 10. For each frequency, determine the eigenvalues of the transverse mode function in the range $\mu_n H < 10$. Also determine the corresponding values of $\kappa_n H$ and the ratio of the plate displacement amplitude to the pressure amplitude in each mode.

Significance

The interaction between different media usually leads to interesting phenomena. In a sense, this is an extension of the analysis of coincidence frequency for an elastic plate in Sect. 5.5.2. The application of symmetry properties to simplify an analysis, which is a general tool, plays a prominent role here.

Solution

The walls are flexible, but they are not locally reacting. Rather, application of a force results in a displacement wave. The consequence it that it is necessary to formulate the equation of motion for the plates. The pressure exerted by the fluid acts as a load. In addition, enforcement of continuity of particle velocity at the walls must be satisfied.

All features are the same above and below the central plane. A general property of a system that has a plane of symmetry is that the response is either symmetric or antisymmetric with respect to that plane. To exploit that feature the origin is placed midway between the plates so that xz is the plane of symmetry. Figure 1 shows the set up. The normal displacement of the plate w is defined to be positive into the fluid.

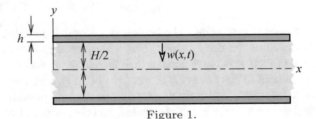

Figure 1.

The frequency-domain representation of the pressure and displacement in a mode are

$$p(x, y, t) = \text{Re}\left(P(x, y)\, e^{i\omega t}\right), \quad w(x, t) = \text{Re}\left(W(x)\, e^{i\omega t}\right) \tag{1}$$

Every point (x, y) above the bisecting xz-plane is matched by a point at $(x, -y)$. A symmetric response has the property that $P(x, -y) = P(x, y)$. In other words P is an even function of y. An antisymmetric response is such that $P(x, -y) = -P(x, y)$, which means that it is an odd function of y. Because P is an analytic function, it must be that $\partial P/\partial y = 0$ at $y = 0$ for a symmetric response and $P = 0$ at $y = 0$ for an antisymmetric response. These, respectively, are the boundary conditions for a rigid and pressure-release wall. Thus, we will consider two configurations of a two-dimensional waveguide whose width is $H/2$: rigid at $y = 0$ and an elastic plate at $y = H/2$ for symmetric modes, and pressure-release at $y = 0$ and an elastic plate at $y = H/2$ for antisymmetric modes. We shall denote which type is under consideration with a superscript "s" or "a", so the effective wall conditions are

$$\frac{\partial P^{(s)}}{\partial y} = 0 \text{ at } y = 0, \quad \bar{n} \cdot \nabla P^{(s)} \equiv -\frac{\partial P^{(s)}}{\partial y} = -i\rho_0\omega\,(-i\omega W) \text{ at } y = H/2$$

$$\frac{}{} \tag{2}$$

$$P^{(a)} = 0 \text{ at } y = 0, \quad \bar{n} \cdot \nabla P^{(a)} \equiv -\frac{\partial P^{(a)}}{\partial y} = -i\rho_0\omega\,(-i\omega W) \text{ at } y = H/2$$

The pressure must be a solution of the Helmholtz equation. Propagation in the axial direction is represented by $\exp(-i\kappa x)$. For the y direction we can immediately take care of the boundary condition at $y = 0$ in Eq. (2) by recognizing that a cosine function gives a zero gradient, whereas a sine function gives a zero value. We leave the argument of each as μy, so the pressure may be written as

$$P^{(s)} = B^{(s)} \cos\left(\mu^{(s)} y\right) e^{-i\kappa^{(s)}x}$$
$$P^{(a)} = B^{(a)} \sin b\left(\mu^{(a)} y\right) e^{-i\kappa^{(a)}x} \tag{3}$$

These forms satisfy the Helmholtz equation if

$$\left(\kappa^{(s)}\right)^2 + \left(\mu^{(s)}\right)^2 = k^2, \quad \left(\kappa^{(a)}\right)^2 + \left(\mu^{(a)}\right)^2 = k^2, \tag{4}$$

The time-domain equation of motion for the plate is Eq. (5.5.27). The pressures p_1 and p_2 appearing there are applied on either side of the plate. In the present situation we shall ignore the fluid loading outside the waveguide. (This is acceptable if there is a liquid within the waveguide and a gas outside, but it might not be a valid assumption if the fluids are similar.) The displacement is defined to be positive if it is into the waveguide, and the pressure pushes the plate away from the fluid. Hence a negative p should cause a positive value of $\partial^2 w/\partial t^2$, which means the pressure term in the equation of motion that is preceded by the negative sign is the one to use. In the two-dimensional problem, there is no variation in the z direction. Thus, the equation of motion is

$$\rho_p h \frac{\partial^2 w}{\partial t^2} + D \frac{\partial^4 w}{\partial x^4} = -\left. p \right|_{y=H/2}$$

where p may be either the symmetric or antisymmetric variable. The parameter D is the flexural rigidity. If the plate is composed of a homogeneous material, rather than a composite, then

$$D = \frac{Eh^3}{12\left(1 - \nu^2\right)}$$

Equation (1) gives the frequency-domain representation of w. To use it we must recognize that the amplitude function $W(x)$ must match the x dependence of the pressure, because it is necessary to match W and P at the plate's surface. Thus, W must be representable as

$$W(x) = \tilde{W} e^{-i\kappa x} \tag{5}$$

where κ and $\tilde{W}$ may be the value for symmetric or antisymmetric modes. Substitution of the representations of W and P into the plate equation of motion leads to

$$\left(D\kappa^4 - \rho_p h\omega^2\right)\tilde{W}^{(s)} = -B^{(s)}\cos\left(\mu^{(s)}\frac{H}{2}\right)$$
$$\left(D\kappa^4 - \rho_p h\omega^2\right)\tilde{W}^{(s)} = -B^{(a)}\sin\left(\mu^{(a)}\frac{H}{2}\right)$$

(6)

The last set of equations result from enforcing the continuity conditions in Eq. (2). These equations are

$$B^{(s)}\mu^{(s)}\sin\left(\mu^{(s)}\frac{H}{2}\right) = \rho_0\omega^2\tilde{W}^{(s)}$$
$$-B^{(a)}\mu^{(a)}\cos\left(\mu^{(a)}\frac{H}{2}\right) = \rho_0\omega^2\tilde{W}^{(s)}$$

(7)

Equations (6) and (7) describe the coupling of the acoustic and structural responses.

Let us begin with the analysis of the symmetric modes. The matrix form of Eqs. (7) and (6) for $B^{(s)}$ and $\tilde{W}$ is

$$\left[D^{(s)}\right]\left\{\begin{array}{c}B^{(s)}\\\tilde{W}^{(s)}\end{array}\right\} = \left\{\begin{array}{c}0\\0\end{array}\right\}$$

(8)

where

$$\left[D^{(s)}\right] = \begin{bmatrix} \mu^{(s)}\sin\left(\mu^{(s)}\dfrac{H}{2}\right) & -\rho_0\omega^2 \\[2ex] \cos\left(\mu^{(s)}\dfrac{H}{2}\right) & \left(D\kappa^4 - \rho_p\omega^2\right) \end{bmatrix}$$

Equation (8) leads to $B^{(s)} = \tilde{W}^{(s)} = 0$ unless the pair of scalar equations is rank-deficient. Thus, we set $\left|\left[D^{(s)}\right]\right| = 0$, which gives

$$\left(D\kappa^4 - \rho_p h\omega^2\right)\mu^{(s)}\sin\left(\mu^{(s)}\frac{H}{2}\right) + \rho_0\omega^2\cos\left(\mu^{(s)}\frac{H}{2}\right) = 0$$

Because the frequency is specified, the value of κ^2 is defined in terms of $\mu^{(s)}$. This means that the characteristic equation for fixed system parameters and frequency is a function of $\mu^{(s)}$. In a nondimensional sense it is a function of $\eta = \mu^{(s)}H$, such that

$$F^{(s)}(\eta) = \left\{ S\frac{\left[(kH)^2 - \eta^2\right]^2}{(kH)^2} - 1 \right\}\eta\sin\left(\frac{\eta}{2}\right) + \frac{H}{h}\frac{\rho_0}{\rho_p}\cos\left(\frac{\eta}{2}\right) = 0 \quad (9)$$

The parameter S is a measure of the dynamic stiffness of the plate. Based on an assumption that Poisson's ratio is $\nu = 0.3$, it is

$$S \equiv \frac{D}{\rho_p h H^2 c^2} = \frac{1}{12\left(1 - \nu^2\right)} \frac{E}{\rho_0 c^2} \left(\frac{\rho_0}{\rho_p}\right) \left(\frac{h}{H}\right)^2 = 0.080$$

The derivation of the equations for the antisymmetric modes follows a similar path. When the pressure is $P^{(a)}$ in Eq. (3), the coefficient equations are like Eq. (8),

$$\left[D^{(a)}\right] \left\{ \begin{array}{c} B^{(a)} \\ \tilde{W}^{(a)} \end{array} \right\} = \left\{ \begin{array}{c} 0 \\ 0 \end{array} \right\}$$

$$\left[D^{(a)}\right] = \begin{bmatrix} \mu^{(a)} \cos\left(\mu^{(a)} \dfrac{H}{2}\right) & \rho_0 \omega^2 \\ \sin\left(\mu^{(s)} \dfrac{H}{2}\right) & \left(D\kappa^4 - \rho_p \omega^2\right) \end{bmatrix} \tag{10}$$

The characteristic equation for the antisymmetric modes is

$$F^{(a)}(\eta) = \left\{ S \frac{\left[(kH)^2 - \eta^2\right]^2}{(kH)^2} - 1 \right\} \eta \cos\left(\frac{\eta}{2}\right) - \frac{H}{h} \frac{\rho_0}{\rho_p} \sin\left(\frac{\eta}{2}\right) = 0 \tag{11}$$

The rank reduction of the Eqs. (8) and (10) at an eigenvalue means that only one equation is independent. We solve the first for the plate displacement in terms of the pressure amplitude, which gives

$$\frac{W_n^{(s)}}{H} = \frac{\mu_n^{(s)} H \sin\left(\mu^{(s)} \dfrac{H}{2}\right)}{(kH)^2} \left(\frac{B^{(s)}}{\rho_0 c^2}\right)$$

$$\frac{W_n^{(a)}}{H} = -\frac{\mu_n^{(s)} H \cos\left(\mu^{(a)} \dfrac{H}{2}\right)}{(kH)^2} \left(\frac{B^{(a)}}{\rho_0 c^2}\right) \tag{12}$$

When the characteristic equations, Eqs. (9) and (11), are evaluated at a fixed kH, they become functions of a single variable, η. These equations may be solved numerically, but doing so requires initial guesses. We shall obtain these by plotting $F^{(a)}(\eta)$ and $F^{(s)}(\eta)$ as functions of η. The eigenvalues are the values of η at which a function crosses the zero axis. Figures 2, 3, and 4 show such plots for $kH = 0.1$, $kH = 1$, and $kH = 10$. The eigenvalues are marked as circles. The curves marked as $F_{ac}^{(s)}$ and $F_{ac}^{(a)}$ denote the second term in each characteristic equation. They are the terms that remain as $\rho_p h / (\rho_0 H) \to 0$. Hence, they represent the purely acoustic effect if the plate did not offer any opposition to an applied pressure, that is, if the walls were pressure-release. The eigenvalues in that case would be $\mu_n H = (2n - 1)\pi$ for the symmetric modes and $\mu_n H = 2n\pi$ for the antisymmetric modes. The contrary case is $\rho_0 H / (\rho_p h) \to 0$. This corresponds to a plate that is extremely massive, and therefore immobile. In such situations the first part of each characteristic equation is dominant. The same behavior is obtained in the limit as $kH \to 0$, that is, low frequencies. The corresponding eigenvalues are $\mu_n H = 2n\pi$ for the symmetric modes and $\mu_n H = (2n - 1)\pi$ for the antisymmetric modes. For the most part, the roots in Fig. 2 match those for the rigid wall limit. The exception are the first two antisymmetric modes. We will see that the axial wavenumber for these modes match the wavenumber at which a flexural wave at $kH = 0.1$ can propagate along the plate in a vacuum. The plate cannot sustain an external load (the pressure) in such conditions.

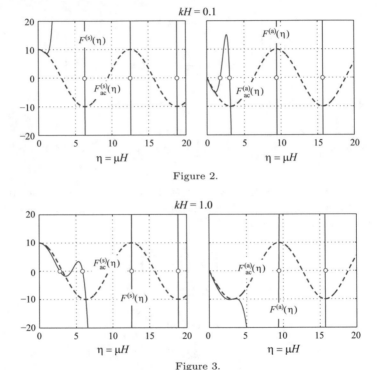

Figure 2.

Figure 3.

At the higher frequency, $kH = 10$, described in Fig. 4, neither the pressure-release nor rigid limits for the wall match the computed eigenvalues. Furthermore, the increment between adjacent eigenvalues is irregular. Both features indicate that the structural dynamics behavior of the wall has an important influence on the acoustic response. The modal properties obtained from numerical solution of both characteristic equations are tabulated below. Perhaps the most obvious feature is the irregularity of the sequence in which symmetric and antisymmetric modes occur at low frequencies. Another manifestation of the structural dynamic effect is the fact that the wall displacement to pressure ratio is much lower at the higher frequency. An important property is that all $\kappa_n H$ values are imaginary for $kH = 0.1$ and 1.0. This means that all of the low-frequency modes are evanescent, so a signal will not propagate in the low-frequency range.

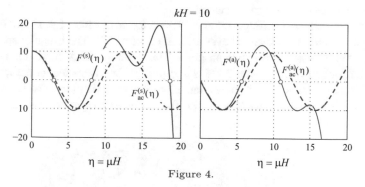

Figure 4.

kH	Mode #n	$\mu_n H$	$\kappa_n H$	$\dfrac{W_n/H}{B_n/(\rho_0 c^2)}$
0.10	Antisymmetric 1	1.7770	−1.7742i	−112.054
	Antisymmetric 2	3.0445	−3.0428i	−14.780
	Symmetric 1	6.2806	−6.2798i	0.8044
	Antisymmetric 3	9.4244	−9.4239i	0.1585
1.0	Symmetric 1	2.7933	−2.6082i	2.7511
	Symmetric 2	4.3647	−4.2486i	3.5736
	Symmetric 3	5.8555	−5.7695i	1.2426
	Antisymmetric 1	9.3891	−9.3357i	0.1673
10.0	Symmetric 1	2.9461	9.5562	0.02932
	Antisymmetric 1	5.6095	8.2785	0.05294
	Symmetric 2	8.1489	5.7961	−0.06546

A graph showing how the eigenvalues depend on kH would be helpful for a closer examination. Such a picture could be obtained by following the preceding analysis for a sequence of closely space kH values. Fortunately, a more efficient method is available. It is based on considering the characteristic equations, Eqs. (9) and (11), to be functions of kH, as well as μH. The value of each function may be considered to form a surface above the $\kappa H, \mu H$-plane. The eigenvalues are the $\kappa H, \mu H$ where the elevation of this surface is zero. Thus, the numerical procedure entails defining a grid of kH and μH values, evaluating $F^{(s)}$ and $F^{(a)}$ at each grid point, and then finding the zero contour of each set of function data. The contours of equal values of $F^{(s)}$ and $F^{(a)}$ are computed. The only contours of interest are those for which $F^{(s)}(\eta) = 0$ or $F^{(a)}(\eta) = 0$. In MATLAB, the operations would be hold on; kH=0:d_k:kH_max; eta=0:d_e:eta_max; C_sym = contour(kH, eta/pi, F_val_sym.', [0 0]); C_anti = contour(kH, eta/pi, F_val_anti.', [0 0]);. In these operations, the contour command graphs the data, the last argument being [0 0] specifies that only the zero contour should be plotted, and hold on at the beginning of the procedure retains the first set of contours when the second set are plotted. A useful aspect is that the $kH, \mu H$ coordinates of the points on each contour are embedded in C_sym and C_anti. This data may be used for further study, for example, to determine the group velocity.

Figure 5 shows the resulting contours. When viewed as plots of $\mu_n H$ as a function of kH, each curve is multivalued over a small range of kH. The irregular spacing of the tabulated data stems from this property. The shading marks the region in which $\mu H > kH$, or equivalently, where κH is imaginary. The associated modes evanesce. Inspection of the contour data shows that all eigenvalues lie in this shaded region if $kH < 2.64$. This means that $kH = 2.64$ is the minimum frequency at which a signal can propagate.

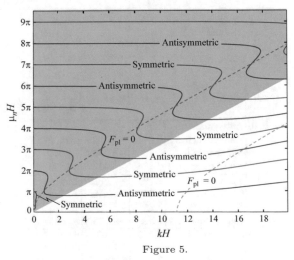

Figure 5.

The dashed curves in Fig. 5 marked $F_{pl} = 0$ are the locus of points at which the term within braces in both characteristic equations vanishes. This term is

$$F_{pl} = \frac{D}{\rho_p h H^2 c^2} \frac{\left[(kH)^2 - \eta^2\right]^2}{(kH)^2} - 1$$

This factor is the nondimensional version of $D\kappa^4 - \rho_p \omega^2$ after substitution of $c^2 \kappa^2 = \omega^2 - c^2 \mu^2$. In other words it is the dispersion equation for waves in an elastic plate that is not loaded by the fluid. The intersection of an F_{pl} curve with one of the eigenvalue contours corresponds to the coincidence frequency for a plate bounding a half-space. When $F_{pl} = 0$, the inertia and stiffness effects of the plate cancel. In the vicinity of the lower curve, the value of F_{pl} is relatively small, so the eigenvalues are relatively insensitive to this term. The upper curve is the contrary condition. The intersection of this F_{pl} curve with any eigenvalue curve is centered in the region where each curve is multivalued, which suggests that these values are very sensitive to the plate's properties. Note that the axial phase speed is ω/κ, so the lower F_{pl} curve describes waves in the plate/wall that are supersonic.

A lesson that can be gleaned from this example is that systems in which there is a strong interaction between an acoustic medium and a structure might exhibit unexpected effects. We also have seen that the cause of these effects can be identified if the properties of the system's response are examined from an appropriate perspective. It is worth noting that the steps leading to the governing equations are typical of all systems in which elastic and acoustic subsystems strongly interact. Continuity of the normal displacement at the interface leads to a boundary condition for the fluid, and the pressure applied to the structure is accounted for as a dynamic load.

9.2.5 Orthogonality and Signal Generation

The primary motivation for identifying the modes of a waveguide is that they simplify an analysis of the signal that is generated by an excitation. This excitation is taken to be a specification of the axial particle velocity on the cross section at $x = 0$. The velocity distribution is $V_s (y)$, so the boundary condition is

$$V_x = -\frac{1}{i\omega\rho_0}\frac{\partial P}{\partial x} = V_s (y) \text{ at } x = 0 \qquad (9.2.42)$$

The other condition is that waves are assumed to be absorbed or perfectly transmitted at a long range, so we are solely concerned with waves that propagate in the positive x direction.

How to satisfy the source condition is the question addressed here. The situation is actually close to that which arose for a spherical source. There the signal propagated to an infinite radial distance, and we used a spherical harmonic series to represent the transverse variation. Here the propagation is in the axial direction. We know how the signal in a mode varies in the transverse y direction, so it is logical to consider a *modal series*,

$$P = \sum_{n=0}^{\infty} P_n \Phi_n (y) e^{-i\kappa_n x} \qquad (9.2.43)$$

In this expression the P_n coefficients are called the *modal amplitudes* of the mode functions. The sum is indicated to begin at $n = 0$, but that is reserved for the plane wave mode in the case of a hard-walled waveguide. Also, if the waveguide has modes whose eigenvalues are purely imaginary, they must be inserted into this series.

The path leading to an expression for the modal amplitudes is like that by which the coefficients of a Fourier series are determined. Indeed, some individuals refer to a modal series, Eq. (9.2.43), as a generalized Fourier series. Both series are a sum of a basis function, modal or trigonometric, multiplied by an unknown coefficient. Each function represents a direction in a generalized linear space and the coefficient represents a vector component in that direction. The similarity extends further, in that the basis functions form an orthogonal set.

The procedure for deriving the orthogonality property is part of Sturm–Liouville analysis in a course on partial differential equations. We begin by observing that the transverse modes satisfy the mixed boundary conditions at $y = 0$ and $y = H$ stated in Eq. (9.2.29),

$$\Phi_n = \frac{\zeta_0}{ik} \frac{d\Phi_n}{dy} \quad \text{at } y = 0$$

$$\Phi_n = -\frac{\zeta_H}{ik} \frac{d\Phi_n}{dy} \quad \text{at } y = H$$

(9.2.44)

Within the domain of the fluid the mode functions satisfy a one-dimensional Helmholtz equation,

$$\frac{d^2}{dy^2} \Phi_n + \mu_n^2 \Phi_n = 0$$

(9.2.45)

The key step is selection of two arbitrary mode indices n and j. The Helmholtz equation for each equation is multiplied by the other mode function. The difference of the two products is

$$\Phi_j \left(\frac{d^2}{dy^2} \Phi_n + \mu_n^2 \Phi_n \right) - \Phi_n \left(\frac{d^2}{dy^2} \Phi_j + \mu_j^2 \Phi_j \right) = 0$$

(9.2.46)

The next operation entails integrating this expression over $0 < y < H$,

$$\int_0^H \left(\Phi_j \frac{d^2}{dy^2} \Phi_n - \Phi_n \frac{d^2}{dy^2} \Phi_j \right) dy + (\mu_n^2 - \mu_j^2) \int_0^H \Phi_j \Phi_n dy = 0$$

(9.2.47)

Integration by parts is applied to the integral containing derivatives. Let us examine one of those terms,

$$\int_0^H \Phi_j \frac{d^2}{dy^2} \Phi_n \, dy \equiv -\int_0^H \frac{d}{dy} \Phi_j \frac{d}{dy} \Phi_n \, dy + \left(\Phi_j \frac{d}{dy} \Phi_n \right) \Big|_{y=H}$$
$$- \left(\Phi_j \frac{d}{dy} \Phi_n \right) \Big|_{y=0}$$

(9.2.48)

If a boundary is rigid, then the normal derivative is zero, whereas the function is zero if a boundary is pressure-release. In either case the terms that are evaluated at the boundaries vanish. Otherwise the impedance boundary conditions in Eq. (9.2.44) relate the y derivative to the function at each wall. Their substitution into the preceding gives

$$\int_0^H \Phi_j \frac{d^2}{dy^2} \Phi_n \, dy = -\int_0^H \frac{d}{dy} \Phi_j \frac{d}{dy} \Phi_n \, dy + \left(\frac{ik}{\zeta_H} \Phi_j \Phi_n \right) \Big|_{y=H}$$
$$- \left(\frac{ik}{\zeta_0} \Phi_j \Phi_n \right) \Big|_{y=0}$$

(9.2.49)

When the same set of operations is performed on the other derivative term in Eq. (9.2.47), the result has the form of this relation with n and j swapped. However, such an interchange does not alter the result, which means that all terms cancel. Thus, regardless of the wall impedances, the integral in Eq. (9.2.47) that contains derivatives vanishes. What remains is the *orthogonality property*,

$$\left(\mu_n^2 - \mu_j^2\right) \int_0^H \Phi_j \Phi_n \, dy = 0 \qquad (9.2.50)$$

If j and n are different, then the integral must vanish. If the indices are equal, then the integral will not evaluate to zero. We can use this fact to set the arbitrary coefficient B_n in Eq. (9.2.33). In particular we will select it such that the integral equals H, which will make the transverse mode functions dimensionless. Let $\tilde{\Phi}_n$ denote a transverse mode function when $B_n = 1$, so that

$$\boxed{\begin{aligned} \Phi_n &= B_n \tilde{\Phi}_n \\ B_n &= \left[\frac{1}{H} \int_0^H \left(\tilde{\Phi}_n\right)^2 dy \right]^{-1/2} \end{aligned}} \qquad (9.2.51)$$

The integral property for any pair of n and j values is captured by using a Kronecker delta, $\delta_{n,j} = 0$ if $j \neq n$ and $\delta_{n,j} = 1$ if $j = n$. Thus,

$$\boxed{\int_0^H \Phi_j \Phi_n \, dy = H \delta_{n,j}} \qquad (9.2.52)$$

From a linear algebra perspective the integral is the scalar product of Φ_n and Φ_j in the functional space of a single Cartesian coordinate. Correspondingly, we say that the mode functions form an *orthonormal* set. Without the above definition of B_n, the mode functions would be the analog of orthogonal vectors that have arbitrary magnitude. With it, the mode functions are analogous to a set of orthogonal unit vectors. A rule that scales the arbitrary coefficient of a mode according to a specified rule is said to be *normalization*. To some extent it would be less confusing to say that the rule "standardizes" a mode because "normal" is synonymous with "orthogonal", which is a property the modes always have. Nevertheless, the term is generally accepted, so we shall use it.

To determine the modal coefficients, we express the given boundary excitation as a series whose basis functions are the transverse modes $\Phi_n(y)$, specifically

$$\boxed{\sum_{n=0}^\infty V_n \Phi_n(y) = V_s(y)} \qquad (9.2.53)$$

An explicit description of the velocity coefficients is obtained from the modal orthogonality property. Thus, the preceding is multiplied by a specific function $\Phi_j(y)$, and both sides are integrated over the cross section. This operation filters the jth term from the sum, with the result that

$$V_j = \frac{1}{H}\int_0^H V_s \Phi_j \, dy \tag{9.2.54}$$

At the same time, Euler's equation states that the axial particle velocity at $x = 0$ corresponding to the modal pressure series is

$$V_s = \frac{i}{\omega\rho_0}\frac{dP}{dx}\bigg|_{x=0} = \frac{1}{\omega\rho_0}\sum_{n=0}^{\infty}\kappa_n P_n \Phi_n(y) \tag{9.2.55}$$

This description of the axial velocity must be the same as Eq. (9.2.53). The alternatives are modal series with the same orthogonal set of basis functions. Equality requires that the coefficients match, which leads to

$$P_j = \rho_0 c \frac{k}{\kappa_j} V_j \tag{9.2.56}$$

The corresponding expression for the pressure field is

$$P(x, y) = \rho_0 c \sum_{j=0}^{\infty}\frac{k}{\kappa_j}V_j \Phi_j(y) e^{-i\kappa_j x} \tag{9.2.57}$$

It is implicit to this expression that evaluation of κ_n is based on a branch cut for the square root that places κ_n in the fourth quadrant of the complex plane.

Equation (9.2.56) tells us that the source might generate an infinite number of waveguide modes. The degree to which the source velocity matches a mode function dictates the magnitude of the velocity coefficient. The other factor affecting the pressure coefficients is κ_n, whose minimum magnitude at a specific frequency occurs in the vicinity of n_{cutoff}. For $n \gg n_{\text{cutoff}}$, κ_n will be much larger than k, which is one reason that the pressure series converges. This property leads to convergence of the modal series at any x. Convergence improves as x increases because the evanescent modes decay increasingly.

The case of a point source situated somewhere on the cross section at $x = 0$ leads to an explicit description. Recall that the volume velocity is the limit as the source's surface area shrinks to zero and the normal velocity of the source increases commensurately. Accordingly, the source is described by a Dirac delta function. For a two-dimensional waveguide with the source situated at $y = y_0$, the volume velocity is per unit width in the invariant direction. In other words, we are considering a line source. Hence, we have

$$V_s = \lim_{\varepsilon\to 0}\left\{\begin{array}{l}(\hat{Q}/\varepsilon)\ \text{if}\ y_0 < y < y_0 + \varepsilon \\ 0\ \text{otherwise}\end{array}\right\} = \hat{Q}\delta(y - y_0) \tag{9.2.58}$$

The velocity coefficients obtained by application of Eq. (9.2.54) are $V_j = \hat{Q}\Phi_j$ $(y_0)/H$. The corresponding pressure field is

$$P_{\text{source}} = \rho_0 c \frac{\hat{Q}}{H} \sum_{j=0}^{\infty} \frac{k}{\kappa_j} \Phi_j (y_0) \Phi_j (y) e^{-i\kappa_j x} \qquad (9.2.59)$$

It is evident that if the source is situated at a location where mode #j is largest, that mode will be excited to the greatest extent possible. Conversely, if the source is situated at a node of the jth mode, that mode will not be excited. Furthermore, if the pressure is observed at a location where the jth mode has a node, there will no evidence of that mode at that location, even though it might be an important mode for the pressure at other points.

Orthogonality of transverse mode functions plays an important role in an analysis of axial power flow. To evaluate the axial component of intensity we require an expression for V_x, which we obtain by applying Euler's equation to the modal series for pressure. This operation yields

$$V_x = \sum_{j=0}^{\infty} V_j \Phi_j (y) e^{-i\kappa_j x} \qquad (9.2.60)$$

The time-averaged power flow past any cross section is the average intensity at fixed x integrated over y,

$$\mathcal{P}_{\text{av}} (x) = \frac{1}{2} \text{Re} \left\{ \int_0^H \left[\rho_0 c \sum_{j=0}^{\infty} \frac{k}{\kappa_j} V_j \Phi_j (y) e^{-i\kappa_j x} \right] \left[\sum_{n=0}^{\infty} V_n^* \Phi_n (y) e^{i\kappa_n^* x} \right] dy \right\}$$

$$= \frac{\rho_0 c}{2} \text{Re} \left\{ \sum_{j=0}^{\infty} \sum_{n=0}^{\infty} \frac{k}{\kappa_j} V_j V_n^* e^{-i(\kappa_j - \kappa_n^*)x} \int_0^H \Phi_j (y) \Phi_n (y) dy \right\}$$

$$(9.2.61)$$

Because the transverse mode functions form an orthogonal set that is normalized to H, the terms in the double sum for which $n \neq j$ integrate to zero. This reduces the expression to

$$\mathcal{P}_{\text{av}} (x) = \frac{\rho_0 c H}{2} \text{Re} \left[\sum_{j=0}^{\infty} \frac{k}{\kappa_j} |V_j|^2 e^{2\,\text{Im}(\kappa_j)x} \right] \qquad (9.2.62)$$

The first notable aspect of this expression is that the modes do not couple, so the power flow is the sum of the contribution of each mode. If both walls are purely reactive, then the κ_j values below cutoff ($j \leq N_{\text{cutoff}}$) will be real. The exponential decay factor in that case is unity, which means that these modal contributions to the power flow are independent of x. The axial wavenumber is negative imaginary for

$j > N_{\text{cutoff}}$. The exponential factor in the summation is real for these modes, but the presence of κ_j in the denominator means that each of these terms is imaginary. The consequence is that evanescent modes do not contribute to the power flow. (The fundamental reason for this property is that the axial velocity in an evanescent mode is $90°$ out of phase from the pressure.) This means that if there is no dissipation the power input to a waveguide at one end is transported downstream by the propagating modes without loss. The situation is different if the impedance of either wall has a resistive part. In that case all axial wavenumbers are complex. We established previously that the κ_j values must lie in the fourth quadrant of the complex plane, which means $\text{Im}\left(\kappa_j\right) < 0$. Thus the power input to each mode at one end of the waveguide is progressively dissipated through the walls(s) as the mode propagates downstream. The decay rate is twice as fast as the decay of the pressure amplitude. If the waveguide is sufficiently long, the pressure and therefore the power, that reaches the far end will be negligible.

EXAMPLE 9.6 The sketch shows a rectangular tank of water that is open to the atmosphere. On the left wall a ribbon transducer is mounted at mid-depth. The transducer executes a uniform translational oscillation in the direction normal to the wall. Its width is sufficiently narrow that it may be approximated as a line whose a volume velocity per unit length is $\text{Re}\left(\hat{Q}_0 e^{i\omega t}\right)$. The bottom and side walls are composed of a thick glass sheet that may be considered to be rigid. The far wall consists of a mosaic of piezoceramic tiles that are actively controlled to achieve a nonreflecting boundary condition. The tank dimensions are $H = 600$ mm and $L = 3$ m, the frequency is 20 kHz, and $\hat{Q}_0 = 8\left(10^{-4}\right)$ m$^2/s$. The wall containing the ribbon transducer is the plane $x = 0$, and y is measured from the bottom. Determine the dependence of $|p|$ on the axial distance x at mid-depth, $y = H/2$, and also determine the depth dependence of $|p|$ at $x = 0$, $L/2$, and L. At each location, identify the portion that is attributable to the modes that are below the cutoff wavenumber. In addition, determine the spatially averaged impedance required of the piezoceramic tiles.

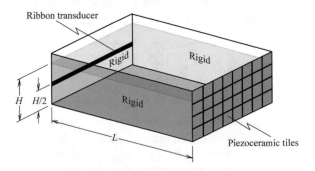

Figure 1.

Significance

The best way to realize the effect of the cutoff phenomenon and the role of the evanescent spectrum, is to carry out a computation. The results will illustrate the general nature of multimodal signals in a waveguide.

Solution

It might not be obvious that this system constitutes a two-dimensional waveguide because the tank is not infinitely wide in any direction parallel to the plane of the excitation. However, the vertical side walls are rigid, which means that $\partial P/\partial z$ must be zero along them. A pressure field that does not depend on z will satisfy this condition. The ribbon transducer executes the same vibration at all z, so a pressure field fitting the description $\mathrm{Re}\left(P\left(x,y\right)e^{i\omega t}\right)$ is fully consistent. Thus, the pressure is described by the modal series in Eq. (9.2.43). The bottom of the tank is rigid, and the top is the free surface, which is well represented as pressure-release. The modes for such a configuration are described by Eq. (9.2.35),

$$\Phi_n = B_n \cos\left(\mu_y y\right), \quad \mu_n = \left(n - \frac{1}{2}\right)\frac{\pi}{H}, \quad n = 1, 2, \ldots \tag{1}$$

The B_n coefficients are normalized according to Eq. (9.2.51), which gives

$$B_n = H^{1/2}\left[\int_0^H \cos\left(\frac{(2n-1)\,\pi y}{2}\frac{}{H}\right)^2 dy\right]^{-1/2} = 2^{1/2} \tag{2}$$

The excitation is a line source having a specified velocity. Aside from the line at $y = H/2$, the wall is rigid, so the velocity boundary condition at $x = 0$ is

$$V_x|_{x=0} = \hat{Q}_0 \delta\left(y - \frac{H}{2}\right), \quad 0 < y < H \tag{3}$$

According to Eq. (9.2.54), the coefficients of a modal series for this velocity distribution are

$$V_n = \frac{1}{H}\int_0^H \hat{Q}_0 \delta(y - H/2)\Phi_n dy = \frac{\hat{Q}_0}{H}\Phi_n(y = H/2) = \frac{\hat{Q}_0}{H}2^{1/2}\cos\left(\frac{(2n-1)\pi}{4}\right) \tag{4}$$

The resulting modal series obtained from Eq. (9.2.57) is

$$P\left(x, y\right) = \rho_0 c \sum_{n=0}^{\infty} \frac{k}{\kappa_n} V_n \Phi_n\left(y\right) e^{-i\kappa_n x} \tag{5}$$

The last consideration before we may evaluate the pressure is selection of the number of terms N to include in the series. The transverse wavenumbers are given by Eq. (1). The cutoff condition is marked by $\mu_n = k = 84.91$ based on $c = 1480$ m/s. This leads to

$$N_{\text{cut}} = \text{floor}\left(\frac{kH}{\pi} + \frac{1}{2}\right) = 16$$

As was noted, convergence of the modal series, Eq. (9.2.43), can result from reduction of $|P_n|$ to a negligible value above some n. This possibility constitutes the convergence criterion for small x, because $|\exp(-i\kappa_n x)|$ will be $O(1)$ regardless of whether κ_n is real or negative imaginary. An important aspect is the fact that κ_n becomes very small near n_{cutoff}. Because κ_n appears in the denominator of Eq. (3), achieving convergence in this manner requires that $N > N_{\text{cut}}$. The other terms in Eq. (3) do not grow with increasing n, so setting N substantially larger than n_{cutoff} should be adequate. We shall use $N = 3N_{\text{cut}}$, which leads to $|P_n| < 0.07 \max |P_n|$ for $n > N$.

The other attribute that leads to convergence of a modal series is the decay of the evanescent modes with increasing x. This decay increases with increasing mode number. Thus, fewer and fewer terms are required as x increases. We could use this attribute to reduce N at each x based on the value of $\exp(-i\mu_n x)$ for $n > n_{\text{cutoff}}$. We shall not do so to avoid complicating the computational program. Furthermore, the processing time required for computations at fixed N is not prohibitive. The only remaining consideration is a caution against a common error. The erroneous procedure evaluates the magnitude of each term in the modal series. The correct procedure evaluates the sum first.

Before we examine data, it is useful to recall the earlier matrix based algorithm for evaluation of a series. Our interest is in the values of P at a set of x and y locations. In Eq. (5) the mode function depends on y, and the exponential function depends on x. Let these locations be x_j, $j = 1, ..., J$ and y_m, $m = 1, ..., M$. The summation over the mode number n can be expressed as a matrix sum. To do so, let $[F]$ be a matrix of transverse mode functions and $[E]$ be a matrix of exponentials, such that $F_{n,m} \equiv \Phi_n(y_m)$ and $E_{j,n} \equiv \exp(-i\mu_n x_j)$. Then $P_{m,j} \equiv P(x_j, y_m)$ is the element of a rectangular array $[P]$ at all locations that is computed from the matrix version of Eq. (4), which is

$$[P]_{J \times M} = \rho_0 c \, [E]_{J \times N} \, [K]_{N \times N} \, [F]_{N \times M} \tag{6}$$

where $[K]$ is a diagonal matrix whose elements are $K_{n,n} = (k/\kappa_n) V_n$. A row of $[P]$ will be the pressure values at all y_m for fixed x_j, whereas a column will be the values at all x_j for fixed y_m. A bonus of this formulation is that changing the modal index to extend from $n = 1$ to $n = N_{\text{cutoff}}$ is the only modification required to exclude the evanescent modes.

The first set of results we shall examine are the pressure amplitudes. According to Eq. (5), they are

$$\hat{P}_n = \rho_0 c \frac{k}{\kappa_n} V_n$$

These values are described by Fig. 2. As was anticipated, the largest coefficients occur in the vicinity of N_{cut}.

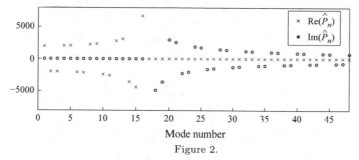

Figure 2.

The axial dependence of the pressure may be computed according to Eq. (6) with x_j being a set of axial locations and $y_1 = H/2$. Figure 3 shows the result. Many individuals anticipate a smooth curve, whereas the actual result shows rounded maxima separated by narrow near-null regions. Capturing these features requires a fine scan; the data for the Fig. 3, as well as for the computations along a cross section used a point spacing set at $\Delta = (2\pi/k)/25$. The cause of this behavior is interference, as it is for a sound beam. The interference occurs here because the multiple modes have different wavelengths. At some locations many are nearly in-phase creating maxima, whereas at other locations the modal contributions nearly annihilate each other. Another noteworthy feature is that the pressure very close to the transducer is 10 dB higher than it is at any downstream maximum.

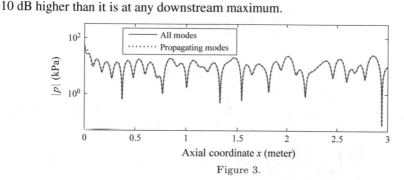

Figure 3.

Figure 3 shows that only very close to $x = 0$ is there significant difference between the pressure obtained from the full series and the series that consists of only propagating modes. This observation is confirmed by the transverse profiles in Fig. 4. To obtain the data for Eq. (6), x_j was set to the axial position and y_m was a sequence of values spaced at Δ. Interference is evident at each cross section. The cross section at the boundary, $x = 0$, shows a sharp pressure peak in the vicinity of $y = 0.3$ m, which is the location of the transducer. With increasing distance from the excitation, that tendency is overcome by the interference effect. Note that the profiles at $x = 1.5$ m and $x = 3$ m are similar in magnitude, as well as spatial dependence, but they are not the same. Further only at $x = 0$ is the contribution of the evanescent modes significant.

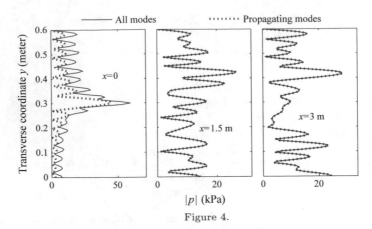

Figure 4.

The last quantity to evaluate is the average impedance at $x = L$. This quantity is merely the average complex pressure amplitude divided by the average particle velocity. There is no variation in the z direction, so averages per unit width are obtained by integrating over y. We use the modal pressure series in Eq. (5) to describe P as a function of y, so that

$$
P_{av}|_{x=L} = \frac{1}{H} \int_0^H P(L, y)\,dy = \frac{\rho_0 c}{H} \sum_{n=1}^{\infty} \frac{k}{\kappa_n} V_n \int_0^H 2^{1/2} \cos\left(\frac{(2n-1)\pi y}{2H}\right) e^{-i\kappa_n L}\,dy
$$

$$
= \rho_0 c \sum_{n=0}^{\infty} \frac{k}{\kappa_n} \left(\frac{2^{3/2} V_n}{(2n-1)\,\pi}\right) (-1)^{n-1} e^{-i\kappa_n L}
$$

$$(7)$$

The average axial velocity is obtained by integrating Eq. (9.2.60) over y,

$$
(V_x)_{av}|_{x=L} = \frac{1}{H} \int_0^H V_x(L, y)\,dy = \frac{1}{H} \sum_{n=1}^{\infty} V_n e^{-i\kappa_n L} \int_0^H 2^{1/2} \cos\left(\frac{(2n-1)\pi y}{2H}\right) dy
$$

$$
= \sum_{n=1}^{\infty} \left(\frac{2^{3/2} V_n}{(2n-1)\,\pi}\right) (-1)^{n-1} e^{-i\kappa_n L}
$$

$$(8)$$

Despite the similarities of the sums in Eqs. (7) and (8), each sum must be evaluated individually. The average specific impedance that results from these expressions is

$$
\zeta_{av} = \frac{Z_{av}}{\rho_0 c} = \frac{P_{av}|_{x=L}}{\rho_0 c \,(V_x)_{av}|_{x=L}} = 1.086 - 0.0043136i
$$

The average impedance is very close to $\zeta = 1$, which is the value for which the reflection coefficient of a plane wave is zero.

Before we leave this system, it is interesting to contemplate an alternative solution based on the method of images. Example 6.8 developed such a solution for the case of a point source in a hard-walled waveguide. The ribbon transducer in the present system effectively is a line source, so all images also are line sources. Images above the free surface, $y = H$, invert images below that surface, while images below the bottom are in-phase replicas of the images above that surface. This creates an infinite sequence of line sources lying in the $x = 0$ cross section centered at $y = H/2$ and separated by distance H. The pressure amplitude radiated by a line source decays as $1/R^{1/2}$ when the distance R transverse to that line is large. If we were to impose a comparable convergence criterion to that employed for the modal series, we would cease a summation of all images when the nth line source above or below has an amplitude that is 1% of the largest contribution. For a field on the axis, the closest image is the actual transducer, which is at $y = H/2, x = 0$. The corresponding source-image distance is $R_0 = x$. The distance to the same field point from the nth image is $R_n = [x^2 + (nH)^2]^{1/2}$. The aforementioned convergence condition suggest that we may omit from the image summation any image for which $1/R_n^{1/2} < 0.01/R_0^{1/2}$. This simplifies to $[x^2 + (nH)^2]^{1/2} > 10^4 x$, which is well approximated as $n > 10^4 x/H$. In other words, the number of images required to obtain a suitably convergent representation of the on-axis pressure must increase drastically in proportion to the distance. The signal arriving at a field point from each image corresponds to a ray. In general, the number of rays required to represent a signal increases greatly with increasing range. In contrast, a fixed, relatively small number of modes usually suffices.

9.3 Three-Dimensional Waveguides

The two-dimensional model of a waveguide encapsulates the conceptual and phenomenological features of other waveguides whose cross section is invariant. This will be evident in the following general formulation and its application to cases of rectangular and circular cross sections.

9.3.1 General Analytical Procedure

As we did previously, we designate the direction in which signals propagate as x. The cross section is $\mathcal{A}$, whose shape and size are constant. The other restriction imposed on this general exploration is that the side wall, which forms the perimeter $\mathcal{C}$, is locally reacting. The specific impedance ζ may vary along the perimeter, but it is independent of x (e.g., a rectangular cross section could have walls different types of sides). Let ξ_2 and ξ_3 denote the transverse coordinates. These would be Cartesian coordinates if the shape is rectangular, or polar coordinates if the shape were circular, but their definition is unimportant at this juncture. The transverse mode functions therefore are $\Phi_n (\xi_2, \xi_3)$, and a waveguide mode is obtained by multiplying these functions by a complex exponential representing the propagation. In other words, we anticipate that the waveguide modes will be described as

$$P = \Phi_n\left(\xi_2, \xi_3\right) e^{-i\kappa_n x} \tag{9.3.1}$$

As always, the wavenumber must lie in the fourth quadrant of the complex plane in order to satisfy the Sommerfeld radiation condition. The transverse mode functions Φ_n are solutions of a Helmholtz equation whose general form is

$$\boxed{\tilde{\nabla}^2\left(\Phi_n\right) + \mu_n^2 \Phi_n = 0, \quad \mu_n^2 = k^2 - \kappa_n^2} \tag{9.3.2}$$

where $\tilde{\nabla}^2$ denotes the portion of the Laplacian that contains derivatives with respect to ξ_2 and ξ_3.

As was the case for a two-dimensional waveguide, values of μ_n will be found from the condition that there be a nontrivial solution consistent with homogeneous boundary conditions at the walls. Application of Euler's equation to describe the particle velocity in the local impedance model, with $\bar{n}$ defined to be into the waveguide, leads to

$$P = \rho_0 c \zeta \left(-\bar{n} \cdot \bar{V}\right) = \frac{\zeta}{ik}\bar{n} \cdot \tilde{\nabla}P \text{ if } \bar{x} \in \mathcal{C} \tag{9.3.3}$$

where $\mathcal{C}$ is the perimeter of the cross section $\mathcal{A}$. Substitution of the representation of P as a waveguide mode leads to a boundary condition for the transverse mode functions,

$$\boxed{\Phi_n = \frac{\zeta}{ik}\bar{n} \cdot \tilde{\nabla}\Phi_n} \tag{9.3.4}$$

In addition to similarities of the equations that must be solved to determine the transverse mode functions, the general problem is like the two-dimensional one because its transverse mode functions are orthogonal. The general orthogonality condition is obtained similarly to the derivation of Eq. (9.2.52). Thus, we consider two eigensolutions and form

$$\begin{aligned}
&\Phi_j\left[\tilde{\nabla}^2\left(\Phi_n\right) + \mu_n^2\Phi_n\right] - \Phi_n\left[\tilde{\nabla}^2\left(\Phi_j\right) + \mu_j^2\Phi_j\right] \\
&\equiv \tilde{\nabla}\cdot\left(\Phi_j\tilde{\nabla}\Phi_n\right) - \tilde{\nabla}\cdot\left(\Phi_n\tilde{\nabla}\Phi_j\right) + \left(\mu_n^2 - \mu_j^2\right)\Phi_j\Phi_n = 0
\end{aligned} \tag{9.3.5}$$

This relation is integrated over a cross section. Application of the divergence theorem, with $\bar{n}$ defined as the normal on the perimeter $\mathcal{C}$ pointing into $\mathcal{A}$, leads to

$$\int_{\mathcal{C}} \bar{n}\cdot\left(\Phi_n\tilde{\nabla}\Phi_j - \Phi_j\tilde{\nabla}\Phi_n\right)d\mathcal{C} + \left(\mu_n^2 - \mu_j^2\right)\iint_{\mathcal{A}} \Phi_j\Phi_n h_2\left(\xi_2\right)h_3\left(\xi_3\right)d\mathcal{A} = 0$$

$$\tag{9.3.6}$$

The terms evaluated on C will vanish for the rigid ($\bar{n} \cdot \tilde{\nabla}\Phi_n = 0$) and pressure-release cases. If the wall is locally reacting, then Eq. (9.3.4) gives

$$\Phi_n \left(\bar{n} \cdot \tilde{\nabla}\Phi_j \right) = \Phi_n \left(\frac{ik}{\zeta}\Phi_j \right) = \left(\frac{ik}{\zeta}\Phi_n \right)\Phi_j = \left(\bar{n} \cdot \tilde{\nabla}\Phi_n \right)\Phi_j \qquad (9.3.7)$$

Thus the integral over C reduces to zero for any local impedance, even if ζ varies along C.

What remains in Eq. (9.3.6) is the integral over $\mathcal{A}$. If the two mode functions correspond to different eigenvalues, $\mu_j \neq \mu_n$, then the integral must vanish. In the case where $\mu_j = \mu_n$, the integral may have any value. We shall use this arbitrariness to normalize the mode functions. Specifically, the Φ_n are scaled such that the integral equals the cross-sectional area. Thus, we will have

$$\iint_{\mathcal{A}} \Phi_j \Phi_n d\mathcal{A} = \mathcal{A}\delta_{j,n} \qquad (9.3.8)$$

Fulfillment of this condition when $j = n$ sets the B_n coefficient of Φ_n, which is arbitrary in regard to satisfying the eigenvalue problem.

Given the similarity of the general relations governing three-dimensional transverse mode functions and those for a two-dimensional waveguide, it is consistent that a series of these functions is used to represent the pressure in the general case. The coefficients of the modes are again designated as P_n, so that

$$P = \sum_{n=1}^{\infty} P_n \Phi_n \left(\xi_2, \xi_3 \right) e^{-i\kappa_n x} \qquad (9.3.9)$$

The axial velocity distribution corresponding to this description of the pressure is

$$V_x = \sum_{n=1}^{\infty} \left(\frac{\kappa_n}{\kappa} \right) \frac{P_n}{\rho_0 c} \Phi_n \left(\xi_2, \xi_3 \right) e^{-i\kappa_n x} \qquad (9.3.10)$$

The procedure for determining the modal amplitudes is the same as it was for the two-dimensional waveguide. Typically, the excitation consists of a specified axial velocity $V_s \left(\xi_2, \xi_3 \right)$ at $x = 0$. This distribution is described by a modal series, specifically

$$V_s \left(\xi_2, \xi_3 \right) = \sum_{n=1}^{\infty} V_n \Phi_n \left(\xi_2, \xi_3 \right) \qquad (9.3.11)$$

To obtain an expression for the velocity coefficients the preceding is multiplied by an arbitrarily selected Φ_j and integrated over $\mathcal{A}$. Orthogonality, Eq. (9.3.8), filters the jth term out of the summation, which leads to

$$V_j = \frac{1}{A} \iint_A \Phi_j \left(\xi_2, \xi_3\right) V_s \left(\xi_2, \xi_3\right) dA \qquad (9.3.12)$$

Equation (9.3.10) evaluated at $x = 0$ must be the same as the modal series of V_s. This requires that the coefficients of Φ_n match, which leads to

$$P_n = \rho_0 c \frac{k}{\kappa_n} V_n \qquad (9.3.13)$$

There is no difference between this expression and Eq. (9.2.56) for the two-dimensional system.

These expressions lead to a general description of the power flow. The time-averaged power flowing across any cross section is the integral over that surface of the time-averaged intensity,

$$P(x) = \iint_A \frac{1}{2} \operatorname{Re} \left(P V_x^*\right) dA = \frac{1}{2} \operatorname{Re} \iint_A \left[\sum_{n=1}^{\infty} P_n \Phi_n \left(\xi_2, \xi_3\right) e^{-i\kappa_n x}\right]$$
$$\times \left[\sum_{j=1}^{\infty} \left(\frac{\kappa_j^*}{\kappa}\right) \frac{P_j^*}{\rho_0 c} \Phi_j \left(\xi_2, \xi_3\right) e^{+i\kappa_j^* x}\right] dA$$
$$(9.3.14)$$

Different indices are used for the summations to assure that all products are included. The only quantities that depends on position are the Φ_n functions. This allows us to sum the integrals of the products of these functions. The result is that

$$P(x) = \frac{1}{2} \operatorname{Re} \sum_{n=1}^{\infty} \sum_{j=1}^{\infty} \left(\frac{\kappa_j^*}{\kappa}\right) \frac{P_n P_j^*}{\rho_0 c} e^{-i\left(\kappa_n - \kappa_j^*\right)x} \iint_A \Phi_n \left(\xi_2, \xi_3\right) \Phi_j \left(\xi_2, \xi_3\right) dA$$
$$(9.3.15)$$

The last step is to apply the orthogonality of the transverse mode functions, which eliminates all coupling between modes. Thus, we find the power flow to be a sum of modal contributions,

$$P(x) = \frac{A}{2\rho_0 c} \operatorname{Re} \sum_{n=1}^{\infty} \frac{\kappa_n^*}{k} |P_n|^2 e^{+2\operatorname{Im}(\kappa_n)x} \qquad (9.3.16)$$

This is a useful relation because it allows for identification of modes that transport the greatest amount of energy. In the case of a purely reactive wall, the axial wave numbers are real for modes that propagate, and imaginary for evanescent modes. The exponential factor for the latter modes is real, but $\operatorname{Re}\left(\kappa_n^*\right) = 0$, so the evanescent modes do not contribute the power transfer. Furthermore, $\operatorname{Im}(\kappa_n) = 0$ for the propagating modes, so the contribution of these modes to the power transfer is independent of x. Thus, $P(x)$ is independent of x if the walls are purely reactive, regardless of the shape of the cross section. The modal contributions to the power flow depend on x only if the wall has a nonzero resistance. In that case the eigenvalues and, therefore the axial wavenumbers, are complex. We selected the branch cut for the square root such that $\text{Im}(\kappa_n) < 0$ in order to satisfy the radiation condition. The consequence is that power flow is attenuated in the downstream direction. The energy that is dissipated is absorbed in the wall.

The relations we have derived for the pressure, particle velocity, and power flow are generally applicable. However, their implementation assumes that we have identified transverse mode functions that satisfy the Helmholtz equation, Eq. (9.3.2), subject to the wall condition in Eq. (9.3.4). The ease of that determination depends on the shape of the cross section. The following studies explore rectangular and circular cross sections, which are the most amenable to formal analysis. Even within that limitation, an analysis using standard methods is not always possible. The derivation allowed the wall impedance to vary along the perimeter. However, the analyses we shall perform require that the boundary condition be constant along any segment of a wall that corresponds to a constant coordinate. In other words, the impedance of each side of a rectangular cross section is required to be constant, and the impedance of a circular wall cannot depend on the circumferential angle. If these conditions are not met, or if the subject is a waveguide whose cross section has an irregular shape, numerical techniques will be required to determine the transverse mode functions. Nevertheless, they do exist for such systems, and they do constitute an orthogonal set.

9.3.2 Rectangular Waveguide

A waveguide whose cross section is rectangular is the least challenging to analyze because we can define a Cartesian coordinate system whose coordinate planes are parallel to the walls. The walls in Fig. 9.13 correspond to $y = 0$, $y = H$, $z = 0$, $z = W$, and the excitation is applied to the end at which $x = 0$. Waves propagate in the positive x direction, and it is assumed that the far end is terminated in a manner that suppresses reflections.

Fig. 9.13 Definition of the Cartesian coordinate system used to analyze propagation in a rectangular waveguide

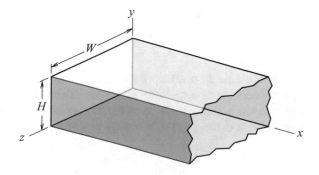

The waveguide mode is

$$\boxed{P\,(x, y, z) = \Phi_n\,(y, z)\exp\left(-i\kappa_n x\right)}$$

(9.3.17)

The differential equation governing a transverse mode function $\Phi\,(y, z)$ is found by substituting this form into the Helmholtz equation, which leads to

$$\left(\frac{\partial^2}{\partial y^2} + \frac{\partial^2}{\partial z^2}\right)\Phi + \left(k^2 - \kappa^2\right)\Phi_n = 0$$

(9.3.18)

Each wall may have a local impedance ζ_j. The normal directions into the fluid are $\bar{e}_2$ for the wall at $y = 0$, $-\bar{e}_2$ for the wall at $y = H$, $\bar{e}_3$ for the wall at $z = 0$, and $-\bar{e}_3$ for the wall at $z = W$, so the specific forms of the general local impedance boundary condition, Eq. (9.3.4), are

$$P = -\rho_0 c \zeta_1 V_y = \frac{\zeta_1}{ik}\frac{\partial P}{\partial y} \text{ at } y = 0$$

$$P = \rho_0 c \zeta_2 V_y = -\frac{\zeta_2}{ik}\frac{\partial P}{\partial y} \text{ at } y = H$$

$$P = -\rho_0 c \zeta_3 V_z = \frac{\zeta_3}{ik}\frac{\partial P}{\partial z} \text{ at } z = 0$$

$$P = \rho_0 c \zeta_4 V_z = -\frac{\zeta_4}{ik}\frac{\partial P}{\partial z} \text{ at } z = W$$

(9.3.19)

We could identify Φ by separation of variables, but it is more direct to guess the form of the general solution Φ and then seek the condition that such a form must satisfy. The analysis of a two-dimensional waveguide began by expressing Φ as a sum of complex exponentials, so it is logical to try the same here. Because either y or z is constant on a boundary, while the other coordinate has a range of values, a solution that is capable of simultaneously solving the boundary conditions on all walls must be a product of functions of y and z. Thus, the appropriate ansatz is

$$\Phi = \phi_y\,(y)\,\phi_z\,(z)$$

$$\phi_y = B_{1y}e^{-i\mu y} + B_{2y}e^{i\mu y}, \quad \phi_z = B_{1z}e^{-i\nu z} + B_{2z}e^{i\nu z}$$

(9.3.20)

The trial solution will satisfy the Helmholtz equation if

$$\mu^2 + \nu^2 + \kappa^2 = k^2$$

(9.3.21)

We substitute the trial form for Φ in Eq. (9.3.20) into the boundary conditions. The ϕ_z function is a common factor in the conditions at constant y, and ϕ_y is a common factor in the conditions on the walls at constant z. Cancelation of these common factors leads to two pairs of boundary conditions, which may be written as

$$\boxed{\begin{aligned}
[D\,(\mu H, kH, \zeta_1, \zeta_2)]\,[B_{1y}\quad B_{2y}]^{\mathrm{T}} &= [0\quad 0]^{\mathrm{T}} \\
[D\,(\nu W, kW, \zeta_3, \zeta_4)]\,[B_{1z}\quad B_{2z}]^{\mathrm{T}} &= [0\quad 0]^{\mathrm{T}}
\end{aligned}} \tag{9.3.22}$$

Let α and β denote the first two variables in the coefficient matrices for each direction, and ζ_a and ζ_b be the corresponding specific impedances. Then these matrices are described by

$$\boxed{[D\,(\alpha, \beta, \zeta_a, \zeta_b)] \equiv \begin{bmatrix} (\zeta_a \alpha + \beta) & (-\zeta_a \alpha + \beta) \\ (-\zeta_b \alpha + \beta)\, e^{-i\eta \ell} & (\zeta_b \alpha + \beta)\, e^{i\alpha} \end{bmatrix}} \tag{9.3.23}$$

In this way the analysis reduces to finding the transverse modes for two two-dimensional waveguides. The wall conditions at constant y set the μ_n eigenvalues, and the conditions at constant z set the ν_n values. The characteristic equations are

$$|[D\,(\mu H, kH, \zeta_1, \zeta_2)]| = 0, \quad |[D\,(\nu W, kW, \zeta_3, \zeta_4)]| = 0 \tag{9.3.24}$$

The representations in Eq. (9.3.20) account for values of μ and ν that have opposite sign, so the only roots for which $\mathrm{Re}\,(\mu) \geq 0$ and $\mathrm{Re}\,(\nu) \geq 0$ should be sought.

The transverse mode function is denoted as $\Phi_{n,j}$ and referred to as the (n, j) mode. Only the ratios B_{2y}/B_{1y} and B_{2z}/B_{1z} can be obtained from Eq. (9.3.22) because the respective $[D]$ matrix is rank-deficient when μH is an eigenvalue. We use the first row of $[D]$ to determine these ratios. Furthermore, when the y and z functions are multiplied, only one arbitrary coefficient, which we denote as B_n, remains. The result is that the transverse mode functions may be written as a product of functions of y and z, which we denote respectively as $\phi_{y,n}$ and $\phi_{z,j}$. Thus, we have established that

$$\Phi_{n,j}\,(y, z) = B_{n,j}\,\phi_{y,n}\,(y)\,\phi_{z,j}\,(z) \tag{9.3.25}$$

where the factors are

$$\begin{aligned}
\phi_{y,n} &= \frac{(\zeta_1 \mu_n H - kH)\,e^{-i\mu_n y} + (\zeta_1 \mu_n H + kH)\,e^{i\mu_n y}}{(\zeta_1 \mu_n H - kH)} \\
\phi_{z,j} &= \frac{(\zeta_3 \nu_j W - kW)\,e^{-i\gamma_j y} + (\zeta_3 \nu_j W + kW)\,e^{i\nu_j y}}{(\zeta_3 \nu_j W - kW)}
\end{aligned} \tag{9.3.26}$$

After the eigenvalues for each direction have been determined, the set of axial wavenumbers are

$$\kappa_{n,j} = \pm \left(k^2 - \mu_n^2 - \nu_j^2\right)^{1/2} \tag{9.3.27}$$

where the alternative sign should be selected such that $\mathrm{Re}\,(\kappa_{n,j}) > 0$ and $\mathrm{Im}\,(\kappa_{n,j}) \leq 0$. If all walls are reactive, then the μ_n and ν_j values are either real or purely imaginary. In such cases cutoff occurs if $\mu_n^2 + \nu_j^2 > k^2$. A corollary is that there is no single cutoff. Rather, for a fixed mode number in one direction there is a mode number for the other direction beyond which the signal evanesces. For example, if $\mu_5 = 0.8k$, then modes are cutoff in the z direction if $\nu_j > 0.6k$. It follows that if $\mu_n > k$, then all modes at that n and higher are evanescent, regardless of the value of ν_j. Similarly, if $\nu_j > k$, then all modes at that j and higher are evanescent. The alternative situation occurs when one or more wall has a resistance, so that its impedance has a real part. In such circumstances all modes decay axially, so that the demarcation between propagating and evanescent modes does not exist. It is reasonable to consider modes to be cutoff if the scale over which they decay, which is $-1/\operatorname{Im}\left(\kappa_{n,j}\right)$, is comparable to or smaller than an axial wavelength, $2\pi/\operatorname{Re}\left(\kappa_{n,j}\right)$. Such a transition will be a noticeable feature of dispersion curves if the resistive part of the wall impedance is small, as it was in Example 9.4. On the other hand, it might be meaningless to characterize a mode in this manner if the resistance is large.

Like a two-dimensional waveguide, a waveguide mode in a rectangular waveguide may be viewed either as a nonuniform plane wave that travels in the x direction with the transverse dependence described by $\Phi_{n,j}$, or as a superposition of a set of plane waves that travel obliquely to the x-axis. Merging Eqs. (9.3.17) and (9.3.20) for a specific (n, j) pair shows that a waveguide mode is a sum of four simple plane waves described by

$$\Phi_{n,j}e^{-i\kappa_{n,j}x} = \sum_{m=1}^{4} C_m e^{-i\bar{k}_m\cdot\bar{x}}$$

$$\bar{k}_m = \kappa_{n,j}\bar{e}_x \pm \mu_{n,j}\bar{e}_y \pm \nu_{n,j}\bar{e}_z$$

(9.3.28)

where the C_m are set by satisfying the boundary conditions at the walls. If $\kappa_{n,j}$ is real, the four plane waves that combine to form a waveguide mode undergo multiple reflections at the side walls. Their interference pattern on a cross section produces the transverse mode function. This is the same phenomenon as the one illustrated in Fig. 9.9, except that each $\bar{k}_m$ has nonzero direction angles in both the xy- and xz-planes.

The general orthogonality property, Eq. (9.3.8), in the case of a rectangular cross section states that

$$\int_0^H \int_0^W \Phi_{n,j}\Phi_{m,\ell}dzdy = HW\delta_{n,m}\delta_{j,\ell}$$

(9.3.29)

The case where $m = n$ and $\ell = j$ sets the coefficient $B_{n,j}$ in Eq. (9.3.25). Substitution of that expression into the orthogonality condition leads to

$$B_{n,j} = (HW)^{1/2}\left(\int_0^H \left(\phi_{y,n}\right)^2 dy\right)^{-1/2}\left(\int_0^W \left(\phi_{z,j}\right)^2 dz\right)^{-1/2}$$

(9.3.30)

We have seen that the factors $\phi_{y,n}$ and $\phi_{z,j}$ in Eq. (9.3.25) are transverse mode functions for a two-dimensional waveguide. This can be a very useful attribute, because it might be quite simple to form the transverse mode functions for a three-dimensional system. For example, consider the three-dimensional transverse mode functions of the tank in Example 9.6. In the y direction, it is a rigid/pressure-release configuration, whereas it is a hard-walled waveguide in the z direction. Multiplying a mode function for each gives

$$\Phi_{n,j} = B_{n,j} \cos\left(\frac{(2n-1)\pi y}{2H}\right) \cos\left(\frac{j\pi z}{W}\right), \quad n = 1, 2, \ldots, \ j = 0, 1, \ldots \quad (9.3.31)$$

A system of particular interest is a hard-walled waveguide, that is, one whose four walls are rigid. The combination of cosine functions in each direction gives

$$\Phi_{n,j}(y, z) = B_{n,j} \cos\left(\frac{j\pi y}{H}\right) \cos\left(\frac{n\pi z}{W}\right), \quad j, n = 0, 1, 2, \ldots \quad (9.3.32)$$

The corresponding eigenvalues and axial wavenumber are

$$\mu_n = \frac{n\pi}{H}, \quad \nu_j = \frac{j\pi}{W}, \quad \kappa_{j,n} = \pm\left[k^2 - \left(\frac{n\pi}{H}\right)^2 - \left(\frac{j\pi}{W}\right)^2\right]^{1/2} \quad (9.3.33)$$

The alternative sign in this expression should be selected such that $\mathrm{Re}\,(\kappa_n) > 0$ and $\mathrm{Im}\,(\kappa_n) \le 0$.

The case where $j = n = 0$ is the plane wave mode, in which $\Phi_{0,0}$ is independent of y and z. The hard-walled waveguide is the only one in which a plane wave mode can exist. This is so for the same reason as a two-dimensional waveguide: In the plane wave mode $\bar{n} \cdot \nabla P$ is identically zero at the walls, which conflicts with the proportionality of pressure and velocity imposed by a finite local impedance.

Examination of Eq. (9.3.33) leads to recognition of an unusual feature. Suppose that the cross section is square, $W = H$. Then any (n, j) mode for which $n \ne j$ has the same axial wavenumber as the (j, n) mode. The same situation can arise even if $W \ne H$. To explore this possibility, let $W = rH$, and let (j_1, n_1) and (j_2, n_2) be the numbers for a pair of modes that have the same κ value. Repeated values of $\kappa_{j,n}$ occur if

$$(n_2)^2 + \left(\frac{j_2}{r}\right)^2 = (n_1)^2 + \left(\frac{j_1}{r}\right)^2 \quad (9.3.34)$$

This condition is met if $n_2 = j_1/r$ and $j_2 = rn_1$. Because the mode indices are integers, repeated $k_{n,j}$ values can arise only if r is a rational number, that is, if W/H is a ratio of integers. Furthermore, unless this ratio consists of small numbers, the overlapping modes will be very high, and possibly evanescent. For example if $r = 3/2$, then $\kappa_{3,2} = \kappa_{2,3}$, $\kappa_{6,4} = \kappa_{4,6}$, etc. In contrast of $r = 20/21$, then $\kappa_{20,21} = \kappa_{21,20}$, etc. Similar situations might arise for other combinations of rigid and pressure-release boundary conditions. Equality of the axial wavenumbers for different modes does not present a complication for the analysis. However, it can cause confusion for an examination of measured data.

After we have evaluated the transverse mode functions, the response to an excitation may be analyzed. The modal summation in Eq. (9.3.9) now must extend over the index for each direction, that is,

$$P = \sum_{n=0}^{\infty} \sum_{j=0}^{\infty} P_{n,j} \Phi_{n,j} e^{-i\kappa_{n,j}x} \tag{9.3.35}$$

As always, a zero subscript applies only if a plane wave mode exists, in which case $\kappa_{n,j} = k$. Similarly, if there is an eigenvalue that is purely imaginary, its contribution must be inserted. The velocity excitation $V_s(x, y)$ extends over the cross section at $x = 0$. Its modal series also is a double sum,

$$V_s = \sum_{n=0}^{\infty} \sum_{j=0}^{\infty} V_{n,j} \Phi_{n,j} \tag{9.3.36}$$

The modes form an orthogonal set according to Eq. (9.3.29). This leads to the specific form of Eq. (9.3.12),

$$V_{n,j} = \frac{B_{n,j}}{HW} \int_0^W \int_0^H V_s \phi_{y,n} \phi_{z,j} dy dz \tag{9.3.37}$$

These coefficients are matched to the modal series for V_x at $x = 0$ obtained from Euler's equation, as was done to obtain Eq. (9.3.13), with the result that

$$P_{n,j} = \rho_0 c \frac{k}{\kappa_{n,j}} V_{n,j} \tag{9.3.38}$$

The considerations regarding convergence for two-dimensional systems are equally applicable here. The largest y mode number N_{cut} for a propagating mode is such that $\mu_{N_{cut}} < k$ and $\mu_{N_{cut}+1} > k$. Similarly, the z mode number for cutoff is such that $\nu_{J_{cut}} < k$ and $\nu_{J_{cut}+1} > k$. In regions that are not close to the excitation at $x = 0$, truncation of the modal series at $n = N_{cut}$ and $j = J_{cut}$ will be adequate. This is so because the modes that are omitted correspond to $\kappa_{n,j} = -i \, \mathrm{Im} \left(\kappa_{n,j} \right)$, so that $\exp(-i\kappa_{n,j}x)$ is negligible. In regions very close to the excitation at $x = 0$, the evanescent modes have not decayed much. Convergence of the pressure series requires inclusion of the evanescent modes. The values of $\left| \kappa_{n,j} \right|$ increase rapidly if $n > N_{cut} + 1$ or $j > J_{cut} + 1$. A truncation at $N = 2N_{cut}$ and $j = 2J_{cut}$ should be adequate, unless x is smaller than a wavelength. For such cases, truncation at $3N_{cut}$ and $3J_{cut}$ will suffice.

EXAMPLE 9.7 Consider the tank in Example 9.6 in the situation where the ribbon transducer is mounted on the vertical center line at $x = 0$, rather than the horizontal center line. All parameters are as stated there, and $W = 0.5$ m. Graph the dependence of $|P|$ on the axial distance along the center line $y = H/2, z = W/2$. Also, graph the profile of $|P|$ for cross sections at $x = 0$ and $x = L$ along the horizontal line $y = H/2$ and the vertical line $z = W/2$.

Significance

The number of modes is the product NJ of the series lengths, which might be quite large. Thus, it is desirable that the computational algorithm be efficient. The need to use two indices to specify a mode complicates development of such an algorithm. Both issues are addressed.

Solution

We shall use the same coordinate system as that in Example 9.6. Unlike the previous arrangement, the ribbon transducer's velocity distribution varies in the z direction. The consequence is that more than one mode in that direction is excited. Furthermore, although the transducer velocity now is independent of y, there is no plane wave mode for this direction. Consequently, all modes in that direction participate, as they did in the previous example. The y dependence of the transverse mode function is the same rigid/pressure-release combination as it was in the previous Example, while the z dependence consists of the modes of a two-dimensional hard-walled waveguide. Thus, the transverse mode functions are

$$\Phi_{n,j} = B_{n,j} \cos\left(\frac{(2n-1)\pi y}{2H}\right) \cos\left(\frac{j\pi z}{W}\right), \quad n = 1, 2, ..., \quad j = 0, 1, 2, ...$$

The corresponding eigenvalues are

$$\mu_n = \frac{(2n-1)\pi}{2H}, \quad \nu_j = \frac{j\pi}{W} \tag{1}$$

The axial wavenumber is either a positive real value or a negative imaginary value,

$$\kappa_{n,j} = \begin{cases} \left(k^2 - \mu_n^2 - \nu_j^2\right)^{1/2} & \text{if } \mu_n^2 + \nu_j^2 < k^2 \\ -i\left(\mu_n^2 + \nu_j^2 - k^2\right)^{1/2} & \text{if } \mu_n^2 + \nu_j^2 > k^2 \end{cases} \tag{2}$$

The value of B_n is set by the normalization in Eq. (9.3.30), which leads to

$$B_{n,j} = \sqrt{2} \text{ if } j = 0, \text{ otherwise } B_{n,j} = 2 \tag{3}$$

Alignment of the transducer in the vertical direction means that the axial velocity everywhere on the cross section at $x = 0$ is zero, except along the line $z = W/2$. Thus, the normal velocity at $x = 0$ is $V_s = \tilde{Q}_0 \delta(z - W/2)$. The corresponding velocity coefficients are indicated by Eq. (9.3.37) to be

$$V_{n,j} = \frac{B_{n,j}\tilde{Q}_0}{H}\phi_{z,j}(z = W/2)\int_0^H V_s \phi_{y,n} dy$$

$$= \frac{2B_{n,j}\tilde{Q}_0}{(2n-1)\pi W}\sin\left(\frac{(2n-1)\pi}{2}\right)\cos\left(\frac{j\pi}{2}\right) \tag{4}$$

The series length is set by the largest cutoff mode numbers for both directions, which are $N_{cut} = \text{floor}((kH/\pi + 1/2)$ and $J_{cut} = \text{floor}(kW/\pi)$. For the present system parameters ($f = 20$ kHz, $c = 1480$ m/s, $H = 0.6$ m, $W = 0.5$ m), these values are $N_{ev} = 16$, and $J_{ev} = 13$. We wish to construct a scan along locations that are close to the boundary, so evanescent modes must be included. Truncation of the series at $n = 3N_{ev}$ and $j = 3J_{ev}$ should be adequate. In principle, we could reduce these numbers as x increases. However, the algorithm that follows is quite efficient, so there is no need to reduce the computational effort.

We have seen in a number of contexts that it is desirable to evaluate a modal-type series by the matrix algorithm employed in Example 9.6. Doing so enables the computation to be carried out in a vectorized manner. To use that algorithm here we must arrange the modes in a sequence described by a single subscript. Let us define an index $m = 1, 2, \ldots$ that is in a one-to-one correspondence to the n, j indices of a transverse mode function. This can be done by incrementing m as n is increased with j fixed, then repeating the process with j increased by one, until the range of n is exhausted. Because $1 \le n \le N$ and $0 \le j \le J$, the rule is

$$m = n + jN \tag{5}$$

Integer arithmetic allows us to identify the n and j values associated with a specific m. If we divide the preceding by N we find that $m/N \ge j$, which leads to

$$j = \text{floor}\left(\frac{m - \varepsilon}{N}\right), \quad n = m - jN \tag{6}$$

where ε is a very small number required to handle round-off error.

This arrangement allows us to implement the algorithm described by Eq. (4) of Example 9.6. The first step is to rewrite the modal series in a single subscript form. The number of modes is $N(J + 1)$, so the series becomes

$$P \approx \rho_0 c \sum_{m=1}^{N(J+1)} \frac{k}{\kappa_m} V_m \Phi_m(y, z) e^{-i\kappa_m x}$$

To convert this to matrix form let the diagonal array $[K]$ hold the factors in the sum that do not depend on position,

$$K_{m,m} = \frac{k}{\kappa_m} V_m \tag{7}$$

The exponential terms are placed in $[E]$, with a row holding the terms for all m at fixed x. The number of rows is the number of x locations at which the pressure will be computed. Thus,

$$E_{q,m} = \exp(-i\kappa_m x_q) \tag{8}$$

The $[F]$ array holds the values of the transverse locations at selected points on a cross section. In any case a column of $[F]$ holds the Φ_m values for all m at a designated y and z. If we are interested in the pressure distribution over a range of y locations at a fixed z, then $[F]$ is formed by adjoining columns for a sequence of y values. Similarly, if our interest is in a range of z locations at fixed y, then the adjoined columns correspond to the successive z values. Thus,

$$F_{m,r} = \Phi_m\,(y_r, z) \text{ or } F_{m,r} = \Phi_m\,(y, z_r) \tag{9}$$

In either case, the pressure evaluation is

$$[P] = \rho_0 c\,[E]\,[K]\,[F] \tag{10}$$

The computational reduction achieved by this algorithm compared to one that actually accumulates sums is quite substantial. Indeed, the computations here, for which the total number of modes is 1920 and the number of points along the axial centerline was 1000, required 0.60 s to execute on a laptop computer with an Intel I7 processor and 8 GB of RAM. Truncating the modal series at $2N_{cut}$ and $2J_{cut}$ reduced this time to 0.37 s.

A few details of the MATLAB program are worth reviewing. The vector $\{m\} = \{1\ 2 \cdots N(J+1)\}$ consists of the single mode indices. Vectors $\{n\}$ and $\{j\}$, which hold the respective mode numbers corresponding to each element of $\{m\}$, were set according to Eq. (6). (The length of $\{n\}$ and $\{j\}$ is the same as the length of $\{m\}$.) The eigenvalues corresponding to $\{n\}$ and $\{j\}$ are placed in arrays $\{\mu\}$ and $\{\nu\}$. The κ_m values also are computed in a vectorized manner. In MATLAB, this operation is kappa=conj(sqrt(k^2+mu.^2-nu.^2)), where ".^" indicates that the square should be done element by element. (The use of the conjugate here is dictated by MATLAB's implementation of the branch cut for a square root.) The vector $\{\kappa\}$ is used to compute the successive rows of $[E]$, with x held constant in each row. The evaluation of $[F]$ is done in a row-wise manner using $\{\mu\}$ and $\{\nu\}$, with the y and z values fixed in a column.

Figure 1 shows $|P|$ as a function of x along the geometric centerline, $y = H/2$, $z = W/2$. The overall pressure amplitude is comparable to the result in Example 9.6. That is, the pattern of relatively broad peaks separated by narrow antinodes is like the preceding result. This is to be expected because Eq. (9.3.28) shows that the signal is a superposition of plane waves, just as it is in the two-dimensional case.

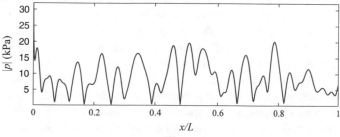

Figure 1.

Profiles of $|P|$ along a cross section appear in Fig. 2. The upper pair describe $x = 0$ and the lower pair are the distributions at $x = 3$ m. The vertical axis of the graphs on the right depicts the y dependence because that is the way the y-axis is physically aligned in Example 9.6. Here too we see interference patterns

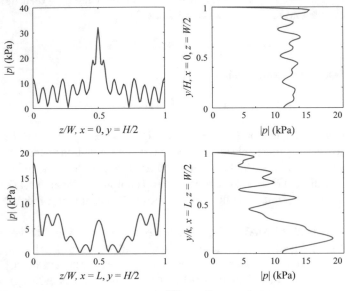

Figure 2.

resulting from superposing the contributions of numerous modes. At $x = 0$ the scan along $z = W/2$ shows the pressure on the transducer. Interference leads to fluctuations of $|P|$ on the face of the transducer, even though V_x is constant. The scan along $y = H/2$ shows a spike of $|P|$ in the vicinity of the transducer. This is like the behavior on the face of a translating piston, where a uniform particle velocity generates a fluctuating face pressure. Neither pattern is evident at the farther cross section because the location is many wavelengths from the transducer.

The symmetry properties of these profiles is important. The waveguide's shape is symmetric with respect to the vertical and horizontal mid-planes, that is, the planes that contain the axial centerline. The pressure does not share this symmetry, because the upper boundary is pressure-release, whereas the lower boundary is rigid. However, it is symmetric with respect to the vertical mid-plane because the walls to the left and right are rigid, and the transducer properties are the same on either side of that plane. This symmetry is manifested in Eq. (4), which indicates that the $V_{n,j}$ velocity coefficients are identically zero if j is odd. As a result, all modes in the vertical direction are excited, but only the even modes for the horizontal dependence are excited. The consequence is that a scan of $|P|$ along a horizontal line is an even function with respect to the middle, whereas a scan along a vertical line shows no pattern repetition. We could have exploited this symmetry by limiting the evaluation of $|P|$ to the range $0 \le z \le W/2$ and only including the even functions of z in the modal series.

9.3.3 Circular Waveguide

Circular waveguides are commonly used in such diverse applications as HVAC and nuclear reactor cooling systems. Cylindrical coordinates, with the axial direction designated as x, are appropriate to this configuration. Propagation in a circular waveguide is the complement of radiation from a vibrating cylinder. Both are governed by the Helmholtz equation in cylindrical coordinates. The difference is that the waves propagate in the transverse direction in the radiation problem, whereas the propagation is axial for a waveguide. Cylindrical Bessel functions are part of the general solution of the Helmholtz equation in cylindrical coordinates. Consequently, much reference will be made to Sect. 7.3.2, where the fundamental properties of these functions are described.

Fundamental Solution for a Waveguide Mode

The pressure is taken to be a wave that propagates in the axial direction with wavenumber κ. The complex amplitude is allowed to depend on the radial distance R and circumferential angle θ. The θ dependence is set by the fact that $(R, \theta \pm 2\pi, z)$ and (R, θ, z) designate the same point, so that any signal must be expressible as a Fourier series in θ with a period of 2π. We shall use a complex exponential to express the θ-dependence of a term in this series. Thus, the ansatz we have identified is

$$p = \mathrm{Re}\left(f\left(R\right)e^{-in\theta}e^{-i\kappa x}e^{i\omega t}\right) \qquad (9.3.39)$$

The choice of κ as the axial wavenumber is consistent with the general analysis in Sect. 9.2.1. However, the variable for the transverse dependence will be μR. This notation is opposite the usage in Sect. 7.3, where the transverse wavenumber of a cylindrical wave was κ and the axial wavenumber was μ. One should bear this in mind when reference is made to prior developments. Another difference from the analysis of radiation is that the axial wavenumber was taken to be known as part of the specification of surface motion, whereas here it is a parameter to be determined.

Substitution of the assumed form of p into the Helmholtz equation in cylindrical coordinates leads to

$$\frac{d^2 f}{dR^2} + \frac{1}{R}\frac{df}{dR} + \left(\mu^2 - \frac{n^2}{R^2}\right) f = 0 \qquad (9.3.40)$$

where

$$\mu^2 = k^2 - \kappa^2 \qquad (9.3.41)$$

This is Bessel's equation, which we previously encountered as Eq. (7.3.8). Its fundamental solutions are the Bessel function $J_n\left(\mu R\right)$ and the Neumann function $N_n\left(\mu R\right)$, so the general solution is

$$f = B_n J_n\left(\mu R\right) + C_n N_n\left(\mu R\right) \qquad (9.3.42)$$

The Neumann function is singular at $R = 0$, so it is discarded by setting $C_n = 0$.

Satisfaction of the boundary condition on the cylindrical wall will be seen to require specific values of μ; these are the eigenvalues. These values will depend on which azimuthal harmonic is being considered. Each μ value will lead to a different κ, so the modes are dispersive. With n restricted to be positive or zero, the waveguide modes and associate transverse mode functions are given by

$$
\begin{aligned}
P &= \Phi\,(R, \theta)\, e^{-i\kappa x} \\
\Phi\,(R, \theta) &= \begin{cases} B J_n\,(\mu R)\, e^{-in\theta} \\ \text{or} \\ B J_n\,(\mu R)\, e^{in\theta} \end{cases}
\end{aligned}
\tag{9.3.43}
$$

Adding and subtracting these fundamental solutions yields sine and cosine functions for the azimuthal dependence. Thus, an alternative form of the waveguide modes is

$$
\begin{aligned}
P &= \Phi\,(R, \theta)\, e^{-i\kappa x} \\
\Phi\,(R, \theta) &= \begin{cases} B J_n\,(\mu R)\cos{(n\theta)} \\ \text{or} \\ B J_n\,(\mu R)\sin{(n\theta)} \end{cases}
\end{aligned}
\tag{9.3.44}
$$

The difference between Eqs. (9.3.43) and (9.3.44) is that the former describes a pair of helical waves that spiral about the x-axis in opposite senses, whereas the latter describes standing wave patterns around the x-axis. The sine and cosine functions are $90°$ out-of-phase and therefore constitute separate solutions. In the special case where $n = 0$, the two helical wave coalesce to a plane wave that propagates in the axial direction, and the sine mode vanishes.

Just as the pattern in the circumferential direction may be viewed as either oppositely traveling waves or standing waves, the same is true for the transverse direction. According to Eq. (7.3.11), the Bessel function is equivalent to the sum of both kinds of Hankel functions,

$$
J_n\,(\mu R) = \frac{1}{2}\left[H_n^{(1)}\,(\mu R) + H_n^{(2)}\,(\mu R) \right]
\tag{9.3.45}
$$

Thus, any of the descriptions of the modes may equivalently be considered to be waves that propagate inward and outward from the center. Each wave individually would be singular at $R = 0$, but their singularities cancel.

In addition to finiteness, any pressure field must be continuous at $R = 0$. This means that the same pressure must be obtained as a field point approaches $R = 0$ along any θ, which is stated as

$$
\lim_{R \to 0} P\,(R, \theta, x) = g\,(x)
\tag{9.3.46}
$$

where $g(x)$ is used to indicate that the limit may depend on x. Continuity of ∇P also is a fundamental requirement. Otherwise, $\nabla^2 P$ will be infinite, which according to the Helmholtz equation corresponds to infinite P. As $R \to 0$, the $\bar{e}_\theta$ direction becomes meaningless, so we may focus on the behavior of $\partial P/\partial R$ at $R = 0$. Because $\bar{e}_R(\theta + \pi) = -\bar{e}_R(\theta)$, the transverse derivative at the origin must change sign when θ is incremented by $180°$,

$$\left. \frac{\partial P}{\partial R} \right|_{R=0, \theta \pm \pi} = - \left. \frac{\partial P}{\partial R} \right|_{R=0, \theta} \tag{9.3.47}$$

Let us assess whether these conditions are satisfied by the waveguide modes. All modes have the form $\Phi_n(R, \theta) \exp(-i\kappa x)$ with $\Phi_n(R, \theta) = f_R(R) f_\theta(\theta)$. The transverse dependence is $f_R(R) = J_n(\mu R) \approx (\mu R/2)^n / n!$ when R is small, and $f_\theta(\theta) = \exp(-in\theta)$. For $n > 0$, $f_R = 0$ at $R = 0$, so the g function is zero independent of x. Thus, the modes for $n \geq 1$ are continuous. For $n = 0$, we have $f_R = 1$ at $R = 0$ and $f_\theta = 1$, so the $n = 0$ modes also are continuous. In regard to continuity of the gradient, for $n \geq 1$, the radial derivative at the center is $\partial \Phi_n/\partial R = (\mu R/2)^{n-1} \exp(-in\theta) / (n-1)!$ evaluated at $R = 0$. This quantity is zero for $n \geq 2$, which satisfies Eq. (9.3.47). For the $n = 1$ modes, we have $\partial \Phi_n/\partial R = \exp(-i\theta)$ at $R = 0$, which also satisfies Eq. (9.3.47). The $n = 0$ modes also are such that $\partial \Phi_0/\partial R = 0$ at $R = 0$. Thus, the Bessel function modes satisfy the continuity requirements at the origin.

Figure 9.14a and b, respectively, shows the behavior of the $n = 0$ and $n = 1$ waveguide modes in the vicinity of $R = 0$. These are contrasted by Fig. 9.14c, which shows a discontinuous pressure field described by $J_0(\mu R) \operatorname{Re}(e^{i\theta})$, and by Fig. 9.14d, where $J_1(\mu R)$ has a discontinuous gradient at $R = 0$.

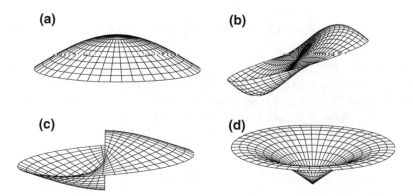

(a)   **(b)**

(c)   **(d)**

Fig. 9.14 Behavior of pressure fields in the vicinity of $R = 0$: **a** The $n = 0$ propagation mode, $P = B J_0(\mu R)$. **b** The $n = 1$ propagation mode, $P = B J_1(\mu R) e^{-i\theta}$. **c** A discontinuous pressure field, $B J_0(\mu R) e^{-i\theta}$. **d** A field having a discontinuous gradient, $P = B J_1(\mu R)$

The radial wavenumber is set by the boundary condition at the wall, where $R = a$. A positive value of V_R corresponds to movement into the wall, so an impedance boundary condition corresponds to

$$P = \rho_0 c \zeta V_R = -\frac{\zeta}{ik} \frac{\partial P}{\partial R} \text{ at } R = a \tag{9.3.48}$$

When we substitute the general solution for a waveguide mode into this condition, the exponential factors cancel, as does the amplitude coefficient B. What remains is the characteristic equation,

$$\boxed{J_n (\mu a) - \frac{i\zeta}{ka} (\mu a) J_n' (\mu a) = 0} \tag{9.3.49}$$

Regardless of the nature of ζ, the characteristic equation for a specified value of the circumferential number n will have an infinite number of roots. These nondimensional roots of the characteristic equation are denoted as $\eta_{n,j}$. The corresponding eigenvalues are $\mu_{n,j} = \eta_{n,j}/a$. The eigenvalues might be imaginary or complex, so we shall arrange them sequentially in increasing order of their real part, $\text{Re}(\mu_{n,1}) < \text{Re}(\mu_{n,2}) < ...$ The manner in which the characteristic equation has been written makes it evident that the $\eta_{n,j}$ values depend only on ζ and ka.

Numerical methods will be required to obtain accurate roots of the characteristic equation. Few software packages offer routines that evaluate derivatives of Bessel functions. A recurrence relation in Eq. (7.3.15) gives

$$\boxed{J_n' (\mu a) = -J_{n+1} (\mu a) + \left(\frac{n}{\mu a}\right) J_n (\mu a)} \tag{9.3.50}$$

This expression may be incorporated as an auxiliary routine, but it is just as easy to incorporate it into the characteristic equation, which becomes

$$J_n (\mu a) - \frac{i\zeta}{ka} (\mu a) \left[-J_{n+1} (\mu a) + \left(\frac{n}{\mu a}\right) J_n (\mu a) \right] = 0 \tag{9.3.51}$$

The methods that are commonly used to solve transcendental equations require initial guesses of the roots $\eta_{n,j} = \mu_{n,j}a$. If ζ is infinite, zero, or purely imaginary, the equation is real. Guesses for the lower roots in that case may be obtained by graphing the characteristic equation as a function of μa. For large μa ($\mu a > 8$ is reasonable), we may employ the asymptotic representation of $J_n (\mu a)$ for large arguments

$$J_n (\mu a) \rightarrow \left(\frac{2}{\pi \mu a}\right)^{1/2} \cos \left(\mu a - \frac{2n+1}{4}\pi\right) \tag{9.3.52}$$

When μa is large, increasing μa changes the square root factor much less than the cosine term. It follows that for large μa, the characteristic equation is approximately periodic in $\Delta (\mu a) = 2\pi$. The consequence of this property is that after we have found a large eigenvalue $\mu_n a \gg 1$, a good starting guess for the next value is $\mu_{n+1}a = \mu_n a + \pi$.

Two indices are required to uniquely specify which eigenvalue is under consideration: the circumferential harmonic number n and the root number j, which gives the sequential placement of the root within the set of eigenvalues at that n. Specification of a transverse mode function requires a third index, denoted as α. This index is used to specify the dependence on θ. We will adopt the convention that if this subscript is $\alpha = 1$ or $\alpha = -1$, we are using helical waves, such that

$$\Phi_{n,j,\alpha} = B_{n,j,\alpha} J_n\left(\mu_{n,j} R\right) e^{i\alpha n\theta} \qquad (9.3.53)$$

Alternatively, if we wish to use real sinusoidal terms to describe the θ dependence, we would set $\alpha = c$ or $\alpha = s$, such that

$$\begin{aligned} \Phi_{n,j,c} &= B_{m,n,c} J_n\left(\mu_{n,j} R\right) \cos\left(n\theta\right) \\ \Phi_{m,n,s} &= B_{n,j,s} J_n\left(\mu_{n,j} R\right) \sin\left(n\theta\right) \end{aligned} \qquad (9.3.54)$$

The discussion thus far covers the generalities. However, examination of specific impedance cases discloses issues that are not yet evident, as well as some significant behaviors.

Rigid Walls

The model of a rigid wall serves as a reference system for discussing other configurations, as well as being useful for its own sake. Satisfaction of Eq. (9.3.49) when ζ is infinite requires that $J_n'\left(\mu a\right) = 0$. Thus, the eigenvalues correspond to locations at which $J_n\left(\mu a\right)$ has an extreme value. Figure 9.15 plots some low-order Bessel functions and shows the smallest roots $\eta_{n,j}$ for each n.

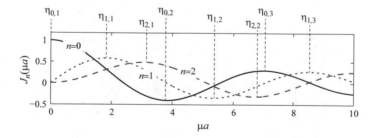

Fig. 9.15 Characteristic equation for a circular waveguide with rigid walls

The first few eigenvalues for a rigid-walled cylinder are listed in Table 9.1.

The axial wavenumbers are defined by Eq. (9.3.41). As was true for the previous situations, a large eigenvalue will lead to cutoff of the axial propagation. The possibilities are

Table 9.1 Roots of the characteristic equation for a rigid-walled cylindrical waveguide

$\eta_{n,j} \equiv \mu_{n,j}a$	$j = 1$	$j = 2$	$j = 3$	$j = 4$
$n = 0$	0	3.8317	7.0156	10.1735
$n = 1$	1.8412	5.3314	8.5363	11.7060
$n = 2$	3.0542	6.7061	9.9695	13.1704
$n = 3$	4.2012	8.0152	11.3459	14.5858

$$\kappa_{n,j} = \begin{cases} \left[k^2 - (\eta_{n,j}/a)\right]^{1/2} & \text{if } \eta_{n,j} < ka \\ -i\left[(\eta_{n,j}/a)^2 - k^2\right]^{1/2} & \text{if } \eta_{n,j} > ka \end{cases} \tag{9.3.55}$$

Because the eigenvalues are roots of a transcendental equation, it is not possible to identify *a priori* how many eigenvalues are below the cutoff value. Nevertheless, the tabulation indicates that the eigenvalues at a fixed root number increase with increasing n. Furthermore, by definition, the eigenvalues at fixed n increase with increasing root number, Thus, it is sufficient to identify the root number J for $n = 0$ at which $\eta_{0,J} > ka$, and the circumferential harmonic N at which $\eta_{1,N} > ka$. Using these as the cutoff numbers assures that all propagating modes are included. Figure 9.16 shows contours of some typical transverse mode functions. Bold contours

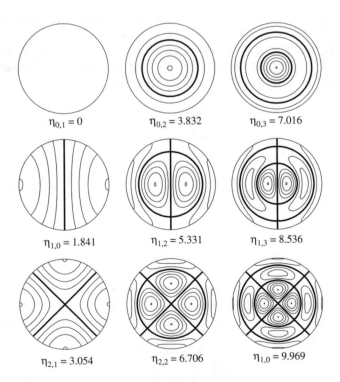

Fig. 9.16 Contours of the transverse mode functions for a rigid-walled circular waveguide. An $\eta_{n,j}$ value is the jth root of the characteristic equation corresponding to circumferential harmonic number n

are zeros, which separate regions of positive and negative pressures. The patterns describe the cosine functions, $\Phi_{n,j,c}$; the sine functions would be rotated by $\pi/(2n)$.

Compliant Walls

Many features exhibited by the modes in the rigid-walled case are typical of those for other wall conditions. Equation (9.3.43) or the alternative form in Eq. (9.3.44) describes any case. A corollary of the occurrence of ka in the characteristic equation, Eq. (9.3.49), is that it will be necessary to recompute its roots if the frequency of interest changes, for example, to perform a frequency sweep in order to determine a dispersion curve. The only exceptions are rigid or pressure-release walls.

If ζ is finite and nonzero, then neither term in the characteristic equation is zero. It follows that unless ζ is purely imaginary, the equation is complex. This means that the eigenvalues must be complex, which is a possibility we will address. The exceptional case is that where ζ is purely imaginary, corresponding to a purely reactive wall. To analyze that case, we set $\zeta = \sigma i$, with $\sigma > 0$ for an inertance and $\sigma < 0$ for a compliance. The corresponding form of the characteristic equation is

$$\boxed{J_n(\mu a) + \sigma \frac{\mu a}{ka} J_n'(\mu a) = 0} \tag{9.3.56}$$

The Bessel function is oscillatory, from which it follows that this equation has an infinite number of roots $\eta_{n,j} = \mu_{n,j} a$, $j = 1, 2, \ldots$

Let us consider the behavior of the eigenvalue for a fixed circumferential harmonic n and root number j as the frequency increases. The nondimensional parameter affecting the eigenvalues is $\sigma/(ka)$. The case of a pressure-release wall, which is the limit as $|\zeta| \to 0$, corresponds to $J_n(\mu a) = 0$. In contrast, $\zeta = \infty$ gives $J_n'(\mu a) = 0$, which is the characteristic equation for a rigid wall. It follows that this eigenvalue will be close to that of a rigid waveguide at very low frequencies, while it will be close to the eigenvalue of a pressure-release configuration for high frequencies. A different trend is that in which n and ka are held fixed. By definition, increasing j corresponds to increasing eigenvalue. Because μa multiplies σ in the characteristic equation, the higher eigenvalues will approach those for a rigid wall regardless of the size of σ. We previously encountered this trend in the analysis of a two-dimensional waveguide whose walls are reactive.

Like a two-dimensional waveguide, a circular waveguide whose wall is purely reactive might have a purely imaginary eigenvalue. To explore this possibility, we set $\mu = i\beta$, $\beta > 0$. Rather than dealing with Bessel functions whose argument is imaginary, it is preferable to convert $J_n(\mu a)$ to the first kind of modified Bessel function, which is given by Eq. (7.3.33) to be

$$J_n(i\beta a) = i^n I_n(\beta a) \tag{9.3.57}$$

Substitution of this expression into the third recurrence relation for a derivative of a
Bessel function in Eq. (7.3.15) gives

$$J'_n(i\beta a) = i^{n-1}\left[I_{n+1}(\beta a) + \frac{n}{\beta a}I_n(\beta a)\right] \tag{9.3.58}$$

Substitution of these relations into the characteristic equation converts it to

$$I_n(\beta a) + \mathrm{Im}(\zeta)\frac{\beta a}{ka}\left[I_{n+1}(\beta a) + \frac{n}{\beta a}I_n(\beta a)\right] = 0 \tag{9.3.59}$$

One feature is immediately apparent. Each of the modified Bessel functions appearing
above is positive for $\beta a > 0$. It follows that a real root of this characteristic equation
can exist only if $\mathrm{Im}(\zeta) < 0$. In other words, a mode whose eigenvalue is purely
imaginary exists only if the wall impedance is a pure compliance. This condition is
consistent with the previous result for a two-dimensional waveguide. We will say
that this is the (n, im) mode, so $\mu_{n,\mathrm{im}} = i\beta_{n,\mathrm{im}}a$.

To get a more detailed picture, let us consider the modal properties for $\sigma = \pm 5$
($\zeta = \pm 5i$), $n = 0$, and $ka = \pi$. The characteristic equation for real eigenvalues,
Eq. (9.3.56), and an imaginary eigenvalue, Eq. (9.3.59), are depicted in Fig. 9.17. The
function that is plotted for imaginary eigenvalues is the above divided by $I_0(\beta a)$.
(This is done because $I_n(\beta a)$ increases exponentially with increasing βa, which
would make it difficult to see the location where the plotted function is zero.) The
intersection of each curve with the zero axis marks a root.

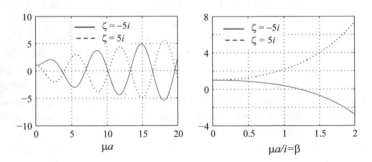

Fig. 9.17 Characteristic equation for a circular waveguide whose wall is compliant, $ka = \pi, n = 0$

Each eigenvalue in the range $\mu a > 3$ for $\sigma = -5$ has a close companion for
positive σ. The exception is the first eigenvalue for $\sigma = 5$, $\mu_1 a = 1.215$, whereas
no real eigenvalue for $\sigma = -5$ that is that small. However, the imaginary eigenvalue
for $\sigma = -5$ also has a small magnitude, which suggests that they might have some
similarities. The first few eigenvalues and corresponding axial wavenumbers are
listed in Table 9.2. In both reactance cases, only one mode is propagative. As ka is
increased, more and more modes enter the spectrum that propagates.

Table 9.2 Eigenvalues of a cylindrical waveguide with a reactive wall, ka = pi

		$j = \text{im}$	$j = 1$	$j = 2$	$j = 3$
$\zeta = -5i$	$\eta_{0,j} \equiv \mu_{0,j}a$	$1.2146i$	3.6657	6.9257	10.1116
	$\kappa_{0,j}a$	3.368	$-1.8888i$	$-6.1722i$	$-9.6112i$
$\zeta = 5i$	$\eta_{0,j} \equiv \mu_{0,j}a$	–	1.0387	3.9910	7.1044
	$\kappa_{0,j}a$	–	2.9649	$-2.4614i$	$-6.3720i$

The transverse mode functions for an inertance and a compliance are depicted in Fig. 9.18. The functions associated with real eigenvalues $\mu_{0,j} = \eta_{0,j}/a$ are like those in Fig. 9.16. This is to be expected because $\zeta = \pm 5i$ is effectively quite close to the rigid condition. However, a finite impedance requires that $d\Phi_{n,j}/dR \neq 0$ at $R = a$, so the maxima, minima, and zeros of the R-dependence are shifted. The single imaginary eigenvalue, $\mu_{n,\text{im}} = i\eta_{n,\text{im}}/a$, corresponds to a mode function that is reminiscent of the plane wave mode. However, the first mode for $\zeta = 5i$ also shows relatively little variation along a transverse lines, so it is too reminiscent of the plane wave mode.

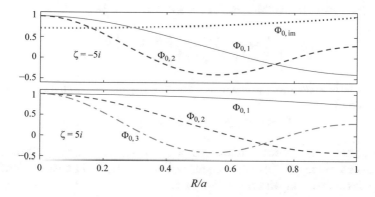

Fig. 9.18 Transverse mode functions for a circular waveguide whose wall is reactive, $ka = \pi$, $n = 0, \zeta = -0.2i$

EXAMPLE 9.8 An annular waveguide consists of the region between two concentric cylinders. The outer radius is a, and the inner radius is b. Consider the case where both cylinders are rigid. Derive the characteristic equation and an expression for the transverse mode functions. Evaluate the $n = 0, n = 1$, and $n = 2$ modes for the case where $b/a = 0.4$. Also, determine the highest frequency for which all nonplanar modes evanesce.

Significance

Coaxial pipes are used in some systems to convey fluid back and forth, especially if the surroundings are narrow. The analysis suggests how the formulation for a complete cylindrical cross section may be generalized. To some extent, it is a precursor of the analyses of a cavity, which is the topic of the next chapter.

Solution

The transverse mode function must satisfy the rigidity condition at both walls, which are

$$\frac{\partial \Phi}{\partial R} = 0 \text{ at } R = a \text{ and } R = b$$

Because $R = 0$ is not a point in the fluid, both the Bessel and Neumann functions in the general solution for R dependence, Eq. (9.3.42), are appropriate. A transverse mode function is the product of that function and the nth harmonic for the θ dependence, so the form is

$$\Phi = [B_n J_n (\mu R) + C_n N_n (\mu R)] e^{-in\theta} \tag{1}$$

The coefficients are found by making this general solution fit the boundary conditions. The matrix form of the result of these operations is

$$\begin{bmatrix} J_n' (\mu a) & N_n' (\mu a) \\ J_n' (\mu b) & N_n' (\mu b) \end{bmatrix} \begin{Bmatrix} B_n \\ C_n \end{Bmatrix} = \begin{Bmatrix} 0 \\ 0 \end{Bmatrix} \tag{2}$$

The characteristic equation, which results from setting the determinant of these equations to zero, is

$$J_n' (\mu a) N_n' (\mu b) - N_n' (\mu a) J_n' (\mu b) = 0 \tag{3}$$

The frequency does not occur in this characteristic equation, which means that the eigenvalues are independent of the frequency. Figure 1 graphs the characteristic equation as a function of μa with $\mu b \equiv (b/a) \, \mu a$.

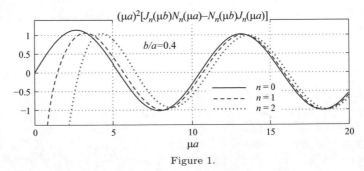

Figure 1.

The zero crossing of the curve for each n is used to initiate a numerical search for the respective eigenvalues. The walls are rigid, so a plane wave mode may exist.

Thus, $n = 0$ is a plane wave, for which $\mu_{0,1} = 0$ and $\Phi_{0,1} = 1$. The result appears in the following tabulation.

	$\mu_{n,j}a$			
n	$j = 1$	$j = 2$	$j = 3$	$j = 4$
0	0	5.3912	10.5577	15.7665
1	1.46178	5.6591	10.6833	15.8481
2	2.8424	6.4160	11.0560	16.0916

When μa is one of the eigenvalues, only one of the scalar equations described by Eq. (2) is independent. Thus, only the ratio of the coefficients is described uniquely. When we use the second coefficient equation, which is the boundary condition at $R = a$, to evaluate C/B, we find that

$$\Phi_{n,j,1} = B_{n,j} \left[J_n \left(\mu_{n,j} R \right) - \frac{J_n' \left(\mu_{n,j}a \right)}{N_n' \left(\mu_{n,j}a \right)} N_n \left(\mu_{n,j} R \right) \right] e^{-in\theta}$$

(Recall that $e^{+in\theta}$, $\cos(n\theta)$, and $\sin(n\theta)$ are equally valid descriptions of the θ-dependence.) Figure 2 shows some transverse mode functions for $b/a = 0.4$. The value of $B_{0,j}$ for each was selected such that the maximum of the respective function is one. If we were to let s be the distance from the inner cylinder, it would not be a terrible approximation to say that these modes fit $\cos(j\pi s/(a - b))$, which are the transverse modes of a hard-walled two-dimensional waveguide.

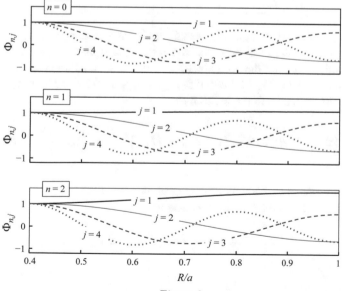

Figure 2.

Signal Generation

Our interest here is the signal generated by the imposition of an axial particle velocity at $x = 0$. If the spatial pattern of the velocity imposed at $x = 0$ does not match a specific mode, all modes are excited. Thus, we consider an arbitrary distribution, $V_x = \mathrm{Re}\left(V_s\left(R, \theta\right) e^{i\omega t}\right)$ at $x = 0$. The procedure for identifying the mix of waveguide modes that is generated by this excitation follows the general outline laid out in Sect. 9.3.1. The first task is to decide how to represent the θ-dependence of the modes. Complex exponentials are suitable for an excitation that rotates about the x-axis, which is typical of noise generation mechanisms in rotating machinery. However, most excitations have a fixed distribution in the θ direction, and trigonometric functions share that property. Hence, we shall use the transverse mode functions in Eq. (9.3.44),

$$\begin{aligned}
\Phi_{n,j,c} &= B_{n,j}\, J_n\left(\mu_{n,j} R\right) \cos\left(n\theta\right) \\
\Phi_{n,j,s} &= B_{n,j}\, J_n\left(\mu_{n,j} R\right) \sin\left(n\theta\right)
\end{aligned} \tag{9.3.60}$$

In the usual situation the range of indices is $n \geq 0$, $j \geq 1$. This scheme is modified if the wall is rigid by designating the first mode for $n = 0$ as $j = 0$ to represent the plane wave mode. In addition, if the wall is purely compliant, the imaginary eigenvalue at any n is designated with the label $j = $ "im". Also, for $n = 0$, $B_{0,j,s}$ are set to zero.

According to the general orthogonality property in Eq. (9.3.8), it must be that

$$\boxed{\int_0^a \int_{-\pi}^{\pi} \Phi_{n,j,\alpha}\Phi_{m,\ell,\beta}\, R d\theta d R = \pi a^2 \delta_{n,m}\delta_{j,\ell}\delta_{\alpha,\beta}} \tag{9.3.61}$$

The Kronecker delta term $\delta_{\alpha,\beta}$ tells us that the sine and cosine modes are orthogonal, even if their eigenvalues are the same. When the subscripts are the same, the orthogonality condition defines the normalization of the $B_{n,j,\alpha}$ coefficients. Substitution of the mode function into Eq. (9.3.60) with $\mu_{n,j} = \eta_{n,j}/a$, followed by integration over θ, reduces the normalization condition to

$$\int_0^a \int_{-\pi}^{\pi} \left(\Phi_{n,j,\alpha}\right)^2 R d\theta d R = \left(1 + \delta_{n,0}\right) \pi \left(B_{n,j}\right)^2 \int_0^a J_n\left(\eta_{n,j}\frac{R}{a}\right)^2 R d R = \pi a^2 \tag{9.3.62}$$

There is no need to evaluate this integral numerically. Changing the integration variable to $\chi = R/a$ converts the equation to

$$\left(1 + \delta_{n,0}\right)\left(B_{n,j}\right)^2 \mathcal{I}_{n,j} = 1 \tag{9.3.63}$$

where

$$\mathcal{I}_{n,j} = \int_0^1 J_n\left(\eta_{n,j}\chi\right)^2 \chi d\chi \tag{9.3.64}$$

Because $\eta_{n,j}$ is a root of the characteristic equation, Eq. (9.3.56), it fits the condition for which a standard formula[2] is valid. The result depends on the nature of the impedance according to

$$
\mathcal{I}_{n,j} = \begin{cases} \dfrac{1}{2} J_n \left(\eta_{n,j} \right)^2 & \text{if } \zeta = 0 \\[3mm] \dfrac{1}{2} \left(1 - \dfrac{\zeta^2 n^2 + (ka)^2}{\zeta^2 \eta_{n,j}^2} \right) J_n \left(\eta_{n,j} \right)^2 & \text{if } \zeta \neq 0 \end{cases}
\tag{9.3.65}
$$

Thus, we have established that the normalizing coefficient is

$$
B_{n,j} = \frac{1}{\left(1 + \delta_{n,0} \right)^{1/2} \left(\mathcal{I}_{n,j} \right)^{1/2}}
\tag{9.3.66}
$$

The incarnation of the general modal series, Eq. (9.3.9), is

$$
P \left(x, R, \theta \right) = \sum_{n=0}^{\infty} \sum_{j=1}^{\infty} \sum_{\alpha=c,s} P_{n,j,\alpha} \Phi_{n,j,\alpha} e^{-i\kappa_{n,j}x}
\tag{9.3.67}
$$

Equation (9.3.13) describes the pressure coefficients in terms of the coefficients of a modal series for the velocity input, $V_s (R, \theta)$. For the case of a cylinder, Eq. (9.3.12), becomes

$$
\begin{aligned}
V_{n,j,c} &= \frac{1}{\pi a^2} \int_0^a \int_{-\pi}^{\pi} V_s (R, \theta) J_n \left(\mu_{n,j} R \right) \cos (n\theta) \, R d\theta d R \\[2mm]
V_{n,j,s} &= \frac{1}{\pi a^2} \int_0^a \int_{-\pi}^{\pi} V_s (R, \theta) J_n \left(\mu_{n,j} R \right) \sin (n\theta) \, R d\theta d R
\end{aligned}
\tag{9.3.68}
$$

The corresponding pressure coefficients are

$$
P_n = \rho_0 c \frac{k}{\kappa_{n,j}} V_{n,j,\alpha}
\tag{9.3.69}
$$

Thus, the modal pressure series is

$$
\begin{aligned}
P \left(x, R, \theta \right) = \rho_0 c \sum_{n=0}^{\infty} \sum_{j=1}^{\infty} \frac{k}{\kappa_{n,j}} \Big[V_{n,j,c} \cos (n\theta) + V_{n,j,s} \sin (n\theta) \Big] \\
\times B_{n,j} J_0 \left(\mu_{n,j} R \right) e^{-i\kappa_{n,j}x}
\end{aligned}
\tag{9.3.70}
$$

---

[2] M.I. Abramowitz and I.A. Stegun, *Handbook of Mathematical Functions*, Dover, (1965) Eq. 11, pp. 485.

The associated axial particle velocity is

$$
V_x\left(x, R, \theta\right) = \sum_{n=0}^{\infty} \sum_{j=1}^{\infty} \left[V_{n,j,c} \cos\left(n\theta\right) + V_{n,j,s} \sin\left(n\theta\right) \right] B_{n,j} J_0\left(\mu_{n,j} R\right) e^{-i\kappa_{n,j}x}
$$

$$(9.3.71)$$

The overall procedure for determining the response begins with solution of the characteristic equation for the eigenvalues. The normalizing coefficients $B_{n,j}$ are evaluated with the aid of Eqs. (9.3.65) and (9.3.66). Evaluation of the integrals for the velocity coefficients in Eq. (9.3.68) might require a numerical algorithm. These items are used to form the modal series, Eq. (9.3.70). The contribution from each circumferential number n may be carried out sequentially. Truncation of the modal series may be done based on the highest n for which $\kappa_{n,1} < k$ and the highest j for which $\mu_{0,j} < k$, as was explained after Eq. (9.3.55).

EXAMPLE 9.9 A circular piston is mounted concentrically at one end of a very long cylindrical tube filled with air. The other end is terminated with a sand barrier that eliminates reflections. The tube is rigid, with a diameter of 100 mm, and the piston diameter is 50 mm. The piston oscillates at 40 kHz. The vibration amplitude is such that the maximum on-axis pressure would be 110 dB//20μPa if the piston were mounted in an infinite baffle in free space. Determine and graph the sound pressure on the axis of the waveguide in a range from 20 mm to 1 m.

Significance

From an operational viewpoint, this example will illustrate some of the facets of constructing a modal series. It also will demonstrate the drastic effect of confining a sound beam.

Solution

Let V_0 be the velocity amplitude of the piston. It is given that the piston is situated concentrically, which means that the axial velocity it imposes is the same at all θ. The specific function is

$$
V_x\left(x = 0\right) = \begin{cases} V_0, & 0 \leq R < b \\ 0, & b < R \leq a \end{cases}
$$

where $b = 0.025$ m is the radius of the piston and $a = 0.05$ m.

Only the $n = 0$ modes are excited because V_x is an axisymmetric distribution. The transverse mode functions of a hard-walled circular waveguide were found in Sect. 9.3.3 to be

$$\Phi_{0,j} = B_{0,j} J_0\left(\eta_{0,j}\frac{R}{a}\right)$$

$$J_0'\left(\eta_{0,j}\right) \equiv -J_1\left(\eta_{0,j}\right) = 0$$

The first four roots $\eta_{0,j}$ for the $n = 0$ modes are listed there in Table 9.1. For the present parameters, $c = 340$ m/s, $\omega = 2.513\left(10^5\right)$ rad/s, $a = 0.05$ m, the highest mode that is not cutoff is the twelfth, for which $\eta_{0,12} = 35.33$. The value of κ for this mode and the first cutoff mode are $\kappa_{0,12} = 216.95$ m^{-1} and $\kappa_{0,13} = -213.73i$ m^{-1}. At $x = 20$ mm, the exponential decay factor for the latter is approximately 1.4%, which clearly demonstrates that we need not be concerned here with evanescent modes.

The scaling coefficients $B_{0,j}$ are given by Eq. (9.3.66). The specific impedance is infinite, and we are only interested in $n = 0$. Therefore, we may set $J_0'\left(\eta_{0,j}\right) = -J_1\left(\eta_{0,j}\right)$, which is squared in the aforementioned equation. Hence, the coefficients are

$$B_{0,1} = 1, \quad B_{0,j} = \frac{2^{1/2}}{J_1\left(\eta_{0,j}\right)} \text{ if } j > 1$$

The velocity coefficients are described in general by Eq. (9.3.68). For the above mode functions, and zero V_x for $R > b$, that expression becomes

$$V_{0,j} = \frac{V_0}{\pi a^2}\int_0^b\int_{-\pi}^{\pi}\Phi_{0,j}Rd\theta dR = \frac{V_0}{\pi a^2}B_{0,j}\int_0^b\int_{-\pi}^{\pi}J_0\left(\eta_{0,j}\frac{R}{a}\right)Rd\theta dR$$

Changing the integration variable to $u = \eta_{0,j}R/a$ leads to a form that is another standard integral.[3] The result is that

$$V_{0,j} = \begin{cases} V_0\dfrac{b^2}{a^2}B_{0,j}, & j = 1 \\[2ex] 2V_0\dfrac{b}{\eta_{0,j}a}B_{0,j}J_1\left(\eta_{0,j}\dfrac{b}{u}\right), & j > 1 \end{cases}$$

The first few terms are $V_{0,j}/V_0 = 0.25, -0.3763, 0.063695, 0.13228, \dots$

The value of V_0 is set by the given pressure in a free space. The maximum on-axis pressure in a sound beam is twice the plane wave value, max $(|P|) = 2\rho_0 c V_0$. The specified value of 110 dB corresponds to max $(|P|) = 8.9443$ Pa, from which we find that $V_0 = 10.961$ mm/s. The Rayleigh distance is $kb^2/2 = 0.231$ m. Hence, a scan out to $x = 1$ m extends into the farfield for the infinite baffle case.

The on-axis pressure in the waveguide is obtained by evaluating the modal series, Eq. (9.3.70) at $R = 0$. Because $J_0(0) \equiv 1$, the modal series reduces to

$$P(x, 0) = \sum_{j=1}^{12} P_{0,j}e^{-i\kappa_{0,j}x}, \quad P_{0,j} = \rho_0 c\frac{k}{\kappa_{n,j}}V_{0,j}$$

---

[3]M.I. Abramowitz and I.A. Stegun, *ibid,* Eq. (11) pp. 484.

Our standard technique for computing a summation like this with x fixed is to define column vector $\{P_j\}$ that holds the coefficients and row vector $[E(x)]$ whose elements are the complex exponentials for each j. To evaluate it at many x, we stack $[E(x)]$ for each location and compute

$$[P] = \begin{bmatrix} [E(x_1)] \\ [E(x_2)] \\ \vdots \end{bmatrix} \{P_j\}$$

The result is shown in Fig. 1. The on-axis pressure for the case where the piston is mounted on an infinite baffle also is shown there. The maximum pressure in the nearfield is much less than it would be in the infinite baffle case, but the pressure in the waveguide does not show spherical decay. Both the large fluctuations over short distance and the weaker signal in the closer ranges have the same explanation. The axial signal for a baffled piston results from constructive and destructive interference between equal strength center and edge waves. In contrast, we found that waveguide modes may be represented as waves that propagate in and out in the transverse direction as they propagate downstream. There are many modes, which means that there are many waves having different amplitudes. There is no location at which all of these waves are in-phase to create a maximum.

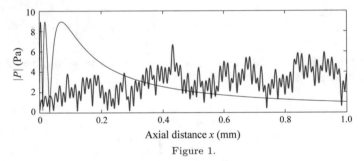

Figure 1.

9.4   Homework Exercises

Exercise 9.1   The sketch shows a conical waveguide that is driven harmonically at its left end by a piston. The right end is terminated by a damping material that eliminates reflections. It is desired to compare the pressure at point B at the wall according to the Webster horn equation to the result obtained by taking the signal to be a radially symmetric wave. For a waveguide that is filled with air at frequencies ranging from 20 Hz to 5 kHz, compare the predictions of both formulations for the magnitude and phase angle of the pressure at point B.

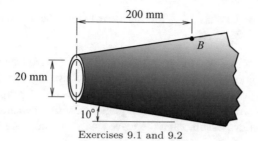

Exercises 9.1 and 9.2

Exercise 9.2 The sketch shows a conical waveguide that is driven harmonically at its left end by a piston. The right end is terminated by a damping material that eliminates reflections. One model for propagation considers the signal to be a radially symmetric wave. Another approximates the radius variation as an exponential expansion, according to which the radius at point B, where $x = 0.2$ m, is related to the end radius by $a_B = a_0 \exp(0.2b)$, where b has units of m^{-1}. Determine the magnitude and phase of the pressure of the pressure at point B according to each model. The frequency varies from 10 Hz to 5 kHz and the fluid is air.

Exercise 9.3 An exponential horn is filled with air. It is 1.2 m long, with end diameters of 10 mm and 100 mm. The pressure at the large end varies harmonically at 800 Hz with an amplitude of 50 Pa. The small end is open to the atmosphere and is surrounded by a large rigid baffle that is flush with the end. (a) Determine the complex amplitudes of the pressure and particle velocity at the left end. Use the end correction on Eq. (3.5.54) for this evaluation. (b) Determine the time-averaged power radiated into the fluid domain to the left of the baffle.

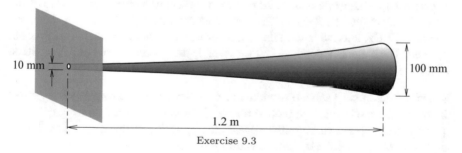

Exercise 9.3

Exercise 9.4 An exponential horn is driven at $x = 0$ by a transducer that imposes a constant velocity amplitude V_0. The specific impedance at $x = L$ is a fixed value ζ independent of the frequency. Derive expressions for the amplitudes of the forward and backward waves as a function of kL when $k > b/2$.

Exercise 9.5 It is desired to design an exponential horn in air to connect to a piston driver. The design requirements are as follows (1): The cross-sectional area at $x = 0$ should be 0.8 m^2. (2) At 200 Hz, the axial particle velocity at $x = 0$ should be 20 dB greater than it is at $x = L$. Equation (3) The cutoff frequency should be 100 Hz. It may be assumed that there are no reflections at the open end. Determine the growth constant b and the length L.

Exercise 9.6 Solve Exercise 9.5 for the case where the undriven end's specific impedance is $\zeta = 0.2 + 0.1i$.

Exercise 9.7 The small end of an exponential horn is driven by a velocity source. The other end is joined to a very long cylindrical waveguide that is terminated by a nonreflecting material. Consequently, only waves that propagate to the right exist within that section. Derive a set of equations for the pressure amplitude in the cylindrical waveguide. Assume that the waves in the cylinder are planar.

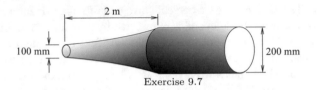

Exercise 9.7

Exercise 9.8 Use the WKB approximation, rather than the exact solution of the Webster horn equation, to carry out the analysis requested in Exercise 9.1.

Exercise 9.9 A horn's radius expands quadratically with axial distance according to $a = \alpha + \beta x^2$. The frequency is ω, and the complex pressure amplitude at $x = 0$ is P_0. (a) Derive expressions describing the WKB approximation for $P(x)$. (b) Evaluate and plot $|P/P_0|$ as a function of x corresponding to the following parameters: $\alpha = 0.25$ m, $\beta = 0.1$ m^{-1}, $f = 1000$ Hz, $\rho_0 = 1.2$ kg/m^3, $c = 340$ m/s.

Exercise 9.10 A seismic wave in the bottom of an ocean channel induces a wave that travels horizontally at 3200 m/s. The dominant harmonic component of this wave occurs at 12 Hz, and the amplitude of the vertical displacement in this component is 2 mm. The channel's depth is 150 m. What is the maximum pressure amplitude within the channel, and at what depth does it occur?

Exercise 9.11 A scan across the width of a two-dimensional waveguide has yielded the transverse mode function plotted below. Use the measured data to identify the specific local impedance of each wall and the eigenvalue of this mode. The fluid is air, the frequency is 400 Hz, and the width of the waveguide is 5 m.

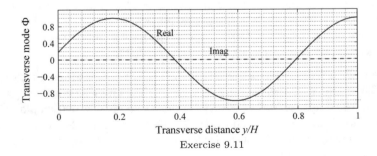

Exercise 9.11

Exercise 9.12 A porous tile is used as the bottom of a two-dimensional water channel whose height is H. A locally reacting model of the bottom sets the specific impedance as $\zeta = i (\beta\omega - \gamma/\omega) + \delta$. The surface of the channel may be considered to be pressure-release. (a) Derive the characteristic equation whose roots are the eigenvalues of the transverse mode functions at a specified value of ω. (b) In the expression for ζ, the parameter β is proportional to a mass per unit surface area, γ is proportional to a stiffness per unit surface area, and δ is a dissipation factor. What is the significance of the frequency being $(\gamma/\beta)^{1/2}$ in the situation where $\delta = 0$? (c) For the ideal case in which $\delta = 0$, derive an equation whose roots are the frequencies at which the various modes are cutoff.

Exercise 9.13 A waveguide is formed by two parallel walls having infinite extent and separated by distance H. One wall is well approximated as pressure-release, while the other is locally reacting with specific impedance ζ. For the case where $kH = 4\pi$ and $\zeta = -2i$, determine the eigenvalues and axial wavenumbers of the first four transverse modes. Plot each mode as a function of the transverse distance y/H, with each mode scaled such that max $(|\Phi_n|) = 1$.

Exercise 9.14 Solve Exercise 9.13 for the case where the specific impedance of the locally reacting wall is $\zeta = 0.9 - 2i$, and all other parameters are as specified there.

Exercise 9.15 The rigid walls of a two-dimensional waveguide are separated by distance H. The far end inhibits reflections. The waveguide is excited by a line source whose volume velocity per unit length is $\hat{Q}$. The source is perpendicular to the xy plane at $y = H/2$ on the cross section $x = 0$. Except for the source, $v_x = 0$ along the cross section at $x = 0$. (a) Derive an expression for the pressure at an arbitrary location for any frequency kH. Which transverse modes contribute to the response? (b) Suppose the value of kH is very close to 2π. What is the nature of the pressure distribution along any cross section at large kx?

Exercise 9.16 A two-dimensional horizontal waveguide has a rigid upper and lower walls. The velocity normal to the vertical wall at $x = 0$ is a translational oscillation over the lower half, while the upper half of the wall is stationary. Thus, the boundary condition at $x = 0$ is $v_x = v_0 \cos(\omega t) [h(y) - h(y - H/2)]$. The fluid is oil, $\rho_0 = 930$ kg/m^3, $c = 1320$ m/s, the height $H = 5$ m, and $\omega = 25000$ rad/s. (a) Determine the value of $|v_0|$ if it is observed that the sound pressure level is 195 dB//1 μPa at a point at $y = H/2$ on the cross section at $x = 8$ m. (b) Repeat the analysis in Part (a) for the case where the sound pressure level is 195 dB at $y = H/4$, $x = 8$ m.

Exercise 9.17 The walls at $y = 0$ and $y = H$ of a rectangular waveguide are rigid, while the walls at $z = 0$ and $z = W$ are pressure-release. The end $x = 0$ is rigid, except that a point source is situated at the corner where $x = y = z = 0$. The termination at the far end of the waveguide prevents reflection of waves. Derive an expression for the pressure at an arbitrary location.

Exercise 9.18 A signal propagates in the horizontal x direction within a long waveguide whose walls at $y = 0$, $y = H$, $z = 0$, and $z = W$ are rigid. The termination at the far end inhibits reflections. The signal is generated by a harmonic vibration at the end $x = 0$ whose spatial pattern is a mix of the plane wave mode and the (2,2) mode, that is, $v_x (x = 0) = [v_0 + v_2 \cos (2\pi y/H) \cos (2\pi z/W)] \cos (\omega t)$. The fluid is air, with $\omega = 500\pi$ rad/s, $H = 3$ m, $W = 4$ m, and $v_0 = v_2 = 400$ mm/s. (a) Evaluate $|P|$ as a function of x along the center line, $y = H/2$, $z = W/2$ for $0 \le x \le 10$ m. (b) Evaluate $|P|$ as a function of y along the transverse line $z = W/2$ at $x = 1$ m and $x = 10$ m.

Exercise 9.19 The walls at $y = 0$, $y = H$, and $z = W$ of a rectangular waveguide are rigid, and the wall at $z = 0$ is pressure-release. The excitation is a uniform velocity distribution over one-quarter of the end $x = 0$, specifically $v_x = v_0 \cos (\omega t)$ for $0 < y < H/2$ and $0 < z < W/2$. The aspect ratio is $W/H = 2.5$, and $kH = 10$. Derive an expression for the pressure at an arbitrary location knowing that the termination at the far end of the waveguide prevents reflection of waves. Then, evaluate this result for the pressure along the lines $y = H/2$ and $z = W/2$ at the cross section $x = 20\pi/k$.

Exercise 9.20 The sketch shows a two-dimensional model of a rectangular waveguide in which a layer of liquid having density ρ_2 and sound speed c_2 overlays a denser liquid whose properties are ρ_1 and c_1. The bottom is rigid, whereas air is above the upper liquid. Consequently, that interface may be considered to be pressure-release from the viewpoint of the liquid. Vibration of the vertical wall at $x = 0$ generates waves that propagate to the right. The far end attenuates these waves, so that there are no return reflections. (a) Derive the characteristic equation governing horizontal propagation modes. (b) Derive an expression for the transverse mode function corresponding to an eigenvalue that would be obtained by solving the characteristic equation. (c) Prove that if $\rho_2 = \rho_1$ and $c_2 = c_1$, then the modal properties reduce to those of two-dimensional waveguide of height H whose walls are rigid and pressure-release. *Hint*: Let μ and η denote the transverse wavenumber of Φ for $0 < y < H/2$ and $H/2 < y < H$, respectively. The pressure and vertical particle velocity derived from a mode function must be continuous. In addition to leading to the conditions that Φ and $d\Phi/dy$ be continuous at $y = H/2$, this requirement tells us that the axial wavenumber κ must be the same for both layers. Thus, it must be that $\kappa = (\omega^2/c_1^2 - \mu^2)^{1/2} = (\omega^2/c_2^2 - \eta^2)^{1/2}$.

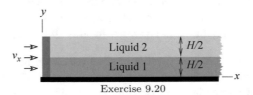

Exercise 9.20

Exercise 9.21 The wall of a cylindrical waveguide is locally reacting, with specific impedance $\zeta = 4i$. Determine the eigenvalues of the first four axisymmetric transverse modes. Plot dependence of these mode functions on the transverse distance R. Carry out the analysis for $ka = 1$ and $ka = 10$.

Exercise 9.22 The wall of a cylindrical waveguide is locally reacting, with specific impedance $\zeta = 4i$. Determine and graph the phase and group veocities of the first four modes from $ka = 15$ down to the cutoff frequency of each mode.

Exercise 9.23 A circular waveguide whose wall is rigid is drive by a vibrating plate at $x = 0$. The spatial pattern of the vibration matches the fundamental nonplanar mode of the waveguide, that is, $v_x = v_0 \Phi_{0,1} \cos(\omega t)$. Determine and graph the nondimensional mean-squared pressure $\left(p^2\right)_{av} / (\rho_0 c v_0)^2$ and nondimensional mean-squared particle velocity $\left(v_x^2\right)_{av} / v_0^2$, averaged over any cross section. Perform the analysis for frequencies ranging from cutoff to $ka = 10$. Compare these dependencies to that of the power transported in that mode.

Exercise 9.24 A cylindrical waveguide whose wall is rigid is excited by a point source situated on the centerline at $x = 0$. Determine the pressure amplitude on-axis at $kx = 40$ as a function of frequency in the range $0.1 \le ka \le 16$.

Exercise 9.25 A cylindrical waveguide whose wall is rigid is excited by a point source situated at $x = 0$. The location of the point source is $R = a/2$ from the centerline. Derive an expression for the pressure field generated by the source.

Exercise 9.26 The region between the inner and outer cylinders is an annular waveguide. A disturbance interior to the inner cylinder induces a harmonic displacement wave that propagates along the inner wall. This wave propagates downstream at sound speed $c_w > c$. In other words, the transverse velocity at $R = b$ is $v_R = V \sin(\omega t - \omega x/c_w)$. The outer cylinder is rigid. (a) Derive an expression for the pressure wave that is generated in the annular region between the cylindrical walls. Does the analysis indicate that there are values of ω that cause a resonance? (b) For the case where $c_w = 3c$, evaluate the pressure at the outer wall as a function of $ka < 20$.

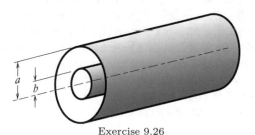

Exercise 9.26

Exercise 9.27 A circular disk is the closure of a very long cylindrical waveguide. The disk is mounted on a shaft that rotates at angular speed Ω. The shaft's centerline is collinear with the centerline of the waveguide. Because of a manufacturing error, the disk is not correctly aligned. Rather, the normal to its surface forms a small constant angle Δ relative to the shaft. The sketch is a side view of the configuration when the plane containing both centerlines is vertical. The effect of the misalignment is to cause a fluctuation in the x coordinate of the boundary. Consider two points at fixed

R and θ. One is on the nominal end, $x = 0$, and the other is on the disk. This distance oscillates as a consequence of the rotation, and there is an axial velocity at (R, θ) at the end. This velocity is $v_x = \Delta\Omega R \sin(\Omega t - \theta)$. The cylindrical wall is rigid, and the far end is terminated in a manner that inhibits reflections. Derive an expression for the pressure within the waveguide. Which transverse modes are excited?

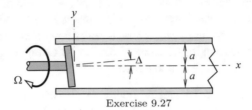

Exercise 9.27

Exercise 9.28 A cylindrical waveguide, whose wall is rigid, is excited at the end $x = 0$. Waves that reach the far end are fully absorbed. The excitation at $x = 0$ consists of a harmonic axial velocity that is uniformly distributed over two nonadjacent quadrants. The velocity in one quadrant is $180°$ out-of-phase from the other, and the velocity in the intervening quadrants is zero. What transverse modes are excited? Derive an expression for the pressure at an arbitrary field point.

Exercise 9.29 A different type of waveguide that features cylindrical waves occurs in an ocean channel. The cylinder in the sketch extends from the ocean floor up to the free surface at the atmosphere. The acoustic domain here is $R > a, 0 < z < H$, $-\pi < \theta < \pi$. The surface of the cylinder undergoes a vibration in the transverse direction, $\bar{e}_R$. (One context in which this configuration might arise is a pipe that brings oil to the surface from a drill hole.) Waves propagate away from the cylinder, as they do for an infinitely long cylinder whose surface vibrates. The fundamental difference is the finite extent of the axial direction. To avoid unnecessary complications, consider an axisymmetric pressure field $p(R, z, t)$. (a) Determine the transverse mode functions for this system. (Note that propagation is in the sense of increasing R, so the transverse modes are functions of z.) Do these modes have a cutoff frequency? (b) Describe the orthogonality and normalization conditions for the mode functions. (c) Describe the modal series for the pressure corresponding to the general situation where the vibration at $R = a$ is a harmonic, with an arbitrary dependence on z, $v_R = \text{Re}(f(z)\exp(i\omega t))$. Derive an expression for the pressure field generated by this vibration.

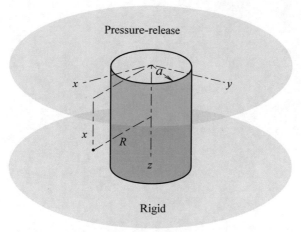

Exercise 9.29

Chapter 10
Modal Analysis of Enclosures

To say that a region containing an ideal fluid is a "cavity" is to imply that it is a void in some solid region. The synonymous term "enclosure" is more descriptive, in the sense that it conveys the notion that the domain is surrounded by a solid boundary. A tube terminated at both ends is the simplest configuration to analyze. Sections 2.5.3 and 3.2.2 addressed planar waves in such a system, but our concern here will extend to higher order waveguide modes. As is true for planar waves, closure results in reflection of the incident waves, thereby setting up a field that features standing waves in all directions. The developments that follow will primarily deal with regular geometries. The chapter will close by developing an approximate method for irregular cavities. The closure will be an approximate method for cavities and elastic structures.

10.1 Fundamental Issues

10.1.1 Wall-Induced Signals

In our study of waveguides in the previous chapter, the fundamental premise was that the far end is terminated in a manner that prevents reflections. We now remove that restriction. Figure 10.1 depicts a rectangular cavity. The length in the x direction is L, and the height in the y direction is H. To avoid unnecessary complications, we begin with a situation in which there is no spatial variation in the z direction. Thus, the frequency-domain response will have the form $p = \text{Re}\left(P\left(x, y\right) e^{i\omega t}\right)$.

Let us consider the situation where a known normal velocity distribution $V_s\left(y\right)$ exists on the face at $x = 0$, and the walls at $y = 0$, $y = H$, and $x = L$ are rigid. We know that if L were infinite, the imposed velocity at $x = 0$ would generate a set of waveguide modes that propagate in the direction of increasing x. The presence

Electronic supplementary material The online version of this chapter (DOI 10.1007/978-3-319-56847-8_10) contains supplementary material, which is available to authorized users.

J.H. Ginsberg, *Acoustics—A Textbook for Engineers and Physicists, Volume II*,
DOI 10.1007/978-3-319-56847-8_10

Fig. 10.1 Coordinate systems for a two-dimensional cavity. An excitation acts at $x = 0$ and the other walls are rigid

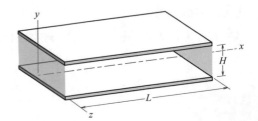

of a wall at finite L, does not alter this fact, but it does result in reflections that are manifested as modes that propagate in the direction of decreasing x. The y dependence of both sets of modes must be the same set of transverse mode functions $\Phi_n (y)$. The transverse mode functions for a hard-walled two-dimensional waveguide are

$$\Phi_n (y) = B_n \cos \left(\frac{n\pi y}{H} \right) \tag{10.1.1}$$

where normalization gives $B_n = 1$ if $n = 0$ and $\sqrt{2}$ otherwise. The axial wavenumber for both the forward and backward propagating modes must be κ_n, but we do not know a priori the modal amplitudes for either set of modes. Hence, the appropriate modal series is

$$
\begin{aligned}
P &= \sum_{n=0}^{\infty} \left[D_{n,1} e^{-i\kappa_n x} + D_{n,2} e^{+i\kappa_n x} \right] B_n \cos \left(\frac{n\pi y}{H} \right) \\
&= \sum_{n=0}^{\infty} \left[D_{n,c} \cos (\kappa_n x) + D_{n,s} \sin (\kappa_n x) \right] B_n \cos \left(\frac{n\pi y}{H} \right)
\end{aligned}
\tag{10.1.2}
$$

where the representation of the x dependence as sinusoidal functions makes it easier to satisfy boundary conditions.

As was noted, this representation satisfies the rigidity conditions at $y = 0$ and $y = H$, and each term in the series individually satisfies the Helmholtz equation because the κ_n wavenumbers are those in Eq. (9.2.10). It remains to satisfy the conditions at $x = 0$ and $x = L$, which are

$$\left. \frac{\partial P}{\partial x} \right|_{x=0} = -i\omega\rho_0 V_s$$

$$\left. \frac{\partial P}{\partial x} \right|_{x=L} = 0 \tag{10.1.3}$$

The second condition is satisfied by each term in the modal series if

$$D_{n,s} \cos (\kappa_n L) = D_{n,c} \sin (\kappa_n L) \tag{10.1.4}$$

We use this relation to eliminate $D_{n,s}$ in Eq. (10.1.2) and then define a new coefficient $D_n \equiv D_{n,c}/\cos(\kappa_n L)$. This converts the series to

$$P = \sum_{n=0}^{\infty} D_n B_n \cos(\kappa_n (L - x)) \cos\left(\frac{n\pi y}{H}\right) \tag{10.1.5}$$

The D_n coefficients are found by satisfying the boundary condition at $x = 0$. To that end, $V_s(y)$ is expanded in a series of the transverse mode functions. Orthogonality of those functions yields

$$V_s(y) = \sum_{n=0}^{\infty} V_n \cos\left(\frac{n\pi y}{H}\right), \quad V_n = \frac{1}{H}\int_0^H V_s B_n \cos\left(\frac{n\pi y}{H}\right) dy \tag{10.1.6}$$

This series and Eq. (10.1.5) are substituted into the first of Eq. (10.1.3). Matching the coefficients of each mode leads to

$$P = -i\omega\rho_0 \sum_{n=0}^{\infty} \frac{B_n}{\kappa_j \sin(\kappa_n L)} V_n \cos(\kappa_n (L - x)) \cos\left(\frac{n\pi y}{H}\right) \tag{10.1.7}$$

This representation of the response to excitation on a boundary is quite similar to the result of the analysis in Sect. 9.2.5. One difference is that the dependence on x here is a superposition of standing waves, rather than propagating waves. Each series term describing the spatial dependence is a *forced cavity mode*. A new feature contained in Eq. (10.1.7) is the possibility of a singularity when any of the axial wavenumbers are such that $\sin(\kappa_n L) = 0$. Because κ_n depends on k, these singularities correspond to resonant frequencies. We shall return to this feature after we consider a different type of excitation.

10.1.2 Source Excitation

Problems pertaining to the response within an enclosure that is induced by motion of a wall are particularly relevant to moving vehicles. However, if the enclosure is a room, the excitation of interest is likely to be a source, such as a person or a musical instrument. Suppose the cavity in question is that in Fig. 10.1, with all walls now taken to be rigid. Because there is no wall that generates propagating modes, there is no conceptual difference between the x and y directions. If we consider the former to be the axial direction, the transverse mode functions are $B_n \cos(n\pi y/H)$, whereas taking the y direction to be axial leads to transverse functions that are $B_j \cos(j\pi x/L)$. It is not reasonable to favor either direction, so we use one of each function to represent P. There is no limit to the values of n and j, so it seems reasonable to try an ansatz for the pressure that is the sum of all possible combinations of the transverse mode functions for each direction. That is, let us try

$$P = \sum_{j=0}^{\infty}\sum_{n=0}^{\infty} B_j B_n D_{j,n} \cos\left(\frac{j\pi x}{L}\right) \cos\left(\frac{n\pi y}{H}\right) \qquad (10.1.8)$$

This expression has the appearance of a Fourier cosine series in each direction, but we soon will find a more profound interpretation.

The above representation clearly satisfies all rigid wall conditions, but there is a fundamental difficulty—none of the terms in the double summation individually satisfy the Helmholtz equation at an arbitrary frequency. Furthermore, there is no indication of the role of a source. Suppose a line source is situated at an arbitrary location (x_0, y_0). It was shown as part of the definition of Green's functions, Sect. 6.4.2, that the field of a harmonic point source is governed by an inhomogeneous Helmholtz equation, specifically Eq. (6.4.22). Thus, the pressure must satisfy

$$\frac{\partial^2 P}{\partial x^2} + \frac{\partial^2 P}{\partial y^2} + k^2 P = -i\omega\rho_0 \hat{Q}\delta\left(x - x_0\right)\delta\left(y - y_0\right) \qquad (10.1.9)$$

The terms in Eq. (10.1.8) will not satisfy this equation individually, but they might do so collectively. Conceptually, we want to find the set of coefficients $D_{j,n}$ for which the discrete, doubly infinite summation satisfies the field equation at a continuum of points within the cavity. Contemplating matching these different types of infinities is bewildering, so let us look at the matching process from a linear algebra perspective. Suppose we were to substitute Eq. (10.1.8) into the inhomogeneous wave equation, with arbitrary values assigned to the $D_{j,n}$ coefficients. The result would be a difference $\Delta(x, y)$ between the left and right side of the equations. Rather than setting Δ to zero at each location, we invoke the interpretation of the series as a sum of unit vectors in a linear functional space (the product of cosines) multiplied by component lengths (the $D_{j,n}$ values). Because all possible solutions of the cavity with rigid walls lie in this linear space, we can assert that the correct set of coefficients is that which causes Δ to be orthogonal to all directions in the linear space. The pressure functions are defined in the domain $0 < x < L$, $0 < y < H$, so orthogonality constitutes an inner product over this space. The functions in the pressure series terms are sinusoidal, which means that the orthogonality property is that of a Fourier series. Hence, a function $F(x, y)$ is orthogonal to the (j, n) term if

$$\int_0^H \int_0^L F(x, y)\left[B_j \cos\left(\frac{j\pi x}{L}\right)\right]\left[B_n \cos\left(\frac{n\pi y}{H}\right)\right] dxdy = 0 \qquad (10.1.10)$$

To implement this concept, we substitute the pressure series into the inhomogeneous wave equation, multiply both sides by a selected function, and then integrate over the domain of the cavity. It is convenient to change the summation indices to r and s, so that we may multiply the series by the (j, n) function. These operations lead to

$$\sum_{r=0}^{\infty}\sum_{s=0}^{\infty}\int_{0}^{H}\int_{0}^{L}\left[B_r B_s D_{r,s}\cos\left(\frac{r\pi x}{L}\right)\cos\left(\frac{s\pi y}{H}\right)\right]\left[k^2-\left(\frac{r\pi}{L}\right)^2-\left(\frac{s\pi}{H}\right)^2\right]$$

$$\times\, B_j B_n \cos\left(\frac{j\pi x}{L}\right)\cos\left(\frac{n\pi y}{H}\right)dx\,dy$$

$$=-i\omega\rho_0\hat{Q}\int_{0}^{H}\int_{0}^{L}\delta\left(x-x_0\right)\delta\left(y-y_0\right)$$

$$B_j b_n \cos\left(\frac{j\pi x}{L}\right)\cos\left(\frac{n\pi y}{H}\right)dx\,dy \qquad (10.1.11)$$

On the left side, we invoke the orthogonality of the functions in each direction. Thus, only the term for which $r=j$ and $s=n$ is nonzero. In that case, inclusion of the B_m in the integrals over x and y, respectively, gives L and H factors. On the right side, we have the integral property of a Dirac delta function, which evaluates the analytical part of the integrand at the location where the delta occurs. The result is

$$LH\left[k^2-\left(\chi_{j,n}\right)^2\right]D_{j,n}=-i\omega\rho_0\hat{Q}B_j B_n \cos\left(\frac{j\pi x_0}{L}\right)\cos\left(\frac{n\pi y_0}{H}\right) \qquad (10.1.12)$$

where

$$\chi_{j,n}=\left[\left(\frac{j\pi}{L}\right)^2+\left(\frac{n\pi}{H}\right)^2\right]^{1/2} \qquad (10.1.13)$$

Substitution of the resulting expression for $D_{j,n}$ into Eq. (10.1.8) leads to

$$P=-i\omega\rho_0\frac{\hat{Q}}{LH}\sum_{j=0}^{\infty}\sum_{n=0}^{\infty}\frac{\left(B_j B_n\right)^2}{k^2-\left(\chi_{j,n}\right)^2}\cos\left(\frac{j\pi x_0}{L}\right)$$

$$\cos\left(\frac{n\pi y_0}{H}\right)\cos\left(\frac{j\pi x}{L}\right)\cos\left(\frac{n\pi y}{H}\right) \qquad (10.1.14)$$

This is a remarkable result because it is the response of the cavity to a point source at an arbitrary location. In other words, it is Green's function for that domain. Observe in this regard that the expression is reciprocal, as is evident by the fact that swapping x and x_0, or y and y_0 does not alter the result.

Thus, we have identified alternative sets of functions that describe the field within the cavity. The Φ_n functions in the previous section were derived from the propagation modes of a waveguide. Changing the excitation changes these functions. We shall refer to them as *forced cavity modes*. These attributes are contrasted by the functions from which Eq. (10.1.14) is constructed. We shall denote them as $\Psi_{j,n}$ and refer to them as *natural cavity modes,* although we often will omit "natural." These functions are independent on the nature of the excitation and therefore represent intrinsic properties of the system. A specific set of wall conditions leads to a specific set of $\Psi_{j,h}$. The functions appearing above are the two-dimensional functions for a hard-walled cavity,

$$\Psi_{j,n} = \cos\left(\frac{j\pi x}{L}\right)\cos\left(\frac{n\pi y}{H}\right) \qquad (10.1.15)$$

Let us substitute this function into the (homogeneous) Helmholtz equation. Doing so leads to recognition that any $\Psi_{j,n}$ is a solution when the acoustic wavenumber is $k = \chi_{j,n}$. These observations are equivalent to saying that $\chi_{j,n}$ are the eigenvalues and $\Psi_{j,n}$ are the eigenfunctions of the cavity. The *natural frequencies* are $\omega_{j,n} = c\chi_{j,n}$. It now is clear why Eq. (10.1.14) is singular if k equals any $\chi_{j,n}$, or equivalently, when $\omega = (\omega_{nat})_{j,n}$. Excitation of any linear system at its natural frequency leads to resonance. This phenomenon occurs because all internal actions balance to sustain a free response, so there is no remaining ability to resist an excitation. The only exception to this statement occurs if the excitation is such that the mode is not excited. For example, suppose that $k = \chi_{2,3}$ ($j = 2$, $n = 3$) in Eq. (10.1.14), but $x_0 = L/4$. Then, $\cos(j\pi x_0/L) = 0$. Mathematically, this gives a term in the pressure series in which zero is divided by zero. The reality is that a true singularity is unlikely because it requires that k exactly equals $\chi_{2,3}$. Furthermore, the analysis excluded dissipation effects that ameliorate resonances. Another aspect to consider is a property identified in Sect. 2.5.3, specifically, that resonance in a linearized analysis results in progressive growth of the transient response. As the response grows, nonlinear effects become increasingly important. Thus, the singularities at resonance merely flag exceptional situations.

An interesting aspect results from rewriting the pressure series, Eq. (10.1.14), in terms of the mode function notation, rather than explicit harmonic functions. Doing so gives

$$P = -i\omega\rho_0 \frac{\hat{Q}}{LH}\sum_{j=0}^{\infty}\sum_{n=0}^{\infty}\frac{\Psi_{j,n}(x_0, y_0)\,\Psi_{j,n}(x.y)}{k^2 - \chi_{j,n}^2} \qquad (10.1.16)$$

This form is descriptive of Green's function for any two-dimensional cavity, with the mode functions appropriate to that domain. Furthermore, three-dimensional situations are described by altering the summation to extend over three indices for the three directions. We will see how this result is obtained in Sect. 10.2.1.

The coexistence of forced and natural cavity modes, which result in different modal series, might cause some degree of confusion, so let us contrast their essential features. The Φ_n in Eq. (10.1.7) represent the transverse dependence of waveguide modes that propagate in opposite senses at any frequency. The $\Psi_{j,n}$ represent the pressure fields that can exist within a closed region without excitation only at the natural frequencies. Some individuals refer to natural cavity modes as *standing wave modes*. Perhaps the most important distinction is that natural modes exist for any enclosed domain, whereas a series of propagation modes obviously assumes that the picture of a waveguide closed at one end is appropriate. For example, although the analysis would not be simple, it is possible to find the cavity modes for an L-shaped room. However, the L-shape does not suggest a conceptual view in which waves propagate back and forth in one direction. Indeed, the direction that might be considered to be axial in one leg of the L-shape would correspondingly be viewed as the transverse direction for the other leg.

Resonance of any dissipationless system is manifested as a singularity at a natural frequency. Regardless of whether the field within a cavity is described as a series of forced or natural modes, resonances will be manifested as an exceptionally large value of a modal coefficient. Resonance occurs in Eq. (10.1.7), which is the forced mode series, when $\sin(\kappa_n L)$ in the denominator is zero for a specific n. This occurs when $\kappa_n L = j\pi$, $j = 0, 1, \ldots$ The axial wavenumber κ_n for a waveguide mode is related to the eigenvalue μ_n of the transverse mode function by $\kappa_n^2 = k^2 - \mu_n^2$, and $\mu_n = n\pi/H$ for a hard-walled waveguide. Hence, the resonances occur when $n^2\pi^2/H^2 + j^2\pi^2/L^2 = k^2$. This condition is identical to the resonance condition $k = \chi_{j,n}$ for the natural cavity mode series in Eq. (10.1.14).

Each type of cavity mode has situations for which it is most useful. A series of forced modes is well suited to situations where the excitation originates on the boundary. However, a forced mode series cannot be implemented directly if the cavity is excited by one or more sources within the fluid. This is so because these modes satisfy the homogeneous Helmholtz equation. We will see in Example 10.4 that in some cases it is possible to use a forced mode analysis to determine the effect of a source. However, the situation addressed there is rather special, and its extension to more general systems would be quite difficult.

The utility of a natural cavity mode formulation is quite opposite. These functions are solutions of the Helmholtz equation that satisfy passive (that is, homogeneous) boundary conditions, so they cannot be used directly to satisfy an inhomogeneous boundary condition associated with an excitation. On the other hand, such a formulation readily handles cases of source excitation.[1]

10.2 Frequency-Domain Analysis Using Forced Cavity Modes

The developments in this section assume familiarity with the properties of propagation modes for waveguides. The various derivations and results in Chap. 9 are not repeated here.

10.2.1 Rectangular Enclosures

A cavity in the shape of a rectangular box, with one wall that undergoes a specified motion, is relatively uncomplicated to analyze, yet it exhibits the same phenomena as

---

[1] It is possible to use the forced mode formulation to determine the field of a source within an enclosed regions. The method follows the development in Sect. 6.4.2, in which a solution is constructed by adding a function $F(\bar{x})$ to the free-space source solution $G(\bar{x})$. Similarly, a natural mode formulation can be used to address boundary excitation by adding to the series a term that satisfies all boundary conditions. Equations for the modal coefficients would be obtained by requiring that the sum of the added term and the natural mode series satisfy the inhomogeneous Helmholtz equation.

any other cavity. The direction normal to the vibrating wall is designated as the axial direction x, with the axial length designated as L. The vibrating wall is defined to be situated at $x = 0$. The velocity distribution at this boundary may be arbitrary, so it is required that $V_x = V_s(y, z)$ at $x = 0$. The sidewalls are the planes $y = 0$, $y = H$, $z = 0$, and $z = W$. Conceptually, the signal within this cavity consists of modes that are radiated from the vibrating wall and propagate in the direction of increasing x. When they arrive at $x = L$, they are reflected, which results in waves that travel back to $x = 0$, where they are reflected, *ad infinitum*. The waves that propagate forward and back in the axial direction are the propagation modes evaluated in Sect. 9.3.2. The amplitudes of the forward and backward modes are quantities to be determined.

All walls other than the vibrating one are taken to be locally reacting, which includes the rigid and pressure-release conditions. The boundary conditions on the sidewalls are the same as those posed in Eq. (9.3.19) for a rectangular waveguide. Thus, the transverse mode functions are as described in Eqs. (9.3.25) and (9.3.26). For a rectangular waveguide, these functions are the product of the transverse modes for two-dimensional waveguides in each direction, specifically

$$\Phi_{j,n} = B_{j,n}\phi_{y,j}(y)\,\phi_{z,n}(z) \tag{10.2.1}$$

Eigenvalues associated with $\Phi_{j,n}$ are μ_j for the y functions, and ν_n for the z functions. It is essential that the modal series account for the effect of all modes. A plane wave mode can exist in the y direction if the walls at $y = 0$ and $y = H$ are rigid. In that case, the sum over the modes for the y direction would begin with $j = 0$, with the first eigenvalue $\mu_0 = 0$. Another possibility is that $y = 0$ and $y = H$ are purely compliant. In that case, there would be a purely imaginary eigenvalue. This mode can be assigned to $j = 0$, so that $\mu_0 = \text{Im}(\mu_{\text{im}})i$. In any other case, the modal sum would begin with $j = 1$. The same considerations apply to the $\phi_{z,n}(z)$ functions, whose eigenvalues are ν_n. A further consideration is that the scaling factors $B_{j,n}$ are required to have been set according to the normalization rule stated in Eq. (9.3.29).

A modal series consists of sums in which the transverse functions multiply complex exponentials representing forward and backward propagation. The amplitudes of these exponentials are denoted as $\alpha_{j,n}$ and $\beta_{j,n}$, respectively, so the ansatz is

$$P = \sum_{n=0}^{\infty}\sum_{j=0}^{\infty}\left[\alpha_{j,n}e^{-i\kappa_{j,n}x} + \beta_{j,n}e^{+i\kappa_{j,n}x}\right]\Phi_{j,n}(y, z) \tag{10.2.2}$$

The axial wavenumbers $\kappa_{j,n}$ must be such that each term in the series individually satisfies the Helmholtz equation, which means that

$$\kappa_{j,n} = \left(k^2 - \mu_j^2 - \nu_n^2\right)^{1/2} \tag{10.2.3}$$

The modal series accounts for forward and backward waves in the x directions, so we may continue the previous practice of placing $\kappa_{j,n}$ in the fourth quadrant of the complex plane.

Because the transverse mode functions satisfy the boundary conditions at the sidewalls, it only remains to make the particle velocity V_x at $x = 0$ match the imposed distribution $V_s (y, z)$ and to make P/V_x be consistent with the locally reacting model at $x = L$. These conditions are

$$\left. \frac{\partial P}{\partial x} \right|_{x=0} = -i\omega\rho_0 V_s (y, z)$$

$$P|_{x=L} = -\frac{\zeta_L}{ik} \left. \frac{\partial P}{\partial x} \right|_{x=L}$$

(10.2.4)

Substitution of the modal series into these boundary conditions gives

$$\sum_{n=0}^{\infty}\sum_{j=0}^{\infty} i\kappa_{j,n} \left(-\alpha_{j,n} + \beta_{j,n}\right) B_{j,n}\Phi_{j,n} (y, z) = -i\rho_0\omega\hat{V}_s (y)$$

$$\sum_{n=0}^{\infty}\sum_{j=0}^{\infty} \left[\alpha_{j,n} \left(1 - \zeta_L\frac{\kappa_{j,n}}{k}\right) e^{-i\kappa_{j,n}L} + \beta_{j,n} \left(1 + \zeta_L\frac{\kappa_{j,n}}{k}\right) e^{+i\kappa_{j,n}L}\right] B_{j,n}\Phi_{j,n}(y, z) = 0$$

(10.2.5)

It might appear that the preceding constitutes two equations for many unknown $\alpha_{j,n}$ and $\beta_{j,n}$ values, but such thinking ignores the orthogonality of the transverse mode functions. Each equation is multiplied by an arbitrarily selected $\Phi_{m,\ell} \equiv B_{m,\ell}\phi_{y,m}\phi_{z,\ell}$ and integrated over the cross section, $0 < y < H, 0 < z < W$. These operations filter out all term in the summations for which $n \neq m$ and $j \neq \ell$ remain. What remains is a pair of equations,

$$\alpha_{m,\ell} - \beta_{m,\ell} = \rho_0 c\frac{k}{\kappa_{m,\ell}}V_{m,\ell}$$

$$\alpha_{m,\ell} \left(1 - \zeta_L\frac{\kappa_{m,\ell}}{k}\right) e^{-i\kappa_{m,\ell}L} + \beta_{m,\ell} \left(1 + \zeta_L\frac{\kappa_{m,\ell}}{k}\right) e^{+i\kappa_{m,\ell}L} = 0$$

(10.2.6)

The $V_{m,\ell}$ coefficients are those of a modal series for V_s defined in Eq. (9.3.37), specifically

$$V_{m,\ell} = \frac{1}{HW}\int_0^W \int_0^H \Phi_{m,\ell} (y, z) V_s (y, z) \, dydz$$

(10.2.7)

Orthogonality has reduced the problem to that of solving two simultaneous equations for $\alpha_{m,\ell}$ and $\beta_{m,\ell}$, which leads to

$$\alpha_{m,\ell} = \frac{\left(1 + \zeta_L\frac{\kappa_{m,\ell}}{k}\right) e^{+i\kappa_{m,\ell}L}}{\Delta (k, j, n)}\rho_0 c\frac{k}{\kappa_{m,\ell}}V_{m,\ell}$$

$$\beta_{m,\ell} = \frac{\left(-1 + \zeta_L\frac{\kappa_{m,\ell}}{k}\right) e^{-i\kappa_{m,\ell}L}}{\Delta (k, j, n)}\rho_0 c\frac{k}{\kappa_{m,\ell}}V_{m,\ell}$$

(10.2.8)

where $\Delta (k, j, n)$ is the determinant of Eq. (10.2.6),

$$\Delta\left(\kappa_{j,n}\right) = \left(1 + \zeta_L \frac{\kappa_{j,n}}{k}\right) e^{i\kappa_{j,n}L} + \left(1 - \zeta_L \frac{\kappa_{j,n}}{k}\right) e^{-i\kappa_{j,n}L} \tag{10.2.9}$$

Substitution of the solution into the modal series, Eq. (10.2.2), leads to

$$P = \rho_0 c \sum_{n=0}^{\infty} \sum_{j=0}^{\infty} \frac{k V_{j,n}}{\kappa_{j,n} \Delta\left(\kappa_{j,n}\right)} \left[\left(1 + \zeta_L \frac{\kappa_{j,n}}{k}\right) e^{i\kappa_{j,n}(L-x)} \right. $$
$$\left. - \left(1 - \zeta_L \frac{\kappa_{j,n}}{k}\right) e^{-i\kappa_{j,n}(L-x)}\right] \Phi_{j,n}\left(y, z\right) \tag{10.2.10}$$

The fact that the variable for the axial dependence in Eq. (10.2.10) is $L - x$ highlights the reflection process at $x = L$. Another manifestation of that attribute comes from factoring $1 + \zeta_L \kappa_{j,n}/k$ out of the bracketed term. Doing so displays the role of a modal reflection coefficient,

$$(R_L)_{j,n} = \frac{\zeta_L \dfrac{\kappa_{j,n}}{k} - 1}{\zeta_L \dfrac{\kappa_{j,n}}{k} + 1} \tag{10.2.11}$$

so that

$$P = \rho_0 c \sum_{n=0}^{\infty} \sum_{j=0}^{\infty} V_{j,n} \frac{k}{\kappa_{j,n}} \left[\frac{e^{i\kappa_{j,n}(L-x)} + (R_L)_{j,n} e^{-i\kappa_{j,n}(L-x)}}{e^{i\kappa_{j,n}L} - (R_L)_{j,n} e^{-i\kappa_{j,n}L}}\right] \Phi_{j,n}\left(y, z\right) \tag{10.2.12}$$

Setting $\Phi_{j,n}\left(y, z\right) = 1$ and $\kappa_{j,n} = k$ would make each term in the summation be like Eq. (3.2.21) for plane waves.

Some individuals prefer a form of Eqs. (10.2.9) and (10.2.10) that results when Euler's identity is used to remove the complex exponentials in favor of sine and cosine functions. The resulting expressions are not displayed here because such forms offer no merit if ζ_L or $\kappa_{j,n}$ is complex.

Resonances correspond to values of k for which $\Delta\left(\kappa_{j,n}\right) = 0$. It can proven that this condition occurs only if the impedance ζ_L at $x = L$ and the impedances of the sidewalls are infinite, zero, or purely imaginary, that is, purely reactive. As was shown earlier, any frequency at which $\Delta = 0$ is a natural frequency of the cavity. If any impedance has a nonzero real part, energy is absorbed by that wall, thereby inhibiting unlimited growth of the pressure amplitude. However, if the resistive part of each impedance is substantially smaller than $\rho_0 c$, a plot of pressure as a function of frequency will exhibit prominent peaks in the vicinity of the natural frequencies associated with zero resistive parts.

When any wall impedance is a finite compliance or inertance, determination of the natural frequencies by solving the eigenvalue problem is not a simple task. The difficult lies in the fact that a transverse eigenvalue, μ_j or ν_n, will depend on the frequency because the associated characteristic equation contains k, see, for example,

Eq. (9.2.32). Concurrently, the axial wavenumber $\kappa_{j,n}$ must satisfy its own characteristic equation, specifically Eq. (10.2.8). In addition, the natural frequency is related to the eigenvalues through Eq. (10.2.3), so that

$$(k_{\text{nat}})_{j,n} = \left(\kappa_{j,n}^2 + \mu_j^2 + \nu_n^2\right)^{1/2} \tag{10.2.13}$$

Thus, solution of the eigenvalue problem in this situation requires simultaneous solution of three characteristic equations and one algebraic equation. The unknowns are the three wavenumbers and the k_{nat}. Such a solution could be obtained by implementing a four variable Newton's method. A simpler, though less efficient, procedure is to sweep through a range of k. For each value, each characteristic equation is solved for the associated wavenumber. If these values are consistent with Eq. (10.2.13), then that value of k marks a natural frequency. Regardless of the procedure that is implemented, care must be exercised to obtain all natural frequencies in the interval of interest.

One of the useful aspects of a series representation of the pressure field is the uncoupled nature of the equations to be solved. That is, we can find the amplitude coefficients α_j and β_j for a mode without finding the coefficients of the other modes. In some situations, it might be that more than one wall moves. In that case, the response may be found by superposing solutions. In each subproblem, one wall moves and the other vibrating walls are held fixed; that is, they become rigid. The axial direction for each subproblem would be normal to the wall that is vibrating. Addition of the pressure from each subproblem yields the pressure field.

EXAMPLE 10.1 All walls of a rectangular cavity are rigid. The dimensions are L in the x direction, H in the y direction, and W in the z direction. The upper half of the wall at $x = 0$ is stationary. The lower half is a rigid plate having mass M that is forced to undergo an oscillation at frequency ω. Thus, the end condition may be written as $v_x = \text{Re}\,(V_0 \exp{(-i\omega t)})\,[h(y) - h(yH/2)]$, where h denotes the step function. Derive an expression for the amplitude of the force that must be applied to the plate. Evaluate this force as a function of the nondimensional frequency kL, for the case where $H/L = 0.2$ and $W/L = 0.1$. The mass of the plate equals the mass of the fluid contained in the cavity.

Significance

This basic operations for the analysis are fairly straightforward. What makes the problem interesting is the interpretation of the nature of the forces acting on the plate, which will afford a different perspective to the meaning of resonances.

Solution

The force acting on the plate must overcome the inertia of the plate and the resultant force exerted by the pressure. This resultant, which acts in the negative x direction if the pressure is positive, is the integral over the plate of the pressure at $x = 0$. Thus, the complex amplitude of the pressure resultant is

$$F_{\text{ac}} = \int_0^W \int_0^{H/2} P|_{x=0}\, dy dz \tag{1}$$

The acceleration of the plate is $\text{Re}\,(i\omega V_0 \exp(i\omega t))$. Thus, the force amplitude required to translate the plate is

$$F = i\omega M V_0 + F_{\text{ac}} \tag{2}$$

Now, we turn our attention to the fluid. The walls that coincide with planes of constant y and z are not forced, so we take them to be the sidewalls of the waveguide. The corresponding transverse mode functions are

$$\Phi_{j,n} = B_{j,n} \cos\left(\frac{j\pi y}{H}\right) \cos\left(\frac{n\pi z}{W}\right), \quad j, n = 0, 1, 2, \ldots \tag{3}$$

Normalization of these functions leads to

$$B_{j,n} = \sqrt{\frac{4}{(1 + \delta_{j,0})(1 + \delta_{n,0})}}$$

The axial wavenumber is

$$\kappa_{j,n} = \left[k^2 - \left(\frac{j\pi}{H}\right)^2 - \left(\frac{n\pi}{W}\right)^2\right]^{1/2} \tag{4}$$

with the usual proviso that $\kappa_{j,n}$ is negative imaginary if the quantity within the brackets is negative.

In Eq. (10.2.4), the axial velocity imposed at $x = 0$ is $V_s\,(y, z)$. Here, this function is specified to be

$$V_s = V_0[h(y) - h(y - H/2)], \quad 0 < y < H,\ 0 < z < W$$

The corresponding velocity coefficients obtained from Eq. (10.2.7) are

$$V_{j,n} = V_0 B_{j,n} \mathcal{I}_{j,n}$$

$$\mathcal{I}_{j,n} = \frac{1}{HW} \int_0^W \int_0^{H/2} \cos\left(\frac{j\pi y}{H}\right) \cos\left(\frac{n\pi z}{W}\right) V_0 dy dz$$

$$= \begin{cases} 0 \text{ if } n > 0 \\[2mm] \dfrac{1}{2} \text{ if } j = n = 0 \\[2mm] \dfrac{1}{j\pi} \sin\left(\dfrac{j\pi}{2}\right) \text{ if } j > 0 \text{ and } n = 0 \end{cases} \tag{5}$$

The $V_{j,n}$ coefficients are zero if $n > 0$ because the imposed velocity is constant in the z direction, and a constant is orthogonal to the z dependence of all transverse mode functions for $n > 0$. The velocity distribution is the excitation, so any velocity coefficient that is zero means that the corresponding mode is not excited. In addition, the factor $\sin(j\pi/2)$ is zero if j is even, so $V_{j,n} \equiv 0$ for even j. Thus, rather than requiring evaluation of a double sum, Eq. (10.2.10) reduces to a single sum over $j = 0, 1, 3, \ldots$ with $n = 0$. The only other adaptation required to use that equation is to let ζ_L be infinite. The resulting series for the pressure field is

$$P = \rho_0 c V_0 \sum_{j=0,1,3,\ldots}^{\infty} \left(B_{j,0}\right)^2 \mathcal{I}_{j,0} \left(\frac{kL}{\kappa_{j,0}L}\right) \left(\frac{e^{i\kappa_{j,n}(L-x)} + e^{-i\kappa_{j,n}(L-x)}}{e^{i\kappa_{j,n}L} - e^{-i\kappa_{j,n}L}}\right) \cos\left(\frac{j\pi y}{H}\right)$$

(6)

The last step is to form the resultant force by substitution of this expression, evaluated at $x = 0$, into Eq. (1). The integral for this evaluation is the same as the one for $\mathcal{I}_{m,\ell}$ in Eq. (5). When the complex exponential sum and difference are replaced by sinusoidal functions, the pressure resultant reduces to

$$F_{\mathrm{ac}} = HW Z_{\mathrm{ac}} V_0$$
$$Z_{\mathrm{ac}} = i\rho_0 c \sum_{j=0,1,3,\ldots}^{\infty} \left(B_{j,0}\right)^2 \left(\mathcal{I}_{j,0}\right)^2 \left(\frac{kL}{\kappa_{j,0}L}\right) \cot\left(\kappa_{j,0}L\right)$$

(7)

An expression for the force that must be applied to the plate is obtained by substituting F_{ac} into Eq. (2), which results in

$$F = HW Z_F V_0$$

(8)

The factor Z_F is the total impedance seen by F. It consists of a contributions from the acoustic pressure and the inertia of the plate,

$$Z_F = Z_{\mathrm{ac}} + Z_M, \quad Z_M = \frac{i\omega M}{HW} \equiv i\rho_0 ckL\left(\frac{M}{\rho_0 LWH}\right)$$

(9)

Because of the simplicity of this system, the value of $Z_F/(\rho_0 c)$ depends on only three parameters: the nondimensional frequency kL, the ratio of the plate's mass to that of the fluid contained within the cavity $M/(\rho_0 LWH)$, and H/L. If j is less than the cutoff value kH/π, then $\kappa_{j,0}$ is real, whereas $\kappa_{j,0}$ is negative imaginary if j exceeds the cutoff value. In either case, the terms inside the summation are real. It follows that both Z_{ac} and Z_M are reactive at any frequency. The corollary is that the time-averaged power input to the cavity by F is zero. This is as it must be, because the walls are rigid and therefore cannot dissipate energy.

The value of j at which the series in Eq. (7) may be truncated is readily identified from Eq. (4), which indicates that $\kappa_{j,0}L \approx -ij\pi(H/L)$ when $j\pi/H \gg k$. For very large j, the terms in the summation are proportional to $\coth(j\pi(H/L))$, which approaches unity. The coefficients $B_{j,0} = 2^{1/2}$ and $\mathcal{I}_{j,0}$ are inversely proportional to $j\pi$. It follows that for large j the terms in the summation are inversely proportional to j^3. Thus, a truncation criterion that $j > kL\,(H/L)\,/\pi$ and $j > 20$ should be adequate. Figure 1 shows the result of computing Z_F for the given parameters with the series truncated at $j = 21$. In order to understand the role of the resultant acoustic force, the frequency dependence of Z_{ac} also is shown in Fig. 1. The second graph gives a finer picture of the behavior in the vicinity of the first resonances.

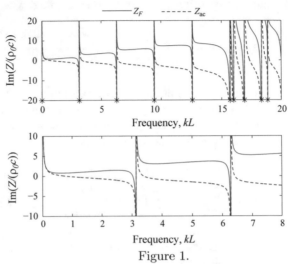

Figure 1.

Resonances occur if the denominator in Eq. (6) vanishes for any j. The denominator may be written as $2i \sin(\kappa_{j,0}L)$. Equating it to zero yields $\kappa_{j,0}L = m\pi$. Then, Eq. (4) yields the natural frequencies,

$$(kL)_{\mathrm{nat}} = \left[(m\pi)^2 + (j\pi)^2 \left(\frac{H}{L}\right)^2 \right]^{1/2}, \quad m = 0, 1, 2, ..., \quad j = 0, 1, 3, ... \quad (10)$$

These values are marked with an asterisk in Fig. 1. The values in ascending order are tabulated below.

j	0	0	0	0	0	0	1	1	1	1	0
m	0	1	2	3	4	5	0	1	2	3	6
$(kL)_{\mathrm{nat}}$	0.00	3.14	6.28	9.42	12.57	15.71	15.71	16.02	16.92	18.32	18.85

Two different j, m pairs give rise to the same natural frequency at $kL = 15.71$, and the spacing between natural frequencies decreases with increasing frequency. Both of these features will be explored in detail when we analyze natural cavity modes.

As expected, $|Z_F|$ is very large in the vicinity of a natural frequency, whereas the value of Z_M is an inertance that is proportional to the frequency. The plots in Fig. 1 indicate that Z_M is the dominant effect away from a resonance. As the frequency is increased beyond a resonance, Z_{ac} transitions to negative imaginary; that is, it is a compliance. Further increase toward resonance results in a sharp increase of this compliance, eventually becoming singular as a resonance is approached from below. The consequence is that at some frequency between resonances the value of $Z_F \equiv Z_M + Z_{ac}$ is zero. This has an important implication.

Let us change our perspective to ask what the plate's displacement would be if we were to drive it with a specified F. The displacement is $U = V_0/(i\omega)$, so Eq. (8) tells us that

$$U = \frac{F}{i\omega HW}\left(\frac{1}{Z_F}\right) \tag{11}$$

From this viewpoint, a large value of Z_F at a resonance leads to a very small displacement. In contrast, a very small value of Z_F leads to a large displacement. This is a resonance of the coupled system formed by the plate and the cavity. We say that it is a *fluid-structure resonance*.

This might seem to be somewhat confusing, but the two situations are readily distinguished if we think of an experiment. The situation implied in the problem statement is that a machine exerts a force amplitude F that can be as large as necessary to maintain a specified value of V_0 regardless of the pressure within the cavity. In this scenario, the pressure will grow greatly when the frequency is close to any $(kL)_{nat}$, so the value of F required to sustain V_0 will also be very large. In another experiment, the machine applies the constant amplitude force F. In that case, Eq. (11) gives the displacement U. Setting $V_0 = i\omega U$ in Eq. (7) gives the corresponding acoustic force F_{ac} applied to the plate,

$$F_{ac} = \frac{Z_{ac}}{Z_M + Z_{ac}}F \tag{12}$$

The value of $F_{ac}/(HW)$ serves as a metric for the pressure within the cavity. In the second experiment, the displacement and pressure within the cavity are very large at frequencies for which $Z_M + Z_{ac} = 0$; this is the fluid-structure resonance condition. The contrasting situation places the frequency at one of the natural frequencies of the cavity, as given in Eq. (10). At these frequencies, Z_{ac} is very large. Equation (11) gives a small value for U in this case, and Eq. (12) gives $F_{ac} = F$. It follows that when we talk about a resonance, we must be unambiguous in specifying how the system is driven. To some extent, we encountered this same issue when we considered closure of one-dimensional waveguides, specifically in Example 3.4.

EXAMPLE 10.2 A rectangular tank of water has width and height $4a$. The bottom and side walls are rigid, and the top is a free surface. A circular piston transducer whose radius is a is flush mounted on the rigid wall at $x = L$, so that $V_x = V_0 \exp(i\omega t)$ on the piston face and zero otherwise. Derive an expression for the pressure at an arbitrary location as a function of ka. Then evaluate $|p|/(\rho_0 c V_0)$ at $x = 0$, $L/2$, and L on the centerline, $y = z = 2a$, for the frequency range $0 < ka < 8$. Also evaluate the face-averaged impedance seen by the transducer. The aspect ratio for the evaluation is $L/a = 5$.

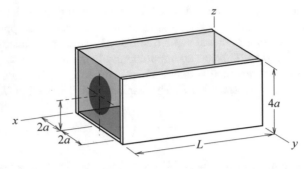

Figure 1.

Significance

From a conceptual viewpoint, the system in this example is very much like the one analyzed in the previous example. The primary difference is the fact that forced cavity modes covering a range of wavenumbers in both the y and z directions are excited here. The other significant difference from the previous example is that numerical methods will be required to evaluate the modal velocity coefficients. A different view of the behavior at resonances will emerge.

Solution

Identification of symmetry properties is a useful starting point. The vertical plane that contains the centerline cuts the tank in half, and the walls have the same properties on either side of this plane. Furthermore, the input velocity field is symmetric with respect to this plane. Consequently, the pressure field must share this symmetry. The normal to the symmetry plane is the y direction, so only forced cavity modes that are even functions with respect to $y = 2a$ will participate in the response. The walls at $y = 0$ and $y = 4a$ are rigid. The full set of two-dimensional modes for rigid walls separated by distance H is $\cos(s\pi/H)$, $s = 0, 1, 2, \ldots$ The symmetric ones correspond to even s. In contrast, there is no symmetry with respect to any horizontal plane, so the full set of rigid/pressure-release two-dimensional modes are required to describe the field's dependence on z. The length of both sides of the cross section is $4a$, so the normalized transverse modes for the analysis are

$$\Phi_{j,n} = B_{n,j} \cos\left(\frac{2j\pi y}{4a}\right) \cos\left(\frac{(2n-1)\pi z}{2(4a)}\right); \quad j = 0, 1, 2, \ldots, n = 1, 2, \ldots$$

$$B_{j,n} = \sqrt{\frac{4}{1 + \delta_{jn,0}}}$$

(1)

The associated axial wavenumbers are

$$\kappa_{j,n} = \left[k^2 - \left(\frac{j\pi}{2a}\right)^2 - \left(\frac{(2n-1)\pi}{8a}\right)^2\right]^{1/2}$$

(2)

As always, the square root should be such that $\kappa_{j,n}$ is either positive real or negative imaginary.

For the sake of variety, the boundary excitation has been situated at $x = L$. The piston's axial velocity is $V_x = V_0$ over its face and $V_x = 0$ for points that are not on the piston. Consequently, the integrand for the modal velocity coefficients is zero outside the piston, so that

$$V_{j,n} = \frac{1}{16a^2} \int_0^{4a} \int_0^{4a} V_x \Phi_{j,n} \, dy \, dz = \frac{V_0}{16a^2} \int\int_{A_{\text{piston}}} \Phi_{j,n} \, dy \, dz \tag{3}$$

In view of the circular shape of the piston's face, it is logical to use polar coordinates centered on the piston to formulate the integral. The coordinate transformation is

$$y = 2a + R \cos \theta, \, z = 2a + R \sin \theta$$

The face of the piston corresponds to $0 < R < a$, $-\pi < \theta < \pi$, which converts the velocity coefficients to

$$\begin{aligned}
V_{j,n} = \frac{1}{16a^2} \int_0^a \int_{-\pi}^{\pi} B_{n,j} &\cos \left(\frac{j\pi}{2} \left(2 + \frac{R}{a} \cos \theta \right) \right) \\
&\times \cos \left(\frac{\pi}{8} (2n - 1) \left(2 + \frac{R}{a} \sin \theta \right) \right) R \, d\theta \, dR
\end{aligned} \tag{4}$$

No analytical result for these integrals is readily available, so they are evaluated numerically. Some software packages contain a double integration routine. (In MATLAB, it is `dblquad`.) Otherwise, the integration may be performed by dividing a circle into a set of small patches subtending $\Delta R'$ and $\Delta \theta$. The double integral is approximately the sum of the value of $\Phi_{n,j}$ at the center point of each patch, multiplied by the area of the patch. Figure 2 plots the values $V_{n,1}$ and $V_{0,j}$. It can be seen that the coefficients decrease, but slowly and in an oscillatory manner.

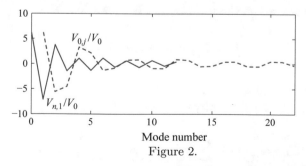

Mode number

Figure 2.

The values of $V_{j,n}$ are the last set of parameters required to form the modal series, Eq. (10.2.10). If we wish to employ the derived formulas, we must recognize that the x direction used for the derivation is reversed from the one we have used for the present system, with the excitation occurring at $x = L$ rather than $x = 0$. We may adjust for this difference by replacing x in the formulas with $L - x$. The undriven end is rigid, so we let ζ_L be infinite in Eq. (10.2.10). This is handled by dropping terms that do not contain ζ_L in the numerator and denominator, which is $\Delta\left(\kappa_{j,n}\right)$ in Eq. (10.2.9). Thus, the modal pressure series is transformed into

$$
P = \sum_{j=0}^{\infty}\sum_{n=1}^{\infty}\frac{k}{\kappa_{j,n}}\rho_0 c V_{n,j}\left(\frac{e^{-i\kappa_{j,n}x} + e^{+i\kappa_{j,n}x}}{e^{+i\kappa_{j,n}L} - e^{-i\kappa_{j,n}L}}\right)\Phi_{n,j}\left(y,z\right)
$$

$$
\equiv \sum_{j=0}^{\infty}\sum_{n=1}^{\infty}\frac{k}{\kappa_{j,n}}\rho_0 c V_{n,j}\frac{\cos\left(\kappa_{j,n}x\right)}{i\,\sin\left(\kappa_{j,n}L\right)}\Phi_{n,j}\left(y,z\right) \tag{5}
$$

Resonances correspond to the zeros of the denominator. This is examined in detail in the discussion.

The face-averaged impedance, which is the average pressure on the piston face divided by the average velocity, may be computed directly from the other parameters. The average velocity amplitude on the piston is V_0. The average pressure is found as an integral over the face of the piston,

$$
P_{\mathrm{av}} = \frac{1}{\pi a^2}\iint_{A_{\mathrm{piston}}} P|_{x=L}\,dydz
$$

The pressure is described by the modal series, Eq. (5), which we evaluate at $x = L$. In order to better understand the magnitude of the terms, we shall replace the complex exponentials with equivalent forms derived from Euler's identity. Thus, we form

$$
P_{\mathrm{av}} = \frac{1}{\pi a^2}\sum_{j=0}^{\infty}\sum_{n=1}^{\infty}\frac{ka}{i\kappa_{n,j}a}\rho_0 c V_{j,n}\cot\left(\kappa_{j,n}L\right)\iint_{A_{\mathrm{piston}}}\Phi_{j,n}\left(y,z\right)dydz \tag{6}
$$

The integral appeared previously in Eq. (3). It follows that

$$
\iint_{A_{\mathrm{piston}}}\Phi_{j,n}\left(y,z\right)dydz = 16a^2\frac{V_{j,n}}{V_0}
$$

The ratio of pressure to velocity is an impedance. Thus, the average acoustic impedance seen by the piston may be determined by evaluating

$$
Z_{\mathrm{ac}} = \frac{P_{\mathrm{av}}}{V_0} = 16\sum_{j=0}^{\infty}\sum_{n=1}^{\infty}\frac{ka}{i\kappa_{j,n}a}\rho_0 c\left(\frac{V_{j,n}}{V_0}\right)^2\cot\left(\kappa_{n,j}L\right) \tag{7}
$$

Evaluation of the pressure and average impedance requires computation of double sums over the n and j mode numbers. Let $J = \max\left(j\right)$ and $N = \max\left(n\right)$ be the numbers at which the summations are halted. One criterion is that J and N may be set such that $V_{j,n}$ is negligible for $j > J$ and $n > N$. The slow oscillatory reduction of $V_{j,n}$ indicated in Fig. 2 suggests that this criterion would require that J and N be quite large. Another possibility is that the series lengths may be set according to the behavior of $\kappa_{j,n}a$. Examination of Eqs. (5) and (7) shows that convergence of the pressure and impedance will be attained if the $\kappa_{n,j}$ values for omitted modes are large in magnitude. This condition will be attained if J is sufficiently large that $\kappa_{J,n}$ is negative imaginary regardless of n, and N is sufficiently large that $\kappa_{j,N}a$ is negative imaginary regardless of j. Equation (2) for $\kappa_{j,n}$ shows that this condition can be met if $J > 2ka/\pi$ and $N > 4ka/\pi + 1/2$. The highest frequency for the computations is $ka = 8$, which gives $J > 6$ and $N > 11$. Doubling these values is more than adequate because $\kappa_{J,1}a \approx -17i$ and $\kappa_{0,N}a \approx -15i$, whereas the smallest values of $\left|\kappa_{n,j}\right|$ are close to zero for modes that are closest to the cutoff condition. A by-product of selecting J and N in this manner is certainty that all modes that might be resonant are included. This is so because $\kappa_{n,j}$ of all omitted modes are negative imaginary, so the exponential terms in the denominators are real.

The computational algorithm developed for a rectangular waveguide in Example 9.7 is equally applicable here. Preliminary to the main frequency loop, all quantities that are independent of frequency are evaluated. This entails setting up an outer loop over $j = 0, ..., J$ and an inner loop over $n = 1, ..., N$. The element index is set at $m = jN + n$. Within these loops, the values of $\tilde{B}_m \equiv B_{j,n}$ are computed according to Eq. (1), and the double integral for $\tilde{V}_m \equiv V_{j,n}/V_0$ is carried out. We are working nondimensionally, so the latter coefficients are given by

$$\tilde{V}_m = \frac{1}{16} \int_0^1 \int_{-\pi}^{\pi} \Phi_{j,n}\,(2 + \cos\theta, 2 + \sin\theta)\,R'd\theta dR'$$

Another item that may be computed independently of the frequency is the value of each transverse mode function on the centerline, $y = z = 2a$, at which the pressure will be computed. These values form $\tilde{\Phi}_m \equiv \Phi_{j,n}\,(2a, 2a)$, which are stored in a diagonal matrix $[\Phi_m]$.

The frequency loop finely increments the ka value between 0 and 8. At each frequency, the loops over $j = 0, ..., J$ and $n = 1, ..., N$ are repeated. The index $m = jN + n$ is set as the subscript for storing arrays as column vectors. The nondimensional axial wavenumber $\tilde{\kappa}_m a$ is obtained from Eq. (2). The pressure at the designated points $x_1 = 0$, $x_2 = L/2$, and $x_3 = L$ and the impedance value are computed by placing the pressure in Eq. (5) and the impedance coefficient in Eq. (7) into matrix form as

$$\left\{ \frac{P}{\rho_0 c V_0} \right\} = [E]\,[D]_{\mathrm{diag}} \left[\tilde{\Phi}_m \right]_{\mathrm{diag}} \left\{ \tilde{V} \right\}$$

$$\frac{Z_{\mathrm{av}}}{\rho_0 c} = \left\{ \tilde{V} \right\}^{\mathrm{T}} [F]_{\mathrm{diag}} \left\{ \tilde{V} \right\}$$

(8)

where

$$D_{m,m} = \frac{ka}{i\kappa_m a}\left(\frac{1}{\sin\left((\tilde{\kappa}_m a)\dfrac{L}{a}\right)}\right)$$

$$E_{s,m} = \cos\left((\tilde{\kappa}_m a)\frac{L}{a}\frac{x_s}{L}\right)$$

$$F_{m,m} = \frac{ka}{i\kappa_m a}\cot\left((\tilde{\kappa}_m a)\left(\frac{L}{a}\right)\right)$$

The pressure amplitude at the designated locations is plotted as a function of frequency in Fig. 3. A casual viewing might lead one to conclude that each plot describes random noise. That this is not so is evidenced by the behavior in the lower frequency range, where distinct resonance peaks can be identified. Each of the high-frequency peaks also is caused by a resonance. The denominators in Eqs. (5) and (7) may be written equivalently as $2i\sin\left(\kappa_{j,n}a\,(L/a)\right)$. Resonance corresponds to this quantity being zero, which occurs if $\kappa_{j,n}a = (a/L)\,s\pi$, $s = 0, 1, 2, \dots$. In view of the definition of $\kappa_{j,n}$ in Eq. (2), the resonance condition is marked by

$$\left(\frac{n\pi}{2}\right)^2 + \left(\frac{(2j-1)\pi}{8}\right)^2 + \left(\frac{a}{L}\pi\right)^2 s^2 = (ka)^2 \tag{9}$$

As the frequency increases, it becomes increasingly likely that there will be one or more n, j pairs that satisfy Eq. (9). For example, the first resonance occurs at $ka = \pi/8 = 0.393$ corresponding to $n = 0$, $j = 1$, and $s = 0$. This is contrasted by the situation for $ka = 7.54$, for which $\left|\sin\left(\kappa_{j,n}a\,(L/a)\right)\right| < 0.05$ for $n = 2$, $j = 4$ $(s = 10)$, $n = 1$, $j = 9$ $(s = 5)$, and $n = j = 4$ $(s = 5)$. The fact that the resonances become very close is further evidenced by the fact that a slight shift to $ka = 7.55$ leads to $\left|\sin\left(\kappa_{j,n}a\,(L/a)\right)\right| < 0.05$ for $n = 0$, $j = 1$ $(s = 12)$, as well as $n = 2$, $j = 4$ $(s = 10)$.

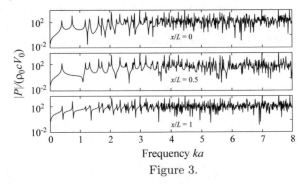

Figure 3.

Figure 4 zooms in on the low-frequency range. With one exception, a resonance frequency is manifested by the pressure at each location showing a peak. The exception occurs at $ka = 0.74$, where the pressure at $x = L/2$ does not show a peak. This is a consequence of the fact that the resonant $(0, 1)$ mode corresponds to

$\cos(\tilde{\kappa}_m (x/L)(L/a)) = 0$ at $x = L/2$. Hence, its does not contribute to the pressure at this location even though the amplitude of this mode is large. Unlike resonances, sharp minima of $|P|$ are not correlated between different locations. These occur because the many terms in the modal sums accumulate to a very small value, and the terms in that sum depend on position.

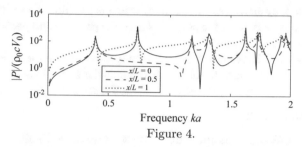

Figure 4.

The $V_{n,j}$ coefficients given in Eq. (4) are real, so Z_{ac} is imaginary. This means that Z_{ac} is reactive, as it was in the previous example. The average impedance on the piston face is depicted in Fig. 5. (The dashed lines describe the discontinuous drop from an infinite inertance to zero compliance as a natural frequency is passed.) The overall appearance of Fig. 5 is like Fig. 1 of the previous example, except that the spacing between natural frequencies is irregular here.

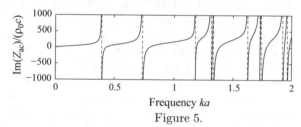

Figure 5.

10.2.2 Spherical Cavities

One seldom encounters a spherical room, although the author does recall visiting a combat flight simulator at an aerospace company in which images were projected on a nearly complete spherical wall. Nevertheless, being able to analyze waves in an enclosed spherical space affords us the opportunity to examine the differences between different types of cavities and thereby understand how curved boundaries affect basic phenomena. We first encountered a spherical cavity in Chap. 6. The earlier treatment was limited to radially symmetric waves. The development of spherical harmonics in Chap. 7 makes it possible to generalize the analysis.

It might seem that a spherical domain is unlike the interior of a box, but there are many similarities. Although a sphere has one physical boundary, whereas a box has six walls, there are additional boundary conditions for the former that play the roles of walls. One we encountered in the study of radially symmetric waves, wherein a primary consideration is that the general solution of the wave equation must not be singular at the center. Other conditions are continuity. If we know the complex amplitude P at a set of spherical coordinates (r, ψ, θ), an increase or decrease of θ by 2π must yield the same values for the pressure and particle velocity. (This condition is identically met by an axisymmetric field, which is the situation we will consider.) Continuity also is an issue at the poles, because the pressure and particle velocity at $\psi = 0$ and $\psi = \pi$ must be finite.

Fortunately, these issues are addressed by a spherical harmonic series representation. Thus, the starting point for our exploration is the general solution for axisymmetric waves in a spherical geometry. This was the subject of Sect. 7.1.

Axisymmetric Excitation

We wish to determine the pressure in the region enclosed by a vibrating spherical wall. The wall's motion is harmonic. A further restriction is that the motion is limited to being axisymmetric. In accord with our standard practice we designate the symmetry axis as z. Axisymmetry of the excitation leads to axisymmetry of the response, so the pressure field is $p(\bar{x}, t) = \text{Re}\left[P(r, \psi) \exp(i\omega t)\right]$. The general solution of the Helmholtz equation in axisymmetric spherical coordinates is a series of spherical harmonics. Each harmonic is a product of a radial function and a polar function.

The polar angle is ψ, but the mathematical variable that was used in Chap. 7 was the transformed variable $\eta = \cos \psi$. Some individuals view η as the projection onto the z axis of a unit radial line at angle ψ. The angular functions are Legendre polynomials, $P_m(\eta)$. When η is replaced by its definition, the result is that the angular functions are $P_m(\cos \psi)$. These are said to be Legendre functions. An analysis of the Helmholtz equation usually will proceed equally well using either ψ or η.

Radiation exterior to a sphere causes waves to propagate outward. A spherical Hankel function $h_m^{(2)}(kr)$ is appropriate to this situation because at long distances it approaches a proportionality to $\exp(-ikr)$. This function is not appropriate by itself for the interior problem because it is singular at the origin, $r = 0$. The singularity of this solution could be canceled by adding the Hankel function $h_m^{(1)}(kr)$ for an inward propagating spherical wave. Such a representation would lead to an analysis that parallels the analysis of a rectangular cavity. Finiteness at $r = 0$ would be the analog of the boundary condition for the wall opposite the driven one in the rectangular geometry.

The merits of seeing this analogy are counterbalanced by the complications associated with dealing with (complex) Hankel functions. The Helmholtz equation is linear and real, so the real and imaginary parts of each Hankel function satisfy that equation individually. The singularity at the origin occurs in the imaginary part, so we discard it. The result is that the radial part of a spherical harmonic for an enclosed region is a spherical Bessel function $j_m(kr)$. The order m is the same as the degree of one of the Legendre functions.

A general solution for the field within a sphere is obtained by multiplying the product of each Legendre function and the corresponding spherical Bessel function by a coefficient whose value must be determined. Thus, the basic ansatz for the complex pressure amplitude is

$$P = \sum_{m=0}^{\infty} C_m j_m (kr) P_m (\cos \psi) \qquad (10.2.14)$$

Before we determine the series coefficients, let us consider the finiteness and continuity conditions that were noted at the beginning of the development. Finiteness was addressed by using only the spherical Bessel function for the radial dependence. This condition also led to truncation of the Legendre polynomial, without which the series would have diverged at $\psi = 0$ and $\psi = \pi$. The fact that the preceding series represents a continuous pressure field results identically from the axisymmetry property. However, there is another finiteness condition that applies to the particle velocity. The velocity components are found from Euler's equation to be

$$V_r = \frac{i}{\rho_0 c} \sum_{m=0}^{\infty} C_m j'_m (kr) P_m (\cos \psi)$$
$$V_\psi = -\frac{i}{\rho_0 \omega r} \sin \psi \sum_{m=0}^{\infty} C_m j_m (kr) (\sin \psi) P'_m (\cos \psi) \qquad (10.2.15)$$

There is nothing irregular about these expressions, but that is only because $P'_m (\cos \psi) \equiv 0$ at $\psi = 0$ and $\psi = \pi$. Suppose this was not true. Points on opposite sides of the polar axis correspond to $\psi = \varepsilon$ at azimuthal positions θ and at $\theta + \pi$. In the limit as $\varepsilon \to 0$, the value of V_ψ would be the same for both points, but the direction of $\bar{e}_\psi$ is opposite. This would correspond to an infinite velocity gradient. According to the continuity equation, such a condition corresponds to an infinite pressure.

Now that we know that the spherical harmonic series satisfies all finiteness and continuity conditions, we may proceed to evaluate the coefficients. They are determined by matching the particle velocity to the vibration of the spherical wall. The normal to the wall is $-\bar{e}_r$. Our convention is that the surface normal points into the fluid, so we consider positive values of the wall's normal velocity to be inward. Thus, we require that

$$v_r = - \mathrm{Re} \left(\hat{V}_S (\psi) e^{i\omega t} \right) \text{ at } r = a \qquad (10.2.16)$$

The amplitude V_S may be complex, in which case it represents a motion of the surface in which not all points vibrate in-phase.

Our task now is to determine the coefficients C_m of the pressure series by satisfying the velocity boundary condition. This is done by using a Legendre function series to describe $V_s (\psi)$. Doing so yields a representation that is consistent with the above expression for V_r of the fluid. Thus, we write

$$\hat{V}_S = \sum_{m=0}^{\infty} V_m P_m (\cos \psi) \qquad (10.2.17)$$

An expression for the coefficients of this series results from application of the orthogonality property of Legendre functions, Eq. (7.1.16). To that end, we multiply the series by $P_j (\cos \psi) \sin \psi$, where j is a specific index, and then integrate over the full range of ψ. Changing the index j back to m gives

$$V_m = \frac{2m + 1}{2} \int_0^\pi \hat{V}_S \, P_m (\cos \psi) (\sin \psi) \, d\psi \tag{10.2.18}$$

We equate the radial velocity at $r = a$, which is obtained from the first of Eq. (10.2.15) to the imposed velocity in Eq. (10.2.17). We could apply the orthogonality property, but a briefer approach argues that the Legendre functions constitute a complete set of linearly independent basis functions. Therefore, it must be that the series coefficients match, which leads to

$$\frac{i}{\rho_0 c} C_m j_m' (ka) = V_m \tag{10.2.19}$$

The resulting description of the pressure is

$$\boxed{P = \rho_0 c \sum_{m=0}^{\infty} V_m \frac{j_m (kr)}{i \, j_m' (ka)} P_m (\cos \psi)} \tag{10.2.20}$$

Earlier we identified resonances as singularities of a frequency-domain description. The singularities of the preceding expression correspond to zeros of the denominator, so we have identified that the mth resonance occurs when

$$\boxed{j_m' (x)\big|_{x=ka} \equiv \frac{m}{ka} j_m (ka) - j_{m+1} (ka) = 0} \tag{10.2.21}$$

This condition can occur for any spherical harmonic. Furthermore, the spherical functions are oscillatory, so there are multiple roots for a specific m. Thus, there is a double infinity of ka values at which the cavity will resonate.

Proper evaluation of the series requires that an adequate number of terms be included. Resonances are an important factor. For a selected value of ka, there are values of m for which the resonance condition is close to being satisfied. However, $j_m (ka (r/a)) / j_m' (ka)$ in Eq. (10.2.20) decreases monotonically with increasing m, provided that m is substantially larger than ka and $r/a \leq 1$. Thus, series truncation at a value of m that is a multiple of ka, such as $\max (m) = 4ka$, should be adequate. Another aspect of series truncation is the behavior of the V_m coefficients. At some value of m, these values will decrease because the corresponding Legendre function series represents a nonsingular V_s function. Usually, truncation based on the behavior of the Bessel functions will assure inclusion of a sufficient number of terms.

In addition to the acoustic pressure, power transfer is an important physical aspect. The instantaneous radial intensity at a point on the wall is

$$I_r = \left. \left(P V_s e^{2i\omega t} + P V_s^* \right) \right|_{r=a} + \text{c.c.}$$

$$= \rho_0 c \sum_{j=0}^{\infty} \sum_{m=0}^{\infty} \left(\frac{V_j V_m e^{2i\omega t} + V_j V_m^*}{i} + \text{c.c.} \right) \frac{j_m(ka)}{j_m'(ka)} P_j(\cos\psi) P_m(\cos\psi)$$

$$(10.2.22)$$

According to this expression, at any point on the wall power might be flowing out of, or into, the wall. The power transfer across a patch surrounding this point might have a nonzero mean value. The total power flow is obtained by integrating this expression over the entire wall. An element of area for this axisymmetric situation is $dS = 2\pi a^2 (\sin\psi) \, d\psi$. Orthogonality of the Legendre functions filters out the cross terms. In addition, $V_m V_m^*$ is a real quantity, so the complex conjugate part cancels this term. The result is that the instantaneous power flow is

$$P = 4\pi a^2 \rho_0 c \sum_{m=0}^{\infty} \left(\frac{V_m^2}{i(m+1)} e^{2i\omega t} + \text{c.c.} \right) \frac{j_m(ka)}{j_m'(ka)} \qquad (10.2.23)$$

From this, we conclude that the average total power transfer is zero. Each term fluctuates about zero at twice the excitation frequency, but all terms are not necessarily in-phase, so there might be no instant at which $P = 0$.

EXAMPLE 10.3 Consider a plane wave that is external to a spherical container. Suppose it induces a radial velocity of the wall that is proportional to the external pressure. Evaluate the pressure field within the container at $ka = 2$ and $ka = 5$.

Significance

The notion that the wall velocity is proportional to the external pressure is a vast simplification. For one, it ignores the fact that the presence of the sphere induces an exterior scattered pressure field. Nevertheless, the stated conditions serve to create a situation from which we may learn much. The analysis will also help us sharpen our computational skills.

Solution

The direction in which the plane wave propagates defines the symmetry of the field, so we let this direction be parallel to the z axis. When the origin of xyz is placed at the center of the cavity, the complex pressure amplitude of the incident wave is $P_0 \exp(-ik\bar{x} \cdot \bar{e}_z) = P_0 \exp(-ikz)$. For a point on the surface, $z = a \cos\psi$, so we begin by setting $V_s = V_0 \exp(-ika \cos\psi)$.

The velocity coefficients for the Legendre series representation of V_s are

$$V_m = \frac{2m+1}{2} V_0 \int_0^\pi e^{-ika\cos\psi} P_m(\cos\psi) \sin(\psi) \, d\psi$$

In general, we cannot expect that a formula for the integral is available, but it is for this case.[2] The result is

$$V_m = (2m + 1) e^{-m\pi i/2} j_m (ka)$$

If this formula were not available, we would evaluate the coefficients numerically. The MATLAB procedure would entail setting up a loop ranging from $m = 0$ to $m = $ max (m). For each m, the integrand would be evaluated with an anonymous function that depends on η, such as `Coeff_int=@(eta) exp(-1i*ka*eta).*leg (m,eta)`, where `leg.m` was described in Example 7.1. The `quadl` function would yield the coefficient according to `V(m + 1) = 0.5*(2*m+1)*quadl (Coeff_int,-1,1,1e-6);`

The V_m values for $ka = 2$ and $ka = 5$ are plotted in Fig. 1. The largest values are centered on a harmonic number that is slightly larger than ka. Beyond that maximum, the coefficients fall off rapidly. According to the above formula, V_m is a real quantity if m is even, and it is imaginary if m is odd. To understand why this is so, consider the integrand in conjunction with writing the complex exponential as $\cos (ka \cos \psi) - i \sin (ka \cos \psi)$. With respect to $\psi = \pi/2$, the first term is an even function, whereas the second is an odd function. At the same time, $P_m (\cos \psi) \sin \psi$ is even with respect to $\pi/2$ for even m and odd for odd m. The integral from 0 to π of the product of even or odd functions is nonzero, whereas the integral of the product of an even and an odd function is zero. This parity behavior will be seen to have an interesting consequence in regard to the pressure distribution.

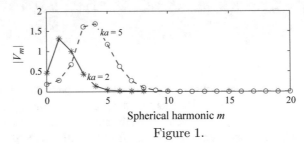

Figure 1.

Figure 1 demonstrates that the V_m values are negligible well before $M = $ max $(m) = 4ka$, which is the suggested rule to truncate the series. The manner in which the pressure field is to be displayed affects the algorithm that is used to evaluate the series. We will construct contours of equal Re $(P (\bar{x}))$ and Im $(P (\bar{x}))$. Doing so entails evaluating the pressure at a set of grid points in axisymmetric spherical coordinates and then calling on the graphics capabilities of our chosen mathematical software. Typically, a contouring graphics routine expects the position coordinates to be Cartesian. Thus, we define a set of r and ψ values at equal increments, $r_n/a = n\Delta_r$ and $\psi_\ell = \ell\Delta_\psi$. The corresponding Cartesian coordinates of the grid points are

---

[2] M.I. Abramowitz and I.A. Stegun, *Handbook of Mathematical Functions*, Dover, p. 440, Eq. 10.1.47 (1965).

$z_{n,\ell}/a = n\Delta_r \cos\left(\ell\Delta_\psi\right)$ and $x_{n,\ell}/a = n\Delta_r \sin\left(\ell\Delta_\psi\right)$. (Because of axisymmetry, it is adequate to view the field in the xz plane.)

Equation (10.2.20) describes P at each grid point. The matrix algorithm for evaluating a modal series at any points is suitable for the present situation. Thus, we define a diagonal matrix $[K]$ whose elements are $K_{m,m} = V_m/j'_m(ka)$, a rectangular matrix $[E]$ whose elements are $E_{n,m} = j_m(kr_n)$, and a rectangular matrix $[F]$ whose elements are $F_{m,\ell} = P_m(\cos\psi_\ell)$. The pressure at all grid points is found by computing

$$[P] = \frac{\rho_0 c}{i}[E][K][F]$$

Figure 2 shows the contours of Re(P) and Im(P) for $ka = 2$. The distribution of Re(P) is antisymmetric with respect to the plane $\psi = \pi/2$, and the distribution of Im(P) is symmetric with respect to that plane. This behavior is a consequence of the real/imaginary property of the V_m coefficients. Both $[E]$ and $[F]$ are real, and $[K]$ is complex solely because its elements are proportional to the V_m coefficients. For even m, V_m is real and $P_m(\cos\psi)$ is even with respect to $\psi = \pi/2$. Division by i results in imaginary quantities that are even. The converse occurs for odd m, because V_m is purely imaginary and $P_m(\cos\psi)$ is odd with respect to $\psi = \pi/2$. Thus, the real pressure contours stem from the odd m terms, whereas the imaginary contours are the contributions of the even m terms.

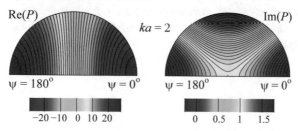

Figure 2

Figure 3 shows the contours for $ka = 5$. The even/odd properties displayed here are the same as those for Fig. 2. There is more fluctuation in the values from point to point, which is to be expected because the frequency is higher. What is unexpected is that the overall pressure level is lower. Also notable is the fact that the real and imaginary parts of P are comparable, whereas Re(P) is much greater than Im(P) at $ka = 2$.

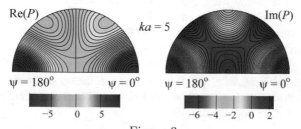

Figure 3.

The magnitude differences are a manifestation of resonance. To explore this aspect, let us return to the pressure series, Eq. (10.2.19). To compensate for differences of the overall Bessel function magnitudes for different order m, let us rewrite that equation in terms of a relative coefficient $(B_m)_{rel}$, such that

$$(B_m)_{rel} = V_m \frac{j_m(ka)}{i\, j'_m(ka)}$$

$$P = \rho_0 c \sum_{m=0}^{\infty} (B_m)_{rel} \frac{j_m(ka(r/a))}{j_m(ka)} P_m(\cos\psi)$$

The ratio of Bessel functions in the summation is always O(1), so $(B_m)_{rel}$ is a measure of the amount each term contributes to the summation. Figure 4 indicates that the dominant contribution at $ka = 2$ comes from the $m = 1$ harmonic. The large value of $(B_1)_{rel}$ stems from $j'_1(ka)$ being very small; $ka = 2.0816$ is a zero of $j'_1(ka)$. Because this is a term for which m is odd, the complex pressure amplitude is essentially a real value everywhere. In other words, it is essentially in-phase. An in-phase pressure distribution, as well as large pressure values, is hallmarks of a cavity resonance. At $ka = 5$, the primary contributors are the $m = 3$ and $m = 4$ spherical harmonics. The values of $(B_3)_{rel}$ and $(B_4)_{rel}$ are comparable in magnitude, which is why Fig. 3 indicates that the real and imaginary parts of $P(\bar{x})$ have similar magnitudes overall. The ratio of these parts is not constant over the field, which means that the pressure field does not consist of an in-phase oscillation. We conclude that $ka = 5$ is not very close to a resonance.

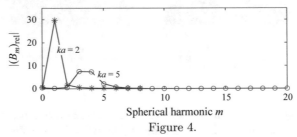

Figure 4.

To close this example, it is useful to take an overview. The excitation of the spherical cavity is a wavelike particle velocity at the wall. Nevertheless, there is no evidence of a comparable wave process in the pressure field. The field at any ka will consist of a complex pressure amplitude that combines real and imaginary parts. The former is antisymmetric with respect to the plane $\psi = \pi/2$, whereas the latter is symmetric with respect to that plane. If ka is close to a resonance, one part will be dominant and the pressure will show an overall maximum magnitude. Off-resonance, the mix of real and imaginary parts results in an instantaneous pressure distribution that fluctuates throughout a period because of the different parities of the real and imaginary parts of the complex amplitude.

Concentric Spheres

Systems in which the acoustic field is the region between concentric spheres are not common. Nevertheless, their analysis serves a number of purposes. The radius of the inner sphere is b, so the acoustic domain is $b < r < a$. The excitation might be applied to either the inner or outer wall as a specified harmonic radial velocity or pressure oscillation. The undriven surface might be rigid or pressure-release. In order to avoid unnecessary complications, we will not consider the alternate possibility that the undriven wall is locally reacting.

It is necessary to satisfy a boundary condition at both walls, but the center, $r = 0$, is not contained in the region occupied by the fluid. This means that we cannot exclude a solution of the spherical Bessel equation that is singular at the center. We could use both types of spherical Hankel functions, $h_m^{(1)}(kr)$ and $h_m^{(2)}(kr)$, to form a spherical harmonic series. These functions, respectively, represent inward and outward propagation. However, as we saw in the previous example, the field in a cavity tends to have a standing wave attribute. Adding and subtracting the two types of Hankel functions lead to the spherical Bessel and Neumann functions, $j_m(kr)$ and $n_m(kr)$, see Eq. (7.1.31). A spherical harmonic series for the complex pressure amplitude results from multiplying each function at a specific m by a different coefficient and then multiplying the sum of these terms by a Legendre function. In other words,

$$P = \sum_{m=0}^{\infty} \left[C_m j_m(kr) + D_m n_m(kr) \right] P_m(\cos \psi) \tag{10.2.25}$$

The corresponding radial particle velocity is

$$V_r = \frac{i}{\rho_0 c} \sum_{m=0}^{\infty} \left[C_m j_m'(kr) + D_m n_m'(kr) \right] P_m(\cos \psi) \tag{10.2.26}$$

The restriction to the passive boundary being rigid or pressure release, accompanied by allowance that the driven sphere is either the inner or outer boundary, admits four possible sets of boundary conditions. We will address the case where the inner sphere is rigid and the outer sphere is driven. The modifications required to address the other cases will be evident. We let $V_a(\psi)$ be the complex amplitude of the radial velocity at $r = a$. Our convention takes a wall velocity to be positive if it points into the fluid, so $V_a > 0$ represents a velocity that is in the direction of $-\bar{e}_r$. Thus, velocity continuity at the boundaries requires that

$$\sum_{m=0}^{\infty} \left[C_m j_m'(ka) + D_m n_m'(ka) \right] P_m(\cos \psi) = i \rho_0 c V_a(\psi)$$

$$\sum_{m=0}^{\infty} \left[C_m j_m'(kb) + D_m n_m'(kb) \right] P_m(\cos \psi) = 0 \tag{10.2.27}$$

As we did for the case of a full spherical cavity, we expand V_a in a Legendre function series,

$$V_a = \sum_{m=0}^{\infty} V_m P_m \left(\cos \psi \right) \tag{10.2.28}$$

The series coefficients are

$$V_m = \frac{2m+1}{2} \int_0^{\pi} V_a P_m \left(\cos \psi \right) \left(\sin \psi \right) d\psi \tag{10.2.29}$$

The Legendre function series for V_a is substituted into the velocity boundary conditions, Eq. (10.2.27). Application of the orthogonality properties of the Legendre functions leads to

$$\begin{aligned} C_m j_m' \left(ka \right) + D_m n_m' \left(ka \right) &= i \rho_0 c V_m \\ C_m j_m' \left(kb \right) + D_m n_m' \left(kb \right) &= 0 \end{aligned} \tag{10.2.30}$$

Solution of these equations yields

$$\begin{aligned} C_m &= i \rho_0 c \frac{n_m' \left(kb \right)}{\Delta \left(ka, kb \right)} V_m \\ D_m &= -i \rho_0 c \frac{j_m' \left(kb \right)}{\Delta \left(ka, kb \right)} V_m \end{aligned} \tag{10.2.31}$$

where

$$\Delta_m \left(ka, kb \right) = j_m' \left(ka \right) n_m' \left(kb \right) - j_m' \left(kb \right) n_m' \left(ka \right) \tag{10.2.32}$$

Resonances correspond to singular values of a C_m and D_m. They occur when $\Delta_m \left(ka, kb \right) = 0$. This condition occurs for each harmonic. As was true for the complete spherical cavity, the oscillatory nature of the Bessel and Neumann functions causes Δ_m to have zeroes at an infinite number of k values. Because there are an infinite number of harmonics, there are a doubly infinite set of resonances.

EXAMPLE 10.4   A concentric inner sphere in a spherical cavity executes a radially symmetric vibration, so that $v_r = v_0 \cos \left(\omega t \right)$ over its entire surface. The outer sphere is rigid. (a) Evaluate the radial dependence of the pressure for $ka = 2$, 8, and 32 in the cases where $b = 0.5$ and $b = 0.1$. (b) Evaluate the dependence of $|P|$ on ka at $r = a$, $r = (a + b)/2$, and $r = b$ for the case where $b = 0.1a$. (c) Consider the limit as $b/a \to 0$ with the source's volume velocity, $\hat{Q} = 4\pi b^2 v_0$, held constant. Derive an expression for the pressure within the sphere.

Significance

The analysis will connect the present development with our investigations of radially symmetric waves and sources. The outcome will be enhanced understanding of acoustic behavior of cavities.

Solution

Radially symmetric situations correspond to the $m = 0$ spherical harmonic. Although the formulas derived in this section are quite general, the simplicity of the $m = 0$ relations allow us to formulate the solution in terms of elementary functions. For $m = 0$, we have $j_0(x) = \sin(x)/x$ and $n_0(x) = -\cos(x)/x$. Thus, the $m = 0$ spherical harmonic for the region between concentric spheres is

$$P = \frac{1}{kr}[C_0 \sin(kr) - D_0 \cos(kr)] \tag{1}$$

The complex radial velocity on the inner sphere is v_0, and the outer sphere is stationary, so the boundary conditions are

$$\left.\frac{dP}{dr}\right|_{r=b} = -i\rho_0\omega v_0, \quad \left.\frac{dP}{dr}\right|_{r=a} = 0$$

Substitution of Eq. (1) into these conditions gives

$$\frac{1}{kb^2}[kb\cos(kb) - \sin(kb)]C_0 + [kb\sin(kb) + \cos(kb)]D_0 = -i\rho_0\omega v_0$$

$$\frac{1}{ka^2}[ka\cos(ka) - \sin(ka)]C_0 + [ka\sin(ka) + \cos(ka)]D_0 = 0$$

Introduction of the identities for the sine and cosine of the difference of two angles reduces the solution of these equations to

$$C_0 = -i\omega\rho_0 kb^2 v_0 \frac{ka\sin(ka) + \cos(ka)}{\Delta_0(ka, b/a)}$$

$$D_0 = i\omega\rho_0 kb^2 v_0 \frac{ka\cos(ka) - \sin(ka)}{\Delta_0(ka, b/a)} \tag{2}$$

where

$$\Delta_0(ka, b/a) = \left[\frac{b}{a}(ka)^2 + 1\right]\sin\left(ka\left(1 - \frac{b}{a}\right)\right) - ka\left(1 - \frac{b}{a}\right)\cos\left(ka\left(1 - \frac{b}{a}\right)\right) \tag{3}$$

An expression for P as a function of r results from substitution into Eq. (1) of the coefficients in Eq. (2). Introduction of $\omega \equiv kc$ and $b \equiv a(b/a)$ yields a nondimensional form that exhibits the fundamental parametric dependencies,

$$P = -i\rho_0 c v_0 \frac{(b/a)^2}{r/a} \frac{ka}{\Delta_0(ka, b/a)}[ka\cos(ka(r/a - 1)) + \sin(ka(r/a - 1))] \tag{4}$$

According to this expression, the complex pressure amplitude everywhere is purely imaginary. This means that the pressure is proportional to $\sin(\omega t)$. The radial velocity of the inner sphere is a cosine, so the pressure is in-phase with the surface acceleration.

To obtain the radial pressure distributions called for in Part (a), we evaluate Eq. (4) for fixed b/a and ka over $b/a < r/a < 1$. The radial profiles of pressure in Fig. 1 are typical. The overall magnitude of each profile is set by the factors preceding the bracketed term in Eq. (4). The value of Δ_0 is especially important in this regard. In most cases, the increase of pressure with decreasing radial distance is evident. However, there is little dependence on r/a at low ka, regardless of the value of b/a. This behavior is consistent with our general expectation for low-frequency oscillations in a cavity. At fixed b/a, the radial interval over which the pressure fluctuates decreases with increasing frequency. Note that dP/dr is zero at $r = a$, as required by that boundary being rigid.

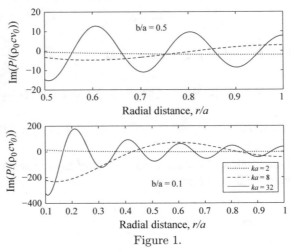

Figure 1.

Frequency response functions offer a different view. Figure 2 shows the result at three locations for the case $b = a/10$. The peaks are resonances at which $\Delta_0 = 0$. Resonance is a system feature, so they are manifested at each of the radial distances. The nulls are frequencies at which the numerator in Eq. (4) vanishes for each r. For a given value of b/a, the occurrence of this condition depends on both ka and r/a, so the nulls are not aligned. The frequencies at which nulls occur at $r = (a + b)/2$ become increasingly close to the resonances as the frequency increases. This behavior is a consequence of b/a being quite small in the present example. When $ka \gg 1$, $b/a \ll 1$, and $r = (a + b)/2$, the numerator in Eq. (4) for $r = (a + b)/2$ is approximately proportional to $ka \cos(ka/2)$ and Δ_0 is approximately proportional to $(ka)^2 \sin(ka) \equiv 2(ka)^2 \sin(ka/2) \cos(ka/2)$. Resonances, at which $\Delta_0 = 0$, occur approximately at $\sin(ka/2) = 0$ and $\cos(ka/2) = 0$. The latter are approximately the same as the zeros of the numerator. When all terms are included, the values of ka at which the numerator for $r = (a + b)/2$ is zero are close to some frequencies at which $\Delta_0 = 0$.

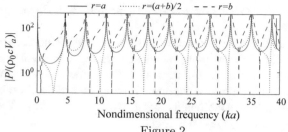

Figure 2.

To evaluate the limit of a very small sphere, we use $v_0 = \hat{Q}/\left(4\pi b^2\right)$ in order to remove v_0 from Eq. (4). Doing so gives

$$P = -i\rho_0 c \frac{\hat{Q}}{4\pi a^2} \frac{a}{r} \frac{ka}{\Delta_0\,(ka, b/a)} \left[ka \cos\left(ka\left(r/a - 1\right)\right) + \sin\left(ka\left(r/a - 1\right)\right)\right]$$

(5)

This expression does not contain b, so it is valid in the limit as $b \to 0$. Note that the limiting form requires that the inner sphere be present, so $r = 0$ is excluded. In effect, Eq. (5) is the solution for a spherical cavity with a point source at its center. The pressure is singular at the source.

10.2.3 Cylindrical Enclosures

The mathematical representation of the pressure field in a rectangular cavity does not depend on which wall vibrates. For a spherical cavity, there is only one mathematical form. In contrast, the mathematical nature of the field in a closed cylinder depends on whether the excitation is an imposed vibration of an end cap or the cylindrical surface. We will explore both possibilities. For the sake of simplicity and brevity, only axisymmetric situations are addressed. More general situations are addressed by a simple adaptation of the present results. This is so because a Fourier series description of an arbitrary θ dependence uncouples due to orthogonality of the circumferential harmonics.

Vibrating End Wall

We consider the end $x = 0$ to execute a specified vibration in an axisymmetric pattern, so that $V_x = V_s\,(R)$. The end $x = L$ is a locally reacting surface with specific impedance ζ_L. The transverse mode functions within a cylinder were derived in Sect. 9.3.3. The axisymmetric modes are

$$\Phi_{0,j}\,(R) = B_{0,j}\, J_0\left(\eta_{0,j}\frac{R}{a}\right), \quad \mu_{0,j} = \frac{\eta_{0,j}}{a}$$

(10.2.33)

The $B_{0,j}$ values in Eq. (10.2.33) are selected to normalize the mode functions according to Eq. (9.3.67).

The $\eta_{0,j}$ quantities are eigenvalues. The characteristic equation from which they are obtained results from satisfying velocity continuity at the cylindrical wall, where the specific impedance is ζ_a. Equation (9.3.57) is the result that was derived in Chap. 9. Because only the $n = 0$ harmonic is being considered, that expression may be simplified by exploiting the identity that $J_0' \left(\eta_{0,j} \right) \equiv -J_1 \left(\eta_{0,j} \right)$. The characteristic equation correspondingly may be written as

$$
J_0 \left(\eta_{0,j} \right) + \frac{i \zeta_a}{ka} \eta_{0,j} J_1 \left(\eta_{0,j} \right) = 0 \qquad (10.2.34)
$$

Evaluation of the modes was discussed in Sect. 9.3.3. The eigenvalues are sequenced in ascending order of their real part. In the case where the wall is rigid, $j = 0$ is the smallest eigenvalue, $\eta_{0,0} = 0$. This is the plane wave mode. It cannot exist if the cylindrical wall is not rigid, because a pressure in the wall would induce a particle velocity in the $\bar{e}_R$ direction, whereas the velocity in a plane wave is solely in the x direction.

If the cylindrical wall impedance is a compliance, that is, if ζ_a is a negative imaginary value, then there is one imaginary eigenvalue. That mode is designated as $j = 0$. The modal series that we will develop will assume that a mode for $j = 0$ exists. If it does not, then the associated modal series coefficient should be set to zero.

In the axial direction, we are presented with the same situation as that for a rectangular enclosure, in that each transverse mode is associated with a pair of waves that propagate forward and back in the x direction. Hence, a modal series will have the same form as Eq. (10.2.2), except that a single summation is used because only the $n = 0$ azimuthal harmonic is excited. The representation of the pressure field is

$$
P = \sum_{j=0}^{\infty} \left[\alpha_{0,j} e^{-i \kappa_{0,j} x} + \beta_{0,j} e^{+i \kappa_{0,j} x} \right] \Phi_{0,j} \left(R \right) \qquad (10.2.35)
$$

The axial wavenumbers are the values that are set by the requirement that each term independently satisfies the Helmholtz equation, which gives

$$
\kappa_{0,j} = \left[k^2 - \left(\eta_{0,j}/a \right) \right]^{1/2} \qquad (10.2.36)
$$

The series accounts for waves that propagate in both directions, so we may retain the requirement that the square root be selected such that $\mathrm{Re} \left(\kappa_{0,j} \right) \geq 0$ and $\mathrm{Im} \left(\kappa_{0,j} \right) \leq 0$. In turn, this requires that the $\eta_{0,j}$ values be in the first or second quadrant of the complex plane, which includes values on the positive real axis.

The steps leading to the determination of the modal coefficients parallel those for analysis of a rectangular cavity. The boundary conditions at the driven end and the opposite passive end are like those in Eq. (10.2.4). Consequently, the results may be derived from those for the rectangular case by modifying any variables associated with the transverse behavior to match the circular geometry. The analog of Eq. (10.2.7) is an inner product over the circular cross section. A differential element of area for an axisymmetric integrand may be taken to be $dA = 2\pi R dR$, and the cross-sectional area is πa^2, so the modal velocity coefficients are

$$V_{0,j} = \frac{1}{\pi a^2} \int_0^a \Phi_{0,j}(R) \, V_s(R) \, 2\pi R dR \qquad (10.2.37)$$

Aside from the fact that the transverse modes are those for a cylindrical waveguide, the pressure field is the same as Eq. (10.2.10),

$$\boxed{\begin{aligned} P = \rho_0 c \sum_{j=1}^{\infty} \frac{k V_{0,j}}{\kappa_{0,j} \, \Delta \left(\kappa_{0,j}\right)} & \left[\left(1 + \zeta_L \frac{\kappa_{0,j}}{k}\right) e^{i\kappa_{0,j}(L-x)} \right. \\ & \left. - \left(1 - \zeta_L \frac{\kappa_{0,j}}{k}\right) e^{-i\kappa_{0,j}(L-x)} \right] \Phi_{0,j}(R) \end{aligned}} \qquad (10.2.38)$$

The quantity $\Delta \left(\kappa_{0,j}\right)$ in the denominator is the same as the determinant in Eq. (10.2.9), except that the axial wavenumbers are those for the cylinder,

$$\boxed{\Delta \left(\kappa_{0,j}\right) = \left(1 + \zeta_L \frac{\kappa_{0,j}}{k}\right) e^{i\kappa_{0,j}L} + \left(1 - \zeta_L \frac{\kappa_{0,j}}{k}\right) e^{-i\kappa_{0,j}L}} \qquad (10.2.39)$$

As was true for a rectangular cavity, resonances correspond to $\Delta = 0$. This condition can occur only if $\mathrm{Re}(\zeta_L) = 0$ and $\kappa_{0,j}$ is real. This feature is evident in the alternate form of Δ obtained by application of Euler's identity. It is

$$\Delta \left(\kappa_{0,j}\right) = 2 \left[\cos \left(\kappa_{0,j}L\right) + i\zeta_L \frac{\kappa_{0,n}}{k} \sin \left(\kappa_{0,j}L\right)\right] \qquad (10.2.40)$$

The procedure by which the natural frequencies may be analyzed when ζ_a or ζ_L is reactive is essentially the same as that discussed in regard to Eq. (10.2.13). The primary difference is that the restriction to axisymmetric modes means that the transverse mode function depends on a single eigenvalue $\eta_{0,j}$, and the characteristic equation for that value features Bessel functions, rather than harmonic functions. The natural frequencies are found by a frequency sweep ranging over k at fixed j. The value $\eta_{0,j}$ is combined with the set of roots $\kappa_{0,j}^{(m)}$ for which $\Delta \left(\kappa_{0,j}\right) = 0$ to evaluate tentative natural frequencies according to

$$(k_{\mathrm{nat}})_{0,j} = \left[\left(\kappa_{0,j}^{(m)}\right)^2 + \left(\frac{\eta_{0,j}}{a}\right)^2\right]^{1/2} \qquad (10.2.41)$$

Any of these values that is very close to the value of k for the computations marks a natural frequency, $\omega_{0,j}^{(m)} = (k_{\mathrm{nat}})_{0,j}/c$.

EXAMPLE 10.5 The concept of an impedance tube developed in Sect. 3.2.4 is that measured properties of the field within a closed waveguide can be used to identify the impedance of a material placed at one end of the tube. The concept was derived on the basis of the field being planar in the transverse direction, but that is not possible in reality. This is so because a planar waveguide mode exists only if the sidewalls are rigid, which cannot be for actual materials. Consider the impedance tube in Example 3.6. As described there, the tube is a cylinder whose diameter is 120 mm, and $L = 2$ m, The pressure field is generated at $x = 0$ by a vibrating piston whose diameter equals that of the tube. The specific impedance of the material that covers the end at $x = L$ was $\zeta_L = 2.763 + 1.168i$. (This is the value used to generate the original data.) The fluid within the impedance tube is water, $\rho_0 = 1000 \, \text{kg/m}^3$, $c = 1480$ m/s. Determine and graph the mean-squared pressure distribution along the centerline when the cylindrical wall is compliant, with $\zeta_a = -5i$. Compare the result to the axial distribution obtained from an idealized rigid wall model. Frequencies to consider are 600 Hz, as stated in the original problem, and 6 kHz.

Significance

Flexibility of the sidewall is a primary issue for the design of impedance tubes. From an instructional standpoint, this example serves to connect the multidimensional phenomena to our earlier studies of plane waves.

Solution

The first task is evaluation of the eigenvalues of the transverse mode functions. The cylindrical wall's impedance is a pure compliance. Consequently, the real eigenvalues $\eta_{0,j}$ are the roots of Eq. (10.2.34). There also is one imaginary eigenvalue, $\eta_{0,\text{im}} = i\beta a$, where β is the root of Eq. (9.3.60) with $n = 0$. This is the type of mode for which $j = 0$ is reserved. The first few eigenvalues for each frequency and corresponding axial wavenumbers are tabulated below. The oscillatory nature of Bessel functions is such that eigenvalues for the higher modes are well approximated by incrementing $\eta_{0,4}$ by multiples of π.

f (Hz)		im	1	2	3	4
600	$\eta_{0,j}$	$0.5022i$	3.7997	6.9981	10.1614	13.3145
	$\kappa_{0,j}$	0.7911	$-3.7502i$	$-6.9714i$	$-10.1430i$	$-13.3005i$
6000	$\eta_{0,j}$	$1.8306i$	3.5107	6.8409	10.0532	13.2319
	$\kappa_{0,j}$	6.3816	5.0048	$-3.0700i$	$-7.9808i$	$-11.7349i$

At 600 Hz, for which $ka = 0.611$, all modes except the one associated with $\eta_{0,\text{im}}$ are cut off. Raising the frequency to 6 kHz only adds one mode to the set of real axial wavenumbers.

The $B_{0,j}$ coefficients that normalize the transverse mode functions are described in Eqs. (9.3.67) and (9.3.66). The result for $n = 0$ is

$$B_{0,j} = \frac{\eta_{0,j}}{J_0 \left(\eta_{0,j}\right)} \left[\left(\frac{ka}{\zeta_a}\right)^2 + \left(\eta_{0,j}\right)^2\right]^{-1/2}$$

To evaluate the coefficients of the modal series for the velocity excitation, we use Eq. (10.2.37) with V_s set to a constant value V_p for a vibrating piston. The resulting integral is a standard one,[3]

$$\frac{V_{0,j}}{V_p} = \frac{1}{\pi a^2} \int_0^a B_{0,j} J_0 \left(\eta_{0,j} R\right) 2\pi R dR = \frac{2B_{0,j}}{\eta_{0,j}} J_1 \left(\eta_{0,j}\right)$$

This expression indicates that the $V_{0,j}$ values decrease with increasing j, both as a consequence of the overall decrease of Bessel functions as their argument increases and as a consequence of the presence of $\eta_{0,j}$ in the denominator. Figure 1 describes these values, with $V_{0,\text{im}}$ assigned to $j = 0$. Beyond $j = 8$, the coefficients are less than 1% of the largest at both frequencies. Thus, the modal series will omit terms for which $j > 8$. (This actually is many more terms than necessary for series convergence, because all modes for $j > 2$ at both frequencies are cut off. The corresponding eigenvalues are negative imaginary, which will lead to $\Delta (k, j)$ being very large.)

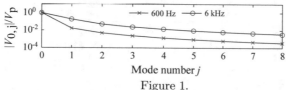

Figure 1.

It is desired to evaluate the pressure at $R = 0$, where the zero-order Bessel function is one. Therefore, the on-axis pressure is obtained by setting $\Phi_{0,j} = B_{0,j}$ in Eq. (10.2.38). Computation of the field at a set of points x_m along this axis may be performed by the matrix algorithm used previously. The $V_{0,j} / V_p$ values form column vector $\{V\}$. A rectangular array $[\Phi]$ holds the terms in Eq. (10.2.38) that depend on axial position, that is,

$$\Phi_{0,n} = \left(1 + \zeta_L \frac{\kappa_{0,n}}{k}\right) e^{i\kappa_{0,n}(L - x_m)} - \left(1 - \zeta_L \frac{\kappa_{0,n}}{k}\right) e^{-i\kappa_{0,n}(L - x_m)}$$

Then, the complex pressure amplitudes at the successive locations are found by evaluating

$$\left\{\frac{P}{\rho_0 c V_p}\right\} = [\Phi] \left[\frac{k}{\kappa_{0,n} \Delta \left(\kappa_{0,j}\right)}\right]_{\text{diag}} \{V\}$$

[3] M.I. Abramowitz and I.A. Stegun, *Handbook of Mathematical Functions*, Dover, (1965) Eq. 11.3.20, p. 484.

where $\Delta\left(\kappa_{0,j}\right)$ is described by Eq. (10.2.39). This computation is performed for both frequencies.

The pressure for planar waves in a closed waveguide (with rigid walls) is given in Eq. (3.2.21). It will be easier to interpret the results if this expression is written in terms of ζ_L rather than the reflection coefficient at $z = L$. Such a form is obtained by multiplying that expression by $\zeta_L + 1$, which leads to

$$\frac{P_{\text{rigid}}(x)}{\rho_0 c V_p} = \frac{(1 + \zeta_L)\, e^{ik(L-x)} - (1 - \zeta_L)\, e^{-ik(L-x)}}{\Delta_{\text{rigid}}}$$

$$\Delta_{\text{rigid}} = (1 + \zeta_L)\, e^{ikL} + (1 - \zeta_L)\, e^{-ikL}$$

The quantity used for an impedance tube calculation is the mean-squared pressure. The axial dependence of $\left(p^2\right)_{\text{av}}$ is depicted in Fig. 2. Clearly, there is a drastic difference between the behavior at the two frequencies, as well as between the rigid waveguide model and the model that account for the compliance of the wall. To understand why this is so, let us first focus on the results at 600 Hz.

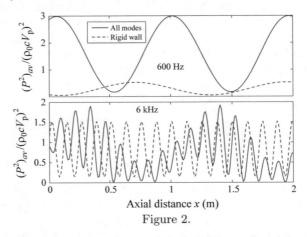

Figure 2.

We found that the zero mode, corresponding to $\eta_{0,\text{im}}$, is the only one that is not cut off at 600 Hz. For all other modes, $\kappa_{0,j}$ is a negative imaginary value. The value of $\Delta(k, j)$ for these modes is very large, which means that these modes contribute little to the modal pressure series. Indeed, the graph of $\left(p^2\right)_{\text{av}}$ would be unchanged if only the "im" mode was used. This explains why the plot of $\left(p^2\right)_{\text{av}}$ at 600 Hz has a sinusoidal appearance.

To explain why this plot differs for that for a rigid wall, we need to compare the $n = 0$ term in Eq. (10.2.38) to the expression for P_{rigid}. The contribution of the "im" mode to $P/\left(\rho_0 c V_p\right)$ is proportional to $k/\kappa_{0,\text{im}}$, which is 0.77, and $V_{0,\text{im}}/V_p$, which is 0.9998. Thus, the different magnitude of the modal description of P and P_{rigid} must stem from differences of the respective Δ values. For the "im" mode, this is $\Delta\left(\kappa_{0,0}\right) = 0.9838 + 2.1789i$, whereas $\Delta_{\text{rigid}} = 2.9133 - 5.1275i$. A resonance corresponds to Δ being zero, so we conclude that the "im" mode is closer at 600 Hz to a resonance than is the rigid model.

The fact that the pressure amplitudes are different does not necessarily invalidate application of the impedance tube concept, which relies on the maximum and minimum values of $(p^2)_{av}$. The procedure derived in Sect. 3.2.4 for extracting ζ_L from measured data should not be applied directly because the axial wavenumber is $\kappa_{0,im}$, rather than k. A comparison of the "im" term in the modal series, Eq. (10.2.38), to the above expression for P_{rigid} suggests that the procedure may be used if we set the wavenumber to $\kappa_{0,im}$ and replace the specific impedance with an effective value given by

$$\zeta_{eff} = \zeta_L \frac{\kappa_{0,im}}{k}$$

Note that in an actual setup we could identify $\kappa_{0,im}$ by measuring the distance between adjacent maxima or minima of $(p^2)_{av}$.

The ratio of the maximum and minimum values of $(p^2)_{av}$ is defined as S^2 in Sect. 3.2.4. The present data for 600 Hz gives $S = 4.2572$. From this, we find the magnitude of the modal reflection coefficient, see Eq. (10.2.11),

$$\left|(R_L)_{0,im}\right| = \frac{S-1}{S+1} = 0.620$$

To find δ, which is the argument of $(R_L)_{0,im}$, we use Eq. (3.2.57) with k replaced by $\kappa_{0,im}$. The value of δ we shall use is the average of the value obtained from each maximum or minimum. The result is $\delta = 0.230$, so the reflection coefficient is $(R_L)_{0,im} = 0.620 \exp(0.230i)$. In turn, the effective impedance is found to be

$$\zeta_{eff} = \frac{1 + (R_L)_{0,im}}{1 - (R_L)_{0,im}} = 3.4727 + 1.5937i$$

The corresponding true impedance is

$$\zeta_L = \frac{k}{\kappa_{0,im}} \zeta_{eff} = 2.6836 + 1.2316i$$

The latter value is quite close to the impedance used to generate the data.

The evaluation of ζ_L at 600 Hz yielded a meaningful result because the concept is based on the axial pressure distribution being sinusoidal. This is not the case at 6 kHz. We previously established that the $j = 1$ mode, as well as the "im" mode, is not cut off at this frequency. Furthermore, $V_{0,1}$ is 20% of $V_{0,im}$. The respective axial wavelengths are $2\pi/\kappa_{0,im} = 0.236$ m and $2\pi/\kappa_{0,1} = 0.301$ m. The combination of the contributions of these modes is the irregular dependence of $(p^2)_{av}$ that appears in Fig. 2. This negates identification of ζ_L at 6 kHz by the scheme developed in Sect. 3.2.4.

Our analysis leads to the conclusion that the given waveguide design is useful at any frequency where the "im" mode is the only one whose eigenvalue is real. As the frequency is increased from 600 Hz, the $j = 1$ mode "cuts on" when $ka = \eta_{0,j}$. This frequency is 3572 Hz. For comparison, if the wall were rigid, the axial wavenumber of the $j = 1$ mode, which is the lowest nonplanar mode, is real above 3760 Hz.

Increasing the stiffness corresponds to making Im (ζ_a) a larger negative number. This causes $\eta_{0,\text{im}}$ to decrease toward zero and each $\eta_{0,j}$ for $j \geq 1$ to approach the respective value for a rigid wall. From this, we may deduce some design objectives for an impedance tube. The first nonplanar mode is cut off below $\omega_{\text{cut}} = \eta_{0,1}c/a$. Thus, a should be minimized. Then, the walls should be made as rigid as possible, so that the cutoff frequencies of the nonplanar modes are close to the ideal values for a rigid wall. Although a locally reacting model is a useful convenience, any wall is part of a dynamic structure that has its own set of natural frequencies. The impedance of the boundary as seen by the fluid within the cavity is greatly reduced at these frequencies. Nevertheless, these characteristics do not alter the basic conclusions, most important of which is that usage of an impedance tube is limited to frequencies below the cutoff frequency of the second mode.

Vibrating Cylindrical Wall

Regardless of how a cylindrical cavity is excited, a series of forced modes must feature a sum of products of a Bessel function of μR and a complex exponential $\exp(\pm i\kappa x)$, because such terms constitute a fundamental solution of the Helmholtz equation for a cylindrical domain. The excitation in the previous section was an imposed vibration at an end, so that waves propagated in the x direction. When the excitation of a cylindrical cavity is an imposed vibration of the cylindrical boundary, a cylindrical wave propagates inward. Its arrival at the centerline generates an outward wave that cancels the singularity of the inward wave. The axial direction x now is transverse to the propagation direction. Correspondingly, the x dependence is set by the need to fit axial waves between the ends. In this view, the significance of R and x is swapped relative to that for vibration at an end.

The wall vibration V_a is a radial velocity, positive into the fluid, that may depend arbitrarily on the axial position. An axisymmetric field occurs when V_a is independent of θ. The ends at $x = 0$ and $x = L$ may be locally reacting, with specific impedances ζ_0 and ζ_L. Thus, we seek a solution of the Helmholtz equation that is finite at $R = 0$ and satisfies impedance boundary conditions at $x = 0$ and $x = L$,

$$P = \frac{\zeta_0}{\imath k}\frac{\partial P}{\partial x} \text{ at } x = 0$$
$$P = -\frac{\zeta_L}{\imath k}\frac{\partial P}{\partial x} \text{ at } x = L \tag{10.2.42}$$

In addition, the (inward) particle velocity at the cylindrical wall must match the input, so it must be that

$$\frac{\partial P}{\partial R} = +i\omega\rho_0 V_s(x) \text{ at } R = a \tag{10.2.43}$$

A comparison of the boundary conditions at the ends, Eq. (10.2.42), to those at the sidewalls of a two-dimensional waveguide, Eq. (9.2.29), shows that they only differ in the labels of parameters and coordinates. Thus, the axial mode functions here will be the same as the transverse functions in Eq. (9.2.33), with y, H, and η_n changed to x, L, and $\eta_{0,j} = \kappa_{0,j}L$, respectively. These functions, therefore, are

$$\Phi_{0,j}(x) = B_{0,j}\left[kL\sin\left(\eta_{0,j}\frac{x}{L}\right) - i\zeta_0\eta_{0,j}\cos\left(\eta_{0,j}\frac{x}{L}\right)\right] \qquad (10.2.44)$$

Like the notation for excitation at an end, using a zero first subscript reminds us that this is an axisymmetric function. The second subscript is the number of the root of the characteristic equation. As always, $j = 0$ is reserved for a zero or an imaginary eigenvalue. Adaptation of Eq. (9.2.32) to the current notation gives the characteristic equation for the eigenvalues $\eta_{0,j}$,

$$-\left[(kL)^2 + \zeta_0\zeta_L\left(\eta_{0,j}\right)^2\right]\sin\left(\eta_{0,j}\right) + i\left(\zeta_0 + \zeta_L\right)(kL)\left(\eta_{0,j}\right)\cos\left(\eta_{0,j}\right) = 0$$

$$(10.2.45)$$

The functions $\tilde{\Phi}_{0,j}$ in Eq. (10.2.44) have a magnitude that varies substantially as the eigenvalue changes. Normalization ensures that the mode functions we use do not share that attribute. The coefficient $B_{0,j}$ is normalized consistently with Eq. (9.2.51). An inner product of the mode function is defined relative to the x direction, so the $B_{0,j}$ values are evaluated according to

$$\left(B_{0,j}\right)^2 = \frac{L}{\displaystyle\int_0^L \left(\Phi_{0,j}\right)^2 dx} \qquad (10.2.46)$$

The other part of the forced modes for this type of excitation describes how waves propagate in the R direction. This dependence must be Bessel functions because the Neumann functions are singular at $R = 0$. A modal series will feature a product of a transverse mode and a Bessel function that depends on R. Specifically, the modal series is taken to be

$$P = \sum_{j=0}^{\infty} P_{0,j} J_0\left(\mu_{0,j}R\right)\Phi_{0,j}(x) \qquad (10.2.47)$$

The transverse wavenumber is denoted $\mu_{0,j}$, just as it was for the case of end excitation. The difference is that previously this wavenumber was obtained by solving a characteristic equation, whereas here it is set by the requirement that each term in the series satisfies the Helmholtz equation. Hence, it must be that the transverse wavenumbers are

$$\mu_{0,j} = \left[k^2 - \left(\frac{\eta_{0,j}}{L}\right)^2\right]^{1/2} \qquad (10.2.48)$$

It remains to match the particle velocity to the imposed velocity of the cylinder wall. Substitution of the modal series into Eq. (10.2.43) and application of the fact that $J_0'(\beta) \equiv -J_1(\beta)$ give

$$- \sum_{j=1}^{\infty} \mu_{0,j} P_{0,j} J_1 \left(\mu_{0,j} a \right) \Phi_{0,j} (x) = i\omega \rho_0 V_a (x) \qquad (10.2.49)$$

The $\Phi_{0,j}$ functions are an orthogonal set, as described by Eq. (9.2.52). We exploit this property by multiplying the preceding relation by an arbitrarily selected $\Phi_{0,n}(x)$ and then integrating the result over $0 < x < L$. The result is that

$$\boxed{P_{0,j} = -\rho_0 c \frac{ik}{\mu_{0,j} J_1 \left(\mu_{0,j} a \right)} V_{0,j}, \quad V_{0,j} = \frac{1}{L} \int_0^L V_a (x) \Phi_{0,j} (x)\, dx} \qquad (10.2.50)$$

The pressure field corresponding to these coefficients is

$$\boxed{P = \rho_0 c \sum_{j=0}^{\infty} \frac{ik V_{0,j}}{\mu_{0,j} J_1 \left(\mu_{0,j} a \right)} J_0 \left(\mu_{0,j} R \right) \Phi_{0,j} (x)} \qquad (10.2.51)$$

Resonances correspond to situations where the denominator of any term in the preceding summation is zero, that is, whenever the transverse wavenumber of the jth mode is a zero of the first-order Bessel function. There are many such zeros; we will use ℓ to denote the root number. Thus, resonances are marked by

$$\mu_{0,j} a = \tilde{\mu}_\ell \text{ where } J_1 \left(\tilde{\mu}_\ell \right) = 0, \ \ell = 1, 2, \dots \qquad (10.2.52)$$

The first few roots are

$$\tilde{\mu}_0 = 0, \ \tilde{\mu}_1 = 3.8317, \ \tilde{\mu}_2 = 7.0156, \ \tilde{\mu}_3 = 10.1735 \qquad (10.2.53)$$

Higher roots are essentially integer multiples of π greater than the last listed value. To find the natural frequencies we recall Eq. (10.2.48),

$$(k_{\mathrm{nat}})_{0,j,\ell} = \left[\left(\frac{\eta_{0,j}}{L} \right) + \left(\frac{\tilde{\mu}_\ell}{a} \right)^2 \right]^{1/2} \qquad (10.2.54)$$

An $\eta_{0,j}$ value is one of the infinite sets of roots of the characteristic equation for the axial direction, Eq. (10.2.45), but k appears in that equation. Thus, unless both ends are either rigid or pressure-release, the value of $\eta_{0,j}$ depends on k. However, the preceding defines the value of k at a natural frequency as a function of $\eta_{0,j}$. Identification of the natural frequencies in the presence of this dilemma was addressed by the discussion of Eq. (10.2.13).

EXAMPLE 10.6 A tentative design of a woofer, which is a loudspeaker that radiates low-frequency sound, consists of a cavity whose cylindrical wall is forced to undergo a uniform expansion at frequency ω. One end of the cavity is rigid, and the other is closed by a loosely tensioned membrane that well approximates a pressure-release condition. The sound generated by this speaker radiates from the vibrating membrane, so the velocity of the membrane is of interest. In particular, at low frequencies, $ka < 1$, the radiated power is proportional to the volume velocity of a source. Therefore, it is desired to derive an expression for the average velocity across the membrane. Evaluate this expression for the range from 20 Hz to 1 kHz for cases where the fluid within the cavity is air and water. The diameter of the cavity is 300 mm and the length is 500 mm.

Significance

The analysis will provide an in-depth exploration of the manner in which a vibrating cylindrical wall influences the interior field. It also will introduce some rudimentary issues for transducer design.

Solution

Formation of the modal series begins with determination of the axial mode functions. It is convenient to let $x = 0$ be the pressure-release end, so that $\zeta_0 = 0$ and ζ_L is infinite. We divide the characteristic equation, Eq. (10.2.45), by ζ_L in order to handle its infinite value, which reduces the equation to

$$(\kappa L) \cos (\kappa L) = 0$$

The root $\kappa L = 0$ leads to $\Phi_{0,j} = 0$, so it is discarded. Thus, κL must be an odd multiple of $\pi/2$,

$$\eta_{0,j} = \kappa_{0,j} L = \frac{2j-1}{2} \pi, \quad j = 1, 2, \ldots \tag{1}$$

The axial mode functions obtained from Eq. (10.2.44) are

$$\Phi_{0,j} = B_{0,j} \sin \left(\frac{(2j-1)\pi x}{2L} \right)$$

To normalize the modes we set

$$\int_0^L \left[B_{0,j} \sin \left(\frac{(2j-1)\pi x}{2L} \right) \right]^2 dx \equiv (B_{0,j})^2 \left(\frac{L}{2} \right) = L$$

so the normalized modes are

$$\Phi_{0,j} = \sqrt{2} \sin \left(\frac{(2j-1)\pi x}{2L} \right) \tag{2}$$

The transverse wavenumbers are given, in Eq. (10.2.48),

$$\mu_{0,j} = \frac{1}{L} \left[(kL)^2 - \left(\frac{2j-1}{2} \right)^2 \pi^2 \right]^{1/2} \tag{3a}$$

The negative imaginary root is taken if $kL < (j - 1/2)\pi$.

It is stated that the radial vibration of the cylindrical wall is uniform, which means that V_a is a constant value. Substitution of Eq. (2) into Eq. (10.2.50) shows that the corresponding velocity coefficients are

$$V_j = \frac{V_a}{L} \int_0^L \Phi_{0,j} \, dx = \frac{2\sqrt{2}}{(2j-1)\pi} V_s \tag{4}$$

Substitution of these quantities into the modal series, Eq. (10.2.51), gives

$$P = -\rho_0 c V_a \sum_{j=1}^{\infty} \frac{4}{(2j-1)\pi} \frac{ik J_0 (\mu_{0,j} R)}{\mu_{0,j} J_1 (\mu_{0,j} a)} \sin \left(\frac{(2j-1)\pi x}{2L} \right) \tag{5}$$

If $(2j-1)\pi > 2kL$, the mode number is above the cutoff value. Then, $\mu_{0,j}$ is negative imaginary, that is, $\mu_{0,j} = -i \hat{\mu}_{0,j}$. We may either directly input this imaginary value to the computer routine for Bessel functions, or alternatively invoke the identity that $J_n(-i\beta) = (-i)^n I_n(\beta)$, see Sect. 9.3.3.

The quantity of interest here is the axial particle velocity at $x = 0$. From Euler's equation, we have

$$V_x = -\frac{1}{i\rho_0\omega} \frac{\partial P}{\partial x} = V_a \sum_{j=1}^{\infty} \frac{2 J_0 (\mu_{0,j} R)}{(\mu_{0,j} L) J_1 (\mu_{0,j} a)} \cos \left(\frac{(2j-1)\pi x}{2L} \right) \tag{6}$$

The volume velocity is the area integral of the particle velocity at $x = 0$. A reference volume velocity is the cylindrical wall's velocity V_a multiplied by the cross-sectional area. Thus,

$$\frac{\tilde{Q}}{\pi a^2 V_a} = \frac{1}{\pi a^2} \int_0^a \frac{V_x|_{x=0}}{V_2} (2\pi R) \, dR$$

An equivalent interpretation of this expression is that it is the ratio of the average axial velocity at $x = 0$ to the wall's velocity. Substitution of Eq. (6) reduces this equation to an integral of $R J_0 (\mu_{0,j} R)$, for which a standard formula[4] is available. The result is that

$$\frac{V_{av}}{V_a} \equiv \frac{\tilde{Q}}{\pi a^2 V_a} = \frac{1}{\pi a^2} \int_0^a V_x|_{x=0} (2\pi R) \, dR = \frac{4}{\pi} V_a \sum_{j=1}^{\infty} \frac{a/L}{(\mu_{0,j} a)^2} \tag{7}$$

Interpretation of the frequency response graphs will be easier if we first compute the natural frequencies. The procedure for this analysis discussed in connection

---

[4]M.I. Abramowitz and I.A. Stegun, *ibid.*

with Eq. (10.2.54) is not needed because the ends are rigid and pressure-release. Consequently, the axial wavenumbers are $\eta_{0,j}/L$ are given by Eq. (1), regardless of the frequency. The resonance condition is $J_1(\mu_{0,j}a) = 0$, whose lower roots are given in Eq. (10.2.53). Substitution of these values into Eq. (10.2.54) yields the natural frequencies. The following tabulation lists the values that occur below 1 kHz.

Air				Water			
Axial #j	ℓ	$\tilde{\mu}_\ell$	$(\omega_{\text{nat}})_{j,\ell}$ (Hz)	Axial #j	ℓ	$\tilde{\mu}_{0,j}$	$(\omega_{\text{nat}})_{j,\ell}$ (Hz)
1	0	0	170	1	0	0	740
2	0	0	510				
1	1	3.8317	711.7				
3	0	0	850				
2	1	3.8317	858.9				

Figure 1 shows the result of evaluating Eq. (7) for the average velocity. When the cylinder is filled with water, the single resonance at 740 Hz matches the tabulated natural frequency. In contrast, for the case where air is the fluid, only the natural frequencies at 170, 510, and 850 Hz are manifested as resonances. (The data was probed to ascertain that the latter is not the adjacent natural frequency of 858.9 Hz.)

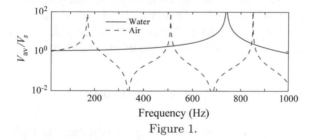

Figure 1.

Do the missing resonances represent an anomaly? Let us look at the pressure on-axis at the midpoint, which is described by Fig. 2. Every natural frequency is manifested by a peak pressure, as expected.

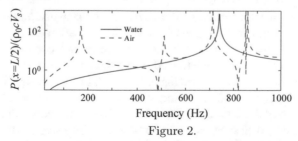

Figure 2.

To explain why the average velocity does not show a resonance at some natural frequencies, we must examine Eqs. (5) and (7) more carefully. The former indicates that the jth pressure term will be singular if at some frequency, $J_1(\mu_{0,j}a) = 0$, which encompasses the alternative that the factor $\mu_{0,j} = 0$. This is the attribute of any natural frequency. The situation for V_{av} in Eq. (7) is different, because a singularity only occurs if the frequency is such that one of the $\mu_{0,j}$ terms in the denominator is zero. The preceding tabulation indicates $\mu_{0,j} = 0$ is associated with the natural frequencies at 170, 510, and 850 Hz. This merely confirms that the graphs are consistent with the analysis, but not why the pressure and average velocity behave differently. The answer lies in Eq. (6) for the axial particle velocity at any location. The denominator of that quantity contains $J_1\left(\mu_{0,j}a\right)$, so it is singular at any natural frequency. However, the numerator indicates that the velocity distribution at $x = 0$ is proportional to $J_0\left(\mu_{0,j}R\right)$. The average value of this function over the area is proportional to $J_1\left(\mu_{0,j}a\right)$, which cancels the singular denominator term. Recall that $\mu_{0,j} = 0$ is the property of a plane wave mode, in which there is no dependence on R. Hence, we have established that if any mode is not constant over a cross section, its contribution to the average velocity will be finite at any frequency, even resonant ones. The average velocity will show a resonance only if the resonant mode is one in which there is no variation in the R direction.

To close this example, let us consider whether this device is useful as a low-frequency loudspeaker. Flatness of the frequency response is a desirable attribute. Filling the cavity with air leads to many resonances in the frequency band, but the average velocity in the water-filled design is essentially flat out to 500 Hz. This might be a desirable attribute, but another feature of the response is not. In the low-frequency range, where the response is flat, the value of V_{av} is nearly equal to V_a. Thus, no advantage is gained by driving the cylindrical surface relative to using a piston that matches the cylinder's cross section. Resonances and amplification factors are two aspects that must be considered in the design of any transducer.

10.3   Analysis Using Natural Cavity Modes

The forced cavity modes depend on the frequency of excitation and what surface is forced to vibrate. In contrast, the natural cavity modes that are the present topic depend on the size and shape of the enclosed region, as well as the nature of the boundaries, but they do not depend on the way in which the field is generated. These modes are properties of the system.

They must be determined prior to analysis of the response to an excitation. If the region has a shape that fits one of the standard coordinate systems, there will be wavenumbers describing the spatial dependence in each coordinate direction. In that case, the mode will be characterized by a subscript whose value marks the set of associated wavenumbers. For example, the modes of a rectangular room would be denoted as $\Psi_{j,m,n}(x, y, z)$. However, many situations feature an enclosed region whose shape is irregular They do not lend themselves to decomposition into identifiable wavenumbers. Such systems are better described by a single subscript denoting the hierarchy of modes. We will begin with an overview, which is more easily presented if we employ the latter representation. Hence, the cavity modes in the general development will be denoted as $\Psi_n(\bar{x})$.

Before we embark, it is necessary to limit the scope of our investigations to situations where the walls fit some combination of the rigid or pressure-release models.

This restriction is imposed because the formulation required to treat any other wall condition is quite complicated, as is evident from a fairly recent analysis.[5] Even under this constraint, a variety of configurations are possible. The development that follows begins with the general aspects of cavity modes that do not depend on the shape and properties of the enclosure. These are the governing equations, orthogonality, and the usage of a modal series to determine the pressure. The general development will be followed by specific analyses of standard shapes. Another limitation is that only the response generated by sources situated within the fluid will be considered. Recall that this is the situation for which a formulation using natural cavity modes is best suited, whereas the field generated by an imposed wall vibration is best described in terms of forced cavity modes. The common practice is to omit "natural" from the terminology. We shall do so because each formulation is used for a different class of excitation. Consequently, there is no context where both natural and forced cavity modes will be encountered.

10.3.1 Equations Governing Cavity Modes

By definition, a cavity mode Ψ_n is a pressure field that can exist within the fluid domain V, even though there is no external excitation. Such a function must be a solution of the Helmholtz equation. In accord with the restrictions stated above, we subdivide the boundary into a rigid portion S_V on which the normal velocity is zero and a pressure-release portion S_P on which the pressure is zero. The pressure field in a cavity mode is $p(\bar{x}, t) = \mathrm{Re}\,(\Psi_n \exp(i\omega t))$, and Euler's equation describes the particle velocity. Hence, the equations governing a cavity mode for this set of surfaces are

$$\nabla^2 \Psi_n + (k_n)^2 \Psi_n = 0, \quad \bar{x} \in V$$
$$\bar{n}(\bar{x}_s) \cdot \nabla \Psi_n = 0 \text{ if } \bar{x}_s \in S_V \qquad (10.3.1)$$
$$\Psi_n = 0 \text{ if } \bar{x}_s \in S_P$$

If we assign an arbitrary value to k_n, the only possible solution of this homogeneous set of equations is $\Psi_n = 0$ everywhere. Thus, we must identify the values of k_n for which there is a nontrivial solution. This is a statement of an eigenvalue problem. The eigenvalues k_n give the natural frequencies as $\omega_n = k_n c$, and the eigenfunction corresponding to each k_n is a cavity mode.

It should be evident that it will be quite difficult, or even impossible, to solve for the eigenvalues k_n unless V has a shape that fits one of the standard coordinate systems. Thus, we shall pursue analytical solutions only for rectangular, spherical, and cylindrical cavities. Any combination of rigid and pressure-release subregions of S is admissible from the viewpoint of general properties. However, we will not be able to solve analytically for the mode functions if the condition on a constant coordinate surface changes. For example, we can use analytical methods if entire

---

[5]J.H. Ginsberg, "Derivation of a Ritz series modeling technique for acoustic cavity-structural systems based on a constrained Hamilton's principle," J. Acoust. Soc. Am. **127**, 2749–2758 (2010).

walls of a box are pressure-release or rigid, but we cannot use analytical methods if any wall of that box is partially rigid and partially pressure-release. Such situations require computational or approximate methods.

The existence and nature of a zero frequency mode has general importance. If the frequency is zero, the Helmholtz equation reduces to Laplace's equation, $\nabla^2 P = 0$. Because $\nabla^2 P \equiv \nabla \cdot \nabla P$, it must be that ∇P is a constant vector. If the entire boundary is rigid, then $\bar{n} \cdot \nabla P$ must be zero at the boundary, which is satisfied only if the constant value of ∇P is zero. In other words, a constant value of P constitutes a nontrivial solution at zero frequency if a fluid is fully enclosed by a rigid boundary. The contrary case is that in which any portion of the boundary is pressure-release. A field whose gradient is constant with a zero value at any location must be zero everywhere. In other words,

> **If a region of fluid is fully enclosed by a rigid boundary, then there exists a zero frequency cavity mode in which the pressure is constant. If any portion of the boundary is pressure-release, a zero frequency mode does not exist.**

10.3.2   Orthogonality

Orthogonality of the cavity modes greatly simplifies determination of the response to source excitation. To derive this property, we consider a pair of modes j and n, whose natural frequencies are ck_j and ck_n. The first step entails multiplying the Helmholtz equation for mode n by mode j and then integrating over $\mathcal{V}$, which gives

$$\iiint_{\mathcal{V}} \Psi_j \nabla^2 \Psi_n d\mathcal{V} + k_n^2 \iiint_{\mathcal{V}} \Psi_j \Psi_n d\mathcal{V} = 0 \qquad (10.3.2)$$

In order to obtain a form in which Ψ_j and Ψ_n appear in the same manner, we exploit the definition that $\nabla^2 \Psi \equiv \nabla \cdot \nabla \Psi$ to convert the first integrand,

$$\iiint_{\mathcal{V}} \left[\nabla \cdot \left(\Psi_j \nabla \Psi_n \right) - \nabla \Psi_j \cdot \nabla \Psi_n \right] d\mathcal{V} + k_n^2 \iiint_{\mathcal{V}} \Psi_j \Psi_n d\mathcal{V} = 0 \qquad (10.3.3)$$

Application of the divergence theorem to the first term leads to

$$- \iint_{\mathcal{S}} \Psi_j \bar{n} \left(\bar{x}_s \right) \cdot \nabla \Psi_n d\mathcal{S} - \iiint_{\mathcal{V}} \nabla \Psi_j \cdot \nabla \Psi_n d\mathcal{V} + k_n^2 \iiint_{\mathcal{V}} \Psi_j \Psi_n d\mathcal{V} = 0$$

$$(10.3.4)$$

where the negative sign preceding the surface integral results from the definition of $\bar{n}$ as the normal oriented into the cavity.

We have limited our consideration to systems in which the wall consists of a rigid portion $\mathcal{S}_V$ and a pressure-release portion $\mathcal{S}_P$. On $\mathcal{S}_V$, we have $\bar{n} \left(\bar{x}_s \right) \cdot \nabla \Psi_n = 0$ and $\Psi_n \left(\bar{x}_s \right) = 0$ on $\mathcal{S}_P$. In either case, one of the factors in the surface integral is zero, so it must be that

$$k_n^2 \iiint_{\mathcal{V}} \Psi_j \Psi_n d\mathcal{V} - \iiint_{\mathcal{V}} \nabla \Psi_j \cdot \nabla \Psi_n d\mathcal{V} = 0 \qquad (10.3.5)$$

If we had begun by multiplying the Helmholtz equation for mode j by mode n, we would have obtained

$$k_j^2 \iiint_{\mathcal{V}} \Psi_j \Psi_n d\mathcal{V} = \iiint_{\mathcal{V}} \nabla \Psi_n \cdot \nabla \Psi_j d\mathcal{V} \qquad (10.3.6)$$

The difference of these two relations is

$$\left(k_n^2 - k_j^2\right) \iiint_{\mathcal{V}} \Psi_j \Psi_n d\mathcal{V} = 0 \qquad (10.3.7)$$

This relation is reminiscent of the one for the transverse mode function of a forced cavity mode. If modes j and n are different, then the factor multiplying the integral is not zero, so the integral must vanish. In other words, two different modes are orthogonal over the space $\mathcal{V}$. Equation (10.3.7) also applies if Ψ_n and Ψ_j are the same function. In that case, the factor multiplying the integral is zero, so the condition is met regardless of the value of the integral. Indeed, we know that the integration will yield a positive value because the mode functions are real, so $(\Psi_n)^2 > 0$ everywhere. This positive value is used to normalize the cavity modes. Specifically, a cavity mode Ψ_n may be multiplied by any constant B_n without altering the fact that they are solutions of the basic governing equations. The arbitrariness of B_n is addressed by setting it such that the integral equals the volume $\mathcal{V}$ of the domain. Doing so yields mode functions that are dimensionless. Thus, we have the first orthogonality condition,

$$\boxed{\iiint_{\mathcal{V}} \Psi_j \Psi_n d\mathcal{V} = \mathcal{V}\delta_{j,m}} \qquad (10.3.8)$$

(For two-dimensional domains, $\mathcal{V}$ would be the area in the plane on which the pressure varies.) When we substitute the first orthogonality condition into Eq. (10.3.5), we obtain the second orthogonality condition,

$$\boxed{\iiint_{\mathcal{V}} \nabla \Psi_j \cdot \nabla \Psi_n d\mathcal{V} = \mathcal{V} (k_n)^2 \delta_{j,m}} \qquad (10.3.9)$$

An exception to this development arises if two modes have the same natural frequency. This situation is common for rectangular-shaped enclosures, as we will see in the upcoming Example 10.7. Furthermore, the repeated natural frequencies always arise for the nonaxisymmetric modes of a cylindrical cavity because of the arbitrariness of how θ is defined. However, in both geometries, the derived cavity modes will be orthogonal, even though $k_j = k_n$. This property stems from the differing spatial patterns, with one being obtainable from the other by rotating the coordinate system by 90° from the other.

10.3.3 Analysis of the Pressure Field

We may now proceed to formulate a modal analysis of the pressure induced by sources. One of the assets of the cavity mode approach is that it can be used to derive directly a time-domain response. We shall develop this type of solution and then use the result to obtain a description of the frequency-domain response.

We begin with a series of cavity modes in which the modal coefficients are allowed to be arbitrary functions of time,

$$p(\bar{x}, t) = \sum_{n=1}^{\infty} \Psi_n(\bar{x}) p_n(t) \qquad (10.3.10)$$

An arbitrary collection of sources are situated within the cavity. The volume velocity of each source is $Q_m(t)$, and the locations are $\bar{x}_m$. The field in the time domain for a single source is governed by the inhomogeneous wave equation, Eq. (6.4.23), with $C = -\rho_0 \dot{Q}_m$. Thus, the field equation for the pressure is

$$\nabla^2 p - \frac{1}{c^2} \frac{\partial^2 p}{\partial t^2} = -\rho_0 \sum_m \dot{Q}_m(t) \delta(\bar{x} - \bar{x}_m) \qquad (10.3.11)$$

Differentiation of the modal series with respect to t operates only on the modal coefficients, and spatial derivatives operate only on the cavity modes. Thus, substitution of the series into the preceding wave equation leads to

$$\sum_{n=1}^{\infty} \left[c^2 \nabla^2 \Psi_n p_n - \Psi_n \ddot{p}_n \right] = -\rho_0 c^2 \sum_m \dot{Q}_m(t) \delta(\bar{x} - \bar{x}_m) \qquad (10.3.12)$$

The procedure mirrors that which led to Eq. (9.2.52). The preceding is multiplied by a specific Ψ_j and integrated over $\mathcal{V}$. Because $\Psi_j \nabla^2 \Psi_n \equiv \nabla \cdot (\Psi_j \nabla \Psi_n) - \nabla \Psi_j \cdot \nabla \Psi_n$, the divergence theorem gives

$$\iiint_{\mathcal{V}} \Psi_j \nabla^2 \Psi_{nn} d\mathcal{V} = \iint_{\mathcal{S}} \Psi_j \bar{n} \cdot \nabla \Psi_n d\mathcal{S} - \iiint_{\mathcal{V}} \nabla \Psi_j \cdot \nabla \Psi_n d\mathcal{V} \qquad (10.3.13)$$

The mode functions satisfy rigid or pressure-release boundary conditions, as prescribed in Eq. (10.3.1), so the surface integral vanishes. Hence, we have found that

$$\sum_{n=1}^{\infty} c^2 p_n \iiint_{\mathcal{V}} \nabla \Psi_j \cdot \nabla \Psi_n dV + \sum_{n=1}^{\infty} \ddot{p}_n \iiint_{\mathcal{V}} \Psi_j \Psi_n dV$$

$$= \iiint_{\mathcal{V}} \Psi_j \left[\rho_0 c^2 \sum_m \dot{Q}_m (t) \, \delta (\bar{x} - \bar{x}_m) \right] dV \qquad (10.3.14)$$

The last step is to invoke both orthogonality properties, Eqs. (10.3.8) and (10.3.9), which serve to filter out of both summations all terms for which $n \neq j$. The remaining terms constitute an ordinary differential equation for the jth modal coefficient,

$$\boxed{\ddot{p}_j + \left(\omega_j\right)^2 p_j = S_j (t), \quad j = 1, 2, ..} \qquad (10.3.15)$$

The S_j excitations are time-dependent *modal source strengths*. They map the spatial distribution of sources into the modal space, according to

$$\boxed{S_j = \frac{\rho_0 c^2}{\mathcal{V}} \iiint_{\mathcal{V}} \Psi_j \sum_m \dot{Q}_m (t) \, \delta (\bar{x} - \bar{x}_m) \, dV = \frac{\rho_0 c^2}{\mathcal{V}} \sum_m \dot{Q}_m (t) \, \Psi_j (\bar{x}_m)}$$

$$(10.3.16)$$

Equation (10.3.15) has the same form as the differential equation of motion for a simple spring-mass system. Its solution is quite accessible. If there is a single source whose time signature is a unit impulse function, that is, $\dot{Q}_1 = \delta (t - t_0)$, then the modal coefficients are the impulse response,

$$p_j = \frac{\rho_0 c^2}{\mathcal{V}} \frac{\Psi_j (\bar{x}_1)}{\omega_j} \sin \left(\omega_j (t - t_0)\right) h (t - t_0) \qquad (10.3.17)$$

By definition, the time-domain Green's function is the pressure due to an impulsive point source. Thus, we have established that for a cavity it is

$$G (\bar{x}, t) = \frac{\rho_0 c^2}{\mathcal{V}} \sum_{j=1}^{\infty} \frac{\Psi_j (\bar{x}_1) \Psi_j (\bar{x})}{\omega_j} \sin \left(\omega_j (t - t_0)\right) h (t - t_0) \qquad (10.3.18)$$

Depending on the nature of the sources' time signatures, the modal differential equations might be solvable by standard methods. If not, they can be solved numerically. One case where an analytical solution is possible is that of steady-state field generated by time harmonic sources. Such an analysis yields the frequency-domain solution. The sources are set to $\dot{Q}_m (x, t) = \delta (\bar{x} - \bar{x}_m) \, \text{Re} \left[i\omega \hat{Q}_m \exp(i\omega t)\right]$. The frequency-domain representation of the pressure coefficients is $p_j = \text{Re} \left(X_j e^{i\omega t}\right)$. The modal source strengths in Eq. (10.3.16) become

$$S_n = \sum_m \Psi_n\left(\bar{x}_m\right) \mathrm{Re}\left(\frac{i\omega\rho_0 c^2 \hat{Q}_m}{\mathcal{V}} e^{i\omega t}\right) \qquad (10.3.19)$$

Synthesis of the modal series, Eq. (10.3.10), using the particular solution for the P_j coefficients yields

$$p\left(\bar{x}, t\right) = \rho_0 c^2 \sum_{n=1}^{\infty} \sum_m \frac{\Psi_n\left(\bar{x}_m\right)\Psi_n\left(\bar{x}\right)}{\omega_n^2 - \omega^2} \mathrm{Re}\left(\frac{i\omega\hat{Q}_m}{\mathcal{V}} e^{i\omega t}\right) \qquad (10.3.20)$$

In the special case of a single source whose strength is a unit value, $\hat{Q}_1 = 1$, the preceding gives the frequency-domain Green's function of the cavity. The similarity of this result and Eq. (10.1.16), which was derived heuristically, is obvious.

The general development is applicable to the cavity modes of any configuration, regardless of the shape of the enclosed region and the nature of the walls. However, Eq. (10.3.20) presumes that the cavity modes have been determined. The reality is that only rectangular and circular shapes may be analyzed without considerable effort.

10.3.4   Rectangular Cavity

To study the field within a closed box, we place the origin of xyz at one corner, with the axes aligned parallel to the edges. Then, the sides of the cavity coincide with the planes $x = 0$, $x = L$, $y = 0$, $y = H$, $z = 0$, and $z = W$. The pressure in the region between two parallel walls was the subject of Sect. 9.2. It was shown there that any combination of rigid and pressure-release surfaces is represented by either a sine or cosine function of the coordinate that is normal to the surfaces. That is the present situation, but there are now three sets of parallel walls. Consequently, there are three sets of harmonic functions: $\Psi_x(x)$ for the walls at $x = 0$ and $x = L$, $\Psi_y(y)$ for the walls at $y = 0$ and $y = H$, and $\Psi_z(z)$ for the walls at $z = 0$ and $z = W$. The wavenumbers are denoted as κ, μ, and ν in the x, y, and z directions, respectively. A product of these harmonic functions will satisfy the Helmholtz equation if the wavenumbers are set properly. Furthermore, because the differential equation and boundary conditions are homogeneous, any solution may be multiplied by a constant B. It follows from these considerations that a mode function is described by

$$\Psi\left(x, y, z\right) = B\Psi_x\left(x\right)\Psi_y\left(y\right)\Psi_z\left(z\right) \qquad (10.3.21a)$$

where the factors are

$$\Psi_x = c_x \cos\left(\kappa x\right) + s_x \sin\left(\kappa x\right)$$
$$\Psi_y = c_y \cos\left(\mu y\right) + s_y \sin\left(\mu y\right)$$
$$\Psi_z = c_z \cos\left(\nu z\right) + s_z \sin\left(\nu z\right) \qquad (10.3.21b)$$

Either a c or s coefficient will be zero, depending on the nature of the side that contains the origin. The nonzero coefficient may be set to one because B serves to collect all arbitrary factors. For example, if the surface $y = 0$ is rigid, then it must be that $\partial P/\partial y = 0$ at $y = 0$, which is met by setting $s_y = 0$ and $c_y = 1$. Alternatively, if $y = 0$ is a pressure-release side, so that $P = 0$ at $y = 0$, then $c_y = 0$ and $s_y = 1$.

The wavenumber in each direction is found by satisfying the boundary conditions at the far walls, $x = L$, $y = H$, and $z = W$. In particular, the condition that there be a nontrivial solution for each direction will require that the argument of the respective sinusoidal function be an integer multiple of $\pi/2$. For example, suppose $P = 0$ on $z = 0$, so that $\Psi_z = \sin(\nu z)$. Then, the rigid condition, $\partial P/\partial z = 0$ on $z = W$, requires that $\cos(\nu W) = 0$, which is satisfied if νW is an odd multiple of $\pi/2$. The other possibility, $P = 0$ at $z = W$, requires $\sin(\nu W) = 0$. This is met if νW is an even multiple of π, but ν cannot be zero. This is so because it would lead to the trivial solution $\Psi = 0$.

Determination of the possible values of the wavenumbers allows us to evaluate the natural frequencies. By definition, Eq. (10.3.21) is a solution of the Helmholtz equation at a natural frequency. Let j, m, n be the sequence number for k, μ, and ν, respectively, so that the corresponding natural frequency is denoted as $\omega_{j,m,m} = ck_{j,m,n}$. Then, it must be that

$$\boxed{\omega_{j,m,m} = c\left(\kappa_j^2 + \mu_m^2 + \nu_n^2\right)^{1/2}} \tag{10.3.22}$$

This leaves only the scaling coefficient $B_{j,m,n}$ to determine. These values are obtained by normalizing the mode functions to satisfy Eq. (10.3.8) when the two modes are the same. The mode function described in Eq. (10.3.21) is products of the functions of each coordinate, and the boundaries consist of constant values of each coordinate. Hence, Eq. (10.3.8) leads to

$$B_{j,m,n} = \left(\frac{LHW}{\displaystyle\int_0^W \Psi_x(x)^2\,dx \int_0^H \Psi_y(y)^2\,dy \int_0^L \Psi_z(z)^2\,dz}\right)^{1/2} \tag{10.3.23}$$

A closed form expression for the coefficient may be found, but it is easier to do so after the functions are specified.

EXAMPLE 10.7 A column of water is used to test vertical propagation in a channel. The height is $4\,\mathrm{m}$, and the dimensions of the horizontal cross section are $0.6\,\mathrm{m} \times 1.2\,\mathrm{m}$. The bottom and sides of the tank are rigid. Determine the natural frequencies of this tank up to $4\,\mathrm{kHz}$.

Significance

The evaluation of the frequencies is straightforward. What makes this example interesting is what the results tell us about the general properties of natural frequencies.

Solution

We shall align the y-axis vertically and set the x-axis horizontally along the short side. This corresponds to $L = 0.6\,\text{m}$, $H = 4\,\text{m}$, and $W = 1.2\,\text{m}$. The surface $y = H$ is the interface between the water and air, so it is well approximated as pressure-release for waves in the water. The vertical sides and bottom are rigid, so it is required that the normal derivatives on them be zero. Satisfaction of this condition at $x = 0$, $y = 0$, and $z = 0$ requires that

$$\Psi_x = \cos(\kappa x), \quad \Psi_y = \cos(\mu y), \quad \Psi_z = \cos(\nu z) \tag{1}$$

The other vertical sides are $x = L$ and $z = W$. Setting the normal derivative on each to zero yields

$$\left.\frac{\partial \Psi_x}{\partial x}\right|_{x=L} = \sin(\kappa L) = 0 \text{ and } \left.\frac{\partial \Psi_z}{\partial z}\right|_{z=W} = \sin(\nu W) = 0$$

These conditions are met if

$$\kappa_j = \frac{j\pi}{L}, \quad j = 0, 1, \ldots; \quad \nu_n = \frac{n\pi}{W}, \quad n = 0, 1, \ldots \tag{2a}$$

Note that $\kappa_0 = 0$ and $\nu_0 = 0$. Both correspond to the plane wave modes of two-dimensional waveguide whose walls are rigid. The horizontal surface at $y = H$ is pressure-release, so setting $\Psi_y = 0$ there leads to

$$\mu_m = \frac{(2m - 1)\pi}{2H}, \quad m = 1, 2, \ldots \tag{2b}$$

The natural frequency corresponding to these wavenumbers is

$$\omega_{j,m,n} = \pi c \left[\left(\frac{j}{L}\right)^2 + \left(\frac{2m - 1}{2H}\right)^2 + \left(\frac{n}{W}\right)^2 \right]^{1/2} \tag{3}$$

The mode functions, which are not needed in the present context, are

$$\Psi_{j,m,n} = B_{j,m,n} \cos\left(\frac{j\pi x}{L}\right) \cos\left(\frac{(2m - 1)\pi y}{2H}\right) \cos\left(\frac{n\pi z}{W}\right) \tag{4}$$

It is specified that we should limit our attention to the range up to $4\,\text{kHz}$. To determine the range of indices to consider, we observe that the natural frequency increases monotonically as either j, m, or n is increased with the other two held fixed.

For a fixed frequency, the largest value of one index corresponds to the smallest value of the other two. The range of indices for the computation is $0 \le j \le J, 1 \le m \le M$, $0 \le n \le N$. Thus, we set the range such that J, M, and N are the smallest values for which $\omega_{J,1,0} > \omega_{max}$, $\omega_{0,M,0} > \omega_{max}$, and $\omega_{0,1,N} > \omega_{max}$. Setting $\omega_{max} = 8000\pi$ rad/s leads to $J = 4$, $M = 23$, and $N = 7$.

A four column list of j, m, n, and $\omega_{j,m,n}$ for $j = 0, .., 3$, $m = 1, .., 22$, and $n = 0, .., 6$ could be sorted by saving them to a spreadsheet. A graphical display is a better way to see the qualitative behavior. Each curve in Fig. 1 describes $\omega_{j,m,n}$ for a fixed j and n as m is increased from one. The lowest frequency, which corresponds to $j = 0$, $m = 1$, and $n = 0$, is 92.5 Hz. This is the cavity's *fundamental natural frequency*.

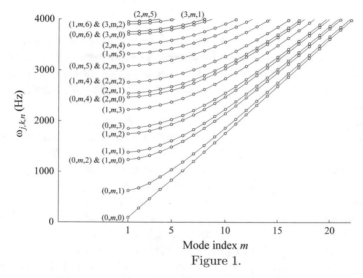

Figure 1.

A few combinations of j and n correspond to overlapping curves, which means the natural frequencies are equal regardless of m. The lowest curve for which this condition occurs is $(0, m, 2)$ and $(1, m, 0)$. Other combinations are $(0, m, 4)$ and $(2, m, 0)$, $(1, m, 4,)$, $(2, m, 2,)$, and so on. This overlap is a consequence of the fact that $W = 2L$. Thus, if a natural frequency corresponds to indices j_1, m, and n_1, the same value will occur for $j_2 = n_1/2$ and $n_2 = 2j_1$, with the same m. This is not to suggest that natural frequencies overlap only if $j_2 n_2 = j_1 n_1$. Other combinations are possible. This is the case for the $(0, m, 5)$ and $(2, m, 3)$ modes. Furthermore, even if a frequency curve does do not overlap others, it is possible that a value on that curve will equal a value on a different curve at a different m. An interesting manifestation of that possibility is a triple value, $\omega_{1,8,3}/(2\pi) = \omega_{1,13,0}/(2\pi) = \omega_{0,13,2}/(2\pi) = 2621$ Hz.

In general, different modes will have the same natural frequency in the low-frequency range if a ratio of cavity dimensions is a small rational fraction. For example, in the present system the ratio of lengths in the x and z directions is $W/L = 2$, so the occurrence of equal frequencies with different (j, n) pairs is rather common. In contrast, $H/L = 20/3$, so equal natural frequencies with different (j, m)

are less common. The same is true for natural frequencies associated with different (m, n) pairs because $H/W = 10/3$.

It was stated in the general discussion that two cavity modes will be orthogonal even if they have the same natural frequency. Let us test that assertion. The orthogonality condition is the volume integral of the product of two functions. In view of the fact that the modes in Eq. (4) are products of x, y, and z functions, the orthogonality integral factorizes to integrals over the range of x, y and z. Each has the same general form,

$$\int_0^\ell \cos{(\beta_r \eta)} \cos{(\beta_s \eta)}\, d\eta \equiv \frac{1}{2} \left[\frac{\sin{((\beta_r - \beta_s)\,\ell)}}{\beta_r - \beta_s} + \frac{\sin{((\beta_r + \beta_s)\,\ell)}}{\beta_r + \beta_s} \right]$$

For the x functions, $\beta_r \ell$ represents $\kappa_j L = j\pi$, while $\beta_r \ell$ represents $\nu_n W = n\pi$ for the z functions. In both cases, the argument of the above sine functions is a multiple of π, and therefore zero. For the y functions, $\beta_r \ell$ represents $\mu_m H = (m - 1/2)\,\pi$, so that $(\beta_r \pm \beta_s)\ell$ also is a multiple of π. It follows that the orthogonality integral vanishes for any combination of two different mode functions, regardless of what their natural frequencies are.

The fact that two or more modes share a natural frequency does have some interesting implications. Suppose a field is established at the natural frequency belonging to two modes. A linear combination of these modes also satisfies the governing equations. For example, in the case of a pair identified in the preceding paragraph, $\Psi = b_1 \Psi_{0,1,2} + b_2 \Psi_{1,1,0}$ satisfies the Helmholtz at $k = \omega_{0,1,2}/c = \omega_{1,1,0}/c$ for any constants b_1 and b_2. We could consider the combined function to be a mode, but shall not do so because it is not independent of the ones we derived. The values of b_1 and b_2 that occur depend on how the field is induced. This phenomenon was first encountered in the vibration of rectangular membranes and plates. Chaladni[6] placed sand on a vibrating square plate. This made it possible to visualize the vibration pattern because the sand migrated toward the nodal lines, at which the plate is quiet. The excitation consisted of a violin bow that was stroked across an edge. Chaladni showed that nodal patterns were not unique. He also showed that they were very difficult to reproduce. This is a consequence of the fact that the factors b_1 and b_2 for the vibration were highly sensitive to small changes in the location where the plate was stroked. In recognition of his contributions, nodal lines are commonly known as *Chaladni lines*.

Even if natural frequencies are not exactly equal, the likelihood that two or more natural frequencies are very close increases drastically as the frequency increases. This attribute may be recognized by drawing a horizontal line through any natural frequency in the graph. Above $\omega_{0,1,1}$, this line will intersect two or more curves. As the frequency increases, the number of curves such as line intersects will increase. Some of those intersections will be close to integer values of m, which mark the natural frequencies. We say that the *modal density* increases with increasing natural

[6]E. Chaladni, *Entdeckungen über die Theorie des Klanges ("Discoveries in the Theory of Sound")* 1787.

frequency. This phenomenon is a general attribute of two- and three-dimensional systems that conserve energy. It can be quite problematic for experimental methods that attempt to identify modal properties from measured data.

EXAMPLE 10.8 Consider a hard-walled chamber whose dimensions are $4 \times 3 \times 6$ m. A line source is situated along one of the 6 m edges. The source is impulsive, with a volume velocity per unit length that changes discontinuously at $t = 0$ from zero to a constant value q_0. This corresponds to a mass acceleration per unit length that is $\rho_0 q_0 \delta(t)$. Determine the signal heard in the middle of the chamber. The fluid is air.

Significance

The solution is a direct application of the modal series formalism. The results will lead to some useful insights regarding our prior study of point sources, and it will highlight some of the difficulties one might encounter in the application of modal series representations.

Solution

The xyz coordinate system in Fig. 1 has the z-axis coincident with the line source and the x-axis coincident with a 4 m edge. The line source has the same properties along its length, and the hard-walls at $z = 0$ and $z = 6$ m allow for the existence of a pressure field that does not vary in that direction. Thus, the pressure field is $p = \text{Re}(P(x, y) \exp(i\omega t))$.

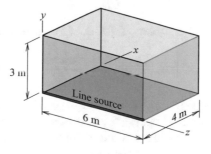

Figure 1.

Despite independence of the z coordinate, it is best to treat the field as three-dimensional, rather than considering it to be two-dimensional. The boundaries in the x and y directions also are rigid, so the modes that contribute to the pressure field are

$$\Psi_{j,n,0} = B_{j,n,0} \cos\left(\frac{j\pi x}{L}\right) \cos\left(\frac{n\pi y}{H}\right), \quad j, n = 0, 1, \ldots \quad (1)$$

where $L = 4\,\text{m}$ and $H = 3\,\text{m}$. The corresponding natural frequencies are

$$\omega_{j,n,0} = c \left[\left(\frac{j\pi}{L} \right)^2 + \left(\frac{n\pi}{H} \right)^2 \right]^{1/2} \tag{2}$$

Normalizing the mode functions to the volume $V = LHW$, in accord with Eq. (10.3.8), leads to

$$B_{j,n,0} = \frac{2}{\left(1 + \delta_{0,j} \right)^{1/2} \left(1 + \delta_{0,n} \right)^{1/2}} \tag{3}$$

Because the source is situated in the corner, its mass flows into one quadrant, rather than into free space. To account for this effect, we multiply the source strength by a factor of four. Another consideration is that Eq. (10.3.16) describes the modal source strength S_j for a discrete set of sources. The line source is a continuous distribution of differential mass accelerations $4\rho_0 q_0 \delta(t)\, dz$, so a summation of sources is replaced by an integral over z. Hence, the first form of Eq. (10.3.16) in this case gives

$$S_{j,n,0} = \frac{\rho_0 c^2}{LHW} \int_0^L \int_0^H \int_0^W \left\{ \int_0^W 4\rho_0 q_0 \delta(t)\, d\xi \right\} \delta(x)\, \delta(y)\, \Psi_{j,n,0}\, dz\,dy\,dx$$

Equation (1) indicates that all modes at the origin are $\Psi_{j,n,0} = B_{j,n,0}$. Thus, the modal source strengths are

$$S_{j,n,0} = \frac{4\rho_0 c^2 q_0}{LH} B_{j,n,0} \delta(t) \tag{4}$$

The modal differential equations, Eq. (10.3.15), corresponding to these coefficients are

$$\ddot{p}_{j,n,0} + \left(\omega_{j,n,0} \right)^2 p_{j,n,0} = \frac{4\rho_0 c^2 q_0}{LH} B_{j,n,0} \delta(t) \tag{5}$$

Quiescent initial conditions are $p_{j,n,0} = 0$ and $\dot{p}_{j,n,0} = 0$ for any t preceding activation of the impulse. After the impulse, the excitation is gone, so only the homogeneous solution remains. The initial conditions for the ensuing response are obtained by considering the infinitesimal interval from $t = 0^-$ before the impulse to $t = 0^+$ afterward. The pressure cannot change suddenly in that interval, because doing so would lead to an infinite particle acceleration. Thus, it must be that $p_{j,n}\left(t = 0^+ \right) = 0$. To determine $\dot{p}_{j,n}\left(t = 0^+ \right)$, we integrate Eq. (5) over $0^- < t < 0^+$. The integral of $p_{n,n}$ over this infinitesimal interval is zero, and the integral of $\ddot{p}_{j,n}$ is $\dot{p}_{j,n}\left(t = 0^+ \right) - \dot{p}_{j,n}\left(t = 0^- \right)$. Thus, we find that

$$\dot{p}_{j,n,0}\left(t = 0^+ \right) = \frac{4\rho_0 c^2 q_0}{LH} B_{j,n,0}$$

The solution of Eq. (5) fitting the initial conditions is the one-degree-of-freedom impulse response,

$$P_{j,n,0} = \frac{4\rho_0 c^2 q_0}{LH} B_{j,n,0} g_{j,n,0}(t)$$

$$g_{j,n,0} = \frac{\sin(\omega_{j,n,0} t)}{\omega_{j,n,0}} h(t) \quad \text{if } j > 0 \text{ or } n > 0 \tag{6}$$

$$g_{0,0,0} = t \, h(t)$$

where the step function serves to remind us that the response is zero for $t < 0$. The special form of the impulse response for the $(0, 0, 0)$ mode is a consequence of the associated natural frequency being zero. The corresponding representation of the pressure at an arbitrary field point is found from Eq. (10.3.10) to be

$$p(\bar{x}, t) = \frac{4\rho_0 c^2 q_0}{LH} \sum_{j=0}^{J} \sum_{n=0}^{N} (B_{j,n,0})^2 \cos\left(\frac{j\pi x}{L}\right) \cos\left(\frac{n\pi y}{H}\right) g_{j,n,0}(t) \tag{7}$$

The dimensionality of q_0 is (volume/time)/length $=$ length^2/time. Therefore, dividing p by $\rho_0 c q_0 / L$ is a convenient way of eliminating dependence on the unspecified value of q_0.

An important aspect is truncation of the series. The impulse response function $g_{j,n,0}(t)$ is the only factor in Eq. (7) whose magnitude varies significantly from term to term. Its magnitude is proportional to $1/\omega_{j,n}$, so it decreases monotonically as j or n increases. Thus, truncation will be based in assuring that any frequencies that are omitted exceed some maximum value. Let us describe this frequency limit as $\sigma \min(\omega_{\text{nat}})$, where σ is a factor to be determined. Because $L > H$, the fundamental nonzero frequency is $\omega_{1,0,0} = \pi c / L$. (The zero frequency is not relevant.) Thus, the task is to identify values of J and N such that $\omega_{J,n,0}$, for any n, and $\omega_{j,N,0}$, for any j, exceed $\sigma(\pi c / L)$. The largest J corresponds to $n = 0$, so we want $\omega_{J,0,0} \geq \sigma(\pi c / L)$, that is $J \geq \text{ceil}(\sigma)$. A similar analysis sets $N \geq \text{ceil}(\sigma H / L)$.

A matrix algorithm may be implemented to evaluate Eq. (7) at specified locations over any time interval. The values of J and N will ultimately be found to be quite large, which means that there are many modes to compute. There are several ways in which the algorithm in Example 10.2 may be adapted to the present formulation. The following was found to be quite efficient relative to others that were tried, yet able to run on a laptop whose memory workspace is limited. Variants that work on more powerful computers might be more rapid, but their workspace requirements could be excessive. The first step creates a row vector of mode numbers $[m] = [1 \ 2 \ 3 \ ...M]$, where $M = (J + 1)(N + 1)$. The pair of wavenumbers in the x and y direction are placed in row vectors $[j]$ and $[n]$, whose length is the same length as $[m]$. Operations to create these quantities are $[j] = \text{floor}(([m] - 1)/(N + 1))$ followed by $[n] = [m] - 1 - [j] * (N + 1)$. These row vectors are used to vectorize computation of the natural frequencies in Eq. (2) and normalizing coefficients in Eq. (3). The MATLAB code for these operations is   `omega=c*sqrt((j*pi/L).^2+(n*pi/H).^2); B=2./sqrt((1+`

(j==0)).*(1+(n==0)));. The result is row vectors [ω] and [B] whose
size matches [m]. To evaluate the modal values, let $\bar{x}_f$, $f = 1, \dots F$, denote the
field points at which we wish to evaluate the pressure, and let t_ℓ be a sequence
of instants at which the pressure is to computed. The values of all mode functions at fixed x are stored in row f of a rectangular array [χ]. Pressure values at several locations may be computed by incrementing f, thereby stacking
rows of [χ]. In MATLAB, the vectorized implementation of this operation defines
two vectors x_f and y_f that hold the coordinates of the set of $\bar{x}_f$ locations.
Then, Chi(f,:)=B.^ 2.*cos(j*pi*x_f/L).*cos(n*pi*y_f/H);.
The pressure in Eq. (7) at the various locations at a specific instant t_ℓ may be computed as the product of [χ] and a column vector of $g_m \equiv g_{j,m,0}$ values, that is,

$$\{p(\{\bar{x}\}, t_s)\} = \frac{\rho_0 c q_0}{L} \frac{4c}{H} [\chi]\{g(t_s)\} \tag{8}$$

A time history may be constructed by adjoining column vectors of {p} values at each
instant, that is,

$$[\{p(\{\bar{x}\}, t_1)\}\{p(\{\bar{x}\}, t_2)\} \cdots] = \frac{\rho_0 c q_0}{L} \frac{4c}{H} [\chi][\{g(t_1)\}\ \{g(t_2)\} \cdots]$$

The highest frequency sets the time step according to the Nyquist criterion. It states
that the sampling interval should be less than half the period of the highest frequency
contained in a signal. The modal series has been truncated J and N. This sets the
highest natural frequency at $\omega_{J,N,0}$. The Nyquist criterion is $\Delta t \leq 0.5 (2\pi/\omega_{J,N,0})$.
The system parameters, $c = 340$ m/s, $L = 4$ m, $H = 3$ m, give $\omega_{J,M,0} = 377 (10^3)$
rad/s, which leads to $\Delta t \leq 8.32$ μs.

For the sake of completeness, the signals received at three points along a diagonal
were computed: $(x, y) = (L/4, 3H/4)$, $(L/2, H/2)$, and $(3L/4, H/4)$. The factor
for modal truncation was set at $\sigma = 100$, which leads to $J = 100, N = 75$. The result
appears in Fig. 2.

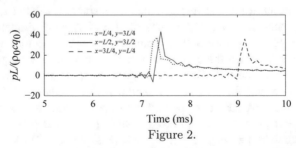

Figure 2.

The plots have a jagged appearance. Calculations with Δt set at one-tenth of
the Nyquist value primarily smoothed out the fluctuations, whereas increasing the
value of σ led to curves whose shape changed from Fig. 2. In other words, a modal
truncation based on $\sigma = 100$ is not adequate. All computations were repeated with
$\sigma = 1000$. This leads to $J = 1000, N = 750$, which corresponds to 751751 modes.

The sampling interval under the Nyquist sampling guideline is $\Delta t = 0.832\,\mu s$. Raising the number of modes and number of instants by a factor of ten relative to Fig. 2 leads to a factor of 100 increase in the computational time. Nevertheless, the effort is still manageable, with an elapsed time of 1.5 min on a laptop computer with an Intel I7 CPU.

The result of the refined computation is the data plotted in Fig. 3. Each plot is much smoother. Some high-frequency wiggles are present. For reasons elucidated below, it is not possible to eliminate them. At each location described in Fig. 3, the pressure shows a sharp peak, followed by a rapid decay. The shape of each signal after this arrival time is reminiscent of the solution to Example 6.5, which explored the field radiated by an impulsive line source in free space. Indeed, that solution can be used in conjunction with the method of images to construct the response of the present system.

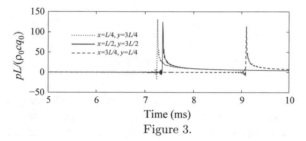

Figure 3.

According to the method of images the walls at constants x and y serve as mirrors for the line source at $x = y = 0$. Figure 4 shows that an infinite number of image sources is required to satisfy the condition that $\bar{n} \cdot \nabla p = 0$ at all walls. A subtle aspect of the image construction is that the walls at $z = 0$ and $z = W$ also serve as mirrors. This leads to a continuous line extending to infinity in the positive and negative z direction. In other words, each image is an impulsive infinite line source, each of whose mass acceleration per unit length is $4\rho_0 q_0 \delta (t)$. Confirmation of the correctness of this model is the fact that the time at which each peak occurs in Fig. 3 is the distance from a field point to one of the images in Fig. 4, divided by c. This is the time required for a signal leaving the image line source at $t = 0$ to arrive at the field point.

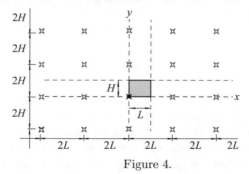

Figure 4.

A corollary is that the waveform received at any point within the cavity begins with arrival of the signal that travels directly from the source, followed by the arrival of pulses resulting from one or more reflections of the source signal. As the distance along the reflected path increases, the arriving signal is weakened by cylindrical spreading, as explained in Example 6.5. This conceptual picture explains the waveforms in Fig. 5, which were obtained by halting computation of Eq. (8) at $t_{max} = 3L/c$. Multiple pulses are seen at each field point. The arrival time for some is sufficiently close that they overlap.

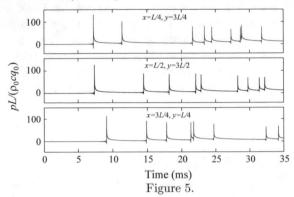

Figure 5.

The availability of an alternate solution makes it possible to recognize some faults with the cavity mode series solution. An exceptionally large number of modes must be included in the series in order to obtain a high-quality waveform. In contrast, the method of images merely requires inclusion of only those images whose signal arrives at the selected field point within the designated time interval. The field of an impulsive line source is singular at the first arrival time. The modal solution cannot replicate this singularity. Furthermore, the modal solution shows high-frequency wiggles prior to the arrival of the signal from each image. This is an impossibility, because early arrival corresponds to a signal that travels supersonically from an image to the field point. Both artifices stem from the fact that the impulsive line source excites all modes equally, as is evident in Eq. (4) for the $S_{j,n}$ values. A very large number of modes is required to reproduce the discontinuity. Increasing the number of modes as we did decreases the scale of the precursor signal in duration and amplitude. However, the precursor cannot be eliminated, and the singularity cannot be replicated, with a finite length modal series.

10.3.5   Cylindrical Cavity

The analysis of cylindrical cavity modes proceeds much like the development for rectangular cavities. We align the x-axis with the centerline, with the origin placed at an end. The mode functions we seek are solutions of the homogeneous Helmholtz equation subject to rigid or pressure-release conditions at the ends $x = 0$ and $x = L$, and on the interior wall of the cylinder, $R = a$. In addition, the mode function must be finite at $R = 0$. This excludes the Neumann function as a possible description of the transverse dependence, so that dependence will be a Bessel function. (If there is a concentric inner cylinder, the finiteness condition is replaced by a boundary condition on the inner cylinder. In that case, a mode function is formed from both the Bessel and Neumann functions.) We designate the transverse wavenumber as μ. The behavior in the axial direction x is like that between parallel walls in a rectangular waveguide. If the field is not axisymmetric, then the dependence on the azimuthal angle θ must be harmonic. Hence, any cavity mode may be described as

$$\Psi = B J_n (\mu R) F (n\theta) [c_x \cos (\kappa x) + s_x \sin (\kappa x)] \qquad (10.3.24)$$

where

$$F (n\theta) = \cos (n\theta), \ \sin (n\theta), \ \text{or} \ \exp (\pm i n\theta) \qquad (10.3.25)$$

Setting $P = 0$ or $\partial P / \partial x = 0$ at $x = 0$ leads to either $c_x = 0$ or $s_x = 0$. Then, satisfying $P = 0$ or $\partial P / \partial x = 0$ at $x = L$ yields an infinite number of κL values. The possible values of μa are determined by satisfying the rigid or pressure-release conditions at $R = a$. This leads to the μa values being either a zero of a Bessel function, or a value at which a Bessel function has an extreme value. The specific conditions are

$$\text{Rigid:} \ \left. \frac{\partial \Psi}{\partial R} \right|_{R=a} = 0 \implies J_n' (\eta_{n,m}) = 0, \ \mu = \frac{\eta_{n,m}}{a}, \ m = 1, 2, \ldots$$

$$\text{Pressure-release:} \ \Psi|_{R=a} = 0 \implies J_n (\eta_{n,m}) = 0, \ \mu = \frac{\eta_{n,m}}{a}, \ m = 1, 2, \ldots$$

$$(10.3.26)$$

Thus, three subscripts are required to denote a cylindrical cavity mode. The first is the order n of the Bessel function, which is the same as the azimuthal harmonic n. The second is the root number m for the radial wavenumber. The third is the root number j for the axial wavenumber κ. Furthermore, as indicated by Eq. (10.3.25), two independent functions describe the dependence on θ for a given triad of n, m, j values. Cavity modes are standing waves, so the sinusoidal functions are more appropriate in the present context. We shall use a subscript $\alpha = c$ or $\alpha = s$ to indicate which sinusoidal function is being considered, with this subscript omitted for the axisymmetric mode, $n = 0$. Correspondingly, the modes $\Psi_{n,m,j,\alpha}$ of a cylindrical cavity are

$$\Psi_{0,m,j} = B_{n,m,0} J_0 \left(\eta_{0,m} \frac{R}{a} \right) [c_x \cos (\kappa_j x) + s_x \sin (\kappa_j x)]$$

$$\Psi_{n,m,j,c} = B_{m,n,j} J_n \left(\eta_{n,m} \frac{R}{a} \right) [c_x \cos (\kappa_j x) + s_x \sin (\kappa_j x)] \cos(n\theta), \ n > 0$$

$$\Psi_{n,m,j,s} = B_{m,n,j} J_n \left(\eta_{n,m} \frac{R}{a} \right) [c_x \cos (\kappa_j x) + s_x \sin (\kappa_j x)] \sin(n\theta), \ n > 0$$

$$(10.3.27)$$

It still remains to determine the natural frequency. This is done by ascertaining the frequency at which a mode function satisfies the Helmholtz equation,

$$\frac{\partial^2 P}{\partial R^2} + \frac{1}{R} \frac{\partial P}{\partial R} + \frac{1}{R^2} \frac{\partial^2 P}{\partial \theta^2} + \frac{\partial^2 P}{\partial z^2} + k^2 P = 0 \qquad (10.3.28)$$

We also have $\partial^2 \Psi_{n,m,j,\alpha}/\partial\theta^2 = -n^2\Psi_{n,m,j,\alpha}$ and $\partial^2\Psi_{n,m,j,\alpha}/\partial x^2 = -\kappa_j^2\Psi_{n,m,j,\alpha}$. In the transverse direction, we know that $J_n\left(\eta_{n,m}R/a\right)$ is a solution of Bessel's equation, which means that

$$\frac{\partial^2\Psi_{n,m,j,\alpha}}{\partial R^2} + \frac{1}{R}\frac{\partial\Psi_{n,m,j,\alpha}}{\partial R} = -\left(\frac{\eta_{n,m}^2}{a^2} - \frac{n^2}{R^2}\right)\Psi_{n,m,j,\alpha} \tag{10.3.29}$$

It follows that $\Psi_{n,m,j}$ is a solution of the Helmholtz equation if

$$\omega_{n,m,j,\alpha} \equiv ck_{j,m,n} = c\left(\kappa_j^2 + \frac{\eta_{n,m}^2}{a^2}\right)^{1/2} \tag{10.3.30}$$

The azimuthal harmonic number n does not appear explicitly in this expression. Nevertheless, inclusion of n in the list of subscripts for the natural frequency is warranted by the fact that the transverse wavenumbers $\eta_{n,m}$ depend on which Bessel function order is under consideration.

An important aspect of the nonaxisymmetric $(n > 0)$ modes is that they occur as a pair $\alpha = c$ and $\alpha = s$ with equal natural frequencies. Despite the equality of natural frequencies, each function is orthogonal to all others. The specific forms of the general modal orthogonality properties and normalization definition, Eqs. (10.3.8) and (10.3.9), for a cylindrical cavity are

$$\int_0^L \int_{-\pi}^{\pi} \int_0^a \Psi_{n,m,j,\alpha}\Psi_{r,s,\ell,\beta} Rd\,Rd\theta dx = \delta_{n,r}\delta_{m,s}\delta_{j,\ell}\delta_{\alpha,\beta}\left(\pi a^2 L\right)$$

$$\int_0^L \int_{-\pi}^{\pi} \int_0^a \left(\frac{\partial\Psi_{n,m,j,\alpha}}{\partial x}\frac{\partial\Psi_{r,s,i,\beta}}{\partial x} + \frac{\partial\Psi_{n,m,j,\alpha}}{\partial R}\frac{\partial\Psi_{r,s,i,\beta}}{\partial R}\right.$$
$$\left. + \frac{1}{R^2}\frac{\partial\Psi_{n,m,j,\alpha}}{\partial\theta}\frac{\partial\Psi_{r,s,i,\beta}}{\partial\theta}\right) Rd\,Rd\theta dz = \delta_{n,r}\delta_{m,s}\delta_{j,\ell}\delta_{\alpha,\beta}\left(\omega_{j,m,n,\alpha}\right)^2\left(\pi a^2 L\right)$$
$$\tag{10.3.31}$$

There is no need to prove these properties because they are covered by the general proof.

The case where all mode numbers match, that is, $n = r$, $m = s$, $j = \ell$, and $\alpha = \beta$, is used to set the normalization coefficients $B_{n,m,j}$. Note that the same $B_{n,m,j}$ value applies to the cosine and sine modes because one is merely rotated about the x-axis relative to the other. An actual determination of the $B_{n,m,j}$ values will entail evaluation of the integral of a product of Bessel functions. Because $\eta_{n,m}$ as an eigenvalue, a formula[7] is available for this integral. The result depends on whether the surface is pressure-release or rigid, specifically

---

[7]M.I. Abramowitz and I.A. Stegun, *Handbook of Mathematical Functions*, Dover, p. 485, Eq. 11.4.5 (1965).

$$\int_0^a J_n\left(\eta_{n,m}\frac{R}{a}\right)^2 R\,dR = \begin{cases} \dfrac{a^2}{2} J'_n\left(\eta_{n,m}\right)^2 & \text{if } J_n\left(\eta_{n,m}\right) = 0 \\[2mm] \dfrac{a^2}{2}\left[1 - \left(\dfrac{n}{\eta_{n,m}}\right)^2\right] J_n\left(\eta_{n,m}\right)^2 & \text{if } J'_n\left(\eta_{n,m}\right) = 0 \text{ and } \eta_{n,m} \neq 0 \end{cases}$$

$$(10.3.32)$$

Upon determination of the mode functions and natural frequencies, an analysis of response in the time domain or frequency domain may proceed by following the steps in the general development provided in Sect. 10.3.1.

EXAMPLE 10.9 A box and a cylinder contain air at standard conditions. The cross section of the box is square, with a side length of w, and the radius of the cylinder is a. Both have equal length L in the axial direction, which is designated as x. The ends at $x = 0$ and $x = L$ are open to the atmosphere and each cross section is sufficiently large to set $P = 0$ at both ends. The other sides are rigid. The parameter that is of interest is the smallest natural frequency corresponding to a mode that is not planar transverse to the x axis. In particular, it is desired that this frequency be a specified value Ω for both configurations. Derive expressions for a and w for which this condition occurs.

Significance

This example will provide quantitative insight into the manner in which dimensions and shape affect natural frequencies.

Solution

Let us begin with a circular cross section. The ends are pressure-release, and the same frequency results from using $\cos(n\theta)$ and $\sin(n\theta)$, so a mode function is described by

$$\Psi_{n,m,j,c} = B J_n\left(\mu_{n,m} R\right) \cos(n\theta) \sin\left(\frac{j\pi x}{L}\right)$$

where j is the root number for the axial eigenvalue, n is the azimuthal harmonic, and m is the root number for the transverse eigenvalue. The latter comes from satisfying the condition $V_R = 0$ at $R = a$, for which the characteristic equation is

$$J'_n\left(\eta_{n,m}\right) = 0, \quad \mu_{n,m} = \frac{\eta_{n,m}}{a}$$

Equation (10.3.30) tells us that the lowest frequency corresponds to minimum values of $\mu_{n,m}$ and κ_j. To identify the appropriate value of the former, we examine Fig. 1, which is a graph of the Bessel function at various orders. The roots $\eta_{n,m}$ are the argument x for which $J_n(x)$ has a maximum or minimum. The first maximum occurs for $n = 0$, but this root gives $J_0(0) = 1$, which is a planar mode. The smallest nonzero value of x leading to an extreme value of $J_n(x)$ occurs at $x \approx 1.8$ for $n = 1$. A more accurate value may be found by invoking the identity $J'_1(\xi) = J_0(x) - J_1(x)/x$ to implement a numerical solution. Doing so yields $\eta_{1,1} = 1.8412$.

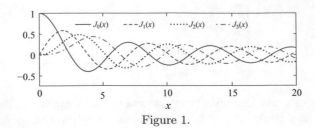

Figure 1.

The axial wavenumber is $\kappa_j = j\pi/L$, with $j > 0$. The minimum occurs for $j = 1$. Thus, the fundamental frequency for nonplanar modes is found from Eq. (10.3.30) to be

$$\min\left(\omega_{j,m,n}\right) = \omega_{1,1,1} = c\left[\left(\frac{\pi}{L}\right)^2 + \left(\frac{\eta_{1,1}}{a}\right)^2\right]^{1/2}$$

The x dependence of the modes of the box is the same as it is for the cylinder. The dependence on y and z must be harmonic, with zero slope at the walls. The corresponding modes are

$$\Psi_{j,m,n} = B\cos\left(\frac{m\pi y}{w}\right)\left(\frac{n\pi z}{w}\right)\sin\left(\frac{j\pi x}{L}\right), \quad m = 0, 1, ..., \quad n = 0, 1, ...$$

The natural frequencies are

$$\omega_{j,m,n} = c\left[\left(\frac{j\pi}{L}\right)^2 + \left(\frac{m\pi}{w}\right)^2 + \left(\frac{n\pi z}{w}\right)^2\right]^{1/2}$$

The lowest natural frequency corresponds to the minimum values of j, n, and m. The minimum value of j is one. Although the smallest values of m and n are zero, both being zero lead to a planar mode. However, setting either m or n to zero and the other to one is acceptable. Thus, we find that

$$\min\left(\omega_{j,m,n}\right) = \omega_{1,1,0} = \omega_{1,0,1} = c\left[\left(\frac{\pi}{L}\right)^2 + \left(\frac{\pi}{w}\right)^2\right]^{1/2}$$

The minimum for each configuration should be Ω, so

$$\left(\frac{\Omega}{c}\right)^2 = \left(\frac{\pi}{L}\right)^2 + \left(\frac{\eta_{1,1}}{a}\right)^2 = \left(\frac{\pi}{L}\right)^2 + \frac{\pi^2}{w^2}$$

This yields

$$a = \frac{1.8412}{\left[\left(\frac{\Omega}{c}\right)^2 - \left(\frac{\pi}{L}\right)^2\right]^{1/2}}, \quad w = \frac{\pi}{\left[\left(\frac{\Omega}{c}\right)^2 - \left(\frac{\pi}{L}\right)^2\right]^{1/2}}$$

Although both dimensions depend on L and Ω/c, their ratio is a constant,

$$\frac{a}{w} = \frac{1.8412}{\pi}$$

The corresponding ratio of the cross-sectional areas is

$$\frac{\mathcal{A}_{cyl}}{\mathcal{A}_{rect}} = \frac{\pi a^2}{w^2} = 1.079\,1$$

The areas are quite close. The associated mode functions also have some attributes in common. Figure 2 shows equal value contours of the modes at $x = L/2$. The thickened contours are nodal lines. The $(1, 1, 1)$ mode with a $\cos\theta$ dependence and the $(1, 1, 0)$ mode of the box are symmetric with respect to the horizontal centerline and antisymmetric with respect to the vertical centerline. Their only nodal line is along the vertical centerline. The patterns are rotated $90°$ for the $(1, 1, 1, \cos\theta)$ mode of a cylinder and the $(1, 0, 1)$ mode of a box.

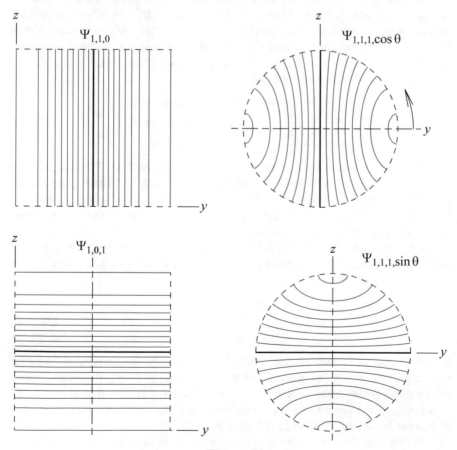

Figure 2.

10.3.6   Spherical Cavity

The objective here is to identify the axisymmetric natural modes of a spherical cavity whose wall is either rigid or pressure-release. In general, the individual terms in a series of forced modes satisfy the Helmholtz equation. Equation (10.2.14) is such a series for an axisymmetric field within a sphere. The individual terms in that series are spherical harmonics. The terms are summed in order to satisfy the boundary condition of an arbitrary excitation on the spherical wall. An analysis of natural cavity modes instead requires that each spherical harmonic individually be consistent with either a zero radial velocity or zero pressure at the boundary. Thus, we begin by considering a natural mode whose form is

$$\Psi = B \, j_m \, (kr) \, P_m \, (\cos \psi) \tag{10.3.33}$$

Although the spherical Neumann function also satisfies the Helmholtz equation, it is excluded by the condition that the pressure at the center must be finite.

As in the previous analyses, the B coefficient will be set by the orthogonality property. The only other parameter not yet set is k, which is the ratio of the natural frequency to the speed of sound. This quantity is found by satisfying the condition at the wall. Rigidity there requires that $\partial \Psi / \partial r = 0$ at $r = a$, whereas a pressure-release condition requires that $\Psi = 0$ at $r = a$. The characteristic equations that result from satisfying these alternative conditions are

$$\left. \begin{array}{l} \text{Rigid: } j'_m \left(\eta_{m,n} \right) = 0 \\ \text{Pressure-release: } j_m \left(\eta_{m,n} \right) = 0 \end{array} \right\}, \quad \eta_{m,n} \equiv k_{m,n} a \equiv \omega_{n,m} \frac{a}{c} \tag{10.3.34}$$

The η quantities, which are the eigenvalues, have two subscripts. The first, n, is the polar harmonic number. Because the spherical Bessel function is oscillatory about a zero value, there are multiple roots at a specified n. The second subscript, m, is the root number. Usually, the roots and therefore the natural frequencies are sequenced in ascending order, so that $\eta_{m,n} < \eta_{m,n+1}$.

Each eigenvalue might lead to a different normalizing coefficient, so those values are denoted as $B_{n,m}$. Thus, we have established that the axisymmetric natural modes of any spherical cavity are

$$\Psi_{n,m} = B_{n,m} \, j_m \left(\eta_{n,m} \frac{r}{a} \right) P_n \, (\cos \psi) \,, \quad \omega_{n,m} = \frac{c}{a} \eta_{n,m} \tag{10.3.35}$$

Despite the common mathematical form of the mode functions, their radial derivative will vanish at $r = a$ for a rigid wall, whereas they will be zero at that location for a pressure-release wall. Because $j_0 \, (0) = 1$, the first root for $m = 0$ in the rigid wall case is $\eta_{0,0} = 0$. Such a mode corresponds to a uniform pressure distribution in the transverse direction. All other eigenvalues are positive.

The first few roots for each case are given in Table 10.1. As is usual for Bessel functions, the higher roots are well approximated by adding multiples of π to these values.

Table 10.1 Nondimensional natural frequencies for the axisymmetric modes of a spherical cavity whose wall is either rigid or pressure-release

Wall condition	Polar harmonic (m)	$k_{m,1}a$	$k_{m,2}a$	$k_{m,3}a$	$k_{m,4}a$
Rigid	0	0	4.4934	7.7253	10.9041
	1	2.0816	5.9404	9.2058	12.4044
	2	3.3421	7.2899	10.6139	13.8461
Pressure-release	0	3.1416	6.2832	9.4248	12.5664
	1	4.4934	7.7253	10.9041	14.0662
	2	5.7635	9.0950	12.3229	15.5146

Example 10.7 showed that the modal density of a rectangular cavity increases as the range of natural frequencies is increased. The spherical Bessel functions are not exactly periodic functions, so there will not be cases where the eigenvalues for different modes are identical. However, they are nearly periodic for large arguments, so modes having different harmonic numbers n might have very close natural frequencies.

The orthogonality conditions for the spherical cavity modes are obtained by specializing the general conditions, Eqs. (10.3.8) and (10.3.9), to an axisymmetric spherical geometry. Thus, we use a differential volume element in the shape of a ring whose radius is $r \sin \psi$, that is, $dV = (2\pi r \sin \psi)(r d\psi) dr$. The result is

$$2\pi \int_0^a \int_0^\pi \Psi_{m,n} \Psi_{r,s} r^2 \sin \psi d\psi dr = \frac{4}{3}\pi a^3 \delta_{m,r} \delta_{n,s}$$

$$2\pi \int_0^a \int_0^\pi \left[\frac{\partial \Psi_{m,n}}{\partial r} \frac{\partial \Psi_{r,s}}{\partial r} + \frac{1}{r^2} \frac{\partial \Psi_{m,n}}{\partial \psi} \frac{\partial \Psi_{r,s}}{\partial \psi} \right] r^2 \sin \psi d\psi dr = \frac{4\pi}{3} a^3 \delta_{m,r} \delta_{n,s} \omega_{m,n}^2$$

$$(10.3.36)$$

As was done for other geometries, the case where the mode numbers match, $m = r$ and $n - s$ are used to set the scaling coefficients $B_{m,n}$. The first of the above equations is used for this purpose. The actual evaluation will entail evaluation of an integral containing products of Bessel functions. We may obtain a formula by introducing into a tabulated integral,[8] the relation of spherical and cylindrical Bessel functions, specifically,

$$j_n\left(\eta_{n,m} \frac{r}{a}\right) = \left(\frac{\pi a}{2\eta_{n,m}}\right)^{1/2} J_{n+1/2}\left(\eta_{n,m} \frac{r}{a}\right) \qquad (10.3.37)$$

This relation is useful for computations, as well as for the normalization integrals. For the latter, it leads to

---

[8]M.I. Abramowitz and I.A. Stegun, *ibid*, p. 485, Eq. 11.4.5.

$$\int_0^a j_n \left(\eta_{n,m}\frac{r}{a}\right)^2 r^2 dr = \begin{cases} \frac{1}{2}a^3 j_{n+1}\left(\eta_{n,m}\right)^2 \text{ if } j_n\left(\eta_{n,m}\right) = 0 \\[2mm] \frac{1}{2}a^3 \left[1 - \left(\frac{n+1/2}{\eta_{n,m}}\right)^2\right] j_n\left(\eta_{n,m}\right)^2 \text{ if } j_n'\left(\eta_{n,m}\right) = 0 \\[2mm] \text{and } \eta_{n,m} > 0 \end{cases}$$

$$(10.3.38)$$

The features unique to the spherical geometry are the specific form of the mode functions, the natural frequencies, and evaluation of the normalizing coefficients. After they have been determined, analysis of the pressure field may follow the general path described in Sect. 10.3.1.

EXAMPLE 10.10 A spherical tank used for chemical processing consists of concentric spheres that separate different fluids. The outer sphere, whose radius is a, is rigid. The inner sphere, whose radius is b, is a permeable membrane that allows the fluids to mix slowly by diffusing through it. This sphere is well approximated as having zero thickness with neither inertia nor stiffness. The density and sound speed within the inner sphere are ρ_1 and c_1, while ρ_2 and c_2 are the properties of the fluid contained between the inner and outer spheres. Derive the characteristic equation and an expression for the cavity modes of this system. Also identify the orthogonality conditions for these modes.

Significance

Our study of propagation through planar layers entailed making solutions of the Helmholtz equation satisfies continuity conditions at interfaces. Those concepts are extended here to spherical waves.

Solution

It is stated that the permeable membrane physically separates the inner and outer regions, but cannot resist any movement. Thus, it must be that the pressure and particle velocity on either side of the inner sphere are equal. In addition, the pressure in both regions must satisfy the Helmholtz equation. The definition of a mode requires that all locations in a system exhibit an oscillation at the same frequency. Thus, the acoustic wavenumbers for the Helmholtz equation are $k_1 = \omega c_1$ and $k_2 = \omega c_2$. The other boundary conditions to be met are zero radial velocity at $r = a$ and finiteness of P at $r = 0$. Hence, we seek pressure fields $P^{(1)}$ and $P^{(2)}$ that satisfy

$$\nabla^2 P^{(1)} + (k_1)^2 P^{(1)} = 0, \quad 0 < r < b$$
$$\nabla^2 P^{(2)} + (k_2)^2 P^{(2)} = 0, \quad b < r < a$$
$$\frac{\partial P^{(2)}}{\partial r} = 0 \text{ at } r = b, \quad P^{(1)} \text{ is finite at } r = 0$$
$$P^{(1)} = P^{(2)} \text{ at } r = a$$
$$\frac{1}{i\omega\rho_1}\frac{\partial P^{(1)}}{\partial r} = \frac{1}{i\omega\rho_2}\frac{\partial P^{(2)}}{\partial r} \text{ at } r = b$$

The continuity conditions at $r = b$ require that the same polar harmonic $P_n (\cos \Psi)$ be associated with the mode in the inner and outer regions. Finiteness at $r = 0$ requires that the radial dependence of $P^{(1)}$ be a spherical Bessel function. The outer region does not contain $r = 0$, so both the spherical Bessel and Neumann functions are required to describe the radial dependence there. It follows that a suitable ansatz for the pressure in each region is

$$P^{(1)} = B^{(1)} \, j_n \, (k_1 r) \, P_n \, (\cos \Psi)$$
$$P^{(2)} = \left[B^{(2)} \, j_n \, (k_2 r) + C^{(2)} \, n_n \, (k_2 r) \right] P_n \, (\cos \Psi) \qquad (1)$$

When these expressions are substituted into the boundary and continuity conditions, the Legendre function is a common factor for each, so it may be canceled. The resulting equations are

$$k_2 \left[B^{(2)} \, j_n' \, (k_2 a) + C^{(2)} \, n_n' \, (k_2 a) \right] = 0$$
$$B^{(1)} \, j_n \, \left(k^{(1)} b \right) = B^{(2)} \, j_n \, (k_2 b) + C^{(2)} n_n \, (k_2 b)$$
$$\left(\frac{1}{\rho_1 c_1} \right) B^{(1)} \, j_n' \, (k_1 b) = \left(\frac{1}{\rho_2 c_2} \right) \left[B^{(2)} \, j_n' \, (k_2 b) + C^{(2)} n_n' \, (k_2 b) \right]$$

The first equation is satisfied by $k_2 = 0$, which leads to $\omega = 0$. Because the outer boundary is rigid, this is the zero frequency mode. The general discussion suggests that such a mode exists for this system. This will be proven after we find the modes whose natural frequency is greater than zero. Thus, k_2 may be factored out of the first equation.

It is convenient to express the coefficient equations in matrix form. In recognition that the frequency sets the acoustic wavenumbers, we replace k_1 and k_2 with ω/c_1 and $k_2 = \omega/c_2$, respectively. Hence, the matrix representation of the coefficient equations is

$$[D (\omega)] \left[B^{(1)} \ \ B^{(2)} \ \ C^{(2)} \right]^\mathrm{T} = \{0\}$$

$$[D (\omega)] = \begin{bmatrix} 0 & j_n' \, (\omega a/c_2) & n_n' \, (\omega a/c_2) \\ j_n \, (\omega b/c_1) & -j_n \, (\omega b/c_2) & -n_n \, (\omega b/c_2) \\ \left(\dfrac{\rho_2 c_2}{\rho_1 c_1} \right) j_n' \, (\omega b/c_1) & -j_n' \, (\omega b/c_2) & -n_n' \, (\omega b/c_2) \end{bmatrix} \qquad (2)$$

At an arbitrary ω, the solution of these equations is $B^{(1)} = B^{(2)} = C^{(2)} = 0$. A non-trivial solution requires that ω be a value that causes the determinant of $[D (\omega)]$ to be zero, thereby making it impossible to invert the matrix. Thus, the characteristic equation is

$$|[D (\omega)]| = 0 \qquad (3)$$

The roots of this (scalar) equation constitute the eigenvalues. Stated differently, they are the natural frequencies corresponding to the axisymmetric cavity modes. We could reduce the determinant to an algebraic expression, but there is little reason to do so because evaluation of its roots would require numerical methods. As was the case for a homogeneous cavity, the oscillatory nature of the Bessel and Neumann functions leads to an infinite number of natural frequencies $\omega_{n,m}$ for a specified polar harmonic n.

When the determinant of $[D]$ vanishes, one of the coefficient equations described in Eq. (2) is not independent of the other two. If we drop the third equation, we may solve the first two equations for $B^{(2)}$ and $C^{(2)}$ as ratios to $B^{(1)}$, that is

$$B^{(2)} = \beta_{n,m} B^{(1)}, \quad C^{(2)} = \gamma_{n,m} B^{(1)}$$

$$\beta_{n,m} = \frac{j_n \left(\omega_{n,m} b/c_1\right) n_n' \left(\omega_{n,m} a/c_2\right)}{j_n \left(\omega_{n,m} b/c_2\right) n_n' \left(\omega_{n,m} q/c_2\right) - n_n \left(\omega_{n,m} b/c_2\right) j_n' \left(\omega_{n,m} a/c_1\right)} B^{(1)}$$

$$\gamma_{n,m} = -\frac{j_n \left(\omega_{n,m} b/c_1\right) j_n' \left(\omega_{n,m} a/c_2\right)}{j_n \left(\omega_{n,m} b/c_2\right) n_n' \left(\omega_{n,m} q/c_2\right) - n_n \left(\omega_{n,m} b/c_2\right) j_n' \left(\omega_{n,m} a/c_1\right)} B^{(1)}$$

The coefficient $B^{(1)}$ is set by the normalization part of the orthogonality condition, so it also will depend on n and m. Therefore, we have established that a mode is described by

$$\Psi_{n,m} = B_{n,m}^{(1)} F_{n,m} (r) P_n (\cos \Psi) \tag{4}$$

where the radial function is

$$F_{n,m}(r) = \begin{cases} j_n \left(\omega_{n,m} r/c_1\right), & 0 < r < b \\ \beta_{n,m} j_n \left(\omega_{n,m} r/c_2\right) + \gamma_{n,m} n_n \left(\omega_{n,m} r/c_2\right), & b < r < a \end{cases} \tag{5}$$

Nowhere in the derivation of the general orthogonality condition in Eqs. (10.3.8) and (10.3.9) was it required that the fluid be homogeneous. It was implicit to the derivation that the mode functions are continuous and that their gradient is continuous. These requirements in the present system are assured by the continuity conditions at $r = b$. The general statements call for an integration over the entire domain, but the modes have different functional dependence in each fluid. Therefore, we decompose the volume integral into integrals over $0 < r < b$ and $b < r < a$. Each integral accounts for the different definitions of the mode function in the inner and outer regions. Because the functions are axisymmetric, we may use $dV = 2\pi (r \sin \Psi) (r d\psi) dr$. The volume of the sphere is $(4/3) \pi a^3$. Thus, the condition tells us that

$$\iiint_{\mathcal{V}} \Psi_{n,m} \Psi_{j,\ell} dV = 2\pi \int_0^\pi \left[\int_0^b \Psi_{n,m} \Psi_{j,\ell} r^2 dr + \int_b^a \Psi_{n,m} \Psi_{j,\ell} r^2 dr \right] \sin \Psi \, d\psi$$

$$= 2\pi B_{n,m}^{(1)} B_{j,\ell}^{(1)} \int_0^\pi P_n (\cos \psi) P_j (\cos \psi) (\sin \psi) \, d\psi \left[\int_0^b F_{n,m} (r) F_{j,\ell} (r) r^2 dr \right.$$

$$\left. + \int_b^a F_{n,m} (r) F_{j,\ell} (r) r^2 dr \right] = \delta_{n,j} \delta_{m,\ell} \left(\frac{4}{3} \pi a^3 \right)$$

If n and j are different, the left side vanishes regardless of m and ℓ because the Legendre functions are orthogonal according to Eq. (7.1.16). The condition that is satisfied by the F function is obtained by setting $j = n$. This property reduces the orthogonality integral to

$$B_{n,m}^{(1)} B_{n,\ell}^{(1)} \left[\int_0^b F_{n,m}(r) F_{j,\ell}(r) r^2 dr + \int_b^a F_{n,m}(r) F_{j,\ell}(r) r^2 dr \right] = \delta_{m,\ell} \frac{2n+1}{3} a^3$$

Normalization of $B_{n,m}^{(1)}$ is obtained by setting $\ell = m$.

To close the analysis, let us verify that there is a zero frequency mode. If $\omega = 0$, then $k_1 = k_2 = 0$. In that case, $j_n(k_1 r)$, $j_n(k_2 r)$, and $n_n(k_2 r)$ are identically zero unless $n = 0$. Thus, a zero frequency mode can only occur for $n = 0$. Furthermore, because $n_0(0)$ is infinite, the zero frequency mode requires that $C^{(2)} = 0$. Substitution of these values, combined with the property that $j_0'(0) = 0$ and $P_0(\cos \Psi) \equiv 1$, leads to recognition that only the second of Eq. (2) is not trivial when $\omega = 0$. The second equation in that case leads to $B^{(2)} = B^{(1)}$. Thus, the zero frequency mode is a constant value that is the same in both regions.

$$\Psi_{0,1}^{(1)} = \Psi_{0,1}^{(2)} = B^{(1)}, \quad 0 < r < b$$

This fits the general property of zero frequency cavity modes.

10.4 Approximate Methods

Several factors limit the types of enclosed domains that can be analyzed mathematically. The primary issues pertain to the shape of the cavity and the nature of the surface bounding the enclosed region. There is an obvious need for a formulation that is more general. Discretizations using either boundary elements or finite elements fit this specification. Their implementations for radiation from a vibrating surfaces, which were described in Chap. 7, are readily modified to address an enclosed region. The fundamental alteration of a boundary element formulation entails reversing the surface normal $\bar{n}$ to point into the domain enclosed by the boundary, rather into the region exterior to the boundary. A merit of this formulation is that concern for the issue of "forbidden frequencies" is not relevant. Finite element implementations also benefit when applied to a cavity because there is no need for infinite elements or any of the other techniques for modeling an infinite domain.

In any event, a model created with either discretization method is likely to feature a large number of variables that are the pressures at points an a spatial grid. The fineness of such a grid must increase as the frequency increases. Consequently, the computational resources required to implement numerical simulation techniques makes them prohibitive for the early stages of the design process, as well as for gaining a preliminary insight when an existing system exhibits anomalous behavior.

Two approximate approaches we shall consider generate relatively small models. The first formulation, which originated with Rayleigh,[9] has two objectives: estimate the lower natural frequencies and find a series solution that best fits the actual cavity modes. The development retains the restriction that the boundary be rigid and/or pressure-release. The second method is known as Dowell's approximation.[10] It is used to model systems in which the walls are flexible, usually because they are part of an elastic structure.

10.4.1   The Rayleigh Ratio and Its Uses

We consider here a limited task: Determine the natural frequencies and modes of an enclosed region whose shape is not amenable to formal mathematical analysis. The concepts we shall explore are applicable to a variety of physical systems. Indeed, they usually are encountered in a course on mechanical and structural vibration. Our implementation is based on the expressions for the acoustic kinetic and potential energy.

Estimation of the Fundamental Frequency

The foundation for the development is the statement of the basic work-energy principle, Eq. (4.5.30). Because we have restricted our attention to cavities whose walls are rigid (no movement) or pressure-release (no traction force), no power is transferred out of the fluid domain. Hence, the acoustical energy $KE + PE$ must be constant. We know that the cavity modes Ψ_n are real functions, and Euler's equation gives $\bar{v}(\bar{x}, t) = \text{Re}\left(i\nabla\Psi_n e^{i\omega t}\right)/(\rho_0\omega)$. This means that the pressure everywhere is in-phase, as is the particle velocity, with the particle velocity leading the pressure by $90°$. Consequently, at instants separated by $T/2 = \pi/\omega$, the kinetic energy is a maximum and the potential energy is zero. Midway between these instants, the potential energy is maximum and the kinetic energy is zero. It follows from Eq. (4.5.30) that $\max(KE) = \max(PE)$. When we use the field properties of the nth mode to form the energy expressions, we find that

$$\max(KE) = \frac{1}{2}\iiint_{\mathcal{V}} \frac{1}{\rho_0(\omega_n)^2}\nabla\Psi_n \cdot \nabla\Psi_n d\mathcal{V}$$

$$\max(PE) = \frac{1}{2}\iiint_{\mathcal{V}} \frac{1}{\rho_0 c^2}(\Psi_n)^2 d\mathcal{V} \tag{10.4.41}$$

[9]J.W. Strutt Lord Rayleigh, *Theory of Sound,* Vol. 2, Dover (1945 reprint) Sect. 88.

[10]Dowell, E.H., and Voss, H.M. (1962). "The effect of a cavity on panel vibration," AIAA J., **1**, 476–477.

Dowell, E.H., Gorman, G.F., and Smith, D.A. (1977). "Acoustoelasticity: General theory, acoustic natural modes and forced response to sinusoidal excitation, including comparisons with experiment" J. Sound Vib., **52**, 519–542.

Equating these expressions yields

$$(\omega_n)^2 = \frac{\iiint\limits_{\mathcal{V}} \frac{1}{\rho_0} \nabla\Psi_n \cdot \nabla\Psi_n d\mathcal{V}}{\iiint\limits_{\mathcal{V}} \frac{1}{\rho_0 c^2} (\Psi_n)^2 d\mathcal{V}} \qquad (10.4.42)$$

This relation is of no direct use, because the analysis leading to a mode function also gives the natural frequency. Its value lies in how Rayleigh generalized it. The numerator and denominator are said to be functionals of Ψ_n. They shall be denoted as $\mathcal{N}[\Psi_n(\bar{x})]$ and $\mathcal{D}[\Psi_n(\bar{x})]$. The *Rayleigh ratio* $\mathcal{R}$ is the same ratio of functionals as above, except that Ψ_n is replaced by a function $\Psi(\bar{x})$ that is not necessarily a cavity mode,

$$\boxed{\mathcal{R} = \frac{\mathcal{N}[\Psi(\bar{x})]}{\mathcal{D}[\Psi(\bar{x})]} \equiv \frac{c^2 \iiint\limits_{\mathcal{V}} \nabla\Psi \cdot \nabla\Psi d\mathcal{V}}{\iiint\limits_{\mathcal{V}} \Psi^2 d\mathcal{V}}} \qquad (10.4.43)$$

The form of this relation is based on the cavity fluid being homogeneous, which is typical for an enclosed region. This restriction allows us to factor ρ_0 and c out of the integrals. Nevertheless, the concepts that will emerge are equally valid for a heterogeneous medium.

The function $\Psi(\bar{x})$ is said to be a *trial function*. It should be selected to satisfy two requirements. The first stems from the fact that the pressure within the fluid must be continuous because the particle velocity must be finite. Therefore, $\Psi(\bar{x})$ must be continuous everywhere. It also is necessary that any function we select be zero on any portion of the surface that is pressure-release. The converse condition is that Ψ should not evaluate to zero on the rigid portion of the boundary. (Satisfaction of the latter condition is not mandatory, but it is advisable.) Why do we need to worry about these alternatives? The mathematical explanation lies in the calculus of variations.[11] A heuristic explanation comes from the observation that the derivation of the Rayleigh ratio required that there be no power transfer at the boundary. We also know that the pressure and normal velocity cannot be specified concurrently anywhere. If $\Psi = 0$ on the pressure-release portion of the boundary, the trial function inherently is consistent with zero power flow across that region, regardless of the value of $\bar{n} \cdot \nabla\Psi$. On the other hand, if Ψ is not zero on the rigid portion of the boundary, then the fact that the principle is based on zero power flow will lead to a result that corresponds to taking $\bar{n} \cdot \nabla\Psi$ to be zero there. This argument does not explain why $\Psi = 0$, rather than $\bar{n} \cdot \nabla\Psi = 0$ is the condition that must be set. Nevertheless, **it is an inviolable requirement that the trial function be zero at any point on the surface that is**

---

[11] Weinstock, *Calculus of Variations*, Dover (1974).

pressure-release. (In the calculus of variations, this is said to be a *geometric* or *imposed* boundary condition.)

To identify some important properties of $\mathcal{R}$, we represent the trial function in terms of the actual mode functions of the cavity. We do not know these functions, but if we did, such a modal series would feature a set of coefficients a_n,

$$\Psi = \sum_{n=1}^{\infty} a_n \Psi_n \tag{10.4.44}$$

It will be noted that no allowance has been made in this series for the $n = 0$ mode, which is the zero frequency mode for an enclosed region whose entire boundary is rigid. We will examine this case after we have completed the derivations.

Although we do not know the mode functions, we know that they satisfy the orthogonality conditions in Eqs. (10.3.8) and (10.3.9). In the latter, k_n is the wavenumber at the nth natural frequency, $k_n \equiv \omega_n/c$. We consider the mode function to be normalized in accord with the first of those conditions. (Normalization merely expedites the derivation.) Thus, substitution of the modal series description of Ψ into the energy functionals leads to

$$\mathcal{N}[\Psi(\bar{x})] = c^2 \sum_{n=1}^{\infty}\sum_{j=1}^{\infty} a_n a_j \iiint_{\mathcal{V}} \nabla\Psi_n \cdot \nabla\Psi_j dV = c^2 \sum_{n=1}^{\infty} (k_n)^2 a_n^2 \mathcal{V}$$

$$\mathcal{D}[\Psi(\bar{x})] = \sum_{n=1}^{\infty}\sum_{j=1}^{\infty} a_n a_j \iiint_{\mathcal{V}} \Psi_n \Psi_j dV = \sum_{n=1}^{\infty} a_n^2 \mathcal{V} \tag{10.4.45}$$

The modes have been ordered in ascending order of their natural frequency, so that $k_n/k_1 > 1$ if $n > 1$. Hence, using the preceding to form $\mathcal{R}$, followed by factoring $(k_1)^2 \equiv (\omega_1/c)^2$ out of the numerator leads to

$$\boxed{\mathcal{R} = \frac{\mathcal{N}[\Psi(\bar{x})]}{\mathcal{D}[\Psi(\bar{x})]} = (\omega_1)^2 \frac{\displaystyle\sum_{n=1}^{\infty} \left(\frac{k_n}{k_1}\right)^2 a_n^2}{\displaystyle\sum_{n=1}^{\infty} a_n^2}} \tag{10.4.46}$$

Except for the $n = 1$ term, each term in the numerator's sum is greater the corresponding term in the denominator's sum. It follows that

$$\mathcal{R} \geq (\omega_1)^2 \tag{10.4.47}$$

with the equality sign being the case only if all a_n other than a_1 are zero, which means that $\Psi = \Psi_1$. In other words, we have shown that **the Rayleigh ratio gives an upper bound to the square of the fundamental frequency of the cavity**.

This does not tell us the quality of the upper bound—is it too high by 10%, 100%, or much more? To examine this, let us suppose that the Ψ function is close to the first mode multiplied by a scaling factor. To indicate this proximity, let a_1 in Eq. (10.4.44) be the scaling factor and let all of the other coefficients be small values by setting $a_n = \varepsilon a'_n, n = 2, 3, ...$, where $|\varepsilon| \ll 1$ and all $|a'_n| \leq 1$. Equation (10.4.46) correspondingly indicates that

$$\mathcal{R} = (\omega_1)^2 \, \frac{a_1^2 + \varepsilon^2 \sum\limits_{n=2}^{\infty} \left(\dfrac{k_n}{k_1}\right)^2 (a'_n)^2}{a_1^2 + \varepsilon^2 \sum\limits_{n=1}^{\infty} (a'_n)^2} = (\omega_1)^2 \left[1 + O\left(\varepsilon^2\right)\right] \qquad (10.4.48)$$

The notation $O\left(\varepsilon^2\right)$ refers to a number whose order of magnitude is ε^2. The preceding tells us that the relative error of the estimate for $(\omega_1)^2$ will be less than the deviation of the trial function from the true fundamental mode, provided that the trial function is a reasonably good estimate, $\varepsilon < 1$.

This leads to a common use of the Rayleigh ratio. To obtain an estimate of the fundamental frequency of a cavity, we may fabricate a trial function that is zero at all locations where the surface is pressure-release. Substitution of that function into the definition of $\mathcal{R}$ can be expected to yield an estimated fundamental natural frequency that is better than the degree to which the trial function approximates the first mode function.

There is one condition that must be satisfied if this expectation is to be met. As was stated, the error in the trial function relative to the true fundamental mode must be $O(\varepsilon)$, but we do not know the true mode. If we blindly select trial functions, we could make a terrible choice. For example, suppose we were to select a trial function for which $a_1 = 0.1$, $a_2 = 1$, and all other $a_n = 0$. This would lead to $\mathcal{R} = 0.9901[(10^{-4})(\omega_1)^2 + (\omega_2)^2]$, which obviously is not close to $(\omega_1)^2$.

This condition usually can be avoided if one is cognizant of a fundamental aspect: the fundamental mode of a system usually exhibits the minimum number of nodal lines consistent with any pressure-release conditions. If we select a trial function having this property, then the estimate for ω_1 can be expected to be less than 40% above the true fundamental frequency, corresponding to $\mathcal{R} < 2\omega_1$. We will take up this issue in the next example.

When all of the boundaries of the cavity are rigid, the fundamental mode is a uniform pressure and the fundamental frequency is zero. We do not need the Rayleigh ratio to tell us this, but that is what Eq. (10.4.48) indicates will be approximated. Can we use our knowledge of the nature of the zero frequency mode to obtain an approximation of the lowest nonzero natural frequency? The answer lies in the observation that if the trial function is orthogonal to the first mode, then the coefficient a_1 will be zero. In that case, Eq. (10.4.46) tells us that $\mathcal{R} \geq (\omega_2)^2$, and Eq. (10.4.48) tells us that the error in the estimate of $(\omega_2)^2$ obtained from $\mathcal{R}$ will be much less than the amount by which the trial function differs from the second mode. Thus, we must modify the trial function in a manner that leads to $a_1 = 0$. The zero frequency mode is a constant value everywhere. We may scale a mode by any factor, so we may set $\Psi_0(\bar{x}) = 1$. (This is what would be obtained from the standard normalization of modes.) A trial function $\Psi(\bar{x})$ is orthogonal to Ψ_0 if

$$\iiint\limits_{\mathcal{V}} \Psi_0 \Psi d\mathcal{V} = \iiint\limits_{\mathcal{V}} (1)\, \Psi d\mathcal{V} = 0 \qquad (10.4.49)$$

This integral is the mean value of Ψ multiplied by the size of the domain.

Thus, if we wish to determine the fundamental nonzero frequency of a rigid-walled cavity, we should employ a trial function whose mean value is zero. If our initial choice for Ψ does not meet this specification, we can obtain a corrected function $\hat{\Psi}(\bar{x})$ by subtracting the mean value of $\Psi(\bar{x})$, that is,

$$\boxed{Rigid - walled\, cavity : \hat{\Psi}(\bar{x}) = \Psi(\bar{x}) - \Psi_{\mathrm{mean}}, \quad \Psi_{\mathrm{mean}} = \frac{1}{\mathcal{V}} \iiint\limits_{\mathcal{V}} \Psi(1)\, d\mathcal{V}}$$

$$(10.4.50)$$

Using $\hat{\Psi}$ to construct the Rayleigh ratio will lead to $R \geq (\omega_2)^2$.

EXAMPLE 10.11 Consider a two-dimensional cavity whose sides at $x = 0$, $x = L$, and $y = 0$ are rigid and whose side at $y = H$ is pressure-release. Identify three alternative trial functions that may be used to estimate the fundamental natural frequency of the system. Then, determine which gives the best estimate when $H = L/4$, $H = L/2$, and $H = 2L$.

Significance

Because the analytical solution for the modal properties is readily obtained, this example will demonstrate how to interpret the performance of the Rayleigh ratio approximation.

Solution

The best way to select a trial function usually is to consider a similar system for which the modes are known. However, we will proceed as though we had no prior experience with a two-dimensional cavity in order to understand how an unfamiliar system might be approached. The boundaries correspond to constant values of x or y, so we consider trial functions whose form is $\Psi = f(x) g(y)$. If $g(y)$ were chosen to be a constant, satisfaction of the pressure-release condition at $y = H$ would require that the constant be zero. The next simplest function is $g(y)$ being linear. A linear function whose value is zero at $y = H$ is $g(y) = H - y$. A constant nonzero value of $f(x)$ would not violate any requirements because $x = 0$ and $x = L$ are rigid, which requires that Ψ not be zero at those locations. The overall magnitude of $f(x)$, as well as $g(y)$, is irrelevant because any scaling factor will cancel when the Rayleigh ratio is formed. We therefore may set $f(x) = 1$, which leads to our first choice for a trial function, $\Psi_1 = H - y$. In general, monomials, that is, x or y to an integer power, are the easiest to work with because they lead to simple integrals. Because Ψ_1 may be written as $x^0 (H - y)$, it seems reasonable to allow for variation in the x direction by raising the power to one, which leads to $\Psi_2 = x (H - y)$. This will test what happens if the trial function is zero on a surface that is not pressure-release. For the third function, let us increase the power of the y dependence, so that $\Psi_3 = x (H - y)^2$. A portion of the boundary is pressure-release, so the issue of eliminating the zero frequency mode is irrelevant.

The analytical steps leading to $\mathcal{R}$ for each trial function, based on integration over $0 < x < L, 0 < y < H$, are tabulated below.

Ψ	$H - y$	$x(H-y)$	$x(H-y)^2$
$\nabla\Psi$	$-\bar{e}_y$	$(H-y)\bar{e}_x - x\bar{e}_y$	$(H-y)^2\,\bar{e}_x - 2x(H-y)\,\bar{e}_y$
$\mathcal{N}[\Psi(\bar{x})]$	$c^2 HL$	$c^2 \dfrac{HL\left(H^2+L^2\right)}{3}$	$c^2\left(\dfrac{H^5L}{5}+\dfrac{4H^3L^3}{9}\right)$
$\mathcal{D}[\Psi(\bar{x})]$	$\dfrac{H^3L}{3}$	$\dfrac{H^3L^3}{9}$	$\dfrac{H^5L^3}{15}$
$\mathcal{R}$	$c^2\dfrac{3}{H^2}$	$c^2\left(\dfrac{3}{H^2}+\dfrac{3}{L^2}\right)$	$c^2\left(\dfrac{20}{3H^2}+\dfrac{3}{L^2}\right)$
$\mathcal{R}\left(\dfrac{L}{c}\right)^2\Big|_{H=L/4}$	**48**	51	109.7
$\mathcal{R}\left(\dfrac{L}{c}\right)^2\Big|_{H=L/2}$	**12**	15	29.67
$\mathcal{R}\left(\dfrac{L}{c}\right)^2\Big|_{H=2L}$	**0.75**	3.75	4.67

The value of $\mathcal{R}$ cannot be less than the square of the fundamental natural frequency. The consequence is the general fact that *the smallest value of R is closest to the true value of ω_1^2*. The lowest value for each H/L is listed in bold type. In each case, the best estimate is obtained with Ψ_1. The actual fundamental frequency is either $\omega_{0,1} = 0.5\pi c/H$ or $\omega_{1,0} = \pi c/L$, depending on the size of H relative to L. These values and the lowest estimate obtained from the Rayleigh ratio are listed below.

H/L	$\omega_{0,1}\left(\dfrac{L}{c}\right)$	$\omega_{1,0}\left(\dfrac{L}{c}\right)$	$\mathcal{R}^{1/2}\left(\dfrac{L}{c}\right)$
0.25	6.283	**3.142**	6.928
0.5	**3.142**	**3.142**	3.464
2	**0.785**	3.142	0.8660

The true fundamental frequency is shown in bold. The error of the Rayleigh ratio estimate for $H/L = 0.5$ and 2 is 10%, but it is greater than 100% for $H/L = 0.25$.

To understand why the estimate for $H/L = 0.25$ is poor, we need to consider the two possible fundamental mode functions, $\Psi_{0,1} = \cos(0.5\pi y/H)$ and $\Psi_{1,0} =$

$\cos(\pi x/L)$. If $H > L/2$, the former is the fundamental mode. In it, there is no variation in the x direction and its value decreases monotonically with increasing y. The first trial function also does not depend on x, and the fact that it increases with increasing y, rather than decreasing, is irrelevant. In contrast, when $H < L/2$, the fundamental mode is $\Psi_{1,0}$, which is independent of y and decreases monotonically with increasing x. None of the trial functions matches this independence on y. Indeed, the results suggest that they are approximating $\Psi_{0,1}$.

There is no simple rule for selecting a trial function. Picking it to match the fundamental mode of a similar system is a good idea. Another aspect is recognizing that the fundamental mode will entail the minimum fluctuation in all directions consistent with the requirement that it be zero at boundaries that are pressure-release. If it is crucial to obtain a good estimate, multiple trial functions may be considered, as was done here. Ultimately, a better approach is to use several trial functions, which is the procedure developed in the next section.

Rayleigh–Ritz Method

The previous example closed with a suggestion is that the Rayleigh ratio could be evaluated for several trial functions. Whichever function led to the smallest value of $\mathcal{R}$ would mark the best estimate of $(\omega_1)^2$. This is a flawed approach because it does not provide a definitive path to an improved answer. Much better results, including estimates of mode functions, may be obtained from another property of $\mathcal{R}$.

Suppose that the trial function is close to the jth cavity mode, where j may be any integer. Proximity of Ψ to Ψ_j does not preclude them differing by a scaling factor. Therefore, this proximity would be manifested in the modal series representation of the trial function, Eq. (10.4.46), by the coefficient a_j being much larger than any other coefficient. We indicate this by setting $a_n = \varepsilon a'_n$, $n \neq j$, where $|\varepsilon| \ll 1$. In other words, the series representation is

$$\Psi = a_j \Psi_j + \sum_{\substack{n=1 \\ n\neq j}}^{\infty} \varepsilon a'_n \Psi_n \tag{10.4.51}$$

When a_n in Eq. (10.4.46) is replaced by $\varepsilon a'_n$, the result is

$$\mathcal{R} = (\omega_j)^2 \, \frac{a_j^2 + \varepsilon^2 \sum_{\substack{n=1 \\ n\neq j}}^{\infty} \left(\dfrac{k_n}{k_j}\right)^2 (a'_n)^2}{a_j^2 + \varepsilon^2 \sum_{\substack{n=1 \\ n\neq j}}^{\infty} (a'_n)^2} \tag{10.4.52}$$

The case where the trial function is exactly proportional to the jth normalized mode corresponds to $\varepsilon = 0$. This is important because differentiation of the above expression with respect to ε leads to

$$\left. \frac{d\mathcal{R}}{d\varepsilon} \right|_{\varepsilon=0} = 0 \tag{10.4.53}$$

To understand this property, let us pretend that we know the cavity modes. Then, we actually could construct the modal series by multiplying any Ψ_j by a coefficient a_j and then adding to that term small contributions of the other modes, with each contribution scaled by a small parameter ε. This construction would allow us to compute the numerator and denominator of the Rayleigh ratio. Altering the value of ε would yield a different value of $\mathcal{R}$. In a plot of $\mathcal{R}$ as a function of ε, the slope would be horizontal, that is, $d\mathcal{R}/d\varepsilon$ would be zero, at $\varepsilon = 0$. This is the property of *stationarity*.

Of course, we do not know the actual mode function. Instead, the concept is to select a trial function that has adjustable parameters. We can use the stationarity property to find the conditions those parameters must satisfy in order that the trial function matches a mode function. It would be problematic to attempt to pursue this concept by selecting a single trial function of position that contains adjustable parameters. It is far easier to use two or more fully prescribed trial functions of position and to let the adjustable parameters be weighting coefficients for their combination.

To see how this concept may be implemented, suppose we have two identified trial functions, Ψ_1 and Ψ_2. Rather than evaluating $\mathcal{R}$ for each function, as we did in the previous example, let us form a new trial function that is a linear combination of Ψ_1 and Ψ_2. That is, let

$$\Psi = b_1 \Psi_1 + b_2 \Psi_2 \tag{10.4.54}$$

Substitution of this expression into the functionals in Eq. (10.4.43) converts $\mathcal{N}$ and $\mathcal{D}$ and therefore $\mathcal{R}$, to algebraic functions of b_1 and b_2. The values of b_1 and b_2 that lead to a Ψ function that most closely fits a mode are marked by $\mathcal{R}$ having an extreme value.

To understand how we may use this property, suppose we were to create a three-dimensional Cartesian plot in which b_1 and b_2 are plotted along two axes and the value of $\mathcal{R}$ as a function of a b_1 and b_2 pair is plotted on the third axis. We wish to identify points in the $b_1 b_2$ plane that lead to extreme values of $\mathcal{R}$. We could do so by searching for a high or low point or any point where it levels out locally. However, there is no reason to actually construct such a plot, because the point we seek is characterized by the gradient with respect to b_1 and b_2 being zero. The components of this gradient are the respective partial derivatives, so we seek the condition for which

$$\frac{\partial \mathcal{R}}{\partial b_1} = 0, \quad \frac{\partial \mathcal{R}}{\partial b_2} = 0 \tag{10.4.55}$$

This does not mean that satisfaction of these conditions will yield the actual mode function. Rather, it states that if we start off with a two-term series representation of a mode, the coefficients b_1 and b_2 that give the closest fit to a mode function are those for which the above derivatives vanish.

We may generalize this notion by considering an N-term series, which requires that we identify N suitable trial functions. They must be a linearly independent set, which means that we cannot obtain one of them as a linear combination of the others. The series is

$$\boxed{\Psi = \sum_{n=1}^{N} b_n \Psi_n (\bar{x})}$$

(10.4.56)

Substitution of this expression into the numerator and denominator functionals converts them to quadratic sums,

$$\mathcal{N} = c^2 \iiint_{\mathcal{V}} \left(\sum_{n=1}^{N} b_n \nabla \Psi_n \right) \cdot \left(\sum_{j=1}^{N} b_j \nabla \Psi_j \right) dV = \sum_{n=1}^{N} \sum_{j=1}^{N} \mathcal{N}_{j,n} b_j b_N$$

$$\mathcal{D} = \iiint_{\mathcal{V}} \left(\sum_{n=1}^{N} b_n \Psi_n \right) \left(\sum_{j=1}^{N} b_j \Psi_j \right) dV = \sum_{n=1}^{N} \sum_{j=1}^{M} \mathcal{D}_{j,n} b_j b_n$$

(10.4.57)

where the coefficients are

$$\boxed{\begin{aligned} \mathcal{N}_{j,n} &= c^2 \iiint_{\mathcal{V}} \nabla \Psi_j \cdot \nabla \Psi_n \, dV \\ \mathcal{D}_{j,n} &= \iiint_{\mathcal{V}} \Psi_j \Psi_n \, dV \end{aligned}}$$

(10.4.58)

This reduces $\mathcal{R}$ to a function of the b_m coefficients, so the stationarity property requires that

$$\frac{\partial \mathcal{R}}{\partial b_m} = 0, \quad m = 1, 2, ..., M$$

(10.4.59)

According to Eq. (10.4.43), $\mathcal{R} = \mathcal{N}/\mathcal{D}$, so we have

$$\frac{\partial \mathcal{R}}{\partial b_m} = \frac{1}{\mathcal{D}^2} \left(\mathcal{D} \frac{\partial \mathcal{N}}{\partial b_m} - \mathcal{N} \frac{\partial \mathcal{D}}{\partial b_m} \right) \equiv \frac{1}{\mathcal{D}} \left(\frac{\partial \mathcal{N}}{\partial b_m} - \mathcal{R} \frac{\partial \mathcal{D}}{\partial b_m} \right) = 0$$

(10.4.60)

Because $\mathcal{D}$ is finite, it must be that

$$\boxed{\frac{\partial \mathcal{N}}{\partial b_m} - \mathcal{R} \frac{\partial \mathcal{D}}{\partial b_m} = 0, \quad m = 1, 2, ..., M}$$

(10.4.61)

These equations are known as the *Rayleigh–Ritz method*. Walther Ritz derived the formulation close to the time of Rayleigh's derivation, but he did so by following a very different route.[12]

---

[12]Rayleigh is widely recognized for this formulation because it appears in his monumental texts on *The Theory of Sound*. Walther Ritz was a pioneering nuclear physicist who independently developed

The Rayleigh–Ritz equation, Eq. (10.4.61), is quite general. To derive its specific form, we differentiate Eq. (10.4.57) with respect to a specific b_m. This operation is

$$\frac{\partial \mathcal{N}}{\partial b_m} = \sum_{n=1}^{N}\sum_{j=1}^{N}\mathcal{N}_{j,n}\frac{\partial}{\partial b_m}\left(b_j b_n\right) = \sum_{n=1}^{N}\sum_{j=1}^{N}\mathcal{N}_{j,n}\left(\delta_{j,m}b_n + b_j\delta_{n,m}\right)$$
$$= \sum_{n=1}^{N}\mathcal{N}_{m,n}b_n + \sum_{j=1}^{N}\mathcal{N}_{j,m}b_j = 2\sum_{n=1}^{N}\mathcal{N}_{m,n}b_n \tag{10.4.62}$$

The last form results from the symmetry of the coefficients, $\mathcal{N}_{j,m} \equiv \mathcal{N}_{m,j}$, which follows from their definition. A similar result applies to the derivative of $\mathcal{D}$. Thus, we obtain the Rayleigh–Ritz equations,

$$\sum_{n=1}^{N}\mathcal{N}_{m,n}b_n - \mathcal{R}\sum_{n=1}^{N}\mathcal{D}_{m,n}b_n = 0, \quad m = 1, 2, ..., N \tag{10.4.63}$$

The nature of the equations to solve becomes evident from their matrix form,

$$\boxed{[[\mathcal{N}] - \mathcal{R}[\mathcal{D}]]\{b\} = \{0\}} \tag{10.4.64}$$

This constitutes a *general (matrix) eigenvalue problem*, whose solution will yield M eigenvalues $\mathcal{R}_m/c$ and corresponding eigenvectors $\{b\}_m$.

The eigensolutions have several remarkable properties. Identification of those properties requires considerable analysis, primarily in linear algebra,[13] so we will only discuss the highlights here. The first is an extension of the upper bound theorem. It asserts that the rth eigenvalue, $\mathcal{R}_r$, is an upper bound for the square of the rth natural frequency,

$$\boxed{\mathcal{R}_r \geq (\omega_r)^2} \tag{10.4.65}$$

This provides no guidelines as to how much larger is the value of $\mathcal{R}_m$. However, unless one makes special effort to find unusual functions, all eigenvalues except the largest can be expected to be within 10% or less of the respective true values, with the error increasing with increasing mode number.

(Footnote 12 continued)
the formulation. He came to it by applying variational calculus concepts he first developed for quantum mechanics. The Rayleigh–Ritz formulation actually is a special application of Ritz' general approach, which is known as the Ritz series method, see J.H. Ginsberg, *Mechanical and Structural Vibrations*, John Wiley and Sons (2000), Chap. 6. This approach has been used for many analyses in physics and engineering. It also is the foundation for the finite element method. Thus, it is remarkable that Ritz wrote only two papers that discussed acoustic-type systems: "On the new method for solving some variational problems of mathematical physics" (1908) and "Theory of the transverse oscillations of a square plate with free boundaries" (1909). These were the first to explain mathematically the occurrence of Chaladni lines in plate vibrations.
[13]L. Meirovitch, *Principles and Techniques of Vibrations,* Prentice-Hall (1997) Chap. 8.

The upper bound property is applicable for any set of trial functions. A much stronger property holds if the series of trial functions is formed in a certain way. Let us denote the basis functions of the N-term series in Eq. (10.4.56) as $\Psi_n^{(N)}$. To this, we add another term consisting of a different trial function F, so that the enlarged set of functions is

$$\Psi_n^{(N+1)} = \Psi_n^{(N)}, \quad n = 1, ..., N, \quad \Psi_{N+1}^{(N+1)} = F \tag{10.4.66}$$

Correspondingly, the eigenvalues obtained from the two series are denoted as $R_r^{(N)}$, $r = 1, 2, ..., N$, and $R_r^{(N+1)}$, $r = 1, 2, ..., N + 1$. The result will satisfy the *separation theorem*, which states that the eigenvalues for the shorter length series fall between the eigenvalues for the longer series, according to

$$R_1^{(N+1)} \le R_1^{(N)} \le R_2^{(N+1)} \le R_2^{(N)} \le \cdots \le R_N^{(N)} \le R_{N+1}^{(N+1)} \tag{10.4.67}$$

In words, this property tells us that as we lengthen the series, the estimate for any natural frequency will decrease or be unchanged. This is a remarkable result in light of the upper bound theorem, because the estimated value must be greater than the true value. Therefore, it must be that lengthening the series will cause the eigenvalues to converge from above to the true natural frequencies.

A corollary of convergence of the eigenvalues is convergence to the true mode functions of the trial function series. This attribute follows from Eq. (10.4.52), which states that if a value of $\mathcal{R}$ is close to $(\omega_j)^2$, then it must be that the corresponding a_j coefficient is much larger than any of the other coefficients. To construct a mode, let $(b_n)_r$ be the nth element of the eigenvector corresponding to $\mathcal{R}_r$. The trial function series corresponding to the rth eigenvector is

$$\Psi_r^{(N)} = \sum_{n=1}^{N} (b_n)_r \, \Psi_n (\bar{x}) \equiv [\Psi_1 (\bar{x}) \quad \Psi_2 (\bar{x}) \quad \cdots \quad \Psi_N (\bar{x})] \{b\}_r \tag{10.4.68}$$

The superscript "N" has been inserted to remind us that we have constructed an N-term series approximation of the actual cavity mode Ψ_r.

EXAMPLE 10.12 It is desired to identify the natural frequencies and mode functions of a two-dimensional cavity in the shape of an ellipse defined as $(x/a)^2 + (y/b)^2 = 1$, where a and b are the semimajor and semiminor axis lengths. The elliptical surface of the cavity is rigid. The mode functions of interest are those that are symmetric with respect to the xz and yz bisecting planes, with a nonzero natural frequency. For the case where $a = 2b$, use series containing two to six terms to determine the natural frequencies and mode functions by the Rayleigh–Ritz method. Assess the frequencies in light of the separation theorem, and compare the results for the first three mode functions.

Significance

Analysis of the mode functions for an elliptical shape is an imposing problem if it is done by seeking eigensolutions of the Helmholtz equation. Thus, it is systems like this for which the Rayleigh–Ritz method is especially useful. The solution will demonstrate how exploiting a system's symmetry properties can improve the analysis without requiring greater computational effort. The results will exemplify the general performance characteristics of the Rayleigh–Ritz method.

Solution

The elliptical wall is rigid, so the trial functions should not vanish on the boundary. Using a power series in each direction is acceptable, and those functions are easy to use. One consideration is the zero frequency mode, which this system possesses. When the Rayleigh ratio is used to estimate the fundamental frequency for such systems, the mean value of the basis function is subtracted from the function in order to estimate the fundamental nonzero frequency. We could do the same here by subtracting from each trial function its mean value over the elliptical cross section. In the Rayleigh–Ritz method, many modes are obtained, so a simpler alternative is to include a trial function that is uniform over the cross section. Suppose we use power series for both the x and y dependencies. We form Ψ as a product of the individual series, and designate the independent coefficients as c_m,

$$\begin{aligned}\Psi &= \left(\alpha_0 + \alpha_1 x + \alpha_2 x^2 + \cdots\right)\left(\beta_0 + \beta_1 y + \beta_2 y^2 + \cdots\right) \\ &= c_1 + c_2 x + c_3 y + c_4 x^2 + c_5 y^2 + c_6 xy + \cdots\end{aligned} \tag{1}$$

It makes sense to halt the series at a stage where all combinations of x and y up to the same degree have been included. Equation (1) corresponds to truncation at quadratic terms.

Equation (1) would be a valid starting point for a general analysis. However, it has been stipulated that we should find the symmetric mode functions. A function is symmetric with respect to the xz plane if $f(x, y) = f(x, -y)$, and symmetry with respect to the yz plane requires $f(x, y) = f(-x, y)$. These properties will be obtained if the x and y polynomials in Eq. (1) are even functions of x and y, which means that they should be polynomials that contain only even powers. Thus, the series we shall use is formed as

$$\begin{aligned}\Psi &= \left(\alpha_0 + \alpha_2 x^2 + \alpha_4 x^4 + \cdots\right)\left(\beta_0 + \beta_2 y^2 + \beta_4 y^4 + \cdots\right) \\ &= c_1 + c_2 x^2 + c_3 y^2 + c_4 x^4 + c_5 y^4 + c_6 x^2 y^2 + \cdots\end{aligned} \tag{2}$$

Note that this series has the same number of terms as Eq. (1), but it is quartic and therefore can be expected to better fit the true modes. The trial functions are the coefficients of the respective c_m in Eq. (2). It is useful to work with their dimensionless equivalents, which are

$$\Psi_1 = 1, \ \Psi_2 = \frac{x^2}{a^2}, \ \Psi_3 = \frac{y^2}{b^2}, \ \Psi_4 = \frac{x^4}{a^4}, \ \Psi_5 = \frac{y^4}{b^4}, \ \Psi_6 = \frac{x^2 y^2}{a^2 b^2} \tag{3}$$

Because the functions are even with respect to x and y, we may implement Eq. (10.4.58) by integrating over one quadrant and then multiplying the result by four. Hence, the domain of integration will be $0 < y < b\left(1 - (x/a)^2\right)^{1/2}$, $0 < x < a$. These functions have the generic form $\Psi_j = (x/a)^{r_j}(y/b)^{s_j}$. Thus, the coefficients for the numerator of $\mathcal{R}$ are described by

$$
\mathcal{N}_{j,m} = 4c^2 \int_0^a \int_0^{b\left(1-(x/a)^2\right)^{1/2}} \left[\frac{r_j}{a}\left(\frac{x}{a}\right)^{r_j-1}\left(\frac{y_j}{b}\right)^{s_j}\bar{e}_x + \frac{s_j}{b}\left(\frac{x}{a}\right)^{r_j}\left(\frac{y}{b}\right)^{s_j-1}\bar{e}_y\right] \cdot
$$
$$
\left[\frac{r_m}{a}\left(\frac{x}{a}\right)^{r_m-1}\left(\frac{y}{b}\right)^{s_m}\bar{e}_x + \frac{s_m}{b}\left(\frac{x}{a}\right)^{r_m}\left(\frac{y}{b}\right)^{s_m-1}\bar{e}_y\right]dydx
$$
$$
= 4abc^2 \int_0^1 \int_0^{\left(1-(x')^2\right)^{1/2}} \left\{\frac{r_j r_m}{a^2}\left(x'\right)^{r_j+r_m-2}\left(y'\right)^{s_j+s_m}\right.
$$
$$
\left. +\frac{s_j s_m}{b^2}\left(x'\right)^{r_j+r_m}\left(y'\right)^{s_j+s_m-2}\right\}dy'dx'
$$
$$
= 4c^2 \int_0^1 \left\{\left(\frac{b}{a}\right)\frac{r_j r_m}{(s_j+s_m+1)}\left(x'\right)^{r_j+r_m-2}\left[1-\left(x'\right)^2\right]^{(s_j+s_m+1)/2}\right.
$$
$$
\left. +\left(\frac{a}{b}\right)\frac{s_j s_m}{(s_j+s_m-1)}\left(x'\right)^{r_j+r_m}\left[1-\left(x'\right)^2\right]^{(s_j+s_m-1)/2}\right\}dx', \quad s_j > 0,\ s_m > 0
$$

(4)

Cases where either $s_j = 0$ or $s_m = 0$ are readily handled separately. The denominator coefficients are

$$
\mathcal{D}_{j,m} = 4 \int_0^a \int_0^{b\left(1-(x/a)^2\right)^{1/2}} \left(\frac{x}{a}\right)^{r_j}\left(\frac{y}{b}\right)^{s_j}\left(\frac{x}{a}\right)^{r_m}\left(\frac{y}{b}\right)^{s_m}dydx
$$
$$
= 4ab \int_0^1 \frac{1}{(s_j+s_m+1)}\left(x'\right)^{r_j+r_m}\left[1-\left(x'\right)^2\right]^{(s_j+s_m+1)/2}dx'
$$

(5)

The most expedient approach is to use numerical methods to evaluate the integrals. A small saving is that $[\mathcal{D}]$ and $[N]$ are symmetric, so it is sufficient to compute only $\mathcal{D}_{j,m}$ and $\mathcal{N}_{j,m}$ for $m \le j$.

To address the fact that only the value of b/a is specified, let us place the eigenvalue problem into nondimensional form. To that end, let $[\mathcal{N}] = c^2\left[\hat{N}\right]$ and $[\mathcal{D}] = a^2\left[\hat{D}\right]$. Doing so converts Eq. (10.4.64) to

$$
\left[\left[\hat{N}\right] - \frac{a^2}{c^2}\mathcal{R}\left[\hat{D}\right]\right]\{b\} = \{0\}
$$

(6)

The eigenvalues are $\left(a^2/c^2\right)\mathcal{R}_n$, which are approximations of the square of nondimensional natural frequencies $k_n a$.

The computer algorithm that was implemented began by defining row vectors for the exponents in Eq. (3) sequenced in the order of the trial functions, such that

$$
[r] = [0\ 2\ 0\ 4\ 0\ 2], \quad [s] = [0\ 0\ 2\ 0\ 4\ 2]
$$

The coefficients in Eqs. (4) and (5) were determined numerically. The MATLAB numerical integration routine that was used is quadl. It requires that the function that is integrated depends only on the integration variable. This is achieved by defining $[r]$ and $[s]$ prior to an outer loop over $j = 1, 2, .., 6$ and an inner loop over $m = 1, .., j$. Inside these loop, anonymous functions defining each integrand are defined with the r and s values referred to explicitly, for example

```
N_integrand=@(x) 4*b_a*r(j)*r(m)*x.^(r(j)+r(m)- 2).*(sqrt(1 - x.^2))...
       .^(s(j) + s(m) + 1)/(s(j) + s(m) + 1);
```

The name of each function is the first argument for the quadl routine. Completion of the loops over m and j fills in $\left[\hat{\mathcal{N}}\right]$ and $\left[\hat{\mathcal{D}}\right]$. The next set of operations is solution of the eigenvalue problem for series lengths ranging from two to six. This is done by creating a loop for N from to six. Within it, submatrices of $\left[\hat{\mathcal{N}}\right]$ and $\left[\hat{\mathcal{D}}\right]$ are input to the eigenvalue solver. In MATLAB, the operation is [b_vals, eigvals] = eig(N_coeff(1 : N, 1 : N), D_coeff(1 : N, 1 : N));. Each column of b_vals is an eigenvector, and the diagonal elements of eigvals are the corresponding values of $\left(a^2/c^2\right) \mathcal{R}_n$. The diagonal elements of eigvals for each N are saved in a matrix for examination upon completion of all operations.

The eigenvectors are needed to generate the mode functions. A useful way in which they may be visualized is with a surface plot, but doing so entails a change in how Eq. (10.4.68) is evaluated. Suppose $[X]$ and $[Y]$ hold the values of x/a and y/b at a grid of points in the ellipse. (These matrices may be created by the meshgrid command in MATLAB.) Let us expand the summation form of the aforementioned equation, such that

$$\Psi_r^{(N)} = \Psi_1\left(\bar{x}\right)\left(b_1\right)_r + \Psi_2\left(\bar{x}\right)\left(b_2\right)_r + \cdots \qquad (10.4.69)$$

where $\Psi_r^{(N)}$ is the rth mode function, the $\Psi_j\left(\bar{x}\right)$ are trial functions evaluated at $\bar{x}$, and $\left(b_j\right)_r$ is the jth element in column r of the eigenvector matrix. (In the present case, this matrix is b_vals.) If we define $\Psi_j\left(\bar{x}\right)$ to be a matrix evaluated at each point in the grid, then the result will be the value of $\Psi_r^{(N)}$ at all grid points. The sum may be evaluated cumulatively. MATLAB code for these operations is Psi_mode_r=zeros(size(X)); for j=1:N; Psi_j= X.^r(j)* (Y/b_a).^s(j);. Psi_mode_r=Psi_mode_r+Psi_j*b_vals(j,r); end;.

Because $\Psi_1 = 1$, so that $\nabla\Psi_1 = 0$, the first row and column of $[\mathcal{N}]$ are zero, which leads to $\mathcal{R}_1 = 0$ for any series length. The corresponding mode function is a constant. The results for the values of $\mathcal{R}_n\,(a/c)^2$ are described in the tabulation. The separation theorem states that any value in a column for $N > 2$ should be less than or equal to the value to the left of it, and greater than or equal to the value above it. The tabulated values are consistent with this theorem. We also see that the last value of $\mathcal{R}$ at any series length is the one that changes most when a term is added to the series. This means that the last mode, whose natural frequency is highest, will be the most in error. The typical overall trend is that at any series length, the error in the eigenvalue, which may be identified by comparison to a solution of the field equations or by examining the trend in a tabulation, increases with increasing mode number. All except the highest may be expected to be reasonably close to the true value.

	$N = 2$	$N = 3$	$N = 4$	$N = 5$	$N = 6$
$\mathcal{R}_1 \, (a/c)^2$	0	0	0	0	0
$\mathcal{R}_2 \, (a/c)^2$	16.0000	15.4534	11.9353	11.8629	11.7446
$\mathcal{R}_3 \, (a/c)^2$		74.5466	70.7570	53.9301	46.5883
$\mathcal{R}_4 \, (a/c)^2$			87.3552	86.8430	69.9076
$\mathcal{R}_5 \, (a/c)^2$				354.1028	177.7429
$\mathcal{R}_6 \, (a/c)^2$					434.1451

The mode functions converge more slowly than the natural frequencies. The changes as the series is lengthened are less readily quantified, so comparisons usually are visual. Figure 1 compares the result of using Eq. (10.4.68) to evaluate Ψ_2 for several series lengths. The differences are not great. Indeed, the high quality of the $N = 2$ result, for which the corresponding value of $\mathcal{R}_2$ is 36% too large, is somewhat surprising.

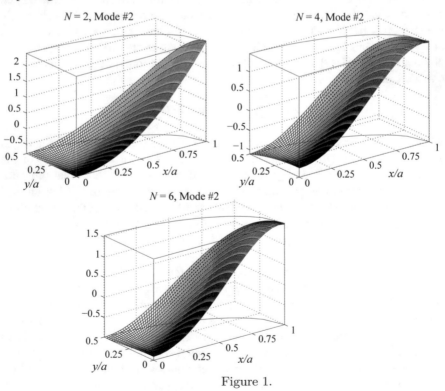

Figure 1.

Length limitations prevent a full comparison of the mode functions for various series lengths. Figure 2 shows some modes for $N = 6$. There seems to be little evidence of the modal distributions for a circular cross section, but Ψ_2 does resemble the $(1, 0)$ mode for a rectangular shape whose aspect ratio is $2 : 1$.

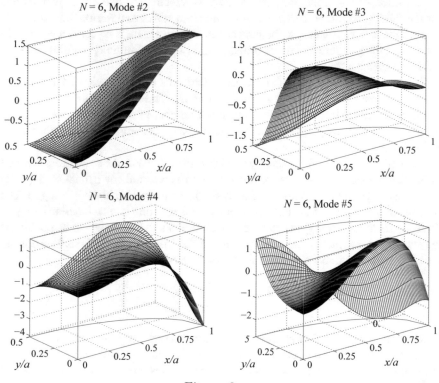

Figure 2.

10.4.2   Dowell's Approximation

Evaluation of the time-domain response within a cavity whose walls are compliant is a core problem for the subject of structural acoustics. Usually, such walls are the surface of an elastic structure. The two media interact in two ways. The pressure exerted by the fluid constitutes a force system that is applied to the structure. Its effect must be incorporated into the structure's equations of motion. The excitation applied to the fluid stems from the requirement that the normal component of its particle velocity and that of the structure be equal everywhere along the surface at which they meet. Analytical solutions of such problems can be found only for very simple circumstances. Numerical modeling techniques often are used, but as noted previously, they might be unwieldy for some applications.

The fundamental formulation of the Rayleigh ratio is valid if the structure is elastic because all that was assumed is that the system is time-invariant and that the sum of kinetic energy and potential energy is conserved in a free vibration. The difficulty with implementation of either the Rayleigh ratio or Rayleigh–Ritz method lies in the selection of trial functions. In addition to selection of one or more functions for the fluid, it would be necessary to formulate a series to describe the structure's displacement. This is where a dilemma arises, because there is no convenient way to assure that the pressure and displacement series satisfy continuity conditions at the fluid-structure interface.

The modeling formulation derived from Dowell's approximation addresses this issue. The analytical method is general. It has the beneficial feature of leading to model equations that feature a relatively small number of variables. However, these benefits are obtained by approximating one of the fundamental fluid-structure inter-action mechanisms.

The portion of the fluid's boundary that is the interface with the structure is denoted as $\mathcal{S}_e$. The displacement of the structure normal to $\mathcal{S}_e$ is an unknown function $w\,(\bar{x}_s, t)$ that must be determined as part of the analysis. The remainder of the bounding surface is denoted as $\mathcal{S}_v$ on which the normal velocity is taken to be a known function $\hat{v}\,(\bar{x}_s, t)$. Thus, $\mathcal{S}_v$ includes any regions that are rigid. Both w and $\hat{v}$ are taken to be positive if they correspond to movement into $\mathcal{V}$, which is the sense in which the surface normal $\bar{n}\,(\bar{x}_s)$ is defined. Hence, continuity at the surface requires that

$$\begin{aligned} \bar{n} \cdot \bar{v} &= \dot{w}, \ \bar{x}_s \in \mathcal{S}_e \\ \bar{n} \cdot \bar{v} &= \hat{v}, \ \bar{x}_s \in \mathcal{S}_v \end{aligned} \tag{10.4.70}$$

The core concept of Dowell's approximation is a series representation of pressure in which the basis functions are the cavity modes for a domain having the same shape, but whose walls are rigid. Analysis of such modes has already been discussed, but their actual determination might represent a significant task if the cavity is irregular in shape. We shall proceed on the assumption that we have found those modes, which we denote as $\Psi_{R,n}$, where n is the mode number and "R" serves to remind us that these are rigid-cavity mode functions. The objective is to derive a set of equations whose solution will describe the pressure field within the cavity and the response of the structure as functions of time. Correspondingly, the coefficients of the cavity mode series are taken to be functions of time to be determined. Thus, the ansatz is

$$\boxed{p\,(\bar{x}, t) = \sum_{n=0}^{N} \Psi_{R,n}\,(\bar{x})\,p_n\,(t)} \tag{10.4.71}$$

where the summation index n begins from zero as a reminder to include the zero frequency mode in which the pressure is uniform throughout the cavity.

The fundamental nature of Dowell's approximation is embedded in the above series. In the time domain, Euler's equation states that the particle acceleration normal to the surface is proportional to the normal component of ∇p. By definition, $\bar{n}\ \cdot$

$\nabla \Psi_{R,n} = 0$ on the boundary. Hence, the series construction gives a zero acceleration on the surface, whereas the time derivative of Eq. (10.4.70) requires that it be $\ddot{w}$. The development proceeds by ignoring this paradox, but the surface continuity conditions will enter the analysis at a later juncture.

In addition to excitation by movement of the boundary, there might be a set of sources at several locations $\bar{x}_m$. Thus, the pressure p is governed by the inhomogeneous wave equation, Eq. (10.3.11), which is

$$\nabla^2 p - \frac{1}{c^2}\frac{\partial^2 p}{\partial t^2} = -\rho_0 \sum_m \dot{Q}_m(t)\, \delta(\bar{x} - \bar{x}_m) \qquad (10.4.72)$$

Equations (10.4.70) are the boundary conditions governing $p(\bar{x}, t)$. The equations satisfied by the rigid-cavity modes are

$$\begin{aligned} \nabla^2 \Psi_{R,n} + (k_{r,n})^2\, \Psi_{R,n} &= 0, \quad \bar{x} \in \mathcal{V} \\ \bar{n} \cdot \nabla \Psi_{R,n} &= 0, \quad \bar{x} \in \mathcal{S}_e \cup \mathcal{S}_v \end{aligned} \qquad (10.4.73)$$

With the aim of employing Green's theorem, we multiply the field equation for a specific rigid body $\Psi_{R,n}$ mode by p, and the field equation for p by $\Psi_{R,n}$. Then, we subtract the latter from the former and integrate over the fluid domain $\mathcal{V}$. The result of these operations is

$$\iiint_{\mathcal{V}} \left\{ p\left[\nabla^2 \Psi_{R,n} + (k_{r,n})^2\, \Psi_{r,n}\right] - \Psi_{r,n}\left[\nabla^2 p - \frac{1}{c^2}\frac{\partial^2 p}{\partial t^2}\right]\right\} d\mathcal{V}$$

$$= \rho_0 \sum_m \dot{Q}_m(t)\, \Psi_{R,n}(\bar{x}_m) \qquad (10.4.74)$$

Green's theorem converts the terms containing the Laplacian to a surface integral according to

$$\iiint_{\mathcal{V}} (p\nabla^2 \Psi_{R,n} - \Psi_{R,n}\nabla^2 p)\, d\mathcal{V} \equiv \iiint_{\mathcal{V}} \left[\nabla \cdot (p\nabla \Psi_{R,n} - \Psi_{R,n}\nabla p)\right] d\mathcal{V}$$

$$= -\iint_{\mathcal{S}} \left[p(\bar{n} \cdot \nabla \Psi_{R,n}) - \Psi_{R,n}(\bar{n} \cdot \nabla p)\right] d\mathcal{S}$$

$$\qquad (10.4.75)$$

The minus sign preceding the surface integral stems from the normal direction $\bar{n}$ being defined to point into $\mathcal{V}$.

The surface integral extends over the entire boundary $\mathcal{S}$. The normal derivative of $\Psi_{R,n}$ is zero everywhere on this surface, whereas $\bar{n} \cdot \nabla p$ is related to the surface normal acceleration by Euler's equation and the time derivative of Eq. (10.4.70). Thus, we have established that

$$\iiint_{\mathcal{V}} \left(p \nabla^2 \Psi_{R,n} - \Psi_{R,n} \nabla^2 p \right) d\mathcal{V} = - \iint_{S_e} \Psi_{R,n} \rho_0 \ddot{w} d\mathcal{S} - \iint_{S_v} \Psi_{R,n} \rho_0 \frac{\partial \hat{v}}{\partial t} d\mathcal{S}$$

$$(10.4.76)$$

It is important to recognized that although the rigid-cavity series cannot give the correct particle velocity at the boundary, the implementation of Green's theorem uses the actual boundary velocity. Thus, the formulation accounts in an average sense for the way movement at the boundary influences the pressure.

Substitution of the preceding relation into Eq. (10.4.74) leads to

$$\iiint_{\mathcal{V}} \Psi_{R,n} \left[\frac{1}{c^2} \frac{\partial^2 p}{\partial t^2} + \left(k_{r,n} \right)^2 p \right] d\mathcal{V} - \iint_{S_e} \Psi_{R,n} \rho_0 \ddot{w} d\mathcal{S} - \iint_{S_v} \Psi_{R,n} \rho_0 \frac{\partial \hat{v}}{\partial t} d\mathcal{S}$$

$$= \rho_0 \sum_m \dot{Q}_m (t) \Psi_{R,n} (\bar{x}_m)$$

$$(10.4.77)$$

The integrands of the surface integrals consist of known quantities, so they belong on the right side of the equation. Furthermore, we have already represented p as a series of rigid-cavity modes according to Eq. (10.4.71). When we substitute that description of p into the preceding, $\Psi_{R,n}$ multiplies every $\Psi_{R,j}$ in the summation. The first orthogonality condition for cavity modes is

$$\iiint_{\mathcal{V}} \Psi_{R,n} \Psi_{R,j} d\mathcal{V} = \mathcal{V} \qquad (10.4.78)$$

Thus, orthogonality filters out of the summation all modes except number n. We thereby obtain a set of uncoupled differential equations for the modal coefficients,

$$\mathcal{V} \left[\ddot{p}_n + \left(\omega_{R,n} \right)^2 p_n \right] = \rho_0 c^2 \iint_{S_e} \Psi_{R,n} \ddot{w} d\mathcal{S} + \rho_0 c^2 \iint_{S_v} \Psi_{R,n} \frac{\partial \hat{v}}{\partial t} d\mathcal{S}$$

$$+ \sum_m \rho_0 \dot{Q}_m (t) \Psi_{R,n} (\bar{x}_m) , \quad n = 0, 1, ..., N$$

$$(10.4.79)$$

where $\omega_{R,n} \equiv c k_{R,n}$ are the natural frequencies of the rigid cavity. Each of the terms in this expression represents a different physical process: $\ddot{p}_n$ represents compressibility, $(\omega_n)^2 p_n$ represents inertia, $\dot{Q}_m$ is an excitation that results from injecting mass into the domain, $\partial \hat{v} / \partial t$ is a surface source distribution resulting from causing the boundary to move, and $\ddot{w}$ is a surface source distribution resulting from movement of the elastic structure at its interface with the fluid.

Equation (10.4.79) constitutes a set of $N + 1$ ordinary differential equations for the $N + 1$ pressure coefficients. If there were no structure, these equations would suffice. Our interest here is the interaction of the acoustic fluid and the structure. In such situations, the surface displacement w is a quantity that must be determined. It is here that the equations of motion for the structure enter the analysis. The displacement of the structure may be described in terms of a set of variables called *generalized coordinates* and denoted as q_j. These variables might be the coefficients of a series, or they might be the mesh displacements in a finite element model. Our interest is in systems that are linear and time-invariant, which means that a reference configuration exists in which the system may remain in a state of static equilibrium. In such cases, it is possible to define the generalized coordinates such that zero values for all place the structure at its reference position. For a given system, there are several methods for formulating the differential equations of motion governing the generalized coordinates. Usually, the derivation of those equations is founded on Lagrange's equations, Hamilton's principle, or the principle of virtual work, which are closely related.[14] The result will be the structure's equations of motion in a canonical form, whose matrix description is

$$[M]\{\ddot{q}\} + [K]\{q\} = \{F\} + \{F_{\text{pressure}}\} \tag{10.4.80}$$

The matrices $[M]$ and $[K]$, which are constant, square, and symmetric, are the *inertia* and *stiffness*. Another structural effect is dissipation, but the actual mechanisms by which energy is lost usually are uncertain. Consequently, dissipation is usually introduced at a later stage.

The elements of the column vectors $\{F\}$ and $\{F_{\text{pressure}}\}$ are *generalized forces*. The former represents the excitation forces acting on the structure, whereas the latter contain the effects of the surface pressure. Determination of $\{F\}$ is part of the standard process by which equations of motion are derived. Determination of $\{F_{\text{pressure}}\}$ is an extension of that process. Identification of the force terms begins with a general description of the surface displacement $w(\bar{x}_s, t)$. A linear structural model leads to a description of the displacement of any point as a series in which each generalized coordinate is multiplied by a spatial function. These functions depend on how the structure is modeled, but the details are not important here. The surface normal displacement is obtained by evaluating that series at an arbitrary surface point $\bar{x}_s$ and then taking a dot product of that vector with the outward normal $\bar{n}(\bar{x}_s)$. The consequence is that w also is represented as a linear series in which the spatial functions depend on $\bar{x}_s$. The general form of this series is

$$w = \sum_{m=1}^{M} \chi_m(\bar{x}_s) q_m \tag{10.4.81}$$

Clearly, the specific form of the χ_m functions depends on how the structure is modeled.

The surface displacement enters the description of the pressure loading through the concept of *virtual work*. This quantity is unlike the actual work done by forces. Rather, it is the work that would result if each generalized coordinate were incremented by an arbitrary differential amount. These increments are denoted as δq_m to distinguish

---

[14]J.H. Ginsberg, *Engineering Dynamics*, Cambridge University Press (2008) Chap. 7

them from actual increments dq_m that occur in the course of a response. Increments of the generalized coordinates lead to a *virtual displacement*. At the surface, this displacement is δw. The functions χ_m are set by the manner in which the structure has been modeled and therefore are unchanged. Thus, the virtual displacement is given by

$$\delta w = \sum_{m=1}^{M} \chi_m (\bar{x}_s) \, \delta q_m \tag{10.4.82}$$

The resultant force exerted by the pressure acting on a patch of the surface is $p(x_s, t) \, dS$ in the direction of $-\bar{n}(\bar{x}_s)$ and δw at that location is positive if it is in the direction of $+\bar{n}(\bar{x}_s)$. Therefore, virtual work done by the pressure acting on S_e is

$$\delta W_{\text{pressure}} = -\iint_{S_e} p(\bar{x}_s, t) \, \delta w \, dS \tag{10.4.83}$$

Substitution of the cavity mode series for p and the preceding expression for δw into the virtual work gives

$$\delta W_{\text{pressure}} = -\sum_{n=0}^{N} \sum_{m=1}^{M} \iint_{S_e} \left[p_n \Psi_{R,n} (\bar{x}_s) \right] \left[\chi_m \, \delta q_m \right] dS \tag{10.4.84}$$

On the other hand, the fundamental definition of the elements of $\{F_{\text{pressure}}\}$ is that they are the coefficients in a canonical description of $\delta W_{\text{pressure}}$, specifically

$$\delta W_{\text{pressure}} = \sum_{m=1}^{M} \left(F_{\text{pressure}} \right)_m \delta q_m \tag{10.4.85}$$

These alternative descriptions of $\delta W_{\text{pressure}}$ must be the same for any set of δq_j values. This can only be true if corresponding coefficients of δq_m are the same, which leads to

$$\left(F_{\text{pressure}} \right)_m = -S_e \sum_{n=0}^{N} p_n \Lambda_{m,n}$$
$$\Lambda_{m,n} = \frac{1}{S_e} \iint_{S_e} \chi_m (\bar{x}_s) \, \Psi_{R,n} (\bar{x}_s) \, dS \tag{10.4.86}$$

It follows that the pressure force vector is

$$\{F_{\text{pressure}}\} = -S_e [\Lambda] \{p\} \tag{10.4.87}$$

In turn, the structure's equations of motion become

$$\boxed{[M] \{\ddot{q}\} + [K] \{q\} + S_e [\Lambda] \{p\} = \{F\}} \tag{10.4.88}$$

The series description of surface displacement, Eq. (10.4.81), is also used to describe the surface acceleration term in the pressure equations. Substitution of the displacement series into Eq. (10.4.79) shows that

$$
\iint\limits_{\mathcal{S}_e} \Psi_{R,n} \ddot{w} d\mathcal{S} = \sum_{m=1}^{M} \left[\iint\limits_{\mathcal{S}_e} \Psi_{R,n}(\bar{x}_s)\,\chi_m(\bar{x}_s)\,d\mathcal{S} \right] \ddot{q}_m = \mathcal{S}_e \sum_{m=1}^{M} \Lambda_{m,n} \ddot{q}_m \quad (10.4.89)
$$

When the preceding expression for the surface displacement is substituted into Eq. (10.4.79), the result is

$$
\mathcal{V}\left[\ddot{p}_n + \left(\omega_{R,n}\right)^2 p_n \right] - \rho_0 c^2 \mathcal{S}_e \sum_{m=1}^{M} \Lambda_{m,n} \ddot{q}_m = \Gamma_n(t), \quad n = 0, 1, ..., N
$$

$$(10.4.90)$$

where the source coefficients are

$$
\Gamma_n = \rho_0 c^2 \iint\limits_{\mathcal{S}_v} \Psi_{R,n}(\bar{x}_s) \frac{\partial}{\partial t}\hat{v}(\bar{x}_s, t)\, d\mathcal{S} + C_0 \sum_m \Psi_{R,n}(\bar{x}_m)\, \dot{Q}_m(t) \quad (10.4.91)
$$

In keeping with the matrix representation of the structural equations of motion, the equations for the acoustical response may be written in that form. Stacking the pressure equations above the structural equations leads to a set of ordinary differential equations whose form is

$$
[\mathcal{M}] \frac{d^2}{dt^2} \left\{ \begin{matrix} \{p\} \\ \{q\} \end{matrix} \right\} + [\mathcal{K}] \left\{ \begin{matrix} \{p\} \\ \{q\} \end{matrix} \right\} = \{\mathcal{F}\}
$$

$$(10.4.92)$$

The system matrices are

$$
[\mathcal{M}] = \begin{bmatrix} \mathcal{V}[I] & -\rho_0 c^2 \mathcal{S}_e [\Lambda]^{\mathrm{T}} \\ [0] & [M] \end{bmatrix}
$$

$$
[\mathcal{K}] = \begin{bmatrix} \mathcal{V}\left[(\omega_R)^2\right] & [0] \\ \mathcal{S}_e [\Lambda] & [K] \end{bmatrix} \quad (10.4.93)
$$

$$
\{\mathcal{F}\} = \left\{ \begin{matrix} \{\Gamma\} \\ \{F\} \end{matrix} \right\}
$$

where $\left[(\omega_R)^2\right]$ is a diagonal matrix whose elements are the rigid-cavity natural frequencies and $[I]$ is an identity matrix. Thus, we have derived a set of coupled equations whose number equals the number of unknown pressure coefficients, $N + 1$, plus the number of unknown displacement coefficients, M. The column vector $\{\Gamma\}$

represents source excitations stemming from injecting mass into the fluid, as well as from imposing a movement of the boundary. The displacement and pressure variables are coupled by the matrix $[\Lambda]$.

These equations cover most situations, but the case where some region in the wall is pressure-release is not addressed. Just as rigid mode series, Eq. (10.4.71), cannot give a nonzero normal velocity on the boundary, it cannot give a zero pressure on the boundary because the rigid-cavity modes are not zero there. An approximation of that condition may be obtained by replacing the pressure-release region with an extremely compliant structure, such as an extremely thin membrane whose surface tension is very low. However, doing so might cause the final equations to be ill-conditioned if the thickness and surface tension are too small.

How to solve the equations derived from Dowell's approximation depends on the type of response we seek. If the excitation is time harmonic, we may obtain the steady-state response from a complex variable representation. We set $p_n = \text{Re}\left(P_n e^{i\omega t}\right)$, $q_n = \text{Re}\left(\xi_n e^{i\omega t}\right)$, $\Gamma_n = \text{Re}\left(\hat{\Gamma}_n e^{i\omega t}\right)$, and $F_n = \text{Re}\left(\hat{F}_n e^{i\omega t}\right)$, which leads to

$$\left[[\mathcal{K}] - \omega^2 [\mathcal{M}]\right] \left\{ \begin{matrix} \{P\} \\ \{\xi\} \end{matrix} \right\} = \left\{ \begin{matrix} \{\hat{\Gamma}\} \\ \{\hat{F}\} \end{matrix} \right\} \tag{10.4.94}$$

Resonances occur at frequencies for which the determinant $\left|[\mathcal{K}] - \omega^2 [\mathcal{M}]\right|$ vanishes.

Solution of the differential equations in the time domain is a little more difficult. Coupled linear ordinary differential equations like Eq. (10.4.92) arise in mechanical and structural vibratory systems. In that area, the equations are commonly solved by modal analysis, in which the modes constitute the eigensolutions of the homogeneous equations. That formulation is not directly applicable here because the coefficient matrices are not symmetric. Consequently, a modal analysis would require consideration of left and right eigenvectors, which is the way in which the equations of motion for rotordynamic systems may be solved.[15] The formulation of such solutions exceeds the scope of the present work.

The differential equations also may be solved by implementing a numerical algorithm, such as one of the Runge-Kutta versions. Most algorithms require that Eq. (10.4.92) be placed in first-order form. This is done by defining a state vector $\{Z\}$ that consists of all unknown variables and their first derivatives,

$$\{Z\} = \left[\{p\}^{\text{T}} \quad \{q\}^{\text{T}} \quad \{\dot{p}\}^{\text{T}} \quad \{\dot{q}\}^{\text{T}}\right]^{\text{T}} \tag{10.4.95}$$

The variables and their first derivatives may be extracted from $\{Z\}$ according to

$$\left\{ \begin{matrix} \{p\} \\ \{q\} \end{matrix} \right\} = [[I] \quad [0]] \{Z\}, \quad \frac{d}{dt}\left\{ \begin{matrix} \{p\} \\ \{q\} \end{matrix} \right\} = [[0] \quad [I]] \{Z\} \tag{10.4.96}$$

---

[15]Ginsberg, Mechanical and Structural Vibrations (2001), Ch. 11.

where the size of the partitions equals the number of unknowns contained in $\{p\}$ and $\{w\}$, which is $N + M + 1$. The *derivative identity* describes the fact that the derivative of the first expression equals the second, from which it follows that

$$[[I] \quad [0]] \frac{d}{dt} \{Z\} = [[0] \quad [I]] \{Z\} \tag{10.4.97}$$

We also use Eq. (10.4.96) to describe the basic differential equations, Eq. (10.4.92), as

$$[[0] \quad [\mathcal{M}]]^{\mathrm{T}} \frac{d}{dt} \{Z\} + [[\mathcal{K}] \quad [0]] \{Z\} = \{F\} \tag{10.4.98}$$

Stacking the derivative identity above this form yields

$$\left[\begin{bmatrix} [I] & [0] \\ [0] & [\mathcal{M}] \end{bmatrix} \frac{d}{dt} \{Z\} = \begin{bmatrix} [0] & [I] \\ -[\mathcal{K}] & -[0] \end{bmatrix} \{Z\} + \begin{Bmatrix} \{0\} \\ \{F\} \end{Bmatrix} \right] \tag{10.4.99}$$

To solve these equations with mathematical software, one merely needs to provide the equation solver with the coefficient matrix multiplying $d\{z\}/dt$ and a function that evaluates the right side of the equality sign at an arbitrary instant based on the current value of $\{z\}$. Initial values required to begin the integration process are the values of $\{p\}$, $\{w\}$, $\{\dot{p}\}$, and $\{\dot{w}\}$ at $t = 0$.

We have seen that Dowell's approximation leads to a set of differential equations that are readily solved and are quite comprehensive in the variety of systems to which the may be applied. However, they do have two flaws. The structural and fluid velocities are not inherently continuous in the normal direction at the interface where the media meet. This was observed at the outset, but another fault was encountered when the differential equations were assembled. The coupled equations are not symmetric. It was shown in Chap. 6 that the laws of acoustics are reciprocal. That is suppose a source is situated at one point and the response is measured at another. The same response would be measured at the original source location if the source were applied at the measurement point. It can be proven that this property requires that the matrix equations for the response be symmetric. This is not the nature of Eq. (10.4.92). Another consequence of the asymmetry of $[\mathcal{M}]$ and $[\mathcal{K}]$ arises in the computation of natural frequencies. These correspond to nonzero responses in the absence of excitation, so they are solutions of an eigenvalue problem,

$$\left[[\mathcal{K}] - \omega^2 [\mathcal{M}] \right] \left[\{p\}^{\mathrm{T}} \quad \{q\}^{\mathrm{T}} \right]^{\mathrm{T}} = 0 \tag{10.4.100}$$

It is possible that some eigenvalues ω^2 will be negative, or they might occur as pairs of complex conjugate values, even though both matrices are real. Neither condition corresponds to a periodic motion. Such a prediction is erroneous, because periodic free vibration is the only possibility in any system that does not dissipate energy, as we saw in the development of the Rayleigh ratio.

Despite these flaws, results obtained from Dowell's method have been found to be quite usable, especially when the fluid is air. Indeed, the formulation has been widely employed to predict cabin noise for aircraft and automobiles. The reason for this success is that the air presents a low impedance to movement of a solid structure. Consequently, the fluid's particle velocity at the interface will be much lower than it is somewhere in the interior of the cavity. (Such an assertion is not likely to be true for a heavy fluid like water.) In addition, issues regarding reciprocity seldom are at the forefront of the questions that the model equations must answer. In any event, the availability of a rather accessible simulation technique makes Dowell's approximation a very useful predictive tool for design and diagnostic tasks.

EXAMPLE 10.13 The sketch depicts a rectangular cavity in which both walls parallel to the plane of the sketch are rigid. At the top, a portion of the wall consists of a plate whose mass is M_1. The wall at the left side is another plate whose mass is M_2. Both plates are extremely stiff, so their movement is essentially a rigid body translation, vertically for the upper plate and horizontally for the side plate. The width of both plates perpendicular to the plane of the diagram is the same as the width of the waveguide, which is 300 mm. The upper plate is supported by spring K and dashpot D, so this subsystem constitutes a one-degree-of-freedom oscillator. A horizontal force applied to the plate on the left is adjusted such that at any frequency ω, it induces a translational velocity of $V_0 \sin(\omega t)$, where $V_0 = 50$ mm/s. All other regions of the wall are rigid. The value of K/M_1 has been selected such that the natural frequency of the oscillator in the absence of fluid loading would equal the lowest nonzero natural frequency of the cavity if all walls were rigid. The value of D is set to give a 0.001 ratio of critical damping for the isolated oscillator. Both M_1 and M_2 are fifty times greater than the mass of the fluid within the cavity. Determine the pressure at three points: $x = 90$, $y = 37.5$ mm; $x = 180$, $y = 75$ mm; and $x = 360/2^{1/2}$, $y = 150/2^{1/2}$ mm, as a function of frequency up to 25% greater than the fourth natural frequency of the system. Also, determine the amplitude of the plate displacement within this range.

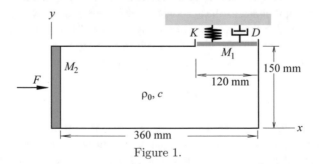

Figure 1.

Significance

The analysis will demonstrate all of the operations required to implement Dowell's approximation without requiring that we formulate a complicated model of the

structure. Features of special interest are the way in which the discontinuous properties of the surface are handle and the influence of the structure on the overall response
characteristics.

Solution

The coordinate system we shall use is shown in the sketch. Each plate is stated
to only translate in the direction normal to its surface. "Translation" means that
all locations execute the same displacement in the z direction. The motion of the
plate at $x = 0$ is the excitation, and a pressure field that is independent of z may
be sustained because the sidewalls are rigid. This observation permits formulation
of a two-dimensional model, in which the pressure field depends only on the x and
y coordinates. Correspondingly, integrals over a surface will feature a width factor,
$W = 0.3\,\text{m}$.

The interface of the horizontal plate and the fluid, which constitutes S_e, occupies
$0.24 < x < 0.36$, $y = 0.15\,\text{m}$. The total area of this portion of the surface is $S_e = W\,(0.12) = 0.036\,\text{m}^2$. All points on this surface displace by the same amount $w\,(t)$,
positive for downward movement (into the fluid). This corresponds to Eq. (10.4.81)
with a single generalized coordinate and a surface displacement function that is one,
that is,

$$w = \chi_1 q, \quad \chi_1 = 1 \text{ for } 0.24 < x < 0.36 \qquad (10.4.101)$$

The plate that is situated at $x = 0$ undergoes a specified motion, and the remainder
of the boundary is rigid. This means that all portions of the boundary excluding the
oscillator's face constitute S_v, with $\hat{v} = \text{Re}\left(V_0 e^{i\omega t}\right)$ on $x = 0$ and $\hat{v} = 0$ elsewhere.
Consequently, integrals containing $\hat{v}$ will have a nonzero contribution from $x = 0$
only.

A rigid-cavity modal series was formulated in Example 10.8. A variant of the
single index sequencing scheme used there is somewhat simpler to formulate. Index
m is associated with cavity mode (j, n), with n being the number for x dependence.
The terms associated with each cavity mode may be defined in a double loop over
$j = 0, ..., J$ and $n = 0, ..., N$. The associated value of m is

$$m = n + 1 + j\,(N + 1), \quad j = 0, 1, ..., J, \quad n = 0, 1, ..., N \qquad (1)$$

where the highest mode numbers in each direction are J and N. Correspondingly,
the rigid-cavity modal properties are

$$\Psi_{R,m} = B_m \cos\left(\frac{j\pi x}{L}\right) \cos\left(\frac{n\pi y}{H}\right), \quad m = 1, 2, ..., (J + 1)\,(N + 1)$$

$$\omega_{R,m} = c\left[\left(\frac{j\pi}{L}\right)^2 + \left(\frac{n\pi}{H}\right)^2\right]^{1/2} \qquad (2)$$

where $H = 0.15\,\text{m}$ and $L = 0.36\,\text{m}$. The modes are normalized according to
Eq. (10.3.8), which gives

$$B_m = \frac{2}{\left(1 + \delta_{0,j}\right)^{1/2} \left(1 + \delta_{0,n}\right)^{1/2}} \tag{3}$$

There are no sources within the fluid, so the source coefficients only contain a contribution from the imposed motion of the plate on the left. Therefore, the surface integral in Eq. (10.4.91) for these coefficients extends only over that plate,

$$
\begin{aligned}
\Gamma_{m,1} &= \rho_0 c^2 \iint_{S_v} \Psi_{R,m}(\bar{x}_s) \frac{\partial}{\partial t} \hat{v}(\bar{x}_s, t) \, dS \\
&= \rho_0 c^2 \int_0^H \Psi_{R,m}(\bar{x}_s)\big|_{x=0} (\omega V_c \cos(\omega t))(W \, dy) \\
&= \rho_0 c^2 W \int_0^H B_m \cos\left(\frac{n\pi y}{H}\right) (\omega V_0 \cos(\omega t)) \, dy = \rho_0 c^2 W H B_m \omega V_0 \cos(\omega t) \, \delta_{n,0}
\end{aligned}
\tag{4}
$$

This tells us that only the modes that are constant in the y direction are directly excited.

The matrix $[\Lambda]$ mathematically couples the modes and the displacement of the upper plate. The coefficients obtained from Eq. (10.4.86) are

$$
\begin{aligned}
\Lambda_{1,m} &= \frac{1}{S_e} \int_{2L/3}^L \left[\chi_1(\bar{x}_s) \Psi_{R,m}(\bar{x}_s) \right]\big|_{y=H} (W \, dx) \\
&= \frac{W}{S_e} \int_{2L/3}^L B_m \cos\left(\frac{j\pi x}{L}\right) \cos(n\pi) \, dx \\
&= \begin{cases} \dfrac{W}{S_e} \left(\dfrac{L}{j\pi}\right) B_m \sin\left(\dfrac{2j\pi}{3}\right) (-1)^{n+1} & \text{if } j \neq 0 \\[2mm] \dfrac{WL}{3S_e} B_\ell (-1)^n & \text{if } j = 0 \end{cases}
\end{aligned}
\tag{5}
$$

Note that each element of $[\Gamma]$ and $[\Lambda]$ is computed within the loops over j and n.

The equations for the acoustical field are supplemented by those for the structure. In the present case, a single differential equation of motion governs the displacement variable q of the oscillator. The only external force acting on this system is the resultant force exerted by the pressure field, which is obtained by integrating the pressure over the face of the piston. Because a positive q has been defined to be positive for downward displacement, a positive pressure represents a negative resultant force. Thus, this force is

$$
\begin{aligned}
F_{\text{pressure}} &= -\iint_{S_e} p(\bar{x}_s, t) \, dS = \sum_{m=1}^{(J+1)(N+1)} \int_{2L/3}^L \Psi_{R,m}(\bar{x}_s)\big|_{y=H} \, P_m (W \, dy), \\
&= -S_e \sum_{m=1}^{(J+1)(N+1)} \Lambda_{1,m} P_m
\end{aligned}
\tag{6}
$$

The equation of motion for the oscillator is $M_1 \ddot{q} + C\dot{q} + Kq = F_{\text{pressure}}$. Substitution of Eq. (6) gives

$$M_1 \ddot{q} + C\dot{q} + Kq + S_e \sum_{m=1}^{(J+1)(N+1)} \Lambda_{1,m} P_m = 0 \tag{7}$$

The mass is specified to be $M_1 = 50\rho_0 \mathcal{V}$, and K is set by the requirement that the isolated natural frequency $(K/M_1)^{1/2}$ equals the fundamental nonzero natural frequency of the rigid-cavity. This frequency belongs to either the $(1, 0)$ or $(0, 1)$ mode. Because $L > H$, the former is the choice, so we set

$$\frac{K}{M_1} = c^2 \left(\frac{\pi}{L}\right)^2$$

The fundamental nonzero frequency of the cavity is 472 Hz. The other parameter is D, which sets the ratio of critical damping to 0.001. This ratio is such that division of Eq. (7) by M_1 converts the coefficient of $\dot{q}$ to $D/M_1 = 2\,(\text{ratio})\,(K/M_1)^{1/2}$, so $D = 2\,(0.001)\,(KM_1)^{1/2}$. The values that result are

$$M_1 = M_2 = 0.972 \text{ kg}, \quad K = 8.57\left(10^6\right) \text{ N/m}, \quad D = 57.7 \text{ N-s/m}$$

The general frequency-domain equations are described in Eq. (10.4.94). To adapt them to the present analysis, we populate $\{P\}$ with the P_m coefficients and replace the structural displacement vector $\{W\}$ with a single coefficient $\hat{q}$, which is the complex amplitude of q. Correspondingly, $[\Lambda]$ is a row vector formed from the coefficients $\Lambda_{1,\ell}$, $[K]$ is the spring stiffness K, and $[M]$ is the piston mass M_2. The structural damping term gives rise to a term $i\omega C$ that adds to $K - \omega^2 M$ in the oscillator's equation of motion. The frequency-domain equations that result are

$$\begin{bmatrix} \mathcal{V}\left[\left[(\omega_R)^2\right] - \omega^2\,[I]\right] & -\rho_0 c^2 S_e \omega^2\,[\Lambda]^T \\ S_e\,[\Lambda] & \left(K + i\omega C - \omega^2 M_2\right) \end{bmatrix} \begin{Bmatrix} \{P\} \\ \xi \end{Bmatrix} = \begin{Bmatrix} \{\hat{\Gamma}\} \\ \{0\} \end{Bmatrix} \tag{8}$$

The computational procedure begins by computing all quantities that do not depend on the frequency. This is done with a pair of nested loops in which j ranges from 0 to J, within which n ranges from 0 to N. Inside these loops, the vector index is $m = n + 1 + j\,(N + 1)$. The mth element of $\left[(\omega_R)^2\right]$, $[\Lambda]$, and $\left\{\tilde{Q}\right\}$ are evaluated. Then, a loop over frequency formulates and solves Eq. (8) over a range of frequencies. The column vector $\{P\}$ of complex pressure amplitudes at each ω is saved as a column of an rectangular matrix $[P_{\text{all}}]$, and the corresponding complex displacement amplitude ζ is saved as an element of a vector.

One response that is requested is the pressure at the midpoint of the cavity. Equation (10.4.71) indicates that at any point (x_f, y_f) the pressure is

$$P = \sum_{m=1}^{(J+1)(N+1)} P_m B_m \cos\left(\frac{j\pi x_f}{L}\right) \cos\left(\frac{n\pi y_f}{H}\right) \tag{9}$$

The upper frequency limit is specified to be 25% greater than the system's fourth nonzero natural frequency. We do not know what that frequency is, so let us assume that the natural frequencies are close to those of the rigid cavity. The lower natural frequencies correspond to $n = 0, 1,$ or 2 and $j = 0, 1,$ or 2. The values are tabulated below.

	$n = 0$	$n = 1$	$n = 2$
$j = 0$	0	7121	14 242
$j = 1$	2967	7714	14 548
$j = 2$	5934	9269	15 429

The fourth nonzero value belongs to the $(1, 1)$ mode. Setting the cutoff to 25% above that value gives 9643 rad/s. We will round it up to 1.6 kHz.

We must decide how many modes to include. One guideline comes from making sure that the highest natural frequency of the model is well above the maximum excitation frequency. This will not be a difficult criterion to meet. A more stringent requirement is that the number of modes should be sufficient to recreate in a modal series the impedance discontinuity of the upper wall, but that provides no definitive guideline. It is reasonable to set the highest mode numbers J and N such that the $(0, N)$ and $(J, 0)$ modes have comparable natural frequencies and that these frequencies are at least twice the highest excitation frequency. We shall try $J = 10$, which leads to $N = \text{ceil}(J (H/L)) = 5$. Because of the vagueness of these guidelines, the computed results were verified by reducing J to 5 and N to 3. Doing so resulted in barely perceptible alterations of the graphed data.

The upper graph in Fig. 2 describes the pressure responses at the prescribed locations. The lower graph shows the corresponding plate displacement. The vertical lines mark the rigid-cavity natural frequencies. The peaks that occur are very close to these values. This attribute is a consequence of the oscillator having a large mass. It appears to the fluid to be a nearly rigid obstacle. Consequently, the resonances in the frequency responses occur close to the natural frequencies of the rigid-cavity modes. However, the pressure at the midpoint is very small when ω is close to the natural frequency of the $(1,0)$ rigid body mode. The pressure at the other locations seems to be nullified at that frequency. The explanation for this feature lies in the concept of a tuned vibration absorber, in which a low-mass oscillator is mounted on a system. The natural frequency of the attachment is set equal to that of the unadorned system. The result is that the previously unadorned part of the system does not respond at that frequency. The piston-spring oscillator on the upper wall serves the same purpose, except that its mass is much greater than the fluid.

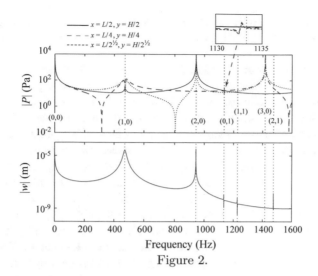

Figure 2.

It still remains to explain why some locations do not show evidence of resonances that are seen at other locations. This behavior is a corollary of the proximity of resonances to the rigid-walled behavior. Consequently, at each resonant frequency $\omega_m \approx (\omega_R)_m$, the pressure coefficient P_m for the resonant mode is much larger than the other coefficients. When the rigid-cavity modal series is evaluated at this frequency, the mth mode will make the dominant contribution. However, if the corresponding mode function $\Psi_{R,m}$ is small at a specific location, then the pressure there will be much smaller than it is elsewhere. In the case of the midpoint, where $x = L/2$ and $y = H/2$, we have $\Psi_{R,m} = B_\ell \cos(j\pi/2) \cos(n\pi/2)$. Thus, if a resonance corresponds to a cavity mode for which j or n is odd, then that resonance will not be observed at the midpoint. Because $x = 360/2^{1/2}$ and $y = 150/2^{1/2}$ are irrational numbers, none of the rigid-cavity mode functions are zero at that point. Consequently, every natural frequency is manifested in a peak of the pressure at that point.

Another interesting feature is the narrow peak and adjacent null that occurs at some frequencies. These all correspond to modes for which $n = 1$. The explanation lies in the source coefficients $\hat{Q}_\ell$, which represent the excitation. According to Eq. (4), these values are nonzero only if $n = 0$. Thus, only the modes that correspond to $n = 0$ are excited directly. If the resonance is of a cavity mode for which $n \neq 0$, the excitation is much weaker than it is for $n = 0$ modes. The drastic change from a maximum to near-null results from the fact that crossing a resonance results in 180° phase shift for the P_ℓ coefficients that are resonant, so instead of adding to the contribution of the nonresonant coefficients, they subtract.

The last effect of note is the fact that the oscillator displacement shows no evidence of the (3, 0) resonance, even though the pressure does. This feature is a consequence of the fact that the length of the piston in the x direction is $L/3$. Equation (5) indicates that the $\Lambda_{1,\ell}$ values are zero if ℓ corresponds to a mode for which j is an integer multiple of 3. This is the mathematical consequence of the fact that $\Lambda_{1,\ell}$ represents the interaction between the piston and a mode and that interaction averages out over the length of the piston if the pressure along it varies as $\cos(3\pi x/L)$, $\cos(6\pi x/L)$, etc

Many of the features discussed in the preceding arise because the structural mass M_2 is large. A system in said to be a case of "light fluid loading" if the structure's mass is large compared to the fluid's. Decreasing the mass of the upper plate leads to system with "heavy fluid loading," although it would be more accurate to say that it is a system with low structural impedance. Dowell's approximation might not be applicable in this parameter range. Nevertheless, carrying out the computation with the plate mass decreased by a factor of 50, so that it equals the mass of the fluid within the cavity and enables us to see how the mass ratio affects the response. Although the plate mass is decreased, the value of K/M_2 is maintained. This has the effect of increasing the importance of the term containing $[\Lambda]$ in the lower part of Eq. (8). As a result, coupling of the P_m coefficients is stronger. Figure 3 shows the pressure responses for this case. The oscillator now is a less substantial obstacle to movement of the fluid, so the cavity's behavior differs considerably from a cavity with rigid walls. One consequence is that the resonant frequencies are noticeably different from the natural frequencies of the rigid-cavity. A by-product is that resonances are more likely to be seen at all points. Stronger coefficient coupling also accounts for the increased occurrence of displacement nulls at various frequencies.

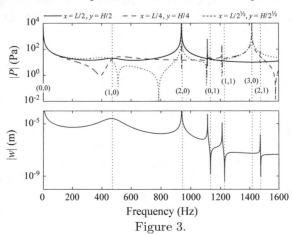

Figure 3.

An overview indicates that Dowell's approximation is readily implemented to describe fairly complicated systems. The most difficult operation in such cases is likely to be determination of the rigid-cavity modes, which are a prerequisite to the formulation. That was not aan issue for the box cavity, nor would it be for any of the rectangular, cylindrical, or spherical cavities we have studied.

10.5   Homework Exercises

Exercise 10.1 An exponential horn whose length is L is closed at its narrow end and open to the atmosphere at its wide end. The cross section is sufficiently small that only plane waves can propagate within it. Derive the characteristic equation whose roots are the horn's natural frequencies. Then, evaluate these frequencies for a horn whose length is $L = \pi/k$ with $b = 0.08/L$. Compare these values to the natural frequencies for plane waves in a uniform waveguide whose length is L with one end open and the other closed.

Exercise 10.2 The sketch depicts a two-dimensional model of a box filled with a liquid. The sides are rigid, and the top is open to the atmosphere. The bottom is bonded to the ground, so any vertical motion of the ground is replicated by the bottom. In the situation of interest, vibration of a nearby machine induces a vertical displacement $w = W \cos (2\pi x/L) \sin (\omega t)$. Derive an expression for the pressure at any location within the tank. Use that expression to evaluate the vertical velocity $v_y/ (W\omega)$ at the midpoint of the free surface for frequencies in the range $0 < kL \le 4\pi$. The aspect ratio is $H/L = 0.6$.

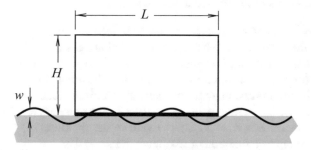

Exercises 10.2, 10.3, and 10.4.

Exercise 10.3 Consider the situation in which a polymer sheet is placed on the top of the liquid of the box in Exercise 10.2. The result is that the specific impedance at this surface $\zeta = 0.8 - 3i$. Evaluate the vertical velocity $|v_z| / (W\omega)$ at the midpoint of the free surface for frequencies in the range $0 < kL \le 4\pi$. Then, compare this response to that which would be obtained with a pressure-release condition, $\zeta = 0$ The aspect ratio is $H/L = 0.6$.

Exercise 10.4 The system of interest is the one in Exercise 10.2, with the sole difference that the vertical wall at $x = L$ is an intertance, with $\zeta = 0.25i$. The aspect ratio is $H/L = 0.6$. The excitation frequency is the lowest value at which the system would resonate if $x = L$ were rigid. Evaluate the vertical velocity $|v_z| / (W\omega)$ at the midpoint of the free surface.

Exercise 10.5 A container whose sides are otherwise rigid is excited by an oscillation in which the vertical wall at $x = 0$ pivots about its horizontal centerline. The result is that its normal velocity is $v_x = v_0 (y/H - 0.5) \cos (\omega t)$. The dimensions of the container are L, H, and W in the x, y, and z directions, respectively. Derive an expression that describes the pressure field within the container.

Exercise 10.6 A rectangular water tank has sides that are composed of very thick glass panels that may be taken to be rigid. The top, $y = H$, is open to the atmosphere, so it is pressure-release. The bottom, $y = 0$, is a thin steel plate that flexes because of ground vibration. The consequence is that the plate undergoes a vibration normal to its surface, specifically, $v_z = v_0 \sin (\pi x/L)$. The dimensions of the bottom plate are $L \times W$. Derive an expression for the pressure $p/ (\rho_0 c v_0)$ at an arbitrary location within the tank.

Exercise 10.7 The normal velocity at the left side of a water tank, $x = 0$, is harmonic with a parabolic dependence on y given by $v_z = v_0(y/b - y^2/b^2) \sin(\omega t)$ for $0 < y < b$. The opposite side, $x = a$, is locally reacting, with local impedance $\zeta_a = 0.6 - 0.3i$. The sides $z = 0$ and $z = d$ are rigid, as is the bottom, $y = 0$, whereas $y = b$ is pressure-release. (a) Derive an expression for the pressure at any location. (b) The dimensions are $a = 10$ mm, $b = 5$ mm, and $d = 3$ mm. Evaluate the pressure at the midpoint $x = a/2$, $y = b/2$, and $z = d/2$. The frequency range is $ka \le 16$. Are resonances evident in this frequency response?

Exercise 10.8 An engineered material has been installed in the ceiling of a closed room. It has the property of behaving like a continuous sheet of bricks, so that it is locally reacting, with an impedance of $0.1kL$. The wall at $x = 0$ undergoes a translational oscillation in which $v_x = v_0 \sin (\omega t)$. The other walls and the floor are rigid. (a) Derive an expression for the pressure at an arbitrary location. (b) Identify the natural frequencies by examining the expression for the pressure. (c) The proportions of the room are $L : H : W = 5 : 3 : 4$. Evaluate the pressure $|p| / (\rho_0 c v_0)$ at the middle of the room as a function of frequency for $ka < 12$.

Exercise 10.9 The walls, floor, and ceiling of a reverberant room are essentially rigid. Its height is H, and its horizontal dimensions are $L \times W$. It is excited by a loudspeaker that emits a constant amplitude at frequency ω. The loudspeaker is situated at the middle of the ceiling. It is sufficiently small that it may be considered to be a point source having volume velocity is $\text{Re} \left(\hat{Q} \exp (i\omega t) \right)$. Derive an expression for the pressure at the middle of the room.

Exercise 10.10 The sound pressure at the middle of the reverberant room in Exercise 10.9 has been measured. At 1 kHz, it is 92 dB//20 μPa. Determine the volume velocity $\hat{Q}$. The room dimensions are 3 m high and 5 m × 4 m horizontally.

Exercise 10.11 A small spherical object has been placed at the center of a spherical cavity filled with water. The outer wall of the cavity executes a radially symmetric vibration whose amplitude is fixed at 0.2 m/s, but the frequency is variable in the range from 1 to 10 kHz. The surface of the spherical object is locally reacting with $\zeta = 0.6 - 0.4i$. The radii are $a = 800$ mm, $b = 10$ mm. The region between the surfaces is filled with water. Determine the frequency at which the pressure on the surface of the object is maximum. What are the pressure amplitudes on the inner and outer surfaces at this frequency?

Exercise 10.12 A transducer is flush mounted on the inner surface of a spherical cavity whose walls otherwise are rigid. The transducer is very small, so it may be approximated as a point source whose volume velocity is $\text{Re}(\hat{Q} \exp{(i\omega t)})$. Derive an expression for the pressure at the center of the cavity.

Exercise 10.13 A cylindrical tank filled with oil has a piston transducer situated concentrically at $x = 0$. The radius of the tank is a, and the radius of the piston is $a/2$. Consequently, the resulting velocity at $x = 0$ is $v_x = v_0 \cos{(\omega t)}$ if $R < a/2$, $v_x = 0$ if $R > a/2$. The other surfaces are rigid. Derive an expression for $p/(\rho_0 c v_0)$ at an arbitrary location. Then, use that expression to obtain an expression for the acoustic force acting on the wall at $x = L$. Evaluate $|p/(\rho_0 c v_0)|$ at $x = L$, $R = 0$ and $|F|/(\pi a^2)$ as functions of $ka < 15$. Parameters are $\rho_0 = 910\,\text{kg/m}^3$, $c = 1460\,\text{m/s}$, $L = 4\,\text{m}$, and $a = 0.25\,\text{m}$.

Exercise 10.14 Exercise 9.27 described the axial velocity at the near end of a cylindrical waveguide when the termination is a misaligned disk that rotates about the centerline at angular speed Ω. The velocity at a point whose cylindrical polar coordinates are $(R, \theta, x = 0)$ is well described as $v_x = \Delta \Omega R \sin{(\Omega t - \theta)}$, where Δ is the angle between the normal to the disk and the centerline. Rather than taking the far end to be nonreflecting, as was done in Exercise 9.27, suppose that the end at $x = L$ is open to the atmosphere, which corresponds to a pressure-release termination. The cylindrical wall is rigid. Derive an expression for the pressure at an arbitrary location.

Exercise 10.15 The fluid contained in a cylinder whose length is L and diameter is $2a$ is excited by harmonic transverse vibration of the cylindrical wall. The velocity amplitude is axisymmetric with a variation in the axial direction as a single lobe of a sine, so that $v_R = v_0 \sin{(\pi x/L)} \sin{(\omega t)}$. The ends of the cavity are rigid. Derive an expression for the pressure amplitude $|P|/(\rho_0 c v_0)$ at an arbitrary location. Then, evaluate this expression at the midpoint, $x = L/2$, $R = 0$, as a function of frequency in the range $ka < 12$. The aspect ratio is $L/a = 5$.

Exercise 10.16 A cylindrical tank was filled with methyl alcohol ($c = 1103\,\text{m/s}$, $\rho_0 = 792\,\text{kg/m}^3$). A small transducer was mounted concentrically at one end. The far end was closed with mylar, which well fits a pressure-release model, whereas the cylindrical wall was lined with a special insulating coating. The radius of the tank was 3 m. The general analysis of the pressure field in a cavity indicates that in a nearly resonant situation the spatial distribution is essentially proportional to the resonant mode. With this in mind, the transducer was swept through a range of frequencies, with the result that 154.3 Hz was identified as the lowest frequency at which the interior field attains a peak amplitude. At this frequency, the pressure amplitude along a transverse line was measured. The data that was collected is described by the following table. The pressure waveforms at these locations were found to be in-phase. (a) Determine the transverse wavenumber μ by fitting the measured pressure to the analytical expression for the pressure field. (b) Determine the dependence of the pressure on the axial distance measured from the driven end. (c) Determine the specific impedance of the liner material. (d) Determine the length of the cylinder.

R/a	0.2	0.4	0.6	0.8
P (kPa)	310	268	203	125

Exercise 10.17 The sketch depicts a rectangular enclosure in which a small source is situated at its midpoint. The source's volume velocity is $Q(t) = \mathrm{Re}(\hat{Q}\exp(i\omega t))$, where $\hat{Q}$ is independent of the frequency ω. The enclosed fluid is oil ($\rho_0 = 910\,\mathrm{kg/m^3}$, $c = 1460\,\mathrm{m/s}$). All sides of the enclosure are rigid, except that the closure at the left end is a lightly tensioned membrane. The fluid on the other side of the membrane is air, so the membrane effectively constitutes a pressure-release termination. Furthermore, the left end is flush mounted with a very large rigid baffle, so the vibration of the membrane causes sound to be radiated into the air. In other words, this system acts like an unconventional type of loudspeaker. The primary property of a low-frequency loudspeaker is its volume velocity. Determine the magnification factor $\left|\hat{Q}_{\mathrm{mem}}/\hat{Q}\right|$ for the vibrating membrane's volume velocity. The frequency range of interest is zero to 2 kHz, and $L = 0.5\,\mathrm{m}$, $W = 0.25\,\mathrm{m}$.

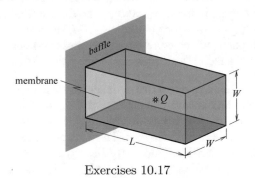

Exercises 10.17

Exercise 10.18 Solve Exercise 10.17 for the case where the cavity is a cylinder. As described there, the closure at the left end is a pressure-release membrane and the other surfaces are rigid. The length is $L = 0.5\,\mathrm{m}$, and the radius is $a = 0.15\,\mathrm{m}$.

Exercise 10.19 The sketch shows two oppositely phased line sources arranged horizontally relative to the middle of the water tank. The sides and bottom are essentially rigid. Consider the situation where the line sources are harmonic, with volume velocities per unit length $\pm\mathrm{Re}(\hat{Q}\exp(i\omega t))$. Derive an expression for the pressure field when the horizontal separation s is finite. (b) Take the limit as $s \to 0$ with $\hat{Q}s$ held constant at D and thereby obtain the pressure field for a dipole line source at the center. (c) Evaluate the pressure distribution (amplitude and phase) along the bottom when $\omega = 0.95\omega_1$ and $\omega = (\omega_1 + \omega_2)/2$, where ω_1 and ω_2 are the two smallest natural frequencies of modes that are excited by the dipole line source. The depth to length ratio is $H/L = 0.4$.

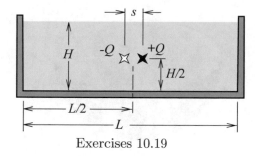

Exercises 10.19

Exercise 10.20 A harmonic point source whose volume velocity is $\text{Re}(Q_0 \exp(i\omega t))$ is situated at the midpoint of a cylindrical cavity. The ends at $x = 0$ and $x = L$ are pressure-release, whereas the cylindrical wall, $R = a$, is rigid. Derive an expression for the pressure amplitude at an arbitrary point on the cylinder's axis. Evaluate the pressure amplitude $|Pa^2/(\rho_0 c Q_0)|$ as a function of distance along this axis in the case where $L = 2a$. Perform this evaluation for cases where the frequency is half the cavity's fundamental natural frequency and 10% greater than that frequency.

Exercise 10.21 The sketch shows a cylinder whose cross section is one-third of a full circle. All surfaces are rigid. Derive expressions for the natural frequencies and mode shapes of the fluid contained within this cavity.

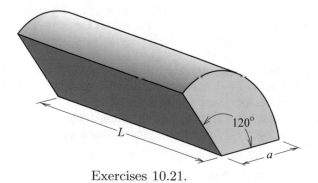

Exercises 10.21.

Exercise 10.22 The wall of a spherical cavity is rigid. The volume of the fluid region is the same volume as the rectangular cavity in Example 10.7. Evaluate all natural frequencies of axisymmetric modes that are below 3 kHz. Graph the data by plotting the frequencies at a fixed polar harmonic against the root number.

Exercise 10.23 The sketch shows a bowl that is filled with water. The hemispherical wall is rigid, whereas the surface is free. Determine the axisymmetric mode functions of the water contained in the bowl.

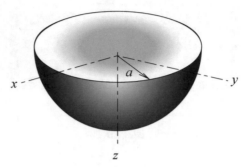

Exercises 10.23

Exercise 10.24 The system of interest is a rigid-walled sphere whose radius is a. A point source whose volume velocity is $\mathrm{Re}(\hat{Q}\exp{(i\omega t)})$ is situated at the center. The pressure field may be described in terms of a series of the cavity's natural modes. An alternative analysis is suggested in Sect. 6.4.2. According to it, the field may be considered to be the superposition of the field generated by the source in free space and a field whose properties are selected to satisfy the boundary conditions, which are violated by the free-space field. In the present case this condition is that the radial particle velocity should be zero at $r = a$. The task here is to formulate both solutions and then compare them by evaluating the dependence of the pressure on the radial distance from the center when $ka = 12$. Part of the comparison should be an assessment of convergence of the modal series as the number of modes is increased.

Exercise 10.25 A harmonic point source whose volume velocity is $\mathrm{Re}(Q_0 \exp(i\omega t))$ is situated at radial distance $r = b$ from the center of a spherical cavity whose radius is a. In order to exploit axisymmetry properties, the radial line through the source is defined to be $\psi = \pi$. The wall of the cavity is rigid. Derive an expression for the pressure amplitude due the point source. Evaluate that expression to determine the radial dependence of the pressure amplitude $\left|Pa^2/\left(\rho_0 c \hat{Q}\right)\right|$ along the radial lines $\psi = 0$ and $\psi = \pi/2$. Parameters for these evaluations are $b = 2a/3$ and $ka = 5$.

Exercise 10.26 An exponential horn is closed at the narrow end and open to the atmosphere at the wide end. Use a series of one, two, and three Ritz basis functions to estimate the natural frequencies $k_n L$ in the case where the exponential growth factor is $\beta = 0.7$. Compare the results to those obtained by solving the Webster horn equation. It may be assumed that the complex pressure amplitude depends only on the axial position x.

Exercise 10.27 The apex angle (side to side) of a cone is $10°$. The cone is open at its a large end, and its wall is rigid. Use a series of one to four basis functions Ritz basis to the natural frequencies $k_n L$ to determine the natural frequencies of modes whose amplitude depends only on the radial distance from the apex.

Exercise 10.28 The apex angle (side to side) of a cone is $30°$. The cone is open at its a large end, and its wall is rigid. The task here is to employ spherical coordinates r, ψ, θ centered at the apex to formulate the Rayleigh–Ritz equations. (a) Identify a set of basis functions that are suitable for identifying the natural frequencies of modes that are axisymmetric relative to the cone's axis, but vary in both the radial and polar direction. (b) Use a series of six of the Ritz basis in Part (a) to identify the nondimensional natural frequencies $k_n L$.

Exercise 10.29 The apex angle (side to side) of a cone is $30°$. The cone is open at its large end, and its wall is rigid. The task here is to employ cylindrical coordinates R, θ, z to formulate the Rayleigh–Ritz equations. The z-axis should coincide with the cone's axis, and the origin should be at the cone's apex. (a) Identify a set of basis functions that are suitable for identifying the natural frequencies of modes that are axisymmetric relative to the cone's axis, but vary in both the radial and polar direction. (b) Use a series of six of the Ritz basis in Part (a) to identify the nondimensional natural frequencies $k_n L$.

Exercise 10.30 The cavity in the sketch is a sector of the region between concentric cylinders. It is filled with water. All walls are rigid. Use the Rayleigh–Ritz method to estimate the cavity's natural frequencies. Carry out the analysis using from one to eight basis functions.

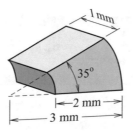

1 mm

$35°$

←2 mm→

3 mm

Exercises 10.30

Exercise 10.31 Suppose that rather than being the interface with an elastic structure, S_e in the development of Dowell's approximation is the interface with a locally reacting material whose impedance is position dependent, $Z(\bar{x}_s)$. What are the corresponding differential equations governing the coefficients of the pressure series?

Exercise 10.32 A piston of mass M, which closes the waveguide in the sketch, is restrained by a spring whose stiffness is K. The termination at the other end is rigid. A harmonically varying force F is applied to the piston. Use Dowell's approximation to determine the pressure amplitude on the face of the piston as a function of the nondimensional frequency $kL < 20$. Parameters for the evaluation are $M/(\rho_0\mathcal{A}L) = 5$ and $(K/M)^{1/2} = 2\pi c/L$. Also determine the lowest five natural frequencies of the system.

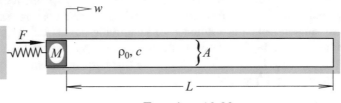

Exercises 10.32

Exercise 10.33 The sketch depicts a closed tank filled with nitrogen, $\rho_0 = 1.165\,\mathrm{kg/m^3}$, $c = 354\,\mathrm{m/s}$. A plate that fully closes the right end is supported by a spring. The mass per unit surface area of the plate is $25\,\mathrm{kg/m^2}$ and the spring stiffness is $500\,\mathrm{MN/m}$. The left end is driven by a plate that extends over half the height of the tank. Both plates translate as rigid bodies, and both extend horizontally across the full width of the waveguide. All other surfaces are rigid. Consequently, the system is two-dimensional in the plane of the sketch. The cavity is excited by harmonic vibration of the left plate at $1.6\,\mathrm{kHz}$, such that $v = 20\sin(\omega t)\,\mathrm{mm/s}$. Determine the pressure distribution along the bottom and the amplitude of the displacement of the plate at the right end.

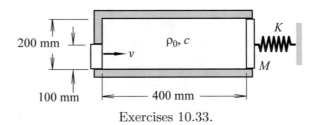

Exercises 10.33.

Exercise 10.34 The region enclosed by the cylinder is filled with an ideal fluid. The left end is terminated by a concentric piston whose radius is $a/2$ and whose mass is M. The remainder of the left end is rigid. Translational displacement w of the piston is resisted by spring K. The cavity is excited by a piston at the right end whose radius is a. As a result, the axial velocity at this end is a specified function $v_L(t)$ for $0 \le R \le a$. (a) Use Dowell's approximation to derive a set of differential equations whose solution would be the pressure coefficients for the acoustic field and the displacement w of the left piston. (b) The parameters are $M/(\rho_0\mathcal{A}L) =$

20, $L/a = 3$, and $(K/M)^{1/2} = 2\pi c/L$. The imposed vibration at the right end is $v_L = v_0 \cos(\omega t)$. Determine and graph the amplitude of the displacement w and the pressure at the center of the right end as a function of frequency for $0 < \omega < 6\pi c/L$. (c) For the parameters in Part (b), determine the first eight natural frequencies of the axisymmetric modes.

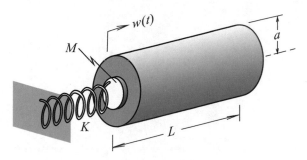

Exercises 10.34

Chapter 11
Geometrical Acoustics

Our explorations thus far have analyzed sound fields from a global perspective, in which field equations and boundary conditions for the entire system were satisfied concurrently. The concepts in this chapter take a localized view to the question of how a signal is modified as it propagates. It constructs rays, which are the paths along which a specific phase propagates through space, and wavefronts, which are surfaces of constant phase at a specific instant. Some aspects of the formulation are an application of differential geometry, so the method is referred to as *geometrical acoustics*. Because ray paths are the primary property that is derived, an alternative name for the formulation is *ray acoustics*. The concept is limited to high-frequency signals, but it is quite general otherwise.

A common application of geometrical acoustics is analysis of propagation through a heterogeneous fluid, in which the ambient density ρ_0 and pressure p_0 are functions of position. However, the ambient properties shall be considered to be independent of time, on the basis that they vary much more slowly than the time scale for the acoustic signal. Analysis of propagation through such a heterogeneous medium is a core task for environmental acoustic studies of the atmosphere and the ocean. The literature on geometrical acoustics and its application to propagation in the ocean and atmosphere is rich. The objective here is to introduce the fundamental concepts. The texts written by Brekhovskikh and Lysanov,[1] Pierce,[2] and Jensen et al.[3] are recommended starting points for further studies.

In order to emphasize development of the basic concepts, we shall add another restriction to the systems we shall consider. Situations where the fluid is not quiescent

[1]L.M. Brekohovskikh and Yu. P. Lysanov, Fundamentals of Ocean Acoustics, Springer-Verlag, 1982.

[2]A.D. Pierce, *Acoustics: An Introduction to Its Physical Principles and Applications*, Acoustical Society of Amer (1989).

[3]F.B. Jensen, W.A. Kuperman, M.B. Porter, and H. Schmidt, *Computational Ocean Acoustics,* 2nd ed (2011).

Electronic supplementary material The online version of this chapter (DOI 10.1007/978-3-319-56847-8_11) contains supplementary material, which is available to authorized users.

© Springer International Publishing AG 2018, corrected publication 2021
J.H. Ginsberg, *Acoustics—A Textbook for Engineers and Physicists, Volume II*,
DOI 10.1007/978-3-319-56847-8_11

in its ambient state are excluded. Winds in the atmosphere might be a reasonably large fraction of the speed of sound. Thus, propagation of sound in the atmosphere is not covered here. In a strict sense, the ocean also is not at rest, but the speed of ocean currents is much less than the speed of sound. Although flow effects are omitted from the present development, the principles and techniques that are derived provide the foundation for study of sound propagation in the atmosphere.

11.1 Basic Considerations: Wavefronts and Rays

We begin with the general representation of a harmonic signal that depends arbitrarily in position, which is $p(\bar{x}, t) = \text{Re}\left(F(\bar{x})\,e^{i\omega t}\right)$. The polar representation of the complex amplitude may be written as $F(\bar{x}) = P(\bar{x})\exp(-i\Delta(\bar{x}))$, where $P(\bar{x})$ and $\Delta(\bar{x})$ are real. The geometrical acoustics approximation defines the nature of $\Delta(\bar{x})$. It considers any signal to be locally planar, with $\bar{e}(\bar{x})$ being the propagation direction and $c(\bar{x})$ being the local sound speed. Such a signal corresponds to $\Delta(\bar{x}) = \bar{x} \cdot \bar{e}(\bar{x})/c(\bar{x})$. (No reference time lag is incorporated because we can define $t = 0$, however, we wish to obtain a pure sine or pure cosine, or any other phase.) Correspondingly, the geometrical acoustics ansatz for the pressure field is

$$p(\bar{x}, t) = P(\bar{x})\,\text{Re}\left(e^{i\omega(t - \bar{x}\cdot\bar{e}(\bar{x})/c(\bar{x}))}\right) \qquad (11.1.1)$$

The fact that the amplitude function $P(\bar{x})$ is real will be crucial to later developments.

To identify how the preceding description is related to wavefronts, we begin with the fundamental definition that *a wavefront is a surface along which the signal at the same instant has the same phase.* A surface that is fixed in space is the locus of points at which a function of the position coordinates has a constant value. A moving surface may be obtained by replacing the constant value by a function of time. Thus, an equation whose form is $f(\bar{x}) = t$ defines a surface that moves through space.

The phase θ of the signal in Eq. (11.1.1) is the coefficient of i in the complex exponential, so we have

$$\frac{\theta}{\omega} = t - \frac{\bar{x}\cdot\bar{e}(\bar{x})}{c(\bar{x})} \qquad (11.1.2)$$

If we fix t, then the locus of points on a wavefront of constant specified θ may be found by solving the preceding for all possible $\bar{x}$. Thus, this relation is the equation defining the wavefront of constant phase θ. It is useful for further developments to collect the position-dependent terms in a single function $\tau(\bar{x})$, whose dimensionality is time,

$$\tau(\bar{x}) \equiv \frac{\bar{x}\cdot\bar{e}(\bar{x})}{c(\bar{x})} \qquad (11.1.3)$$

The description of a wavefront provided by Eq. (11.1.2) actually has two interpretations. The values of $\bar{x}$ on a wavefront, and therefore wavefront's shape and overall position, depend of the value of $t - \theta/\omega$. If we hold the phase at a constant θ and vary t, we will see the wavefront flow through space as time evolves. Alternatively, we can fix t and construct a family of wavefronts at that instant corresponding to different values of the θ. In other words, wavefronts may be considered to be a set of surfaces at a specific instant or a single surface that moves through space. Both views are equivalent, because only the value of $t - \theta/\omega$ is important. This duality tells us that everything we need to know may be found by following the wavefront corresponding to zero phase. Therefore, we take the wavefront equation to be

$$\boxed{\tau(\bar{x}) = t} \tag{11.1.4}$$

After we have described the zero-phase wavefront at an instant t, the wavefront along which the phase is $\theta \neq 0$ will occupy the same location at the instant $t' = t + \theta/\omega$.[4]

The description on Eq. (11.1.1) is quite general. The geometrical acoustic approximation is based on a fundamental assumption regarding spatial dependence of the amplitude and phase. For a simple plane wave, the phase is $\theta = \omega \bar{x} \cdot \bar{e}/c = kx$ so that $\nabla \theta = k\bar{e} = (2\pi/\lambda)\bar{e}$. In view of the definition of θ in Eq. (11.1.2), we may consider $\nabla \theta$ to be proportional to the reciprocal of a local wavelength. Thus, variation of the phase over a small spatial scale implies that $\nabla \theta$ is relatively large. With this observation in mind, let us differentiate Eq. (11.1.1) to determine the pressure gradient,

$$\nabla p \equiv [\nabla P - iP\nabla\theta] \, \mathrm{Re}\left(e^{i(\omega t - \theta(\bar{x}))}\right) \tag{11.1.5}$$

An overview of the various signals we have considered indicates that in many cases, the wavelength scale is much smaller than the spatial scale over which the pressure amplitude varies. A geometric acoustics analysis is based on the *a priori* assumption of this behavior. More definitively, it is required that $|\nabla P| \ll |P\nabla\theta|$. When this condition is met, we say that the amplitude $P(\bar{x})$ is a *slowly varying* function of position. A further assumption comes from differentiation of Eq. (11.1.2), which gives

$$\nabla\theta = -\omega\nabla\left(\frac{\bar{x} \cdot \bar{e}(\bar{x})}{c(\bar{x})}\right) = -\frac{\omega\bar{e}}{c(\bar{x})} + \theta\frac{\nabla c}{c(\bar{x})} \tag{11.1.6}$$

Unless the fluid is homogeneous, ∇c is not zero. A geometrical acoustic analysis assumes that the sound speed varies slowly relative to the local wavelength, in which case $\nabla c \approx \bar{0}$. The combination of both assumptions leads to the geometrical acoustics assumption of the pressure gradient as

---

[4]This assertion that the set of wavefronts at a specific instant may be identified by following the evolution of a specific wavefront is only valid in the context of linear acoustics. For example, it assumes that the signal does not alter the properties of the fluid, even though the speed of sound depends on the pressure. If nonlinear effects are included, wavefronts and rays might also depend on the pressure and particle velocity associated with a specific phase. This condition will be encountered in Sect. 13.4.2.

$$\boxed{\nabla p \approx - \mathrm{Re}\left(i \frac{\omega \bar{e}}{c(\bar{x})} P e^{i(\omega t - \theta(\bar{x}))} \right)} \tag{11.1.7}$$

Another requirement, which will be relaxed later, is that the radius of curvature of the wavefront should be large compared to the wavelength. When both conditions are met, the signal may be considered to be *locally planar*. Conversely, ray theory might require correction, or it might not be valid at all, if either property is not obtained in some region.

The extension of Eq. (11.1.1) to an arbitrary signal is

$$\boxed{p(\bar{x}, t) = F(\bar{x}, t'), \quad t' = t - \tau(\bar{x})} \tag{11.1.8}$$

The essence of the geometrical acoustics approximation is that the explicit dependence of F on $\bar{x}$ is taken to be very gradual relative to the manner in which $\tau(\bar{x})$ varies spatially. This leads to the pressure gradient being represented by

$$\nabla p \approx - \frac{\partial F}{\partial t'} \nabla \tau \tag{11.1.9}$$

The nature of the wavefronts is defined by Eq. (11.1.4). An examination of Eqs. (11.1.1) and (11.1.8) indicates that knowledge of the wavefronts is fundamental to describing the pressure field. This observation should cause the reader to wonder how rays enter the picture? That question is addressed by Fig. 11.1, which follows a zero-phase wavefront. At instant t, a specific signal is located at $x = \bar{\xi}$. The location $\bar{\xi}$ shifts in time by the amount necessary to keep it on the wavefront. The locus of positions at which this signal is located is a curved line whose equation is $\bar{x} = \bar{\xi}(t)$. This is a parametric description of a ray, in which time is the parameter. Of course, at this stage we do not know the functional dependence.

Fig. 11.1 Relationship of rays and wavefronts in the geometrical acoustics approximation. The normal $\bar{e}$ to the wavefront at any location is the direction in which the signal at that location propagates

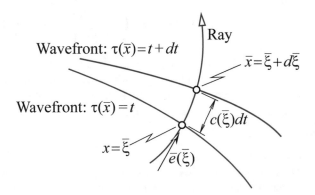

If we know the wavefront at any instant t and the sound speed at any location, there is a graphical construction that will lead to wavefronts and rays at later times. We consider a single step from t to $t + dt$. The planar nature of the propagation means that

the phase velocity at $\bar{\xi}$ is $c\left(\bar{\xi}\right)$ in the direction that is normal to the wavefront at $\bar{\xi}$. The unit vector for this normal direction is $\bar{e}\left(\bar{\xi}\right)$ in Fig. 11.1. Given the current wavefront, we can draw this normal, either manually or with a CAD program. In a time interval dt, the signal propagates through distance $c\left(\bar{\xi}\right)dt$, which we lay out along $\bar{e}\left(\bar{\xi}\right)$. This gives the next point on the ray as $\bar{x}=\bar{\xi}\left(t+dt\right)=\bar{\xi}\left(t\right)+\bar{e}\left(\bar{\xi}\right)c\left(\bar{\xi}\right)dt$. If we apply this construction to many points on the wavefront at t, we will establish where the wavefront is at $t+dt$. Thus, we have updated the location of the wavefront. The process may be repeated as long as necessary. If we track the position of a particular point on the initial waveform as time evolves, we will obtain the ray path for the signal that was at that initial point at that initial instant.

There are several reasons why this graphical procedure is only used conceptually. Most significantly, it is approximate because it can only be implemented for a finite time step. In addition, any method we use to graphically identify $\bar{e}\left(\bar{\xi}\right)$ will inherently be approximate. There is no way to quantify either type of error. Beyond considerations of accuracy, the procedure is tedious, because many time steps would be required to identify the rays and wavefronts over a reasonably large interval.

As we progress through this chapter, we will encounter two approaches by which geometrical acoustics analyses are implemented. The first, which is limited to fluids whose properties depend on a single spatial coordinate, directly determines the rays. The second approach is suitable for an arbitrary heterogeneity. It simultaneously determines wavefronts and rays. Both formulations are used primarily as the basis for computational algorithms that we will explore. However, before we can address either formulation, it is necessary that we identify the governing field equation.

11.1.1 Field Equations for an Inhomogeneous Fluid

The ambient density ρ_0 and pressure p_0 are known function of position, and the fluid is taken to be at rest in this state. We seek to determine the acoustic perturbations, $\rho'\left(\bar{x},t\right)$, $p'\left(\bar{x},t\right)$, and $\bar{v}\left(\bar{x},t\right)$. (It is convenient to temporarily revert to the use of a prime to denote increments of state variables from their ambient value.) A body force acting on the fluid may depend on the density, so it is denoted as $\bar{b}\left(\bar{x},\rho_0+\rho'\right)$. The body force adds a term to the momentum equation, Eq. (4.1.7), but the general continuity equation, Eq. (4.1.5), is unmodified.

A linearized set of equations is obtained by introducing the small signal approximations,

$$\rho=\rho_0\left(\bar{x}\right)+\rho'\left(\bar{x},t\right),\quad p=p_0\left(\bar{x}\right)+p'\left(\bar{x},t\right),\quad \bar{v}=\bar{v}'\left(\bar{x},t\right) \tag{11.1.10}$$

in which $\left|\rho'\right|\ll\rho_0$, $\left|p'\right|\ll p_0$, and $\left|\bar{v}\right|$ is much less than the speed of sound at ambient conditions. When we introduce this decomposition into the continuity equation and drop all terms that are products of primed quantities, we find that

$$\boxed{\frac{\partial\rho'}{\partial t}+\nabla\cdot\left(\rho_0\bar{v}'\right)=0} \tag{11.1.11}$$

Substitution of the small-signal representation into the momentum equation gives

$$\frac{\partial \bar{v}'}{\partial t} = -\left(\frac{\nabla p_0 + \nabla p'}{\rho_0}\right) + \frac{1}{\rho_0}\left[\bar{b}\,(\bar{x}, \rho_0) + \left(\frac{\partial \bar{b}}{\partial \rho}\right)_0 \rho'\right] \qquad (11.1.12)$$

Note that we have used the smallness of ρ' to replace $\bar{b}$ by its first two terms in a Taylor series relative to the ambient state. All primed variables are zero in the ambient state, so the ambient pressure must satisfy

$$\nabla p_0 = \bar{b}\,(\bar{x}, \rho_0) \qquad (11.1.13)$$

Hence, the linearized momentum equation is

$$\boxed{\frac{\partial \bar{v}'}{\partial t} = -\frac{\nabla p'}{\rho_0} + \left(\frac{\partial \bar{b}}{\partial \rho}\right)_0 \frac{\rho'}{\rho_0}} \qquad (11.1.14)$$

Three state variables appear in the continuity and momentum equations. The third relation is the equation of state. The assumption that heat flow is negligible remains valid. However, because the entropy now may be position-dependent, the pressure must be taken to depend on both the density and entropy, $p = p\,(\rho, S)$. Negligible heat flow between particles corresponds to taking the entropy of a specific particle to be constant as the particle moves through space. The material derivative describes how a property of a moving particle changes as it moves. Thus, we seek a relation between p' and ρ' based on $dS/dt = 0$. The total pressure is $p = p_0 + p'$, but $dp_0/dt = 0$ because the ambient pressure is taken to vary only with position. Thus, the material derivative of the pressure equation of state is

$$\frac{dp'}{dt} = \frac{\partial p}{\partial \rho}\frac{d\rho}{dt} + \frac{\partial p}{\partial S}\frac{dS}{dt} = \frac{\partial p}{\partial \rho}\left(\frac{\partial \rho}{\partial t} + \bar{v}' \cdot \nabla \rho\right) \qquad (11.1.15)$$

This expression does not invoke the small-signal approximation. To do so we introduce $\rho = \rho_0\,(\bar{x}) + \rho'$ and drop higher order terms. Furthermore, the small-signal approximation indicates that dp'/dt and $\partial p'/\partial t$ differ by second-order terms, which are dropped. Thus, the equation of state for a heterogeneous fluid is a linear differential equation,

$$\boxed{\frac{\partial p'}{\partial t} = c^2\left(\frac{\partial \rho'}{\partial t} + \bar{v}' \cdot \nabla \rho_0\right), \quad c^2 = \left(\frac{\partial p}{\partial \rho}\right)_0} \qquad (11.1.16)$$

Although this definition of c is the same as it is for a homogeneous fluid, c now is a function of position.

We have derived three differential equations that govern p', ρ', and $\bar{v}'$, but we seek a single differential equation for p'. Toward that end, we substitute $\partial \rho'/\partial t$ from

Eq. (11.1.11) into the equation of state, which leads to

$$\frac{\partial p'}{\partial t} = -\rho_0 c^2 \nabla \cdot \vec{v}' \tag{11.1.17}$$

Next, we take the time derivative of this relation. Substitution of $\partial \vec{v}/\partial t$ from Eq. (11.1.14) removes the particle velocity. The result is

$$\frac{1}{\rho_0 c^2} \frac{\partial^2 p'}{\partial t^2} = -\nabla \cdot \frac{\partial \vec{v}'}{\partial t} = \nabla \cdot \left[\frac{\nabla p'}{\rho_0} - \left(\frac{\partial \bar{b}}{\partial \rho} \right)_0 \frac{\rho'}{\rho_0} \right] \tag{11.1.18}$$

A rearrangement of terms yields

$$\rho_0 \nabla \cdot \left(\frac{\nabla p'}{\rho_0} \right) - \frac{1}{c^2} \frac{\partial^2 p'}{\partial t^2} = \rho_0 \nabla \cdot \left[\left(\frac{\partial \bar{b}}{\partial \rho} \right)_0 \frac{\rho'}{\rho_0} \right] \tag{11.1.19}$$

If the term in the right side were not present, this would be a field equation for p'. Indeed, most references do not list this term because they ignore the body force at the outset. To see why it is allowable to do so, consider a plane wave propagating in direction $\bar{e}$ in the ocean. The body force per unit volume is gravitational attraction, so $\bar{b} = \rho g \bar{e}_z$ and $(\partial \bar{b}/\partial \rho)_0 = g \bar{e}_z$. Let us use the relations for a homogeneous fluid to compare the two terms whose divergence is taken in the above. Thus, we set $p' = \text{Re} \left(\hat{P} \exp (i\omega t - i k \bar{e} \cdot \bar{x}) \right)$ and $\rho' \approx p'/c^2$. This leads to estimates of the two terms as

$$\left| \frac{\nabla p'}{\rho_0} \right| \approx \frac{k |p'|}{\rho_0}, \quad \left| \left(\frac{\partial \bar{b}}{\partial \rho} \right)_0 \frac{\rho'}{\rho_0} \right| \approx \frac{g |p'|}{\rho_0 c^2} \tag{11.1.20}$$

The ratio of the second term to the first is $g/(\omega c)$, which is extremely small even at $\omega = 1$ rad/s, unless the fluid is remarkably slow. For this reason, we shall proceed without considering the body force effect. (It might not be allowable to do in some situations, such as a strong magnetic field confining a plasma).

In the case of a gas, notably the atmosphere, dropping the body force term in Eq. (11.1.19) is the sole fundamental simplification of the field equation. The equation for a heterogeneous gas therefore is

$$\boxed{\nabla^2 p' - \frac{1}{\rho_0(\bar{x})} \nabla \rho_0 \cdot \nabla p' - \frac{1}{c(\bar{x})^2} \frac{\partial^2 p'}{\partial t^2} = 0} \tag{11.1.21}$$

If the density changes gradually, the preceding field equation may be simplified further by setting $\nabla \rho_0$ to zero. To determine the conditions for which this approximation is reasonable, let us again make use of the description of a plane wave in a homogeneous fluid to estimate the effect. The density gradient is approximated as the difference $\Delta \rho'$ between the minimum and maximum values divided by the distance ℓ over which these extreme values are observed. For ρ_0, we use the average value. For a harmonic plane wave propagating in direction $\bar{e}$, we have $\nabla p' = -ik\bar{e}p'$. The maximum magnitude of $\nabla \rho_0 \cdot \nabla p'$ occurs if $\bar{e}$ is parallel to the density gradient. Thus, we have the estimates that

$$\nabla^2 p' \approx -k^2 p', \quad \frac{\nabla \rho_0}{\rho_0} \cdot \nabla p' \approx -ik\frac{\Delta \rho'}{\rho_{av}\ell}p' \tag{11.1.22}$$

With k replaced by $2\pi/\lambda$, the order of magnitude of the ratio of these terms is $O\left((\Delta\rho/\rho_{av})/(2\pi\ell/\lambda)\right)$.

Density variations in the atmosphere can be quite large, but that is not true for water. To demonstrate this assertion consider standard formulas for the density of water provided by Pierce.[5] The value of ρ_0 depends on the ambient pressure and temperature, and salinity in the case of the ocean. The reference value for seawater is $\rho_0 = 1027\,\text{kg/m}^3$ at $10°C$. Raising the ambient pressure to $2\,\text{MPa}$, which corresponds to a $200\,\text{m}$ depth, raises the density by less than 0.1%. A temperature increase to $30°C$ decreases the density by less than 2%. Distilled water exhibits similar tendencies. The density of seawater also depends on salinity, whose reference value is 35 parts per thousand (ppt). It is very rare that the salinity varies by more than 2 ppt from the reference, and such a fluctuation changes the density by 0.15%. Furthermore, it is unusual that these extreme changes of ambient conditions occur over less than a kilometer. In contrast, the wavelength at $10\,\text{Hz}$ is approximately $150\,\text{m}$, and the wavelength decreases with increasing frequency. In other words, $2\pi\ell/\lambda$ will be large. Thus, neglecting the contribution of the density gradient to the field equation is well warranted if the medium is a large body of water or most other liquids. One exception occurs at the interface between distinct layers, which was studied in Chap. 5. The geometrical acoustic formulation addresses discontinuous changes of density and sound speed as separate considerations.

When the simplifications that ignore the effects of a body force and of a spatially varying density are invoked, what remains of the field equation is

$$\boxed{\nabla^2 p' - \frac{1}{c(\bar{x})^2}\frac{\partial^2 p'}{\partial t^2} = 0} \tag{11.1.23}$$

This equation resembles the linear wave equation for a homogeneous fluid. However, the spatial dependence of c has a fundamental effect.

It should be apparent at this juncture that any analysis we pursued in previous chapters probably will not be valid for a heterogeneous fluid because the field equation is a partial differential equation with variable coefficients. Indeed, any direct mathematical analysis would be challenging. Nevertheless, it is important that we have derived this equation because it validates considering the pressure to behave

---

[5]A.D. Pierce, *Acoustics*, McGraw-Hill (1981), reprinted by ASA Press (1989) pp. 33–34.

locally like a plane wave. In addition, Eq. (11.1.21) will be crucial for the analysis of the pressure amplitude in an fluid whose heterogeneity is an arbitrary function of position.

11.1.2 Reflection and Refraction of Rays

The geometrical acoustic approximation that a ray behaves locally as though it were planar has some important corollaries. Suppose a wave is incident on a locally reacting solid surface. The reflected wave may be identified by adapting the laws for reflection of a plane wave at oblique incidence to a planar boundary. In keeping with the restrictions of geometrical acoustics, the surface's principal radii of curvature are required to be much greater than a typical wavelength. When this restriction is fulfilled, the surface may be replaced by the tangent plane at the point of incidence. Other than the dependence of the normal $\bar{n}$ on the point of incidence, the reflection process is the same as that of planar waves from a flat surface. Thus, the normal $\bar{n}$ and the incident ray form a plane. The reflected ray lies in this plane at the same angle as the angle of incidence. A consequence of this behavior is that rays that were parallel prior to reflection will not be parallel after reflection, as is illustrated by Fig. 11.2.

Fig. 11.2 Reflection of rays from a curved surface

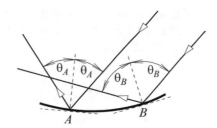

At very high frequencies, surfaces that fit a local impedance model tend to be very stiff. For that reason, solid surfaces are typically considered to be rigid in the geometrical acoustics approximation. Then, the amplitude of the reflected wave at the point of incidence equals that of the incident wave. If the impedance Z is finite, the reflection coefficient in Eq. (5.2.15) may be used,

$$P_R = RP_I, \quad R = \frac{Z - \rho_0 c/\cos\theta}{Z + \rho_0 c/\cos\theta} \tag{11.1.24}$$

As is true for plane waves, a description using a reflection coefficient should be limited to frequency-domain analyses, unless Z is real.

It is unusual that two fluids meet at a curved interface. A possibility is that two fluids are separated by a highly compliant solid sheet. Like the case of reflection from a solid surface, the reflection and transmission of an incident ray is defined relative to the normal at the point of incidence. The process is illustrated in Fig. 11.3. The reflected and refracted rays lie in the plane formed by the incident ray and the normal at the point of incidence. The angle of reflection equals the angle of incidence. The angle of transmission is governed by Snell's law, which states that

$$\frac{\sin \theta_2}{c_2} = \frac{\sin \theta_1}{c_1} \qquad (11.1.25)$$

Fig. 11.3 Transmission and reflection of a ray incident on the curved interface of two fluids

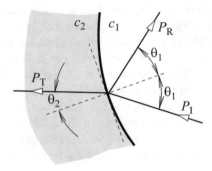

The amplitudes of the signal on the reflected and transmitted rays are taken to be those derived for a plane wave. The reflection and transmission coefficients are as given in Eq. (5.3.8),

$$R = \frac{Z_2 - Z_1}{Z_2 + Z_1}, \quad T = \frac{2Z_2}{Z_2 + Z_1}, \quad Z_j = \frac{\rho_j c_j}{\cos \theta_j} \qquad (11.1.26)$$

The angle of transmission is measured relative to the local normal. Consequently, even if incident rays are parallel, the amplitudes of transmitted and reflected rays will vary with position on the interface.

We saw in Sect. 5.3 that supercritical incidence occurs if $\cos \theta_1 > c_1/c_2$, which obviously can only occur if $c_2 > c_1$. Supercritical incidence leads to a wave in the receiving fluid that evanesces with increasing distance perpendicularly the interface. In the case of a truly planar interface, such a wave propagates parallel to the surface. In a strict sense such a wave does not fit the geometrical acoustic model.

EXAMPLE 11.1 A plane wave propagates parallel to the axis of symmetry of the cap of a sphere whose radius is a. The wave is reflected from the inner concave surface of the cap, which is rigid. Determine and graph the paths of the reflected rays.

Significance

This system, which is the acoustical analog of the mirror in a reflecting optical telescope, has an acoustic application as part of an extremely sensitive microphone. The only mathematical tools required for the analysis are those of trigonometry, so the basic aspects of ray paths will be easy to identify. The results will exhibit the phenomenon of a caustic, at which rays are concentrated.

Solution

The system is axisymmetric, so the picture is the same for any slicing plane that contains the axis of the spherical cap. The range of polar angles $\psi \geq 0$ for one slice is depicted in Fig. 1. The spherical radius of the cap is a. The sketch shows a ray at an arbitrary distance R from the axis of symmetry. This defines the polar angle ψ to the point where the ray is incident on the mirror according to $\psi = \sin^{-1}(R/a)$. The normal direction $\bar{n}$ coincides with the radial line. Because the angle of reflection is ψ, the reflected ray, the radial line, and the axis of symmetry form an isosceles triangle. The distance where the reflected ray intersects the axis of symmetry is $s = (a/2)/\cos\psi$ inward from the center.

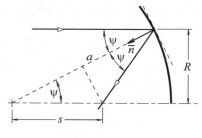

Figure 1.

A full picture requires consideration of many rays, so we shall consider a range of R values. If the maximum ψ is too large, a reflected ray will go on to be incident at a second point on the cap. This situation may be avoided if the cap is such that the maximum polar angle is less than 45°. We will consider a cap whose maximum polar angle is 30°.

Construction of the rays is assisted by defining a coordinate system whose x-axis coincides with the axis of symmetry, with the origin at the center. We construct a set of incident rays, spaced at equal increments of the offset distance. Because of the axisymmetric nature of the system, it is sufficient to consider positive polar angles. For each R, we draw an incident ray from $(0, R)$ to $(a\cos\psi, R)$, and a reflected ray from $(a\cos\psi, R)$ to $(s, 0)$, where s is described above. In Fig. 2, the reflected ray are extended beyond the axis of symmetry.

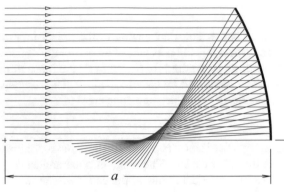

Figure 2.

Every cutting plane features the same picture. Furthermore, Fig. 2 only shows half the rays in a cutting plane because there are additional rays for $\psi < 0$. The full set for a cross section are depicted in Fig. 3. The reflected rays extend beyond the window that is plotted. The collection of rays corresponding to all cutting planes represents an axisymmetric distribution. Thus, the locus of straight rays at offset distance R is a cylinder, and the locus of reflected rays corresponding to a specified R is a cone whose apex is situated at distance s.

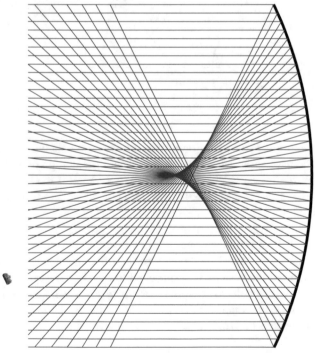

Figure 3.

There is a region near the z-axis where the rays are closely spaced. Our later investigations will show that the amplitude grows in regions where the spacing of rays decreases. Another interesting aspect is this region is delimited by a curve beyond which there are no neighboring reflected rays. To better understand this feature Fig. 4 zooms in on the reflected rays in this region of concentration. The decreased spacing between rays and the absence of rays beyond a bounding curve are produced by nearly parallel rays intersecting. The minimum s from such an intersection to the center of the spherical cap is the ray associated with $R \to 0$. This corresponds to $\psi \to 0$, for which $s \to a/2$. The bounding curve generated by the intersecting rays is called a *caustic*, and $s = a/2$ marks the *arrête*, which is the term given to the location where a caustic terminates.

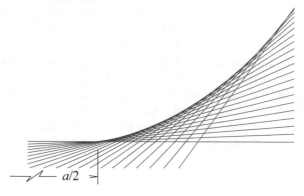

Figure 4.

In some applications, it is desirable that all rays intersect in a very compact region. Convergence at a point, which is the definition of a *focus*, can be attained with a mirror that is the interior of a paraboloid of revolution. (This is the subject of Homework Problem 11.1.) All rays for the case of a spherical cap nearly converge at a point if the largest polar angle is relatively small. However, there is no size that leads to a perfect focus. In optics, this is known as *spherical aberration*.

11.2 Propagation in a Vertically Stratified Medium

An important model in environmental acoustics considers the fluid to have properties that vary with vertical elevation, which is designated as z. The ambient density, pressure, and temperature are taken to depend solely on z, and the fluid is quiescent in the ambient state. The analysis that follows parallels the manner in which a classic textbook on underwater acoustics[6] begins. The development has the attractive feature of emphasizing the physical properties of rays. Modifications to account for flow

---

[6]L. Brekohovskikh and Yu. Lysanov, *Fundamentals of Ocean Acoustics*, Springer-Verlag (1982).

in the ambient state, such as winds in the atmosphere, are best deferred until the basic concepts of geometrical acoustics are understood. The text by Pierce[7] offers an excellent portal to this subject. For underwater acoustics flow is less of an issue because the flow rates are lower than in the atmosphere, and the sound speed is much higher.

11.2.1 Snell's Law for Vertical Heterogeneity

We begin with the case where finite thickness layers are stacked vertically. Propagation in the ocean typically is analyzed by measuring z downward with $z = 0$ at the free surface. That is the convention used here. A ray starting in layer n behaves like a plane wave, so it is transmitted into layer $n + 1$. The direction of the transmitted ray is governed by Snell's law. Consequently, the ray will remain in a constant vertical plane. Whereas the treatment of transmission in Sect. 5.3 used the angle of incidence, the developments that follow use the *grazing angle* χ of the ray, which is measured from the horizontal to the ray. The arrangement is depicted in Fig. 11.4. It will be necessary to distinguish between propagation that is downward (increasing z) and upward (decreasing z). This can be done by letting χ be in the range $-\pi/2 \le \chi \le \pi/2$, with negative values corresponding to upward propagation.

Fig. 11.4 Definition of the grazing angle for propagation in a vertically stratified fluid

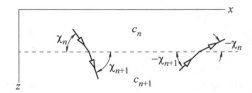

Let "I" denote the conditions at the location where the ray was generated. Then, Snell's law requires that

$$\frac{c_{n+1}}{\cos \chi_{n+1}} = \frac{c_n}{\cos \chi_n} = \cdots = \frac{c_I}{\cos \chi_I} \tag{11.2.1}$$

The situation where the sound speed $c(z)$ is a continuous function corresponds to layers having differentially small thickness. This condition is described by replacing c_n with $c(z)$, and χ_n with $\chi(z)$. Thus, the grazing angle at any depth is defined by

$$\boxed{\cos \chi = \frac{1}{\mu(z)} \cos \chi_I, \quad \mu = \frac{c_I}{c(z)}} \tag{11.2.2}$$

The quantity μ is the *index of refraction*. It captures the manner in which the depth dependence affects any ray, whereas the $\cos \chi_I$ factor is specific to one ray. It is

---

[7]Pierce, *ibid.*

evident that the grazing angle changes continuously if c is a continuous function of depth. This means that each ray path will be curved, so that χ measures the angle between the tangent to this curve and the horizontal.

It is worth noting that no consideration has been given to a reflected ray. The reflection coefficient depends on the difference of the $\rho c / \sin \chi$ values on either side of the interface. A differential difference of these value associated with a continuous variation of ρ and c leads to a reflection coefficient that is infinitesimal.

The observation that χ defines the local tangent to the ray path is the basis for identification of these paths. Consider Fig. 11.5, which depicts a differential arc $d\ell$. Its slope is dz/dx, so we have

$$dx = \frac{dz}{\tan \chi} = \frac{\cos \chi}{\left[1 - (\cos \chi)^2\right]^{1/2}} dz, \quad d\ell = \frac{dz}{\sin \chi} = \frac{1}{\left[1 - (\cos \chi)^2\right]^{1/2}} dz$$

$$(11.2.3)$$

Snell's law and trigonometry lead to descriptions of both differential distances in terms of $\cos \chi_I$,

$$dx = \frac{\cos \chi_I}{\left[\mu^2 - (\cos \chi_I)^2\right]^{1/2}} dz$$

$$(11.2.4)$$

$$d\ell = \frac{\mu}{\left[\mu^2 - (\cos \chi_I)^2\right]^{1/2}} dz$$

Fig. 11.5 Pythagorean theorem for the arclength along a ray

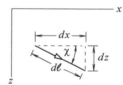

The first of Eq. (11.2.4) defines x as a function of z. Because μ is a function of z, the relation $x(z)$ may be found as a definite integral starting from the first location (x_I, z_I). At the location where the ray is generated, the sense of the ray might be downward. In that case $z > z_I$. Upward propagation leads to $z < z_I$. In either case, x increases monotonically. Both cases are captured by taking the absolute value of the integral, so that

$$x = x_I + \left| \int_{z_I}^{z} \frac{\cos \chi_I}{\left[\mu(\xi)^2 - (\cos \chi_I)^2\right]^{1/2}} d\xi \right|$$

$$(11.2.5)$$

The second of Eq. (11.2.4) is useful because the local propagation speed in geometrical acoustics is $c(z)$. Thus, the time required for a signal to propagate distance $d\ell$ along a ray is $dt = d\ell/c(z) \equiv \mu d\ell/c_I$. The arclength is described by the second of Eq. (11.2.4), so

$$dt = \frac{\mu(z)}{c_I} d\ell = \frac{\mu(z)^2}{c_I \left[\mu^2 - (\cos \chi_I)^2\right]^{1/2}} dz \tag{11.2.6}$$

If a signal departs from the first location at t_1, the time at which it arrives at another point on a specified ray is

$$t = t_1 + \left| \int_{z_1}^{z} \frac{\mu(\xi)^2}{c_I \left[\mu(\xi)^2 - (\cos \chi_I)^2\right]^{1/2}} d\xi \right| \tag{11.2.7}$$

The absolute value of the integrand is used here because the elapsed time increases monotonically, regardless of whether the propagation is upward or downward.

Evaluation of Eq. (11.2.5) for a range of z gives the ray path, and evaluation of Eq. (11.2.7) gives the time at which the signal will arrive at a point on that path. Propagation into a region where the sound speed decreases in the direction of propagation leads to increasing values of $|\chi|$. This is the situation depicted Fig. 11.6a and d. There is no limit to the vertical propagation in these cases, other than incidence at the bottom. The behavior of rays in cases where the sound speed increases in the direction of propagation are depicted in Fig. 11.6b and c. There the grazing angle progressively decreases.

Fig. 11.6 Influence of the gradient of the sound speed on the curvature of a ray

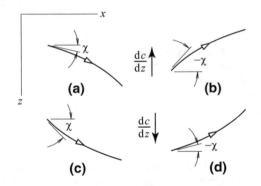

A simple explanation of the tendency for a ray to bend may be gained by considering a wave that initially is planar, with a wavefront that lies in the vertical plane. If the sound speed increases with depth, so that the signal propagates faster as the depth increases, the wavefront will advance progressively farther as the depth increases. The direction of propagation is perpendicular to the wavefront, therefore the ray begins to point upward. The case where $dc/dz < 0$ leads to the opposite effect, wherein the wavefront advances more as the depth decreases.

It might ultimately happen that the grazing angle is zero. The sound speed at which this condition occurs will be denoted as c_0. Setting $\chi = 0$ in Snell's law gives

$$c_0 = \frac{c_I}{\cos \chi_I} \tag{11.2.8}$$

This value is dependent on the initial grazing angle. If a ray reaches a depth at which $c(z) = c_0$ for its associated χ_I, then it will attain a horizontal tangency, which is the limit of its vertical propagation. This condition occurs only if there is a depth at which $c(z)$ equals the value of c_0 for the ray. For instance, rays that are launched vertically remain straight. After a ray attains this horizontal tangency, the vertical sense of the propagation reverses, so the horizontal tangency is said to be a *turning point*.

It might happen that there are turning points above and below the launch point. In Fig. 11.7a $c(z_B)$ and $c(z_D)$ equal $c(z_A)/\cos \chi_I$. The ray is launched downward, but it curves upward because $dc/dz > 0$. Beyond turning point B it continues to curve upward until it reaches a depth at which $dc/dz < 0$. This results in downward curvature. Upward propagation ceases at turning point D, and beyond that point the propagation direction is downward. It follows that that this ray is confined to the range $z_D \le z \le z_B$. Depending on the sound speed profile $c(z)$ and the value of χ_I, other rays will be confined differently or not at all.

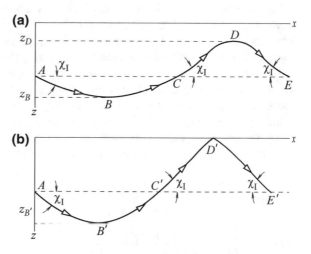

Fig. 11.7 Turning points and the surface limit the vertical propagation of rays

An exceptional situation occurs if a turning point occurs at the depth where the sound speed is a maximum. In the region above that depth, a ray curves upward, whereas a ray in the region below that depth curves downward. At the turning point the propagation direction is horizontal. If c is a maximum at that depth the tendency for the horizontal ray to curve downward is equal to the tendency for it to curve upward. The consequence is that the horizontal ray splits in two, with one part going up and the other going down.

Another possible condition is that a ray arrives at the free surface or the horizontal bottom of a channel. Such a ray is reflected from the boundary. Because of the locally planar nature of the signal, the reflected ray obeys the law for reflection of plane waves at a free surface. Thus, χ for the reflected ray is $-\chi$ for the incident one. This is the situation in Fig. 11.7b. The launch angle χ_I is larger, so the lower turning point occurs at a greater depth $z_{B'}$. Reflection at the surface will occur if there is no turning point at a lesser depth than z_I, or if the depth at which $c(z) = c_I / \cos \chi_I$ is negative.

An interesting feature of a ray path results from the observation that $|\chi|$ for a specific ray depends only on z. Thus, the grazing angle at a given depth beyond a turning point is the negative of the value at that depth prior to the turning point. This means that a ray is symmetric with respect to a vertical line through a turning point. This feature is exhibited in Fig. 11.7 which shows that the grazing angle at the original depth z_A is either χ_I or $-\chi_I$, depending on whether the propagation is downward or upward. This attribute also applies relative to a surface reflection point. Furthermore, the situation at points E and E' in Fig. 11.7 is the same as point A, except that the horizontal range is different. Thus, the pattern beyond these points is a periodic repetition of the pattern from point A.

It is useful to consider how to handle the integrals in Eqs. (11.2.5) and (11.2.7) in view of the nature of the ray paths in Fig. 11.7. From Snell's law, we know that $0 \le |\cos \chi_I| \le \mu(z)$. From the starting location to the lower turning point, where $z = z_B$ and $x = x_B$, the integral's upper limit progressively increases. Consequently, the integral leads to a value of x that increases monotonically with increasing z. Beyond the turning point, the upper limit decreases. The consequence is that both integrals decrease from their values at the turning point. This cannot be correct because it leads to x values beyond the turning point that are less than x_B, and elapsed times required for the signal to propagate beyond the turning point that are less than the time to reach the turning point. This seeming paradox is actually a consequence of ignoring the existence of a singularity. The condition for a turning point is $\mu(z) = \cos \chi_I$, so the integrands are singular at these points. To go beyond this point, we must perform piecewise evaluation of the integrals. That is, if x_{tp}, z_{tp} are the coordinates of a turning point, then we must perform the evaluations as

$$
x = x_I + \left| \int_{z_I}^{z_{tp}} \frac{\cos \chi_I}{\left[\mu(\xi)^2 - (\cos \chi_I)^2 \right]^{1/2}} d\xi \right| + \left| \int_{z_{tp}}^{z} \frac{\cos \chi_I}{\left[\mu(\xi)^2 - (\cos \chi_I)^2 \right]^{1/2}} d\xi \right|
$$

$$
t = t_I + \left| \int_{z_I}^{z_{tp}} \frac{\mu(\xi)^2}{c_I \left[\mu(\xi)^2 - (\cos \chi_I)^2 \right]^{1/2}} d\xi \right| + \left| \int_{z_{tp}}^{z} \frac{\mu(\xi)^2}{c_I \left[\mu(\xi)^2 - (\cos \chi_I)^2 \right]^{1/2}} d\xi \right|
$$

$$(11.2.9)$$

A better alternative is to recognize that the first term in the right side of these expressions gives x_{tp} and t_{tp}. This allows us to write these equations as

$$
x = x_{tp} + \left| \int_{z_{tp}}^{z} \frac{\cos \chi_I}{\left[\mu(\xi)^2 - (\cos \chi_I)^2 \right]^{1/2}} d\xi \right|
$$

$$
t = t_{tp} + \left| \int_{z_{tp}}^{z} \frac{\mu(\xi)^2}{c_I \left[\mu(\xi)^2 - (\cos \chi_I)^2 \right]^{1/2}} d\xi \right|
$$

$$(11.2.10)$$

The same modifications are required to evaluate the ray path and elapsed time beyond a reflection point. The integrand is not singular at that point. Thus, the integration may be carried without special consideration up to the reflection point, then using the values of x and t at the reflection point to proceed onwards. For example, in Fig. 11.7b, we may use the above procedure to find x and t for any point on segments $A'B'$ and $B'C'D'$. To go beyond the reflection point, and thereby find segment $D'E'$, we would start with $x_{D'}$ and $t_{D'}$ and integrate over $0 < z \le z_{E'}$. Beyond point E', periodic replication applies. The symmetry property of the waveform relative to turning points and waveforms could be used to reduce the integrations. In any event, this piecewise method of handling slope reversals at turning points and reflections is troublesome to incorporate into an autonomous computer program. An alternative approach is developed in the next example. It evaluates the integrals in a stepwise manner.

The curvature of rays is related to the gradient of the sound speed. To identify this relation, we again consider a differential arc $d\ell$, but now account for the change of slope along it. Thus, at the upper end of the element in Fig. 11.8 the grazing angle is χ, while at the lower end it is $\chi + d\chi$. The center of curvature C is the point at which lines perpendicular to the tangents intersect. The radius of curvature $\mathcal{R}$ is the distance from the center of curvature to a point on the arc at the specified z.

Fig. 11.8 Geometric construction of the radius of curvature $\mathcal{R}$ using the properties of a ray

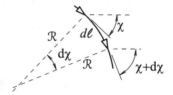

The angle between these perpendiculars is the difference of the grazing angles, so the differential arclength is $d\ell = \mathcal{R}d\chi$. The grazing angle on a specific ray may be considered to be a function of the arclength, so $d\chi = (\partial\chi/\partial\ell)\,d\ell$. Implicit differentiation of Snell's law, Eq. (11.2.2), gives

$$- (\sin\chi)\,\frac{\partial\chi}{\partial\ell} = \frac{\partial}{\partial\ell}\left(\frac{\cos\chi_I}{c_I}c(z)\right) = \frac{\cos\chi_I}{c_I}\left(\frac{dc}{dz}\frac{dz}{d\ell}\right) \qquad (11.2.11)$$

We found from Fig. 11.5 that $dz = (\sin\chi)\,d\ell$. Assembling these relations leads to

$$\boxed{\frac{1}{\mathcal{R}} = \frac{\partial\chi}{\partial\ell} = -\frac{\cos\chi_I}{c_I}\frac{dc}{dz}} \qquad (11.2.12)$$

Most treatments regard the radius of curvature as a positive number, so they take the absolute value of the preceding relation. This is not done here because the sign of $\mathcal{R}$ describes the direction in which the arc is curving. As shown in Fig. 11.6a and b, increasing z leads to decreasing c. Therefore, the slope increases with depth, $d\chi/dz > 0$ and $\mathcal{R} > 0$. In Fig. 11.6c and d, the value of c increases with increasing z. In that case, $d\chi/dz > 0$ and $\mathcal{R} > 0$. A different view is that the center of curvature lies below the ray path if $\mathcal{R} > 0$, whereas it lies above the ray path if $\mathcal{R} < 0$.

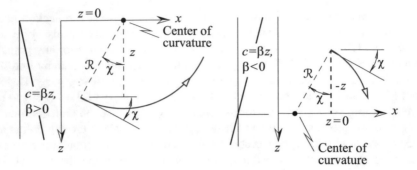

Fig. 11.9 Construction of circular ray paths for a fluid whose sound speed varies linearly with depth. The reference depth $z = 0$ need not be situated in the fluid

Knowledge of the radius of curvature is most useful when the sound speed depends linearly on z, so that dc/dz is constant. It is evident that $\mathcal{R}$ is constant in this case, but more information is embedded in Eq. (11.2.12). Let us define $z = 0$ to be the depth at which $c = 0$, so that $c = \beta z$. (It is not important that this reference depth probably is outside the fluid.) Correspondingly, we have

$$\mathcal{R} = -\frac{c_{\mathrm{I}}}{(\cos \chi_{\mathrm{I}})\,\beta} = -\frac{c}{(\cos \chi)\,\beta} = -\frac{z}{\cos \chi} \qquad (11.2.13)$$

The center of curvature is at distance $\mathcal{R}$ in the direction of the inward normal to the ray. This distance is constructed in Fig. 11.9 for cases where dc/dz is positive, $\beta > 0$, and negative, $\beta < 0$. In either case, the radius of curvature is the hypotenuse of a right triangle whose vertical side is z. The means that the center of curvature is situated at the virtual depth where the sound speed vanishes. This knowledge may be used for a graphical construction. Given the launch location and the initial grazing angle, a line perpendicular to the initial ray tangent is constructed. The intersection of this line with the $z = 0$ reference elevation marks the fixed center of curvature. The ray path is a circle of radius $\mathcal{R}$ centered on this point.

It has been suggested that a case of arbitrary sound speed dependence may be analyzed by approximating it as a sequence of layers in which $c(z)$ depends linearly on z. This is equivalent to replacing $c(z)$ with a pointwise linear interpolation. Mathematical software enables direct evaluation of the integrals in Eqs. (11.2.5) and (11.2.7), so there is little reason to pursue an approximate procedure.

EXAMPLE 11.2 The sound speed profile in a certain region of the ocean varies parabolically with depth according to $c(z) = c(z_{\min}) + [c(0) - c(z_{\min})]\,(z/z_{\min} - 1)^2$, with $z_{\min} = 200\,\mathrm{m}$, $c(z_{\min}) = 1460\,\mathrm{m/s}$, and $c(0) = 1520\,\mathrm{m/s}$. Consider two rays emitted by a source at a depth of $150\,\mathrm{m}$, both of which are launched toward the surface. One departs at a grazing angle of $12°$, whereas the other is such that it has a turning point at the surface.

Determine and plot the path and the travel time along each ray as a function
of the horizontal range x.

Significance

The numerical algorithm developed herein handles turning points without the
requirement of user intervention. The results will shed some light on the overall
significance of sound speed variation on propagation in the ocean.

Solution

The first task is to determine the range of z covered by each ray. For the first,
the initial grazing angle is $\chi_I = -12°$ and the sound speed at $z_I = 150$ m is
$c_I = 1463.8$ m/s. The corresponding sound speed for a turning point is $c_0 \equiv$
$c_I/\cos \chi_I = 1496.5$ m/s. There are two depths at which $c(z) = c_0$: $z_B = 44.11$ m and
$z_D = 355.87$ m. The initial grazing angle is upward, opposite the case in Fig. 11.7a.
Thus, the ray first travels upward from z_I to z_B, then downward from z_B to z_D, then
upward from z_D to z_B, and so on. We only need to follow one cycle, after which we
may invoke the periodicity property. To determine the initial angle for the second
ray, we compute c at $z = 0$, which is 1520 m/s. The grazing angle at this depth is
zero in order to have a turning point. Snell's law states that the initial angle should
be $\chi_I = \cos^{-1}(c(0)/c_I) = -15.636°$. The depth of the lower turning point corresponding to these values is 400 m.

Determination of a path requires evaluation of Eqs. (11.2.5) and (11.2.7) at a
sequence of depths z_n. The computational effort will be reduced greatly if the integrals
in Eqs. (11.2.5) and (11.2.7) are carried out in an incremental manner from one depth
to the next. If the values of x_{n-1} and t_{n-1} at a specific x_{n-1} have been determined,
and the values at the next point are given by

$$
\begin{aligned}
x_n &= x_{n-1} + \left| \int_{z_{n-1}}^{z_n} \frac{\cos \chi_I}{\left[\mu(\xi)^2 - (\cos \chi_I)^2\right]^{1/2}} d\xi \right| \\
t_n &= t_{n-1} + \left| \int_{z_{n-1}}^{z_n} \frac{\mu(\xi)^2}{c_I \left[\mu(\xi)^2 - (\cos \chi_I)^2\right]^{1/2}} d\xi \right|
\end{aligned}
\tag{1}
$$

In most cases, the integrals may be evaluated with standard software. Indeed, if
$z_n - z_{n-1}$ is extremely small, a crude strip rule would be adequate. However, turning
points are such that at some depth x_j it happens that $\mu(x_j) = \cos(\chi_I)$, so both
integrands are singular at such points. This condition may be addressed by using
Eq. (1) only if the value of z at a turning point does not fall in the interval between
z_{n-1} and z_n. If it does, an alternative evaluation based on a series expansion of the
denominator will allow us to pass the turning point. To see how to proceed, consider
an upper turning point. Let J be the index of this point, so that $z_J = z_B$. Because this
is an upper point, adjacent points are at a greater depth. Thus, $z_{J-1} = z_{J+1} = z_J + \Delta$,

with $\Delta \ll 1$. The integration variable is expressed as $\xi = z_B + \varepsilon$. Series expansions based on the smallness of ε in this interval give

$$c(z) = c(z_J + \varepsilon) \approx c(z_J) + \varepsilon \left(\frac{dc}{dz}\right)_J,$$

$$\left[\mu(\xi)^2 - (\cos \chi_\mathrm{I})^2\right]^{-1/2} = \left[\frac{c_\mathrm{I}^2}{c(z_J + \varepsilon)^2} - (\cos \chi_\mathrm{I})^2\right]^{-1/2}$$

$$\approx \left[\frac{c_\mathrm{I}^2}{c(z_J)^2}\left[1 - \frac{2\varepsilon}{c(z_J)}\left(\frac{dc}{dz}\right)_J\right] - (\cos \chi_\mathrm{I})^2\right]^{-1/2}$$

Because z_J is a turning point, it must be $c_\mathrm{I}/c(z_J) = \cos \chi_\mathrm{I}$. This observation leads to

$$\left[\mu(\xi)^2 - (\cos \chi_\mathrm{I})^2\right]^{-1/2} \approx \left[-\frac{c(z_J)^3}{2c_\mathrm{I}^2 \left(\dfrac{dc}{dz}\right)_J}\right]^{1/2} \varepsilon^{-1/2} \qquad (11.2.14)$$

Note that dc/dz is negative at the upper turning point, so the square root yields a real value. Substitution of this approximation and $d\xi = d\varepsilon$ into Eq. (1) leads to

$$x_J - x_{J-1} = x_{J+1} - x_J = (2\Delta)^{1/2}\left[\frac{c(z_J)}{\left|\left(\dfrac{dc}{dz}\right)_J\right|}\right]^{1/2}$$

$$t_J - t_{J-1} = t_{J+1} - t_J = \frac{(2\Delta)^{1/2}}{c_\mathrm{I}\mu(z_J)}\left[\frac{c(z_J)}{\left|\left(\dfrac{dc}{dz}\right)_J\right|}\right]^{1/2} \qquad (2)$$

The values of x_{J-1} and t_{J-1} presumably have been determined by invoking Eq. (1). The above expressions yield the corresponding values of x_J and t_J, then x_{J+1} and t_{J+1}. These expressions also apply to the case of a lower turning point, for which $(dc/dz)_J > 0$. In that case, the points neighboring the turning point will be at a slightly smaller depth, so that $z_{j-1} = z_{J+1} = z_j - \Delta$.

The actual ray path has two turning points. Figure 1 describes the scheme by which the depth z is discretized. It is based on the recognition that a selected sampling interval Δ_1 between the launch point and the first turning point might not fit the interval between the z_I and the lower turning point. Therefore, two intervals are used: $\Delta_1 = (z_\mathrm{I} - z_B)/N_1$ for points at $z \leq z_\mathrm{I}$ and $\Delta_2 = (z_D - z_\mathrm{I})/N_2$ for points at $z \geq z_\mathrm{I}$. This leads to assignment of the depths according to $z_n = z_\mathrm{I} - (n-1)\Delta_1$ for $n = 1$ to $N_1 + 1$; $z_n = z_B + (n - N_1 - 1)\Delta_1$ for $n = N_1 + 2$ to $2N_1 + 1$; $z_n = z_\mathrm{I} + (n - 2N_1 - 1)\Delta_2$ for $n = 2N_1 + 2$ to $2N_1 + N_2 + 1$; and $z_n = z_C - (n - 2N_1 - N_2 - 1)\Delta_2$ for $n = 2N_1 + N_2 + 2$ to $2N_1 + 2N_2 + 1$.

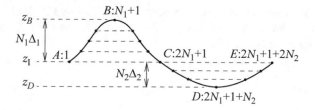

Figure 1.

The initial values are $x_1 = 0$ and $t_1 = 0$ at $x_1 = 0$. The values of x_j and t_j at regular points are found by using a numerical integration routine to evaluate Eq. (1). A typical step in MATLAB is `x(j)=x(j-1)+abs(quadl(F_x, z(j-1), z(j)))`. This scheme is used to evaluate points 2 to N_1, then the upper turning point is passed by using Eq. (2) to evaluate the z and t values at points $N_1 + 1$ and $N_1 + 2$. The next phase is evaluation of the regular points $N_1 + 3$ to $2N_1 + 1$. From there, the procedure is replicated, with Eq. (2) used to evaluate points $2N_1 + N_2 + 1$ and $2N_1 + N_2 + 2$. For points beyond $2N_1 + 2N_2 + 1$, the data is replicated according to $x_n = x_{2N_1+2N_2+1} + x_{n-2N_1-2N_2-1}$ and $t_n = t_{2N_1+2N_2+1} + t_{n-2N_1-2N_2-1}$.

The ray path in Fig. 2 is constructed by plotting (z_n, x_n) pairs such that z is displayed as a function of x. The first graph shows that the path is moderately curved, with a depth variation of approximately 310 m over a range of more than 4 km. The result for travel time along the ray appears to depend linearly on the range. This property stems from the facts that the grazing angle is small along the entire path and the phase velocity $c(z)$ varies little. Consequently, the horizontal component of the phase velocity which is $c(z)\cos(\chi)$ is nearly constant.

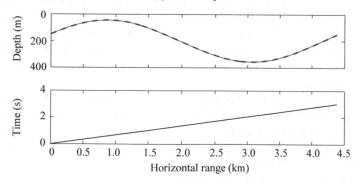

Figure 2.

The path appears to be sinusoidal. Whether this is the true nature of the path is tested by plotting a sine curve relative to the average value of z, which is $z_{av} = [\max(z) + \min(z)]/2$. The fitted curves are $z = z_{av} + A\sin[2\pi(x - x_0)/L]$ and $t = \max(t)b$, where $A = (z_D - z_B)/2$, $L/2$ is the horizontal distance between adjacent locations at which $z = z_{av}$, and x_0 is the minimum range at which $z = z_{av}$ with $dz/dx > 0$. This sine curve is indistinguishable from the computed path.

The ray corresponding to $\chi_I = -15.636°$ is described by Fig. 3. Launching a ray at an upward or downward initial angle that exceeds this value will generate a ray that reflects from the free surface.

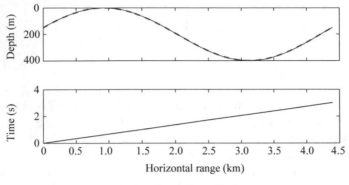

Figure 3.

11.2.2 Intensity and Focusing Factor

Thus far, we have not examined how the pressure magnitude is altered as a signal propagates. Although we applied Snell's law for an interface between different fluids to derive equations describing the ray paths, we cannot extend the analogy by applying the reflection and transmission coefficients. The reason for this is that the fluid properties vary continuously. Because $\rho c / \sin \chi$ changes by an infinitesimal amount across a layer of depth dz, the reflection coefficient is essentially zero and the transmission coefficient is essentially one. Our need is to determine how much the transmission differs from one, so we shall pursue a different approach. The analysis examines the manner in which power flows along rays in order to obtain an expression for intensity.

A common measurement in the ocean is mapping of the field from a small source. Typically the source radiates axisymmetrically relative to the vertical, which means that the field is independent of the circumferential angle of a cylindrical coordinate system whose axial direction is vertical. This situation is depicted in Fig. 11.10. The ray path in this picture is described by giving the transverse distance R as a function of z. The fact that the properties are the same for any vertical plane containing the z-axis, combined with the locally planar nature of the geometrical acoustics approximation, requires that R depends on z in the same manner as x depends on z for two-dimensional Cartesian coordinates. In addition, a ray path is altered if the depth z_0 of the source or the grazing angle at the source are altered. Thus, we may consider the ray path to be defined functionally as $R = R(z, z_0, \chi_I)$.

The analysis of intensity is founded on the concept of a *ray tube*, which is the region enclosed by a group of rays that leave a source at orientations that are infinitesimally different. A typical tube appears in Fig. 11.10. The sides of this ray tube lie in vertical

Fig. 11.10 A ray tube consisting of all rays emanating from a point at grazing angles between χ_I and $\chi_I + d\chi_I$, and between azimuthal angles between θ and $\theta + d\theta$

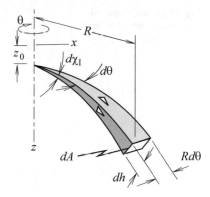

planes whose circumferential angle differ by $d\theta$. The upper and lower bounding surfaces correspond, respectively, to initial grazing angles χ_I and $\chi_I + d\chi_I$. The transverse distance to a cross section $d\mathcal{A}$ of the tube is R. The cross section is a rectangle whose sides are $Rd\theta$ horizontally and dh in the vertical plane.

The significance of a ray tube stems from the fact that the particle velocity, and therefore the intensity, is parallel to a ray. Consequently, there is no power flow across the sides, so the time-averaged power that flows across any cross section of the ray tube must be constant. The particle velocity is parallel to each ray, so the local intensity is perpendicular to the cross section at which it occurs. It follows that the constant power in a ray tube is such that

$$dP = \frac{|P|^2}{2\rho_0 c} d\mathcal{A} \text{ is constant along a ray} \tag{11.2.15}$$

An equivalent way of stating this principle is that the time-averaged power that flows across any two cross sections of the same ray tube must be the same. A corollary is that the power radiated by a source into the tube must flow across the cross section at the far end of the tube. Our objective is to describe this principle in terms of the properties of the rays in order to obtain an expression for the intensity.

The cross section of a ray tube in the vicinity of a specified depth is depicted in Fig. 11.11. Both ray paths in this Figure have been terminated at the same depth, so the transverse distance to the end of the upper ray is $R(z, z_0, \chi_I)$, while it is $R(z, z_0, \chi_I + d\chi_I)$ for the lower ray.

Because power is the area integral of intensity, we seek an expression for the cross-sectional area $d\mathcal{A} = (Rd\theta)(dh)$. The side dh forms one side of a right triangle whose hypotenuse is the line connecting the points in the bounding rays at depth z. The horizontal distance between the ends is solely a consequence of the dependence of R on the initial grazing angle. These angles are χ_I and $\chi_I + d\chi$ for the rays in the sketch, from which it follows that the horizontal distance between points on the rays is $\left| R(z, z_0, \chi_I + d\chi) - R(z, z_0, \chi_I) \right|$. Therefore, we have

Fig. 11.11 Side view of a ray tube generated by a source at depth z_0

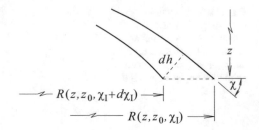

$$dh = \left| R\left(z, z_0, \chi_I + d\chi_I\right) - R\left(z, z_{0,} \chi_I\right)\right| \left| \sin \chi \right| = \left| \frac{\partial R}{\partial \chi_I} d\chi_I \right| \left| \sin \chi \right| \quad (11.2.16)$$

Thus, the cross-sectional area of the ray tube at the output end is

$$d\mathcal{A} = (Rd\theta) \, dh = \left| \frac{\partial R}{\partial \chi_I} \sin \chi \right| Rd\chi_I d\theta \quad (11.2.17)$$

The normal to $d\mathcal{A}$ is parallel to the rays within the tube, and therefore parallel to the particle velocity on that cross section. Hence, the time-averaged power flowing across $d\mathcal{A}$ is

$$d\mathcal{P} = I_{av} d\mathcal{A} = I_{av} \left| \frac{\partial R}{\partial \chi_I} \sin \chi \right| Rd\chi_I d\theta \quad (11.2.18)$$

The preceding is the time-averaged power that passes the cross section of the ray tube at depth z. This must be the same as the power that is input to the ray tube by the source. To describe this quantity, we observe that heterogeneous effects are significant only over distances that are many wavelengths. Thus, if we surround the source with a small sphere of radius ε, the pressure on this sphere will be the same as it would be if the fluid was homogeneous. If a sphere radiates omnidirectionally, its power $\mathcal{P}$ is distributed evenly over the surface area $4\pi\varepsilon^2$, so the intensity would be a constant value, $\mathcal{P}/\left(4\pi\varepsilon^2\right)$. The radial intensity for a source that is axisymmetric relative to the vertical axis is defined by multiplying this constant reference value by a directivity factor $\mathcal{D}_I$. This factor depends on the polar angle, which is the complement of the grazing angle, $\psi = \pi/2 - \chi_I$. Thus, the intensity at radial distance ε from the source is given by

$$I_0\left(\varepsilon, \chi_I\right) = \frac{\mathcal{P}}{4\pi\varepsilon^2} \mathcal{D}_I\left(\pi/2 - \chi_I\right) \quad (11.2.19)$$

The power input $d\mathcal{P}_0$ to the ray tube is the product of I_0 and the area of a patch of the surrounding sphere, whose sides are $\varepsilon d\chi_I$ in the vertical plane and $(\varepsilon \cos \chi_I)\, d\theta$ horizontally. Thus, we have

$$d\mathcal{P}_0 = I_0\left(\varepsilon, \chi_I\right) \varepsilon^2 \left(\cos \chi_I\right) d\chi_I d\theta = \frac{\mathcal{P}}{4\pi} \mathcal{D}_I\left(\pi/2 - \chi_I\right) \left(\cos \chi_I\right) d\chi_I d\theta \quad (11.2.20)$$

Equating this quantity to the power flowing across an arbitrary cross section in Eq. (11.2.18) yields

$$
I_{av} = \left(\frac{P}{4\pi} \right) \frac{D_I \, (\pi/2 - \chi_I) \cos \chi_I}{R \left| \left(\dfrac{\partial R}{\partial \chi_I} \right) \sin \chi \right|}
\tag{11.2.21}
$$

The same concept may be employed to examine the intensity when a field is generated by some agent other than a small source. If we know the intensity I_1 along a cross section at one location, and we wish to determine the value at another cross section, equality of the power flows requires that

$$
d\mathcal{P} = (I_{av})_2 \, dA_2 = (I_{av})_1 \, dA_1
\tag{11.2.22}
$$

When we use Eq. (11.2.18) to describe each term, we obtain

$$
(I_{av})_2 = (I_{av})_1 \, \frac{R_1 \left| \left(\dfrac{\partial R}{\partial \chi_I} \right)_1 \sin \chi_1 \right|}{R_2 \left| \left(\dfrac{\partial R}{\partial \chi_I} \right)_2 \sin \chi_2 \right|}
\tag{11.2.23}
$$

If we know the ray path, we can evaluate all terms in this expression other than the value at the input end. Thus, we can use this relation to monitor the intensity as the signal propagates along the ray path provided that we have a starting value for the intensity.

The *focusing factor* is used to compare the intensity to the hypothetical value that would be obtained at a specified location if the same source radiated into a homogeneous fluid. The radial distance would be $r = \left(R^2 + x^2 \right)^{1/2}$. Equation (11.2.19) describes the intensity for a homogeneous fluid, so we may find the reference intensity by replacing R in that relation by r. Thus, the focusing factor is

$$
\frac{I_{av}}{I_0 \, (r, \chi_I)} = \frac{r^2 \cos \chi_I}{R \left| \dfrac{\partial R}{\partial \chi_I} \sin \chi \right|}
\tag{11.2.24}
$$

The condition where $\partial R / \partial \chi_I = 0$ marks a location at which the cross-sectional area of a ray tube narrows to zero. The actual pressure must be finite, so this condition marks a condition where ray theory requires some adjustment. Depending on the specific nature of the signal, it might be that the singularity may be corrected by using a higher order approximation within the context of a linear acoustic theory, or it might be appropriate to incorporate a nonlinear correction. In any event the pressure will be greatly enhanced in the vicinity of a location where $\partial R / \partial \chi_I = 0$.

Extreme narrowing of ray tubes is manifested in either of two ways. One is a *focus*, like that encountered in an optical lens. A focus is marked by convergence

(a)   **(b)**

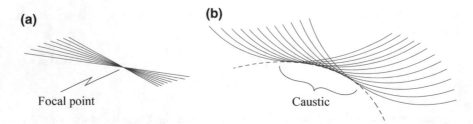

Focal point   Caustic

Fig. 11.12 Two situations in which geometric acoustics predicts a singularity of intensity: **a** a focus, **b** a caustic

of a group of rays on a single point. According to geometrical acoustics, the rays emerge from the focus continuously, unless the focus is at an interface between media. Figure 11.12a displays a focus at which a cone of rays intersect. The quantity $\partial R/\partial \chi_I$ is zero because there is no change in the transverse distance R for rays associated with initial grazing angles that differ by an infinitesimal amount. This is different from the situation where two rays associated with finitely different χ_I values intersect. In that case, the pressure at the intersection is the (finite) sum of the pressure associated with each ray.

Figure 11.12b illustrates a different singularity. In it, there is a curve in two dimensions, or a surface in three dimensions, on which rays for infinitesimally different χ_I arrive tangentially. We previously encountered this phenomenon in Example 11.1. The line or surface that is the locus of points on which R does not change when χ_I is changed infinitesimally is a *caustic* . When the fluid properties and the starting location of a ray are specified, the transverse distance R to a point on a ray path depends only on the depth and the initial grazing angle, that is, $R = f(z, \chi_I)$. Hence, the occurrence of a caustic in the axisymmetric field in a vertically stratified fluid is marked by

$$R = f(z, \chi_I) \text{ and } \frac{\partial f}{\partial \chi_I} = 0 \qquad (11.2.25)$$

In principle, we could identify this locus by solving the second equation for z over a range of χ_I values, then using the first equation to find the corresponding R values. In practice, the ray path usually is determined numerically, so the function f will not be available. A caustic is marked visually by a greatly reduced spacing between adjacent rays, with all arriving rays being on one side of the caustic. The absence of rays on the other side of the caustic is an artifice of the geometrical acoustics approximation. The actual behavior is that the field evanesces very rapidly in the direction normal to the caustic. Nevertheless, geometrical acoustics is quite useful even here because it informs us that there is a region where the pressure is greatly enhanced. Furthermore, Eq. (11.2.23) may be used to evaluate the intensity beyond the caustic. (This extension is a topic in Sect. 11.3.2.)

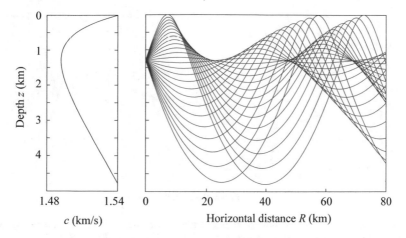

Fig. 11.13 Ray paths for a point source in a channel modeled by Munk in the Journal of the Acoustical Society of America, vol. 55, pp. 220–226 (1974). The dependence of sound speed on depth is $c = c_1 \left(1 + \varepsilon \left(\xi + e^{-\xi} - 1\right)\right)$ where $\xi = 2 \left(z - z_1\right)/z_1$. The coefficients are $c_1 = 1492$ m/s, $\varepsilon = 0.0074$, and $z_1 = 1300$ m

A classic demonstration of the occurrence of caustics may be found in a paper by Munk.[8] The sound speed profile described in Fig. 11.13 was suggested there to be a good prototype for studies of phenomena in the ocean at moderate latitudes. The sound speed profile has a minimum at $z_1 = 1300$ m. An omnidirectional source is located at this depth, so that $z_I = z_1$. Setting $dc/dz = 0$ at z_I results in the ray paths having an inflection point at the depth of the source. However, the sound speed profile is not symmetric relative to this depth. Consequently, the appearance of a ray path below z_I does not mirror that appearance above.

The rays appearing in Fig. 11.13 correspond to initial grazing angles ranging from $14.40°$ upward to $14.40°$ downward. This is the range of angles for which rays are not reflected from the surface. The rays described here extend to a greater range than those computed by Munk, whose results were not shown in their entirety beyond a range of 25 km.

The most evident feature of the ray paths is that there are regions where there are none. These are *shadow zones*. Usually, one thinks of a shadow as the blocking effect of a solid object. Here, a caustic obstructs the sound field. Regions of enhanced intensity correspond to decreased spacing between adjacent rays. They appear as darkened regions in the figure. The source is at $R = 0$, $z = 1300$ m, so the enhancement near that location is not surprising. The next enhanced region occurs in the range from 10 to 20 km at depths between 400 m–1.3 km. This appears to be a caustic. Other regions of interest are close to the source depth at ranges in the vicinity of 50 and 70 km. These seem to be somewhat like a focus, but they also seems to resemble a caustic. Figure 11.14 zooms in on the regions where rays are concentrated.

---

[8]W. Munk, "Sound channel in an exponentially stratified ocean, with application to SOFAR," Acoustical Society of America, Vol. 55 (1974) pp. 220–226.

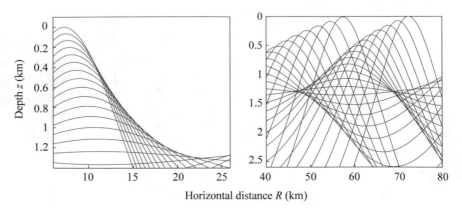

Horizontal distance R (km)

Fig. 11.14 Zoomed views of regions of enhanced intensity in the sound channel studied by Munk in the Journal of the Acoustical Society of America, vol. 55, pp. 220–226 (1974)

The caustic in the left figure is a boundary beyond which the pressure is zero (or very small if the geometrical acoustics approximation is corrected). The low point of this feature occurs at the depth of the source. The concentration of rays in the vicinity of $R = 47$ km and $R = 70$ km in the right figure also occurs in the vicinity of the source depth z_I. The zoomed views show that these concentrations are not foci, because only rays at nearly equal grazing angles intersect. This is a fundamental aspect of caustics. Each has a distinct arrete, which is the minimum distance for the caustic. The rays become more diffuse at the other end of each caustic. Additional computations would show that caustics like these will occur even if z_I is not the same as depth z_1 for minimum $c(z)$. However, in that case no caustics would be observed at z_I or z_1.

EXAMPLE 11.3 A radially symmetric point source is situated at $z_I = 150$ m. The sound speed dependence on depth is as described in Example 11.2. Identify the regions where the pressure is enhanced. Then explain how this data may be used to evaluate the focusing factor at a designated location.

Significance

In addition to showing how to map rays, some interesting phenomena, including the role of reflections, will be explored.

Solution

We begin with an evaluation of the ray paths that remain below the surface. For this, we follow the procedure developed in the previous example. Stated briefly, the initial grazing angle is set in a range whose limits are the values extracted from Eq. (11.2.8), specifically $\chi_I = \cos^{-1}(c_I/c(z=0)) = \pm 15.363°$. The turning point depths min (z) and max (z) for each angle are computed, and the depth is discretized into intervals of Δ_1 for min $(z) \leq z \leq z_I$ and Δ_2 for $z_I \leq z \leq$ max (z). Starting from the source location, the transverse value of R at each depth up to the first turning point is computed by applying the first of Eq. (1) in Example 11.2. The turning point and the next are evaluated by applying Eq. (2). Then, Eq. (1) is used to evaluate points up to the second turning point, followed by Eq. (2) for that point and the point beyond. The final part of evaluation uses Eq. (1) to return to z_I, after which the path is periodically replicated. Figure 1 is the result of this computation.

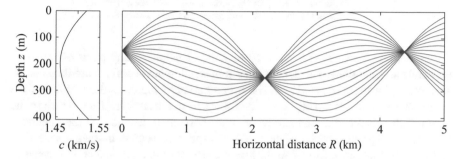

Figure 1. Rays radiated by the source that do not reflect from the surface.

It can be seen that all rays have the same periodicity horizontally, and that they all intersect at a common point. These intersections are foci. Both properties are a consequence of $c(z)$ being an even function relative to the depth z_{min} at which it is a minimum. Another consequence of the symmetric nature of the sound speed profile is that the depth of foci are either z_I or max $(z) - z_I$.

In addition to the occurrence of foci, there also appears to be large shadow zones. However, Fig. 1 does not give a complete picture, because it does not account for rays that reflect from the free surface. Such a condition is marked by one of the turning points being above the surface, that is, $z_0 < 0$. Thus, reflected rays in the present case correspond to $15.363° < \chi_I < 90°$ and $-90° < \chi_I < -15.363°$. Rays that reflect are computed by using Eq. (1) in Example 11.2 to evaluate the values of R and x at the reflection point. If we let J equal to the index of the reflection point, then the properties for the next point are set to place its properties symmetrically, so that $R_{J+1} = R_J + (R_j - R_{J-1})$.

The result of including rays that reflect is Fig. 2. It appears that only one shadow zone remains. (Even that zone would disappear if there were a bottom at a finite depth, in which case there would also be reflections at that boundary.) The ray paths are far apart in the regions that were quiet in the previous figure. This tells us that the intensity is low in those regions. There appears to be a caustic that touches the surface at a range of 2.5 km formed from a group of rays for which $\chi_I > 0$, and another at a range of 2.8 km formed from a group for which $\chi_I < 0$.

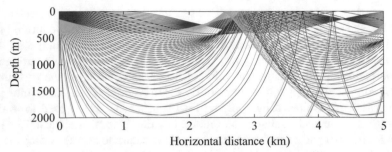

Figure 2. All rays radiated by the point source. Blue paths are rays that
do not relect from the surface, red rays radiate upward from
the source ($\chi_I < 0$), and green rays radiate downward ($\chi_I > 0$).

To examine these features more closely, Fig. 3 displays only paths of rays that
reflect from the surface. The details remain masked in the overview, but they can be
seen clearly in the zoomed views. The broad darkened region in the vicinity of the
source is due to nearly vertical rays that reflect from the surface. They return to the
depth of the source at a comparatively small horizontal distance, so they enhance
the already large pressure in the nearfield of a spherical source. In the zoomed view
of the horizontal range from 2 to 3.2 km, we see that there are indeed two caustics
formed from rays that are incident at the surface. We also see that reflection of the
rays associated with each caustic seems to result in each forming another caustic.

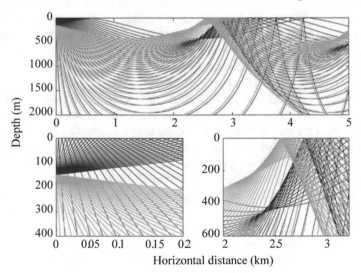

Figure 3. Rays that reflect from the free surface when
the sound speed depends parabolically on
depth.

Graphs such as these give a qualitative picture, but the data also may be used to calculate the focusing factor. The quantity in Eq. (11.2.24) that is not readily available is $\partial R/\partial \chi_1$. Because z is held constant in this derivative, we could estimate it with a finite difference if we had values of R corresponding to a discretization of z that is common to the ray paths. The problematic aspect is that our algorithm for evaluating ray paths leads to the type of data depicted in Fig. 4. The two ray paths correspond to slightly different initial grazing angles χ_1 and χ_2. Points marked a and b on each curve are the result of computations, so the R and z coordinates of each point are known. The algorithm that we use sets an equal depth increment for the computation of a ray, but that increment is not constant from ray to ray. Thus, $z_a(\chi_2) \neq z_a(\chi_1)$ and $z_b(\chi_2) \neq z_b(\chi_1)$.

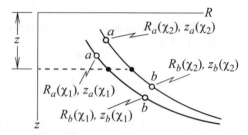

Figure 4.

We wish to determine the intensity at the indicated depth z midway between these ray paths. To that end we first evaluate the transverse distance to the point on each ray path at depth z. This is done by interpolating between the coordinates of the computed point. A linear interpolation gives

$$R\left(z, \chi_j\right) = \left(\frac{z_b - z}{z_b - z_a} R_a + \frac{z - z_a}{z_b - z_a} R_b\right)\Bigg|_{\chi_j}$$

Then a first-order finite difference approximation gives

$$\frac{\partial R}{\partial \chi_1}\bigg|_{z,(R_a+R_b)/2} = \frac{R(z, \chi_2) - R(z, \chi_1)}{\chi_2 - \chi_2}$$

Application of this procedure to create an intensity map for the entire field would require effort. One approach would establish a spatial grid at which the algorithm would be implemented. A search algorithm would identify the two rays that are closest to a selected grid point, and the data points on each ray that are directly above and below the selected point. Section 11.3.2 will develop a formulation that is readily programed and more accurate.

11.3 Arbitrary Heterogeneous Fluids

The model of a vertically stratified fluid is useful for environmental acoustic studies. However, it might be that the sound speed depends on the horizontal range, as well as depth, especially at very long ranges. The analytical approach in the previous section has the attractive feature of being pictorial, and therefore readily understood. The method that follows is mathematical and more abstract, but it is general.

11.3.1 Ray Tracing Equations

The formulation begins with the general equation for the zero-phase wavefront, which was stated previously in Eq. (11.1.4),

$$\tau\left(\bar{x}\right) \equiv \frac{\bar{x} \cdot \bar{e}\left(\bar{x}\right)}{c\left(\bar{x}\right)} = t \tag{11.3.1}$$

The corresponding pressure is given by Eq. (11.1.1) for a harmonic signal or Eq. (11.1.8) for an arbitrary time signature. In either case, a constant value of τ locates the wavefront on which the phase is zero when t is that value of τ.

The geometrical acoustics approximation of propagation was depicted in Fig. 11.1, where a specific signal is observed at position $\bar{x} = \bar{\xi}$ at time t. At a slightly later instant $t + dt$, it is situated on the same ray at $\bar{x} = \bar{\xi} + d\bar{\xi}$. Because $\bar{\xi}$ marks the position of a specific phase, it must be that $d\bar{\xi}/dt$ is the phase velocity. A different view is that the local sound speed is $c\left(\bar{\xi}\right)$, and the propagation is in the direction $\bar{e}\left(\bar{\xi}\right)$ tangent to the ray path. Both views are correct, so it must be that

$$\frac{d\bar{\xi}}{dt} = c\left(\bar{\xi}\right)\bar{e}\left(\bar{\xi}\right) \tag{11.3.2}$$

Another view of the propagation is that the phase moves along a ray such that it remains on the same wavefront. The wavefront at time t is described by Eq. (11.3.1), and its shape at time $t + dt$ is described by $\tau\left(\bar{\xi} + d\bar{\xi}\right) = t + dt$. A Taylor series expansion this expression gives

$$\tau\left(\bar{\xi}\right) + \nabla\tau \cdot d\bar{\xi} = t + dt \tag{11.3.3}$$

The fact that $\tau\left(\bar{\xi}\right) = t$ reduces this relation to

$$\nabla\tau \cdot \frac{d\bar{\xi}}{dt} = 1 \tag{11.3.4}$$

A fundamental property of the gradient of a function is that it is normal to the surface on which the function is constant. Along a wavefront τ is constant, and $\bar{e}\left(\bar{\xi}\right)$ is the normal, so we have $\nabla \tau = |\nabla \tau| \, \bar{e} \, (\bar{\xi})$. Substitution of this description and Eq. (11.3.2) into the preceding leads to $|\nabla \tau| = 1/c \, (\bar{\xi})$. An equivalent description of this property is

$$\nabla \tau \cdot \nabla \tau = \frac{1}{c \, (\bar{\xi})^2} \tag{11.3.5}$$

This is the *eikonal equation*. Because the magnitude of $\nabla \tau$ is inversely proportional to the local sound speed, and it is in the sense of the propagation, it is called the *wave slowness*. We shall denote this quantity as $\bar{s} \, (\bar{\xi})$, so we have

$$\bar{s} \, (\bar{\xi}) \equiv \nabla \tau = \frac{\bar{e} \, (\bar{\xi})}{c \, (\bar{\xi})} \tag{11.3.6}$$

Manipulation of the relations developed thus far will lead to a set of first-order differential equations in which the dependent variables are $\bar{\xi}$ and $\bar{s}$. The derivation begins by evaluating the rate at which $\bar{s}$ changes. One way in which that derivative may be found is by the chain rule based on considering $\bar{s}$ to be a function of position along a ray, which leads to

$$\frac{d\bar{s}}{dt} = \nabla \bar{s} \cdot \frac{d\bar{\xi}}{dt} = \nabla \bar{s} \cdot (c\bar{e}) \tag{11.3.7}$$

(The quantity $\nabla \bar{s}$ is the gradient of a vector, which we usually do not encounter. In Cartesian coordinates, each component of $\bar{s}$ may depend on x, y, and z, and the unit vector parallel to each axis is constant. Correspondingly, we would find that $\nabla \bar{s} = \nabla s_x \bar{e}_x + \nabla s_y \bar{e}_y + \nabla s_z \bar{e}_z$.) We use the definition of $\bar{s}$, Eq. (11.3.6), which indicates that $\bar{e} = c\bar{s}$, to eliminate the unit vector,

$$\frac{d\bar{s}}{dt} = c^2 \nabla \bar{s} \cdot \bar{s} \equiv \frac{1}{2} c^2 \nabla \, (\bar{s} \cdot \bar{s}) \tag{11.3.8}$$

The eikonal equation gives $\bar{s} \cdot \bar{s} = 1/c^2$. The gradient of this quantity is $\nabla \, (\bar{s} \cdot \bar{s}) = (-2/c^3) \, \nabla c$. Substitution of this relation converts the preceding equation to

$$\frac{d\bar{s}}{dt} = -\frac{\nabla c}{c \, (\bar{\xi})} \tag{11.3.9}$$

In addition, substituting $\bar{e} = c\bar{s}$ into Eq. (11.3.2), which describes the phase velocity, leads to

$$\frac{d\bar{\xi}}{dt} = c \, (\bar{\xi})^2 \, \bar{s} \, (\bar{\xi}) \tag{11.3.10}$$

Equations (11.3.9) and (11.3.10) are coupled ordinary differential equations that govern $\bar{\xi}$ and $\bar{s}$ as functions of time. They are known as the *ray tracing equations*. Except for a few special cases, their solution requires computational methods. The numerical algorithms typically implemented in software solve first-order equations that have been placed into a matrix form,

$$\boxed{\frac{d}{dt}\{X\} = \{F\}}$$
(11.3.11)

The ray tracing equations are expressed in this form by placing the components of $\bar{s}$ and $\bar{\xi}$ in the column vector $\{X\}$. The components of the right side of each equation are placed in the same sequence in $\{F\}$. The result is a set of six differential equations,

$$\{X\} = \begin{bmatrix} s_x & s_y & s_z & \xi_x & \xi_y & \xi_z \end{bmatrix}^{\mathrm{T}}$$
$$\{F\} = \begin{bmatrix} -\dfrac{1}{c}\begin{bmatrix} \dfrac{\partial c}{\partial \xi_x} & \dfrac{\partial c}{\partial \xi_y} & \dfrac{\partial c}{\partial \xi_z} \end{bmatrix} & c^2 \begin{bmatrix} s_x & s_y & s_z \end{bmatrix} \end{bmatrix}^{\mathrm{T}}$$
(11.3.12)

The sound speed is a function of position. Hence, it is implicit that embedded in the computation of $\{F\}$ are routines that evaluate c and the components of its gradient as functions of the position coordinates ξ_x, ξ_y, and ξ_z.

To start the numerical solution, the initial conditions must be stated. The initial instant may be set as $t = 0$ because time does not appear explicitly in the ray tracing equations. The analysis of rays in a vertically stratified fluid began by setting the starting position and the grazing angle. The analogous information here is a specification of where a ray is situated, $\bar{\xi} = \bar{\xi}_0$, and the direction in which it is aimed, $\bar{e} = \bar{e}_0$, at $t = 0$. The corresponding initial wave slowness is $\bar{s}_0 = \bar{e}_0/c\left(\bar{\xi}_0\right)$. The result of the numerical solution is a set of $\bar{\xi}$ and $\bar{s}$ values at a set of t value for a single ray. Depending on how the signal is generated, we might find families of rays by changing the starting point or the initial tangent direction.

It generally is a good idea to have an external check to monitor the error when differential equations are solved numerically. For example, if energy is conserved in a dynamic system, the mechanical energy contained in the response at each instant can be verified to be constant. A conservation-like principle is the eikonal equation, $\bar{s} \cdot \bar{s} = 1/c^2$. Solution of the differential equations gives the values of $\bar{\xi}$ and $\bar{s}$ at some instant t. Knowledge of these values makes it possible to compute $c\left(\bar{\xi}\right)^2 \bar{s} \cdot \bar{s} - 1$. The degree to which the computed value differs from zero is a fractional measure of the error.

One situation where the ray tracing equations may be solved analytically is that of a homogeneous fluid, for which $\nabla c \equiv 0$. Equation (11.3.9) states that s is constant in this case, so it never changes from its initial value, $\bar{s}_0$. In turn, this means that the right side of Eq. (11.3.10) is constant at the initial value $\bar{s}_0$. The solution of that equation is

$$\bar{\xi} = \bar{\xi}_0 + c^2 \bar{s}_0 t = \bar{\xi}_0 + \bar{e}\left(\bar{\xi}_0\right) ct$$
(11.3.13)

This is the equation of a straight line. Thus, the ray tracing equations for a homogeneous fluid yield rays that are straight, which is a property we already knew. Furthermore, the signal at instant t is at distance ct from the starting point, as it must for plane wave propagation.

EXAMPLE 11.4 Consider a fluid whose sound speed depends linearly on the depth. Start with an ansatz for a ray path that is the general equation for a circle. Show that the ray tracing equations are satisfied by this representation if the center of the circle is situated at the extrapolated elevation where the value of c is zero.

Significance

The process of working backward to fit the trial solution to the ray tracing equations will clarify what the forward solution process actually entails. It also serves to tie the graphical approach for vertically stratified media to the mathematical formulation for arbitrary heterogeneity.

Solution

The mathematical operations are somewhat simplified by defining $z = 0$ to be the depth at which $c = 0$. (This location might be outside the region in which the fluid resides.) This definition of z leads to $c = \beta z$, where β is an arbitrary constant. There are several ways in which a circular path may be represented mathematically. We will use a parametric form in which the grazing angle χ is the parameter. Figure 1 shows that χ also is the angle of the radial line to a point on the ray. The coordinates of the center are (x_0, z_0). Neither these coordinates nor the radius $\mathcal{R}$ is presumed to be known.

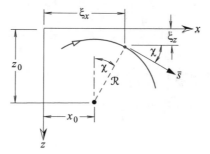

Figure 1.

The diagram shows that the coordinates of a point on the circular arc are

$$\xi_x = x_0 + \mathcal{R} \sin \chi, \quad \xi_z = z_0 - \mathcal{R} \cos \chi \qquad (1)$$

Thus, the sound speed at $\bar{x} = \bar{\xi}$ is

$$c\left(\bar{\xi}\right) = \beta \left(z_0 - \mathcal{R}\cos\succ\right) \tag{2}$$

The slowness vector is tangent to the ray, and its magnitude is $1/c$, so we find from the sketch that

$$\bar{s} = \frac{1}{c\left(\bar{\xi}\right)} \left(\bar{e}_x \cos\succ + \bar{e}_z \sin\chi\right) \tag{3}$$

The last aspect of the trial solution is the introduction of time. A signal travels along a ray at the local speed of sound. This is the rate at which point $\bar{\xi}$ moves along the arc. The path is circular, so the corresponding angular speed is

$$\frac{d\chi}{dt} = \frac{c\left(\bar{\xi}\right)}{\mathcal{R}} \tag{4}$$

Our approach is to substitute these representations into the ray tracing equations. Because all quantities have been defined in terms of χ, derivatives with respect to time may be evaluated by the chain rule. We begin with Eq. (11.3.10), for which we have

$$\frac{d\bar{\xi}}{dt} = \left(\frac{d\xi_x}{d\chi}\bar{e}_x + \frac{d\xi_z}{d\chi}\bar{e}_z\right)\frac{d\chi}{dt} = \mathcal{R}\left(\bar{e}_x\cos\succ + \bar{e}_z\sin\succ\right)\left(\frac{c}{\mathcal{R}}\right) \equiv c^2\bar{s}$$

This shows that Eq. (1) identically satisfies Eq. (11.3.10) independently of the radius and coordinates of the center. In other words, any circular path is consistent with that equation.

Consideration of Eq. (11.3.9) proceeds similarly. An expression for ds/dt may be found by differentiating Eq. (3). The key observation is that Eq. (2) gives c as a function of χ, which means that Eq. (3) actually is a description of s as a function of χ. This calls for the chain rule for differentiation, which leads to

$$
\begin{aligned}
\frac{d\bar{s}}{dt} &= \left(\frac{d\chi}{dt}\right)\left(\frac{d\bar{s}}{d\chi}\right) = \frac{c}{\mathcal{R}}\left[\frac{d}{d\chi}\left(\frac{1}{c}\right)\left(\bar{e}_x\cos\chi + \bar{e}_z\sin\chi\right)\right.\\
&\quad \left. + \frac{1}{c}\frac{d}{d\chi}\left(\bar{e}_x\cos\chi + \bar{e}_z\sin\chi\right)\right]\\
&= \left(\frac{c}{\mathcal{R}}\right)\left[-\frac{1}{c^2}\frac{dc}{d\chi}\left(\bar{e}_x\cos\chi + \bar{e}_z\sin\chi\right) + \frac{1}{c}\left(-\bar{e}_x\sin\chi + \bar{e}_z\cos\chi\right)\right]\\
&= \frac{1}{\mathcal{R}}\left[-\left(\frac{\beta\mathcal{R}\sin\chi}{\beta z_0 - \mathcal{R}\cos\chi}\right)\left(\bar{e}_x\cos\chi + \bar{e}_z\sin\chi\right) + \left(-\bar{e}_x\sin\chi + \bar{e}_z\cos\chi\right)\right]
\end{aligned}
$$

Collection of like components reduces this expression to

$$\frac{d\bar{s}}{dt} = \frac{\beta}{\beta\left(z_0 - \mathcal{R}\cos\chi\right)\mathcal{R}}\left[z_0\left(-\bar{e}_x\sin\chi + \bar{e}_z\cos\chi\right) - \mathcal{R}\bar{e}_z\right] \tag{5}$$

This expression constitutes the left side of Eq. (11.3.9). The right side is found by using the original description $c = \beta z$ to evaluate the gradient. Thus,

$$-\frac{\nabla c}{c} = -\frac{1}{c}\left(\frac{\partial c}{\partial z}\bar{e}_z\right)\bigg|_{z=\xi_x} = -\frac{1}{\beta(z_0 - \mathcal{R}\cos\chi)}(\beta\bar{e}_z) \tag{6}$$

When we equate Eqs. (5) and (6), we see that the $\bar{e}_x$ and $\bar{e}_z$ components match only if $z_0 = 0$. Because z_0 is the depth of the center of curvature, the finding that $z_0 = 0$ places the center of the circular ray path at the depth where c vanishes. As is true for the first ray tracing equation, Eq. (11.3.9) is satisfied for any $\mathcal{R}$. This quantity and the actual location of the center are set by the initial position and angle at which the ray is launched. The circular nature of the ray path in a linearly varying fluid was identified previously. Solution of the ray tracing equations gives us a concrete illustration of what the ray tracing equations represent.

11.3.2   Amplitude Dependence

Transport Equation

Suppose we know the waveform of a signal at the starting location $\bar{\xi}_0$ of a ray. We wish to know the waveform at a downstream point $\bar{\xi}_A$ on that ray. Part of such a determination requires that we find the travel time for the signal. The discussion of Eq. (11.1.4) informs us that we can extract this property directly from the solution of the ray tracing equations. The solution algorithm sets the time base such that $t = 0$ for the starting point $\bar{\xi}_0$. The wavefronts that are computed correspond to zero phase. This means that the signal that departed from $\bar{\xi}_0$ at $t = 0$ will be observed at the computed location $\bar{\xi}_A$ at time t_A. According to the aforementioned discussion, a signal that departs from $\bar{\xi}_0$ at some instant $t' > 0$ will arrive at $\bar{\xi}_A$ at time $t_A + t'$. Consequently, if we retain the time value associated with each $\bar{\xi}$ value obtained in a numerical solution of the ray tracing equations, then we may use that time to determine the time delay of the waveform function at each location. A different route leading to the same conclusion comes from the observation that the independent variable for the ray tracing equations is t, but it does not appear explicitly anywhere in those equations other than as the rate variable for derivatives. Thus, shifting the time at $\bar{\xi}_0$ from $t = 0$ to $t = t'$ merely increments the time value at any point $\bar{\xi}_A$ from t_A to $t_A + t'$.

Knowledge of the time delay allows us to determine when a specific phase arrives at any location along a computed ray. However, this knowledge sheds no light on the amplitude dependence of the signal. Here, we will develop a procedure by which this dependence may be identified if the signal is harmonic. The basic notion is to seek conditions that must be satisfied in order that the geometrical acoustics description of pressure be a solution of the position-dependent wave equation. For this analysis, we use the field equation that accounts for density variation, which is Eq. (11.1.21). Doing so yields a result that is valid for gases, as well as liquids.

We begin with the basic ansatz in Eq. (11.1.1). In term of the wavefront function $\tau(\bar{x})$, it is

$$p(x, t) = P(\bar{x}) \, \text{Re} \left(e^{i\omega(t - \tau(\bar{x}))} \right) \tag{11.3.14}$$

From this, we find the pressure gradient to be

$$\nabla p = \text{Re} \left((\nabla P - i\omega P \nabla \tau) \, e^{i\omega(t - \tau(\bar{x}))} \right) \tag{11.3.15}$$

The divergence of this expression is

$$\nabla^2 p = \text{Re} \left(\left(\nabla^2 P - 2i\omega \nabla P \cdot \nabla \tau - i\omega P \nabla^2 \tau - \omega^2 P \nabla \tau \cdot \nabla \tau \right) e^{i\omega(t - \tau(\bar{x}))} \right)$$

The dependence of p on t is harmonic. Thus, substitution of the preceding into Eq. (11.1.21) leads to

$$\nabla^2 p - \frac{1}{\rho_0} \nabla \rho_0 \cdot \nabla p - \frac{1}{c^2} \frac{\partial^2 p}{\partial t^2} = \text{Re} \left\{ \left[\nabla^2 P - 2i\omega \nabla P \cdot \nabla \tau - i\omega P \nabla^2 \tau \right. \right.$$
$$\left. -\omega^2 P \nabla \tau \cdot \nabla \tau \ - \ \frac{1}{\rho_0} \nabla \rho_0 \cdot (\nabla P - i\omega P \nabla \tau) \right.$$
$$\left. \left. +\frac{\omega}{c^2} P \right] e^{i\omega(t - \tau(\bar{x}))} \right\} = 0$$

$$\tag{11.3.16}$$

This equation must be satisfied for any combination of $\bar{x}$ and t values, so the complex exponential may be canceled. The fact that both P and τ are real assists decomposition of the factor of this exponential into its real and imaginary parts. The real part is

$$\nabla^2 P + \frac{1}{\rho_0} \nabla \rho_0 \cdot \nabla P - \omega^2 \left(\nabla \tau \cdot \nabla \tau - \frac{1}{c^2} \right) P = 0 \tag{11.3.17}$$

Geometrical acoustics considers ω to be very large, which means that terms that are proportional to ω^2 are dominant. This attribute makes it permissible to drop the terms that not contain ω. What remains is the factor in parentheses multiplying P. The relation must be true for any P. Therefore, the term inside the parentheses must be zero. The result is a restatement of the eikonal equation, so it provides no new information.

The imaginary part of Eq. (11.3.16) is the one that is useful. All of these terms are proportional to ω, so none may be dropped as being insignificant relative to others. Factoring out $-\omega$ converts the imaginary part to the *transport equation*,

$$\boxed{2\nabla P \cdot \nabla \tau + P \nabla^2 \tau - \frac{1}{\rho_0} P \nabla \rho_0 \cdot \nabla \tau = 0} \tag{11.3.18}$$

This expression features three terms, each of which is a derivative of one variable in a triple product. The term $-(\nabla \rho_0)/\rho_0$ is identically $\rho_0 \nabla (1/\rho_0)$ and the two factor for the first term suggests that the underlying term is P^2. Consequently, let us multiply the transport equation by P/ρ_0. Doing so gives

$$2\frac{P}{\rho_0}\nabla P \cdot \nabla\tau + \frac{P^2}{\rho_0}\nabla^2\tau + P^2\nabla\left(\frac{1}{\rho_0}\right)\cdot\nabla\tau \equiv \nabla\cdot\left(\frac{P^2}{\rho_0}\nabla\tau\right) = 0 \qquad (11.3.19)$$

This relation is further simplified by the definition that $\nabla\tau \equiv s \equiv \bar{e}/c$. Substitution of this relation reduces the transport equation to

$$\boxed{\nabla\cdot\left(\frac{P^2}{\rho_0 c}\bar{e}\right) = 0} \qquad (11.3.20)$$

The particle velocity in the geometrical acoustic approximation fits the relation for a planar wave, according to which $\bar{V} = \bar{e}P/(\rho_0 c)$. Therefore, the quantity inside the parentheses is twice the time-averaged intensity tangent to the ray. If we already have solved the ray tracing equations, then the intensity is the sole unknown in this version of the transport equation. In principle, we could employ a numerical differential equation solver to determine how the intensity varies along the ray path. However, there is a simpler way of extracting P along a ray. It does not require solution of a differential equation.

Relation to Ray Tube Area

The occurrence of intensity as the variable governed by the transport equation suggests that this relation is fundamentally like the result for intensity in a depth-dependent heterogeneity in Sect. 11.2.2. To examine this observation, we consider a ray tube, such as the one in Fig. 11.15. The starting point for the ray tube is $\bar{x} = \bar{\xi}_0$, and the cross-sectional area at any location $\bar{\xi}$ is the infinitesimal value $dA\left(\bar{\xi}\right)$.

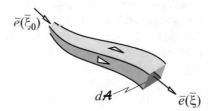

Fig. 11.15 Ray tube in an arbitrary heterogeneous fluid. The edges are rays whose initial tangent corresponds to four sets of direction angles that differ by differential amounts

Green's (divergence) theorem is applied to the transport equation, with the domain $\mathcal{V}$ being the region contained within the ray tube. The intensity is in the direction $\bar{e}$ tangent to the rays, and the normal to the side walls is perpendicular to $\bar{e}$. Consequently, there is no contribution to the integral from the side walls. It follows that

$$\iiint_{\mathcal{V}}\nabla\cdot\left(\frac{P^2}{\rho_0 c}\bar{e}\right)d\mathcal{V} = \iint_{dA(\bar{\xi})}\bar{e}\cdot\left(\frac{P^2}{\rho_0 c}\bar{e}\right)dA + \iint_{dA(\bar{\xi}_0)}(-\bar{e})\cdot\left(\frac{P^2}{\rho_0 c}\bar{e}\right)dA = 0$$

$$(11.3.21)$$

The ray tube cross section is infinitesimal in extent, so the pressure may be considered to be constant along it. (The behavior at caustics will be addressed separately.) The direction $-\bar{e}$ in $d\mathcal{A}_I$ corresponds to an inflow. Hence, the above relation states that the time-averaged power that flows into the ray tube at its beginning must equal the power that flows out at the end, that is

$$\frac{P\left(\bar{\xi}\right)^2}{\rho_0\left(\bar{\xi}\right)c\left(\bar{\xi}\right)}d\mathcal{A}\left(\bar{\xi}\right) = \frac{P\left(\bar{\xi}_0\right)^2}{\rho_0\left(\bar{\xi}_0\right)c\left(\bar{\xi}_0\right)}d\mathcal{A}\left(\bar{\xi}_0\right) \tag{11.3.22}$$

The explicit solution for the pressure amplitude at position $\bar{\xi}$ on a ray is

$$P\left(\bar{\xi}\right) = \left[\frac{\rho_0\left(\bar{\xi}\right)c\left(\bar{\xi}\right)}{\rho_0\left(\bar{\xi}_0\right)c\left(\bar{\xi}_0\right)}\frac{d\mathcal{A}\left(\bar{\xi}_0\right)}{d\mathcal{A}\left(\bar{\xi}\right)}\right]^{1/2}P\left(\bar{\xi}_0\right) \tag{11.3.23}$$

Equations (11.3.23) and (11.2.23) are equivalent. Indeed, Eq. (11.3.22) could have been obtained from the principle that the time-averaged power that flows into the ray tube must equal the power that emerges. This was the basis for the analysis of vertically stratified media. In view of this observation, it is reasonable to ponder why the present derivation derived a transport equation, and then used it to determine the amplitude dependence. The answer is that the energy conservation argument does not hold if there is a flow in the ambient state, whereas a modified transport equation would be available.

The preceding developments indicate that we can determine the amplitude of the signal anywhere along a ray if we know the amplitude on that ray at some starting point. But to do so we must evaluate the ray tube area. With that determination as the objective, let us return to the situation addressed in Sect. 11.2, specifically, an axisymmetric field in a depth-dependent fluid. The solution of the ray tracing equations gives position along a ray as a function of time. Different initial values of the grazing and azimuthal angles generate different rays, so position on a ray is a function of t, χ_I, and θ_I. However, by virtue of axisymmetry, the range R and depth z of a point depend only on t and χ_I because the azimuth angle is invariant along a ray. Rather than considering t to be the independent variable, it is advantageous for this discussion to use the arclength ℓ measured along a ray from the starting position. This quantity is in a one-to-one correspondence to t. Indeed, the differential position increment in a time t is $d\ell = c\left(\bar{\xi}\right)dt$, so we could determine ℓ by a direct integration along the ray path. With this change of variables, the transverse distance and depth are described functionally as $R\left(\ell, \chi_I\right)$ and $z\left(\ell, \chi_I\right)$. This is the situation depicted in Fig. 11.16. (The difference between this Figure and Fig. 11.11 is that R previously was considered to depend explicitly on z.)

A ray tube cross section consists of all points at the same t. Because all rays within the tube are very close, the rays intersect the cross section at the same arclength ℓ. Thus, the coordinates of the end point on all upper rays in Fig. 11.16 are $R\left(\ell, \chi_I\right)$ and $z\left(\ell, \chi_I\right)$, while the coordinates at the end of all lower rays are $R\left(\ell, \chi_I + d\chi_I\right)$ and $z\left(\ell, \chi_I + d\chi_I\right)$.

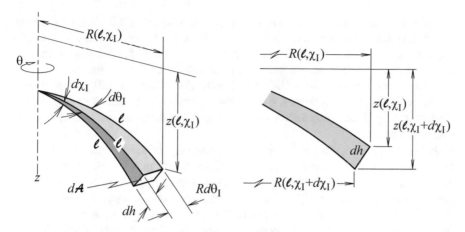

Fig. 11.16 Evaluation of ray tube area for the axisymmetric field radiated by a point source. The initial grazing angle is χ_I, the azimuthal angle is θ, and ℓ is the arclength along a ray measured from the source

The Pythagorean theorem gives the distance in the vertical plane between these points,

$$dh = \left\{ [R(\ell, \chi_I + d\chi_I) - R(\ell, \chi_I)]^2 + [z(\ell, \chi_I + d\chi_I) - z(\ell, \chi_I)]^2 \right\}^{1/2}$$

$$= \left[\left(\frac{\partial R}{\partial \chi_I} \right)^2 + \left(\frac{\partial z}{\partial \chi_I} \right)^2 \right]^{1/2} d\chi_I \qquad (11.3.24)$$

The distance in the horizontal plane between the rays bounding the tube is $Rd\theta_I$. The corresponding cross-sectional area is

$$d\mathcal{A}(\bar{\xi}) = R \left[\left(\frac{\partial R}{\partial \chi_I} \right)^2 + \left(\frac{\partial z}{\partial \chi_I} \right)^2 \right]^{1/2} d\chi_I d\theta_I \qquad (11.3.25)$$

This expression would be useful if we knew R and z as analytical functions, but numerical solution of the ray tracing equations gives a digitized representation. In that case if two adjacent rays n and $n+1$ are launched at grazing angles that differ by a very small value $\Delta\chi_I$, then the derivatives may be approximated as first-order finite differences,

$$\frac{\partial R}{\partial \chi_I} \approx \frac{R_{n+1} - R_n}{\Delta \chi_I}, \quad \frac{\partial z}{\partial \chi_I} \approx \frac{z_{n+1} - z_n}{\Delta \chi_I} \qquad (11.3.26)$$

where R_n and z_n are the coordinates on ray n at the instant of interest, and R_{n+1} and z_{n+1} are the coordinates for ray $n+1$ at that instant.

A hint as to how to proceed when the field is not axisymmetric comes from the observation that the bracketed term in Eq. (11.3.25) describes dh. This distance is measured in the vertical plane between points on adjacent rays that are at the same arclength. Thus, dh represents the magnitude of the change of position when

ℓ and θ_I are held constant. An equivalent statement is that dh is derived from the Jacobian of the transformation from the (R, z, θ) position variables to the $(\ell, \chi_\mathrm{I}, \theta_\mathrm{I})$ variables of a ray path when θ_I and ℓ are held constant. A description of ray tube area in terms of a Jacobian is the basic approach described by Jensen et al.[9] for handling an arbitrary situation. Ultimately, evaluation of this quantity would be done via numerical methods because the ray paths are known as discrete data sets. Rather than following that approach, we will develop a formulation that relies solely on vector algebra.

In Fig. 11.17, a ray tube is formed by a selected ray, designated as number 1, and two adjacent rays, numbered 2 and 3. The adjacent rays must be independent of ray 1. For example, if ray 1 corresponds to launch angles χ_I and θ_I, then ray 2 could correspond to $\chi_\mathrm{I} + \Delta\chi_\mathrm{I}$ and θ_I, while ray 3 could correspond to χ_I and $\theta_\mathrm{I} + \Delta\theta_\mathrm{I}$, where the angle increments are extremely small. This analysis of ray tube area requires that the evaluation of points on each ray use a common time base. (This would be the case if we prescribe the same output instants when the ray tracing equations are solved numerically.) This scheme allows us to identify points $\bar\xi_1$, $\bar\xi_2$, and $\bar\xi_3$ on the respective rays that correspond to any selected instant t_n. Therefore, these three points lie on the same wavefront $\tau = t_n$.

Fig. 11.17 Vectors used to evaluate the cross-sectional area of a ray tube

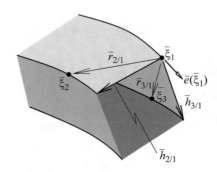

From $\bar\xi_1$ we construct relative position vectors to the other points,

$$\bar h_{2/1} = \bar\xi_2 - \bar\xi_1, \quad \bar h_{3/1} = \bar\xi_3 - \bar\xi_1 \tag{11.3.27}$$

The cross section area of this ray tube is $\Delta\mathcal{A} = |\bar h_{2/1}|\, |\bar h_{3/1}| \sin\left(\mathrm{angle}\left(\bar h_{2/1}, \bar h_{3/1}\right)\right)$. This is the magnitude of a scalar triple product, with the tangent vector as the third vector. Furthermore, although this area is finite, the fact that the rays correspond to initial conditions that are only slightly different allows us to consider the area to be an approximation of $d\mathcal{A}$, so we have found that

$$d\mathcal{A} \approx \left|\left(\bar h_{2/1} \times \bar h_{3/1}\right) \cdot \bar e_1\left(\bar\xi_1\right)\right| \equiv c\left(\bar\xi_1\right)\left|\left(\bar h_{2/1} \times \bar h_{3/1}\right) \cdot \bar s_1\left(\bar\xi_1\right)\right| \tag{11.3.28}$$

[9]F.B. Jensen et al., W.A. Kuperman, M B. Porter, and H. Schmidt, *Computational Ocean Acoustics*, 2nd ed., Chap. 3, Springer (2011).

If we wish to ascertain how the amplitude varies along a specific ray, which has been designated as ray 1, we would employ the preceding description of dA to formulate Eq. (11.3.23) for each computed point on that ray. Alternatively, we can apply the computation of dA to the point on each ray at a common instant. The outcome of such an evaluation would be the distribution of pressure along a wavefront. Repetition of either calculation for all rays and all wavefronts would yield a map of the pressure field.

Reflections and Caustics

The procedure we have developed entails computing a set of rays by solving the ray tracing equations. Each ray corresponds to a different initial condition at $t_1 = 0$, which may be a different initial position $\bar{\xi}_I$ and/or a different initial tangent direction $\bar{e}_I$. The differential equation solution marches forward incrementally. Upon reaching the farthest location, the solver is halted. After all rays have been computed, the properties of the rays may be used to determine how the pressure varies along any ray, provided that the computation has described a sufficient number of rays to approximate ray tubes.

The evaluation of each ray may be carried without interruption, unless a ray encounters a boundary. In that case, the ray is reflected. In view of the locally planar nature of a signal in the geometrical acoustics approximation, the rules for reflection of a plane wave that is incident on an infinite plane may be employed here. In Fig. 11.18, a ray is incident on a solid boundary at $\bar{\xi}_B$. The normal to the surface, pointing into the fluid, at this location is $\bar{n}(\bar{\xi}_B)$, and $\bar{e}(\bar{\xi}_{B-})$ is the tangent to the ray path immediately before it touches the boundary. The tangent to the reflected ray immediately after it leaves the surface is $\bar{e}(\bar{\xi}_{B+})$. The angle of incidence ψ_B equals the angle of reflection.

Fig. 11.18 Unit vectors describing the incidence and reflection of a ray at position $\bar{\xi}_B$ in a solid surface

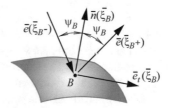

We wish to describe $\bar{e}(\bar{\xi}_{B+})$ in a manner that does not require defining a coordinate system. To that end, let $\bar{e}_t(\bar{\xi}_B)$ denotes the tangent to the surface at the point of incidence in the plane formed by $\bar{e}(\bar{\xi}_{B-})$ and $\bar{n}(\bar{\xi}_B)$. It follows that the component representation of the tangent directions is

$$\bar{e}(\bar{\xi}_{B-}) = \bar{e}_t(\bar{\xi}_B)\sin\psi_B - \bar{n}(\bar{\xi}_B)\cos\psi_B, \quad \bar{e}(\bar{\xi}_{B+}) = \bar{e}_t(\bar{\xi}_B)\sin\psi_B + \bar{n}(\bar{\xi}_B)\cos\psi_B$$

$$(11.3.29)$$

Subtracting the first equation from the second yields the desired relation. The angle of incidence may be found from a dot product, so we have

$$\boxed{\bar{e}(\bar{\xi}_{B+}) = \bar{e}(\bar{\xi}_{B-}) + 2\bar{n}(\bar{\xi}_B)\cos\psi_B, \quad \psi_B = \cos^{-1}\left(-\bar{e}(\bar{\xi}_{B-})\cdot\bar{n}(\bar{\xi}_B)\right)}$$ (11.3.30)

Subsequent to reflection, solution of the ray tracing equations may resume. The initial conditions for that computation would be the values of $\bar{\xi}_B$ and $\bar{e}\left(\bar{\xi}_{B+}\right)$ at t_B.

The computation of the amplitude variation along a ray also must account for a reflection. Because the angles of incidence and reflection are equal, the ray tube area immediately prior to incidence is the same as the area after reflection. Hence, the only reason the amplitude might be altered by the reflection is that the reflection coefficient is not unity. Suppose the surface is locally reacting with impedance Z. Then, the reflection coefficient is

$$\mathcal{R} = \frac{Z\left(\bar{\xi}_B\right) - \rho_0 c / \cos \psi_B}{Z\left(\bar{\xi}_B\right) + \rho_0 c / \cos \psi_B} \tag{11.3.31}$$

Neither the local impedance nor the angle of incidence is required to be constant along the surface. If that is the situation, some rays will be attenuated more than others as a result of reflection. However, this effect often is ignored by taking the surface to be rigid, $\mathcal{R} = 1$, based on the fact that at very high frequencies most materials have a large impedance. In any case, the amplitude immediately after reflection is $\mathcal{R}$ times the amplitude before. Using the properties at the reflection point as the input for the reflected ray tube leads to a description of the reflected signal as

$$p\left(\bar{\xi}, t\right) = \operatorname{Re}\left(\left[\frac{\rho_0\left(\bar{\xi}\right) c\left(\bar{\xi}\right)}{\rho_0\left(\bar{\xi}_B\right) c\left(\bar{\xi}_B\right)} \frac{d\mathcal{A}\left(\bar{\xi}_B\right)}{d\mathcal{A}\left(\bar{\xi}\right)}\right]^{1/2} \mathcal{R} P\left(\bar{\xi}_{B-}\right) e^{i\omega(t-\tau(\bar{x}))}\right) \tag{11.3.32}$$

The occurrence of a caustic is the consequence of the global behavior of a family of rays, rather than a special feature of a single ray. Thus, there is no need to interrupt solution of the ray tracing equations upon arrival at a caustic. (Indeed, one would be unlikely to recognize the existence of a caustic until the full set of rays has been computed.) The occurrence of a caustic does alter the amplitude dependence. We shall merely describe how to account for that effect because the analysis is beyond the present scope. The interested reader is referred to Pierce,[10] which also contains references to the original analyses.

Mathematical analysis would reveal that a harmonic signal undergoes a $\pi/2$ phase delay as it propagates along its ray through a caustic. Specifically, if ℓ_A and ℓ_B, respectively, are the arclengths to points on a ray prior to, and following, the caustic at ℓ_{caustic}, then we have

$$p\left(\bar{\xi}_B, t\right) = \operatorname{Re}\left(\left[\frac{c\left(\bar{\xi}_B\right)}{c\left(\bar{\xi}_A\right)} \frac{d\mathcal{A}\left(\bar{\xi}_A\right)}{d\mathcal{A}\left(\bar{\xi}_B\right)}\right]^{1/2} P\left(\bar{\xi}_A\right) e^{i\omega(t-\tau(\bar{x}))}\right) \quad \text{if } \ell_B < \ell_{\text{caustic}}$$

$$p\left(\bar{\xi}_B, t\right) = \operatorname{Re}\left(\left[\frac{c\left(\bar{\xi}_B\right)}{c\left(\bar{\xi}_A\right)} \frac{d\mathcal{A}\left(\bar{\xi}_A\right)}{d\mathcal{A}\left(\bar{\xi}_B\right)}\right]^{1/2} P\left(\bar{\xi}_A\right) e^{-i\pi/2} e^{i\omega(t-\tau(\bar{x}))}\right) \quad \text{if } \ell_B > \ell_{\text{caustic}}$$

$$\tag{11.3.33}$$

---

[10] A.D. Pierce, *Acoustics,* Sect. 9-4, ASA Books reprint of 1981 edition.

If it should be that two caustics are situated on a ray between $\bar{\xi}_A$ and $\bar{\xi}_B$, then the phase delay would be $2 (\pi/2)$, and so on. Note that the ray tube area decreases to zero at a caustic, which means that the linear acoustics approximation requires correction. In the vicinity of the caustic, the amplitude will be greatly enhanced.

A heuristic explanation of this phase shift may be obtained by considering the ray tube area beyond a caustic to be negative. This view stems from the description of ray tube area as a proportionality to the Jacobian. The rays cross over as they pass the caustic. The Jacobian is positive ahead of the caustic, zero at the caustic, and negative beyond it. The amplitude is proportional to the square root of the area, and the square root of a negative number gives a $\pm i = \exp(\pm i\pi/2)$. The root that applies is $-i$ for a delay of the signal. This aspect may be viewed as a manifestation of the principle of causality. A factor $+i$ as the signal passes through a caustic corresponds to a phase lead. If that were the case, it would mean that a feature is observed at a location infinitesimally beyond the caustic prior to its observation at a location that infinitesimally precedes the caustic.

EXAMPLE 11.5 Air is contained within a rectangular waveguide whose side walls are composed of tempered glass, which is well approximated as being rigid. As sketched below, circular plates of radius a are attached to the horizontal walls. The plates are cooled, thereby altering the sound speed in the region between them. Because the depth H is small and the walls are rigid, it is reasonable to consider the sound speed to vary only in the horizontal directions. Hence, the cooled region may be taken to constitute a vertical cylinder of lower sound speed. The specific dependence is $c = c_0 + (c_1 - c_0) \left(1 - R^2/a^2\right)$, where R is the transverse distance from the centerline connecting the plates. Outside this region, the sound speed is c_0. The wall at $x = 0$ generates a 200 kHz harmonic plane wave, and the wall at $x = L$ is actively controlled to eliminate reflections. Draw the rays and wavefronts. Then determine the dependence of pressure amplitude on distance along the wavefront that touches the front of the cylindrical region, where $x = 2.5a$. Parameters for the evaluation are $c_0 = 330$ m/s, $c_1 = 280$ m/s, and $a = 250$ mm.

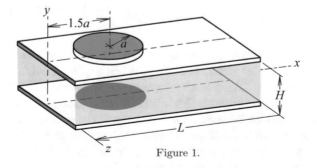

Figure 1.

Significance

Application of numerical methods to solve the ray tracing and transport equations is a primary thrust, and the results will greatly enhance our understanding of the behavior of rays and wavefronts, including a much deeper insight to the behavior near caustics.

Solution

The sound speed depends on the transverse distance R, but the waveguide is best described in terms of Cartesian coordinates. Because the sound speed is independent of the vertical position y and the top and bottom walls are rigid, the pressure field will be independent of y, so $\bar{s}$ and $\bar{\xi}$ may be described solely in terms of x and z components. The column vector of variables for a numerical solution is

$$\{X\} = \begin{bmatrix} s_x & s_z & \xi_x & \xi_z \end{bmatrix}^{\mathrm{T}}$$

The corresponding vector driving $d\{X\}/dt$ is

$$\{F\} = \left[-\frac{1}{c} \begin{bmatrix} \dfrac{\partial c}{\partial \xi_x} & \dfrac{\partial c}{\partial \xi_z} \end{bmatrix} \quad c^2 \begin{bmatrix} s_x & s_z \end{bmatrix} \right]^{\mathrm{T}}$$

The transverse distance from the vertical centerline of the cooling plates to $\bar{\xi}$ is $R = \left| \bar{\xi} - 1.5a\bar{e}_x \right|$, which is used to evaluate the given expression for $c(\bar{\xi})$. The corresponding gradient of c is $\nabla c = (dc/dR)\,\nabla R$. We know that $\nabla R = \bar{e}_R$, where $\bar{e}_R$ is the unit vector from the center to point $\bar{\xi}$. Thus,

$$\nabla c = \frac{dc}{dR}\bar{e}_R = \begin{cases} -\dfrac{2R}{a^2}(c_1 - c_0)\,\bar{e}_R & \text{if } R < a \\ 0 & \text{if } R \geq a \end{cases}$$

This gradient must be described in term of x and z components, with ξ_x and ξ_z as the variables. The required expressions are obtained by a transformation to Cartesian coordinates, which sets

$$\bar{\xi} = \xi_x \bar{e}_x + \xi_z \bar{e}_z, \quad R = \left| \bar{\xi} - 1.5a\bar{e}_x \right|, \quad \bar{e}_r = \frac{\bar{\xi} - 1.5a\bar{e}_x}{R}$$

The numerical solver routine must be selected. Matlab® offers several. The one that was used to generate the data is ODE45, which implements a fourth/fifth-order Runge–Kutta algorithm. The input arguments for this routine are the name of a function F_rhs that will evaluate $\{F\}$, followed by a column or row vector t_span holding t values at which the solution will be output. The next input argument is the initial value of $\{X\}$, which will be addressed below. The last input argument for ODE45 is an optional set of parameters setting error limits. The results here were obtained by setting the relative error tolerance to 10^{-6}.

There are two alternative procedures for stepping through time with ODE45. If t_span holds all of the time instants, the output from a single call to ODE45 would be a rectangular array, which we will denote as $[X_{\text{out}}]$. The nth row of $[X_{\text{out}}]$ would be the solution $\{X\}$ at time t_n. The alternative procedure explicitly steps through time by setting t_span as a two-element vector $[t_{n-1} \; t_n]$. The value of $\{X\}$ at t_n is the last row of $[X_{\text{out}}]$ in that case. (The other rows are $\{X\}$ values at intermediate instants listed in $\{t_{\text{out}}\}$.) The time history for all variables in this procedure may be constructed by adjoining the transpose of the column vectors $\{X\}$. In either treatment, the time values at which data is output are stored in a column vector $\{t\}$. Clearly, the single-call procedure is easier to implement, and it was found to require much less time to execute. However, the explicit stepping procedure was found to be significantly more accurate.

The numerical integration must be performed for each ray. This was done within a loop over the ray number. The solution for all rays should use a common base of t values. Doing so aligns the data such that the computed $\bar{\xi}$ values for all rays lie on a common set of wavefronts. The time increment for the evaluation was taken to be the time required to travel a distance $a/25$ at speed c_0, and the maximum time was set as the time to travel $6a$ at speed c_0. Thus, the number of t_n values is $N = (6a/c_0) / (a/25c_0) = 150$. Each pass through the ray number loop corresponds to a ray that departs from the plane $x = 0$ at an incremented distance z_j in the z direction. Each ray initially is a straight line, so $\bar{e} = \bar{e}_x$ at the start, which is defined to be $t = 0$. Let J be the number of rays. The diameter of the cooled region is $2a$, so the rays were initialized at a uniform spacing to cover a width of $3a$, which gives a separation of $\Delta = 3a/ (J - 1)$ between rays. The corresponding initial conditions for the solution for ray j were $s_x = 1/c_0$, $s_z = 0$, $\xi_x = 0$, $\xi_z = -1.5a + \Delta (j - 1)$.

A single pass through the outer loop over ray numbers gives $[X_{\text{out}}]$ for ray j. The four rows of $[X_{\text{out}}]$ consist, respectively, of s_x, s_z, ξ_x, ξ_z at each t_j. The row of ξ_x data was stored as column j of a rectangular array $[\xi_x]$, and the ξ_z row was likewise stored in column j of $[\xi_x]$. The eikonal error metric for each ray, numbered j, was obtained by evaluating $c^2[(s_x)^2 + (s_z)^2] - 1$ for each instant t_n, and then storing the error values at all instants as column j of the rectangular array $[\varepsilon_{\text{eik}}]$. The largest eikonal errors were found to occur along rays 21 and 25. Figure 2 displays the variation of this quantity along the respective rays. The largest error for any ray was found to be less than 0.08%.

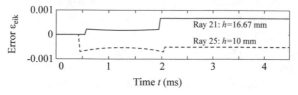

Figure 2.

When the data is stored as described above, a ray may be drawn by plotting a column of $[\xi_x]$ against the corresponding column of $[\xi_y]$, while wavefronts are obtained by plotting a row of $[\xi_x]$ against the corresponding row of $[\xi_y]$. The rays and wavefronts appear in Fig. 3. For the sake of clarity, only every tenth of the evaluated wavefronts is shown there. There are many interesting facets to this picture. The rays that pass through the cooled region are bent inward toward the x-axis. This is consistent with the property observed for vertical stratification. It results from a wavefront being retarded where the sound speed is lower, thereby having the effect of turning the wavefront. There appears to be regions where there is no ray, but these are not shadow zones. Rather the rays are sparse there. If we were to greatly increase the number of computed rays, we would see that rays do pass through those regions. Regions where the rays are sparse correspond to low, but not zero, pressure.

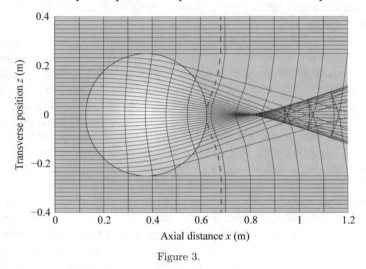

Figure 3.

Figure 3 shows that there is a region in which many rays converge. Figure 4 zooms in on this region. The concentration is not a focus because the rays do not intersect at a single point. Another feature leading to the same conclusion is that the wavefronts beyond the convergence are not circles, as they would be if the rays were radiated from a focus. Rather, the concentration of rays is a caustic. Inspection of a region beyond the arrete shows that at any point there are three rays. Two arrive at comparatively large angles from either side of the x-axis, whereas the third comes at a small angle from the opposite side of the x-axis. This results in folding of a wavefront. The part closer to the arrete is a curved arc. It is the locus of constant τ values for the shallow rays. The portions of a folded wavefront that intersect at the x-axis correspond to the large angle rays.

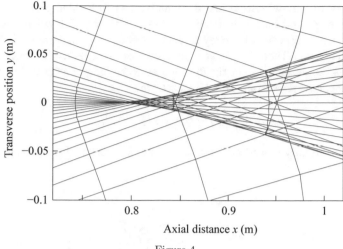

Figure 4.

The location of the caustic can be shifted by adjusting the value of c_1 with c_0 fixed. Decreasing c_1 increases the heterogeneity, thereby increasing the curvature of the rays within the cooled region. This brings the caustic closer to the origin. Indeed, if c_1 is sufficiently large, the caustic will occur inside that region.

It still remains to evaluate the pressure along a wavefront. For this analysis, there are several items we must determine. First, we must determine where the computed ray paths intersect the wavefront that was stipulated. The point where this wavefront tangentially touches the cooled region is $\xi_x = 2.5a = 0.625\,\mathrm{m}$, $\xi_z = 0$. The ray passing through this point is the straight one along the center line. Its ray number is $j = (J + 1)/2 = 31$. None of the computed ξ_x values for this ray exactly equals 0.625 m. We can use interpolation to reach the specified wavefront. First, we scan the ξ_x values for ray 31 to identify the points that bracket $\xi_x = 0.625$ m. These are $(\xi_x)_{69,31} = 0.6225$ and $(\xi_x)_{70,31} = 0.6325$. A linear interpolation between these points should give $(\xi_x)_{\mathrm{wf},31} = 0.625$ m, that is,

$$0.625 = 0.6225 + \beta\,(0.6325 - 0.6225) \implies \beta = 0.25$$

The interpolation fraction β is used to find the value of τ for the wavefront, according to

$$\tau_{\mathrm{wf}} = \tau_{69} + \beta\,(\tau_{70} - \tau_{69}) = 2.0681 \text{ ms}$$

This interpolation also is employed to determine the x and y coordinates at which any ray j intersects the wavefront, according to

$$\left\{\begin{array}{c}(\xi_x)_{\mathrm{wf},j} \\ (\xi_y)_{\mathrm{wf},j}\end{array}\right\} = \left\{\begin{array}{c}(\xi_x)_{69,j} \\ (\xi_y)_{70,j}\end{array}\right\} + \beta\left\{\begin{array}{c}(\xi_x)_{70,j} - (\xi_x)_{69,j} \\ (\xi_y)_{70,j} - (\xi_y)_{70,j}\end{array}\right\}, \quad j = 1, ..., J$$

Now that the ray-wavefront intersections have been located, the next task is to determine the ray tube areas. Reference to Fig. 5 will assist following this development. The depth perpendicular to ray diagrams is constant, and we are only interested in the ratio of areas. Therefore, an area ratio reduces to the ratio of distances between intersection points. Let $h_{wf,j}$ be the distance between the points at which rays j and $j+1$ intersect the wavefront. These distances are

$$h_{wf,j} = \left| \bar{\xi}_{wf,j+1} - \bar{\xi}_{wf,j} \right|, \quad j = 1, ..., J-1$$

The rays are equally spaced at distance Δ when they depart from $x = 0$, so $h_{1,j} = \Delta$ for all j.

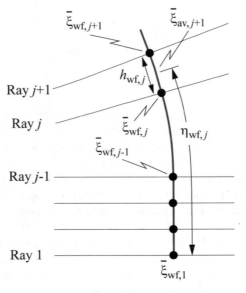

Figure 5.

The last quantity required to evaluate the pressure amplitude according to Eq. (11.3.23) is the characteristic impedance. The stipulated wavefront does not lie in the cooled region, so the ratio of $\rho_0 c$ along it to the $\rho_0 c$ value at the initial point is unity. Nevertheless, we shall incorporate this effect into the computation because doing so would allow us to probe inside the cooled region. The value of ρ_0 is not specified for any location, but we can deduce it from the given ambient properties. In the ambient state, the fluid everywhere is quiescent, so there can be no pressure gradient. It follows that the pressure within the cooled region must be the same as it is outside this region. We shall assume that the ambient pressure is atmospheric, so $P_0 = 1$ atm $= 101.32$ kPa. We know that $c = (\gamma P_0/\rho_0)^{1/2}$ for an ideal fluid, with $\gamma = 1.4$ for air, and the value of $c\left(\bar{\xi}\right)$ at any location is given. Solving this relation for the density gives $\rho_0\left(\bar{\xi}\right) = \gamma P_0/c\left(\bar{\xi}\right)^2$.

The ray tube numbered j covers the distance between $\bar{\xi}_{wf,j}$ and $\bar{\xi}_{wf,j+1}$. It is reasonable to use the impedance midway between those points to formulate Eq. (11.3.23). These values are $(\rho_0 c)_{av,j} \equiv \rho_0 (\bar{\xi}_{av,j}) c (\bar{\xi}_{av,j})$, where $\bar{\xi}_{av,j}$ is the midpoint position,

$$\bar{\xi}_{av,j} = \frac{1}{2} \left(\bar{\xi}_{wf,j} + \bar{\xi}_{wf,j+1} \right)$$

The pressure at the midpoint corresponding to these parameters is indicated by Eq. (11.3.23) to be

$$\frac{P_{wf,j}}{P_I} = \left(\frac{(\rho_0 c)_{av,j}}{(\rho_0 c)_{x=0}} \right)^{1/2} \left(\frac{\Delta}{h_{wf,j}} \right)^{1/2}$$

where P_I is the pressure amplitude in the plane wave at $x = 0$.

These pressure ratios are associated with the midpoint between rays. To plot them, we will use as the abscissa the distance along the wavefront measured from the point where the wavefront intersects the first ray. This distance, which we shall denote as $\eta_{wf,j}$, is shown in Fig. 5 to be approximately the accumulation of the ray tube widths preceding it, according to.

$$\eta_{wf,1} = \frac{1}{2} h_{wf,1}, \quad \eta_j = \frac{1}{2} h_{wf,j} + \sum_{n=1}^{j-1} h_{wf,n}$$

Figure 6 plots these the pressure ratio against distance along the wavefront, with the abscissa shifted such that $\eta = 0$ at the centerline. The number of rays for this graph was raised to 121 to give the graphs a less jagged appearance. The wavefront that was specified contains the point $x = 0.625\,\text{m}$, $y = 0$. Its equation is $\tau = 2.0681\,\text{ms}$. The pressure distribution along two other wavefronts also appear in this graph. The point $x = 0.5\,\text{m}$, $y = 0$ lies on the wavefront $\tau = 1.6633\,\text{ms}$. This wavefront passes through the cooled region. The third wavefront is $\tau = 2.4469\,\text{ms}$. It contains the point $x = 0.75\,\text{m}$, $y = 0$, which places it slightly to the left of the arrete. Reference to Fig. 3 shows that the intervals where $|P| / P_I$ is less than one correspond to regions where the rays are sparse, while the rays are closely spaced in regions where the pressure ratio is enhanced. This enhancement becomes more prominent as the wavefront comes closer to the caustic.

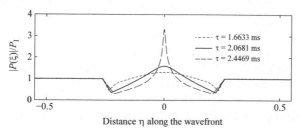

Figure 6.

Evaluation of the Signal at a Field Point – Eigenrays

The operations described thus far entail evaluating ray paths by solving the ray tracing equations. The pressure along a ray is then determined by solving the transport equation. In this procedure, we cannot specify *a priori* the points where the ray falls. This leads to the question of how can we find the pressure at a specific point? Stated differently, the question is how can we determine the ray path that connects a source and a specific field point?

The situation is depicted in Fig. 11.19 where the source is at $\bar{\xi}_I$, and the receiver's location is the designated field point $\bar{\xi}_f$. Only direct and surface-reflected rays are considered. It is permissible to ignore bottom-reflected rays if the water is very deep compared to the depths of the source and field point, because signals that propagate along bottom-reflected rays to $\bar{\xi}_f$ will have traveled over very large arclengths. Unless a caustic occurs, the ray tube area of this set of rays will be greatly enlarged, so the amplitude will be greatly reduced. (Bottom reflections are an important aspect of propagation in a shallow channel.)

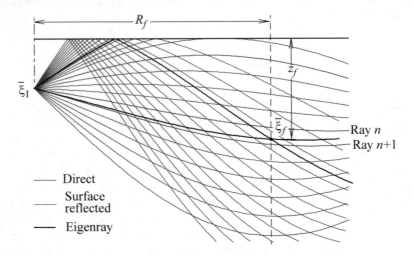

Fig. 11.19 Identification of eigenray connecting a point source and a selected field point

Contributors to the signal at $\bar{\xi}_f$ are a ray that arrives directly from the source, and a ray that arrives after it is reflected from the free surface. Both are referred to as *eigenrays,* which is the term given generally to a ray that connects the source and a selected field point. Solution of the ray tracing equations presumably has given the two families of rays in Fig. 11.19, where each ray corresponds to a different grazing angle at the source. How can we identify the eigenrays given that none of the rays in the Figure actually intersects $\bar{\xi}_f$? This is not a trivial questions, and it is much more difficult if some rays undergo many surface and bottom reflections.

The simplest approach is to use interpolation. To do so, we first identify the pair of direct rays n and $n+1$ that bracket $\bar{\xi}_f$, that is, the rays that come closest to $\bar{\xi}_f$ above and below. Let R_f be the horizontal range to $\bar{\xi}_f$, and let z_n and z_{n+1} be the depths of the points on the respective rays that are at R_f. A linear interpolation of the initial grazing angles gives

$$\frac{(\chi_1)_f - (\chi_1)_n}{z_f - z_n} = \frac{(\chi_1)_{n+1} - (\chi_1)_n}{z_{n+1} - z_n} \tag{11.3.35}$$

The value of $(\chi_1)_f$ is used to find the eigenray. To do so, the ray tracing equations are solved with $(\chi_1)_f$ as the initial grazing angle. The eigenray for surface reflection is found by repeating these steps using the rays in the reflected set that bracket $\bar{\xi}_f$.

The total pressure at $\bar{\xi}_f$ is the sum of the contributions from all eigenrays connecting the source and field points. In the case of Fig. 11.19, let τ_d and τ_r be the values of the wavefront functions for the direct and reflected eigenrays, respectively. Also, let P_d and P_r be the corresponding amplitudes. Both amplitudes are obtained by evaluating the cross-sectional areas of the respective ray tubes, with the appropriate corrections for reflection coefficients and phase delays at caustics. Then the total pressure at $\bar{\xi}_f$ is

$$p\left(\bar{\xi}_f, t\right) = \text{Re}\left[P_d e^{i\omega(t-\tau_d)} + P_r e^{i\omega(t-\tau_r)}\right] \tag{11.3.36}$$

If more than two eigenrays intersect the selected field point, than the contribution of each is additive.

EXAMPLE 11.6 A radially symmetric point source is situated at depth H below the free surface of a deep body of water. Density and sound speed variations are negligible. Use geometrical acoustics to derive an expression for the pressure at range R_F and depth z_F. Sound speed variations over this range are negligible. Compare that result to the result of an analysis that uses the method of images.

Significance

The main feature of the analysis is that it explains how the Jacobian is used to evaluate analytically the ray tube area. The juxtaposition of geometrical acoustics and the method of images illustrates their different perspectives.

Solution

The fluid is homogeneous, so all rays are straight lines. There are only two eigenrays because reflections from the bottom are deemed to be negligible. The direct eigenray is the line from the source to the field point, and the reflected eigenray intersects the surface somewhere between the source and the receiver. The arrangement is described in Fig. 1.

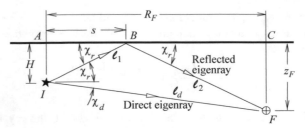

Figure 1.

A geometrical acoustics analysis begins with determination of the ray paths. Because they are formed from straight lines, we may use geometry for the determination, rather than solving the ray tracing equations. The angle of incidence equals the angle of reflection. Consequently, right triangles IAB and BCF are similar. Equating the ratio of sides yields an expression for the horizontal distance s to the reflection point,

$$\frac{s}{H} = \frac{R_F - s}{z_F} \implies s = \frac{HR_F}{z_F + H}$$

We must evaluate ray tube areas for each eigenray. We could do so by considering each eigenray to be one limit of its ray tube with rays at initial grazing angles $\chi_d + d\chi$ and $\chi_r + d\chi$ as the other limit. However, doing so would be tedious. An analytical approach employs the Jacobian in Eq. (11.3.25). This formulation requires description of the (R, z) coordinates of a field in terms of the initial grazing angle for the ray path through that point and the distance ℓ along that ray path to the field point. The source depth H is fixed in that evaluation

For the direct path, the arclength is ℓ_d, and the initial grazing angle is χ_d. In terms of these variables, the coordinates of the field point F are

$$R_F = \ell_d \cos \chi_d, \quad z_F = H + \ell_d \sin \chi_d$$

We differentiate these expressions with respect to χ_d, and substitute the results into Eq. (11.3.25). This gives

$$dA_d = R_F \left[\left(\frac{\partial R_F}{\partial \chi_d} \right)^2 + \left(\frac{\partial z_F}{\partial \chi_d} \right)^2 \right]^{1/2} d\chi_d d\theta_d = (\ell_d)^2 (\cos \chi_d) \, d\chi_d d\theta \quad (1)$$

The presence of a reflection complicates the evaluation of ray tube area for reflected rays. We seek a description of R_F and z_F in terms of χ_r and the total arclength along the reflected ray. Figure 1 indicates that this arclength is $\ell_r = \ell_1 + \ell_2$. Because H is fixed, the distance ℓ_1 depends solely on χ_r. Using this relation to form $\ell_2 = \ell_r - \ell_1$ gives ℓ_2 as a function of χ_r and ℓ_r. We employ this relation to describe R_F and z_F for the reflected ray,

$$\ell_1 = \frac{H}{\sin \chi_r}, \quad \ell_2 = \ell_r - \ell_1 = \ell_r - \frac{H}{\sin \chi_r}$$

$$R_F = \ell_1 \cos \chi_r + \ell_2 \cos \chi_r = \ell_r \cos \chi_r$$

$$z_F = \ell_2 \sin \chi_r = \ell_r \sin \chi_r - H$$

The ray tube area obtained from Eq. (11.3.25) is

$$dA_r = R_F \left[\left(\frac{\partial R_F}{\partial \chi_r} \right)^2 + \left(\frac{\partial z_F}{\partial \chi_r} \right)^2 \right]^{1/2} d\chi_1 d\theta_1$$

$$= R_F \left[(\ell_r \sin \chi_r)^2 + (\ell_r \cos \chi_r)^2 \right]^{1/2} d\chi_r d\theta = (\ell_r)^2 (\cos \chi_r) d\chi_r d\theta$$

(2)

Equation (11.3.22) equates twice the intensity at an arbitrary location to the value at the start of the ray tube. In the present situation, the input to both ray tubes is a radially symmetric point source. The time-averaged intensity is the source's radiated power divided by the surface area of a surrounding sphere. For both ray tubes, this sphere has a very small radius ε. Therefore, for both the direct and reflected eigenvalues the starting value is

$$\left(\frac{P^2}{\rho_0 c} \right)_{\mathrm{I}} = \frac{2P}{4\pi\varepsilon^2}$$

(3)

The cross-sectional area for both ray tubes at the input end consists of a small patch of the surrounding sphere at polar angle $\pi/2 - \chi$ covering a polar angle $d\chi$ and azimuthal angle $d\theta$. Therefore, we have

$$(dA_{\mathrm{I}})_d = (\varepsilon d\chi_d)(\varepsilon \cos \chi_d) d\theta, \quad (dA_{\mathrm{I}})_r = (\varepsilon d\chi_r)(\varepsilon \cos \chi_r) d\theta$$

(4)

Application of Eq. (11.3.22) in the case where the input to a ray tube is a radially symmetric point source leads to

$$\frac{P_d^2}{\rho_0 c} dA_d = \left(\frac{P^2}{\rho_0 c} \right)_{\mathrm{I}} (dA_{\mathrm{I}})_d \quad \text{and} \quad \frac{P_r^2}{\rho_0 c} dA_r = \left(\frac{P^2}{\rho_0 c} \right)_{\mathrm{I}} (dA_{\mathrm{I}})_r$$

Substitution of Eqs. (1)–(4) into these relations gives

$$\frac{P_d^2}{\rho_0 c} \ell_d^2 (\cos \chi_d) d\chi_d d\theta = \left(\frac{2P}{4\pi\varepsilon^2} \right) \varepsilon^2 (\cos \chi_d) d\chi_d d\theta$$

$$\frac{P_r^2}{\rho_0 c} (\ell_r)^2 (\cos \chi_r) d\chi_r d\theta = \left(\frac{2P}{4\pi\varepsilon^2} \right) \varepsilon^2 (\cos \chi_r) d\chi_r d\theta$$

$$P_d = \frac{1}{\ell_d} \left(\frac{\rho_0 c P}{2\pi} \right)^{1/2}$$

From this we find that the pressure amplitudes are

$$P_d = \frac{1}{\ell_d}\left(\frac{\rho_0 c \mathcal{P}}{2\pi}\right)^{1/2}, \quad P_r = \frac{1}{\ell_r}\left(\frac{\rho_0 c \mathcal{P}}{2\pi}\right)^{1/2} \tag{5}$$

For a wave that departs from the source at zero phase and propagates at constant phase speed c along a ray, the phase variable is $\tau\left(\bar{x}\right) = \ell/c$. The reflection coefficient at the free surface may be taken to be -1. Adding the signal received at the field point from each eigenray therefore gives

$$p\left(\bar{\xi}_f, t\right) = \mathrm{Re}\left[P_d e^{i\omega(t-\ell_d/c)} + P_r\left(-1\right)e^{i\omega(t-\ell_r/c)}\right]^{1/2}$$

$$= \left(\frac{c\mathcal{P}}{2\pi}\right)^{1/2}\mathrm{Re}\left[\left(\frac{e^{-i\omega\ell_d/c}}{\ell_d} - \frac{e^{-i\omega\ell_r/c}}{\ell_r}\right)e^{i\omega t}\right]$$

Evaluation of the pressure by the method of images was addressed in Chap. 6. The source 1 and its image 2 are depicted in Fig. 2. Because the free surface is pressure-release, the strength of the image is the negative of the source's strength. It is evident that the radial distance r_1 from the source to the field point equals ℓ_d and that $r_2 = \ell_1 + \ell_2 = \ell_r$.

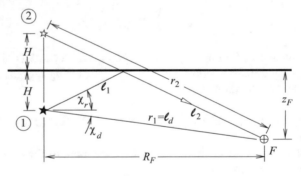

Figure 2.

The complex pressure amplitude radiated by a point source in a homogenous fluid is $[\rho_0 c \mathcal{P}/(2\pi)]^{1/2}\exp\left(-ikr\right)/r$, where $\mathcal{P}$ is the time-averaged power. The pressure received at the field point from the image is inverted, so the combination of the source and image pressures is

$$p\left(\bar{\xi}, t\right) = \left(\frac{\rho_0 c \mathcal{P}}{4\pi}\right)^{1/2}\mathrm{Re}\left[\left(\frac{e^{-ikr_1}}{r_1} - \frac{e^{-ikr_2}}{r_2}\right)e^{i\omega t}\right]$$

Because $r_1 = \ell_d$ and $r_2 = \ell_r$, this is the same result as that obtained from ray theory. Although the analysis using the method of images is much easier, seeing how ray theory is implemented, particularly the analytical evaluation of ray tube area, makes some of the abstract concepts more understandable.

11.4 Fermat's Principle

The developments thus far rest on Snell's law, which ultimately is founded on the basic principles that underlie other acoustic phenomena. A different formulation for analyzing ray paths is available. Fermat stated his principle in 1662 without a mathematical derivation. It postulates that *the ray connecting two points is such that the path minimizes the travel time between the points.* We will see that it leads to Snell's law, as well as a differential equation that is equivalent to the ray tracing equations. However, that is not the reason we study it. Rather it offers a different perspective regarding the nature of rays.

It is possible to justify Fermat's principle with a heuristic argument based on an analogy. This might be the thought process that led Fermat to state the principle. Suppose a large number of people have gathered at the base of a mountain. Each individual is made to follow a different path as they travel to a specified point at the other side of the mountain. Some paths might be easier to walk, so individuals following such paths can travel more quickly. However, these fast paths might be long. Other paths might be much shorter, thereby counterbalancing the fact that those paths are difficult to traverse. If we consider when individuals arrive at the destination, the first arrival will be the individual who followed the optimum path. Individuals who follow paths that are very close to the optimum will arrive slightly later, whereas individuals who follow nonoptimum paths will arrive at a wide range of later times. In the acoustical analogy, signals that follow the optimum and near optimum rays will arrive nearly synchronously, thereby reinforcing each other. Signals following nonoptimal paths will be unsynchronized, and therefore destructively interfere with each other.

Any derivation of basic principles must start from a set of postulates. We could begin with the ray tracing equations as the foundation for a derivation of Fermat's principle. Instead, we will derive the ray tracing equations from it. One benefit of proceeding in this manner is that it will introduce a general mathematical tool called the *calculus of variations*, which is very useful for a wide range of subjects, especially classical and relativistic mechanics.

In Fig. 11.20 several paths connect selected points A and B. One of these paths is optimal in the sense that the time for a signal to travel along it is minimal. Some other paths are far from optimal, but one path in the Figure is very close. This is a *variational path*. The general approach is to compare the travel time along the optimal path to the time along a variational path. The variational path is defined to be separated from the optimal path by an infinitesimal difference $\delta \bar{\xi}$ of their positions. More specifically, if $\bar{x} = \bar{\xi}$ is a point on the optimal path, then the corresponding point on the variational path is placed at $\bar{x} = \bar{\xi} + \delta \bar{\xi}$. The reason for taking $\delta \bar{\xi}$ to be infinitesimal is that doing so allows us to truncate a Taylor series at the second term.

Although we do not know the optimal path, that is, the eigenray, let us pretend that we do. The minimum travel time along this ray between points A and B may be found by evaluating

Fig. 11.20 The optimal ray path corresponds to the minimum time between two designated points A and B

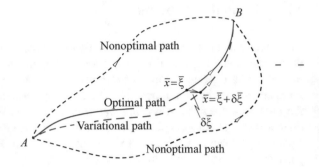

$$T_{\min} = \oint_A^B \frac{d\ell}{c\left(\bar{\xi}\right)} \tag{11.4.1}$$

The differential arclength $d\ell$ is the distance between two adjacent points on the optimal path. We shall use the symbol "d" to indicate differential increments pertaining to the same path. Thus, the positions of the adjacent points are $\bar{\xi}$ and $\bar{\xi} + d\bar{\xi}$, as shown in Fig. 11.21. It follows that

$$d\ell = \left(d\bar{\xi} \cdot d\bar{\xi}\right)^{1/2} \tag{11.4.2}$$

The variational path is obtained by shifting from position $\bar{x} = \bar{\xi}$ to position $\bar{x} = \bar{\xi} + \delta\bar{\xi}$. This corresponds to infinitesimal amount $\delta\bar{\xi}$. The symbol δ is used to indicate that the increment is infinitesimal, but it pertains to a difference between a quantity on the optimal and variational paths. Thus, any quantity that contains the symbol δ is not the result of a physical process. For this reason, it is said to be a *virtual increment*. Thus, $\delta\bar{\xi}$ is a virtual displacement, whereas $d\bar{\xi}$ is an actual displacement. Other than being required to be infinitesimal, the direction and magnitude of the virtual displacement are arbitrary. As shown in Fig. 11.21, $d\bar{\xi}$ is tangent to the optimal path, whereas $\delta\bar{\xi}$ may be in any direction. Because $\delta\bar{\xi}$ is arbitrary, there is no implication that the displaced point, whose position is $\bar{x} = \bar{\xi} + \delta\bar{\xi}$, is on the same wavefront as $\bar{x} = \bar{\xi}$.

Fig. 11.21 Relation between a displacement $d\bar{\xi}$ along minimum-time path and a variational displacement $\delta\bar{\xi}$ to a neighboring sub-optimal path

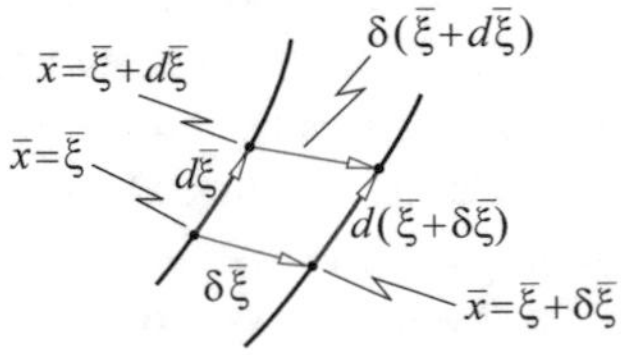

We need to compare the propagation time in Eq. (11.4.1) to the time for the variational path. This requires a description of the differential arclength along the variational path, which in turn requires that we describe the position of the displaced point on the variational path. Figure 11.21 shows that this position may be regarded as the result of a virtual displacement from the point $\bar{\xi} + d\bar{\xi}$ on the optimal path, or the result of an actual displacement from point $\bar{\xi} + \delta\bar{\xi}$ on the variational path. Both constructions must lead to the same point. An equivalent statement is that the same new position must be obtained, regardless of whether the virtual displacement precedes or follows the actual displacement. Thus, it must be that

$$\left(\bar{\xi} + d\bar{\xi}\right) + \delta\left(\bar{\xi} + d\bar{\xi}\right) = \left(\bar{\xi} + \delta\bar{\xi}\right) + d\left(\bar{\xi} + \delta\bar{\xi}\right) \implies \delta\left(d\bar{\xi}\right) = d\left(\delta\bar{\xi}\right)$$

(11.4.3)

This property will soon be found to be quite useful.

The arclength along the variational path is the magnitude of the differential increase in the position of a point on the variational path, that is, $d\ell' = \left|d\left(\bar{\xi} + \delta\bar{\xi}\right)\right|$. A dot product may be used to construct this distance. Higher order differentials are negligible, so we have

$$\begin{aligned} d\ell' &= \left[\left(d\bar{\xi} + \delta\left(d\bar{\xi}\right)\right) \cdot \left(d\bar{\xi} + \delta\left(d\bar{\xi}\right)\right)\right]^{1/2} \\ &= \left[d\bar{\xi} \cdot d\bar{\xi} + 2d\bar{\xi} \cdot \delta\left(d\bar{\xi}\right)\right]^{1/2} \equiv \left[(d\ell)^2 + 2d\bar{\xi} \cdot \delta\left(d\bar{\xi}\right)\right]^{1/2} \\ &= d\ell + \frac{d\bar{\xi}}{d\ell} \cdot \delta\left(d\bar{\xi}\right) \end{aligned}$$

(11.4.4)

We use this expression to construct the propagation time along the variational path. The sound speed along the variational path is $c\left(\bar{\xi} + \delta\bar{\xi}\right)$, so the time required for a signal to travel along this path is

$$T_{\text{var}} = \oint_A^B \frac{1}{c\left(\bar{\xi} + \delta\bar{\xi}\right)}\left[d\ell + \frac{d\bar{\xi}}{d\ell}\,\delta\left(d\bar{\xi}\right)\right]$$

(11.4.5)

It is easier to proceed if we replace the sound speed with its reciprocal

$$\boxed{U\left(\bar{\xi}\right) \equiv \frac{1}{c\left(\bar{\xi}\right)}}$$

(11.4.6)

(Some treatments use the index of refraction μ in Eq. (11.2.2), but that would require the unnecessary introduction of a reference sound speed.)

At this juncture, we call on Fermat's principle. According to it, the travel time along the variational path can differ at the most from the minimal time by second-order differentials. Therefore, we express T_{var} in terms of the lowest order differentials that distinguish it from T_{min} in Eq. (11.4.1). A Taylor series is used to treat the integrand, so that

$$
\begin{aligned}
T_{\text{var}} &= \oint_{A}^{B} U\left(\bar{\xi} + \delta\bar{\xi}\right)\left[d\ell + \frac{d\bar{\xi}}{d\ell} \cdot \delta\left(d\bar{\xi}\right)\right] \\
&= \oint_{A}^{B} \left[U\left(\bar{\xi}\right) + \delta\bar{\xi} \cdot \nabla U\left(\bar{\xi}\right)\right]\left[d\ell + \frac{d\bar{\xi}}{d\ell} \cdot \delta\left(d\bar{\xi}\right)\right] \\
&= \oint_{A}^{B} U\left(\bar{\xi}\right) d\ell + \oint_{A}^{B}\left[U\left(\bar{\xi}\right)\frac{d\bar{\xi}}{d\ell} \cdot \delta\left(d\bar{\xi}\right) + \delta\bar{\xi} \cdot \nabla U\left(\bar{\xi}\right) d\ell\right]
\end{aligned}
\tag{11.4.7}
$$

The first integral is T_{min}. Thus, the statement that the lowest order representation of T_{var} must be the same as T_{min} leads to the condition that

$$
\oint_{A}^{B}\left[U\left(\bar{\xi}\right)\frac{d\bar{\xi}}{d\ell} \cdot \delta\left(d\bar{\xi}\right) + \delta\bar{\xi} \cdot \nabla U\left(\bar{\xi}\right) d\ell\right] = 0
\tag{11.4.8}
$$

The ray tracing differential equations consider the position of a point on a ray to be a function of time, but none of the terms in the integral contains time. This suggests an alternative description of position on a ray path, in which it is defined by the arclength ℓ to a point. In other words, we consider ℓ to be the independent variable, so that the position of a point on the optimal path is described functionally as $\bar{x} = \bar{\xi}\left(\ell\right)$. This is where Eq. (11.4.3) enters, because we can assert that

$$
\delta\left(d\bar{\xi}\right) = d\left(\delta\bar{\xi}\right) = \left[\frac{d}{d\ell}\left(\delta\bar{\xi}\right)\right]d\ell
\tag{11.4.9}
$$

This step converts the statement of Fermat's principle in Eq. (11.4.8) to an integral over ℓ, specifically,

$$
\oint_{A}^{B}\left[U\left(\bar{\xi}\right)\frac{d\bar{\xi}}{d\ell} \cdot \frac{d}{d\ell}\left(\delta\bar{\xi}\right) + \delta\bar{\xi} \cdot \nabla U\right]d\ell = 0
\tag{11.4.10}
$$

The last operation is based on the fact that although the dependence of $\delta\bar{\xi}$ on ℓ is arbitrary, it is a function that we select. Its selection sets $d\left(\delta\bar{\xi}\right)/d\ell$, so this derivative is not an independent quantity. It may be eliminated with an integration by parts, which gives

$$
U\left(\bar{\xi}\right)\frac{d\bar{\xi}}{d\ell} \cdot \left(\delta\bar{\xi}\right)\Bigg|_{\ell_A}^{\ell_B} - \oint_{A}^{B}\left[\frac{d}{d\ell}\left(U\left(\bar{\xi}\right)\frac{d\bar{\xi}}{d\ell}\right) - \nabla U\right] \cdot \left(\delta\bar{\xi}\right) d\ell = 0
\tag{11.4.11}
$$

At this juncture, we take an overview. It was stated that $\delta\bar{\xi}$ is an arbitrary function of ℓ, but that is not exactly true. The variational path is an alternative between point A and B, so the variational path must intersect these points. This means that $\delta\bar{\xi}\left(\ell_A\right) = \delta\bar{\xi}\left(\ell_B\right) = 0$. The consequence is that the terms in the preceding relation that are evaluated at the ends of the path are zero. In turn, this requires that the integral must evaluate to zero for any $\delta\bar{\xi}$ function we select. If we wish that the integral of

$\bar{a} \cdot \bar{b}$ be zero for any function $\bar{b}$, it must be that $\bar{a}$ is identically zero. Thus, we find

$$\boxed{\frac{d}{d\ell}\left(U\left(\bar{\xi}\right)\frac{d\bar{\xi}}{d\ell}\right) - \nabla U = 0} \qquad (11.4.12)$$

Minimization of integral quantities arise in a variety of areas, notably analytical dynamics. The mathematical process we have followed is known as the calculus of variations. Equation (11.4.12) in a general analysis is known as the *Euler-Lagrange equation*. It is a second-order differential equation governing $\bar{\xi}$, which is a fundamental difference with the (first order) ray tracing equations. Another fundamental difference is the absence of the wave slowness $\bar{s}$.

Our intent at the outset was to derive the ray tracing equations from Fermat's principle, but the appearance of Eq. (11.4.12) does not suggest how to meet that objective. Let us try to work backward, by eliminating $\bar{s}$ from the ray tracing equations. The second ray tracing equation, Eq. (11.3.10), states that

$$\frac{d\bar{\xi}}{dt} = c^2\bar{s} \qquad (11.4.13)$$

A different view is that $\bar{\xi}$ depends on the arclength ℓ, which leads to a description of the preceding derivative according to $d\bar{\xi}/dt = \left(d\bar{\xi}/d\ell\right)\left(d\ell/dt\right)$. The local sound speed is the rate at which the arclength changes, so that

$$c = \frac{d\ell}{dt} \qquad (11.4.14)$$

Matching the two descriptions of the rate of change of the position on a ray leads to

$$\frac{d\bar{\xi}}{dt} - \frac{d\bar{\xi}}{d\ell}\frac{d\ell}{dt} \equiv \frac{d\bar{\xi}}{d\ell}c - c^2\bar{s} \qquad (11.4.15)$$

Replacing c with $1/U$ thereby converts the second ray tracing equation to

$$\boxed{\bar{s} = U\frac{d\bar{\xi}}{d\ell}} \qquad (11.4.16)$$

Now we turn to the first ray tracing equation, Eq. (11.3.9), which is

$$\frac{d\bar{s}}{dt} = -\frac{\nabla c}{c} \qquad (11.4.17)$$

On the right side, we set $c = 1/U$, so that

$$\frac{\nabla c}{c} = U\nabla\left(\frac{1}{U}\right) = -\frac{1}{U}\nabla U \qquad (11.4.18)$$

We introduce the chain rule to handle $d\bar{s}/dt$ on the left side of Eq. (11.4.17) because our view now is that $\bar{s}$ is a function of ℓ. The result is that the first ray tracing equation becomes

$$\frac{d\bar{s}}{d\ell}\frac{d\ell}{dt} \equiv \frac{d\bar{s}}{d\ell}c = \frac{1}{U}\nabla U \qquad (11.4.19)$$

Replacement of c with $1/U$ in this equation leads to

$$\boxed{\frac{d\bar{s}}{d\ell} = \nabla U} \qquad (11.4.20)$$

The last step is to substitute $\bar{s}$ from Eq. (11.4.16) into this expression. The result is the Euler-Lagrange equation, Eq. (11.4.12). In other words, the ray tracing equations and the Euler-Lagrange equation are equivalent specifications of a ray path. The differences are that the ray tracing equations give position as a function of time, with additional information about the tangent to the path in the form of $\bar{s}$. If we wish to also determine the arclength to a point on a ray, we can do so by adding ℓ to the set of unknowns $\{X\}$ in Eq. (11.3.12) and inserting into the scalar differential equations another equation: $d\ell/dt = c\left(\xi_x, \xi_y, \xi_z\right)$. The Euler-Lagrange equation describes the position in terms of ℓ. If we wish to know a local tangent or the propagation time, we must extract that information after the equation has been solved.

Some mathematical analyses are best performed with a single differential equation. However, realistic situations usually require numerical methods. In that case, the ray tracing equations are preferable. The reason for this assertion is that most numerical algorithms require that the equations be first order, which is how the ray tracing equations are posed. Of course, the Euler-Lagrange equation could be transformed into first-order form, but as we have seen, such equations would be the ray tracing equations or a variant of them.

We have demonstrated that Fermat's principle correctly describes the rays in a fluid whose sound speed is a continuous function of position. It also is valid at discontinuities. To demonstrate this, consider Fig. 11.22. We wish to identify the horizontal distance x_O where the ray from A to B crosses the interface between two fluids. Points A and B are fixed. A coordinate system whose z-axis measures depth into the second fluid is situated at the horizontal position of A.

Fig. 11.22 Construction of a ray path for transmission across an interface between two fluids

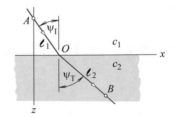

The travel times along each segment of the ray are ℓ_1/c_1 and ℓ_2/c_2. These distances are

$$\ell_1 = \left[z_A^2 + (x_O)^2\right]^{1/2}, \quad \ell_2 = \left[z_B^2 + (x_B - x_O)^2\right]^{1/2} \tag{11.4.21}$$

We seek the value of x_O that minimizes the sum of the travel time along each segment. This requires that

$$\frac{d}{dx_O}\left(\frac{\ell_1}{c_1} + \frac{\ell_2}{c_2}\right) = \frac{d}{dx_O}\frac{\left[z_A^2 + (x_O)^2\right]^{1/2}}{c_1} + \frac{d}{dx_O}\frac{\left[z_B^2 + (x_B - x_O)^2\right]^{1/2}}{c_2}$$
$$= \frac{x_O}{c_1\ell_1} - \frac{(x_B - x_O)}{c_2\ell_2} = 0 \tag{11.4.22}$$

According to Fig. 11.22, $\sin\psi_1 = x_O/\ell_1$ and $\sin\psi_2 = (x_B - x_O)/\ell_2$, so the preceding is a restatement of Snell's law.

There is a degree of elegance in having a general principle that leads to the equations for ray paths in inhomogeneous fluids as well as at interfaces. Indeed, the generality of Fermat's principle probably is its most important attribute. For example, it can be used to derive equations describing the rays in the presence of an ambient flow.[11] It also can link geometrical acoustics to optics and the Hamiltonian formulation of analytical mechanics.

EXAMPLE 11.7 A point source is located at position A in an ocean channel and a hydrophone is situated at point B. The sound speeds are c_1 for the water, $c_2 < c_1$ for the air, and $c_3 > c_1$ for the sediment. Describe the eigenrays for point B.

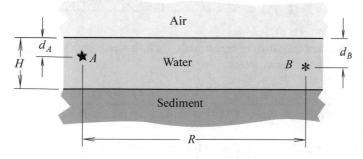

Figure 1.

Significance

This application of Fermat's principle to the determination of eigenrays is quite straightforward, but the results have a surprising feature. Thus, this example affirms that the perspective offered by Fermat's principle might cause us to re-examine our initial assumptions.

---

[11]O. Bühler, *Waves and Mean Flows*, Cambridge University Press (2009) pp. 81–83.

Solution

We begin with a sketch of possible paths between points A and B. Paths 1, 2, and 3 in Fig. 2 are familiar, but path 4 might be considered to be unexpected. In addition to these, others exist, but we will consider them later. Path 1 is the direct path. Because the sound speed along it is c_1, the minimum travel time is obtained for the shortest path. Thus, Fermat's principle confirms that this path is a straight line.

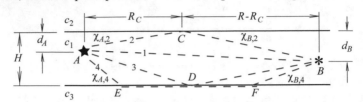

Figure 2.

Paths 2 and 3 feature a single reflection. Let us determine whether Fermat's principle yields ray paths for which the angle of reflection equals the angle of incidence. In the case of the surface-reflected wave, points A and B are fixed and we know that the elevation of point C at which it is incident. This ray path is set if the horizontal location of point C is specified. Therefore, we select the horizontal range R_C to this point as the unknown to be determined by Fermat's principle. Trigonometric equations describe the length of the two straight segments that form path 2. Division of those lengths by the sound speed gives the travel time along each. Fermat's principle applies to the sum of these contributions, which is

$$T_2(R_C) = \frac{\left(d_A^2 + R_C^2\right)^{1/2}}{c_1} + \frac{\left[d_B^2 + (R - R_C)^2\right]^{1/2}}{c_1}$$

Fermat's principle states that the correct value of R_C is that which gives $dT_2/dR_C = 0$. Evaluation of the derivative gives

$$\frac{dT_2}{dR_C} = \frac{1}{c_1} \frac{R_c}{\left(d_A^2 + R_C^2\right)^{1/2}} - \frac{1}{c_1} \frac{(R - R_C)}{\left[d_B^2 + (R - R_C)^2\right]^{1/2}} = 0 \tag{1}$$

The length ratios in Eq. (1) are the cosines of the respective grazing angle depicted in the sketch. Hence, Eq. (1) reduces to

$$\frac{\cos \chi_{2,A}}{c_1} - \frac{\cos \chi_{2,B}}{c_1} = 0$$

Cancelation of c_1 shows that the grazing angles are equal. Thus, Fermat's principle is consistent with the familiar reflection law. There is no need to repeat this analysis for path 3, because the arrangement is the same as path 2.

Path 4 involves two fluids. It consists of a straight segment from the source to point E on the interface, then a horizontal path in the sediment to point F, from which a straight path brings the signal to point B. The location of points E and F each depend on the horizontal range, so this path is a function of two parameters. For the sake of variety, let us use the grazing angles $\chi_{A,4}$ and $\chi_{B,4}$. The corresponding expression for the travel time is

$$T_4\left(\chi_{A,4}, \chi_{B,4}\right) = \frac{1}{c_1} \frac{H - d_A}{\sin \chi_{A,4}} + \frac{1}{c_3}\left[R - \left(\frac{H - d_A}{\tan \chi_{A,4}}\right) - \left(\frac{H - d_B}{\tan \chi_{B,4}}\right)\right] + \frac{1}{c_1}\frac{H - d_B}{\sin \chi_{B,4}}$$

The value of T_4 must be stationary with respect to both grazing angles, so we set

$$\frac{\partial T_4}{\partial \chi_{A,4}} = \left(\frac{H - d_A}{c_1}\right)\frac{\left(-\cos \chi_{A,4}\right)}{\left(\sin \chi_{A,4}\right)^2} - \left(\frac{H - d_A}{c_3}\right)\frac{(-1)}{\left(\sin \chi_{A,4}\right)^2} = 0$$

$$\frac{\partial T_4}{\partial \chi_{B,4}} = -\left(\frac{H - d_B}{c_2}\right)\frac{(-1)}{\left(\sin \chi_{B,4}\right)^2} + \left(\frac{H - d_B}{c_3}\right)\frac{\left(-\cos \chi_{B,4}\right)}{\left(\sin \chi_{B,4}\right)^2} = 0$$

(2)

These expressions reduce to

$$\cos \chi_{A,4} = \frac{c_1}{c_3} = \cos \chi_{B,4}$$

The angles of the line segments relative to the normal to the interface are $\psi_{A,4} = \pi/2 - \chi_{A,4}$ and $\psi_{B,4} = \pi/2 - \chi_{B,4}$, so Eq. (2) leads to

$$\sin \psi_{A,4} = \sin \psi_{B,4} = \frac{c_1}{c_3}$$

(3)

This is the expression for the critical angle of incidence. It is a real value only if the sound speed in the receiving fluid is greater than the speed in the transmitting fluid. Thus, Fermat's principle tells us that the alternative path exists if c_3 for the sediment is greater than c_1 for the water. If that is the case, then the ray arrives and departs from the interface at the critical angle. One way to view this is that the source at point A emits many rays. Rays that arrive at the sediment at an angle of incidence less than the critical value are reflected upward and transmitted downward. However, the ray that arrives at the critical angle of incidence generates a transmitted ray that is parallel to the interface. This ray runs along the interface on the sediment side. The water now becomes the receiving fluid. The ray emerges from the sediment when the angle from the interface to the receiving point is the critical angle for transmission from the fluid into the sediment.

Paths 2 and 3 entail a single reflection. Multiple reflection paths also are possible. A few are shown in Fig. 3. In general, there the two paths that feature N reflections, with the their difference being whether the first reflection is at the top or bottom. We will analyze ray 5, from which the procedure for other cases should be apparent.

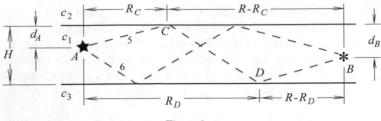

Figure 3.

The reflection points are defined by the ranges R_C and R_D, so the travel time is a function of two parameters,

$$T_5(R_C, R_D) = \frac{\left[(d_A)^2 + (R_C)^2\right]^{1/2}}{c_1} + \frac{\left[H^2 + (R_D - R_C)^2\right]^{1/2}}{c_1}$$
$$+ \frac{\left[(H - d_B)^2 + (R - R_D)^2\right]^{1/2}}{c_1}$$

For an extreme value, the derivative of T_5 with respect to both parameters must be zero,

$$\frac{\partial T_5}{\partial R_C} = \frac{1}{c_1} \frac{R_C}{\left[(d_A)^2 + (R_C)^2\right]^{1/2}} - \frac{1}{c_1} \frac{(R_D - R_C)}{\left[H^2 + (R_D - R_C)^2\right]^{1/2}} = 0$$

$$\frac{\partial T_5}{\partial R_D} = \frac{1}{c_1} \frac{(R_D - R_C)}{\left[H^2 + (R_D - R_C)^2\right]^{1/2}} - \frac{1}{c_1} \frac{(R - R_D)}{\left[(H - d_B) + (R - R_D)^2\right]^{1/2}} = 0$$

(4)

These equations may be solved for R_C and R_D. To interpret these equations we observe that the grazing angle of the segments of ray 5 in the Figure are

$$\chi_{AC} = \cos^{-1}\left(\frac{R_C}{\left[(d_A)^2 + (R_C)^2\right]^{1/2}}\right)$$

$$\chi_{CD} = \cos^{-1}\left(\frac{R_D - R_C}{\left[H^2 + (R_D - R_C)^2\right]^{1/2}}\right)$$

(11.4.23)

$$\chi_{AC} = \cos^{-1}\left(\frac{R - R_D}{\left[(H - d_B)^2 + (R - R_D)^2\right]^{1/2}}\right)$$

Thus, the values of R_C and R_D that extremize T_5 according to Eq. (4) correspond to equal grazing angles for the arriving and departing rays at points C and D. In other words, the angle of reflection equals the angle of incidence at points C and D. There is nothing new in this.

The fact that Fermat's principle identifies the critical angle path through the sediment is another demonstration of its generality. One of the paradoxes of this path is that the concept of a ray tube having an identifiable cross-sectional area breaks down. This is so because grazing angles less than the critical value in Eq. (3) do not lead to real rays, while grazing angles greater than critical lead to rays that are transmitted into the sediment, and therefore cannot reach the vicinity of point B. This is one of many phenomena that do not fit the geometric acoustic model. A few others are the fact that a sound is heard in a shadow zone, and that sound is heard behind a rigid barrier. Phenomena that do not fit into the geometrical acoustic model are referred to as *diffraction*. Analyses of diffraction phenomena generally require a level of mathematics that is beyond the scope of this book.

11.5 Homework Exercises

Exercise 11.1 The mirror in the sketch is a surface of revolution $z = f(R)$, with $f(R)$ increasing monotonically. A plane wave propagating in the negative z direction is reflected. Consider a ray that is incident at arbitrary R. The angle of reflection equals the angle of incidence, so $\tan \psi = df/dR$. Derive an expression in terms of $R, f(r)$, and df/dR for the coordinate z_f where the reflection of this ray intersects the z-axis. Then consider the specific case of a paraboloid of revolution, for which $f(R) = h_z (R/a)^2$. Show that z_f is a constant in this case, which means that all reflected rays intersect at a true focus.

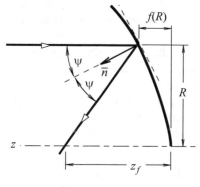

Exercise 11.1

Exercise 11.2 The cross section of the prism in the sketch is an isosceles triangle with a vertical axis of symmetry. Assume that the prism is a fluid with sound speed c_b, and set c_a as the sound speed in the surrounding fluid. (a) Derive an expression for the angle of transmission ψ for rays that emerge from the prism. (b) If $c_a = 200\,\text{m/s}$ and $c_b = 400\,\text{m/s}$, what is the largest value of ϕ for which a plane wave emerges from the prism? (b) If $c_a = 400\,\text{m/s}$ and $c_b = 200\,\text{m/s}$, what is the largest value of ϕ for which a plane wave emerges from the prism? (c) It is given that $c_a \neq c_b$. Qualitatively explain why there is no value of the apex angle $2\phi > 0$ for which the plane wave that emerges propagates horizontally, regardless of whether $c_a > c_b$ or $c_a < c_b$.

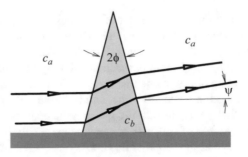

Exercise 11.2

Exercise 11.3 The reflection from a flat surface of the signal radiated by a point source is describable by the method of images if the surface is rigid or pressure-release. In the case of a locally reacting surface, geometrical acoustics also leads to the conclusion that the reflected rays seem to emanate from a point source below the surface. It also leads to the conclusion that the image source has a directivity factor. Consider an omnidirectional point source situated at height H above a surface whose local impedance is Z. The source is harmonic at frequency ω. Determine the location of the image and the directivity of that image. Does the field of the image source actually satisfy the Helmholtz equation in the actual fluid?

Exercise 11.4 The sound speed profile in a very deep ocean channel consists of a lower layer of depth H in which the sound speed is constant at c_1, and an upper layer in which the sound speed decreases linearly with depth. The sound speed is continuous across the interface of these layers. It is convenient to describe c in terms of the height z' above the bottom. Thus, the sound speed is $c = c_1 + m\left(z' - H\right)h\left(z' - H\right)$, with $m > 0$ and $z' = 0$ at the bottom. Any ray path within the upper layer must be a downward curving circular arc. Consequently, a ray departing from the bottom at any upward grazing angle must eventually return to the bottom. Derive an expression for the distance between the points of departure and return.

Exercise 11.5 At the surface of an ocean channel, the sound speed is 1470 m/s. It decreases linearly with depth, down to 500 m, where the sound speed is 1420 m/s. Downward from that depth the sound speed increases linearly, with a value of 1470 m/s occurring again at 1000 m. Determine the path of a ray launched horizontally at a depth of 200 m. Base the analysis on the fact the path is circular if the sound speed is linearly dependent on the depth.

Exercise 11.6 External heating of a waveguide has resulted in a sound speed that increases linearly from axial position $x = 0$ to position $x = L$, but the sound speed is constant transversely. The result is that $c = c_0\left(1 - x/L\right) + c_1 x/L$. A point source is situated in the middle of the waveguide, at $x = 0$, $y = H/2$, where $H = 2$ m is the waveguide's width. Consider rays launched at grazing angle of 15°, 45°, and 75°. (Note that the heterogeneity varies in the x direction, so the grazing angle are measured relative to the y direction.) Do any of these rays intersect a wall? If so, what is the location of this intersection, and what is the propagation time from the instant the ray is launched to the instant when it arrives at that wall? Parameters are $c_0 = 300$ m/s, $c_1 = 400$ m/s, $H = 3$ m.

Exercise 11.7 The sound speed in a hard-walled two-dimensional HVAC duct varies parabolically according to $c(z) = c_0 + c_1(1 - 2z/H)^2$, where z is the transverse distance. The walls are at $y = 0$ and $y = H$. Determine the path of two rays that are launched downward from $z = H/2$: χ_{Ia} is such that a wall is a turning point, and $\chi_{Ib} = 0.5\chi_{Ia}$. Parameters are $c_0 = 300$ m/s, $c_1 = 50$ m/s, $H = 2$ km.

Exercise 11.8 The sound speed in a hard-walled two-dimensional HVAC duct varies parabolically according to $c(z) = c_0 + c_1(1 - 2z/H)^2$, where z is the transverse distance. The walls are at $z = 0$ and $z = H$. Determine the paths of rays launched downward from $z = H/2$ resulting from an initial grazing angle that is $1.25\chi_{Iw}$, where χ_{Iw} is such that a wall is a turning point. Parameters are $c_0 = 300$ m/s, $c_1 = 50$ m/s, $H = 2$ km.

Exercise 11.9 The sound speed in a large body of water varies sinusoidally with depth relative to a mean value, $c(z) = c_0 + \varepsilon \sin(2\pi z/H)$. A radially symmetric point source is situated at $z_0 = 10H$. This depth is sufficiently large that there is no need to consider reflections from the surface or bottom. (a) Consider a ray that is launched horizontally. Does it have a turning point? If so, at what depth? (b) Determine the path followed by this ray. (c) Determine the travel time along this ray. Plot this property as a function of the horizontal distance from the source, and also as a function of the depth. Parameters are $c_0 = 1500$ m/s, $\varepsilon = 25$ m/s, $H = 100$ m.

Exercise 11.10 The sound speed in a large body of water varies sinusoidally with depth relative to a mean value, $c(z) = c_0 + \varepsilon \sin(2\pi z/H)$. A radially symmetric point source is situated at $z_0 = 10H$. This depth is sufficiently large that there is no need to consider reflections from the surface or bottom. Determine the paths of rays launched at initial grazing angles ranging from $15°$ to $45°$ in a $1°$ increment.

Exercise 11.11 Consider the waveguide in Exercise 11.10. (a) Evaluate the paths of rays that are launched from the source at grazing angles of $2.99°$ and $3.01°$. (b) Use the result of Part (a) to evaluate the ray tube area corresponding to rays that depart from in the source range $2.99° \leq \chi \leq 3.01°$ in an interval of azimuthal angles subtending $0.02°$. (c) Determine the intensity in this ray tube at a horizontal distance of 500 m from the source.

Exercise 11.12 Evaluate the ray tube area $\mathcal{A}$ for two incident rays in Exercise 11.2 that are infinitesimally separated by elevation dh. From that expression determine the amplitude $|P_T|$ of the wave that emerges from the prism if the incident wave was harmonic with amplitude $|P_I|$. Graph $d\mathcal{A}/dh$ and $|P_T|/|P_I|$ as functions of θ for two cases: $c_b/c_a = 2$, $\rho_b/\rho_a = 1.2$, and $c_b/c_a = 0.5$, $\rho_b/\rho_a = 0.8$.

Exercise 11.13 A radially symmetric source is situated at depth z_0 in a vertically stratified fluid. The sound speed is an arbitrary function $c(z)$, where z is the depth.

It is desired to determine the intensity within a ray tube that surrounds the vertical ray path. The first step in the analysis considers a ray that departs from the source slightly off vertical, so the initial grazing angle is $\chi_I = \pi/2 - \Delta$, with $\Delta \ll 1$. (a) Show that the slope of this ray at an arbitrary depth is given, to leading order in powers of Δ, by

$$\frac{dR}{dz} \approx \Delta \frac{c(z)}{c_I}$$

where R is the transverse distance from the z-axis. (b) This approximation is valid at any depth for which $c(z)/c_I$ is $O(1)$. Subject to this limitation derive an expression for the cross section area of the ray tube that emanates from the source as a cone whose apex angle is 2Δ. Then use that area to determine the depth dependence of the time-averaged intensity along the downward ray.

Exercise 11.14 Consider the waveguide in Exercise 11.6. The task here is to determine the pressure at the wall when the radially symmetric source, which is situated at $x = 0, y = H/2$, is time-harmonic, with time-averaged power output $\mathcal{P}$. Derive an expression for the distance x_f at which a ray that departs from the source at grazing angle χ_I intersects a wall. Then use that expression to determine the pressure amplitude at the location where rays intersect the wall. Consider initial grazing angles of $\chi_I = 0°, 30°$, and $60°$. (Note that the heterogeneity varies in the x direction, so the grazing angle are measured relative to the y direction.)

Exercise 11.15 Suppose that the reference sound speeds in Example 11.5 are reversed, so that $c_0 = 280\,\text{m/s}$, $c_1 = 330\,\text{m/s}$. All other parameters are as stated there. Compute the ray paths and wavefronts. Do caustics form? Are there shadow zones?

Exercise 11.16 Due to extraordinary circumstances the sound speed in the atmosphere has been found to depend on the horizontal distance x and altitude y measured from a point that is 1000 m above the ground. The dependence $c = 350 + xy/200\,\text{m/s}$, where the units of x and y are meters. A weather balloon transmits an omnidirectional point source from that location. (a) Compute the paths of rays that are launched in the xy plane in all directions covering the full 360° range in 10° increments. (b) Determine the wavefronts corresponding to the rays in Part (a).

Exercise 11.17 In the region $x < 0$, a fluid is homogeneous, with sound speed c_0. To the right, $x > 0$, the sound speed varies sinusoidally in both the x and y directions, according to $c = c_0 - \Delta \sin(\pi x/L) \cos(\pi y/H)$ for $0 < x < L$. For $x > L$ the sound speed returns to $c = c_0$. The parameters are $c_0 = 1500\,\text{m/s}$, $\Delta = 60\,\text{m/s}$, $L = 3\,\text{m}$, $H = 1.5\,\text{m}$. An initially plane wave whose propagation direction is $\bar{e} = \bar{e}_x$ is incident at $x = 0$ on the heterogeneous region. Determine the ray paths and wavefronts in the region $0 > x > 6L$.

Exercise 11.18 Consider the fluid in Exercise 11.17. A harmonic plane wave having amplitude B propagates at 8° above the x-axis in the region $x < 0$. Consider the ray that is incident on the heterogeneous region at $x = 0, y = H/4$. (a) Determine the path of this ray. (b) What is the location of this ray and the direct of the tangent to that ray when it emerges from the heterogeneous region at $x = L$. (c) Determine the amplitude of the signal described by this ray at the location where the ray emerges from the heterogeneous regions

Exercise 11.19 A source at point A emits rays in all directions. Use Fermat's principle to prove that the ray that arrives at field point B after reflection from the corner is parallel to the ray that reflects from the vertical wall.

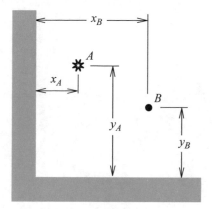

Exercise 11.19

Exercise 11.20 The sketch shows two points at opposite sides of a rigid cylinder. The fluid in which it is immersed is homogeneous, with constant sound speed c. Suppose there is a ray connecting these points. Because path segments within the fluid must be straight, the only way a connecting ray might exist is if it wraps around the surface of the cylinder, as shown in the sketch. Given distances L_1 and L_2 use Fermat's principle to determine the angles ϕ_1 and ϕ_2 that mark the limits of this contact. Draw a sketch of the result. Also determine the corresponding propagation time for a signal that follows this path.

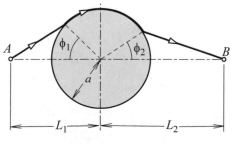

Exercise 11.20

Chapter 12
Scattering

Have you ever wondered why you can hear sound even though there is a building between you and the source, or why the sky is blue, or how dolphins can identify objects underwater? These phenomena are manifestations of scattering. At a fundamental level, scattering is the reflection of a signal from a finite-sized object. Knowledge of the scattering properties of a body probably is of the greatest concern for underwater acoustics, not only for its relevance to naval applications like sonar, but also for animal bioacoustics. Scattering phenomena often also are important for studies of atmospheric propagation.

We will see that sometimes the analysis of scattering is closely related to the description of acoustic radiation. Like that problem, scattering analyses are usually performed in the frequency domain. The key parameter is the wavelength relative to the size of the scattering region. The primary tools for low frequencies are derived from the Kirchhoff–Helmholtz integral theorem, whereas geometrical acoustics is the approach we will employ for high frequencies. Both formulations entail the judicious application of well-justified approximations.

The problematic situation is the mid-frequency range, in which the object's dimensions are comparable to the wavelength. There are few objects that are conducive to an analytical study of the scattered field at an arbitrary frequency. The sphere is one that we shall take up. Spherical harmonics will play an important role in that study. As is true for many simplified models, the availability of this solution will help us understand the validity of the low- and high-frequency approximations. Nevertheless, we seldom encounter spheres as actual acoustical systems. Realistic shapes require computational methods.

Electronic supplementary material The online version of this chapter
(DOI 10.1007/978-3-319-56847-8_12) contains supplementary material, which is
available to authorized users.

J.H. Ginsberg, *Acoustics—A Textbook for Engineers and Physicists, Volume II*,
DOI 10.1007/978-3-319-56847-8_12

12.1   Background

The general situation has two finite-sized bodies that are surrounded by a fluid whose extent is infinite. One generates a signal that encounters the other body as an obstacle to its propagation. The modification of the incident signal is the scattered signal. Reflection is one aspect of the scattering process, but other phenomena are involved. In the most accurate model, the pressure distribution on the surface of the scattering body causes that body to move. Such motion modifies the scattered signal in comparison with what it would be if the body were immobile. Incorporation of the effect of that motion might require consideration of the laws of structural dynamics. Another complication is that the scattered signal returns to the generating body. Consequently, that body also is a scatterer, so its surface response is different from that associated with generation of the incident signal. For example, if the generating body is an electromechanical transducer, the pressure and particle velocity corresponding to a specified voltage are modified. This alters the signal that is incident on the second body that was considered at the outset to be the scatterer, and so on. This is the phenomenon of *mutual scattering*. The extended version features many bodies whose scattered fields interact.

An analysis of mutual scattering is quite complicated and requires prior understanding of single-body scattering. The latter is a very good model if the distance between the two bodies is much greater than the acoustic wavelength and the largest dimension of either body, because the farfield of either scattered signal will eventually decay inversely with the distance from that body. The consequence is that the signal radiated by the generating body will be little affected by the signal returning from the scatterer.

Thus, the general problem we shall address is that body S_0 generates the signal $\mathrm{Re}\left(P_\mathrm{I}\left(\bar{x}\right)e^{i\omega t}\right)$ that is incident on the scattering body S. The scattered signal is $\mathrm{Re}\left(P_\mathrm{s}\left(\bar{x}\right)e^{i\omega t}\right)$. The total pressure is the sum of these contributions,

$$P\left(\bar{x}\right) = P_\mathrm{I}\left(\bar{x}\right) + P_\mathrm{s}\left(\bar{x}\right) \tag{12.1.1}$$

As noted, because S and S_0 are widely separated, the field generated by S_0 is essentially the same as it would be if S were not present. Thus, P_I is a solution of the Helmholtz equation that matches the normal velocity on S_0, and obeys the Sommerfeld radiation condition at large radial distances r_0 from S_0,

$$\nabla^2 P_\mathrm{I} + k^2 P_\mathrm{I} = 0$$
$$\bar{n} \cdot \nabla P_\mathrm{I} = -i\omega\rho_0 V_0, \quad \bar{x} \in S_0$$
$$\lim_{r_0 \to \infty} r_0 \left(\frac{\partial P_\mathrm{I}}{\partial r_0} + ik P_\mathrm{I}\right) = 0 \tag{12.1.2}$$

In principle, the incident field can be any field consistent with these equations. However, we have imposed the restriction that S is very distant from S_0, which means that it is reasonable in most cases to consider the field that is incident on S to be locally planar.

If the scattering body is capable of motion, either as a rigid body, or as a consequence of its deformability, then the surface velocity on S is $\bar{V}(\bar{x})$. The normal velocity in the acoustic signal must match the normal component of $V(\bar{x})$. Also, the total pressure must satisfy the Helmholtz equation. Because P is the superposition in Eq. (12.1.1) and P_{I} individually satisfies the Helmholtz equation, the scattered pressure must satisfy

$$\boxed{\begin{array}{c} \nabla^2 P_{\mathrm{s}} + k^2 P_{\mathrm{s}} = 0 \\ \bar{n} \cdot \nabla P_{\mathrm{s}} = -\bar{n} \cdot \nabla P_{\mathrm{I}} - i\omega\rho_0\bar{n} \cdot \bar{V}, \quad \bar{x} \in S \end{array}} \tag{12.1.3}$$

In addition, the scattered field at large radial distances r from S must satisfy the Sommerfeld radiation condition relative to that body,

$$\lim_{r \to \infty} r \left(\frac{\partial P_{\mathrm{s}}}{\partial r} + ik P_{\mathrm{s}} \right) = 0 \tag{12.1.4}$$

These equations may be interpreted as a statement that the scattered field is the pressure radiated by S when the normal velocity of its surface is $\bar{n} \cdot \bar{V} + (1/i\omega\rho_0)\,\bar{n} \cdot \nabla P_{\mathrm{I}}$.

By definition, the scattered pressure is the field created by the insertion of S into the fluid domain. The pressure on that surface is the total quantity $P(\bar{x})$. By analogy with a radiation problem, the Kirchhoff–Helmholtz integral theorem (KHIT) describes P_{s} at any location $\bar{x}_0$ exterior to S. Equation (7.4.14) states that

$$\boxed{P_{\mathrm{s}}(\bar{x}_0) = \iint\limits_{S} [P(\bar{x}_s)\,\bar{n}(\bar{x}_s) \cdot \nabla G(\bar{x}_0, \bar{x}_s) - \bar{n}(\bar{x}_s) \cdot \nabla P(\bar{x}_s)\,G(\bar{x}_0, \bar{x}_s)]\,dS}$$

$$(12.1.5)$$

Although KHIT is valid for any Green's function, the one we will use is the free-space version,

$$G(\bar{x}_0, \bar{x}_s) = \frac{e^{-ik\hat{r}}}{4\pi\hat{r}}, \quad \hat{r} = |\bar{x}_0 - \bar{x}_s| \tag{12.1.6}$$

The gradient operator is applied at $\bar{x}_s$, so that

$$\nabla G(\bar{x}_0, \bar{x}_s) = \frac{\bar{x}_s - \bar{x}_0}{\hat{r}} \frac{d}{d\hat{r}} G(\bar{x}_0, \bar{x}_s) = \frac{1}{4\pi\hat{r}} \left(ik + \frac{1}{\hat{r}} \right) e^{-ik\hat{r}} \left(\frac{\bar{x}_0 - \bar{x}_s}{\hat{r}} \right) \tag{12.1.7}$$

The viewpoint in Eq. (12.1.5) is that the scattered pressure is the field radiated from S by the total pressure field. An alternative form stems from the perspective of Eqs. (12.1.3) and (12.1.4), which holds that P_{s} is a solution of the Helmholtz equation that satisfies the Sommerfeld radiation condition. Thus, P_{s} *by itself* must be consistent with the KHIT,

$$P_{\mathrm{s}}(\bar{x}_0) = \iint\limits_{\mathcal{S}} [P_{\mathrm{s}}(\bar{x}_s)\,\bar{n}(\bar{x}_s) \cdot \nabla G(\bar{x}_0, \bar{x}_s) - \bar{n}(\bar{x}_s) \cdot \nabla P_{\mathrm{s}}(\bar{x}_s)\,G(\bar{x}_0, \bar{x}_s)]\,d\mathcal{S}$$

$$(12.1.8)$$

In this view, the scattered pressure at a field point is the field radiated from $\mathcal{S}$ due to the scattered pressure and particle velocity on that surface.

Some individuals are puzzled by the availability of two forms of the KHIT for P_{s}, so let us examine this issue. By definition, the total pressure is $P = P_{\mathrm{s}} + P_{\mathrm{I}}$. Hence, if we subtract the above version from Eq. (12.1.5), we find that

$$\iint\limits_{\mathcal{S}} [P_{\mathrm{I}}(\bar{x}_s)\,\bar{n}(\bar{x}_s) \cdot \nabla G(\bar{x}_0, \bar{x}_s) - \bar{n}(\bar{x}_s) \cdot \nabla P_{\mathrm{I}}(\bar{x}_s)\,G(\bar{x}_0, \bar{x}_s)]\,d\mathcal{S} = 0 \quad (12.1.9)$$

This identity may be derived in a different way. The incident field satisfies the KHIT for the surface $\mathcal{S}_0$ from which it emanates, that is,

$$P_{\mathrm{I}}(\bar{x}_0) = \iint\limits_{\mathcal{S}_0} [P_{\mathrm{I}}(\bar{x}_s)\,\bar{n}(\bar{x}_s) \cdot \nabla G(\bar{x}_0, \bar{x}_s) - \bar{n}(\bar{x}_s) \cdot \nabla P_{\mathrm{I}}(\bar{x}_s)\,G(\bar{x}_0, \bar{x}_s)]\,d\mathcal{S}$$

$$(12.1.10)$$

At the same time, we could form the KHIT for $P_{\mathrm{I}}(\bar{x}_0)$ by including the scattering body in the domain. Then, the surface integral would consist of contributions from $\mathcal{S}$ and $\mathcal{S}_0$. According to the preceding, the contribution from $\mathcal{S}_0$ gives $P_{\mathrm{I}}(\bar{x}_0)$, so the contribution from $\mathcal{S}$ must be zero. This is the condition stated in Eq. (12.1.9). Whether it is preferable to use Eq. (12.1.5) or Eq. (12.1.8) sometimes depends on what is known about the field.

12.2 Scattering by Heterogeneity

In Example 11.5 of the previous chapter, we used ray theory to examine how a signal passes through a region in which the fluid properties vary spatially. In the situation considered there, the density and sound speed were continuous functions of position, so there was no impedance mismatch. Consequently, sound was transmitted into and out of the heterogenous region without reflections. In the situation, we consider here $\mathcal{S}$ is the surface that bounds a region in which the density ρ' and sound speed c' are different from the ambient properties exterior to $\mathcal{S}$. Thus, the incident signal is scattered by $\mathcal{S}$, as well as being transmitted into the interior, then out again. Both ρ' and c' are taken to be constant, so the scattering region is inhomogeneous from the perspective of the overall domain, although it is internally homogeneous.

12.2.1 General Equations

We cannot determine the scattered pressure directly from the KHIT integral because we do not know the surface pressure and particle velocity. Perhaps we can gain some insight if we examine the field in the homogeneous region V interior to S. A prime will denote that a quantity is associated with the interior, so the position of a point in V is labeled in Fig. 12.1 as $\bar{x}'$. Point C is central to V, such as the centroid, and $\bar{x}_0$ is an arbitrary field point exterior to V. The z-axis has any convenient reference orientation. It serves as the axial direction for a set of spherical coordinates, r, ψ_0, θ_0 that locates the field point.

Fig. 12.1 Definition of positions and coordinates relative to fluid region V whose properties differ from those of the surrounding fluid

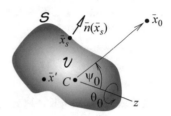

The pressure and normal particle velocity must be continuous along the interface S between the inner and outer fluids, so it must be that

$$P'\left(\bar{x}_s\right) = P\left(\bar{x}_s\right) \text{ and } \frac{1}{i\omega\rho'}\bar{n}\left(\bar{x}_s\right) \cdot \nabla P'\left(\bar{x}_s\right) = \frac{1}{i\omega\rho_0}\bar{n}\left(\bar{x}_s\right) \cdot \nabla P\left(\bar{x}_s\right) \quad (12.2.1)$$

Upon substitution of these relations, Eq. (12.1.5) becomes

$$P_s\left(\bar{x}_0\right) = \iint_S \bar{n}\left(\bar{x}_s\right) \cdot \left[P'\left(\bar{x}_s\right)\nabla G\left(\bar{x}_0, \bar{x}_s\right) - \frac{\rho_0}{\rho'}\nabla P'\left(\bar{x}_s\right) G\left(\bar{x}_0, \bar{x}_s\right)\right] dS$$

$$(12.2.2)$$

The divergence theorem converts the integral to one that extends over the interior domain. Note that $\bar{n}\left(\bar{x}_s\right)$ points out of the interior domain, so no sign change is entailed in this operation. Thus, we have

$$P_s\left(\bar{x}_0\right) = \iiint_V \nabla \cdot \left[P'\left(\bar{x}'\right)\nabla G\left(\bar{x}_0, \bar{x}'\right) - \frac{\rho_0}{\rho'}\nabla P'\left(\bar{x}\right) G\left(\bar{x}_0, \bar{x}'\right)\right] dV \quad (12.2.3)$$

where ∇ is the gradient at interior point $\bar{x}'$.

The fact that $\bar{x}_0$ is exterior to V simplifies this integral because nowhere does an interior point $\bar{x}'$ coincide with $\bar{x}_0$. Consequently, $G\left(\bar{x}_0, \bar{x}'\right)$ is a solution of the Helmholtz equation with no additional source term, that is, $\nabla^2 G\left(\bar{x}_0, \bar{x}'\right) = -k^2 G\left(\bar{x}_0, \bar{x}'\right)$. In addition, $P'\left(\bar{x}'\right)$ is a solution of the Helmholtz equation for the interior domain, where the sound speed is c'. It follows that $\nabla^2 P'\left(\bar{x}'\right) = -\left(\omega/c'\right)^2 P'\left(\bar{x}'\right)$. These relations convert the KHIT to

$$P_{\rm s}\left(\bar{x}_0\right) = \iiint\limits_{\mathcal{V}} \left[\left(1 - \frac{\rho_0}{\rho'}\right) \nabla P'\left(\bar{x}'\right) \cdot \nabla G\left(\bar{x}_0, \bar{x}'\right) \right.$$
$$\left. -k^2 \left(1 - \frac{\rho_0 c^2}{\rho'\left(c'\right)^2}\right) P'\left(\bar{x}'\right) G\left(\bar{x}_0, \bar{x}'\right) \right] d\mathcal{V} \qquad (12.2.4)$$

Further simplifications result if we limit consideration to cases where $\bar{x}_0$ is in the farfield. The farfield approximations apply if the smallest distance from a point within $\mathcal{V}$ to $\bar{x}_0$ is much greater than the largest dimension of the region. The farfield approximations use some central point C as the reference point, so that $\bar{x}_C \equiv \bar{0}$. The length of the line from point C to $\bar{x}_0$ is the radial distance r, and the direction is $\bar{e}_r$. The line from any point $\bar{x}'$ to $\bar{x}_0$ is considered to be parallel to $\bar{e}_r$, so the distance between $\bar{x}'$ and $\bar{x}_0$ differs from r by the projection of the vector connecting those points. The relations are

$$r = |\bar{x}_0|, \quad \bar{e}_r = \frac{\bar{x}_0}{r}, \quad \hat{r} \equiv |\bar{x}_0 - \bar{x}'| \approx r - \bar{x}' \cdot \bar{e}_r \qquad (12.2.5)$$

This leads to a simplified form of the Green's function in Eq. (12.1.6),

$$G\left(\bar{x}_0, \bar{x}'\right) \approx \frac{e^{-ikr}}{4\pi r} e^{ik\bar{x}' \cdot \bar{e}_r} \qquad (12.2.6)$$

Furthermore, $\nabla G\left(\bar{x}_0, \bar{x}'\right)$ is the gradient at $\bar{x}'$, with $\bar{x}_0$ held fixed, so that

$$\nabla G\left(\bar{x}_0, \bar{x}'\right) = \nabla \hat{r} \frac{d}{d\hat{r}} G\left(\bar{x}_0, \bar{x}'\right) \approx \bar{e}_r \frac{ike^{-ikr}}{4\pi r} e^{ik\bar{x}' \cdot \bar{e}_r} \qquad (12.2.7)$$

The orientation of $\bar{x}_0$ is described by the polar angle ψ_0 and azimuthal angle θ_0. Hence, substitution of these expressions into Eq. (12.2.4) leads to recognition that farfield scattering from a heterogeneity is described by two directivity factors. The general representation is

$$\boxed{P_{\rm s}\left(\bar{x}_0\right) = \frac{e^{-ikr}}{4\pi r} \left[\left(1 - \frac{\rho_0}{\rho'}\right) F_1\left(\psi_0, \theta_0\right) + \left(1 - \frac{\rho_0 c^2}{\rho'\left(c'\right)^2}\right) F_2\left(\psi_0, \theta_0\right) \right]} \qquad (12.2.8)$$

where the directivities are

$$\boxed{\begin{aligned} F_1\left(\psi_0, \theta_0\right) &= ik\bar{e}_r \cdot \iiint\limits_{\mathcal{V}} e^{ik\bar{x}' \cdot \bar{e}_r} \nabla P'\left(\bar{x}'\right) d\mathcal{V} \\ F_2\left(\psi_0, \theta_0\right) &= -k^2 \iiint\limits_{\mathcal{V}} e^{ik\bar{x}' \cdot \bar{e}_r} P'\left(\bar{x}'\right) d\mathcal{V} \end{aligned}} \qquad (12.2.9)$$

Without further development, these relations merely tell us that the strength of the scattered field depends on the mismatch of densities and of the bulk moduli, with the former being attributable to the particle velocity within the scattering volume and the latter due to the pressure in that region. Evaluation of the directivity factors requires judicious approximations of the interior field. Two that have been used consider different situations. The Born approximation is based on an assumption that the properties of the scattering region do not drastically differ from those of the surrounding homogeneous fluid. An approximation due to Rayleigh addresses objects whose size is much smaller than an acoustic wavelength.

12.2.2 The Born Approximation

If the region inside S was the same as the fluid in the outer region, the pressure inside S would be the incident value at that location. The Born approximation addresses the situation in which the difference between the fluids is not substantial, that is, when $\rho' \approx \rho_0$ and $\rho' (c')^2 \approx \rho_0 c^2$. In that case, it is reasonable to assume that the inside pressure differs little from P_I. Setting $P'(\bar{x}') = P_I(\bar{x}')$ in Eq. (12.2.4) converts the directivity factors in Eq. (12.2.9) to computable functions.

The typical situation is a plane wave that propagates in direction $\bar{e}_I$. Let $\hat{P}_I$ be the amplitude at the origin, so that

$$P_I(\bar{x}') = \hat{P}_I \exp\left(-ik\bar{x}' \cdot \bar{e}_I\right), \quad \nabla P'(\bar{x}') = -ik\bar{e}_I \hat{P}_I \exp\left(-ik\bar{x}' \cdot \bar{e}_I\right) \quad (12.2.10)$$

Substitution of these representations into Eq. (12.2.9) gives

$$\begin{aligned}
F_1(\psi_0, \theta_0, \psi_I, \theta_I) &= k^2 \hat{P}_I \bar{e}_r \cdot \bar{e}_I \mathcal{L}(\psi_0, \theta_0, \psi_I, \theta_I), \\
F_2(\psi_0, \theta_0, \psi_I, \theta_I) &= -k^2 \hat{P}_I \mathcal{L}(\psi_0, \theta_0, \psi_I, \theta_I) \\
\mathcal{L}(\psi_0, \theta_0, \psi_I, \theta_I) &= \iiint_{\mathcal{V}} e^{ik(\bar{x}' \cdot \bar{e}_r - \bar{x}' \cdot \bar{e}_I)} d\mathcal{V}
\end{aligned} \quad (12.2.11)$$

Because the integrals for the directivity factors F_1 and F_2 extend over the domain of the scattering region, it makes sense to align the coordinate system consistently with the shape of the body. Then, $\bar{e}_I$ and $\bar{e}_r$ to the field point have arbitrary orientation relative to the coordinate system. We use spherical angles ψ_I and θ_I defined relative to the z-axis to describe $\bar{e}_I$, while angles ψ_0 and θ_0 describe $\bar{e}_r$.

$$\begin{aligned}
\bar{e}_I &= \sin \psi_I \cos \theta_I \bar{e}_x + \sin \psi_I \sin \theta_I \bar{e}_y + \cos \psi_I \bar{e}_z \\
\bar{e}_r &= \sin \psi_0 \cos \theta_0 \bar{e}_x + \sin \psi_0 \sin \theta_0 \bar{e}_y + \cos \psi_0 \bar{e}_z
\end{aligned} \quad (12.2.12)$$

The presence of ψ_I and θ_I as variables affecting the directivity factors emphasizes that the scattered pressure depends on the propagation direction of the incident wave relative to the body, as well as the direction to the field point, which is defined by ψ_0 and θ_0. The position $\bar{x}'$ and differential volume element dV may be described in terms of Cartesian coordinates, cylindrical coordinates, or spherical coordinates, whichever is most convenient.

This description of Born scattering is not limited to a certain frequency range. However, its utility is limited by the condition that the scattering region is homogeneous with properties that are very much like those of the surrounding fluid.

EXAMPLE 12.1 A plane wave is incident on a region in the shape of a cylinder whose radius is a and length is L. The sound speed and density in this region are very close to the ambient properties, so the Born approximation may be applied. Derive an expression for the scattered pressure in the form $(r/a) \left| P_s \left(\bar{x}_0 \right) / \hat{P}_1 \right|$ for arbitrary $\bar{e}_1$ and $\bar{e}_r$. Then, specialize this expression to the case of backscatter, for which $\bar{e}_r = -\bar{e}_1$. Use that expression to evaluate the dependence of the backscattered pressure on the orientation of the incident wave for $ka = 0.1$, 1, and 10. The aspect ratio is $L/a = 20$, and the fluid properties are $\rho_0/\rho\prime = c/c' = 0.95$.

Significance

The solution demonstrates the way in which scattering integrals are formulated in general. The results that are obtained are a specific case of backscatter, which later will be addressed in greater detail. Most importantly, the results are illustrative of general scattering properties.

Solution

The starting point is Fig. 1, which illustrates the scattering region, the unit vectors, and the xyz coordinate system for the description of vectors. The origin is situated at the centroid of the cylinder, and the z-axis is aligned with the centerline. The axisymmetry of the system makes it permissible to situate the x-axis in the plane containing $\bar{e}_1$ and $\bar{e}_z$. This means that we can set $\theta_1 = 0$ without losing generality of the analysis.

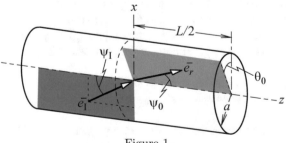

Figure 1

The shape is best described in terms of cylindrical coordinates defined relative to xyz, so that $dV = R d\theta\, dR\, dz$. The relevant vectors are

$$\bar{e}_I = \sin\psi_I \bar{e}_x + \cos\psi_I \bar{e}_z$$
$$\bar{e}_r = \sin\psi_0 \cos\theta_0 \bar{e}_x + \sin\psi_0 \sin\theta_0 \bar{e}_y + \cos\psi_0 \bar{e}_z \qquad (1)$$
$$\bar{x}' = R\cos\theta \bar{e}_x + R\sin\theta \bar{e}_y + z\bar{e}_z$$

When these representations are substituted into the integral in Eq. (12.2.11), the result is

$$\mathcal{L}(\psi_0, \theta_0, \psi_I, \theta_I) = \int_{-L/2}^{L/2}\int_0^a\int_{-\pi}^{\pi} \exp[ikR(\sin\psi_0\cos\theta_0 - \sin\psi_I)\cos\theta$$
$$+ ikR(\sin\psi_0\sin\theta_0)\sin\theta + ikz(\cos\psi_0 - \cos\psi_I)]R d\theta\, dR\, dz$$

Several simplifications are available. The first two terms in the integrand are independent of z, whereas the last term is independent of R and θ. The integral may be factorized accordingly. Furthermore, the integral over z may be evaluated analytically. The integral over the range of θ may be split into integrals over positive and negative θ. Replacing θ by $-\tilde{\theta}$ in the latter interval allows the integral to extend only over positive θ. The result is that

$$\mathcal{L}(\psi_0, \theta_0, \psi_I, \theta_I) = a^2 L \mathcal{G}(\psi_0, \theta_0, \psi_I)\, \mathcal{H}(\psi_0, \psi_I) \qquad (2)$$

The $\mathcal{H}$ factor is the result of integration over z,

$$\mathcal{H}(\psi_0, \psi_I) = \int_{-1/2}^{1/2} \exp\left[ikL\hat{z}(\cos\psi_0 - \cos\psi_I)\right] d\hat{z}$$

It may be integrated analytically, which leads to

$$\mathcal{H}(\psi_0, \psi_I) = \begin{cases} \dfrac{\sin\left[(kL/2)(\cos\psi_0 - \cos\psi_I)\right]}{(kL/2)(\cos\psi_0 - \cos\psi_I)} & \text{if } \psi_0 \neq \psi_I \\ 1 \text{ if } \psi_0 = \psi_I \end{cases} \qquad (3)$$

The $\mathcal{G}$ factor comes from integration over θ and R. It is

$$\mathcal{G}(\psi_0, \theta_0, \psi_I) = 2\int_0^1\int_0^{\pi} \exp\left[ika\hat{R}(\sin\psi_0\cos\theta_0 - \sin\psi_I)\cos\theta\right]$$
$$\times \cos\left[ka\hat{R}(\sin\psi_0\sin\theta_0)\sin\theta\right]\hat{R} d\theta\, d\hat{R} \qquad (4)$$

Numerical methods are required to evaluate this function for arbitrary orientation angles. However, a closed form result is available for the backscatter case.

After the value of $\mathcal{L}(\psi_0, \theta_0, \psi_I, \theta_I)$ is determined for a specific set of angles, the directivity factors in Eq. (12.2.11) are known. Their substitution, and $\bar{e}_r \cdot \bar{e}_I$ from Eq. (1), into Eq. (12.2.8) gives the scattered pressure. Collecting the parameters into nondimensional groups leads to

$$P_s(\bar{x}_0) = \frac{\hat{P}_I}{4\pi} \left(\frac{a}{r}\right) e^{-ikr} (ka)(kL) \left[\left(1 - \frac{\rho_0}{\rho'}\right) (\sin\psi_0 \cos\theta_0 \sin\psi_I + \cos\psi_0 \cos\psi_I)\right.$$

$$\left. - \left(1 - \frac{\rho_0 c^2}{\rho'(c')^2}\right)\right] \mathcal{G}(\psi_0, \theta_0, \psi_I) \mathcal{H}(\psi_0, \psi_I) \tag{5}$$

Evaluation of the $\mathcal{G}$ function for arbitrary ψ_I, ψ_0, and θ_0 requires numerical integration, which is easier if the available software has a routine for double integration. In MATLAB, it is dblquad, whose input arguments are the name of a function that evaluates the integrand, as well as the lower and upper integration limits for both variables.

Equation (5) offers little insight to the nature of the scattered field. The results for an arbitrary set of angles could be displayed as a spherical plot for fixed ψ_I, in which ψ_0 and θ_0 are the spherical angles and $(r/a)\left|P_s/\hat{P}_I\right|$ is the radial distance. Many such plots would describe the dependence on ψ_I. Fortunately, the information that is requested sets $\bar{e}_r = -\bar{e}_I$. This corresponds to setting $\theta_0 = \pi$ and $\psi_0 = \pi - \psi_I$. For these angles, the preceding expression reduces to

$$\left(\frac{r}{a}\right) \frac{P_s(\bar{x}_0)}{\hat{P}_I} = \frac{e^{-ikr}}{4\pi} (ka)(kL) \left[\frac{\rho_0}{\rho'}\left(1 + \frac{c^2}{(c')^2}\right) - 2\right]$$
$$\mathcal{G}(\pi - \psi_I, \pi, \psi_I) \mathcal{H}(\pi - \psi_I, \psi_I) \tag{6}$$

where

$$\mathcal{G}(\pi - \psi_I, \pi, \psi_I) = 2\int_0^1 \int_0^\pi \exp\left(-2ika\hat{R}\sin\psi_I \cos\theta\right) \hat{R}d\theta d\hat{R}$$
$$\mathcal{H}(\pi - \psi_I, \psi_I) = \frac{\sin(kL\cos\psi_I)}{kL\cos\psi_I}$$

Analytical evaluation is possible by invoking general formulas provided in Abramowitz and Stegun's compendium.[1] The eventual result is

$$\mathcal{G}(\pi - \psi_I, -\pi, \psi_I) = \pi\frac{J_1(2ka\sin\psi_I)}{ka\sin\psi_I} \tag{7}$$

There are different ways in which parametric trends may be viewed. We can consider the value of P_s that would be measured at a fixed location, which sets r.

[1]M.I. Abramowitz and I.A. Stegun, *Handbook of Mathematical Functions*, Dover, (1965), p. 360 for Eq. (9.1.210), and p. 484 for Eq. (11.3.20).

Because $a^2 L = \mathcal{V}/\pi$, that view considers P_s to be proportional to the volume of the heterogenous medium, as well as the square of the frequency. However, changing the dimensions or the frequency also has a significant effect on the angular dependence. The first of Eq. (6) indicates that $\mathcal{G}$ is a function of ka, as well as ψ_{I}. The value of $\mathcal{G}$ for $\psi_{\mathrm{I}} = 0$ is π for any ka, as well as for any ψ_{I} if $ka \ll 1$. Furthermore, $\mathcal{G} = \pi$ is the maximum value for any ka and ψ_{I}. These properties mean that the $\mathcal{G}$ function leads to enhanced directivity in the axial direction as ka increases. In the case of $\mathcal{H}$, its maximum is $\mathcal{H} = 1$. This value occurs at any ψ_{I} if $kL \ll 1$, and at $\psi_{\mathrm{I}} = \pi/2$ for any kL. Thus, $\mathcal{H}$ leads to greater directivity in the transverse direction as the frequency increases. The low-frequency limit is examined in detail in the next section.

Figure 2 displays polar plots of $(r/a)\,|P_s/P_{\mathrm{I}}|$ at the designated frequencies. Backscatter at broadside incidence is somewhat larger at the lowest frequency, and it is strongly directional in the broadside direction at the middle frequency. At the highest frequency, backscatter is comparable for end-on and broadside incidence, and both are confined to a smaller range of angles.

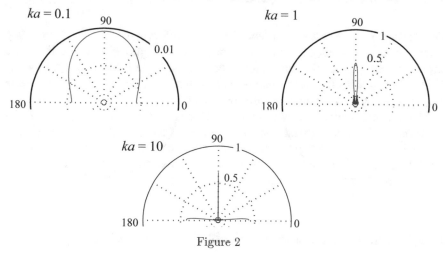

Figure 2

To gain insight to this shifting trend, let us consider individually $\mathcal{G}\,(\pi - \psi_{\mathrm{I}}, -\pi, \psi_{\mathrm{I}})$ and $\mathcal{H}\,(\pi - \psi_{\mathrm{I}}, \psi_{\mathrm{I}})$, whose product describes the backscatter directivity. Examination of Eq. (7) reveals that the maximum value of the former is $\mathcal{G} = \pi$ at end-on incidence and that $\mathcal{G} = \pi J_1\,(2ka)\,/\,(ka)$ at broadside incidence. This value decreases in an oscillatory manner with increasing ka. In contrast, the maximum value of $\mathcal{H}$ broadside is $\mathcal{H} = 1$, and $\mathcal{H} = \sin(kL)\,/\,(kL)$ for end-on incidence. This too decreases in an oscillatory manner with increasing frequency. With these trends in mind, consider the graphs in Fig. 3. Each function is large in a narrow interval. The directivity of P_s is the product of the functions, which explains why it is very small away from $\psi_{\mathrm{I}} = 0°$, $90°$, and $180°$. Increasing ka increases kL, so $J_1\,(2ka)$ and $\sin(kL)$ will grow and decrease. If either term is near zero, the value of p_s will be very small in the respective direction. Consequently, whether backscatter is greater at broadside or end-on incidence strongly depends on the frequency.

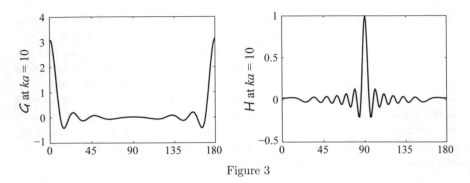

Figure 3

12.3 Rayleigh Scattering Limit

The *Rayleigh scattering limit* is a general term that refers to the field scattered by any object whose size is much less than the acoustic wavelength. Specifically, if a is the largest dimension, then the Rayleigh scattering limit is the behavior when $ka \ll 1$. This condition leads to several analytical simplifications. It also will disclose some qualitative aspects of the field that are not evident from the general equations.

The Rayleigh limit is sometimes referred to as a quasistatic approximation. This terminology stems from the fact that because of the small size of the body, the phase of any long-wavelength signal will be essentially constant over the extent of the body. This means that any complex exponential representing a phase shift within the scattering body or on its surface may be approximated as having unit value.

12.3.1 The Rayleigh Limit of the Born Approximation

The Born approximation is not limited by the size of the heterogeneous region. Its associated directivity factors must be evaluated for a specific body, and a specific frequency. The Rayleigh limit of the Born approximation yields a formula for the scattered field that does not require evaluation of integrals. At the same time, it is less general because it considers the scattering region to be much smaller than a wavelength.

The previous example indicates how we should proceed to obtain the low frequency version. As before, we shall only consider the case of an incident plane wave, which is described by Eq. (12.2.10). Let a be the largest dimension of the scattering region. If the origin is interior to the region, then any position within this region is such that $|\bar{x}'| \le a$. Furthermore, we are interested in the situation when $2\pi/k \gg a$. It follows that $k\,|\bar{x}'| \ll 1$, so that the integrand in Eq. (12.2.11) is essentially one, which gives $\mathcal{L} = \mathcal{V}$. The corresponding directivity factors are

$F_1 = k^2 \hat{P}_1 \bar{e}_r \cdot \bar{e}_1 \mathcal{V}$ and $F_2 = -k^2 \hat{P}_1 \mathcal{V}$. The only directional factor that arises is $\bar{e}_r \cdot \bar{e}_1$. Therefore, we may define the z-axis to be parallel to $\bar{e}_1$. This reduces Eq. (12.2.8) to

$$P_s\left(\bar{x}_0\right) = k^2 \hat{P}_1 \frac{e^{-ikr}}{4\pi r} \mathcal{V}\left[\left(\frac{\rho_0 c^2}{\rho'\left(c'\right)^2} - 1\right) - \left(\frac{\rho_0}{\rho'} - 1\right)\cos\psi_0\right] \qquad (12.3.1)$$

where ψ_0 is the angle between $\bar{e}_1$ and the radial direction to the field point. This expression tells us that the strength of the scattered field at a fixed location is proportional to the volume of the scatterer. This attribute surprises some individuals who intuitively think that the cross-sectional area represents the amount the sound is blocked. Another important observation is that the field consists of a monopole whose strength is proportional to the difference of the bulk moduli, combined with a dipole component that is proportional to the difference of densities. The polar angle to the field point is measured from the incident direction $\bar{e}_1$. Thus, the dipole is aligned in the direction of the incident wave. Both the monopole and dipole strengths are proportional to the square of the frequency.

It is instructive to compare this result to the solution of Example 12.1, which is specific to a cylindrical region. That result is valid for any frequency, but it is only descriptive of a specific shape. The Rayleigh limit described Eq. (12.3.1) is valid only for low frequencies, but it is descriptive of scattering from any region, regardless of its shape. Another difference is that the example was limited to situations where the scattered signal is observed on the same line as that from the planar source to the region. This corresponds to $\psi_1 = 180°$ in the above expression. The result for a cylindrical region when $ka \ll 1$ is identical to Eq. (12.3.1) when $\mathcal{V} = \pi a^2 L$ and $\cos\psi_1 = -1$. Despite the generality of the present result, the assumptions embedded in the model lead to a description of the scattered pressure that depends on the location of the field point, but is independent of the orientation of the body. In contrast, such dependence is displayed in the Born approximation.

Viewing the dependence of the scattered pressure in Eq. (12.3.1) in terms of nondimensional variables offers a different perspective. The largest dimension has been denoted as a, so the volume can be described as βa^3, where β is a shape factor. Then $r/a \left|P_s/\hat{P}_1\right|$ depends solely on β, $(ka)^2$, and ψ_1.

This trend was used by Rayleigh[2] to explain why the sky is blue. At midday, the light rays that are incident at our location from the sun essentially are white, which means that they are composed of all colors in the visual spectrum in nearly equal amounts. These rays constitute the image of the sun. If there were no scattering, these are the only rays that would travel from the sun to an observer on the ground. The rest of the sky would appear to be black, as it does on the Moon. What actually happens is that the light rays that propagate through the atmosphere in other directions are scattered by dust and water particles in the atmosphere. Some of these rays reach an observer on the ground. The Rayleigh description of scattering indicates that the

---

[2]J.W. Strutt, Lord Rayleigh, "On the light from the sky, its polarization and colour," Philosophical Magazine, series 4, vol. 41 (1871) pp. 107–120.

blue spectrum, which comprises the short-wavelength/high-frequency portion of the visual spectrum, scatters more strongly than the red spectrum. Thus, the contribution of the blue spectrum to what people see when they view the sky is stronger than the contribution of the red spectrum.

12.3.2 Mismatched Heterogeneous Region

If the properties of a small region of fluid differ significantly from those of the surrounding fluid, the Born approximation is not valid. A different analysis, also based on ka being small, is available in that case. Let us multiply the Helmholtz equation by a^2 and define a nondimensional gradient $\hat{\nabla} = a\nabla$. Smallness of ka means that $\hat{\nabla} P$ is the dominant term in the Helmholtz equation within the heterogeneity. Hence, the Helmholtz equation for the interior field reduces to the Laplace equation. Smallness approximations do not apply to the exterior scattered field, which must satisfy the Helmholtz equation. The boundary conditions on the surface $\mathcal{S}$, which is the interface between the media, are continuity of pressure and normal particle velocity. Let $P'(x)$ be the pressure within the heterogeneity, and $P_s(\bar{x})$ be the scattered field. The governing equations are

$$
\left.
\begin{aligned}
\nabla^2 P' &= 0, \quad \bar{x} \in \mathcal{V} \\
\nabla^2 P_s + k^2 P_s &= 0, \quad \bar{x} \notin \mathcal{V} \\
P' &= P_s + P_1 \\
\frac{\rho_0}{\rho'}\bar{n} \cdot \nabla P' &= \bar{n} \cdot (\nabla P_s + \nabla P_1)
\end{aligned}
\right\} \; \bar{x} \in \mathcal{S}
\tag{12.3.2}
$$

In addition the scattered field must satisfy the radiation condition in Eq. (12.1.4).

To see how the limitation to small ka simplifies the analysis, we shall consider a specific situation. Suppose $\mathcal{V}$ is a sphere and the incident signal is a plane wave propagating in the axial direction, so that $\bar{e}_1 = \bar{e}_z$. Let $\hat{r} = r/a$ be the nondimensional radial position, and set $\bar{n} = \bar{e}_r$. Then, the incident pressure and its normal derivative on the surface of the sphere are

$$
P_1|_{\hat{r}=1} = \hat{P}_1 e^{-ika\cos\psi}
$$
$$
\left.\frac{\partial P_1}{\partial \hat{r}}\right|_{\hat{r}=1} = -ika\cos\psi \hat{P}_1 e^{-ika\cos\psi}
\tag{12.3.3}
$$

Smallness of ka allows us to expand the complex exponential in a series that is truncated at the lowest order frequency-dependent term, so that

$$
P_1|_{\hat{r}=1} = (1 - ika\cos\psi)\,\hat{P}_1
$$
$$
\left.\frac{\partial P_1}{\partial \hat{r}}\right|_{\hat{r}=1} = -ika\cos\psi \hat{P}_1
\tag{12.3.4}
$$

We may use a spherical harmonic expansions to represent P_s and $\tilde{P}$. Both series may be truncated at the $m = 1$ harmonic, because that is the highest harmonic appearing in the incident pressure and pressure gradient on the surface, both of which constitute the excitation. The series for the scattered pressure is

$$P_s = \sum_{m=0}^{1} \hat{B}_m P_m (\cos\theta) h_m^{(2)} (ka\hat{r}) = \hat{B}_0 h_0^{(2)} (ka\hat{r}) + \hat{B}_1 h_1^{(2)} (ka\hat{r}) \cos\psi \quad (12.3.5)$$

This description is valid everywhere, but it may be simplified in the vicinity of the sphere based on $ka \ll 1$ and $\hat{r} = O(1)$. This form is

$$P_s \approx \frac{B_0}{\hat{r}} + \frac{B_1}{\hat{r}^2} \cos\psi, \quad ka\hat{r} \ll 1$$
$$B_0 = \frac{i}{ka} \hat{B}_0, \quad B_1 = \frac{i}{2(ka)^2} \hat{B}_0 \quad (12.3.6)$$

The value of $ka\hat{r}$ is much less than one everywhere within $\mathcal{V}$, so the spherical harmonic representation of the pressure in the scattering region may be represented by a similar form,

$$P' = \sum_{m=0}^{\infty} A_m P_m (\cos\theta) j_m (kr) \approx A_0 + A_1 \hat{r} \cos\psi \quad (12.3.7)$$

These series are substituted into the boundary conditions in Eq. (12.3.2), and like dependencies on ψ are matched. This operation yields

$$A_0 = \hat{P}_1 + B_0$$
$$A_1 = -ika\hat{P}_1 + B_1$$
$$0 = -B_0 \quad (12.3.8)$$
$$\frac{\rho_0}{\rho'} A_1 = -ika\hat{P}_1 - 2B_1$$

Solution of these equations leads to an explicit description of the pressure and its gradient within $\mathcal{V}$,

$$P' = \hat{P}_1 \left(1 - \frac{3ik}{2 + \rho_0/\rho'} r \cos\psi \right)$$
$$\nabla P' = -\frac{3ik}{2 + \rho_0/\rho'} \hat{P}_1 r \cos\psi \quad (12.3.9)$$

These expressions are used to evaluate the volume integrals in Eq. (12.2.9). The result is

$$\boxed{P_s (\bar{x}_0) = \hat{P}_1 k^2 \frac{e^{-ikr}}{4\pi r} \frac{4\pi a^3}{3} \left[\left(\frac{\rho_0 c^2}{\rho' (c')^2} - 1 \right) - \frac{3(\rho_0/\rho' - 1)}{2 + \rho_0/\rho'} \cos\psi_0 \right]} \quad (12.3.10)$$

The above has several features that are like the behavior predicted by the Born approximation. The magnitude of the scattered pressure increases as the square of the frequency. It consists of a monopole whose strength is proportional to the difference of the bulk moduli, and a dipole whose strength is proportional to the difference of densities. The amplitude of the monopole term matches the result of the Born approximation, Eq. (12.3.1), but the dipole moment is different. In the case where $\rho_0/\rho' \approx 1$, the dipole moment derived here is 50% greater than the previous result. The Born approximation was introduced empirically, whereas the present result stems from an analysis that only assumes that ka is sufficiently small.

Although the above formula was derived for a sphere, we know that at very low frequencies, geometric details that are much smaller than the wavelength are unimportant. Thus, it is reasonable to use Eq. (12.3.10) with $(4/3)\,\pi a^3$ replaced by $\mathcal{V}$, rather than Eq. (12.3.1), to describe scattering from a small embedded region of fluid, regardless of its shape.

12.3.3   Scattering from a Rigid Body

Analysis of the field scattered by an immovable body is quite complicated if the wavelength is comparable to the size of the body. Restricting our consideration to situations where the body is small enables us characterize the general nature of the scattered field. The development is founded on the Kirchhoff–Helmholtz integral theorem in Eq. (12.1.8), which states that P_s is the field radiated from S due to the scattered pressure and particle velocity on that surface. We begin by considering the situation where the rigid body is immobilized by some means. After that, we will formulate a description that accounts for the effect of removing restraints against movement.

Stationary Bodies

When the surface S is stationary, the normal component of the particle velocity must vanish everywhere on S. The total signal is the sum of the incident and scattered parts, so it must be that

$$\bar{n} \cdot \bar{V}_s = -\bar{n} \cdot \bar{V}_I, \quad \bar{x} \in S \tag{12.3.11}$$

Specific results may be obtained for low-frequency scattering of an incident plane wave,

$$P_I = \hat{P}_I e^{-ik\bar{e}_I \cdot \bar{x}} \implies \bar{V}_I = \bar{e}_I \frac{\hat{P}_I}{\rho_0 c} e^{-ik\bar{e}_I \cdot \bar{x}} \tag{12.3.12}$$

where $\bar{e}_I$ is the propagation direction. It will be useful for the analysis if we situate the origin at the geometric centroid C of the body, and set the phase lag of the wavefront through point C to zero. This is the arrangement depicted in Fig. 12.2.

Fig. 12.2 Plane wave is
incident on a rigid body

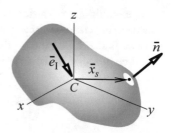

The definition of the low-frequency range is that the wavelength is much larger
than the size of the scattering body. In that case, it must be true that $k\,|\bar{x}| \ll 1$ for
any point in the body. Consequently, $\bar{V}_{\mathrm{I}}$ will be adequately described by the first few
terms in a Taylor series,

$$\bar{V}_{\mathrm{I}} = \bar{e}_{\mathrm{I}}\frac{\hat{P}_{\mathrm{I}}}{\rho_0 c}\,[1 - ik\bar{e}_{\mathrm{I}} \cdot \bar{x} + \cdots] \tag{12.3.13}$$

The surface normal velocity in the scattered signal must cancel the incident contribution, so we have

$$\bar{n} \cdot \bar{V}_{\mathrm{s}} = \bar{n} \cdot \bar{e}_{\mathrm{I}}\frac{\hat{P}_{\mathrm{I}}}{\rho_0 c}\,[-1 + ik\bar{e}_{\mathrm{I}} \cdot \bar{x} + \cdots]; \quad \bar{x} \in \mathcal{S} \tag{12.3.14}$$

A useful way to interpret this expression is that the scattered pressure is equivalent
to the pressure radiated by the body when the normal velocity on the surface is the
quantity on the right side. This viewpoint allows us to invoke the multipole expansion
of radiation from a rigid body, which was the topic in Sect. 7.4.2. The first contribution
to $\bar{n} \cdot \bar{V}_{\mathrm{s}}$ is the component of the particle velocity $\bar{V}_{\mathrm{I}}$ in the direction of the inward
normal. In a rigid body translation, all points have the same velocity. Therefore, the
first term represents a rigid body translational oscillation whose complex velocity
amplitude is $-\bar{V}_{\mathrm{I}}$. Such a motion leads to dipole radiation, but does not contribute
to the monopole amplitude. The second term in Eq. (12.3.14) may be written as
$i\bar{n} \cdot \bar{V}_{\mathrm{I}}\,(\bar{e}_{\mathrm{I}} \cdot k\bar{x})$. This term contributes to both the monopole amplitude and dipole
moment. However, $|\bar{x}|$ can be no larger than the largest dimension a, and we have
restricted our consideration to situations where $ka \ll 1$. This observation makes it
permissible to neglect the contribution of this term to the dipole moment. Hence, by
these considerations, we have recognized that the monopole amplitude A accounts for
the volume velocity associated with a normal velocity that is $i\bar{n} \cdot \bar{V}_{\mathrm{I}}\,(\bar{e}_{\mathrm{I}} \cdot k\bar{x})$, whereas
$\bar{D}$ is the dipole moment for a rigid body oscillation whose complex amplitude is $-\bar{V}_{\mathrm{I}}$.

The multipole expansion is given in Eq. (7.4.23). We let $G\left(\bar{x}_0, \bar{x}_C\right)$ be the free-space Green's function and ignore higher order terms on the basis that they are weak if $2\pi/k$ is much larger than the body's size. The corresponding expression for the scattered pressure in the farfield is

$$P\left(\bar{x}_0\right) = \left(A + ik\bar{D}\cdot\bar{e}_r\right)\frac{e^{-ikr}}{r} \tag{12.3.15}$$

The monopole amplitude is stated by the first of Eq. (7.4.24) to be

$$
A = \frac{i\omega\rho_0}{4\pi}\iint\limits_{S}\left[\left(\bar{n}\cdot\bar{e}_\mathrm{I}\frac{\hat{P}_\mathrm{I}}{\rho_0 c}\right)\left(ik\bar{e}_\mathrm{I}\cdot\bar{\xi}\right)\right]d\mathcal{S} = -\frac{k^2\hat{P}_\mathrm{I}}{4\pi}\iint\limits_{S}\left(\bar{n}\cdot\bar{e}_\mathrm{I}\right)\left(\bar{e}_\mathrm{I}\cdot\bar{\xi}\right)d\mathcal{S}
$$
$$
= -\frac{k^2\hat{P}_\mathrm{I}}{4\pi}\iiint\limits_{\mathcal{V}}\nabla\cdot\left[\bar{e}_\mathrm{I}\left(\bar{e}_\mathrm{I}\cdot\bar{\xi}\right)\right]d\mathcal{V}
$$
$$\tag{12.3.16}$$

where the divergence theorem is applied with no sign change because $\bar{n}\left(\bar{\xi}\right)$ is oriented outward from $\mathcal{V}$. To evaluate the gradient, let ℓ_j denote the direction cosines of $\bar{e}_\mathrm{I}$, so $\bar{e}_\mathrm{I}\cdot\bar{\xi} = \ell_1 x + \ell_2 y + \ell_3 z$. Because $\bar{e}_\mathrm{I}$ is a constant, the integrand reduces to $\bar{e}_\mathrm{I}\nabla\left(\bar{e}_\mathrm{I}\cdot\bar{\xi}\right) \equiv \bar{e}_\mathrm{I}\cdot\bar{e}_\mathrm{I} = 1$. Correspondingly, the monopole amplitude reduces to

$$A = -\frac{k^2\mathcal{V}}{4\pi}\hat{P}_\mathrm{I} \tag{12.3.17}$$

Determination of the dipole moment does not require evaluation of a surface integral. The requisite analysis has been performed in Sect. 7.4.2, which treated the field radiated by an oscillating rigid body. Thus, we may employ Eq. (7.4.39), with the translational velocity sets at $-\bar{V}_\mathrm{I} \equiv -\bar{e}_\mathrm{I}P_\mathrm{I}/\left(\rho_0 c\right)$. This equation is written in matrix form to facilitate describing the possibility that the dipole moment might not be aligned with the translational velocity. Thus, we find that

$$\{D\} = \frac{ik}{4\pi}P_\mathrm{I}\left[\mathcal{V}\left[I\right] + \left[W\right]\right]\{e_\mathrm{I}\} \tag{12.3.18}$$

The quantity $\rho_0\left[W\right]$ is the virtual mass, defined in Eq. (7.4.35).

The scattered pressure is obtained by adding the monopole and dipole contributions according to Eq. (12.3.15). In keeping with the matrix form of the relation for the dipole moment, this pressure field is

$$P_s\left(\bar{x}_0\right) = -\frac{k^2\hat{P}_\mathrm{I}}{4\pi}\left[\mathcal{V} - \{e_r\}^\mathrm{T}\left[\mathcal{V}\left[I\right] + \left[W\right]\right]\{e_\mathrm{I}\}\left(1 - \frac{i}{kr}\right)\right]\frac{e^{-ikr}}{r} \tag{12.3.19}$$

where r is the distance from the centroid of the scattering body to the field point. The farfield version is obtained by dropping the i/kr term. The corresponding description of the force exerted by the scattered pressure is stated in Eq. (7.4.35). Replacing $\bar{V}_C$ in that expression with $\bar{e}_I P_I / (\rho_0 c)$ gives

$$\boxed{\{F\} = ik\hat{P}_I[W]\{e_I\}} \tag{12.3.20}$$

Both $\mathcal{V}$ and $[W]$ are shape-dependent quantities that scale as the cube of the size of the body, which is measured by dimension a. Therefore, the scattered pressure observed at a fixed field point depends on the cube of the size, the shape of the body, the square of the frequency, and the orientation of $\bar{e}_I$ and $\bar{e}_r$ relative to the body. Nondimensionally, it is proportional to $\left|\hat{P}_I\right|$, a/r, and $(ka)^2$, and the direction angles for both directions. These attributes, as well as the fact that the field is a superposition of a monopole and dipole, are the same as those found for Rayleigh scattering from a region of heterogeneity.

Evaluation of the scattered pressure when $[\Lambda]$ is known requires a description of $\{e_i\}$ and $\{e_r\}$. For this, we may use spherical angles defined relative to the z-axis, with ψ_I and ψ_r being the respective polar angles. The representation is

$$\boxed{\begin{aligned} \{e_I\} &= [\sin\psi_I\cos\theta_I \quad \sin\psi_I\sin\theta_I \quad \cos\psi_I]^{\mathrm{T}} \\ \{e_r\} &= [\sin\psi_r\cos\theta_r \quad \sin\psi_r\sin\theta_r \quad \cos\psi_r]^{\mathrm{T}} \end{aligned}} \tag{12.3.21}$$

The expression for the scattered pressure simplifies considerably when the body is axisymmetric. We shall designate z as this axis. A consequence of axisymmetry is that we may take $\bar{e}_I$ to lie in the zx, so that $\theta_I = 0$. Setting the off-diagonal terms of $[\Lambda]$ to zero reduces the farfield scattered pressure in Eq. (12.3.19) to

$$\boxed{\begin{aligned} P_s(\bar{x}_0) = -\frac{k^2\hat{P}_I}{4\pi}&\left[\mathcal{V} - \left(\mathcal{V} + W_{1,1}\right)\sin\psi_I\sin\psi_r\cos\theta_r\right. \\ &\left. - \left(\mathcal{V} + W_{3,3}\right)\cos\psi_I\cos\psi_r\right]\frac{e^{-ikr}}{r} \end{aligned}} \tag{12.3.22}$$

For a sphere, $W_{1,1} = W_{3,3} = (2/3)\pi a^3$, which corresponds to the added mass being half the mass of the displaced fluid. For a thin disk, $W_{1,1} = 0$ and $W_{3,3} = (8/3)a^3$. In the thin disk approximation, the displaced volume is essentially zero, so the added mass is not related to the displaced mass. It is important to recognize that although $\mathcal{V}$ is very small, the monopole term should not be taken to be zero because the dipole contribution might also be very small. If the scattering body is homogeneous with mass density ρ_b, then we can set $\mathcal{V} = M/\rho_b$.

EXAMPLE 12.2 The sketch shows an arrangement in which projector A and hydrophone B are situated along orthogonal lines relative to the center of a scattering body that is rigid and stationary. The distance from the source to the body is sufficiently large to consider the incident wave to be planar, and the wavelength of the incident wave is much greater than the body's largest dimension. The body is axisymmetric relative to the z-axis, which is coplanar with $\bar{e}_I$ and $\bar{e}_r$. The virtual mass matrix is described relative to the volume of the body, according to $W_{1,1} = W_{2,2} = 5\mathcal{V}$, $\Lambda_{3,3} = \mathcal{V}$. Determine the dependence on the angle of incidence ψ of the scattered pressure at point B, normalized according to $4\pi r \, |P_s \, (\bar{x}_A) \, / P_I| \, / \left(k^2 \mathcal{V} \right)$.

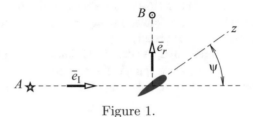

Figure 1.

Significance

Considering a specific case of scattering from a rigid body gives a clearer picture of the angular dependence, as well as some understanding to the influence of the virtual added mass.

Solution

To employ Eq. (12.3.22), we must define the direction angles. We define the x-axis to lie in the plane formed by $\bar{e}_I$ and $\bar{e}_r$, which also contains the z-axis. This is allowable because it does not matter how much the body is rotated about its axis of symmetry. The formula was derived on the basis that $\bar{e}_I$ has a positive component in the x direction, which means that the x-axis is down and to the right in the sketch. Thus, the unit vectors are

$$\bar{e}_I = \cos\psi\bar{e}_z + \sin\psi\bar{e}_x, \quad \bar{e}_r = \sin\psi\bar{e}_z - \cos\psi\bar{e}_x \qquad (1)$$

A comparison of these expressions to the general description in Eq. (12.3.21) leads to recognition that $\psi_I = \psi$, $\theta_I = 0$, $\psi_r = \pi/2 - \psi$, and $\theta_r = \pi$. The result is that

$$\frac{4\pi r}{k^2 \mathcal{V}} |P_s \, (\bar{x}_0)| = 1 + \frac{1}{\mathcal{V}} \left(W_{1,1} - W_{3,3} \right) \cos\psi \sin\psi = 1 + \frac{\left(W_{1,1} - W_{3,3} \right)}{2\mathcal{V}} \sin\left(2\psi \right) \qquad (2)$$

This function is depicted in Fig. 2. Note that this is not a directivity. Rather, a radial line at a specific ψ indicates the magnitude of $|P_s|$ when the axis of symmetry, which is x, is aligned at that angle.

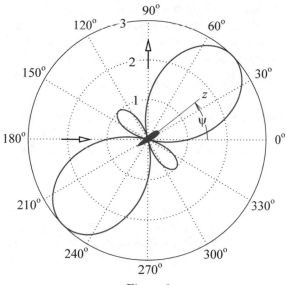

Figure 2.

The plot has interesting features, some of which might be considered to be counter-intuitive. At $\psi = 0, 90°, 180°$, and $270°$, the normalized scattered pressure is one. This is the monopole effect. But it is one also at $\psi = 135°$ and $\psi = 315°$. At these angles, the dipole contribution associated with the amplitude of the dipole contribution is twice the monopole value, but it is $180°$ out-of-phase from the monopole. In contrast, the scattered pressure arriving at hydrophone B has a maximum at $\psi = 45°$ and $\psi = 225°$ because the two contributions are in-phase. The nulls which occur at $\psi = 105°, 165°, 285°$, and $245°$ result from the two effects having equal magnitude by opposite phase.

Many individuals think of scattering as a mirror-like effect, which is the viewpoint of geometrical acoustics. In such thinking, $\psi = 45°$ and $\psi = 225°$ are like placing a mirror in the xy-plane. The angle of incidence of $\bar{e}_I$ on this plane equals the angle of reflection of $\bar{e}_r$, which should lead to a maximum pressure at hydrophone B. This expectation is confirmed by Fig. 1, but that does not mean that the underlying reasoning is correct. If it were, then $\psi = 135°$ and $\psi = 270°$ would be nulls because the corresponding angle of reflection is downward. In general, low-frequency scattering is not a process of reflection.

Effect of Rigid Body Motion

We now turn to the fact that the scattering body often is immersed in the fluid without support, as in the case of a rain drop. Such movement generates a radiating pressure field that adds to the pressure field obtained when the body is stationary. Section 7.4.2 developed a multipole expansion that describes low-frequency radiation from an oscillating rigid body. That development led to the observation that rotation of a body at low frequencies generates a very weak quadrupole that is negligible if the dipole contribution is nonzero. This is the present situation, so we only need to consider translational motion, which is governed by Newton's second law.

The most expedient way to proceed is to return to Eq. (12.1.3), which is the velocity continuity condition for any scattering problem,

$$\bar{n} \cdot \nabla P_s = -\bar{n} \cdot \nabla P_I - i\omega \rho_0 \bar{n} \cdot \bar{V}, \quad \bar{x} \in \mathcal{S} \tag{12.3.23}$$

This relation allows us to regard the scattered field as the superposition from two incident plane waves: the actual one, and one whose particle velocity is $-\bar{V}$. The pressure amplitude of the latter is $\rho_0 c V$. In other words, the effective incident pressure is

$$(P_I)_{\text{effective}} = \hat{P}_I e^{-ik\bar{e}_I \cdot \bar{x}} - \rho_0 c V e^{-ik\bar{e}_v \cdot \bar{x}} \tag{12.3.24}$$

The first place we introduce this replacement is the analysis of the monopole part of the scattered pressure, Eq. (12.3.17). This effect does not depend on the direction in which the incident wave propagates, but the phase lag between the incident velocity and $\bar{V}$ is 180°. Therefore, the monopole amplitude is

$$A = -\frac{k^2 \mathcal{V}}{4\pi} A = -\frac{k^2 \mathcal{V}}{4\pi} \left(\hat{P}_I - \rho_0 c |V| \right) \tag{12.3.25}$$

The modification of the dipole moment entails replacement of $P_I \bar{e}_I$ with $P_I \bar{e}_I - \rho_0 c \bar{V}$. Doing so converts the dipole moment in Eq. (12.3.18) to

$$\{D\} = \frac{ik}{4\pi} [\mathcal{V}[I] + [W]] \left\{ \hat{P}_I \{e_I\} - \rho_0 c \{V\} \right\} \tag{12.3.26}$$

The last step in the analysis is determination of $\bar{V}$. With that as our objective, we formulate Newton's Second Law for the object. It equates the inertia of the body to the resultants of the incident and scattered pressure distributions. Let M be the body's mass. The complex amplitude of its acceleration is $i\omega \bar{V}$, so it must be that

$$M \left(i\omega \bar{V} \right) = \bar{F}_I + \bar{F} \tag{12.3.27}$$

The scattered pressure resultant is described by (12.3.20), with the incident pressure set to $P_I \bar{e}_I - \rho_0 c \bar{V}$,

$$\{F\} = ik [W] \left\{ \hat{P}_I \{e_I\} - \rho_0 c \{V\} \right\} \tag{12.3.28}$$

The resultant of the incident pressure is

$$\bar{F}_I = \iint\limits_S (-P_I \bar{n}) \, d\mathcal{S} \tag{12.3.29}$$

The sense of $\bar{n}$ is outward from the region $\mathcal{V}$ enclosed by $\mathcal{S}$, so the divergence theorem converts this to

$$\bar{F}_I = -\iiint\limits_{\mathcal{V}} (\nabla P_I) \, d\mathcal{V} \tag{12.3.30}$$

The scale over which P_I varies is an acoustic wavelength, which is restricted here to being much larger than the size of the body. Therefore, we may consider ∇P_I to be a constant in this expression. For a plane wave, $\nabla P_I = -ik\hat{P}_I\bar{e}_I \exp(-ik\bar{e}_I \cdot \bar{x})$. We evaluate this gradient at the origin and invoke the smallness of $k\bar{x}$, so that

$$\bar{F}_I = ik\mathcal{V}\hat{P}_I\bar{e}_I \Longleftrightarrow \{F_I\} = ik\mathcal{V}\hat{P}_I \{\bar{e}_I\} \tag{12.3.31}$$

Substitution of $\{F\}$ and $\{F_I\}$ into Newton's law gives

$$Mi\omega \{V\} = ik [W] \left\{ \hat{P}_I \{e_I\} - \rho_0 c \{V\} \right\} + ik\mathcal{V}\hat{P}_I \{\bar{e}_I\} \tag{12.3.32}$$

A rearrangement of terms yields an expression that may be solved for $\{V\}$, specifically

$$\boxed{ \left[\frac{M}{\rho_0} [I] + [W] \right] \{V\} = \frac{\hat{P}_I}{\rho_0 c} [\mathcal{V} [I] + [W]] \{\bar{e}_I\} } \tag{12.3.33}$$

This expression confirms that it was correct to say in Chap. 7 that $\rho_0 \lfloor W \rfloor$ is an added mass.

The closure of the analysis uses the value of $\{V\}$ found from the preceding to form the dipole moment in Eq. (12.3.26). Then, the expressions for the monopole amplitude and dipole moment are substituted into the mulipole expansion, Eq. (7.4.23). Letting $G(\bar{x}_0, \bar{x}_C)$ be the free-space Greens function ultimately leads to

$$\boxed{ \begin{aligned} P_s(\bar{x}_0) = -\frac{k^2}{4\pi} &\left[\mathcal{V}\left(\hat{P}_I - \rho_0 c \,|V| \right) - \{e_r\}^T [\mathcal{V}[I] \right. \\ &\left. + [W]\,] \left\{ \hat{P}_I \{e_I\} - \rho_0 c \{V\} \right\} \left(1 - \frac{i}{kr} \right) \right] \frac{e^{-ikr}}{r} \end{aligned} } \tag{12.3.34}$$

As is true for scattering from a stationary rigid body, a quantitative evaluation of this expression requires that we know $[W]$. The values for a sphere and thin circular disk are provided in Eq. (7.4.38) and at the concluding paragraph of the preceding section.

The special case of a neutrally buoyant object, such as a submarine, is interesting. The mass is the same as that of the displaced fluid, so $M/\rho_0 = \mathcal{V}$. According to Eq. (12.3.33), this situation leads to $\rho_0 c\bar{V} = \hat{P}_I\bar{e}_I$, from which it follows that $P_s = 0$. This situation occurs because the neutral buoyant body moves in unison with the fluid, so it does not impede the propagation of the incident wave. This phenomenon assumes that the acoustic wavelength is much larger than any dimension of the object.

It also ignores the fact that no object is truly rigid. The role of structural flexibility is examined in the last section of this chapter.

EXAMPLE 12.3 A thin aluminum disk is suspended in a large vat of water in a manner that allows it to be rotated about its vertical diameter without otherwise restraining the disk. A plane wave traveling horizontally is incident on the disk, with the angle between $-\bar{e}_{\mathrm{I}}$ and the normal to the disk being ψ_{I}. The mass of the disk is 1.7 g, and its radius is 20 mm. The frequency of the incident wave is 100 Hz, and its amplitude is 10 kPa. (a) Determine the magnitude and direction of the disk's complex velocity relative to the z direction as a function of ψ_{I}. (b) Determine the amplitude of the backscattered pressure as a function of ψ_{I} and compare the result to backscatter for the case of a stationary disk.

Significance

The focus is on the tensorial nature of the added mass effect and how it affects the backscattered pressure. The quantitative results illuminate the general effect of allowing an object to move.

Solution

The coordinate system and geometry for this system are shown in Fig. 1. The given parameters correspond to $ka = 8.491 \left(10^{-3}\right)$, which certainly is sufficiently small to warrant application of Rayleigh scattering theory. We begin by assembling the basic quantities. According to Eq. (7.4.38), the virtual mass tensor is a single element, $\Lambda_{3,3} = (8/3)\, a^3$. It is not surprising that only the 3,3 element is nonzero, because incidence of a wave traveling parallel to the surface of the disk will not lead to significant scattering. Another consideration is that $\mathcal{V}$ is not given. We could take it be zero, based on thinness of the disk, but the monopole contribution is dominant at orientations for which the dipole contribution is very small. Therefore, we determine it from the mass, which gives $\mathcal{V} = M/\rho_{\mathrm{al}}$.

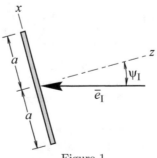

Figure 1

Axisymmetry allows us to situate $\bar{e}_{\mathrm{I}}$ in the xz-plane, so

$$\{e_{\mathrm{I}}\} = [\sin\psi_{\mathrm{I}} \quad 0 \quad -\cos\psi_{\mathrm{I}}]^{\mathrm{T}}$$

All terms in Eq. (12.3.33) have been characterized, so we may solve that equation for the complex velocity amplitude. Doing so leads to

$$
\{V\} = \frac{\hat{P}_I}{\rho_0 c}
\left[
\begin{bmatrix}
\dfrac{M}{\rho_0} & 0 & 0 \\[2mm]
0 & \dfrac{M}{\rho_0} & 0 \\[2mm]
0 & 0 & \dfrac{M}{\rho_0} + \dfrac{8}{3}a^3
\end{bmatrix}
\right]^{-1}
\begin{bmatrix}
\dfrac{M}{\rho_{al}} & 0 & 0 \\[2mm]
0 & \dfrac{M}{\rho_{al}} & 0 \\[2mm]
0 & 0 & \dfrac{M}{\rho_{al}} + \dfrac{8}{3}a^3
\end{bmatrix}
\begin{Bmatrix}
\sin\psi_I \\[2mm]
0 \\[2mm]
-\cos\psi_I
\end{Bmatrix}
\tag{1}
$$

The matrix being inverted is diagonal, so we find that

$$
V_x = \left(\frac{\rho_0}{\rho_{al}}\right)\frac{\hat{P}_I}{\rho_0 c}\sin\psi_I, \quad V_y = 0
$$

$$
V_z = -\left(\frac{M + (8/3)\,\rho_{al}a^3}{M + (8/3)\,\rho_0 a^3}\right)\left(\frac{\rho_0}{\rho_{al}}\right)\frac{\hat{P}_I}{\rho_0 c}\cos\psi_I
\tag{2}
$$

Both complex velocity components are real, so their relative phase is zero. This corresponds to a temporal velocity vector that is oriented in a constant direction. It lies in the xz-plane, so we may describe $\bar{V}$ in terms of its magnitude $|V|$ and angle θ relative to the z-axis as $V_x = |\bar{V}|\sin\theta$, $V_z = |\bar{V}|\cos\theta$. (Both components are required to assure that θ is assigned to the proper quadrant.) The velocity components, as well as $|V|$ and θ, are graphed in Fig. 2. Perhaps the most unexpected feature is the fact that $\bar{V}$ is parallel to $\bar{e}_z$ when $\psi_I = 90°$. Although the disk is thin, it moves at this incidence because its volume is finite.

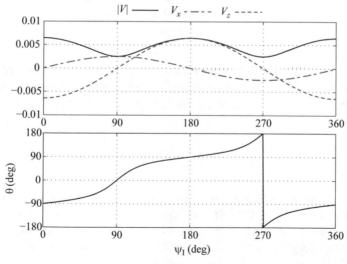

Figure 2

The complex velocity amplitude is used to compute $(r/a)\,|P_I|\,/(\rho_0 c)$ according to Eq. (12.3.34). We are interested in the farfield backscatter, so we set $\bar{e}_r = -\bar{e}_I$ and drop the term that is the reciprocal of kr. Thus, we have

$$
\left(\frac{r}{a}\right)\frac{|P_s(\bar{x}_0)|}{\rho_0 c} = -\frac{(ka)^2}{4\pi}\left(\frac{M}{\rho_{al}a^2}\right)\left(\hat{P}_I - \rho_0 c\,|V|\right)
$$

$$
+\frac{(ka)^2}{4\pi}\left\{\begin{array}{c}\sin\psi_I\\0\\-\cos\psi_I\end{array}\right\}^{\mathrm{T}}\begin{bmatrix}\dfrac{M}{\rho_{al}}&0&0\\[2mm]0&\dfrac{M}{\rho_{al}}&0\\[2mm]0&0&\dfrac{M}{\rho_{al}}+\dfrac{8}{3}a^3\end{bmatrix} \tag{3}
$$

$$
\times\left\{\hat{P}_I\left\{\begin{array}{c}-\sin\psi_I\\0\\\cos\psi_I\end{array}\right\} - \rho_0 c\left\{\begin{array}{c}V_x\\0\\V_z\end{array}\right\}\right\}
$$

Figure 3 is constructed by substituting $\{V\}$ at each ψ_I into Eq. (3). The value of $|P_s|$ for the case of a stationary disk is obtained by setting $\{V\} = \{0\}$.

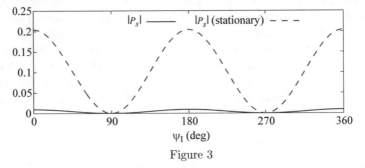

Figure 3

It can be seen that the backscatter is much greater if the disk is immobilized. An explanation of this behavior may be found in an evaluation of the ratio of V_x and V_z to the respective components of $\bar{V}_I$, which are $V_I \sin\psi_I$ and $-V_I \cos\psi_I$. Equation (2) indicate that

$$
\frac{V_x}{\bar{V}_I\cdot\bar{e}_x} = \frac{\rho_0}{\rho_{al}} = 0.370,\quad \frac{V_z}{\bar{V}_I\cdot\bar{e}_z} = \left(\frac{M+(8/3)\,\rho_{al}a^3}{M+(8/3)\,\rho_0 a^3}\right)\left(\frac{\rho_0}{\rho_{al}}\right) = 0.954 \tag{12.3.35}
$$

Movement of the disk in the $\bar{e}_x$ direction leads to little scattering, so the mismatch between V_x and $\bar{V}_I \cdot \bar{e}_x$ has little effect on the scattering. In contrast, movement of the disk in the $\bar{e}_z$ direction is important. This velocity component is very nearly equal to $\bar{V}_I \cdot \bar{e}_z$. If they truly were equal, there would be no scattering. This phenomenon is comparable to ocean waves. If they are incident on a stationary object, there is a large splash. In contrast, a floating object induces little ripple. Another aspect to note is that the scattered pressure is essentially zero at $\psi_I = 90°$ and $270°$. At these angles, the contribution associated with $[W]$ is zero. The remaining terms are negligible because $M/\rho_{al} = \mathcal{V}$ is much less than a^2.

12.4 Measurements and Metrics

Acoustic scattering is the core phenomenon for a number of applications, ranging from biomedical ultrasonics to fishing to naval sonar. Thus, it is undoubtedly true that scattered signals are measured much more frequently than they are evaluated analytically. This bias is even stronger if we include marine mammals and bats, who have remarkable capabilities to "see" with sound.

Sonar has two implementations. Passive sonar is a process of listening, in that it detects sound radiated by a body. If the object that is to be located is stationary and does not vibrate, it will emit no sound and therefore cannot be located. Active sonar does not suffer from this shortcoming. It is used recreationally and commercially to locate fish. Typically, this entails broadcasting a signal from the boat, then measuring the signal returned to the boat. Suitable processing of the returned signal can determine whether it is scattered by fish, and if so, where the fish are located. This application is a *monostatic scattering measurement*. The "mono" prefix is used because signal generation and measurement are at a single location. Many of the sonar concepts follow those for radar, wherein "static" refers to the apparatus being stationary. Monostatic scattering is the primary concept for biomedical ultrasound, wherein the device that generates the probing signal also detects the signal scattered by a heterogeneous region, such as a tumor.

Naval sonar systems applications rely on both monostatic and *bistatic scattering measurements*. In a bistatic measurement, the incident signal is generated at one location and the scattered signal is detected at another. This can be done from an airplane or helicopter by dropping sonobuoys, or it can be done from ships and submarines by using a towed array.

Ultrasonic imaging for medical diagnostics is the acoustical analog of our visual sense, in that they rely on contrast. The field scattered by a heterogeneity is scanned spatially. Strong returns correspond to greater impedance mismatch, so they are displayed as lighter regions. Weaker returns correspond to little or no heterogeneity, and therefore are displayed as dark regions. The field properties of the scattered field, such as its multipole resolution, typically is not an issue in biomedical applications.

The underlying properties of the scattered field are of vital importance to sonar applications. The general objective there is to deduce the nature of the scattering body from measurements. We shall not delve into how the measured signal is used to identify the fundamental nature of the scattering body, which is a process of *target identification*. However, if we consider the result for the field scattered by a rigid body, we can gain a hint at the possibility of using scattering data to characterize the scatterer. The virtual mass matrix $\rho_o\,[W]$ in Eq. (12.3.19) is symmetric, so it contains six unknown elements. Another unknown parameter is the volume of the scattering body. Thus, if we have several measurements, corresponding to angles of incidence and/or different measurement locations, it is conceivable that we could infer these unknown parameters.

Obviously, this notion relies on several assumptions. Several sources and/or receivers are required to obtain bistatic measurements. It is necessary to determine the location of the body. Range information can be determined by using a pulse, rather than a continuous wave. The simplest measurement uses the time τ for a signal to return to the source location in a monostatic measurement. Because the travel time to the body is $\tau/2$, the range will be $c\tau/2$. Orientation of the line from the source/receiver location to the body can be determined by rotating a highly directive source until the scattered signal is largest, or by using a phased array. Another assumption is that the orientation of the scattering body does not change between measurements. This can be addressed by taking the measurements over a short interval. The most fundamental assumption buried in the notion of identifying V and $[\Lambda]$ is that the analysis is valid. If the body is flexible, then the scattered field will be much more complicated than a simple superposition of monopole and dipole fields. Indeed, in that case, the nature of the field is likely to be highly dependent on the frequency. An efficient way of performing scattering measurements over a frequency range is to alter the pulse from a pure tone burst. The shape of the envelope can be adjusted (amplitude modulation), and the frequency can be varied (frequency modulation) within the pulse's interval. Then, FFT processing of the scattered signals, which also will be pulses, will yield frequency spectra that can be converted to transfer functions.

Unless the duration of the pulse is very long relative to the range divided by the sound speed, the scattered signal will be measured distinctly from the incident. The logical question is what to do with it? Other than travel time and location, the most fundamental metric is based on a concept developed for radar. It originates from a hypothetical situation. Suppose a planar harmonic wave is normally incident on a rigid stationary disk. Let us further suppose that the planar wave is a confined beam whose cross section exactly matches the disk. This is the situation depicted in Fig. 12.3. In the high-frequency limit of ray acoustics, the scattered signal P_s is the regular reflection of the incident signal, corresponding to a reflection coefficient of one. In that case, the power transported in the reflected signal is the same as that of the incident signal. The magnitude of the time-averaged intensity in the incident wave is $I_I = |P_I|^2 / (2\rho_0 c)$, so the power in the scattered signal is $(\mathcal{P}_s)_{av} = I_I \mathcal{A}$, where $\mathcal{A}$ is the cross-sectional area of the side of the disk that is irradiated by the incident signal. Division of the time-averaged power in the reflected wave by the intensity of the incident wave yields a measure of the cross-sectional area that blocks the incident wave.

Fig. 12.3 Model of scattering from a disk used to define the scattering cross section. The picture generally is not descriptive of the actual scattering process

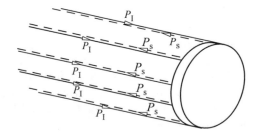

To extend this idea to the actual field scattered by an arbitrary body, the time-averaged power transported in the scattered signal is computed by integrating over a sphere S_0 that surrounds the scatterer. The spherical coordinates of a point on the sphere's surface are r_0, ψ_0, θ_0. The radius r_0 of this sphere is taken to be much greater than $2\pi/k$, as well as the largest dimension of the body. Then, the radial velocity on the sphere's surface is $V_r = P_s(r_0, \psi_0, \theta_0)/(\rho_0 c)$. The general definition of the *scattering cross section* is that it is the time-averaged power transported by the scattered signal divided by the intensity of the incident wave,

$$\sigma_s = \frac{(P_s)_{av}}{I_I} = \iint\limits_{S_0} \frac{I_s(r_0, \psi_0, \theta_0)}{I_I} dS_0, \quad I_s = \frac{|P_s|^2}{2\rho_0 c} \qquad (12.4.1)$$

The intensity in the farfield is proportional to $1/r_0^2$, so setting $dS_0 = r_0^2 (\sin \psi_0) d\theta_0 d\psi_0$ will yield a value for σ_s that is independent of r_0.

In the hypothetical model of scattering from a disk, $\sigma_s = A$. This leads to interpretation of σ_s as the apparent area of the scattering object that blocks the incident wave. Because the field scattered by an arbitrary body depends on the orientation of $\bar{e}_I$ relative to the body, the scattering cross section depends on that orientation. The value of σ_s might exceed the area of the body's projection on to the plane normal to the $\bar{e}_I$, or it might be much smaller.

Consider the case of scattering from a stationary rigid body. Equation (12.3.19) describes the scattered field as the sum of a monopole whose strength is proportional to the displaced mass $\rho_0 V$, and a dipole whose moment is proportional to $[V[I] + [W]]\{e_I\}$. It is useful to identify the direction of the dipole moment corresponding to a known $[W]$ because scattering associated with the dipole will be strongest in that and the opposite direction. Toward that objective, we write the vectorial part of the dipole moment in terms of its magnitude Φ and a unit vector $\{e_D\}$ that is the direction of the vector. In other words, we set $\Phi\{e_D\} = [V[I] + [W]]\{e_I\}$. It follows that

$$\Phi = |[V[I] + [W]]\{e_I\}|, \quad \{e_D\} = \frac{[V[I] + [W]]\{e_I\}}{\Phi} \qquad (12.4.2)$$

This transformation yields a description of $\{e_D\}$ in terms of components relative to the xyz coordinate system associated with the known components of $[W]$. We also represent the direction $\{e_r\}$ from the origin to a field point on the surface of the surrounding sphere S_0 in terms of components relative to the same coordinate system. Doing so allows us to evaluate product $\{e_r\}^T [[V[I] + [W]]]\{e_I\}$, which now is $\Phi\{e_r\}^T\{e_D\}$. In other words, the directivity of the dipole is the dot product of $\bar{e}_r$ and $\bar{e}_D$, and the strength of the dipole is proportional to Φ. Let ψ_0 be the angle between the vectors, that is,

$$\psi_0 = \cos^{-1}\left(\{e_r\}^T\{e_D\}\right) \qquad (12.4.3)$$

This corresponding representation of the farfield pressure scattered by a stationary rigid body is

$$P_s(\bar{x}_0) = -\frac{k^2 \hat{P}_I}{4\pi} [\mathcal{V} - \Phi \cos \psi_0] \frac{e^{-ikr_0}}{r_0} \tag{12.4.4}$$

Integration of this expression over the surface of the sphere yields the radiated power. We define the polar direction for a set of spherical coordinates to be $\bar{e}_D$. The field is axisymmetric with respect to an axis aligned with $\{e_D\}$, so the differential area element is set to $2\pi r_0^2 \sin \psi_0 d\psi_0$. Division of $\mathcal{P}_s$ by $|I_I| = \left|\hat{P}_I\right|^2 / (2\rho_0 c)$ yields the scattering cross section. The result is

$$\boxed{\sigma_s = \frac{k^4}{4\pi} \left(\mathcal{V}^2 + \frac{1}{3} |[\Phi]\{e_I\}|^2\right)} \tag{12.4.5}$$

It should be noted that the integration over $\mathcal{S}_0$ has removed any dependence on the orientation of the dipole moment, so there is no need to actually evaluate $\{e_D\}$. Specific results are

$$\text{Sphere: } \sigma_s = \frac{7}{9} (\pi a^2) (ka)^4$$

$$\text{Thin circular disk: } \sigma_s = \frac{16}{27\pi^2} (\pi a^2) (ka)^4 (\cos \psi_I)^2 \tag{12.4.6}$$

where ψ_I is the angle between the normal to the disk and the incident wave's direction.

In the case of a sphere, the factor πa^2 is the physical area that blocks the incident wave at any angle. For a thin disk, the maximum blocking area is πa^2, which corresponds to normal incidence. At arbitrary incidence, the physical blocking area is $\pi a^2 \cos \psi_I$, so the above expression indicates that the decrease of σ when ψ_I is increased is due to more than the decrease of the blocking area. The Rayleigh scattering limit describes cases where $ka \ll 1$, so the dependence of σ_s for each shape on $(ka)^4$ tells us that the bodies appear to be much smaller than their cross-sectional area.

The scattering cross section represents a spatial average of what is seen at various field points. A different metric is used to convey what is seen at a specific location. This metric is defined in terms of a solid angle, which is a geometric parameter we encountered in Sect. 7.4.1. A solid angle will arise in the next section, so it is helpful to review the concept. Consider the normal $\bar{n}(\bar{x}_s)$ to the surface. The intersection of the surface and any plane that contains this normal is a curve. We are interested in the differential arc of this curve that is centered at $\bar{x}_s$. The center of curvature of this arc is the intersection of $\bar{n}(\bar{x}_s)$ and any other normal to the arc, such as that at either end. The distance from the center of curvature to the surface is the radius of curvature of the arc. The plane containing $\bar{n}(\bar{x}_s)$ is defined by its rotation about $\bar{n}(\bar{x}_s)$. At some rotation angle, the arc has the largest radius of curvature. A further rotation by 90°

forms an arc with the smallest radius. These planes and arcs are depicted in Fig. 12.4. The radii r_1 and r_2 are the *principal radii of curvature*. (For a sphere $r_1 = r_2 = a$, whereas $r_1 = a$ and r_2 is infinite for a cylinder.) The arclengths are $d\ell_1 = r_1 d\theta_1$ and $d\ell_2 = r_1 d\theta_2$ where $d\theta_n$ is the angle between the normals to the surface at the ends of $d\ell_n$. The area of the patch that is formed is $dS = d\ell_1 d\ell_2$. The definition of a solid angle is

$$d\Gamma = \frac{dS}{r_1 r_2} \qquad (12.4.7)$$

where Γ is the symbol we shall employ for the solid angle. Thus, the solid angle subtended by the patch is $d\Gamma = d\theta_1 d\theta_2$.

Fig. 12.4 Surface patch formed from arcs whose radii of curvature are the principal values at a point

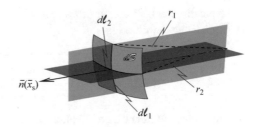

The scattering cross section is defined in Eq. (12.4.1) as an integral over a sphere. Both principal radii of that surface are r_0, so the differential element of area may be written as $dS_0 = r_0^2 d\Gamma$. Concurrently, any integral may be represented as the antiderivative of its integrand. Thus, the definition of σ_s may be written in two ways, as

$$\sigma_s = \iint_{S_0} \frac{I_s(r_0, \psi_0, \theta_0)}{I_I} \left(r_0^2 d\Gamma\right) = \iint_{S_0} \frac{d\sigma_s}{d\Gamma} d\Gamma \qquad (12.4.8)$$

The quantity $d\sigma_s/d\Gamma$ is the *differential scattering cross section*. Matching the above integrands gives

$$\frac{d\sigma_s}{d\Gamma} = \frac{I_s(r_0, \psi_0, \theta_0)}{I_I} r_0^2 = \frac{r_0^2 |P_s(r_0, \psi_0, \theta_0)|^2}{|P_I|^2} \qquad (12.4.9)$$

In other words, $d\sigma_s/d\Gamma$ is proportional to the directivity of the farfield intensity. It is implicit to this expression that the scattering cross section is defined for a specific orientation of $\bar{e}_I$ relative to the body, so that ψ_0 and θ_0 are defined relative to a body-fixed direction. If we define that direction to be $\bar{e}_r$, then backscatter corresponds to $\bar{e}_r = -\bar{e}_I$ ($\psi_0 = \pi$). A measurement at an arbitrary field point gives the differential scattering cross section in a bistatic measurement. The special case where $\bar{e}_r = \bar{e}_I$ ($\psi_0 = \pi$) is said to be *forward scatter*.

A quantity that causes confusion for some is the *backscatter cross section*, which is denoted as σ_{back}. Despite the name, it is not derived from the actual scattering cross section. Rather, it is the value of the scattering cross section that would be obtained if the scattering was isotropic at the backscatter value. Accordingly, if $d\sigma_s/d\Gamma$ were constant at its value for $\bar{e}_r = -\bar{e}_I$, integrating it over S_0 would give a 4π factor, so that

$$\sigma_{\text{back}} = 4\pi \left(\frac{d\sigma}{d\Gamma} \right)_{\bar{e}_r = -\bar{e}_k} = 4\pi r_0^2 \frac{I_s(r_0, \pi, 0)}{I_I} = 4\pi r_0^2 \left| \frac{P_s(r_0, \pi, 0)}{P_I} \right|^2 \quad (12.4.10)$$

Like the other metrics, σ_{back} depends on the orientation of the incident signal relative to the body. Because $|P_s|$ is proportional to $1/r_0$, the value of σ_{back} is independent of r_0.

Typically, the backscatter cross section is described logarithmically as the *target strength.* The definition is

$$TS = 10 \log \left(\frac{\sigma_{\text{back}}}{4\pi r_{\text{ref}}^2} \right) \quad (12.4.11)$$

where r_{ref} is a reference distance; the standard value is $r_{\text{ref}} = 1$ m. An expression for the target strength in terms of measured pressure values results from substitution of the definition of σ_{back} and Eq. (12.4.9) into the above. Let $\mathcal{L}_{\text{back}}$ be the sound pressure level at the source/receiver location in a monostatic measurement, and let $\mathcal{L}_I$ be the sound pressure level in the incident wave at that location. Then

$$TS = \mathcal{L}_{\text{back}} - \mathcal{L}_I + 20 \log \left(\frac{r_0}{r_{\text{ref}}} \right) \quad (12.4.12)$$

Measurement of backscatter calls for a monostatic setup, in which the source and receiver are collocated, and possibly the same device. The usage of the above to evaluate target strength requires that the incident signal not be included in $\mathcal{L}_{\text{back}}$. This condition will be attained if the incident signal is a pulse that terminates before the scattered signal returns to the receiver.

12.5 High-frequency Approximation

Low-frequency approximations of scattering phenomena are associated with a great acoustician, Rayleigh. It is appropriate that another pioneer, Kirchhoff, should be associated with high-frequency scattering. Kirchhoff approximations are essentially applications of ray acoustics to the scattering problem. We will restrict our attention to the usual case wherein the incident wave is planar.

We begin with an analysis of backscatter from a rigid, stationary body. The first task is to locate the point on the surface at which the normal is opposite $\bar{e}_I$. This location depends on the shape of the body and the orientation of $\bar{e}_I$ relative to the body. This point is labeled as C in Fig. 12.5, which is a side view in the first plane of principal curvature. The intensity of the backscattered wave is found by analyzing the reflection of the incident ray tube that arrives at a differential element dS of surface area that is centered at point C. The edges of dS are the arcs $d\ell_1 = r_1 d\theta_1$ depicted in Fig. 12.5 and $d\ell_2 = r_2 d\theta_2$ in the second plane of principal curvature. The angles $d\theta_j$ are the angles subtended by each relative to the respective centers of curvature.

Fig. 12.5 Description of backscatter as a process of specular reflection

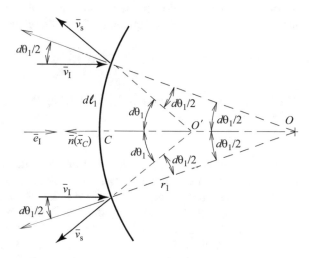

The intensity of the incident plane wave is $\bar{I}_I = \bar{e}_I |P_I|^2 / (2\rho_0 c)$ everywhere, so the power transported by this ray tube is

$$dP_I = I_I dS = I_I r_1 r_2 d\theta_1 d\theta_2 \qquad (12.5.1)$$

In the geometrical acoustics approximation , the reflection of each ray is like it would be if the surface was an infinite plane with normal $\bar{n}$. The angle of incidence at the extremities of the ray tube in Fig. 12.5 is $d\theta_1/2$. The angle of reflection equals the angle of incidence, so the reflected rays are confined to a tube that is bounded by rays that are at angle $d\theta_1$ relative to $-\bar{e}_I$. The consequence is that the reflected rays in this bundle represent a spherical wave that radiates from a virtual source at point O'. (There is a single center for all rays because the angle of incidence for all is infinitesimal. If one were to consider incident rays at large distances from the backscatter centerline, the reflected rays would appear to radiate from centers distributed along the centerline behind point C. This is a manifestation of spherical aberration, which we first encountered in Example 11.1.)

The reflected rays in the plane of Fig. 12.5 are centered at point O'. At a radial distance r_0, the arc subtended between the extremities of $d\ell_1$ is $r_0\,(2d\theta_1)$. A similar description applies to the other principal plane of curvature, so the arc subtended by the rays in that plane is $r_0\,(2d\theta_2)$. Thus, the reflected rays at distance r_0 cross a differential surface patch whose area is $d\mathcal{S}_0 = 4r_0^2 d\theta_1 d\theta_2$. All rays contained within the scattered ray tube have the same amplitude, so the intensity along $d\mathcal{S}_0$ is constant at I_s. The power transported across this patch in the scattered wave is

$$dP_s = I_s d\mathcal{S}_0 = I_s \left(4r_0^2 d\theta_1 d\theta_2\right) \tag{12.5.2}$$

The surface is rigid, so the reflection coefficient for each incident ray is one. It follows that the power flow in the tube of scattered rays must equal the power flow in the tube of incident rays. Equating dP_s to dP_I yields

$$I_s = I_I \frac{r_1 r_2}{4r_0^2} \tag{12.5.3}$$

The backscatter cross section is found by using this intensity to form Eq. (12.4.9), which leads to

$$\sigma_{\text{back}} = \pi r_1 r_2 \tag{12.5.4}$$

A review of the derivation will reveal that it assumes that there is a single point C on the surface at which the normal is exactly opposite to $\bar{e}_I$. If there are multiple points that meet this criterion, the conservation of energy argument may still be applied, provided that it accounts for the contribution from each scattering site. Each contribution is a tube centered on direction $-\bar{e}_I$. Because the scattering sites are separated by a finite distance, the tubes would not intersect. Hence, the intensity of each tube may be added as scalars to find the backscatter cross section. It also is possible that there is no point at which there is normal incidence. An example of this condition is a thin disk whose normal is not aligned with $\bar{e}_I$. Another assumption is implicit, specifically, that the surface is convex. The situation where the surface is concave at the normal incidence point can be analyzed by merging the ray tube approach here with the analysis of focusing carried out in Example 11.1.

The preceding analysis of backscatter may be generalized to predict bistatic scattering. The first step in doing so is to locate point C at which the normal to the surface is such that the angle of reflection to a designated field point equals the angle of incidence. From there, the derivation invokes concepts from differential geometry. The interested reader is referred to Pierce's presentation.[3]

It is much easier to evaluate the scattering cross section σ_s for a rigid body in the high-frequency limit. Figure 12.6 depicts all rays that are incident on the illuminated side of the body. They are confined to a (general) cylinder. The projection of the body's shape onto a plane perpendicular to $\bar{e}_I$ defines the cross section $\mathcal{S}_I$ of this bundle.

---

[3] A.D. Pierce, *Acoustics*, Mc-Graw-Hill (1981), ASA reprint, (1989), Sect. 8-8.

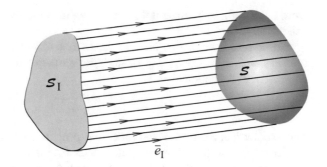

Fig. 12.6 Bundle of rays of a plane wave that is incident on body S from a cylinder whose cross section is S_I

The power flow in this bundle is $I_I S_I$. Although the incident rays are reflected over a wide range of directions, their reflection is lossless. Thus, the power scattered on the illuminated side of the body equals the power contained in the incident ray tube. On the shadow side, the total pressure is essentially zero, but the total pressure is the sum of incident and scattered contributions. Thus, the power in the scattered field on that side equals the power flow in the bundle of incident rays that appear to have passed through the body. This is the same ray tube as that which is incident on the illuminated side, so the power in the scattered field on the shadow side equals the power flow in the incident side, which is $I_I S_I$. The total power flow in the scattered field, therefore, is twice this value. Hence, we find from Eq. (12.4.9) that

$$\boxed{\sigma_s = 2 S_I} \tag{12.5.5}$$

It is interesting to consider the geometrical acoustics results for a sphere and a thin circular disk. Both principal radii for a sphere equal the radius a, and the perpendicular area is $S_I = \pi a^2$. If the angle between the normal to the disk's surface and the incident direction is χ, the cross section blocked by the disk is an ellipse whose semi-major radii are a and $a \cos \chi$. Therefore, the incident ray tube area is $S_I = \pi a^2 \cos \chi$. The principal radii of curvature for the disk are infinite because the surface is flat. It follows that

$$\text{Sphere: } \sigma_s = 2\pi a^2, \quad \sigma_{\text{back}} = \pi a^2$$
$$\text{Disk: } \sigma_s = 2\pi a^2 \cos \chi, \quad \sigma_{\text{back}} = \infty \tag{12.5.6}$$

The infinite value might seem surprising, but its meaning becomes obvious when we recall the general interpretation of a scattering cross section as the area that blocks the incident wave. Consider the reflection from an infinite plane surface. An incident wave at normal incidence is reflected backward, so that $\bar{I}_s = -\bar{I}_I$ everywhere. To evaluate the scattered power in this case, we could define S_0 to be a hemisphere of radius r_0 centered on the surface, then let $r_0 \to \infty$. The result would be an infinite value of $\mathcal{P}_s$ because $|\bar{I}_s|$ is independent of r_0. Thus, the scattering cross section σ of an infinite wall is infinite. The backscatter cross section takes the backscatter differential cross section to be representative of the scatterer's properties in all directions. In other words, it is infinite because it does not recognize the fact that the surface has finite extent.

The preceding result may be extended to treat incidence on a surface that is locally reacting, with impedance $\rho_0 c \zeta$. Backscatter corresponds to normal incidence, so the coefficient of reflection is

$$R = \frac{\zeta - 1}{\zeta + 1} \tag{12.5.7}$$

The geometrical features of the reflection process are unaltered. Thus, the tube of scattered rays has the same extent as it does for the case where the surface is rigid. From the analysis in Sect. 5.2.2, we know that the power flow in the reflected ray tube as it leaves the surface is $d\mathcal{P}_s = |R|^2 d\mathcal{P}_I$. The time-averaged intensity in the tube when it crosses the surrounding sphere still is $I_s = d\mathcal{P}_s / d\mathcal{S}_0$. The result is that the previous expression for the farfield scattered intensity becomes

$$I_s = I_I |R|^2 \frac{r_1 r_2}{4 r_0^2} \tag{12.5.8}$$

The corresponding expression for the backscatter cross section is

$$\boxed{\sigma_{\text{back}} = \pi r_1 r_2 |R|^2} \tag{12.5.9}$$

We know that $|R| < 1$ for any passive nonideal material, which means that rigid and pressure-release conditions lead to the largest backscatter at high frequencies.

The geometrical acoustics approximation may also be invoked to analyze bistatic scattering. The general approach entails formulating the KHIT with the surface response described as though the surface is locally planar. Thus, the reflected signal constitutes the scattered field on the surface. The reflection coefficient for the ray that is incident at surface point $\bar{x}_s$ is

$$R(\bar{x}_s) = \frac{\zeta \bar{e}_I \cdot \bar{n}(\bar{x}_s) - 1}{\zeta \bar{e}_I \cdot \bar{n}(\bar{x}_s) + 1} \tag{12.5.10}$$

In the geometrical acoustics limit, rays are incident on the illuminated side only. Correspondingly, the pressure and the particle velocity are taken to be zero on the shadowed side. This region is identifiable by $\bar{e}_I$ being outward from the surfaces. Thus, the total pressure and normal velocity on the surface are

$$\bar{e}_I \cdot \bar{n}(\bar{x}_s) < 0 \implies \begin{cases} P(\bar{x}_s) = [1 + R(\bar{x}_s)] P_I(\bar{x}_s) \\ \bar{V}_s(\bar{x}_s) \cdot \bar{n}(\bar{x}_s) = \frac{1}{\rho_0 c} [-1 + R(\bar{x}_s)] P_I(\bar{x}_s) \bar{e}_I \cdot \bar{n}(\bar{x}_s) \end{cases}$$
$$\bar{e}_I \cdot \bar{n}(\bar{x}_s) > 0 \implies P(\bar{x}_s) = 0 \ \& \ \bar{V}_s(\bar{x}_s) \cdot \bar{n}(\bar{x}_s) = 0$$
$$\tag{12.5.11}$$

To implement Eq. (12.1.5), we could define $\bar{e}_z$ to align with $\bar{e}_1$, and place the origin at the centroid of the body. Then, if we set the phase of the incident plane wave to zero at the origin, we would have

$$P_I(\bar{x}_s) = \hat{P}_I e^{-ik\bar{x}_s \cdot \bar{e}_z} \qquad (12.5.12)$$

Substitution of these representations into the KHIT would define all of the terms in the integrand, but that does not mean we can directly proceed to evaluating the integral numerically. One difficulty that awaits is the necessity to formulate a way of describing $\bar{x}_s$ and $\bar{n}(\bar{x}_s)$ in a way that fits the selected integration scheme. This is a standard issue for any formulation of KHIT. The more difficult feature stems from the fact that this is a high-frequency theory. Thus, $k\bar{x}_s \cdot \bar{e}_3$ will be much greater than unity over most of the surface. This means that the integrand of Eq. (12.1.5) would be a rapidly oscillating function of position. It is difficult to obtain a convergent result in such situations without expending a great deal of computational effort. Asymptotic methods, such as the method of stationary phase,[4] might be useful for such efforts.

12.6 Scattering from Spheres

We have analyzed scattering from a variety of bodies in the low-frequency limit of the Rayleigh approximation and the high-frequency limit of the Kirchhoff approximation. However, we have no idea of the actual frequency range for which either approximation is valid, nor of how accurate either is in its range of validity. Answering these questions for an arbitrary scattering object would require considerable computational or experimental work. Fortunately, the spherical shape is amenable to analytical investigations. As was noted at the outset, a basic concept for an analysis of scattering is to convert the problem to an analogous radiation problem. Radiation from spheres was the subject of Sect. 7.2.

12.6.1 Stationary Spherical Scatterer

For scattering from a sphere, there is no special orientation of the polar axis. Therefore, we align the z-axis with the propagation direction of the incident plane wave, $\bar{e}_z = \bar{e}_1$. The origin of the coordinate system is situated at the center of the sphere, and the phase of the incident wave is defined to be zero at that location. Thus, the incident signal is $\hat{P}_I \exp(-ikz)$. In spherical coordinates, the axial distance is $z = r \cos\psi$.

---

[4] C.M. Bender & S.A. Orszag, *Advanced Mathematical Methods for Scientists and Engineers* (1978).

An identity known as an addition theorem[5] converts the exponential to a spherical harmonic series,

$$P_{\mathrm{I}} = \hat{P}_{\mathrm{I}} e^{-ikz} = \hat{P}_{\mathrm{I}} \sum_{m=0}^{\infty} (2m+1)\,(-i)^m\, j_m\,(kr)\, P_m\,(\cos\psi) \tag{12.6.1}$$

Rigidity requires that the surface velocity be zero. The surface normal is the radial direction. Correspondingly, continuity of the normal velocity on the surface requires that

$$\left.\frac{\partial P_{\mathrm{s}}}{\partial r}\right|_{r=a} = -\left.\frac{\partial P_{\mathrm{I}}}{\partial r}\right|_{r=a} = -k\hat{P}_{\mathrm{I}} \sum_{m=0}^{\infty} (2m+1)\,(-i)^m\, j_m'\,(ka)\, P_m\,(\cos\psi) \tag{12.6.2}$$

A spherical harmonic series for P_{s} satisfies the Helmholtz equation and the Sommerfeld radiation condition. The incident field is independent of the azimuthal angle θ, which means that the scattered field will be axisymmetric relative to the z-axis. Therefore, the scattered pressure may be represented as a series of axisymmetric spherical harmonics, see Eq. (7.2.1),

$$P_{\mathrm{s}} = \sum_{m=0}^{\infty} B_m h_m\,(kr)\, P_m\,(\cos\psi) \tag{12.6.3}$$

Recall that the spherical Hankel functions appearing here are the second kind, which correspond to waves that propagate outward when the convention is to use $\exp(i\omega t)$. The series coefficients are determined by using this form to satisfy the surface boundary condition. Doing so leads to

$$B_m = -(2m+1)\,(-i)^m\, \frac{j_m'\,(ka)}{h_m'\,(ka)}\, \hat{P}_{\mathrm{I}} \tag{12.6.4}$$

The corresponding expression for the scattered pressure at any location is

$$P_{\mathrm{s}} = -\hat{P}_{\mathrm{I}} \sum_{m=0}^{\infty} (2m+1)\,(-i)^m\, \frac{j_m'\,(ka)}{h_m'\,(ka)}\, h_m\,(kr)\, P_m\,(\cos\psi) \tag{12.6.5}$$

The series expansion may be specialized to the farfield by applying the asymptotic expansion in Eq. (7.1.35), which is

$$h_m\,(kr) \to \frac{i^{m+1}}{kr} e^{-ikr} \tag{12.6.6}$$

---

[5]M.I. Abramowitz and I.A. Stegun, *Handbook of Mathematical Functions*, Dover, (1965), p. 440, Eq. 10.1.47.

This limiting form is valid if $kr \gg m$. This attribute might seem to be problematic because the spherical harmonic series sums over all m. Fortunately, the B_m coefficients decrease rapidly as m increases. (To verify this behavior, consider Eq. (7.1.36), which describes the behavior of the spherical functions as m increases.) Thus, for any kr, there always is a value $m = M$ for which B_M is sufficiently small to permit truncation of the series. We may use Eq. (12.6.6) to describe terms for $m \leq M$. We thereby obtain

$$
P_s = P_1 \frac{a}{r} e^{-ikr} \chi(\psi, ka), \quad \chi(\psi, ka) = -i \sum_{m=0}^{M} (2m+1) \frac{j_m'(ka)}{ka h_m'(ka)} P_m(\cos \psi)
$$

$$(12.6.7)$$

One approach to evaluate the series is to accumulate the contributions of increasing m until there is negligible change in the value of the sum. However, modern software is most efficient when the operations are vectorized, which requires that the range of the summation be set prior to the actual computation. A viable approach sets the maximum m as an integer multiple of ka. If that multiple is too large, the value of $j_m'(ka)/h_m'(ka)$ will cause an underflow; the results reported here were obtained using $\max(m) = \mathrm{ceil}(5ka)$. Vectorization is achieved by defining a column vector $\{F\}$ that consists of the values of $(2m+1) j_m'(ka)/\left(ka h_m'(ka)\right)$ for $0 \leq m \leq \max(m)$. If the Legendre function values $P_m(\cos \psi)$ are stored in a column vector $\{L\}$, then the summation reduces to $\{F\}^{\mathrm{T}}\{L\}$.

The function $\chi(\psi, ka)$ is analogous to the directivity function for radiation. From it, the differential scattering cross section Eq. (12.4.9) may be evaluated as

$$
\frac{d\sigma}{d\Gamma} = a^2 |\chi(\psi, ka)|^2
$$

$$(12.6.8)$$

The backscatter cross section is obtained from the time-averaged intensity at $\psi = 180°$, for which $P_m(-1) = (-1)^m$, so that

$$
\chi(\pi, ka) = \sum_{m=0}^{\infty} (-1)^m (2m+1) \frac{j_m'(ka)}{ka h_m'(ka)}
$$

$$(12.6.9)$$

The backscatter cross section is

$$
\sigma_{\mathrm{back}} = 4\pi a^2 |\chi(\pi, ka)|^2
$$

$$(12.6.10)$$

For comparison, the value of σ_{back} in the Rayleigh scattering limit may be found from Eq. (12.3.22) by setting $W_{3,3} = 2\pi a^3$, $\psi_1 = 0$, and $\psi_0 = -\pi$, whereas the geometrical acoustics limit is independent of ka,

$$\sigma_{back} = \frac{25}{9}\pi a^2 (ka)^4, \quad ka \ll 1$$
$$\sigma_{back} = 2\pi a^2, \quad ka \gg 1 \tag{12.6.11}$$

In Fig. 12.7, the computed backscatter cross section is compared to the low- and high-frequency approximations. The lower graph shows that the Rayleigh limit is indistinguishable from the spherical harmonic series up to $ka = 0.6$, while the geometrical acoustic approximation essentially is close to the mean about which the computed value fluctuates when $ka > 10$. The amplitude of this fluctuation decreases as ka increases, with an excursion that is less than 5% of the mean value above $ka = 20$.

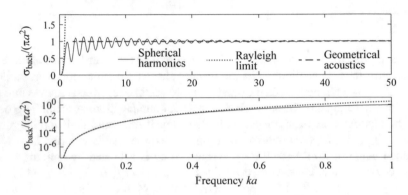

Fig. 12.7 Frequency dependence of the backscatter cross section of a rigid, stationary sphere

Evaluation of the scattering cross section requires integration of the time-averaged bistatic intensity. This quantity in the farfield is $\bar{e}_r |P_s|^2 / (\rho_0 c)$. It contains all cross products of spherical harmonics. However, an evaluation of radiated power entails integration over the surrounding sphere, and the Legendre functions are orthogonal over that domain, see Eq. (7.1.16). Orthogonality was shown in Eq. (7.1.17) to decouple the spherical harmonics in regard to evaluation of radiated power. In the present context, the radiated power in the scattered field is

$$\mathcal{P}_s = \int_0^\pi \frac{P_s P_s^*}{2\rho_0 c} \left(2\pi r^2 \sin\psi d\psi\right)$$
$$= \frac{|P_I|^2}{2\rho_0 c} \sum_{m=0}^\infty 4\pi a^2 (2m+1) \left(\frac{j_m'(ka)}{ka \left|h_m'(ka)\right|}\right)^2 \tag{12.6.12}$$

The factor preceding the summation is the intensity of the incident wave, so the scattering cross section of a rigid sphere is

$$\sigma_s = \frac{\mathcal{P}_s}{I_1} = 4\pi a^2 \sum_{m=0}^\infty 2(2m+1) \left(\frac{j_m'(ka)}{ka \left|h_m'(ka)\right|}\right)^2 \tag{12.6.13}$$

Figure 12.8 shows that unlike the backscatter cross section, σ_s monotonically increases toward the geometrical acoustics limit, $\sigma_s = 2$, without any fluctuation. The value of σ_s in the Rayleigh limit, see Eq. (12.4.6), closely matches the computed function for $ka < 0.45$.

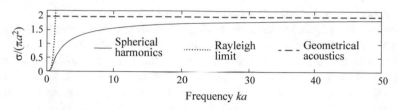

Fig. 12.8 Frequency dependence of the scattering cross section of a rigid, stationary sphere

Before we examine why σ_{back} and σ_s behave differently, let us examine the scattering response from the perspective of the scattering directivity function $\chi\,(\psi, ka)$. Graphs like those in Fig. 12.9 were published by Stenzel in 1938,[6] which is quite impressive given the nonexistence of electronic computing. At low frequencies, the angular dependence of χ fits the Rayleigh limit in Eq. (12.3.22), with the monopole and dipole contributions being in-phase for $\psi > 90°$ and oppositely phased for $\psi < 90°$. As the frequency increases, side lobes become increasingly numerous,

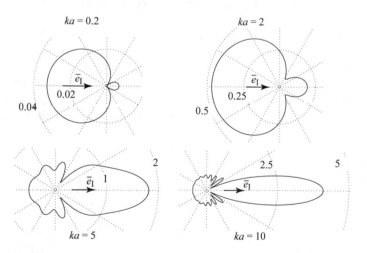

Fig. 12.9 Scattering directivity function $\chi\,(\psi, ka)$ as a function of the observation angle relative to the propagation direction $\bar{e}_I$ of the incident wave

---

[6]H. Stenzel, "On the perturbation of a sound field caused by a rigid sphere," *Electr. Nachrichtentech.*, vol. 15, (1938) pp. 71–28.

which is a trend for acoustic radiation problems. An unexpected trend that grows as the frequency increases is strong forward scattering, in which the maximum value of χ increases and the beamwidth decreases as ka increases. This behavior is not predicted by ray theory, according to which the incident signal should be blocked in the shadow of the sphere. Because $P = P_s + P_I$, such reasoning leads to the expectation that forward scattering should be such that $|P_s| = |P_I|$.

The inability of geometrical acoustics to predict forward scattering is not a failure of the theory. This phenomenon is another diffraction effect. Such effects arose in our study of geometrical acoustics, which does not predict that sound will be heard on the quiet side of a caustic. It also arose there as the fact that a ray at critical incidence to the interface of two fluid might propagate along the interface, then re-emerge into the transmitting fluid. The term "*diffraction*" refers to any high-frequency phenomenon that cannot be described by ray acoustics. The process that ray acoustics missed in scattering is *creeping waves*. Their existence would be difficult to detect from direct evaluations of the spherical harmonic description of the pressure field. At the same time, availability of a series solution is crucial, because it is the beginning point for a mathematical proof of the existence of creeping rays.

Watson in 1918[7] addressed the fact that series such as those derived here require an increased number of terms to converge as the frequency increases. Only rudimentary computational tools were available, so he needed to be "smart". He developed a technique that transforms the series to a contour integral. Evaluation by the methods of complex analysis produces a solution in the form of a series of residues. Each residue corresponds to a pair of waves that pass a point on the surface. Figure 12.10 provides a qualitative picture of the first pair, which are the dominant contributors to the signal.

The incident rays depicted in Fig. 12.10 touch the sphere at $\psi = 90°$, where they are tangent to the sphere. In other words, they are at grazing incidence. They launch rays that propagate along meridian circles (constant θ) in the sense of decreasing ψ, which is the reason they are said to creep over the surface. Ultimately, they focus at the shadowed pole ($\psi = 0$). They emerge from there as waves that travel in the sense of increasing ψ along the diametrically opposite meridian, for which the azimuthal

Fig. 12.10 Incident rays P_I that are at grazing incidence curve around the sphere as creeping waves (*the dotted lines*). They circumnavigate the sphere along meridians

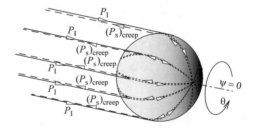

[7]G.N. Watson, "The Diffraction of Electric Waves by the Earth; the Transmission of Electric Waves Around the Earth," Proc. Royal Soc. London **A95** (1918), pp. 83–88, 546–563.

angle is $-\theta$. In other words, an individual ray of the creeping wave follows a great circle that contains the polar axis. Each ray circumnavigates the sphere along its great circle. Eventually, they return to the location where they were launched. From there, the rays execute another circumnavigation, then another, and so on. Each passage is described in Watson's solution by another residue in the complex integration.

The reason it is sufficient to only account for the first circumnavigation is that the wavenumber for these rays is complex, so they attenuate as they propagate. Because this is a high-frequency phenomenon, the behavior in the vicinity of any surface patch is like what it would be if that patch laid in a planar boundary. Thus, the surface ray launches a tangential ray. This signal is said to be *shed* from the surface. The shedding at a few surface locations is depicted in Fig. 12.11, but it occurs continuously along the surface. These shed waves are the agents that illuminate the shadow zone.

Fig. 12.11 Rays shed from creeping wave on the surface of a sphere

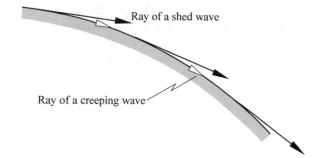

The wave that is shed in the backscattering direction combines with the reflected signal predicted by geometrical acoustics. The latter is referred to as the *specularly reflected wave*. The fluctuations of σ_{back} displayed in Fig. 12.7 stem from constructive and interference between the contribution of the creeping waves and the specularly reflected wave. As the frequency increases, the decay rate for the creeping waves increases. Above $ka = 20$, this attenuation is sufficiently strong to make shedding in the backscatter direction negligible. This leaves only the contribution of specular reflection, which explains why the computed σ_{back} differs little from the geometrical acoustic limit at very high frequencies.

Diffraction is responsible for acoustical phenomena in other systems. Sometimes, we can identify the mechanism directly from a mathematical analysis. For example, the side lobes of a sound beam are a diffraction effect associated with the discontinuity of the velocity at the edge of a piston in a baffle, which leads to center and edge waves for the on-axis signal. In other situations, specialized analyses based on ka being very large are required. This is the situation regarding the Watson transformation. The mathematical sophistication required to carry out such analyses exceeds the level set for this text. An excellent introduction is available in the text by Junger and Feit.[8] Pierce[9] explores the same phenomena, as well as diffraction around wedges

---

[8]M.C. Junger and D. Feit, "Sound, Structures, and Their Interaction," 2nd ed., MIT Press, ASA reprint, Chap. 12 (1986).

[9]A.D. Pierce, *Acoustics*, Mc-Graw-Hill (1981), ASA reprint, (1989), Chap. 9.

and in ocean acoustics, and Keller's geometrical theory of diffraction[10] offers much
in terms of comprehensiveness and generality.

12.6.2   Scattering by an Elastic Spherical Shell

Acoustic radiation from a spherical shell was shown in Sect. 7.2.4 to be profoundly
dependent on frequency. At the resonance frequencies, the sole agent sustaining
any excitation applied to the shell is the resistive part of the acoustic loading. Now,
consider the implications for scattering from a spherical shell. The surface pressure
in the incident signal may be regarded conceptually as the excitation, so the pressure
in the scattered signal is analogous to the radiated field. Thus, if the frequency is
close to one of these fluid-loaded resonances, it is reasonable to expect that the field
will be quite different from what it would be away from these frequencies. It also is
reasonable to expect that if the shell is composed of a material whose density and
sound speed are much greater than the fluid's properties, the scattered signal away
from resonances would be like that for a rigid target. Our objective is to explore the
correctness of these expectations.

We will analyze scattering from an elastic spherical shell by merging the analyses
of scattering from a rigid sphere and radiation from a spherical shell. Toward that
end, let us decompose the scattered field in two parts. The first is that which would be
found if the shell was rigid and stationary. It is called the *blocked pressure*, denoted
as $p_b = \mathrm{Re}(P_b \exp{(i\omega t)})$. The second contribution is induced by the motion of the
shell resulting from its elasticity. Correspondingly, we will denote this part of the
scattered field as $p_e = \mathrm{Re}(P_e \exp{(i\omega t)})$. Thus, we have

$$P_s = P_b + P_e \qquad (12.6.14)$$

This decomposition is substituted into Eq. (12.1.3), which govern the scattered field.
When each term is associated with its respective effect, we obtain

$$\nabla^2 P_b + k^2 P_b = 0, \quad \frac{\partial P_b}{\partial r} = -\frac{\partial P_1}{\partial r} \text{ at} r = a \qquad (12.6.15a)$$

$$\nabla^2 P_e + k^2 P_e = 0, \quad \frac{\partial P_e}{\partial r} = \omega^2 \rho_0 W \text{ at } r = a \qquad (12.6.15b)$$

In the boundary condition for P_e, the parameter W is the complex amplitude of the
radial displacement of the shell, that is, $w = \mathrm{Re}(W \exp{(i\omega t)})$. The corresponding
radial velocity is $v_r = \mathrm{Re}(i\omega W \exp{(i\omega t)})$. In addition to satisfying the Helmholtz
equation and a surface boundary condition, both P_b and P_e must satisfy the Sommerfeld radiation condition, Eq. (12.1.4).

---

[10]J.B. Keller, "Geometric Theory of Diffraction," *J. Opt. Soc. Am.* **52**, (1962) pp. 116–130.

The conditions that the blocked pressure must satisfy are identical to those for the pressure scattered by a rigid sphere. Our interest is the case of an incident plane wave, which was analyzed in the previous section. Consequently, we may employ Eqs. (12.6.3) and (12.6.4) to describe the blocked pressure, so that

$$P_b = -\hat{P}_I \sum_{m=0}^{\infty} (2m+1)\,(-i)^m\, \frac{j_m'(ka)}{h_m'(ka)} h_m(kr)\, P_m(\cos\psi) \qquad (12.6.16)$$

The corresponding farfield description is provided by Eq. (12.6.7).

Now, let us analyze P_e. The radial displacement w represents the response of the elastic shell to the total acoustic pressure applied at its surface. This is the fully coupled problem solved in Sect. 7.2.4. That analysis was limited to excitations that are axisymmetric, which is true for an incident plane wave. It was shown that the spherical harmonic expansions of the radial displacement w and meridional displacement u in Eq. (7.2.35) will satisfy the equations of motion if the coefficients are selected properly. The specific expressions are

$$w = \mathrm{Re}\left(We^{i\omega t}\right), \quad W = \sum_{m=0}^{\infty} W_m P_m(\cos\psi)$$
$$u = \mathrm{Re}\left(Ue^{i\omega t}\right), \quad U = \sum_{m=0}^{\infty} U_m \frac{d}{d\psi} P_m(\cos\psi) \qquad (12.6.17)$$

Furthermore, the pressure induced by the shell's motion constitutes a radiated field, so it too must be expressible as a spherical harmonic series,

$$P_e = \sum_{m=0}^{\infty} C_m P_m(\cos\psi)\, h_m(kr) \qquad (12.6.18)$$

Satisfaction of the velocity boundary condition, Eq. (12.6.15b), requires that

$$k C_m h_m'(ka) = \omega^2 \rho_0 W_m \qquad (12.6.19)$$

Thus, P_e will be determined when we find an expression for W_m.

The total pressure acting on the surface of the shell is the sum of the incident, blocked, and structural contributions. In the basic differential equations of motion for a spherical shell, the surface pressure is decomposed into two parts: an applied loading that does not depend on the shell's motion, and an acoustic part that consists of the pressure induced by the motion. In the present context, the part that is generated by the shell's motion is P_e, and the effective loading is the sum of P_I and P_b. The series for P_e is the same as the description of the acoustic pressure obtained from Eq. (7.2.41), so we may determine P_e directly from the radiation analysis. To do so, we equate the applied pressure q_{applied} to the negative of the sum of the incident and blocked pressures, where the negative sign applies because $q_{applied}$ was taken to be positive if it is outward. Equation (12.6.1) is a spherical harmonic series for a plane wave propagating in the direction of increasing z. When it is combined with Eq. (12.6.16) for P_b, the total pressure loading is

$$q_{applied} = - (P_I + P_b)|_{r=a}$$

$$= -\hat{P}_I \sum_{m=0}^{\infty} (2m + 1) (-i)^m \left[j_m (ka) - \frac{j'_m (ka)}{h'_m (ka)} h_m (ka) \right] P_m (\cos \psi)$$

$$(12.6.20)$$

When the fraction in the bracketed term is cleared, the numerator is a quantity called a Wronskian. Application of the definition that $h_m (ka) = j_m (ka) - i n_m (ka)$ gives a form that is described by a simplifying identity,[11]

$$j_m (ka) h'_m (ka) - j'_m (ka) h_m (ka) \equiv -i \left[j_m (ka) n'_m (ka) \right.$$

$$\left. - j'_m (ka) n_m (ka) \right] = \frac{-i}{(ka)^2} \quad (12.6.21)$$

The result is that the applied load reduces to

$$q_{applied} = -\hat{P}_I \sum_{m=0}^{\infty} \frac{(-i)^{m+1} (2m + 1)}{(ka)^2 h'_m (ka)} P_m (\cos \psi) \quad (12.6.22)$$

The radiation analysis derived algebraic expressions for the displacement coefficients W_m and U_m in terms of the coefficient F_m of a series representation of the applied pressure. The preceding equation is such a description, so the force coefficients are

$$\boxed{F_m = -\hat{P}_I \frac{(-i)^{m+1} (2m + 1)}{(ka)^2 h'_m (ka)}} \quad (12.6.23)$$

The relation between F_m and W_m is described by Eq. (7.2.50) in terms of a structural stiffness $(Z_e)_m$ and an acoustic impedance $(Z_f)_m$, such that

$$\boxed{\left[(Z_e)_m + (Z_f)_m \right] (i k_e a) \frac{W_m}{a} = \frac{a}{h} \frac{F_m}{\rho_e c_e^2}} \quad (12.6.24)$$

where ρ_e and c_e are the density and a certain wave speed for the shell's material, and $k_e = \omega/c_e$. The structural impedance describes the combination of inertial and elastic effects,

---

[11] M.I. Abramowitz and I.A. Stegun, *ibid.*, p. 439. Use Eq. (10) to eliminate $n_{n-1} (z)$ and $j_{n-1} (z)$ in Eq. 10.1.31.

$$(Z_e)_m = \frac{k_e a}{i} \frac{\left[\left(\kappa_m^{(1)} \right)^2 / (k_e a)^2 - 1 \right] \left[\left(\kappa_m^{(2)} \right)^2 / (k_e a)^2 - 1 \right]}{\left(\kappa_m^{(0)} \right)^2 / (k_e a)^2 - 1}$$

(12.6.25)

In this expression $\kappa_m^{(j)} = \Omega_m^{(j)} a / c_e$, where $\Omega_m^{(1)}$ and $\Omega_m^{(2)}$ are the natural frequencies for axisymmetric-free vibration of the shell in the mth spherical harmonic. Also, $\kappa_m^{(0)}$ is a nondimensionalization of the frequency $\Omega_m^{(0)}$ at which there is no radial displacement, regardless of the excitation. The determination of these parameters is described in Sect. 7.2.4. The parameter $k_e = \omega / c_e$, where c_e is the phase speed of an extensional wave of plane strain. The value $(Z_e)_m$ is purely imaginary, so it is a reactance. Inclusion of the effects of internal dissipation would add a resistive part. However, this resistance would be much less than the reactance, except in the vicinity of either natural frequency, where the reactive part is zero.

The fluid impedance is

$$(Z_f)_m = -i \frac{\rho_0 c}{\rho_e c_e} \left(\frac{a}{h} \right) \frac{h_m (ka)}{h'_m (ka)}$$

(12.6.26)

The real part is positive, representing resistance associated with radiation damping. The imaginary part also is positive, and therefore an inertance, corresponding to an added mass effect. Figure 7.7 indicates that the reactive part has a broad peak that occurs at a frequency that increases monotonically with increasing m. In contrast, the resistive part of $(Z_f)_m$ approaches a constant value with increasing frequency.

Equation (7.2.41) describes the coefficients of a spherical harmonic series for the pressure. Substitution of the expressions for W_m and F_m, followed by synthesis of the series, leads to

$$P_e - \hat{P}_1 \frac{\rho_0 c}{\rho_e c_e} \frac{a}{h} \sum_{m=0}^{\infty} \frac{(-i)^m (2m+1)}{[(Z_e)_m + (Z_f)_m] (ka)^2 h'_m (ka)^2} \frac{h_m (kr)}{} P_m (\cos \psi)$$

The farfield version is obtained by applying Eq. (12.6.6), which is the asymptotic property of the Hankel function. The result is

$$(P_e)_{ff} = i \hat{P}_1 \frac{\rho_0 c}{\rho_e c_e} \frac{a}{h} \frac{a}{r} e^{-ikr} \sum_{m=0}^{M} \frac{(2m+1)}{[(Z_e)_m + (Z_f)_m] (ka)^3 h'_m (ka)^2} P_m (\cos \psi)$$

(12.6.27)

where M is the largest index for which the factor of the Legendre function is significant.

The amplitude of the mth spherical harmonic is maximized when the magnitude of $(Z_e)_m + (Z_f)_m$ is a minimum. Here, an analogy with standard vibration theory comes into play. The portion $(Z_e)_m$ represents the effect of inertia and elasticity. If internal dissipation is ignored, $(Z_e)_m$ is imaginary. The fluid effect is contained in $(Z_f)_m$. It is complex, with the imaginary part acting as a virtual mass that adds to $(Z_e)_m$. The real part represents radiation damping, in which power is radiated into the fluid. In a standard spring-mass-dashpot system, a resonance is associated with the magnitude of the combined impedance being a minimum. However, if the damping is not too large, the resonance condition will be close to that for which the imaginary part is zero. Similarly, although the true maximum of a spherical harmonic corresponds to the frequency that minimizes $|(Z_e)_m + (Z_f)_m|$, it is simpler to approximate the frequency at which a *fluid-loaded resonance* occurs as the value for which $\text{Im}((Z_e)_m + (Z_f)_m) = 0$.

These resonances affect the selection of the harmonic M at which the series may be truncated. In the analysis of scattering from a rigid sphere, the criterion for selection of M was found to be $M \gg ka$, whose satisfaction assures that $h'_m (ka)^2$ is very large. That criterion applies here also, but there is another consideration. It is imperative that any harmonic $\tilde{m}$ that is close to resonance be included in the summation. That is, we should set $M > \tilde{m}$. Figure 7.7 describes the impedances as a function of frequency for fixed m, but the issue here is the dependence of $\text{Im} \left[(Z_e)_m + (Z_f)_m \right]$ on m at fixed ka.

The natural frequencies as functions of m are described in Fig. 7.5, and $\Omega_m^{(0)}$ falls between these frequencies. If M is set well above the intersection of the line at the current frequency ka with the lower branch, then $\text{Im} (Z_e)_m$ will be large negatively. This accelerates the convergence rate of the series for P_e, beyond what is attributable to the behavior of the Hankel function. The next example uses $M = \text{ceil}(10ka)$, which is much larger than necessary.

Elastic effects occur in a number of ways in Eq. (12.6.27), so let us evaluate P_e in the case where $\text{Im} \left[(Z_e)_{\tilde{m}} + (Z_f)_{\tilde{m}} \right]$ actually is zero. Doing so will lead to the pressure at a fluid-loaded resonance. The frequency for this resonance is denoted as $\tilde{\omega} \equiv \tilde{k}c$. The total impedance reduces to $\text{Re} (Z_f)_{\tilde{m}}$ at $\tilde{k}a$. Equation (12.6.26) describes the fluid impedance, so we have

$$\left[(Z_e)_{\tilde{m}} + (Z_f)_{\tilde{m}} \right]\Big|_{\tilde{k}a} = \text{Re} (Z_f)_{\tilde{m}}|_{\tilde{k}a} = \frac{\rho_0 ca}{\rho_e c_e h} \text{Re} \left[-i \frac{h_{\tilde{m}} \left(\tilde{k}a \right)}{h'_{\tilde{m}} \left(\tilde{k}a \right)} \right] \qquad (12.6.28)$$

The Hankel function of the second kind is defined in terms of the real spherical Bessel and Neumann functions to be $h_{\tilde{m}}^{(2)} \left(\tilde{k}a \right) = j_{\tilde{m}} \left(\tilde{k}a \right) - i n_{\tilde{m}} \left(\tilde{k}a \right)$. With this, the impedance becomes

$$\left[(Z_e)_{\tilde{m}} + (Z_f)_{\tilde{m}}\right]\big|_{\tilde{k}a} = \frac{\rho_0 c a}{\rho_e c_e h} \, \text{Re}\left[-i \frac{h_{\tilde{m}}\left(\tilde{k}a\right) h_{\tilde{m}}'\left(\tilde{k}a\right)^*}{h_{\tilde{m}}'\left(\tilde{k}a\right) h_{\tilde{m}}'\left(\tilde{k}a\right)^*} \right]$$

$$= \frac{\rho_0 c a}{\rho_e c_e h} \left(\frac{-n_{\tilde{m}}\left(\tilde{k}a\right) j_{\tilde{m}}'\left(\tilde{k}a\right) + j_{\tilde{m}}\left(\tilde{k}a\right) y_{\tilde{m}}'\left(\tilde{k}a\right)}{\left| h_{\tilde{m}}'\left(\tilde{k}a\right) \right|^2} \right)$$

$$(12.6.29)$$

The numerator is the Wronskian in Eq. (12.6.21), so the total impedance at a fluid-loaded resonance reduces to

$$\left[(Z_e)_m + (Z_f)_{\tilde{m}}\right]\big|_{\tilde{k}a} = \frac{\rho_0 c a}{\rho_e c_e h} \frac{1}{\left(\tilde{k}a\right)^2 \left| h_{\tilde{m}}'\left(\tilde{k}a\right) \right|^2} \qquad (12.6.30)$$

Presumably, the $\tilde{m}$ term in $(P_e)_{\text{ff}}$ is dominant. Substitution of the impedance at resonance into Eq. (12.6.27) yields

$$(P_e)_{\text{ff}} \approx \hat{P}_1 \frac{e^{-i\tilde{k}r}}{\tilde{k}r} (2\tilde{m} + 1) \, e^{-2i\,\arg(h_{\tilde{m}}'(\tilde{k}a))} P_{\tilde{m}}(\cos\psi) \qquad (12.6.31)$$

The surprising aspect of this result is that the peak magnitude of $(P_e)_{\text{ff}} / P_1$ appears to depend only on the harmonic number and the acoustic wavenumber at which the resonance occurs. However, this is a deceptive view, because both $\tilde{m}$ and $\tilde{k}$ depend strongly on all system parameters.

EXAMPLE 12.4 A spherical shell ($h/a = 40$) composed of aluminum ($\rho_e = 2100$ kg/m^3, $c_e = 6410$ m/s) is submerged in water. Determine the backscatter cross section σ_{back} and total scattering cross section σ in the range $ka < 10$. Compare this to the frequency dependence of each parameter if the sphere was rigid and stationary. Then, repeat the analysis for the case where the sphere is suspended in the air.

Significance

The dependence on a number of system parameters, as well as the interplay between spherical harmonic number and frequency, can best be understood by performing quantitative analyses like the one to be performed here. The evaluation itself will provide a useful review of some fundamental concepts.

Solution

The total scattered field is the sum of the blocked and structural contributions. Because P_b is the same as the scattered field for the rigid case, we shall compute it, and

P_e as separate entities. Our strategy is rather straightforward. The *in-vacuo* natural frequencies $\Omega_m^{(1)}$ and $\Omega_m^{(2)}$, as well as the zero-radial-displacement frequency $\Omega_m^{(0)}$, are independent of the excitation frequency. The truncation criterion is $M = 10ka$, and the maximum frequency to be considered is $ka = 10$. Therefore, we begin by computing the modal properties for azimuthal harmonics from zero to 100. This computation entails solving the characteristic equation, Eq. (7.2.44). The K coefficients to use are those in Eq. (7.2.48), which describe membrane and flexural deformation effects. The roots of this quadratic equation are the values $\kappa_m^{(j)} = \Omega_m^{(j)} a / c_e$, $j = 1, 2$, and $\kappa_m^{(0)}$ in Eq. (7.2.46).

The remaining operations are performed for a succession of ka value. The values of $h_m(ka)$ for $0 \le m \le M + 1$ are computed and used in conjunction with a recurrence relation to evaluate $h'_m(ka)$ for $0 \le m \le M$. These values are required to evaluate both P_b and P_e, and the latter also requires $j'_m(ka)$, which is the real part of $h'_m(ka)$. Let us denote the coefficients of the spherical harmonic series for P_b/P_I in the farfield as $C_{b,m}$ and those for P_e/P_I as $C_{e,m}$. Thus, from Eqs. (12.6.7) and (12.6.27), we have

$$\frac{P_b}{P_I} = \frac{a}{r} e^{-ikr} \sum_{m=0}^{M} C_{b,m} P_m(\cos\psi), \quad \frac{P_e}{P_I} = \frac{a}{r} e^{-ikr} \sum_{m=0}^{M} C_{e,m} P_m(\cos\psi)$$

$$C_{b,m} = -i(2m+1) \frac{j'_m(ka)}{ka h'_m(ka)} \tag{1}$$

$$C_{e,m} = i \frac{\rho_0 c a}{\rho_e c_e h} \frac{(2m+1)}{\left[(Z_e)_m + (Z_f)_m\right](ka)^3 h'_m(ka)^2}$$

All coefficients for a specific ka may be computed in a single vectorized step.

The backscatter cross section was defined in Eq. (12.4.10). The backscatter direction is $\psi = \pi$, for which $P_m(-1) = (-1)^m$. Thus,

$$(\sigma_{back})_e = 4\pi a^2 \left| \sum_{m=0}^{M} (-1)^m \left(C_{b,m} + C_{e,m}\right) \right|^2 \tag{2}$$

$$(\sigma_{back})_b = 4\pi a^2 \left| \sum_{m=0}^{M} (-1)^m C_{b,m} \right|^2$$

Evaluation of the scattering cross section would be more difficult were it not for the orthogonality property of Legendre functions. Because of it, there is no coupling of harmonics in the scattered power. The steps for this analysis are

$$\sigma_s = \frac{P_s}{I_I} = \int_0^\pi \left(\frac{P_b}{P_I} + \frac{P_w}{P_I} \right) \left(\frac{P_b^*}{P_I^*} + \frac{P_w^*}{P_I^*} \right) (2\pi r^2 \sin \psi d\psi)$$

$$= 2\pi a^2 \sum_{j=0}^M \sum_{m=0}^M \int_0^\pi (C_{b,j} + C_{e,j}) (C_{b,m}^* + C_{e,m}^*) P_j (\cos \psi) P_m (\cos \psi) \sin \psi d\psi$$

$$= \pi a^2 \sum_{m=0}^M \frac{4}{2m+1} \left[|C_{b,m}|^2 + |C_{e,m}^*|^2 + 2 \,\mathrm{Re}\, (C_{b,m} C_{e,m}^*) \right] \tag{3}$$

The scattering cross section for the rigid sphere is obtained by setting $C_{e,m} = 0$.

Figure 1 shows the backscatter and total scattering cross sections. The spikes correspond to fluid-loaded resonances, whose occurrence was anticipated in the earlier discussion. The puzzling aspect is their occurrence only in the low-frequency band.

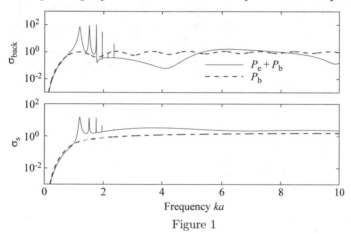

Figure 1

An examination of this behavior begins by considering the resonance condition in the absence of fluid loading. In that case, $(Z_f)_{\tilde{m}}$ is identically zero, so a resonance corresponds to $(Z_e)_m$, which is imaginary, being zero. At a randomly selected ka, a plot of $\mathrm{Im}\,(Z_e)_m$ as a function of m would show that it crosses the zero axis at a value of m that is not an integer, so it is not a resonant frequency. Increasing ka eventually will bring on a condition where $\mathrm{Im}\,(Z_e)_m = 0$ at an integer value of m. Because we are ignoring the role of fluid loading, the corresponding value of $(C_e)_{\tilde{m}}$ would be infinite. Further increase of ka increases the m value at which $(Z_e)_m = 0$, until it reaches another integer value of m, and so on.

A crucial aspect of the effect of fluid loading is that $(Z_f)_m$ is zero only at $ka = 0$. The two quantities that affect the magnitude of $(C_e)_m$ are $|(Z_e)_m + (Z_f)_m|$ and $|h'_m (ka)|$, both of which are in the denominator of the above expression. A fluid-loaded resonance was defined as a frequency $\tilde{k}a$ at which $\mathrm{Im}((Z_e)_{\tilde{m}} + (Z_f)_{\tilde{m}}) = 0$. However, $\mathrm{Re}\,(Z_f)_{\tilde{m}}$ is not zero, so it is possible that $|(Z_e)_m + (Z_f)_m|$ is not close to a minimum. Furthermore, even if it is, the denominator of $(C_e)_{\tilde{m}}$ might not be minimized because the magnitude of its denominator is $|(Z_e)_m + (Z_f)_m| \, |h'_m(ka)|$; a small impedance might be multiplied by a large Hankel function.

To see how these factors work together, Fig. 2 examines them at $ka = 1.188$, at which the first peak in σ_{back} and σ_s occur. This value of ka is very close to the first natural frequency of the $m = 2$ spherical harmonic, $ka \approx \kappa_2^{(1)}$. The uppermost graph indicates that the reactance at $m = 2$ is zero, and the resistance is small. The actual values are $(Z_e)_2 + (Z_f)_2 = 0.13341 - 0.02512i$. The second graph shows the total impedance is minimized at $m = 2$. The general trend for $h'_m(ka)$ is that it rises rapidly as m increases beyond ka. However, the third graph indicates that this increase is not drastic for $m = 2$, for which $h'_2(1.188) = 0.12829 + 4.8275i$. The logarithmic scale in the fourth graph obscures relative magnitudes, but it is evident that $(C_e)_2$ is the largest coefficient.

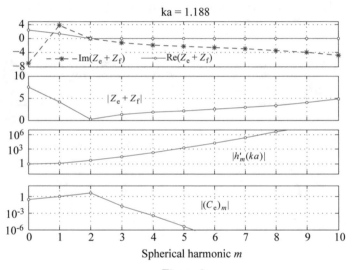

Figure 2

As ka is increased, the plots of $(Z_e)_m + (Z_f)_m$ and $h'_m(ka)$ as functions of m maintain their general shape, except that $\mathrm{Im}[(Z_e)_m + (Z_f)_m]$ crosses the zero axis at an increasing m. Figure 3 describes the behavior at $ka = 4$, which does not correspond to a peak in either scattering cross section. The plot of $\mathrm{Im}[(Z_e)_m + (Z_f)_m]$ crosses the zero axis between $m = 12$ and $m = 13$, and the minimum value of $|(Z_e)_m + (Z_f)_m|$ occurs at $m = 11$, where it is 0.03865. However, $h'_{12}(4) = 4.4811\,(10^{-6}) + 2.0608\,(10^4)\,i$, which is quite large, and this quantity is squared in the denominator of $(C_e)_m$. The consequence is that the largest value of $(C_e)_m$ occurs at $m = 3$, but it is not especially large relative to the values for adjacent m. Hence, this frequency does not lead to enhanced scattering.

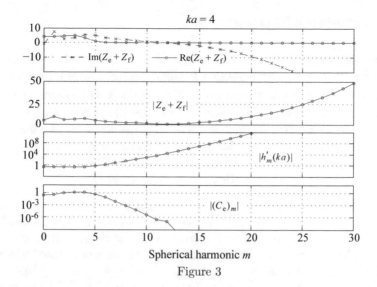

Figure 3

These observations suggest that the increase of $|h'_m(ka)|$ for $m > ka$ is responsible for the suppression of high-frequency resonances. An example of this behavior is $ka = 8.710$ at $\tilde{m} = 19$, for which $(Z_e)_{\tilde{m}} + (Z_f)_{\tilde{m}} = 8.8890\left(10^{-10}\right) - 6.7896\left(10^{-5}\right)i$. For comparison, at $m = 18$ the value is $(Z_e)_{18} + (Z_f)_{18} = 1.7893\left(10^{-8}\right) + 0.85717i$. However, at $m = 19$, $\left|h'_{19}(8.71)\right| = 3.2851\left(10^4\right)$, so the relative smallness of the impedance term is overwhelmed by the largeness of $h'_{19}(ka)$. The largest coefficient value at $ka = 8.710$ is $(C_e)_m = 1.4761$ at $m = 7$, whereas $(C_e)_{19} = 6.5216\left(10^{-5}\right)$. Thus, the absence of high-frequency resonances in the scattering cross sections may be attributed to the fact that the spherical harmonic excitation becomes much weaker with increasing m, as well as the low radiation efficiency of the higher spherical harmonics of the surface displacement. (A spherical harmonic whose number is large corresponds to a short wavelength/subsonic surface motion.)

Although resonances are not present in scattering cross sections at high frequencies, the shell's flexibility does influence the scattering properties at all frequencies. As shown in Fig. 1, the value of the total scattering cross section is considerably increased by the flexibility of the shell, and the backscatter cross section shows considerable variability.

There is no need to display the cross sections for the case where the surrounding fluid is air, because they are identical to those in Figs. 12.7 and 12.8 for a rigid, stationary sphere. The flexibility of the shell has little effect on the scattering because there is an enormous difference in the properties of the shell relative to air and water. The parameter that governs the overall magnitude of P_e relative to P_b is $\rho_0 c / (\rho_e c_e)$, which is $3(10^{-5})$ for air and aluminum, and 0.11 for water and aluminum. Thus, P_e is very a small contributor to the scattered field. Even resonances are insignificant relative to the blocked pressure.

12.7   Homework Exercises

Exercise 12.1 A solid plastic sphere immersed in an ideal liquid has density and sound speed that are slightly larger than the liquid. A plane wave is incident on this object. Specialize the Born approximation of $(r/a) \left| P_s(\bar{x}_0) / \hat{P}_1 \right|$ to fit this spherical configuration. Doing so will yield an integral representation of the scattered pressure at a field point in the farfield as a function of ka, ρ/ρ_0, and c/c_0, and the direction angles to the field point. Evaluate this integral for the case of backscatter corresponding to $\rho'/\rho_0 = 1.02$ and $c'/c_0 = 1.05$. Plot the backscatter amplitude in the interval $ka < 2$, and compare it to the Rayleigh limit in Eq. (12.3.1).

Exercise 12.2 A submerged object is surrounded by eight equally spaced hydrophones, as shown in the sketch. Projector A emits an omnidirectional signal at 20 Hz. This signal is measured to be 210 dB//1 μPa at one meter from the source. The table provides the amplitude of the scattered signal measured at each hydrophone. The ambient properties for the water are $\rho_0 = 1004$ kg/m^3 and $c = 1470$ m/s. A visual inspection of the object indicates that its volume is 0.12 m^3. (a) Assuming that the object's physical properties are comparable to water, use the tabulated data to deduce the density and sound speed within the object. Use the Rayleigh limit of the Born approximation to analyze the data. (b) To explore the effect of measurement error, contaminate the tabulated data by adding to each value a random error that ranges over ± 0.4 Pa with a uniform probability distribution. Repeat the calculation in Part (a) using the error-contaminated data. What does the result say about the significance of measurement error?

Hydrophone #	1	2	3	4	5	6	7	8
$|P_s|$ (mPa)	1.5296	1.3526	1.1022	0.92514	0.92514	1.1022	1.3526	1.5296

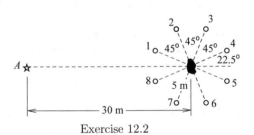

Exercise 12.2

Exercise 12.3 A transducer capable of transmitting and receiving is situated at point A, and a receiving hydrophone is situated 400 m away at point B. It is desired to use this arrangement to measure the physical properties of a bottle-nosed dolphin, based on the notion that it does not differ much from water. The scheme uses transducer A to generate a tone burst consisting of three cycles of a 50 Hz harmonic. The scattered signal is measured by both transducers. The received signals are plotted below. (The pulse received at position B directly from the source at point A has been removed.) The timescale is based on $t = 0$ being the instant when the tone burst is initiated at source A. The measured signals allow for determination of the geometrical properties required to evaluate the Rayleigh limit of the Born approximation. It is known that the transducer A generates a radially symmetric signal that is 49 kPa at 1 m from the center, and a measurement of the displacement of the dolphin indicates that its volume is 0.987 m^3. The density and sound speed of water are the nominal values, $c = 1480$ m/s and $\rho_0 = 1000$ kg/m^3. Based on the assumption that the dolphin is approximately homogeneous, determine its average density and sound speed.

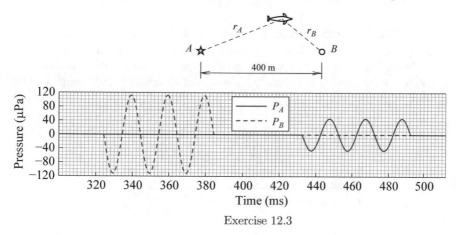

Exercise 12.3

Exercise 12.4 Consider the cylinder in Example 12.1 when the incident wave arrives broadside, $\psi_I = 90°$. Determine the directivity of the scattered pressure in the plane formed by the incident wave's propagation direction and the axis of the cylinder. Frequencies of interest are $ka = 0.1$, 1, and 10.

Exercise 12.5 A plane wave is incident at direction $\bar{e}_I$ on the stationary rigid rectangular box in the sketch. The box, whose walls have negligible stiffness and mass, is filled with a liquid for which $\rho' = 0.92 \rho_0$ and $c' = 0.94$. The box sides are a, b, and d. (a) Use the Born approximation to derive an expression for the low-frequency scattered pressure in the form $(r/a) \left| P_s\left(\bar{x}_0\right) / \hat{P}_I \right|$ for arbitrary $\bar{e}_I$ and scattering direction $\bar{e}_r$. For the development, describe $\bar{e}_I$ and $\bar{e}_r$ in terms of spherical angles described relative to the z-axis. (b) Specialize the result in Part (a) to the backscatter case, $\bar{e}_r = -\bar{e}_I$ when a plane wave is incident in the xz-plane. (c) Let $\ell = \mathcal{V}^{1/3}$, where $\mathcal{V}$ is the volume, be a representative length scale. Compare the backscatter directivity for two shapes having the same volume: (1) $a = b = d = \ell$, (2) $a = b = 0.5\ell$, $d = 4\ell$.

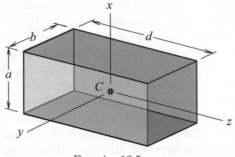

Exercise 12.5

Exercise 12.6 The cylinder in the sketch is rigid and stationary. It is insonified by a plane wave at frequency ω whose propagation direction is parallel to the centerline of the cylinder. The radius a is extremely small compared to the length L, and $ka \ll 1$. Derive an expression for the scattered pressure at an arbitrary field point $\bar{x}$ in the farfield.

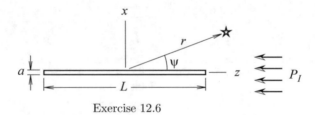

Exercise 12.6

Exercise 12.7 The sketch shows a mass-spring oscillator consisting of a sphere whose mass is M that is supported by a spring whose stiffness is K. A harmonic plane wave, which is traveling parallel to the direction in which the spring is oriented, causes the sphere to oscillate. The sphere is rigid, so the only motion that it executes is a rigid body translation $u = \mathrm{Re}\,(U\exp{(i\omega t)})$. It may be assumed that $ka \ll 1$. (a) Derive an expression for U as a function of ka. (b) Derive an expression for the scattered pressure P_s at an arbitrary field point in the farfield. (c) Determine the value of ka that will maximize $|P_s|$.

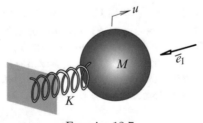

Exercise 12.7.

Exercise 12.8 A 20 mm diameter ball bearing is insonified by a 1 kHz plane wave. Evaluate its scattering cross section in situations where it is restricted from moving, and where it is suspended in a manner that permits it to translate. Compare the results for a ball compose of lead ($\rho = 11340$ kg/m³, $c = 1322$ m/s), to those for an aluminum ball ($\rho = 2100$ kg/m³, $c = 6410$ m/s).

Exercise 12.9 A plane wave at 60 Hz in air is obliquely incident on a stationary rigid disk whose radius is 200 mm. Its direction of propagation is 20° off the normal to the disk. Determine the scattering cross section, the backscatter cross section, and the target strength corresponding to this angle of incidence.

Exercise 12.10 A land mine is buried in sand, $\rho_0 = 1140$ kg/m³, $c = 105$ m/s, which may be regarded as a liquid. It is insonified by a 15 Hz plane wave whose amplitude is 500 Pa. The mine is buried sufficiently deeply that reflections of the scattered wave from the surface may be ignored. It is a 200-mm-long cylinder whose radius is 150 mm, and both ends are closed with hemispherical caps. The mass of the cylinder is 120 kg and the virtual mass coefficients are $\Lambda_{1,1} = \Lambda_{2,2} = 10\mathcal{V}$, $\Lambda_{3,3} = 2.5\mathcal{V}$. Determine its scattering cross section and target strength in the situation where $\bar{e}_{\mathrm{I}} = -\bar{e}_x$. Do these values depend strongly on whether the mine is stationary or movable?

Exercise 12.11 A plane harmonic wave in the atmosphere is incident on an object that is symmetric about the z-axis, with $\bar{e}_{\mathrm{I}} = -\bar{e}_z$. The frequency is 500 Hz. The scattered pressure at a point in the farfield is described in terms of spherical harmonics as

$$P_s = P_{\mathrm{I}} \frac{e^{-ikr_0}}{kr_0} \sum_{m=0}^{\infty} B_m P_m (\cos \psi_0) \, e^{i(m+1)\pi/2}$$

The coefficient values are $B_0 = 0.5$, $B_1 = 0.2 - 0.4i$, $B_2 = 1.5 + 0.3i$, $B_3 = -0.2 + 0.2i$ Pa, while all higher values are zero. The forward scattered pressure is 2 Pa at $r = 10$ m. (a) What is the backscattered pressure at $r = 10$ m. (b) What are the differential scattering cross section values at $\psi = 0$, 90°, and 180°? (c) What is the scattering cross section of this object? (d) What is the target strength?

Exercise 12.12 The axisymmetric object in the sketch is a body of revolution. It is generated by rotating about the centerline a circular arc of radius $4a$ whose center is situated at $5a$ from the centerline. A plane harmonic wave traveling in the direction $-\bar{e}_x$ is incident on the object. According to geometrical acoustics, what is the backscatter cross section of this object? Would this value change if the incident wave's direction deviated slightly from broadside incidence?

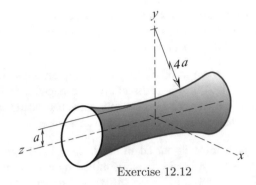

Exercise 12.12

Exercise 12.13 A crude model for scattering from a rough surface considers normal incidence of a plane wave on a surface that is nominally flat, but actually has a sinusoidal variation in its elevation. Such a situation is depicted in the sketch, where the elevation is $h = h_0 \cos (2\pi x / L)$. If the acoustic wavelength $2\pi / k$ is much smaller than the spatial period L of the surface, it is permissible to invoke geometrical acoustics. Construct a sketch of the manner in which the incident rays scatter. What general conclusions can be drawn from this construction?

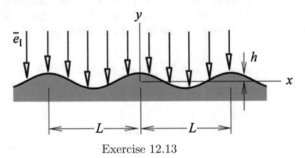

Exercise 12.13

Exercise 12.14 A plane harmonic wave propagating horizontally in water is incident on a solid sphere. The sphere is suspended by a cable that permits it to move in the horizontal plane. Derive expressions for the backscattered pressure and the velocity of the center of the sphere as a function of $ka < 10$ for the case where the sphere's mass is $5\pi a^3 \rho_0$.

Exercise 12.15 Exercise 12.7 entailed an analysis of scattering from a spring-supported rigid sphere in the low-frequency regime, $ka \ll 1$. (a) Perform the investigation requested there, but do not assume that ka is small. (b) Evaluate $\rho_0 c \, |V_C| \, / \, |P_1|$ as a function of $ka < 5$ for four parametric combinations: $M = \sigma \rho_0 \pi a^3$ and $K = \chi M \, (c/a)^2$ with $\sigma = 2$ and 8, $\chi = 1$ and 2. Identify the frequency at which $|V_C|$ is maximized.

Exercise 12.16 A 3-m diameter steel shell is neutrally buoyant in a very large vat of ethyl alcohol ($\rho_0 = 785$ kg/m^3, $c = 1144$ m/s). It is insonified by a plane harmonic wave. Determine the scattering directivity, that is, $(r/a) \, |P_s / P_1|$ as a function of ψ.

Identify the portion of this function that is attributable to the elasticity of the shell. Carry out this evaluation for cases where the frequency of this wave is 99% of the upper natural frequency of the $m = 0$, 1, and 2 in-vacuo modes.

Exercise 12.17 The sketch shows a plane wave that is incident on a rigid hemisphere mounted on an infinite rigid baffle. The sum of the incident and scattered fields must have zero particle velocity normal to the baffle, as well as the normal to the hemisphere's surface, that is, $(\bar{v}_I + \bar{v}_s) \cdot \bar{e}_z = 0$ for $r > a$, $\psi = \pi/2$ and $(\bar{v}_I + \bar{v}_s) \cdot \bar{e}_r = 0$ for $r = a$, $0 \le \psi < \pi/2$. A suitable representation draws on the development in Sect. 7.2.3, which showed that the field radiated by a vibrating hemisphere on a rigid baffle can be represented as a series of the even-numbered spherical harmonics. At the same time, if the hemisphere was not present, the incident wave would be reflected with the reflection coefficient being one. Thus, an ansatz for the field scattered by the hemisphere and baffle is

$$P = P_I + P_R + P_h$$

$$P_I = \hat{P}_I e^{+ikz}, \quad P_R = \hat{P}_I e^{-ikz}, \quad P_h = \sum_{m=0}^{\infty} B_m h_{2m}(kr) P_{2m}(\cos \psi)$$

Derive an expression that describes the B_m coefficients.

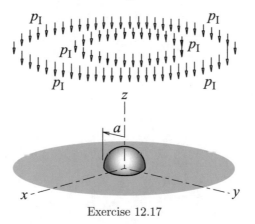

Exercise 12.17

Exercise 12.18 Consider the hemisphere-baffle configuration in Exercise 12.17. The scattered pressure is $P_s = P_R + P_h$. The frequency of interest is $ka = 10$. (a) Evaluate the scattered pressure as a function of distance along the z-axis. Compare the result to the geometrical acoustics approximation for the backscatter. (b) Evaluate the dependence of the scattered pressure along a line parallel to the baffle at $z = 4a$. (c) Evaluate the dependence of the scattered pressure on the polar angle ψ at $r = 4a$.

Chapter 13
Nonlinear Acoustic Waves

When told that a system's response is quite large, a knowledgeable individual will anticipate that nonlinear effects are important. This certainly is the case in acoustics, as exemplified by an explosion, wherein peak pressures might be well above atmospheric. However, even if the pressure is not large, there is another parameter whose largeness might require consideration of nonlinear effects. That parameter is the propagation distance. The signal radiated by a transducer might not seem to be especially large, but nonlinear effects grow with increasing distance. We might observe harmonic and intermodulation distortion of the waveform, decreased signal strength, or the formation of shocks (discontinuous changes of the waveform). Although these effects would seem to be deleterious, there are some applications that rely on them.

Analysis of nonlinear effects encounters two fundamental difficulties relative to linear theory. The most obvious is the inapplicability of the principle of superposition. But there is another. In our studies thus far, conversion to the frequency domain made it possible to analyze many systems. The concept of using a complex exponential to remove the explicit dependence on time ceases to be valid in the presence of nonlinearity. The reason becomes evident when we consider a nonlinear term such as p^2. If p oscillates at frequency ω, then p^2 contains a term whose frequency is 2ω. The occurrence of such a term in the field equation for p means that p cannot be a pure harmonic. The consequence is that we will need to work in the time domain, so far fewer systems are amenable to analysis.

In many treatises, the term *finite amplitude wave* is used to refer to a signal whose amplitude is sufficiently large that nonlinear effects cannot be ignored. The notion behind this terminology is that ignoring terms like p^2 is equivalent to considering p to be infinitesimal. Whether it is allowable to do so depends on several factors, most notable of which are the amplitude of the signal and the propagation distance. If either is large, one might need to consider nonlinear effects.

We will not pursue a historical survey of the numerous approaches that have been used. Rather, the methods we shall explore have been selected either for their

Electronic supplementary material The online version of this chapter
(DOI 10.1007/978-3-319-56847-8_13) contains supplementary material, which is
available to authorized users.

J.H. Ginsberg, *Acoustics—A Textbook for Engineers and Physicists, Volume II*,
DOI 10.1007/978-3-319-56847-8_13

generality or for their directness. We will follow Riemann's seminal analysis,[1] which is based on the fundamental properties of differential equations. One of its beneficial attributes is that it does not require specification of the excitation. Another approach we will develop applies Fourier series to analyze systems in which the excitation is temporally periodic. The idea is not to convert the problem to the frequency domain. Rather, the formulation yields differential equations for the Fourier coefficients that are easier to solve. A third approach for our consideration is perturbation analysis, which decomposes the nonlinear problem to a sequence of linearized problems. This method cannot proceed if the analogous linear problem cannot be solved.

13.1   Riemann's Solution for Plane Waves

It is remarkable that Riemann's solution was published in 1859, and that it was contemporaneous with Earnshaw's analysis[2] of the same problem. The two papers followed very different paths. Riemann used the continuity and momentum principles to obtain equations governing the fundamental state variables: pressure, particle velocity, and density. Earnshaw formulated the analysis in terms of displacement of a specific particle in what is called a Lagrangian kinematical description. Most importantly for us, Riemann's analysis proceeds in a logical manner, whereas Earnshaw's requires a degree of intuition. The results of the analyses are equivalent. Both may be found in treatises by Rayleigh[3] and Lamb.[4] Poisson[5] actually was the first to solve this problem, so some refer to the final result as the Poisson solution. However, that treatment assumed isothermal conditions. We will follow Riemann's analysis because of its directness and generality.

13.1.1   Analysis

The fluctuating acoustic part of the pressure is p, so the absolute pressure in the current state is $p_0 + p$. The corresponding density is ρ. The direction of propagation is designated as x. The particle velocity, positive in the sense of increasing x, is v. The basic equations of continuity and momentum for a one-dimensional wave are

---

[1]B. Riemann, "Über die Fortpflanzung ebener Luftwellen von endlicher Schwingungsweite," Abhandlungen der Königlichen Gesellschaft der Wissenschaften zu Göttingen, Vol. 8 (1859) pp. 245–264.

[2]S. Earnshaw. "On the mathematical theory of sound," Trans. Royal Soc. London (1860) pp. 133–148.

[3]J.W. Strutt, Lord Rayleigh, "Theory of Sound," Vol 2, (1877), Dover reprint (1945) Articles 252 and 253.

[4]Lamb, *Hydrodynamics*, (1916), Dover reprint, (1945). Articles 260 and 261.

[5]S.D. Poisson, "Memoire sr lat théorie de son," J. l'École Polytechnique Paris, Vol 7 (1808) pp. 319–392.

Eqs. (2.1.6) and (2.1.10). They are

$$\frac{\partial \rho}{\partial t} + \frac{\partial}{\partial x}(\rho v) = 0$$

$$\rho \left(\frac{\partial v}{\partial t} + v \frac{\partial v}{\partial x} \right) = -\frac{\partial p}{\partial x} \tag{13.1.1}$$

In addition, we retain the assumption of adiabatic compression and expansion, based on the expectation that the pressure varies at a rate that is too high to permit significant heat transfer.

The notion motivating the Riemann solution is that there is a family of curves in space-time along which p and v are constant. If this is true, then as long as we stay on one of these curves, we can consider p to be a function of t and x, and v to be a function of p. A curve t as a function of x (or x as a function of t) having this property is a *characteristic*. Thus, along a characteristic, the functional dependence is $v = v(p)$. As a consequence of the adiabatic assumption, the equation of state may be taken to be $\rho = \rho(p)$.

The task is to solve the first-order conservation equations for the relation between x and t consistent with the assumed functional dependence on p. The derivatives of ρ and v appearing in the basic equations are converted by the chain rule to

$$\frac{\partial \rho}{\partial t} = \frac{1}{dp/d\rho} \frac{\partial p}{\partial t}, \quad \frac{\partial}{\partial x}(\rho v) = \frac{d}{dp}(\rho v) \frac{\partial p}{\partial x}$$

$$\frac{\partial v}{\partial t} = \frac{dv}{dp} \frac{\partial p}{\partial t}, \quad \frac{\partial v}{\partial x} = \frac{dv}{dp} \frac{\partial p}{\partial x} \tag{13.1.2}$$

In linear theory, the square of the sound speed is defined to be $dp/d\rho$ in the ambient state. In the nonlinear formulation, the relevant value is this derivative at the current state. Many references define c^2 to be this value and denote c_0 as the linear speed of sound. It is less confusing if c is retained as the linear speed of sound, so $\tilde{c}^2$ is $dp/d\rho$ at the current state,

$$\boxed{\tilde{c} = \left(\frac{dp}{d\rho} \right)^{1/2} \equiv \left(\frac{1}{d\rho/dp} \right)^{1/2}} \tag{13.1.3}$$

When Eq. (13.1.2) is substituted into the basic continuity and momentum equations, the result is

$$\frac{1}{\tilde{c}^2} \frac{\partial p}{\partial t} + \left(\frac{1}{\tilde{c}^2} v + \rho \frac{dv}{dp} \right) \frac{\partial p}{\partial x} = 0$$

$$\rho \frac{dv}{dp} \frac{\partial p}{\partial t} + \left(\rho v \frac{dv}{dp} + 1 \right) \frac{\partial p}{\partial x} = 0 \tag{13.1.4}$$

Because of the assumed interdependence of variables, the only quantities that depend explicitly on x and t are the derivatives of p. Thus, the preceding may be considered to be a pair of first-order equations for $\partial p / \partial t$ and $\partial p / \partial x$ in which the coefficients are functions of p. The matrix form of the equations is

$$
\left[\begin{array}{cc} \dfrac{1}{\tilde{c}^2} & \left(\dfrac{1}{\tilde{c}^2} v + \rho \dfrac{dv}{dp} \right) \\[2mm] \rho \dfrac{dv}{dp} & \left(\rho v \dfrac{dv}{dp} + 1 \right) \end{array} \right] \left\{ \begin{array}{c} \partial p / \partial t \\[2mm] \partial p / \partial x \end{array} \right\} = \left\{ \begin{array}{c} 0 \\[2mm] 0 \end{array} \right\}
\tag{13.1.5}
$$

These equations are homogeneous. If the relation between p and v were arbitrary, their solution would be that both $\partial p / \partial t$ and $\partial p / \partial x$ are zero. These conditions lead to the trivial solution that p is constant. A nontrivial solution is obtained only if the pair of equations cannot be solved. In other words, a nontrivial solution along a characteristic curve exists only if the determinant of the coefficient matrix vanishes. This condition reduces to

$$
\frac{1}{\tilde{c}^2} - \left(\rho \frac{dv}{dp} \right)^2 = 0
\tag{13.1.6}
$$

Thus, we find that

$$
\boxed{ \frac{dv}{dp} = \pm \frac{1}{\rho \tilde{c}} }
\tag{13.1.7}
$$

The equation of state relates ρ and p, and $\tilde{c}$ is obtained by differentiating that relation. Hence, the preceding constitutes a first-order ordinary differential equation that relates p and v on a characteristic. It may be integrated, subject to the condition that the fluid is quiescent in the ambient state, $v = 0$ if $p = 0$. Thus, the integrated relation is

$$
\boxed{ v = \int_0^p \frac{dp'}{\rho \left(p' \right) \tilde{c} \left(p' \right)} }
\tag{13.1.8}
$$

We shall proceed on the assumption that we have evaluated this integral.

It is useful at this juncture to recall the relation between the particle velocity and the pressure fluctuation in a linear plane wave, $\bar{v} = \bar{e} p / (\rho_0 c)$. The propagation here may be in either the positive or negative x direction, so $\bar{e} = \pm \bar{e}_x$, and $\bar{v} = v \bar{e}_x$. Thus, if we were to differentiate the relation for a linear wave, we would obtain $dv/dp = \pm 1 / (\rho_0 c)$. The present nonlinear analysis tells us that the effective characteristic impedance is the value at the current state. An analogous relationship occurs when the force in a spring depends nonlinearly on the strain, which is equivalent to the strain being a function of the force. Then, the incremental force dF required to increase the strain by $d\varepsilon$ depends on the value of the current force. In other words, $dF = K(F) d\varepsilon$, where $K = 1(d\varepsilon/dF)$. For a linear spring, K is the (constant) stiffness, so $K(F)$ is said to be the *tangential stiffness*. If we were to use similar terminology for a nonlinear plane wave, we would call $\rho\tilde{c}$ the tangential characteristic impedance of the fluid.

A corollary of the coefficient matrix in Eq. (13.1.5) having a zero determinant is that the two differential equations are equivalent. When Eq. (13.1.7) applies, both require that

$$\boxed{\frac{\partial p}{\partial t} + (v \pm \tilde{c})\frac{\partial p}{\partial x} = 0}\qquad(13.1.9)$$

Note that the correspondence of alternative signs must be maintained, so a negative sign corresponds to a wave that propagates in the direction of decreasing x.

The preceding tells us that anywhere along a characteristic, $\partial p/\partial t$ is proportional to $\partial p/\partial x$, where the proportionality factor is $-(v \pm \tilde{c})$, whose constancy follows from the fact that both v and $\tilde{c}$ depend on the p value for that characteristic. Constancy of p along a characteristic may be expressed alternatively by equating p at any two (x, t) pairs on the same characteristic. The points we select are differentially apart, so we require that $p(x + dx, t + dt) = p(x, t)$. Application of a Taylor series leads to

$$\left(\frac{\partial p}{\partial x}dx + \frac{\partial p}{\partial t}dt\right)\Bigg|_{\text{constant } p(x,t)} = 0\qquad(13.1.10)$$

When we substitute Eq. (13.1.7), the result is

$$\boxed{\left[\frac{dx}{dt} - (v \pm \tilde{c})\right]\Bigg|_{\text{constant } p(x,t)} = 0}\qquad(13.1.11)$$

Because v and $\tilde{c}$ are constant in this condition, it follows that dx/dt is constant. In other words, the slope dx/dt defining where a specific p occurs in space-time is constant at $\pm\tilde{c} + v$. Stated differently, the characteristics are straight lines! If $\tilde{p}$ is known at position x_0 at time t_0, then the same pressure will occur at position x at time t, with

$$x - x_0 = (v \pm \tilde{c})(t - t_0)\qquad(13.1.12)$$

This simple property has some of profound implications, which we shall now examine.

13.1.2   Interpretation

From a viewpoint of phenomenology, the most important aspect is that the phase velocity of a plane wave that propagates in the direction of increasing x is $(\tilde{c} + v)\bar{e}_x$. The difference between $\tilde{c}$ and the linear speed of sound, c, is a consequence of the variability of the equation of state. Consider an ambient state in which the fluid is at rest and the pressure is the sum of the actual ambient pressure p_0 plus the acoustic perturbation p. Then, $\tilde{c}$ would be the linear sound speed c. The phase speed also differs from c because the particle velocity is nonzero. Suppose that in this ambient state, the fluid flows uniformly at $v\bar{e}_x$. In this condition, the fluid is at rest relative to a reference frame that translates in unison with the ambient flow. That reference frame is inertial, so sound propagates relative to the moving reference frame at speed $\tilde{c}\bar{e}_x$. The propagation velocity as seen by a fixed observer is the velocity relative to the moving reference frame plus the velocity of the reference frame, that is, $\tilde{c}\bar{e}_x + v\bar{e}_x$. In the terminology of nonlinear mechanics, the difference between $\tilde{c}$ and c is said to be a consequence of *material nonlinearity*. The fact that v adds to the speed of sound is said to be a *convective nonlinearity*. Although the two sources of nonlinearity are unrelated in regard to their cause, they affect the propagation speed similarly.

The (linear) d'Alembert solution for plane waves actually is quite similar to the present result. It states that constant values of p propagate along lines for which either $t - x/c$ or $t + x/c$ is constant. These lines in space-time are the characteristics of the linear wave equation. Figure 2.3 describes the characteristics as graphs of t as a function of x. Similar graphs are extremely useful for nonlinear waves. Let us begin with a boundary value problem, in which v at $x = 0$ is a specified function of t.

The propagation is taken to be in the sense of increasing x, so the positive sign preceding $\tilde{c}$ applies in Eq. (13.1.11) and everywhere else. The abscissa is x, so the slope of a characteristic line is $dt/dx = 1/(\tilde{c} + v)$. This slope is less than $1/c$ if $p > 0$. To see why, note that Eq. (13.1.7) in this case gives $dv/dp > 0$, which means that $v > 0$ if $p > 0$. Furthermore, the equation of state for ideal gases and liquids is such that $\tilde{c} > c$ if $p > 0$. Because the slope is less than $1/c$, a characteristic along which $v > 0$ describes propagation that is faster than the linear speed of sound. The converse is true for characteristics associated with $v < 0$. They describe propagation at a speed that is less than c. (It is best to avoid using terms like subsonic or supersonic here.)

To visualize these properties, let us suppose we know the particle velocity at $x = 0$ as a function of time. (This is a simplification of the actual boundary condition for a source, such as a vibrating wall. The proper description of this condition may be found in the next section.) The case of a single period of a sine pulse starting at $t = 0$ is plotted along the t-axis in Fig. 13.1.

Fig. 13.1 Characteristics lines for propagation of the nonlinear wave generated by a sinusoid velocity pulse at $x = 0$

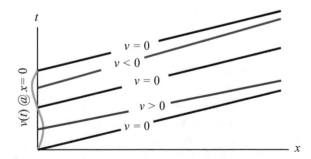

The characteristics through the instants when $v = 0$ at $x = 0$ represent propagation at the linear speed of sound because v and p are zero along them. Because v at the boundary is zero for $t < 0$ and $t > 2\pi/\omega$, the pressure is zero in the region below the lowest characteristic and above the uppermost characteristic. Signals for which $v > 0$ cover the same distance in a shorter elapsed time, whereas signals for which $v < 0$ take longer to travel a given distance. It follows that the shape of a waveform will change as the wave propagates. A linear wave, which is described by the d'Alembert solution, is said to be dispersionless, because all waveforms maintain their shape. In a waveguide, nonplanar modes have a phase speed that depends on the frequency, which means that the waveforms undergo frequency dispersion. The changing shape of the waveform for a nonlinear plane wave is said to be *amplitude dispersion*.

Because the characteristics for different v values are not parallel, it is evident that these lines will eventually intersect. At such intersections, the analysis indicates that two signals having different values of v will exist at the same x and t. This is an impossible condition. What actually happens is that discontinuities, which are called *shocks*, form. We will explore this phenomenon later. For now, we will consider the intersection of characteristics as a limit beyond which the Riemann solution is not valid.

A consequence of p being constant on a characteristic is that lines that originate at instants when p is maximum will retain this property. Similarly, lines that originate from minimum or zero p remain so. We could use this attribute to construct a sketch of a waveform at fixed x or a profile at fixed t. This is one approach described in the next section. However, before we address the quantitative evaluation of a signal, let us consolidate what we have learned.

The pressure and particle velocity are constant along a characteristic, and a characteristic is the locus of points in space-time at which $t - x/(\tilde{c} + v)$ is a specified constant. This constant value is a phase variable for the wave. We shall use τ to denote it, so that

$$p = F(\tau), \quad \tau = t - \frac{x}{(\tilde{c} + v)} \tag{13.1.13}$$

The logical question at this juncture is what is $F(\tau)$? The answer lies in matching the preceding general form to the manner in which the signal is generated.

13.1.3 Boundary and Initial Conditions

Consider the situation where the plane wave is generated by movement of a large piston or a wall that translates as a rigid body. Let $x = X(t)$ denote the (known) position of the wall at any instant. Then, continuity of particle velocity requires that the velocity of the fluid *at the current location* of the piston face equals the piston's velocity, that is,

$$v = \dot{X}(t) \text{ at } x = X(t) \tag{13.1.14}$$

This is referred to as a *moving boundary condition*. Because of the intricate manner in which the state variables are related, we shall take a somewhat reverse approach, in which relations that we seek are taken to be known.

Because the boundary condition imposes a particle velocity, the first step is to remove pressure from the functional dependencies. Toward that end, we invert the expression of v in terms of p obtained from Eq. (13.1.8). This step allows us to consider the relationship to be $p = p(v)$. The implication of this switch of dependencies is that $\tilde{c}$ is known if we know v. This feature is made explicit by letting $\tilde{C}$ be the value of $\tilde{c}$ at a specified v, so that

$$\tilde{C}(v) = \tilde{c}(p(v)) \tag{13.1.15}$$

It follows that the slope of a characteristic is $dx/dt = \tilde{C}(v) + v$.

Given this, let us track a signal that departs from the boundary at time τ. The particle velocity everywhere along the characteristic for this signal is $\dot{X}(\tau)$, so its slope is $\tilde{C}(\dot{X}(\tau)) + \dot{X}(\tau)$. Also, the starting point of the characteristic is $(X(\tau), \tau)$. The time required for this signal to arrive at an arbitrary point (x, t) is $t - \tau$, and the propagation distance to this point is $x - X(\tau)$. Equating $[x - X(\tau)]/(t - \tau)$ to dx/dt yields

$$\frac{x - X(\tau)}{t - \tau} = \tilde{C}(\dot{X}(\tau)) + \dot{X}(\tau) \tag{13.1.16}$$

A minor rearrangement of terms yields a solution whose form is like the one derived by Earnshaw, specifically

$$\boxed{\tau = t - \frac{x - X(\tau)}{\tilde{C}(\dot{X}(\tau)) + \dot{X}(\tau)}, \quad v = \dot{X}(\tau)} \tag{13.1.17}$$

These relations fully describe the signal. To realize this, suppose we select a value of τ. Given a function $X(t)$, we may evaluate X and $\dot{X}$ at $t = \tau$, which sets the value of $\tilde{C}(\dot{X}(\tau))$. Then, the first of the above relations gives the locus of x, t values for passage of this signal, that is, it defines the characteristic along which the particle velocity is $\dot{X}(\tau)$. The value of p along this characteristic may be found from the relation $p = p(v)$ that we would determine from Eq. (13.1.8). Note that the derived solution is consistent with Eq. (13.1.13), because it merely entails measuring the propagation distance from the location of the piston, rather than the reference location at which $x = 0$.

Despite the availability of the general solution in Eq. (13.1.17), it seldom is important to satisfy the moving boundary condition. To see why this is so, consider Fig. 13.2, which depicts the position of the wall in space-time as the curve $x = X(\tau)$ that oscillates about the t-axis. The specific function is one cycle of $X = -(\varepsilon c/\omega)\cos\omega\tau$, which leads to $\dot{X} = \varepsilon c \sin(\omega\tau)$. The parameter ε is the *acoustic Mach number*. Later examination will show that ε is a very small number, with values exceeding 0.001 being extraordinarily uncommon.

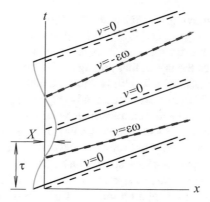

Fig. 13.2 Characteristics originating from the instantaneous position $X(\tau)$ of a moving boundary whose nominal position is $x = 0$. The *solid lines* are the result of satisfying velocity continuity at the instantaneous position of the boundary, whereas the *dashed lines* are the characteristics if the boundary's velocity is approximated as occurring at $x = 0$

Each point on the $x = X(\tau)$ curve is the starting point for a characteristic, where τ is the instant at which the associated signal was generated. The slope of any characteristic is $\Delta x/\Delta t = \tilde{C}\left(\dot{X}(\tau)\right) + \dot{X}(\tau)$. The characteristics described in Fig. 13.2 are those for values of τ at which $\dot{X}$ is either zero, or a maximum, or a minimum. These characteristics are compared in the figure to the nominal characteristic that would result if the value of $\dot{X}(\tau)$ occurred at $x = 0$. For the selected X function, the maximum and minimum values of $\dot{X}$ occur when $X = 0$. These characteristics are the same as the corresponding nominal ones. This arrangement is contrasted with those associated with instants when X is a maximum or minimum. The value of $\dot{X}$ is zero at these instants, which means that the phase speed is c, but these characteristics are shifted by the maximum amount relative to the nominal value.

The key aspect is that satisfying the moving boundary condition shifts a characteristic parallel to the x-axis by $X(\tau)$. This effect is quite different from the effect of the dependence of phase speed on particle velocity, which changes the slope of the characteristics. Consequently, the associated time shift increases proportionally to x. In the vicinity of the boundary, these time shifts might be comparable. However, with increasing x, the velocity dependence of the phase speed overwhelms the moving boundary condition effect.

It follows that if a signal's amplitude is within the range of common experience, the error resulting from applying velocity continuity at the undisplaced position of a boundary is insignificant. This allows us to replace the actual boundary condition in Eq. (13.1.14) with an approximate stationary boundary condition that sets $v = \dot{X}(t)$ at $x = 0$. Correspondingly, the particle velocity anywhere is described by removing $X(\tau)$ from the numerator of Eq. (13.1.17). Thus, the solution reduces to

$$\boxed{v = \dot{X}(\tau), \quad \tau = t - \frac{x}{\tilde{C}(v) + v}} \tag{13.1.18}$$

As in the case of the moving boundary formulation, the corresponding value of p may be found by inverting Eq. (13.1.18). This relation requires specification of the equation of state, which is the topic for the next section.

Figure 13.1 describes the signal that propagates in the positive x direction when the excitation at $x = 0$ is a known function of time. How would the analysis be altered if the domain was $x < 0$? In that case, the negative sign in Eqs. (13.1.7) and (13.1.11) would apply. Hence, the slope of a characteristic would be $dt/dx = -1/(\tilde{c} - v)$. As before $p > 0$ corresponds to $\tilde{c} > c$. However, now $dv/d\rho < 0$, so $p > 0$ corresponds to $v < 0$. Thus, as was true for the previous case, the phase speed of a positive pressure is $\tilde{c} + |v| > c$, and the phase speed of a negative pressure is $\tilde{c} - |v| < c$.

When we studied linear wave propagation, we satisfied arbitrary initial conditions by combining the d'Alembert solution for waves that propagate in either direction. This process cannot be implemented for nonlinear waves. A general reason is that solutions cannot be superposed. If we add solutions for forward and backward propagation, the result will generate product terms that do not fit the general scheme. A more specific reason is that each characteristic line describes a constant pressure value. If we contemplate combining waves that propagate in both directions, characteristic curves for propagation in each direction would intersect. The pressure at any (x, t) would be the sum of the value for the forward and backward characteristics that intersect there. Thus, any attempt to combine forward and backward waves would contradict the condition that the pressure is constant along a characteristic.

The consequence is that we may only use the Riemann solution to satisfy a limited initial value problem. Suppose we know that a forward propagating wave has been established somehow. (The argument is equally applicable to a backward propagating wave.) Further suppose that at some instant, which we designate as $t = 0$, we have measured the profile v as a function of x. This situation is depicted in Fig. 13.3. Other than where they originate from, this diagram shows that there is no difference between the characteristics for this initial value problem and those in the previous figure for a boundary value problem.

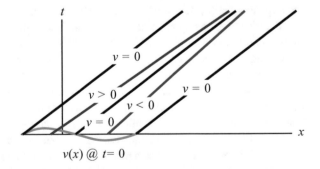

Fig. 13.3 Characteristics for a nonlinear wave that propagates in the sense of increasing x. The properties of this wave are taken to be known at $t = 0$

The task of actually satisfying this type of initial value problem is expedited by a minor modification. The phase variable for a boundary value problem, Eq. (13.1.13), is $\tau = t - x/(\tilde{c} + v)$. This definition facilitates matching the general solution to whatever function is specified at the boundary. For the initial value problem, we wish to match the general solution to whatever x function is specified at $t = 0$. This process is assisted by redefining the phase variable to be $\theta = x - (\tilde{c} + v)t$, so that $\theta = x$ at $t = 0$. The corresponding form of the general solution is

$$\boxed{p = f(\theta), \quad \theta = x - (\tilde{c} + v)t} \tag{13.1.19}$$

This is the form obtained by Poisson.[6]

13.1.4 Equations of State

If we wish to construct waveforms (p as a function of t at specified x) and profiles (p as a function of x at specified t), it is necessary to evaluate the integral in Eq. (13.1.8), which relates the values of p and v along a characteristic. Doing so requires specification of the equation of state, which relates p and ρ. We begin with the case of an ideal gas that is compressed and expanded adiabatically,

$$\boxed{\frac{p}{p_0} + 1 = \left(\frac{\rho}{\rho_0}\right)^\gamma \iff \frac{\rho}{\rho_0} = \left(\frac{p}{p_0} + 1\right)^{1/\gamma}} \tag{13.1.20}$$

The linear speed of sound in an ideal gas is $c^2 = \gamma p_0/\rho_0$. We use this relation to eliminate the ambient pressure, which converts the preceding relations to

$$\frac{p}{\rho_0 c^2} = \frac{1}{\gamma}\left(\frac{\rho}{\rho_0}\right)^\gamma - 1 \iff \frac{\rho}{\rho_0} = \left(\frac{\gamma p}{\rho_0 c^2} + 1\right)^{1/\gamma} \tag{13.1.21}$$

From this, we find that

$$\boxed{\frac{\tilde{c}}{c} = \left(\frac{\gamma p}{\rho_0 c^2} + 1\right)^{\frac{\gamma-1}{2\gamma}}} \tag{13.1.22}$$

To determine the relation between p and v, we substitute ρ from Eq. (13.1.21) and $\tilde{c}$ from the preceding into Eq. (13.1.8), which becomes

$$v = \frac{1}{\rho_0 c} \int_0^p \left(\frac{\gamma p'}{\rho_0 c^2} + 1\right)^{-\left(\frac{\gamma+1}{2\gamma}\right)} \left(\frac{\rho_0 c^2}{\gamma}\right) d\left(\frac{\gamma p'}{\rho_0 c^2}\right) \tag{13.1.23}$$

---

[6]S.D. Poisson, " Memoire sr la théorie de son," J. l'École Polytechnique Paris, Vol. 7 (1808) pp. 319–392.

The result is

$$v = c\left(\frac{2}{\gamma-1}\right)\left[\left(\frac{\gamma p}{\rho_0 c^2}+1\right)^{\left(\frac{\gamma-1}{2\gamma}\right)}-1\right] \tag{13.1.24}$$

This equation may be solved for p as a function of v, which is

$$\frac{p}{\rho_0 c^2}=\frac{1}{\gamma}\left[\left(1+\frac{\gamma-1}{2}\frac{v}{c}\right)^{2\gamma/(\gamma-1)}-1\right] \tag{13.1.25}$$

We can gain some insight into the significance of nonlinearity by examining graphs of $\tilde{c}$ and v as functions of p; Fig. 13.4 describes both functions nondimensionally for the case where the gas is air, $\gamma = 1.4$. The dotted lines are the relations for linear acoustics, $\rho = \rho_0 + p/c^2$, $\tilde{c} = c$, $v = p/(\rho_0 c)$.

Fig. 13.4 Nonlinear dependence of ρ, $\tilde{c}$, and v on the pressure in an ideal gas for $\gamma = 1.4$. The analogous relation for a linearized equation of state is shown as a *dotted line*

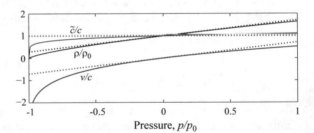

The linear acoustic approximations are very close to the corresponding exact curve in the range $|p|/p_0 < 0.2$. This range corresponds to an acoustic amplitude that is 20% of the ambient pressure. In air, for which p_0 is 1 atm, this amplitude is approximately 20 kPa, which corresponds to a harmonic signal at 217 dB//20 μPa. Such a signal is beyond anything that would be encountered in a conventional acoustic application.

The fact that these relations are close to those of linear theory does not mean that nonlinearity is unimportant, because the phase variable τ is not simply $t - x/c$. One could say that the linearized analysis gives the correct solution, except that it places that solution in the wrong space-time location.

The logical question at this juncture is: What modifications are required if the fluid is a liquid? The equation of state for an adiabatic process in a liquid is given by Eq. (2.3.18),

$$\frac{p}{\rho_0 c^2}=\left(\frac{\rho}{\rho_0}-1\right)+\frac{1}{2}\frac{B}{A}\left(\frac{\rho}{\rho_0}-1\right)^2+\cdots \tag{13.1.26}$$

This series relation is derived experimentally. The value of A is the bulk modulus, $A = K = \rho_0 c^2$. It is convenient to consider B/A to be a single coefficient, because that is the manner in which B arises in nonlinear waves. The value of B/A for water ranges from 4.2 for distilled water at atmospheric pressure and 0°C to 6.2 for distilled water between 200 and 4000 atm at 30°C. A nominal value $B/A = 5$ may be used if environmental conditions are unspecified. It also is the value for seawater at atmospheric pressure, 20°C, and 35 parts per thousand salinity.[7]

The equation of state for a liquid contains only two terms because the bulk modulus is very large. For example, $\rho_0 c^2 = 2.2$ GPa for water. Hence, large values of $\rho/\rho_0 - 1$ lead to very large values of p that are outside the usual realm of acoustics. Indeed, it is challenging to generate pressures that are sufficiently large to measure the coefficients of the cubic and higher terms in the equation of state. To illustrate this fact, a harmonic signal at 260 dB//1 µPa in water corresponds to $|\rho/\rho_0 - 1| = 0.0065$.

A corollary of halting the equation of state at quadratic terms is that any constitutive parameters that are derived from it may be truncated at a consistent level. To formulate Riemann's solution, we need the density as a function of pressure. A Taylor series expansion of the quadratic equation gives

$$\frac{\rho}{\rho_0} = 1 + \frac{p}{\rho_0 c^2} - \frac{1}{2}\frac{B}{A}\left(\frac{p}{\rho_0 c^2}\right)^2 + \cdots \tag{13.1.27}$$

Differentiation of Eq. (13.1.26) gives

$$\frac{\tilde{c}}{c} = \left[1 + \frac{B}{A}\left(\frac{\rho}{\rho_0} - 1\right)\right]^{1/2} = 1 + \frac{B}{2A}\left(\frac{p}{\rho_0 c^2}\right) + \cdots \tag{13.1.28}$$

In turn, the differential equation relating v to p becomes

$$\frac{dv}{dp} = \frac{1}{\rho \tilde{c}} = \frac{1}{\rho_0 c \left(1 + \dfrac{p}{\rho_0 c^2} + \cdots\right)\left[1 + \dfrac{B}{2A}\left(\dfrac{p}{\rho_0 c^2}\right) + \cdots\right]} \tag{13.1.29}$$

A consistent series truncation of a binomial series simplifies the differential equation to

$$\frac{dv}{dp} = \frac{1}{\rho_0 c}\left[1 - \left(\frac{B}{2A} + 1\right)\left(\frac{p}{\rho_0 c^2}\right) + \cdots\right] \tag{13.1.30}$$

Separating variables with the lower limits of the integral set to give $v = 0$ if $p = 0$ yields

$$\frac{v}{c} = \left(\frac{p}{\rho_0 c^2}\right) - \frac{1}{2}\left(\frac{B}{2A} + 1\right)\left(\frac{p}{\rho_0 c^2}\right)^2 + \cdots \tag{13.1.31}$$

---

[7]R.T. Beyer, "The Parameter B/A," M.F. Hamilton & D.T. Blackstock, *Nonlinear Acoustics*, eds., Acoustical Society of America (2008) p. 34.

The inverse relation is

$$\frac{p}{\rho_0 c^2} = \frac{v}{c} + \frac{1}{2}\left(\frac{B}{2A} + 1\right)\left(\frac{v}{c}\right)^2 + \cdots \tag{13.1.32}$$

The expressions for all quantities derived from the equation of state for a liquid have been truncated on the basis that $|p|/(\rho_0 c^2)$ is very small. Although the bulk modulus of any gas is much less than that of water, it nevertheless is large, so $|p|/(\rho_0 c^2)$ for a signal in an ideal gas typically is much less than one. Series expansions of the basic quantities based on taking $|p|/(\rho_0 c^2)$ to be much less than one are said to constitute *small-signal approximations*. Waves that have this feature are said to be *weakly nonlinear*. Acoustical studies seldom need to consider strong nonlinearity.

In such circumstances, the relations for ideal gases and liquid may be combined. To see how this is achieved, consider a Taylor series in powers of $p/(\rho_0 c^2)$ of the equation of state for an ideal gas, Eq. (13.1.21). When this series is truncated at the first term beyond the expression for linear theory, the result is

$$\frac{\rho}{\rho_0} = 1 + \frac{p}{\rho_0 c^2} - \frac{\gamma - 1}{2}\left(\frac{p}{\rho_0 c^2}\right)^2 + \cdots \tag{13.1.33}$$

This relation and Eq. (13.1.27) are power series. Changing B/A in the latter to $\gamma - 1$ yields the expression for a gas. A description that covers both media is obtained by *coefficient of nonlinearity* defining a β such that

$$\boxed{\beta = \begin{cases} \dfrac{1}{2}(\gamma + 1) : \text{Ideal gas} \\ \dfrac{B}{2A} + 1 : \text{Liquid} \end{cases}} \tag{13.1.34}$$

Because the equations of state now match, the expressions for $\tilde{c}$ and v in terms of p derived from them also will match. The unified constitutive relations that result are

$$\boxed{\begin{aligned} \frac{\rho}{\rho_0} &= 1 + \frac{p}{\rho_0 c^2} - (\beta - 1)\left(\frac{p}{\rho_0 c^2}\right)^2 + \cdots \\ \frac{\tilde{c}}{c} &= 1 + (\beta - 1)\left(\frac{p}{\rho_0 c^2}\right) + \cdots \\ \frac{p}{\rho_0 c^2} &= \frac{v}{c} + \cdots \end{aligned}} \tag{13.1.35}$$

The coefficient of nonlinearity is defined as it is in Eq. (13.1.34) because the difference between the nonlinear and linear phase speeds is proportional to β, specifically

$$\tilde{c} + v = c + \beta v = c + \beta \frac{p}{\rho_0 c} \qquad (13.1.36)$$

Smallness of $p/(\rho_0 c^2)$ has the further implication that it usually is acceptable to use $v = p/(\rho_0 c)$, which is the same as the relation according to linear theory. In that case, we may shorten the analysis slightly. Instead of solving Eq. (13.1.18) for v and then evaluating the corresponding p, we may convert that relation to one that governs p directly,

$$\boxed{p = \rho_0 c \dot{X}(\tau), \quad v = \dot{X}(\tau), \quad \tau = t - \frac{x}{c\left(1 + \beta p/\left(\rho_0 c^2\right)\right)}} \qquad (13.1.37)$$

where the boundary condition is taken as $v = \dot{X}(t)$ at $x = 0$.

In summary, four fundamental aspects collectively constitute the small-signal approximation:

- Continuity of particle velocity at a vibrating surface is satisfied at the static reference position of that boundary.
- Liquids and gases are governed by the same equations.
- The pressure and particle velocity for a plane wave propagating in the direction of increasing x are related by $p = \rho_0 c v$.
- The phase speed along a characteristic is $c + \beta p/(\rho_0 c)$.

Each of these simplifies some aspect of an analysis, but none is essential.

13.1.5 Quantitative Evaluations

Now that we know $\tilde{c}$ and v as functions of p, the next task it to extract profiles and waveforms from the Riemann solution. We will address the boundary value solution in Eq. (13.1.13), but the developments are readily modified to treat the initial value solution in Eq. (13.1.19). It is convenient to nondimensionalize the velocity excitation at $x = 0$ by setting

$$\boxed{\dot{X}(\tau) = \varepsilon c V(t)} \qquad (13.1.38)$$

The parameter ε is the acoustic Mach number we encountered earlier. If we define $V(t)$ such that $\max(|V|) = 1$, then ε is the ratio of the maximum particle velocity to the linear speed of sound. In the small-signal approximation, ε is very small.

Graphical Construction

A useful approximation in some situations is a representation of $V(t)$ as a piecewise linear function, that is, a sequence of straight lines with a few locations at which the slope changes. As the signal propagates, waveforms and spatial profiles of pressure retain that attribute if the small-signal approximations apply. Consequently, it is only necessary to follow the propagation of the phases at which the slope of the initial waveform changes. After those phases are located, the intervals between them remain straight lines.

Suppose τ is an instant at which the slope of $V(\tau)$ changes. The particle velocity for this phase is $\varepsilon c V(\tau)$, and the small-signal approximation of the pressure is $p(\tau) = \varepsilon \rho_0 c^2 V(\tau)$. To construct a waveform, the value of x is fixed. According to Eq. (13.1.36), the phase speed is $c[1 + \beta \varepsilon V(\tau)]$. Hence, the time required for this signal to propagate from the origin to the specified x is $x/[c + \beta \varepsilon c V(\tau)]$. By definition, this signal was generated at $x = 0$ when $t = \tau$, so the corresponding arrival time of $p(\tau)$ is $\tau + x/[c + \beta \varepsilon c V(\tau)]$. A plot of the waveform observed at x would show this time as the abscissa and $p(\tau)$ as the ordinate. This process is repeated for each value of τ at which $V(\tau)$ changes slope, after which the waveform is obtained by joining the plotted points with straight lines. The procedure is illustrated in Fig. 13.5 for a triangular pulse whose duration is T.

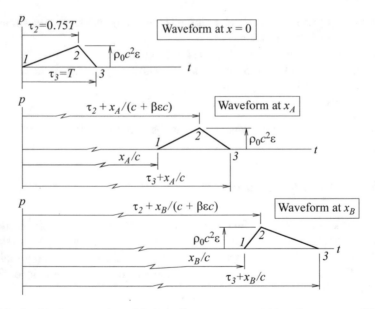

Fig. 13.5 Graphical construction of the waveforms at two spatial locations $x_A < x_B$. The phase speed of the signal at each instant of slope discontinuity is constant

There are three instants along which the slope changes discontinuously, $\tau_1 = 0$, $\tau_2 = 0.75T$, and $\tau_3 = T$. The particle velocity at τ_2 is the maximum value εc. The first plot is the waveform at $x = 0$, which is obtained by setting $p = \rho_0 c \varepsilon V(t)$. Because $v(\tau_1) = v(\tau_3) = 0$, the phase speed of these signals is the linear speed of sound, c. The time required for $p(\tau_1)$ and $p(\tau_3)$ to propagate to a position $x_A > 0$ is x_A/c. The pressure along the characteristic line associated with τ_2 is $p = \rho_0 c^2 \varepsilon$. The phase speed along this characteristic is constant at $c + \beta \varepsilon c V(\tau)$. The propagation time required for $p(\tau_2)$ to arrive at x_A is $x_A/(c + \beta\varepsilon c V(\tau))$, which is less than the travel time for $p(\tau_1)$ and $p(\tau_3)$. Thus, the waveform seems to begin to lean in the sense of earlier time. (If the waveform has an interval in which p is negative, that portion of the waveform will seem to lean in the sense of later time.)

To obtain the waveform at a farther location x_B distance, we can evaluate the arrival time for each phase by adding the departure time τ to x_B divided by the phase speed for that pressure, as is done in Fig. 13.5. An alternative is to use the waveform at x_A as the reference. The propagation time from x_A to x_B is $x_B - x_A$ divided by the phase speed corresponding to the pressure on that characteristic.

At a certain x, the arrival time of $p(\tau_2)$ will equal that of $p(\tau_1)$. Such a condition marks the formation of a shock. Equating the two arrival times gives $\tau_2 + x/(c + \beta\varepsilon c) = \tau_1 + x/c$. This value of x is the *shock formation distance*,

$$x_{\text{shock}} = \frac{c + \beta\varepsilon c}{\beta\varepsilon c} c (\tau_2 - \tau_1) \qquad (13.1.39)$$

(This expression could be simplified by approximating the numerator as c based on $\varepsilon \ll 1$.) The value of x_{shock} depends on the magnitude and shape of the initial waveform. For now, we regard this as the limiting range for our evaluations.

Evaluation of a spatial profile follows a similar procedure. We select τ as the phase of a significant feature we wish to follow. The phase speed of this feature is $c + \beta\varepsilon c$. It departed the boundary at time τ. At a designated instant t, the propagation time is $t - \tau$, so the feature will be situated at $x = (c + \beta\varepsilon c)t$.

Figure 13.6 depicts this construction for the initial waveform in Fig. 13.5. The first profile describes $t_A = \tau_3$. This is the instant when $p(\tau_3)$ leaves the boundary, so $x_3 = 0$. Because $p(\tau_1) = 0$, its phase speed is c and its location is $x_1 = ct_A$. The propogation time for $p(\tau_2)$ is $t_A - \tau_2$. According to the Riemann solution, its location

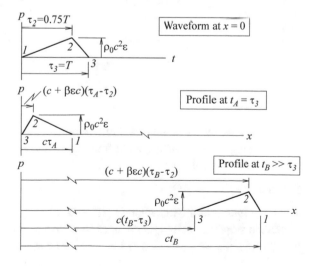

Fig. 13.6 Construction of spatial profiles at various instants

is $x_2 = c(t_A - \tau_2) + \beta \varepsilon c(t_A - \tau_2)$. However, the pulse is very brief. Consequently, $c(t_A - \tau_2)$ is small, and $\beta \varepsilon c(t_A - \tau_2)$ is a negligible addition. Thus, at this early instant the spatial profile differs little from the prediction of linear theory. The second profile is for a much later instant t_B. The phase speed of $p(\tau_1 = 0)$ and $p(\tau_3)$ is c, so the spatial extent of the pulse remains at $c\tau_3$. However, the profile changes shape because the phase speed of $p(\tau_2)$ is greater than c. Because $p(\tau_2)$ propagates farther than the beginning and end of the pulse, the profile seems to lean in the sense of increasing x. (Regions where the pressure is negative would seem to lean in the sense of decreasing x.)

With increasing time, the profile propagates farther, and the $p(\tau_2)$ signal increasingly advances relative to the $p(\tau_1)$ and $p(\tau_3)$ signals. Eventually, it catches up to $p(\tau_1)$, which marks the advent of a shock. The instant t_{shock} when this condition occurs may be found by equating the locations where $p(\tau_1)$ and $p(\tau_2)$ occur. The shock formation distance $x_{\text{shock}} = ct_{\text{shock}}$ is the same as Eq. (13.1.39).

EXAMPLE 13.1 The waveforms described in the sketch are alternative velocity inputs at $x = 0$. Their only difference is phase-reversal. For each case, determine the shock formation distance, x_{shock}. Then, evaluate each waveform at $x = x_{\text{shock}}/2$ and $x = x_{\text{shock}}$. The value of v_0 is sufficiently small to permit considering the propagation to be weakly nonlinear.

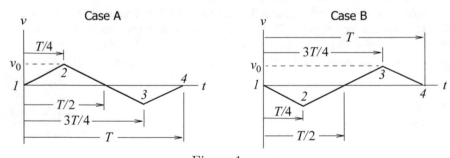

Figure 1.

Significance

Graphical constructions enhance intuitive understanding, which is further enhanced by examining the similarities and differences between the propagation properties of the alternative waveforms.

Solution

The input functions are the velocity at $x = 0$, where $\tau = t$. There are four instants at which the slope of each input changes. For case A, $v_1^{(A)} = 0$ at $\tau = 0$, $v_2^{(A)} = v_0$ at $\tau = T/4$, $v_3^{(A)} = -v_0$ at $\tau = 3T/4$, and $v_4^{(A)} = 0$ at $\tau = T$. For case B, the velocities at τ_2 and τ_3 are the negative of the value for case A. According to the weakly nonlinear approximations, $p/(\rho_0 c^2) = v/c$, so the equations of the relevant characteristics are

$$
\begin{array}{cc}
\text{Case A} & \text{Case B} \\[4pt]
t_1 = \dfrac{x}{c} & t_1 = \dfrac{x}{c} \\[10pt]
t_2 = \dfrac{T}{4} + \dfrac{x}{c + \beta v_0} & t_2 = \dfrac{T}{4} + \dfrac{x}{c - \beta v_0} \\[14pt]
t_3 = \dfrac{3T}{4} + \dfrac{x}{c - \beta v_0} & t_3 = \dfrac{3T}{4} + \dfrac{x}{c + \beta v_0} \\[14pt]
t_4 = T + \dfrac{x}{c} & t_4 = T + \dfrac{x}{c}
\end{array}
$$

A shock forms at the location where characteristics for different values of τ intersect. The denominator of the x term for t_2 in case A is bigger than the denominator for t_1, so there is a value of x at which t_2 and t_1 are equal. Similar reasoning suggests that it is possible that $t_4 = t_3$ in case A, whereas in case B the only possibility is that $t_3 = t_2$. These conditions occur at

$$
\text{Case A: } t_2 = t_1 \implies \frac{x}{c} - \frac{x}{c + \beta v_0} = \frac{T}{4} \implies \beta \frac{v_0}{c} x = (c + \beta v_0) \frac{T}{4}
$$

$$
\text{Case A: } t_4 = t_3 \implies \frac{x}{c - \beta v_0} - \frac{x}{c} = \frac{T}{4} \implies \beta \frac{v_0}{c} x = (c - \beta v_0) \frac{T}{4}
$$

$$
\text{Case B: } t_3 = t_2 \implies \frac{x}{c - \beta v_0} - \frac{x}{c + \beta v_0} = \frac{T}{2} \implies 2\beta \frac{v_0}{c} x = \frac{c^2 - (\beta v_0)^2}{c} \frac{T}{2}
$$

The value of v_0/c is stated to be sufficiently small to justify using weak shock approximation, so the expression for x in each case reduces to

$$
x_{\text{shock}} = \frac{cT}{4\beta (v_0/c)}
$$

To plot the waveforms at $x = x_{\text{shock}}/2$, we substitute $x_{\text{shock}}/2$ into the expressions for each t_j and then plot each time value against the corresponding pressure $\rho_0 c v (\tau_j)$. The time values are simplified by using the smallness of v_0/c to approximate $1/(c \pm \beta v_0)$ by the first two terms of a binomial series. For example,

$$
\text{At } x = \frac{x_{\text{shock}}}{2}, \quad t_2 \approx \frac{T}{4} + \frac{\left(1 - \beta \dfrac{v_0}{c}\right)}{c} \frac{cT}{8\beta (v_0/c)} = \frac{T}{8} + \frac{x_{\text{shock}}}{2c}
$$

The full set of values is

$$\text{Case A } (x = x_{\text{shock}}/2) \quad \text{Case B } (x = x_{\text{shock}}/2)$$

$$t_1 = \frac{x_{\text{shock}}}{2c} \qquad\qquad t_1 = \frac{x_{\text{shock}}}{2c}$$

$$t_2 = \frac{T}{8} + \frac{x_{\text{shock}}}{2c} \qquad t_2 = \frac{3T}{8} + \frac{x_{\text{shock}}}{2c}$$

$$t_3 = \frac{7T}{8} + \frac{x_{\text{shock}}}{2c} \qquad t_3 = \frac{5T}{8} + \frac{x_{\text{shock}}}{2c}$$

$$t_4 = T + \frac{x_{\text{shock}}}{2c} \qquad t_4 = T + \frac{x_{\text{shock}}}{2c}$$

The waveform corresponding to these instants is shown in Fig. 2.

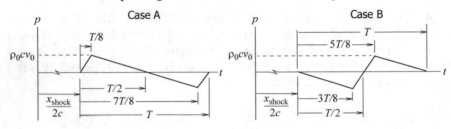

Figure 2.

The same procedure with x set to x_{shock} leads to

$$\text{Case A } (x = x_{\text{shock}}) \quad \text{Case B } (x = x_{\text{shock}})$$

$$t_1 = \frac{x_{\text{shock}}}{c} \qquad\qquad t_1 = \frac{x_{\text{shock}}}{c}$$

$$t_2 = \frac{x_{\text{shock}}}{c} \qquad\qquad t_2 = \frac{T}{2} + \frac{x_{\text{shock}}}{c}$$

$$t_3 = T + \frac{x_{\text{shock}}}{c} \qquad t_3 = \frac{T}{2} + \frac{x_{\text{shock}}}{c}$$

$$t_4 = T + \frac{x_{\text{shock}}}{c} \qquad t_4 = T + \frac{x_{\text{shock}}}{c}$$

Figure 3 shows that both waveforms have indeed developed a shock at x_{shock}.

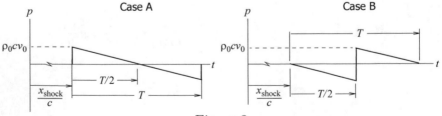

Figure 3.

Although it might seem as though there is a fundamental difference between the signals generated by the alternative inputs, they actually are quite similar. Rather than considering case B to be the negative of case A at $x = 0$, the similarities become evident if we compare the behavior of the positive and negative phases. In comparison with case A, the positive phase in case B is retarded in time by $T/2$, and the negative phase is advanced by $T/2$. From this perspective, the corresponding phases in each case are altered in the same way as the signal propagates. Both waveforms have the shape of teeth of a sawblade. We will see that this is a tendency that is shared by all oscillatory waveforms.

Computation of Characteristics Graphical construction of waveforms and profiles is impractical if the input is not piecewise linear. It also is inappropriate if we believe that the acoustic Mach number is too large to consider the signal to be weakly nonlinear. The method developed here is a computational version of the graphical algorithm. It allows for evaluation of as many points as necessary to resolve the input waveform, so it can be used in any situation.

We begin with the weakly nonlinear signal. In the graphical method, a value of the phase variable τ is selected. The initial velocity $v(\tau)$ is used to evaluate the associated pressure $p(\tau)$, from which either x or t is determined by solving the equation for the nonlinear phase speed of that signal. The concept is to use a computer algorithm to perform these operations for many values of τ in situations where the input function is complicated.

The first step is to discretize the interval during which the input function is nonzero. At $x = 0$, τ and t are equal, so we let τ_n denote the sampled instants. (Usually we select a uniform increment, so that $\tau_n = n\Delta$, but doing so is not essential.) The corresponding set of particle velocities is $v_n = \varepsilon c V(\tau_n)$, and the pressures are $p_n = \varepsilon \rho_0 c^2 V(\tau_n)$.

If we wish to construct the waveform at some point x_A, we solve Eq. (13.1.36) for the time t_n corresponding to the selected values of x_A and τ_n, which gives

$$t_n = \tau_n + \frac{x_A}{c(1 + \beta \varepsilon V(\tau_n))} \tag{13.1.40}$$

Plotting p_n as the ordinate and t_n as the abscissa for all n yields the waveform. To obtain the profile at a specific instant t_A, we solve the characteristic's equation for the location x_n,

$$x_n = (1 + \beta \varepsilon V(\tau_n)) c(t_A - \tau_n) \tag{13.1.41}$$

The desired profile results from plotting p_n versus x_n.

Unlike the graphical method, the procedure is readily modified to handle a situation where $\varepsilon \equiv \max(v)/c$ is sufficiently large that the weakly nonlinear approximations might not be valid. It is seldom necessary to consider this possibility for a liquid, so Eq. (13.1.25) for an ideal gas would be the governing equation of state. Rather than using $p_n = \rho_c c v_n$ to evaluate the pressure associated with characteristic τ_n, we would use Eq. (13.1.25). The corresponding value $\tilde{c}_n$ would be found from Eq. (13.1.22). Characteristic lines in this case are described by the original expression, Eq. (13.1.13), so the relation between x and t for the signal that departed the boundary at time τ is

$$x = (t - \tau_n)(\tilde{c}_n + v_n) \qquad (13.1.42)$$

In all other respects, the evaluation would proceed as it does when the signal is weakly nonlinear.

EXAMPLE 13.2 A plane wave in air is induced by a harmonically varying velocity at $x = 0$. The motion begins at $t = 0$, and the velocity amplitude is $0.002c$. Evaluate the spatial distribution of the pressure signal at $t = 50T$, where T is the period of the boundary velocity. Also determine the waveform at $x = 50\lambda$, where λ is the wavelength according to linear theory.

Significance

Application of the general procedure for tracing signals along characteristic lines serves to enhance our ability to implement the method for other problems. Furthermore, harmonic signals are the fundamental building blocks of linear acoustics, so using such a signal as the exemplar will bring to the fore some important phenomena.

Solution

The excitation is $v = \varepsilon c \sin(\omega \tau) h(t)$ at $x = 0$ with $\varepsilon = 0.002$ as specified. For reference, $\varepsilon = 2\left(10^{-3}\right)$ corresponds to $140\,\mathrm{dB//20\,\mu Pa}$. The fluid is specified to be air, so $\gamma = 1.4$. The sound pressure level is quite high. For this reason, we should assess whether it is appropriate to use the small-signal approximations. To do so, let us consider the signal at the instant when the velocity is εc. According to Eq. (13.1.25), $v/c = 0.002$ corresponds to $p = 0.0020024\rho_0 c^2$, whereas the small-signal approximations give $p = 0.002\rho_0 c^2$. The error for the latter value is 0.1%, so application of the small-signal approximations is justified.

There is no need to specify the frequency if we work nondimensionally. To that end, we write the equation for the characteristic as

$$\omega \tau_n = \omega t - \frac{kx}{(1 + \beta v_n/c)}$$

In other words, we take the characteristic value to be $\omega \tau_n$ and consider the coordinates in space-time to be kx and ωt.

Before we can select the τ_n values, we must decide the interval to be discretized. The velocity input at $x = 0$ begins at $t = 0$, at which instant $v = 0$. This corresponds to $p = 0$, so the characteristic line through the origin is $t - x/c = 0$. Thus, at any instant t, the signal will extend in space from $x = 0$ to $x = ct$. Standard terminology is to say that $x = ct$ is the *leading edge*. We seek the profile for $t = 50(2\pi/\omega)$, so the range to be sampled is $0 \le \omega \tau \le 100\pi$. To assure that the sampling leads to smooth curves, the sampling rate is selected to be 20 per period, so $\omega \tau_n = n(2\pi//20)$,

$n = 0, 1, \ldots 1000$. In regard to evaluation of waveforms, the acoustic wavelength is $\lambda = 2\pi/k$, so we must determine the waveform at $kx = 100\pi$. It will be proven later that a periodic excitation at the boundary leads to waveforms having the same period. Consequently, we only need to see one period of the waveform. The sampling of $\omega\tau_n$ for the evaluation of profiles is adequate for this purpose.

The algorithm is implemented by defining a column vector of τ_n values. The column vector of velocity values at the sampled instants is $v_n/c = \varepsilon \sin(\omega\tau_n)$. The pressure corresponding to each v_n is $p_n/(\rho_0 c^2) = v_n/c$. To find a spatial profile at time ωt, we solve the characteristic definition for the kx_n value associated with p_n at this ωt, so that $kx_n = (1 + \beta\varepsilon v_n/c)(\omega t - \omega\tau_n)$. Waveforms result from solving Eq. (1) for ωt_n at location kx_n corresponding to pressure p_n, which gives $\omega t_n = \omega\tau_n + kx/(1 + \beta v_n/c)$.

Figure 1 describes the profile at $\omega t = 100\pi$. (Division of kx_n by 2π gives distance as a fraction of a linear wavelength. Division of ωt_n by 2π gives time as a fraction of the period.) The first plot displays the full extent of the signal. The compressed nature of the x scale makes it difficult to see specific features, so the second graph shows the profile from the boundary outward to five wavelengths. Nothing remarkable is evident—the distribution seems to be the sine curve predicted by linear theory. The third plot shows the region five wavelengths back from the wavefront, $x = ct$. The wave in this region seems to be leaning forward in space. This is a consequence of the positive pressure phase propagating faster than c, while the negative phase propagates slower than c.

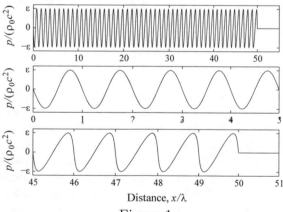

Figure 1.

What would happen if we were to perform the same construction for a much larger time? The leaning effect will be so exaggerated that the computation would yield a multivalued profile. This behavior is exhibited in Fig. 2, which describes the forward part of the spatial profile at $\omega t = 300\pi$. Only one

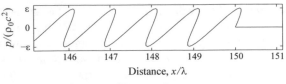

Figure 2.

pressure can exist at a specific location and instant, so *it is evident that there is something wrong*. What is missing is the formation of a shock, at which the pressure changes discontinuously. The instant at which the shock forms is that at which a multivalued condition first occurs. Because the positive lobe at the leading edge, $x = ct$, leans forward the most, occurrence of a shock is first manifested at the leading edge as infinite $\partial p / \partial x$. We will investigate shocks in a later section, where we will find that an infinite gradient is a general criterion marking shock formation.

The time required for a positive pressure to propagate to a specific x is less than it is for a negative pressure. Therefore, the leaning appearance of waveforms is in the opposite sense of a spatial profile. This feature is evident in Fig. 3. With increasing x, the leaning tendency increases, until a shock forms at a certain distance. The first occurrence of a shock in a waveform is manifested by a vertical tangent at some instant, at which $\partial p / \partial t$ is infinite.

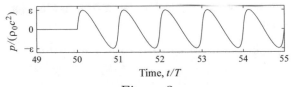

Figure 3.

The waveform has other interesting properties. It is zero until $t = x/c$ and then, it is periodic at the period T of the boundary excitation. Periodicity follows from the fact that the characteristic lines in x, t space corresponding to a periodic $v(\tau)$ form a repeated pattern. This attribute will be proven to be general. A periodic waveform generated by a periodic excitation may be represented as a Fourier series. Thus, another effect of nonlinearity is to create harmonics whose amplitude is dependent on the propagation distance. It might happen that the some harmonics are depleted, while others are generated or enhanced. In any event, propagation leads to an ever increasing level of distortion. Distortion is another effect that will be explored in greater detail.

Solution of a Nonlinear Algebraic Equation

The method of constructing profiles and waveforms by computing x or t along a characteristic is reliable and readily programmed, so why seek another method? One reason is that analyses that include dissipation or nonplanar effects might be posed in a different form. A more common reason lies in the fact that the interval between the instants at which a waveform is computed is not uniform, that is, a constant increment

$\tau_{n+1} - \tau_n$ does not lead to a constant increment $t_{n+1} - t_n$. The consequence of the waveform being known discretely at instants that are not uniformly spaced is that standard FFT routines cannot be used to decompose a waveform. This is a serious shortcoming, because the harmonic content of the distorted wave often is of primary interest.

The concept underlying the method we shall now develop is that the Riemann solution may be assembled as a nonlinear equation for p as a function of x and t. Such an equation may be solved with the aid of numerical methods. We will derive the method for the boundary value problem, but the procedure may be modified with little effort to address the initial value problem. As before, the boundary condition is $v = \varepsilon c V(t)$ at $x = 0$.

Let us begin by considering the procedure when ε is sufficiently small to warrant considering the signal to be weakly nonlinear. Correspondingly, we set $p = \rho_0 c^2 \varepsilon V(t)$ at $x = 0$. The same approximation is used to replace v/c in the equation for the characteristics. Thus, the nonlinear plane wave matching the boundary velocity is defined by

$$p = \varepsilon \rho_0 c^2 V \left(t - \frac{x/c}{1 + \beta p / (\rho_0 c^2)} \right) \qquad (13.1.43)$$

It is evident that Eq. (13.1.43) is a function relating x, t, and p. Cases where p and either x or t are specified are equivalent to the previous method, which used constancy of the phase speed along a characteristic. Our interest here is in finding the value of p at specified x and t. The equation cannot be solved analytically, so we use numerical methods. It is useful to nondimensionalize the pressure relative to the bulk modulus, such that

$$q = \frac{p}{\rho_0 c^2} \qquad (13.1.44)$$

Correspondingly, the equation to be solved is

$$G(q, x, t) \equiv q - \varepsilon V \left(t - \frac{x/c}{1 + \beta q} \right) = 0 \qquad (13.1.45)$$

Many methods have been developed to find the roots of a nonlinear algebraic equation. One that works well for the function described above is Newton's method, which is sometimes referred to as the method of tangents. Let $q^{(j)}$ denote the jth iteration. The next iteration is described by

$$q^{(j+1)} = q^{(j)} - \left(\frac{G(q, x, t)}{\frac{\partial}{\partial q} G(q, x, t)} \right) \Bigg|_{q = q^{(j)}} \qquad (13.1.46)$$

The derivative of the G function in Eq. (13.1.45) is

$$\frac{\partial}{\partial q} G(q, x, t) = 1 - \varepsilon \dot{V} \left(t - \frac{x/c}{1 + \beta q} \right) \frac{\beta x/c}{(1 + \beta q)^2} \tag{13.1.47}$$

where $\dot{V}()$ denotes the time derivative of the excitation function.

A straightforward modification allows for evaluation of the signal in situations where the signal is strongly nonlinear. In that case, the phase speed as a function of pressure is $\tilde{c}(p) + v(p)$, which are the functions given by Eqs. (13.1.22) and (13.1.24), respectively. The corresponding function representing the Riemann solution is

$$G(q, x, t) \equiv q - \varepsilon V \left(t - \frac{x}{\tilde{c}(\rho_0 c^2 q) + v(\rho_0 c^2 q)} \right) = 0 \tag{13.1.48}$$

Carrying out a numerical solution by Newton's method requires the value of $\partial G/\partial q$ corresponding to the latest iterative value of $q^{(j)}$. The expression for that derivative is complicated by the intricate nature of the expressions for $\tilde{c}(p)$ and $v(p)$. Fortunately, it is adequate to use Eq. (13.1.47) for a weakly nonlinear wave to evaluate $\partial G/\partial q$, because a small error in this quantity merely alters the increment of $q^{(j)}$ in Eq. (13.1.46).

The Newton solver algorithm requires an initial guess, $q^{(0)}$. The fact that $|q| \ll 1$ suggests that a good guess would be the value according to linear theory, which is

$$q^{(0)} = \varepsilon V \left(t - \frac{x}{c} \right) \tag{13.1.49}$$

This usually will lead to a root in a few iterations, but this assertion only holds if x is not too close to the shock formation distance. A more robust approach is to evaluate $G(q, x, t)$ for a set of values q_n. Then, the best starting guess is that for which $G(q_n, x, t)$ is closest to zero. A computational operation would implement

$$\min \left(|G(q_n, x, t)| \right) \Longrightarrow q^{(0)} = q_n \tag{13.1.50}$$

In MATLAB, if G is a vector array of these values, one can find n by writing `[n,min_G]=min(abs(G))`. The appropriate range of q_n values is that which matches the range of $\varepsilon V(t)$ for all possible t. For example, if $V(t) = \sin(\omega t)$, then the range of q_n would be $-\varepsilon$ to $+\varepsilon$.

To illustrate this procedure, consider the signal in Example 13.2. Let us determine the value of q at $kx = 100\pi$ at the instant $\omega t = 101.98\pi$, which is slightly less than one period after the wavefront arrives at that location. The range of q_n values is $-\varepsilon$ to ε, but the range is extended in Fig. 13.7 in order to view the behavior of the G function with greater clarity. The function is sampled at $\Delta q_n = \varepsilon/50$. A scan of the computed values of G shows that it crosses the zero axis between $q_{91} = -4 \left(10^{-4} \right)$, where $G = -2.747 \left(10^{-6}\right)$, and $q_{92} = -3.6 \left(10^{-4}\right)$, where $G = 7.623 \left(10^{-6}\right)$. The first q_n gives the smaller magnitude of G, so the solution algorithm would begin with the initial value $q^{(0)} = -4 \left(10^{-4}\right)$. The root is found after one iteration based on a convergence criterion that $\left|q^{(j)} - q^{(j-1)}\right| < \varepsilon/10^5$; it is $q = -3.8944 \left(10^{-4}\right)$.

Figure 13.7 also shows a situation where the concept of finding q by solving a nonlinear equation becomes problematic. When $\omega t = kx + 1.98\pi$ at $kx = 200\pi$, which is twice as far from the boundary, there are three values of q for which $G = 0$ at the selected x and t. This means that three values of q are consistent with the Riemann solution. Hence, the occurrence of multiple roots for G at a selected x and t is another way that shocks are manifested. The Riemann solution is not valid in the presence of shocks. Nevertheless, the method by which shocks are incorporated into an analysis will require the multi-valued portion of the Riemann waveform. Such an evaluation could implement the root search procedure, with the scan initiated near each q value at which the G function changes sign.

Fig. 13.7 Graph of the function $G\,(q, x, t)$ whose root gives the nondimensional pressure $q = p/\left(\rho_0 c^2\right)$ at $\omega t = kx + 1.98\pi$, $\varepsilon = 0.002$

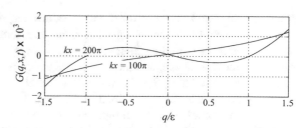

EXAMPLE 13.3 The particle velocity at $x = 0$ is $\varepsilon V (t) = \varepsilon c \left[\sin(\omega t) + 0.5 \sin(2\omega t)\right]$ for $t > 0$, and it was zero for $t < 0$. The fluid is air. Consider the case where $\varepsilon = 9 \left(10^{-5}\right)$. Compare the pressure waveforms at $kx - 800\pi$ and $kx = 1600\pi$ to the waveform predicted by linear theory. Further, because the excitation is periodic at $T = 2\pi/\omega$, it may be represented as a Fourier series. Construct a graph showing the dependence of the Fourier pressure coefficients on the propagation distance.

Significance

The evaluation of the signal is unremarkable, but the growth or decay of harmonics is a primary issue for nonlinear acoustics. This evaluation is a useful reminder of spectral techniques that were developed in Chap. 1.

Solution

We will implement the Newton solver algorithm in nondimensional form because the value of ω is not given. The value of ε is sufficiently small to justify considering the wave to be weakly nonlinear. The function to be solved is

$$G\left(q, kx, \omega t\right) \equiv q - \varepsilon\left[\sin\left(\omega t - \frac{kx}{1 + \beta q}\right) + 0.5 \sin\left(2\omega t - \frac{2kx}{1 + \beta q}\right)\right] h\left(t\right)$$

The derivative of this function with respect to q is

$$\frac{\partial}{\partial q} G(q, x, t) = 1 - \varepsilon\left[\cos\left(\omega t - \frac{kx}{1 + \beta q}\right) + \cos\left(2\omega t - \frac{2kx}{1 + \beta q}\right)\right] \frac{(kx)\,\beta}{\left(1 + \beta q\right)^2} h\left(t\right)$$

The value of β for air is 1.2.

The value of q predicted by linear theory is the most efficient way to initiate Newton's method. If doing so fails to lead to a convergence, then the search algorithm described by Eq. (13.1.50) may be invoked. (This alternative procedure was not required to obtain the present results.) Thus, the root search is initiated with

$$q^{(0)} = \varepsilon c\left[\sin\left(\omega t - kx\right) + 0.5 \sin\left(2\omega t - 2kx\right)\right] h\left(t\right)$$

We select an even increment for the time values, $\omega t_n = 2\pi n/N$, $n = 0, ..., N - 1$, in order to exploit FFT technology. The nondimensional fundamental frequency is $\omega \tau = 2\pi$, and the sampling interval is $\omega \Delta t = 2n/N$. Thus, the fundamental frequency of the data is one, and its highest frequency is $N/2$. Although we will only examine harmonics $m \le 5$, it is crucial that the highest frequency be greater than the significant frequencies contained in the data. The waveform at $x = 0$ contains only two harmonics, so the sampling criterion would be met with a small value of N. However, as the signal propagates its steepens in a portion of a period. This raises the Nyquist requirement. Consequently, the evaluations will be carried out with $N = 128$.

The procedure followed for each kx is to perform a program loop in which t_n is set, and the corresponding value of q_n is found by Newton's method. After completion of that loop, the set of q_n values is input to an FFT routine. The output of that routine is a set of DFT coefficients that usually must be scaled to obtain Fourier series coefficients P_m. (In MATLAB, this factor is $2/N$.) The P_m values obtained at each kx are preserved for plotting. A simple check of the computations uses the fact that the mean value of $v\left(t\right)$ at $x = 0$ is zero. This attribute should be preserved as the signal propagates. (The first value returned by most FFT routines is the mean value, P_0.)

The waveforms at the requested locations appear in Fig. 1. The abscissa is the retarded time, that is, the elapsed time after the wavefront arrives at $\omega t = kx$. Two periods appear there, but the data for the second period merely periodically replicates that of the first, with $\omega t_{n+N} = \omega t_n + 2\pi$ and $p_{n+N} = p_n$ for $n = 0, ...N - 1$.

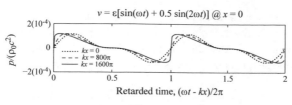

$v = \varepsilon[\sin(\omega t) + 0.5 \sin(2\omega t)] \ @ \ x = 0$

Figure 1.

The waveform at $kx = 0$ is the same as the linear solution. Although it has the appearance of being advanced for positive p and retarded for negative p, this quality is a consequence of the manner in which the harmonics combine, not the effect of nonlinearity. The distortion of the waveform enhances this initial appearance. At $kx = 1600\pi$, the effect is sufficiently large that the slope of the waveform seems to be nearly vertical at the start of each period.

Waveforms at two values of kx were requested, but such a sample is too sparse for a description of the position dependence of the harmonic amplitudes. Therefore, the waveforms at $kx_m = 400\pi$, 800π, 1200π, and 1600π were processed. The result is Fig. 2. Only the imaginary parts of the amplitudes are shown because the real parts remain zero. None of the amplitudes are constant. The first and second harmonics are the ones contained in the input, whereas the third and higher harmonics are generated by nonlinearity. There are no dissipation mechanisms, so the total energy in a period is conserved. Typical discussions of the generation of harmonics state that energy is transferred from lower harmonics into higher harmonics as the wave propagates. This graph shows this description to be only partially true. The harmonics that are not directly generated are $n \geq 3$. They do indeed grow as the wave propagates. The directly generated harmonics are $n = 1$ and $n = 2$, but only the former is depleted, whereas $m = 2$ grows slightly.

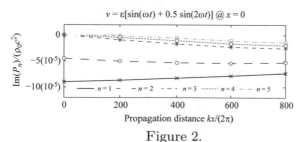

$v = \varepsilon[\sin(\omega t) + 0.5 \sin(2\omega t)] \ @ \ x = 0$

Figure 2.

The interchange of energy between harmonics and the resulting distortion of the waveform are sensitive to the content of the excitation. To demonstrate this, Fig. 3 shows the waveforms that result from shifting the second harmonic of the input by 180°, so that $v = \varepsilon [\sin(\omega t) - 0.5 \sin(2\omega t)]$ at $x = 0$. The interference of the harmonics of $p(t)$ at $x = 0$ is such that the starting waveform has less of the leaning appearance than it did in the previous case. The nonlinear advancement and retardation are the same as in the previous case. However, the overall leaning appearance of the initial waveform is less noticeable than it was in the previous case. The consequence is that the maximum slope of the waveform at $kx = 1600\pi$ is reduced, which suggests that shocks will form at a larger distance.

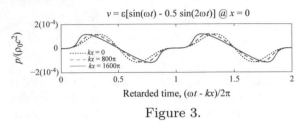

Figure 3.

The harmonic content of these signals is described in Fig. 4. Reversal of the second harmonic at $x = 0$ has a rather drastic effect on the growth and decay of harmonic amplitudes. Increasing propagation distance leads to depletion of the $m = 2$ harmonic. The energy lost from that harmonic is transferred into the other harmonics, including the first. The growth of the higher harmonics is similar to the previous case, but now the fourth harmonic grows more rapidly than the third.

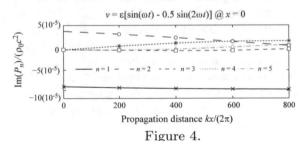

Figure 4.

13.2   Effects of Nonlinearity

The examples disclosed that some interesting phenomena arise in nonlinear plane waves. Two of particular interest are harmonic distortion and the occurrence of shocks. Both features are encountered in other types of waves. Both are sufficiently important to warrant a more detailed analysis, which is what we shall now do.

13.2.1   Harmonic Generation

A periodic excitation at the boundary leads to a periodic waveform. This is so because the characteristics form a periodic pattern parallel to the t-axis. Periodicity is readily proven. Two phases separated by a period are τ and $\tau + T$. The arrival time of these phases at a specified x is t and $t + \Delta t$, where

$$t = \tau + \frac{x}{c + \beta\varepsilon V(\tau)}, \quad t + \Delta t = (\tau + T) + \frac{x}{c + \beta\varepsilon V(\tau + T)} \tag{13.2.1}$$

Because $V(\tau)$ is periodic, $V(\tau + T) = V(\tau)$, so it must be that $\Delta t = T$. A corollary of periodicity is that the signal may be expressed as a Fourier series. Rather than using an FFT to process discretized data, as we did in Example 13.3, let us consider a formal analysis of the Fourier coefficients.

It is convenient to measure time relative to the signal's arrival at position x. Because the pressure at the wavefront is zero, the arrival time is x/c. Thus, the time variable is the *retarded time t'*, where

$$t' = t - \frac{x}{c} \tag{13.2.2}$$

We consider the weakly nonlinear wave generated by a periodic boundary vibration $v = \varepsilon c V(t)$ at $x = 0$. We further require that $V(t) = 0$ for $t < 0$, in order that the velocity be continuous at the initial instant. This condition is added because a discontinuity in the waveform means the existence of a shock, which invalidates using the Riemann solution throughout a period.

A periodic function that is zero at $t = 0$ may be represented as a Fourier sine series. The period of the excitation is T, so the fundamental frequency is $\omega = 2\pi/T$. A Fourier sine series whose independent variable is t' is used,

$$p = \sum_{n=1}^{\infty} P_n \sin(n\omega t'), \quad t' > 0 \tag{13.2.3}$$

Concurrently, the Riemann solution in Eq. (13.1.37) also describes p, so it must be that

$$\sum_{n=1}^{\infty} P_n \sin(n\omega t') = \rho_0 c^2 \varepsilon V(\tau), \quad \tau = t' + \frac{x}{c}\beta\varepsilon V(\tau) \tag{13.2.4}$$

The relation for τ differs from the earlier form by application of a binomial series expansion to clear the denominator, based on the fact that $|\varepsilon| \ll 1$.

The Fourier pressure coefficients are found by applying the orthogonality property of sine functions to Eq. (13.2.4). The first period following $t' = 0$ covers the interval $0 < t' < T$. Therefore, the coefficients are given by

$$P_n = \rho_0 c^2 \varepsilon \frac{2}{T} \int_0^T V(\tau) \sin(n\omega t') \, dt' \tag{13.2.5}$$

Because $V(\tau)$ is an arbitrary function, we use Eq. (13.2.4) to change the integration variable from t' to τ. Doing so gives

$$P_n = \rho_0 c^2 \varepsilon \frac{2}{T} \int_0^T V(\tau) \sin\left[n\omega\left(\tau - \frac{x}{c}\beta\varepsilon V(\tau)\right)\right]\left[1 - \frac{x}{c}\beta\varepsilon \dot{V}(\tau)\right] d\tau \quad (13.2.6)$$

The factor of $V(\tau)$ is a perfect differential, which suggests an integration by parts. Because $V(0) = V(T) = 0$, this gives

$$P_n = \rho_0 c^2 \varepsilon \frac{2}{n\omega T} \int_0^T \dot{V}(\tau) \cos\left[n\omega\left(\tau - \frac{x}{c}\beta\varepsilon V(\tau)\right)\right] d\tau \quad (13.2.7)$$

The periodic nature of the integrand allows the integration range to be shifted to $-T/2 < \tau < T/2$. Furthermore, because $V(\tau)$ is an odd function relative to $\tau = 0$, $\dot{V}(\tau)$ is even, so that the product of $\dot{V}(\tau)$ and the cosine term also is even. Therefore, the integration interval may be reduced to $0 < \tau < T/2$ with the result of the integral doubled. The result is

$$\boxed{P_n = \rho_0 c^2 \varepsilon \frac{2}{n\pi} \int_0^{T/2} \dot{V}(\tau) \cos\left[\frac{2n\pi}{T}\left(\tau - \frac{x}{c}\beta\varepsilon V(\tau)\right)\right] d\tau} \quad (13.2.8)$$

It might be desirable to compare these P_n values to those obtained from FFT analysis of a waveform, which is the procedure in the previous example. To do so, it is necessary to be cognizant of the fact that the FFT gives the coefficients of a complex Fourier series, whereas Eq. (13.2.8) gives the coefficients of a Fourier sine series.

Equation (13.2.8) may be used to describe the propagation of any initial waveform for which $p = 0$ at the beginning of a period. The expression seems to be unpromising for an analytical integration, but it can be evaluated numerically. Harmonic excitation is one case where an analytical result can be obtained. If $V(\tau) = \sin(\omega\tau)$, then an identity for the product of cosines, accompanied by the change of variables $\xi = \omega\tau$, leads to

$$P_n = \rho_0 c^2 \varepsilon \frac{2}{n\pi} \int_0^\pi \cos(\xi) \cos[n(\xi - kx\beta\varepsilon \sin(\xi))] d\xi$$

$$\equiv \rho_0 c^2 \varepsilon \frac{1}{n\pi} \int_0^\pi \{\cos[(n-1)\xi - nkx\beta\varepsilon \sin(\xi)]$$

$$+ \cos[(n+1)\xi - nkx\beta\varepsilon \sin(\xi)]\} d\xi \quad (13.2.9)$$

Both parts of the integrand match a fundamental formula for Bessel functions.[8] From it, we find that

$$\frac{P_n}{\rho_0 c^2} = \frac{\varepsilon}{n}\left[J_{n-1}(n\beta\varepsilon kx) + J_{n+1}(n\beta\varepsilon kx)\right] \equiv \frac{2}{n\beta kx}J_n(n\beta\varepsilon kx) \quad (13.2.10)$$

---

[8]M.I. Abramowitz and I.A. Stegun, *Handbook of Mathematical Functions*, Dover (1965) Eq. (9.1.21).

where the last form stems from a recurrence relation for the Bessel functions. A more meaningful form results when distance is described relative to the shock formation distance of an initially harmonic plane wave, $(x_{shock})_{sine}$. Shock formation is the subject of the next subsection. It will be found that $(x_{shock})_{sine} = 1/(\varepsilon\beta k)$, so that

$$\boxed{\frac{P_n}{\rho_0 c^2} = \frac{2\varepsilon}{n\sigma} J_n(n\sigma)} \tag{13.2.11}$$

The symbol σ is the distance ratio,

$$\boxed{\sigma = \frac{x}{(x_{shock})_{sine}} = \beta\varepsilon kx} \tag{13.2.12}$$

The corresponding series representation is

$$\boxed{p = \rho_0 c^2 \varepsilon \sum_{n=1}^{\infty} \frac{2}{n\sigma} J_n(n\sigma) \sin(n\omega t')} \tag{13.2.13}$$

Because this analysis began with the Riemann solution, the basic limitation to regions where $x < (x_{shock})_{sine}$, or $\sigma < 1$, applies equally to the Fourier series.

The Fourier series analysis was first worked out by Fubini-Ghiron in 1935.[9] A remarkable aspect is the simplicity of the result, for it tells us that the harmonic amplitudes depend on only two parameters: the harmonic number and distance relative to the position where a shock first forms. However, this property is specific to a sinusoidal excitation function; other excitations will lead to dependence on additional parameters.

Figure 13.8 plots the amplitudes. We see that the fundamental harmonic amplitude lessens with increasing distance, whereas the higher harmonic amplitudes grow. Another important property of a sinusoidal excitation is that the amplitudes at any x decrease as the harmonic number increases. This brings up an ambiguity regarding the description of harmonics. The terminology used here is to refer to $n = 1$ as the fundamental or first harmonic, $n = 2$ as the second harmonic, and so on. However, one can find many technical works wherein $n = 1$ is referred to as the fundamental, $n = 2$ is the first harmonic, and $n = 3$ is the second harmonic. This usage is consistent with terminology for music, but we shall avoid it.

Simpler expressions for the range dependence of the amplitudes may be obtained if we restrict our attention to the region close to the source at $x = 0$. Application to Eq. (13.2.13) of the series approximation of Bessel functions gives

[9] I.E. Fubini-Ghiron. "Anomalies in acoustic wave propagation of large amplitudes," Alta Freq. Vol. 4 (1935) pp. 530–581.

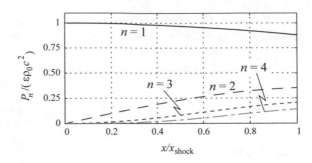

Fig. 13.8 Plot of the Fourier series coefficients for the signal generated by a sinusoidal excitation at $x = 0$. The distance x_{shock} is the location where a shock forms, and the analysis is not valid beyond that distance

$$\frac{P_1}{\rho_0 c^2 \varepsilon} = 1 - \frac{1}{8}\left(\frac{x}{x_{shock}}\right)^2 + \ldots$$

$$\frac{P_2}{\rho_0 c^2 \varepsilon} = \frac{1}{2}\left(\frac{x}{x_{shock}}\right) + \ldots, \quad \frac{P_3}{\rho_0 c^2 \varepsilon} = \frac{3}{8}\left(\frac{x}{x_{shock}}\right)^2 + \ldots \tag{13.2.14}$$

This confirms quantitatively the trend that is evident in Fig. 13.8 that the primary effect near the source, $x = 0$, is growth of the second harmonic linearly with propagation distance. The depletion of the fundamental, as well as the growth of third harmonic, depends quadratically on the propagation distance. All of these effects are weak until x is a substantial fraction of x_{shock}.

In view of the fact that we could have obtained the data in Fig. 13.8 from FFT analysis, as we did in Example 13.3, one might think that Fubini-Ghiron's development is important only from a historical perspective. This is far from the truth. For one, many evaluations would be required before we realized that the harmonic amplitudes depend on only two parameters. In addition, the availability of a formula for the amplitudes allows us to delve further into basic properties. One of particular interest is energy conservation.

Consider a temporally periodic wave in the region between two planes at x_1 and x_2, both of which are less than x_{sh}. As a consequence of periodicity of the pressure and particle velocity, the sum of the kinetic energy and potential energy per unit volume at any location x is the same at the end of any period as it was at the beginning. Thus, the total energy in the region between the two planes is conserved over any period. It follows that the average energy transported into this region at x_1 in an interval of a period must equal the average energy that flows out of this region at x_2. The cross sectional areas at x_1 and x_2 are identical and the locations are arbitrary, so it must be that the time-averaged intensity in the propagation direction is constant.

The time-averaged intensity is the same basic expression as it is in linear theory,

$$(I_x)_{av} = \frac{1}{T}\int_{t_0 - T/2}^{t_0 + T/2} pv\, dt \tag{13.2.15}$$

where t_0 is arbitrary. The order of magnitude of p and v is ε, so the intensity is $O\left(\varepsilon^2\right)$. However, as we proceed some higher-order terms might arise. To assure that we consistently account for all such terms, we shall employ the small-signal approximations, but retain the quadratic terms in the relation between v and p. Equation (13.1.35) describes this relation, so we have

$$(I_x)_{\mathrm{av}} = \frac{c}{T} \int_{t_0 - T/2}^{t_0 + T/2} p \left[\frac{p}{\rho_0 c^2} - \frac{1}{2}\beta \left(\frac{p}{\rho_0 c^2} \right)^2 \right] dt \qquad (13.2.16)$$

The series is defined in terms of the delayed time t', Correspondingly, the intensity may be computed by integrating over $T/2 < t' < 3T/2$. (This is the first full interval centered on an instant when $p = 0$.) This changes the preceding expression to

$$(I_x)_{\mathrm{av}} = \frac{\rho_0 c^3}{T} \int_{T/2}^{3T/2} \left[\left(\frac{p}{\rho_0 c^2} \right)^2 - \frac{1}{2}\beta \left(\frac{p}{\rho_0 c^2} \right)^3 \right] dt' \qquad (13.2.17)$$

The pressure is an odd function with respect to $t' = T$, which means that $p\left(t'\right) = -p\left(2T - t'\right)$. Therefore $p\left(t'\right)^3$ also is an odd function. This means that its contribution to the integral from $T/2 < t' < T$ will be the negative of its contribution from $T < t' < 3T/2$. The consequence is that the contribution of p^3 vanishes. Thus the time-averaged intensity is given by the same expression as it is in linear theory. Periodicity permits adjusting the interval to $0 < t' < T$, so that

$$(I_x)_{\mathrm{av}} = \frac{1}{T\rho_0 c} \int_0^T p^2 dt' \qquad (13.2.18)$$

We wish to prove that the Riemann solution for any excitation leads to a time-averaged intensity that is independent of x. Doing so is much easier if we limit consideration to weakly nonlinear signals. In that case the pressure is related to the boundary velocity by $p\left(\tau\right) = \rho_0 c^2 \varepsilon V\left(\tau\right)$, and τ is related to t' by Eq. (13.2.4). The latter relation is used to convert the integration variable from t' to τ, which leads to

$$\begin{aligned} (I_x)_{\mathrm{av}} &= \frac{1}{T\rho_0 c} \int_0^T p^2 d\left[\tau - \frac{x}{c}\beta\varepsilon V\left(\tau\right) \right] \\ &= \frac{\rho_0 c^3}{T}\varepsilon^2 \int_0^T V\left(\tau\right)^2 \left[1 - \frac{x}{c}\beta\varepsilon \dot{V}\left(\tau\right) \right] d\tau \\ &= \frac{\rho_0 c^3}{T}\varepsilon^2 \left[\int_0^T V\left(\tau\right)^2 d\tau - \frac{x}{3c}\beta\varepsilon V\left(\tau\right)^3 \Big|_0^T \right] \end{aligned} \qquad (13.2.19)$$

Because $V\left(\tau\right)$ is periodic, the cubic term vanishes. Furthermore, the integrand depends only on the nature of $V\left(\tau\right)$, so it is independent of x. This proves that $(I_x)_{\mathrm{av}}$ is independent of x, being

$$\boxed{(I_x)_{\text{av}} = \frac{\rho_0 c^3}{T} \varepsilon^2 \int_0^T V(\tau)^2 \, d\tau}$$

(13.2.20)

An alternative form of this expression results from Parseval's theorem, Eq. (1.4.12), which describes the mean-squared pressure corresponding to a Fourier series. The fact that the integration is performed over a period of the τ variable, not time t, is irrelevant. Thus, the time-averaged intensity is related to the Fourier series coefficients of $V(\tau)$ by

$$(I_x)_{\text{av}} = \frac{1}{2} \rho_0 c^3 \varepsilon^2 \sum_{n=1}^{\infty} |V_n|^2$$

(13.2.21)

Because the derivation was restricted to weakly nonlinear waves, this expression actually only is accurate to $O(\varepsilon^2)$. An analysis that does not invoke the weak nonlinearity would confirm that the Riemann solution describes a signal in which power flows constantly to the location where a shock forms, regardless of the magnitude of the signal.

13.2.2 Shock Formation

We have seen three manifestations of the occurrence of a shock: intersection of characteristic lines, multivaluedness of the function whose roots give the pressure, and vertical tangency of a waveform or spatial profile. Any of these three features may be used as the basis for a quantitative analysis of the distance where a shock forms. The result from each is the same, but the differences in the analysis demonstrate different techniques for analyzing other systems.

We begin with an excitation at $x = 0$ that is a piecewise linear waveform like the one in Example 13.1. Whereas we previously tracked the propagation of the phases at which p is maximum and zero, we now consider signals that depart the boundary at two arbitrary instants τ_1 and $\tau_2 > \tau_1$; the corresponding pressures are p_1 and p_2. The phase speeds associated with these signals are $\tilde{c}_1 + v_1$ and $\tilde{c}_2 + v_2$. The equations for the two characteristics are

$$\tau_1 = t - \frac{x}{\tilde{c}_1 + v_1}, \quad \tau_2 = t - \frac{x}{\tilde{c}_2 + v_2}$$

(13.2.22)

At the intersection of these lines, the values of x and t are the same. Because $\tau_2 > \tau_1$, such an intersection can occur only if the slope $dx/dt = 1/(\tilde{c}_2 + \beta v_2)$ of the τ_2 characteristic is less than the slope $1/(\tilde{c}_1 + \beta v_1)$ of the τ_1 characteristic. We have seen that both $\tilde{c}$ and v increase monotonically with increasing p. It follows that the characteristic lines will intersect only if $p_2 > p_1$. This is another way of saying that a shock forms because a later high-pressure phase catches up to an earlier lower pressure phase. To solve for x, we take the difference of the two equations, which leads to

$$x = \frac{(\tilde{c}_1 + v_1)(\tilde{c}_2 + v_2)(\tau_2 - \tau_2)}{(\tilde{c}_2 + v_2) - (\tilde{c}_1 + v_1)} \qquad (13.2.23)$$

The shock formation distance, x_{shock}, in general is the minimum value of x at which a shock occurs at any instant. Rather than solving this relation for a variety of τ_1, τ_2 pairs, we can identify directly which pair of characteristics intersect at the smallest x. Specifically, the τ_2 characteristic should be the one for which $\tilde{c}_2 + v_2$ is the largest value, and the τ_1 characteristic should be the one for which $\tilde{c}_1 + v_1$ is the smallest value. Equation (13.1.39) is the result for a simple triangle waveform because in that case, p_2 is the maximum and $p_1 = 0$.

Equation (13.2.23) applies if $p(\tau)$ at $x = 0$ is piecewise linear. If it is a continuously differentiable function, it is more useful to consider two adjacent instants by letting $\tau_1 = \tau$ and $\tau_2 = \tau + d\tau$. Then, $(\tilde{c}_2 + v_2) - (\tilde{c}_1 + v_1)$ becomes $d(\tilde{c} + v)$. Furthermore, $\tilde{c} + v$ is a function of p, and p is a function of x and τ. These observations lead to identification of the condition for formation of a shock,

$$\boxed{x = \frac{(\tilde{c} + v)^2}{\left[\dfrac{d}{dp}(\tilde{c} + v)\right]\dfrac{\partial p}{\partial \tau}}} \qquad (13.2.24)$$

We know that $\tilde{c} + v$ increases monotonically with increasing p, so that $d(\tilde{c} + v)/dp > 0$. Therefore, we conclude that a shock occurs in the portion of the waveform for which p increases with increasing t.

Equation (13.2.24) is valid for a signal at any amplitude (within the limits of the assumptions of adiabatic propagation in an inviscid fluid). In acoustic applications, the acoustic Mach number ε is small, so Eq. (13.1.36) may be used to describe the phase speed. Doing so gives

$$x = \frac{c\left(1 + \beta\dfrac{p}{\rho_0 c^2}\right)}{\beta\dfrac{d}{d\tau}\left(\dfrac{p}{\rho_0 c^2}\right)} \qquad (13.2.25)$$

In the small-signal approximation $p/(\rho_0 c^2) = v/c = \varepsilon V(\tau)$. Thus, when terms whose order of magnitude is ε or smaller are dropped, the preceding becomes a simple expression for x. This value depends on the value of τ. The shock formation distance is defined to be the smallest x at which characteristics intersect because we must assume that the basic formulation ceases to be valid in a region where a shock exists at any instant. The minimum distance, x_{shock}, corresponds to the positive maximum of $dV/d\tau$. Thus,

$$x_{\text{shock}} = \frac{c}{\beta\varepsilon \max\left(\dfrac{dV}{dt}\right)} \qquad (13.2.26)$$

This expression confirms that the shock forms at the phase of the waveform that changes most rapidly. It also tells us that the distance at which a shock first forms is inversely proportional to the magnitude of the excitation.

Our second encounter with shock formation occurred when we developed a numerical procedure to evaluate the pressure. This entailed finding the value $q \equiv p/(\rho_0 c^2)$ that satisfies $G(q, x, t) = 0$. Figure 13.7 describes this function at $x < x_{\text{shock}}$, in which case there is one root, and at $x > x_{\text{shock}}$, in which case there are three roots. As x is increased from the first value, the slope of G vs. q at the root decreases. The transition from one root to three is marked by a horizontal tangent at the root. This suggests that the condition leading to x_{shock} is marked by

$$\frac{\partial}{\partial q} G(q, x, t) = 0 \qquad (13.2.27)$$

This derivative is given by Eq. (13.1.47), so we set

$$1 - \varepsilon \frac{dV}{d\tau}\bigg|_{\tau = t - \frac{x/c}{1+\beta q}} \left[\frac{x}{c} \frac{\beta}{(1+\beta q)^2}\right] = 0 \qquad (13.2.28)$$

Because the order of magnitude of q is ε, it is acceptable to approximate the denominator terms as one. The smallest value of x at which this condition occurs corresponds to the largest value of $dV/d\tau$, regardless of when it occurs. The resulting expression for x_{shock} is the same as Eq. (13.2.26).

Another way in which we may locate a shock is by explicitly searching for a vertical tangency of the waveform or spatial profile. This property is recognizable by the fact that the interval in space for a profile, or in time for a waveform, between the characteristics associated with τ and $\tau + d\tau$ decreases as they come closer to intersecting. We shall pursue this analysis for a waveform and leave the corresponding analysis of a profile for Exercise 13.13. The first step is to derive an expression for $\partial p/\partial t$. The substitution $p = \rho_0 c^2 q$ somewhat simplifies differentiation of the small-signal approximation in Eq. (13.1.43), which gives

$$\frac{\partial q}{\partial t} \equiv \varepsilon \frac{dV}{d\tau}\bigg|_{\tau = t - \frac{x/c}{1+\beta q}} \left[1 + \frac{x/c}{(1+\beta q)^2} \beta \frac{\partial q}{\partial t}\right] \qquad (13.2.29)$$

from which we find that

$$\left[\frac{x/c}{(1+\beta q)^2} \beta\varepsilon \frac{dV}{d\tau}\bigg|_{\tau = t - \frac{x/c}{1+\beta q}} - 1\right] \frac{\partial q}{\partial \tau} = -\varepsilon \frac{dV}{d\tau}\bigg|_{\tau = t - \frac{x/c}{1+\beta q}} \qquad (13.2.30)$$

For a vertical tangency, we seek the condition that leads to $\partial q/\partial t$ being infinite. This occurs if the bracketed term is zero. The minimum x at which this condition occurs corresponds to the maximum value of dV/dt, so a vertical tangent first occurs at

$$\frac{x}{c} = \frac{(1 + \beta q)^2}{\beta \varepsilon \max \left(\dfrac{dV}{d\tau} \right)} \tag{13.2.31}$$

Dropping βq in comparison with 1 yields Eq. (13.2.26).

Identification of the location where a shock forms is possible by three alternative methods because of the manner in which the Riemann solution is posed. For other systems, these approaches might not be equally viable. Indeed, if the solution is the result of a numerical analysis, the only suitable approach might be a numerical evaluation of the slope of a waveform or profile.

The value of x_{shock} depends on the nature of the input function $V(\tau)$. The typical concern is a sine function, $V(\tau) = \sin(\omega\tau - \phi)$. In that case, $\max(dV/d\tau) = \omega$, so that

$$\boxed{(x_{shock})_{sine} = \frac{1}{\varepsilon\beta k}} \tag{13.2.32}$$

For reference values, a harmonic excitation at 1 kHz in air at 140 dB//20µPa corresponds to $\varepsilon = 0.0020$ and $x_{shock} = 23$ m, while a 200 dB//1 µPa signal at 1 kHz underwater corresponds to $\varepsilon = 6.5 \left(10^{-5}\right)$ and $x_{shock} = 1.04$ km.[10] These values may be scaled for other signal amplitudes and frequencies, so that a 1 MHz signal in water at $\varepsilon = 6.5 \left(10^{-5}\right)$ shocks at 1.04 m, while the same signal at $\varepsilon = 6.5 \left(10^{-6}\right)$ forms a shock at 10.4 m.

13.2.3 Propagation of Weak Shocks

Propagation of a wave beyond the location where a shock first forms causes its spatial profile or waveform to develop an extremely steep slope over a very small interval. This condition is depicted in Fig. 13.9. Within this interval, shear stresses grow, which means the some accounting must be given to the effects of viscosity and heat conduction. Model field equations that account for the role of viscosity and other nonideal effects would be much more difficult to derive and solve. Furthermore, the specific details of the small-scale region within which the signal changes drastically might be of little interest if we can find a simpler way to describe how this phenomenon affects the overall propagation features. The formulation that meets this objective replaces the actual velocity dependence by a model in which the velocity

---

[10] A 200 dB signal in water has an amplitude that is 1.4 atm. Hence, generating such a signal without causing cavitation requires that the depth be sufficiently large that the hydrostatic pressure exceeds the amplitude.

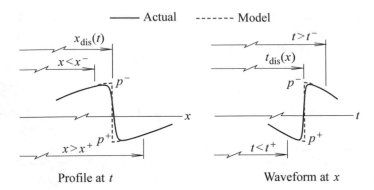

Profile at t Waveform at x

Fig. 13.9 Actual pressure profile and waveform at a shock and the idealized model. The latter replaces extremely steep gradients and high accelerations with discontinuous changes in space and time

changes discontinuously. Figure 13.9 shows that the actual pressure profile changes over a distance that is much less than a wavelength in space. It also shows that the actual waveform changes over an interval that is much shorter than a period. If we do not closely inspect the signal, the model of a discontinuous change is a good approximation. A discontinuity of the pressure in either space or time is an artifice, but one implies the other. This is so because Euler's equation and the equation of state require that $\partial p / \partial t$ is infinite if ∇p is infinite.

The term "shock" is synonymous with discontinuity. We have already used x_{shock} to denote the distance at which the wave first develops a vertical tangency, so $x_{\text{dis}}(t)$ will denote the location of the discontinuity in a spatial profile at an arbitrary instant. For a wave propagating in the sense of increasing x, it must be that $x_{\text{dis}} \geq x_{\text{shock}}$. Similarly, $t_{\text{dis}}(x)$ is the instant at which a shock is observed in the waveform at an arbitrary location. Other parameters are described in Fig. 13.9. The position infinitesimally behind the discontinuity is denoted as x_{dis}^- and the corresponding pressure is p^-, whereas the position infinitesimally ahead of the discontinuity is x_{dis}^+ and the corresponding pressure is p^+. A minor confusion results from the usage of $+$ and $-$ to denote when and where a feature occurs, rather than relative size. Specifically, we have $p^- > p^+$, $t^- > t^+$, and $x^- < x^+$. In terms of the characteristic variable, because p^+ is ahead of p^- in space, it must be that p^+ departed from the boundary at an earlier time, so that $\tau^- > \tau^+$. Multiple discontinuities may exist, but each is treated individually.

Rankine–Hugoniot Jump Conditions

If the model of a discontinuous pressure change at a shock is to be useful, we must determine how the discontinuity moves in space. This analysis modifies the derivation of the conservation principles to include a discontinuous change of the state variables. The domain we consider is a control volume contained between two fixed locations, $x_1 < x_{\text{dis}} < x_2$. Figure 13.10 shows this control volume at two instants t and $t + dt$. The position of the discontinuity is x_{dis} at each instant. The field is independent of

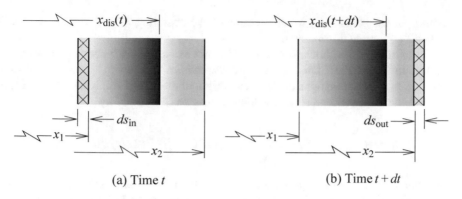

(a) Time t (b) Time $t + dt$

Fig. 13.10 A stationary control volume surrounding a moving shock

position transverse to the propagation distance, so the cross section may be considered to be a unit area.

In Fig. 13.10(a), a chunk of fluid is situated to the left of the control volume. This chunk contains the fluid that will enter the control volume during the interval from t to $t + dt$. The width of this chunk is ds_{in}. Because this width is infinitesimal, the velocity of all particles contained in it is essentially $v(x_1, t)$. By definition, the left face of this chunk arrives at x_1 in an elapsed time of dt, so $ds_{\text{in}} = v(x_1, t) \, dt$. In Fig. 13.10(b), the chunk of fluid to the right of the control consists of the particles that exited in the interval from t to $t + dt$. As a consequence of ds_{out} being infinitesimal, the velocity of these particles is essentially $v(x_2, t)$. Hence, $ds_{\text{out}} = v(x_2, t) \, dt$.

The principles we shall consider are conservation of mass, momentum and impulse, and energy and work. The formulation of each balance principle is essentially the same. Let $\mathcal{H}$ denote the quantity that is the subject of the principle. Thus, the mass per unit volume is $\mathcal{H} = \rho$, the momentum per unit volume is $\mathcal{H} = \rho v$, and the energy per unit volume is $\mathcal{H} = (1/2) \rho v^2 + U$, where U is the potential energy per unit volume stored within the fluid when it is compressed and expanded. Equation (4.5.7) describes U in terms of the pressure and the equation of state.

There is an external agent $\mathcal{F}$ that changes $\mathcal{H}$. For momentum, it is the impulse of the pressure acting on the control volume, which is positive for $p(x_1)$ and negative for $p(x_2)$. The impulse is the net force times dt, so we define $\mathcal{F} = \left[p(x_1) - p(x_2) \right] dt$ corresponding to a unit cross section. The external agent changing the energy is the net work done by the pressure at the two ends. This work may be evaluated as the instantaneous power multiplied by dt, so we set $\mathcal{F} = \left[p(x_1) v(x_1) - p(x_2) v(x_2) \right] dt$. Mass is conserved, so $\mathcal{F} = 0$ for conservation of mass.

The description of the contributions from the chunks of fluid that enter and leave the control volume is based on their infinitesimal extent. According to the central limit theorem, $\mathcal{H}(x, t)$ within each chunk is the same as the value at the adjacent end of the control volume. Hence, $\mathcal{H}(x_1, t) \, ds_{\text{in}}$ enters the control volume and $\mathcal{H}(x_1, t) \, ds_{\text{out}}$ exits. We also must account for the amount of the $\mathcal{H}$ quantity contained within the control volume at instant t, which is obtained by integrating $\mathcal{H}(x, t)$ over $x_1 < x < x_2$. Let $\mathcal{I}(t)$ denote this amount.

The existence of a shock means that $\mathcal{H}$ is not a continuous function of x, so the integration to describe $\mathcal{I}$ must be done in a piecewise manner from x_1 to x_{dis} and then from x_{dis} to x_2. The positions to the left and right of the discontinuity are x^- and x^+, so the integration over the control volume at time t is described by

$$\mathcal{I}(t) \equiv \int_{x_1}^{x_2} \mathcal{H}(x, t)\, dx = \int_{x_1}^{x^-} \mathcal{H}(x, t)\, dx + \int_{x^+}^{x_2} \mathcal{H}(x, t)\, dx \qquad (13.2.33)$$

We could follow similar arguments to describe the integral at time $t + dt$, with the shock situated at $x_{dis}(t + dt)$. A simpler procedure employs Leibnitz' rule for differentiating an integral with a variable limit. The difference between x^- and x^+ is infinitesimal, so $dx^-/dt = dx^+/dt = dx_{dis}/dt \equiv v_{dis}$. The result is that

$$\begin{aligned} \mathcal{I}(t + dt) &= \mathcal{I}(t) + dt \frac{d}{dt}\mathcal{I}(t) \\ &= \mathcal{I}(t) + \left\{ \int_{x_1}^{x^-} \frac{\partial}{\partial t}\mathcal{H}(x, t)\, dx + \int_{x^+}^{x_2} \frac{\partial}{\partial t}\mathcal{H}(x, t)\, dx \right. \\ &\quad \left. + v_{dis}(t) \left[\mathcal{H}(x^-, t) - \mathcal{H}(x^+, t) \right] \right\} dt \end{aligned} \qquad (13.2.34)$$

Each balance principle states that $I(t + dt)$ must equal $I(t)$ plus the net amount of $\mathcal{H}$ that enters the control volume plus the increment due to the $\mathcal{F}$ quantity, that is,

$$\mathcal{I}(t + dt) = \mathcal{I}(t) + \mathcal{H}(x_1, t)\, v(x_1, t)\, dt - \mathcal{H}(x_2, t)\, v(x_2, t)\, dt + \mathcal{F} dt \qquad (13.2.35)$$

This relation applies in any situation. Substitution of the representations of $\mathcal{I}$ at t and $t + dt$ into the conservation equation reduces it to

$$\begin{aligned} &\int_{x_1}^{x^-} \frac{\partial}{\partial t}\mathcal{H}(x, t)\, dx + \int_{x^+}^{x_2} \frac{\partial}{\partial t}\mathcal{H}(x, t)\, dx + v_{dis}(t) \left[\mathcal{H}(x^-, t) - \mathcal{H}(x^+, t) \right] \\ &\quad = \mathcal{H}(x_1, t)\, v(x_1, t) - \mathcal{H}(x_2, t)\, v(x_2, t) + \mathcal{F} \end{aligned}$$
$$(13.2.36)$$

The last step is to shrink the control volume by letting x_1 approach x_{dis} from below, $x_1 \to x^-$, and letting x_2 approach x_{dis} from above, $x_2 \to x^+$. This operation reduces the integrals to zero. In addition, because x_1 and x_2 now are at the shock, we have $v(x_1, t) \to v_{dis}$ and $v(x_2, t) \to v_{dis}$. To simplify the notation, let $\mathcal{H}^- \equiv \mathcal{H}(x^-, t)$ and $\mathcal{H}^+ \equiv \mathcal{H}(x^+, t)$. A minor rearrangement of terms lead to a general description of each balance principle as

$$\mathcal{H}^+ \left(v^+ - v_{dis} \right) = \mathcal{H}^- \left(v^- - v_{dis} \right) + \mathcal{F} \qquad (13.2.37)$$

Replacing the symbols $\mathcal{H}$ and $\mathcal{F}$ with their definition for each balance principle leads to the *Rankine–Hugoniot (jump) conditions*. The mass equation is

$$\rho^+\left(v^+ - v_{\text{dis}}\right) = \rho^-\left(v^- - v_{\text{dis}}\right)$$

(13.2.38)

The momentum equation is

$$\rho^+ v^+\left(v^+ - v_{\text{dis}}\right) + p^+ = \rho^- v^-\left(v^- - v_{\text{dis}}\right) + p^-$$

(13.2.39)

The energy condition is

$$\left[\frac{1}{2}\rho^+\left(v^+\right)^2 + U^+\right]\left(v^+ - v_{\text{dis}}\right) + p^+ v^+ = \left[\frac{1}{2}\rho^-\left(v^-\right)^2 + U^-\right]$$
$$\left(v^- - v_{\text{dis}}\right) + p^- v^-$$

(13.2.40)

These relations are general and therefore apply to nonacoustic applications such as blast waves and combustion in ramjet engines, as well as shocks formed by supersonic aircraft.

Our objective is to determine where the discontinuity is located and how fast it is moving. The analysis is simplified if we limit consideration to weak nonlinearity, in which case the pressure on either side of the shock is much less than the bulk modulus. Nevertheless, the operations are intricate, see for example Pierce's treatment,[11] We shall only take an overview of the analysis. A key step is to form a linear combination of the Rankine–Hugoniot equations in order to obtain an expression for the enthalpy per unit mass, $h = (U + pv)/\rho$. This quantity is important because the differential increment of entropy per unit mass in a process is related to the change of h according to $TdS = dh - (dp)/\rho$. The order of magnitude of $p/\left(\rho_0 c^2\right)$ is taken to be $O(\varepsilon)$. Simplification of the recast Rankine-Hugoniot equations based on keeping the largest terms eventually leads to recognition that the order of magnitude of dS is $O\left(\varepsilon^3\right)$. In other words, entropy is essentially constant across a weak shock.

The implication of constancy of entropy is that the nonlinear acoustic relations derived from the Riemann solution remain valid everywhere other than x_{dis}. This property leads to a relation between the velocity of the shock and state of the fluid behind and ahead of the shock. Let $v_{\text{av}} = \left(v^- + v^+\right)/2$ and $\tilde{c}_{\text{av}} = \tilde{c}\left(v_{\text{av}}\right)$. The weakly nonlinear properties lead to

$$v^- = \frac{p^-}{\rho_0 c}, \quad v^+ = \frac{p^+}{\rho_0 c}, \quad v_{\text{av}} = \frac{p^- + p^+}{2\rho_0 c}, \quad \tilde{c}_{\text{av}} = \frac{1}{2}\left[\tilde{c}\left(v^-\right) + \tilde{c}\left(v^+\right)\right]$$

(13.2.41)

The velocity of the discontinuity is found to be

$$v_{\text{dis}} = \tilde{c}_{\text{av}} + v_{\text{av}}$$

(13.2.42)

---

[11] Pierce, *ibid.*, pp. 576–577.

These relations describe a situation in which the shock is propagating in the positive x direction. Because a shock is characterized by a forward-leaning of the spatial profile, p^- will be greater than p^+. Thus, the shock's location will move slower than the phase speed behind the shock and faster than the phase speed ahead. If it should happen that $p^+ = -p^-$, then $\tilde{c}_{av}$ will be c. In other words, the shock will move at the linear speed of sound.

Equal-Area Rule

It still remains to determine how x_{dis} depends on time. One approach solves a differential equation based on setting $v_{dis} = dx_{dis}/dt$ in Eq. (13.2.42). Figure 13.11 describes the parameters required to carry out this analysis. In it, the multivalued pressure profile is the Riemann solution beyond the shock formation distance. Suppose the value of x_{dis} is known at instant t. We can use the method in Sect. 13.1.5 to determine the characteristics τ_1 and τ_3, and the associated pressures corresponding to the values of x_{dis} and t. (The pressure at τ_2 is irrelevant to this discussion.) The unshocked portion of the waveform adjacent to τ_3 is farther advanced than x_{dis}, so $p^+ = p(\tau_3)$. Similarly, the unshocked portion of the waveform adjacent to τ_1 is behind the shock, so $p^- = p(\tau_1)$. Because τ_1 and τ_3 are functions of x_{dis} and t, so too are p^- and p^+. This means that Eqs. (13.2.41) and (13.2.42) combine to give v_{av} as a function of x_{dis} and t. Therefore, Eq. (13.2.42) is a statement that $dx_{dis}/dt = f(x_{dis}, t)$—it is a differential equation. The initial condition is the value of x_{dis} at the instant when the profile first develops a vertical tangency.

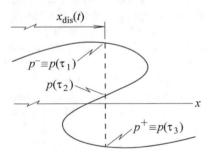

Fig. 13.11 Relationship of the parameters for a shock to the pressure waveform obtained from the Reimann solution

Although this procedure is feasible, numerical methods would be required to solve the differential equation, and the need to find multiple roots of the Riemann solution at each integration step would require considerable effort. Furthermore, the method provides little understanding of the underlying process. Fortunately, there is an alternative that provides a graphical interpretation and determines x_{dis} directly. Figure 13.12 shows a multivalued profile obtained from the Riemann solution. The difference between it and the previous figure is that the portions of this waveform that are advanced beyond x_{dis} and retarded behind x_{dis} are shaded. The ultimate result will be that x_{dis} is such that the total shaded area, positive for the region where $x > x_{dis}$

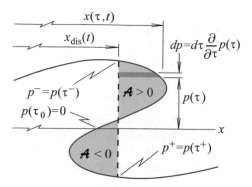

Fig. 13.12 Spatial pressure profile obtained by evaluation of the Riemann solution beyond the location where a shock begins to form. This profile is a multivalued function of x for characteristic values in the range $\tau^- < \tau < \tau^+$. The position of the area of the profile at $x > x_{\text{dis}}$ is positive, while the area for $x < x_{\text{dis}}$ is negative

and negative for $x < x_{\text{dis}}$, is zero. We shall prove this property, because doing so leads to a method by which x_{dis} can be determined at any instant.

If t is specified, then p and x are unique functions of the characteristic variable τ. The values of τ corresponding to p^- and p^+ are τ^- and τ^+, respectively. The lower portion of the profile after the discontinuity is farther advanced than the discontinuity, so p^+ is the smallest value of p obtained from the Riemann solution when $x = x_{\text{dis}}$ and t is the time for the profile. Similarly, the upper portion of the profile preceding the discontinuity is less advanced than the discontinuity, so p^- is the largest value of p obtained from the Riemann solution at $x = x_{\text{dis}}$ when t is the designated time. If follows that the range of characteristic variables associated with the shaded area is $\tau^+ < \tau < \tau^-$. The small-signal approximation of the equation for a characteristic tells us that the location of a specific $p(\tau)$ at instant t is

$$x(\tau, t) = (t - \tau)(c + \beta v) = (t - \tau)\left(c + \frac{\beta}{\rho_0 c} p(\tau)\right) \qquad (13.2.43)$$

As suggested by Fig. 13.12, the total area $\mathcal{A}$ is best evaluated by accumulating $(x - x_{\text{dis}}) \, dp$, so that

$$\mathcal{A}(t) = \int_{p^+}^{p^-} [x(\tau, t) - x_{\text{dis}}(t)] \, dp \qquad (13.2.44)$$

If $\mathcal{A}(t)$ is constant, then it should be that $d\mathcal{A}/dt = 0$. Leibnitz' rule is used to differentiate the integral,

$$\frac{d\mathcal{A}}{dt} = \int_{p^+}^{p^-} \left[\frac{d}{dt} x \left(\tau, t\right) - v_{\text{dis}} \right] dp + \left[x \left(\tau^-, t\right) - x_{\text{dis}} \right] \frac{\partial p^-}{\partial t} \tag{13.2.45}$$

$$- \left[x \left(\tau^+, t\right) - x_{\text{dis}} \right] \frac{\partial p^+}{\partial t}$$

The definitions of τ^- and τ^+ are that $x \left(\tau^-, t\right) = x \left(\tau^+, t\right) = x_{\text{dis}} \left(t\right)$, so both terms obtained from the integration limits vanish. We know from the boundary condition that $p \left(\tau\right) = \epsilon \rho_0 c V \left(\tau\right)$, so p is a single-valued function of τ. Accordingly, we change the integration variable from p to τ, so that $dp = \left(\partial p / \partial \tau\right) d\tau$. Also, differentiation of Eq. (13.2.43) with respect to t, with τ held fixed, gives

$$\frac{dx}{dt} = c + \frac{\beta}{\rho_0 c} p \left(\tau\right) \tag{13.2.46}$$

The result is that

$$\frac{d\mathcal{A}}{dt} = \int_{\tau^+}^{\tau^-} \left(c + \frac{\beta p \left(\tau\right)}{\rho_0 c} - v_{\text{dis}} \right) \frac{\partial}{\partial \tau} p \left(\tau\right) d\tau$$

$$= \left[\left(c - v_{\text{dis}}\right) p \left(\tau\right) + \frac{\beta}{2\rho_0 c} p \left(\tau\right)^2 \right] \Bigg|_{\tau=t^+}^{\tau=t^-} \tag{13.2.47}$$

A rearrangement of terms lead to

$$\frac{d\mathcal{A}}{dt} = \left(p^- - p^+\right) \left[c - v_{\text{sh}} + \frac{p^- - p^+}{2\rho_0 c} \right] \tag{13.2.48}$$

The term in the bracket is zero according to Eq. (13.2.42), which was derived from the Rankine-Hugoniot relations. Thus, we have proven that $d\mathcal{A}/dt = 0$. This means that the total area $\mathcal{A}$ is constant at its value at the instant when the shock first formed, which is $\mathcal{A} = 0$. Because the total area bounded by a graph of $p \left(x.t\right)$ versus x and the line $x = x_{\text{dis}}$ is zero, then the area to the left of the discontinuity must equal the area to the right. This is the *equal-area rule*.

Evaluation of Spatial Profiles

Enforcement of the equal-area rule leads to a method for determining the location of the discontinuity at any instant. There is a common situation in which little analysis is required. Let τ_0 be a characteristic along which the pressure is zero. Suppose that $p \left(\tau\right)$ is a continuous odd function relative to τ_0, that is, $p \left(\tau_0 + \delta\right) = -p \left(\tau_0 - \delta\right)$. It follows that $p^+ = -p^-$ regardless of t, so that $v_{\text{dis}} = c$ and $x_{\text{dis}} \left(t\right) = x \left(\tau_0, t\right) = c \left(t - \tau_0\right)$.

The application of the equal-area rule when $p \left(\tau\right)$ is an arbitrary function begins by selecting a trial value for the discontinuity's location; let us denote this choice as x_{tr}. Obviously, x_{tr} must be somewhere in the interval where the profile is multivalued. The area $\mathcal{A}$ created in this manner will depend on x_{tr}, as well as the value of t for the profile. The basic notion is to identify which x_{tr} in a range of values yields $\mathcal{A} = 0$.

To do so, we need a procedure for evaluating A. To that end, the integration variable in Eq. (13.2.44) is changed from p to τ, which leads to

$$A(x_{tr}, t) = \int_{\tau^+}^{\tau^-} [x(\tau, t) - x_{tr}] \frac{\partial}{\partial \tau} p(\tau) d\tau \qquad (13.2.49)$$

This expression can be specialized to the situation where the velocity at $x = 0$ is $\varepsilon c V(t)$. Then, the weakly nonlinear pressure is $\rho_0 c^2 \varepsilon V(\tau)$. In view of the position dependence in Eq. (13.2.43), the area function is

$$A(x_{tr}, t) = \rho_0 c^2 \varepsilon \int_{\tau^+}^{\tau^-} \{c(t - \tau)[1 + \beta \varepsilon V(\tau)] - x_{tr}\} \dot{V}(\tau) d\tau \qquad (13.2.50)$$

Several terms are exact differentials. Integration by parts is used to handle the terms that contain τ as a factor. The result is

$$\boxed{\begin{aligned} A(x_{tr}, t) &= \rho_0 c^2 \varepsilon \left\{ c(t - \tau - x_{tr}) V(\tau) + \frac{1}{2} \beta \varepsilon c(t - \tau) V(\tau)^2 \right\} \Big|_{\tau^+}^{\tau^-} \\ &+ \rho_0 c^2 \varepsilon \int_{\tau^+}^{\tau^-} c \left[V(\tau) + \frac{1}{2} \beta \varepsilon V(\tau)^2 \right] d\tau \end{aligned}} \qquad (13.2.51)$$

A determination of x_{dis} for the spatial profile at a specified t may be performed by following a set procedure. It begins by letting x_{tr} be the smallest x in the multi-valued range, solving for the smallest and largest roots, τ^+ and τ^-, respectively, of Eq. (13.2.43) when $x = x_{tr}$ and t is the instant of interest, then evaluating A for that choice of x_{tr}. Progressively incrementing x_{tr} will reduce the value of A. The value of x_{dis} lies between the values of x_{tr} for which A changes sign.

Evaluation of Waveforms

We have established that the equal-area rule can be used to determine where a discontinuity occurs in a pressure profile at an arbitrary instant. It also can be used to determine the instant at which the discontinuity occurs in the waveform at a designated x. The multivalued waveform in Fig. 13.13 is the result obtained directly from the Riemann solution. The instant at which the discontinuity is observed at this location is $t_{dis}(x)$, which leads to a total area $A'(x, t_{dis})$. The interval in which the waveform is multivalued corresponds to $p^+ < p < p^-$. According to the discussion of profiles, the value of τ is a unique function of p, so it is best to us $dA' = (t_{dis} - t) dp$. Therefore,

$$A'(x, t_{dis}) = \int_{p^+}^{p^-} [t_{dis}(x) - t(\tau, x)] dp \qquad (13.2.52)$$

To verify that the equal-area rule applies to waveforms, we evaluate dA'/dx. Application of Leibnitz' rule to the preceding gives

Fig. 13.13 Construction of the area bounded by a multivalued waveform and a the location of the discontinuity at an arbitrary location x

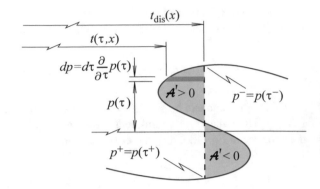

$$\frac{d}{dx}\mathcal{A}'(x, t_{\text{dis}}) = \int_{p^-}^{p^+}\left[\frac{d}{dx}t_{\text{dis}}(x) - \frac{\partial}{\partial x}t(\tau, x)\right]dp + \left[t_{\text{dis}}(x) - t\left(\tau^-, x\right)\right]\frac{\partial}{\partial x}p^-$$

$$- \left[t_{\text{dis}}(x) - t\left(\tau^+, x\right)\right]\frac{\partial}{\partial x}p^+ \qquad (13.2.53)$$

By definition, $t\left(\tau^-, x\right) = t\left(\tau^+, x\right) = t_{\text{dis}}(x)$, so the terms from the integration limits vanish. Also, $(d/dx)\,t_{\text{dis}}(x) \equiv 1/v_{\text{dis}}(x)$. To continue the analysis, it is convenient to rewrite Eq. (13.2.43), which is the small-signal approximation of the equation for a characteristic, as

$$\boxed{t(\tau, x) = \tau + \frac{x}{c + \beta v(\tau)}} \qquad (13.2.54)$$

Differentiation of this relation with respect to x, with τ held fixed, yields

$$\frac{\partial}{\partial x}t(\tau, x) = \frac{1}{c + \beta v(\tau)} \qquad (13.2.55)$$

The result is that the spatial rate of change of A' becomes

$$\frac{d}{dx}\mathcal{A}'(x) = \int_{p^+}^{p^-}\left[\frac{1}{v_{\text{dis}}(x)} - \frac{1}{c + \beta v(\tau)}\right]dp \qquad (13.2.56)$$

The analysis is restricted to weakly nonlinear signals, so $v = p/(\rho_0 c)$ and $c \gg v(\tau)$. These approximations lead to

$$\frac{d}{dx}\mathcal{A}'(x) = \int_{p^+}^{p^-}\left[\frac{c + \beta p/(\rho_0 c) - v_{\text{dis}}(x)}{c v_{\text{dis}}(x)} + \cdots\right]dp$$

$$= \frac{1}{c v_{\text{dis}}(x)}\left[c + \frac{1}{2}\beta\frac{p}{\rho_0 c} - v_{\text{dis}}(x)\right]p\bigg|_{p^+}^{p^-} \qquad (13.2.57)$$

The bracketed term is identically zero according to Eq. (13.2.42), so we have proven that the gradient of $A'(x)$ is identically zero. Consequently, $A'(x)$ is constant at the value when the discontinuity began to form, so $A'(x) = 0$. Thus, we have proven that the equal-area rule applies to waveforms, as well as spatial profiles.

We may use the equal-area rule to determine when the discontinuity occurs in the waveform at a designated x. If the excitation function $V(\tau)$ is odd relative to a phase τ_0 at which $V(\tau) = 0$, then symmetry places the shock at $t_{\text{dis}} = \tau_0 + x/c$. The procedure when $V(\tau)$ is arbitrary is a modification of the procedure for profiles.

To carry out the analysis, we need an expression for A' in terms of the trial instant t_{tr} for the discontinuity. To that end, we use the equation for a characteristic, Eq. (13.2.54), to replace t in Eq. (13.2.52) and then set $v = \varepsilon c V(\tau)$ and $p = \varepsilon \rho_0 c^2 V(\tau)$, and change the integration variable to τ. These operations give

$$A'(x, t_{\text{tr}}) = \varepsilon \rho_0 c^2 \int_{\tau^+}^{\tau^-} \left[t_{\text{tr}} - \tau - \frac{x}{c + \beta \varepsilon c V(\tau)} \right] \frac{dV(\tau)}{d\tau} d\tau \qquad (13.2.58)$$

A binomial series simplifies the fraction, so that

$$A'(x, t_{\text{tr}}) \approx \varepsilon \rho_0 c^2 \int_{\tau^+}^{\tau^-} \left[t_{\text{tr}} - \tau - \frac{x}{c}(1 - \beta \varepsilon V) \right] \frac{dV(\tau)}{d\tau} d\tau \qquad (13.2.59)$$

We apply integration by parts to the term for which τ is a coefficient. All other terms are perfect differentials, so that

$$A'(x, t_{\text{tr}}) = \rho_0 c^2 \varepsilon \left\{ \left[\left(t_{\text{tr}} - \tau - \frac{x}{c} \right) V(\tau) + \frac{1}{2} \beta \varepsilon \frac{x}{c} V(\tau)^2 \right] \Big|_{\tau^+}^{\tau^-} + \int_{\tau^+}^{\tau^-} V(\tau) \, d\tau \right\}$$

$$(13.2.60)$$

The value of t_{dis} is the value of t_{tr} that gives $A' = 0$ when x is set. Carrying out the search for this value of t_{tr} requires determination of the values of τ^+ and τ^- corresponding to specified values of x and t_{tr}. These values respectively are the smallest and largest roots extracted from Eq. (13.2.54). The procedure for carrying out the analysis is described in the next example.

EXAMPLE 13.4 In Example 13.2, the excitation at $x = 0$ is a harmonically varying velocity starting at $t = 0$. Determine the waveforms at $x/(x_{\text{shock}})_{\text{sine}} = 1.2, 1.5, 2,$ and 4.

Significance

A primary focus is development of a general computational algorithm for fitting shocks into the Riemann solution. The resulting waveforms will shed light on the trend for the signal to approach a sawtooth waveform.

Solution

The starting point for the analysis is the evaluation of the Riemann solution. The direct computational method used to solve Example 13.2 may be used for this purpose because there is no need to have the pressure data on a base of equally spaced t values. The nondimensional shock formation distance is $kx_{shock} = 1/(\varepsilon\beta)$, so we seek the waveform at locations where $kx = \sigma kx_{sh} = \sigma/(\beta\varepsilon)$. The acoustic Mach number is $\varepsilon = 0.002$. The Riemann solution at a set of equally spaced $\omega\tau_n$ values starting at $\omega\tau_1 = 0$ yields a set of $q_n \equiv p/(\rho_0 c^2)$ values corresponding to instants ωt_n, where

$$q_n = \varepsilon \sin(\omega\tau_n)\, h(\omega\tau_n)$$
$$\omega t_n = \omega\tau_n + \frac{\sigma}{\beta\varepsilon}\,\frac{1}{1 + \beta\varepsilon \sin(\omega\tau_n)\, h(\omega\tau_n)} \tag{1}$$

The evaluation of the Riemann solution yields a correlated set of $\omega\tau_n$, ωt_n, and q_n data at fixed σ in which the constant interval for the $\omega\tau$ values is $2\pi/N$.

The Riemann solution beyond x_{shock} is typified by the waveform at $\sigma = 4$ in Fig. 1. Two different conditions are seen to exist. At the leading edge, which corresponds to $\tau = 0$, there is no prior signal. The multivalued waveform has no special symmetry properties in this interval, so fitting a discontinuity will require an analysis of the waveform area. Subsequent to arrival of the shock at the leading edge, the multivalued intervals are centered on $\omega\tau = 2j\pi$, which is the phase at which $p = 0$ and $dp/d\tau$ is the maximum positive value. With respect to these instants, p is an odd function. Thus, the shock will occur at the time values for which $\omega(t_{dis} - x/c) = 2j\pi$.

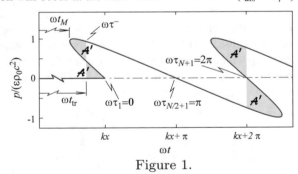

Figure 1.

We begin by modifying the computed waveform in the periods after the first. Let τ_J be any instant for which $p(\tau_J) = 0$. In view of the periodic nature of $p(\tau)$, and the fact that $p(0) = 0$, the second and subsequent discontinuities correspond to $J = N + 1, 2N + 1, \ldots$. Therefore, $\omega t_{dis} = \omega\tau_J + kx$ for these discontinuities. Now that we know the instant at which a discontinuity occurs, we can correct the multivalued pressure. Smaller τ values that occur later than ωt_J, and subsequent τ values that occur earlier are physically impossible. These phases belong to the discontinuity. This condition can occur for no more than $N/2$ points on either side of ωt_J. Thus, the procedure is to set $\omega \tau_{J-n} = \omega \tau_J$, $\omega t_{J-n} = \omega t_{dis}$, and $p_{J-n} = 0$ if $\omega \tau_{J-n} > \omega \tau_J$, and $\omega \tau_{J+n} = \omega \tau_J$, $\omega t_{J-n} = \omega t_{dis}$, and $p_{J+n} = 0$ if $\omega \tau_{J+n} < \omega \tau_J$. These operations are implemented for $n = 1, ..., N/2$.

Fitting the shock at the leading edge requires explicit examination of the waveform area according to Eq. (13.2.60) for each of the Riemann data values. The excitation function is $V(\tau) = \sin(\omega \tau) h(\omega t)$. Furthermore, the area of the waveform is situated in $p \geq 0$, so the lower limit in that expression corresponds to $\tau^+ = 0$ and $V(\tau^+) = 0$. Thus, the waveform area corresponding to a trial instant ωt_{tr} for the discontinuity is

$$\omega \mathcal{A}'(kx, \omega t_{tr}) = \rho_0 c^2 \varepsilon \left\{ \left[(\omega t_{tr} - \omega \tau^- - kx) \sin(\omega \tau^-) + \frac{1}{2} \beta \varepsilon kx \sin(\omega \tau^-)^2 \right] + \left[1 - \cos(\omega \tau^-) \right] \right\}$$

(1)

The true value of ωt_{dis} is obtained by setting $\omega t_{tr} = \omega t_n$ for a range of n and then searching for the value of n that best fits the criterion that $\omega \mathcal{A}'(kx, \omega t_n) = 0$. Explanation of the procedure is facilitated by considering Fig. 2, which zooms in on the multivalued interval at the leading edge. The dots mark points on the waveform that are the computed data. The first point, $\omega \tau_1 = 0$, corresponds to the largest t in the multivalued interval. The earliest time in the data set is ωt_M. The value of M is determined by seeking the index corresponding the minimum of the ωt_n data.

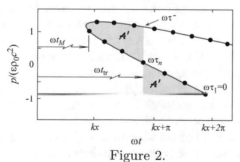

Figure 2.

The search procedure entails setting ωt_{tr} equal to a computed ωt_n value in the range $1 \leq n \leq M$. By definition, $\omega \tau^-$ is the characteristic value on the upper part of the multivalued segment when the time is ωt_{tr}. In Fig. 2, this point is the intersection of the vertical line through point n and the upper branch. Because a constant increment of the $\omega \tau$ values leads to uneven spacing of the ωt values, this intersection will not fall on a point that was computed along the upper branch. Therefore, the value of $\omega \tau^-$ is not yet known, but evaluation of the area in Eq. (1) requires that we locate this intersection. There are several options for this determination. The most accurate one uses the method in Sect. 13.1.5 to determine the largest of the three values of p

given by the Riemann solution at kx and this ωt_{tr}. A simpler alternative interpolates between data points on the upper branch on either side of ωt_{tr}. The simplest to implement selects the computed point on the upper branch that is closest to ωt_n. The latter procedure will be adequate if the value of N is sufficiently large. Regardless of which method is used, the concept is to use the value of $\omega \tau^-$ to compute $\omega \mathcal{A}'$. This is done beginning at $n = 1$, and ending at the most at $n = M$. The search may be halted when the sign of $\mathcal{A}'$ changes.

The resulting values of $\omega \mathcal{A}'$ in the full range of possible ωt_{tr} are shown in Fig. 3 for $x = 1.2 x_{sh}$. The value of $\mathcal{A}'$ is zero at $\omega t_{tr} = 199.9464$, so this is the value of ωt_{dis}. The computations used $N = 512$, which corresponds to $\Delta(\omega \tau) = 2\pi/512$. This is a reasonably fine resolution, but the nearly linear nature of the plotted curve suggests that it would have been adequate to use a smaller N.

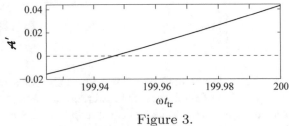

Figure 3.

After ωt_{dis} has been identified, the waveform may be corrected. The first step is to identify the instant that is closest to ωt_{dis} on the middle branch. We denote this instant as ωt_K. The index of points that lie on the middle branch is $1 < K < M$. The points in the multivalued interval that are delayed too much are adjusted by assigning them to the bottom of the discontinuity, so that $\omega t_n = \omega t_{dis}$ and $p_n = 0$ if $\omega t_n > \omega t_{dis}$ and $n < K$. Data values that occur too early are shifted to the top of the discontinuity by setting $\omega t_n = \omega t_{dis}$ and $p_n = p_K$ if $\omega t_n < \omega t_{dis}$ and $n > K$.

These procedures yield a set of $(\omega t, p)$ data that is single-valued function of time. Figure 4 is the result at the four specified locations. At 20% farther than the location where the shock begins to occur, the discontinuity is barely noticeable. Nevertheless, there is a visible similarity of the waveform to the teeth of a saw. Propagation to farther distances enhances this appearance. Indeed, at $\sigma = 4$, there is no evidence of a sinusoidal function. Another important aspect is the effect of the discontinuity on the peak pressure. At $\sigma = 1.2$ and $\sigma = 1.5$, the maximum pressure in the Riemann solution occurs slightly after each of the multivalued intervals. In contrast, at $\sigma = 2$ and at $\sigma = 4$ the maximum and minimum pressures in the Riemann solution occur in the multivalued interval. Thus, fitting a discontinuity to those waveforms results in reduction of the peak pressure. From this, we may deduce that the shock wave dissipates energy.

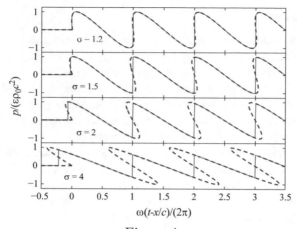

Figure 4.

An interesting question is how the discontinuities propagate relative to the wave-form. The waveform is an odd function relative to the second and subsequent discontinuities. These discontinuities propagate at the linear speed of sound, so their relative spacing is maintained at cT, and the interval between their occurrence in a waveform will be T. This is not true for the leading discontinuity because for it, p^- is positive and $p^+ = 0$. The average pressure across this shock is $p_{av} = p^-/2$. The Rankine-Hugoniot relation, Eq. (13.2.42), states that this discontinuity will propagate at $\tilde{c} \left(p^- / (2\rho_0 c) \right) + p^- / (2\rho_0 c) > c$. Therefore, the distance from the shock at the leading edge to the next shock will decrease with increasing t. For the same reason, the leading edge shock will precede arrival of the next shock by more than T. Eventually, the value of p^- progressively decreases as the discontinuity cuts off an increasing portion of the multivalued waveform. In terms of nondimensional variables, the discontinuity's arrival at a specified x is advanced by $\omega t_{dis} - kx$. Figure 5 is the result of computing this quantity over a broad range of σ values. Plotting the time advancement as a ratio to 2π allows us to see it as a fraction of a period. The rate of this advancement decreases with increasing distance due to the decrease of p^-. These features are examined more closely in the next section.

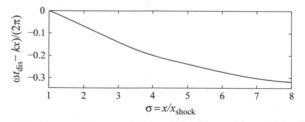

Figure 5.

Old-Age Behavior

The previous example demonstrates the drastic ways that a signal is affected by the creation of shocks. These effects grow as a signal propagates well beyond the location where a shock first forms. A square wave provides a convenient means for exploring these phenomena because the equal-area rule can be enforced without requiring integration to evaluate the waveform area. Although this is the only type of waveform we will consider, the observations are widely applicable because of the tendency of all waves to evolve to a sawtooth appearance.

The excitation is a particle velocity at $x = 0$ that is a periodic square wave beginning at $t = 0$. The peak particle velocity is set at εc, with $\varepsilon \ll 1$. The small-signal approximation tells us that $p = \rho_0 c v$, so the pressure also is a square wave, as shown in Fig. 13.14, where the maximum value is $\hat{p} = \rho_0 c^2 \varepsilon$. The presence of discontinuities in which the pressure increases tells us that the signal at $x = 0$ already contains shocks.

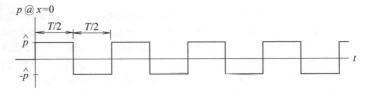

Fig. 13.14   A periodic square wave generated at $x = 0$. The pressure is zero for $t < 0$

The positive pressures propagate faster and the negative pressures propagate slower than the linear speed of sound. Because $|\varepsilon| \ll 1$, the Riemann solution has the property that straight segments of the waveform remain straight as the wave propagates. Thus, according to the Riemann solution, the square lobes become trapezoids that lean forward in time as the signal propagates. Figure 13.15 shows this waveform at a propagation distance that is not too large. The phase speed of the maximum positive pressures is $c (1 + \beta \varepsilon)$, whereas the phase speed at zero pressure is c. Thus,

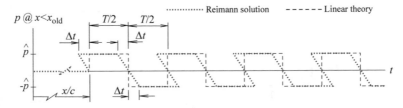

Fig. 13.15   Comparison of the Riemann solution and linear theory for propagation of a periodic square wave. The location is not too distant from the source

the time advancement Δt of the positive pressure plateaus relative to the arrival time of the preceding $p = 0$ phases is

$$\Delta t = \frac{x}{c} - \frac{x}{c\,(1 + \beta\varepsilon)} \approx \beta\varepsilon\frac{x}{c} \qquad (13.2.61)$$

The phase speed of the negative pressure plateaus is $c\,(1 - \beta\varepsilon)$. To $O(\varepsilon)$, these signals are retarded in time relative to $p = 0$ by the same Δt. As part of the analysis, we will identify the largest distance, labeled x_{old}, for which this picture is appropriate.

According to the Riemann solution, $p = 0$ and $\partial p/\partial \tau > 0$ when t equals x/c plus an integer multiple of the period. The multivalued portion of the Riemann solution contains these instants. They are corrected by application of the equal-area rule.

The piecewise linear nature of the Riemann waveform simplifies this adjustment. After the first cycle, the waveform in the multivalued intervals are odd functions relative to the instant when $p = 0$. Figure 13.16 shows a vertical line through $p = 0$ within these intervals. The waveform area for negative p is the negative of the area for positive p. It follows that these lines define the time t_{dis} of discontinuities. The discontinuity values are $p^- = \hat{p}$ and $p^- = -\hat{p}$. The Rankine-Hugoniot relation states that shocks propagate at the linear speed of sound plus the average of the particle velocities ahead of and behind the discontinuity. This construction in Fig. 13.16 is consistent with this principle.

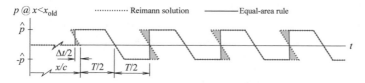

Fig. 13.16 Application of the equal-area rule to a periodic square wave. The propagation distance is relatively small

The situation at the leading edge is different, because no signal precedes the $p = 0$ phase. A vertical line that splits Δt leads to a waveform area that consists of identical triangles to the left and right, so the discontinuity at the leading edge is advanced by $\Delta t/2$. It follows that the leading edge arrives at position x when $t = x/c - \Delta t/2 = (x/c)\,(1 - \beta\varepsilon/2)$. The corresponding propagation speed of this shock is $x/t = c/\,(1 - \beta\varepsilon/2) \approx c\,(1 + \beta\varepsilon/2)$. This too is consistent with the Rankine–Hugoniot relations because it is the mean of the sound speeds for $p^- = \hat{p}$ and $p^+ = 0$. Increasing the propagation distance leads to an increase of Δt. At some location, the Riemann solution predicts that the last pressure on a positive plateau occurs at the instant of a discontinuity, and that the first pressure on a negative plateau also occurs at that instant. This condition, which is characterized by $\Delta t = T/2$, is described by Fig. 13.17. Only the leading edge is exempt from this behavior.

Beyond the location described by Fig. 13.17, the waveform is altered in a fundamentally different manner that is ongoing. This position marks the advent of the *old-age stage,* so it is labeled as x_{old}. Setting $\Delta t = T/2$ gives

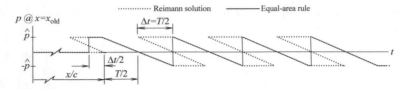

Fig. 13.17 Distortion of an initial square wave at the onset of old age. The time advancement of the positive pressure equals half a period

$$x_{\text{old}} = \frac{cT}{2\beta\varepsilon} = \frac{cT}{2\beta} \left(\frac{\rho_0 c^2}{\hat{p}} \right) \tag{13.2.62}$$

In the present case of a periodic square wave, the old-age stage begins when a plateau has advanced sufficiently that its entire occurrence falls in the interval where the Riemann solution is multivalued. Beyond that occurrence, the discontinuity intersects the Riemann solution on a sloping portion. A negative plateau also is cut off because it is delayed into the multivalued interval. This is the situation in Fig. 13.18. The equal-area rule maintains all discontinuities, other than the leading edge, at the location where the Riemann solution gives $p = 0$. The consequence is that the peak maximum and minimum pressure amplitudes are reduced. Any wave that propagates beyond its shock formation distance will ultimately exhibit this phenomenon. It is the hallmark of the *old-age stage*.

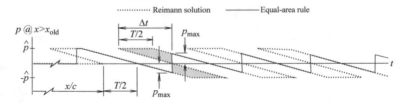

Fig. 13.18 Waveform of an initially square wave at old-age. Distortion results in a decrease of the peak pressure

An expression for the peak pressure in the old-age stage is obtained from similar triangles formed by a discontinuity and the Riemann solution, specifically

$$\frac{p_{\text{max}}}{T/2} = \frac{\hat{p}}{\Delta t} \tag{13.2.63}$$

Substitution of Δt from Eq. (13.2.61) and $\hat{p} = \rho_0 c^2 \varepsilon$ leads to

$$p_{\text{max}} = \frac{\rho_0 c^2}{2\beta} \left(\frac{cT}{x} \right) = \hat{p} \frac{x_{\text{old}}}{x}, \quad x > x_{\text{old}} \tag{13.2.64}$$

The first form is remarkable because it indicates that the peak pressure is independent of $\hat{p}$. In other words, if all other parameters are held fixed, increasing the amplitude of the input will not alter the pressure observed at a specified location in the old-age region. Why is this so? If we increase the amplitude, the value of x_{old} decreases. Consequently, the signal enters old-age at a closer location, so x_{old}/x decreases. Thus, the increase in the input amplitude is offset by the greater old-age reduction. This phenomenon is called *saturation*. It limits the loudness of the sound that can be heard. To see why, suppose we select a location x. At very small values of $\hat{p}$, this location is much smaller than the value of x_{old} given by Eq. (13.2.62). Increasing the amplitude of the input proportionally raises the amplitude of the Riemann solution. However, the increase also reduces the value of x_{old}. Further increase eventually reduces x_{old} to x at the field point, which occurs when $\hat{p} = \rho_0 c^2 (cT) / (2\beta x)$. This is the saturation pressure. Further increase of $\hat{p}$ leads to $x > x_{old}$, which means that Eq. (13.2.64) now describes the peak pressure. As an example, consider a signal in air for which $1/T = 1$ kHz. The saturation amplitude at $x = 100$ m is 197 Pa. The RMS pressure in one cycle of the sawtooth waveform is $p_{max}/3^{1/2}$, so the sound pressure level at saturation is 135 dB. The only way in which a louder sound at that location can be obtained with a plane wave is to raise T, that is, lower the fundamental frequency.

The reduction of peak pressure in the old-age stage brings to the fore the issue of mechanical energy. Because we are dealing with simple plane waves, and the particle velocity is $p/(\rho_0 c)$ in the small-signal approximation, the time-averaged intensity has the same form as it does for a linear signal whose waveform is arbitrary. For the present purpose, we may use any period subsequent to the first, so that

$$I_{av}(x) = \frac{1}{T} \int_{x/c+nT}^{x/c+(n+1)T} \frac{p(x,t)^2}{\rho_0 c} dt \qquad (13.2.65)$$

For the initial waveform in Fig. 13.14, we have $I_{av}(0) = \hat{p}^2/(\rho_0 c)$. The area under the square of trapezoidal waveform in Fig. 13.15 may be found by adding the contribution of the plateau and the linear decrease, which gives

$$I_{av}(x) = \frac{2}{\rho_0 cT}\left[\hat{p}^2\left(\frac{T}{2} - \Delta t\right) + \frac{1}{3}\hat{p}^2 \Delta t\right] = \frac{\hat{p}^2}{\rho_0 c}\left(1 - \frac{2}{3}\frac{x}{x_{old}}\right), \quad x < x_{old} \qquad (13.2.66)$$

In the region where old-age behavior has taken over, integration of the square of the triangular waveform in Fig. 13.18 leads to

$$I_{av}(x) = \frac{2}{\rho_0 cT}\left(\frac{1}{3}p_{max}^2\frac{T}{2}\right) = \frac{1}{3}\frac{\hat{p}^2}{\rho_0 c}\left(\frac{x_{old}}{x}\right)^2, \quad x > x_{old} \qquad (13.2.67)$$

At x_{old}, both expressions give $I_{av} = (\hat{p})^2/(3\rho_0 c)$. We see that the time-averaged intensity decreases with increasing x at all locations. This means that some energy is lost, rather than flowing downstream. Before the onset of old-age, the loss is linear. In old-age, the power flow decreases as the inverse square of distance, so the energy loss is much more drastic.

The loss of energy is a consequence of the thermodynamic mechanisms captured by the Rankine–Hugoniot relations. This effect arises even if nonideal effects are negligible. However, dissipative effects like viscosity and molecular relaxation are always present. Attenuation of harmonics composing a periodic signal typically increases with increasing harmonic number, as will be seen in Sect. 13.3.2. This effect is not necessarily large, depending on the properties of the fluid. However, it combines with the old-age behavior, which reduces all harmonics. Ultimately, what remains is a fundamental harmonic that has been reduced sufficiently to permit the usage of linear acoustic theory.

One of the surprising attributes of shocks is that they affect a pulse differently from a steady-state signal. We encountered this difference in the case of continuous square wave excitation, where the discontinuity at the leading edge was unlike those embedded in the subsequent response. To explore this further, we consider a pulse at $x = 0$ that is the first cycle of the square wave, after which it is cut off. In Fig. 13.19, the pulse has been propagating for a while, but it has not yet reached old-age.

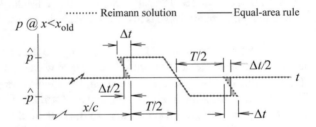

Fig. 13.19 Early stage in the propagation of single period of a square wave

The Riemann solution for the pulse is the same as its prediction for the first period of the continuous wave. As a consequence, the shock at the leading edge behaves in the same way as the shock for the continuous wave. Termination of the waveform after one cycle leads to $p^+ = -\hat{p}$ and $p^- = 0$. The multivalued waveform at the trailing edge is not what it was for a continuous wave. Rather, it is an image of the waveform at the leading edge. Therefore, the retardation of the discontinuity at the trailing edge is the same as the advancement at the leading edge. The consequence is that the pulse duration is extended to $x/c - \Delta t/2 < t < x/c + T + \Delta t/2$, whereas it is $x/c < t < x/c + T$ in linear theory.

Further propagation eventually brings the pulse to the transition where old-age behavior sets in. This occurs when the last pressure on the positive plateau catches up to the discontinuity, which is the same instant as that at which the first pressure on the negative plateau is delayed to the trailing discontinuity. This occurs when $\Delta t = T$, which leads to x_{old} being twice as distant as the value in Eq. (13.2.62). Figure 13.20 shows the waveform beyond the old-age location.

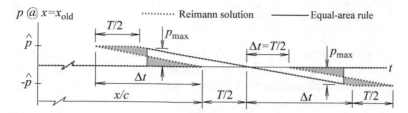

Fig. 13.20 Distortion of a square wave pulse beyond the distance where old-age behavior begins

Application of the equal-area to locate the discontinuity in the old-age signal would be a little more difficult than it was for the continuous wave because the waveform area has a different shape on either side of the discontinuity. Nevertheless, it is evident that the waveform preserves its sawtooth shape with an ever decreasing amplitude, and that its duration increases progressively as the propagation distance increases.

It is reasonable to ponder whether the old-age behavior of a square wave pulse is representative of general trends. Evidence that it may be found in the results of Example 13.4, where the waveform strongly resembles a sawtooth at a distance as small as $x = 2x_{shock}$. Indeed, if we set $T = 2\pi/\omega$ for the fundamental frequency, the expression in Eq. (13.2.62) gives $x_{old} = \pi (x_{shock})_{sine}$. A comparison of the waveforms at $x = 2x_{shock}$ and $x = 4x_{shock}$ in Fig. 3 of the aforementioned example shows that old-age amplitude reduction has occurred at the larger distance. The old-age behavior of any signal will be like that of the square wave, but the details of where it sets in, where the discontinuities are situated, and how the amplitude decreases with increasing distance will depend on the nature of the input function.

13.3 General Analytical Techniques

We began our study of linear acoustics with the d'Alembert solution. Like the Riemann solution, it is obtained by analyzing the characteristics. To extend the scope of linear systems that could be investigated, we invoked other methods to solve the wave equation. In the same way, if we are to expand our exploration of nonlinear acoustic waves, we will need an appropriate field equation. The nonlinear wave equation we will derive contains no approximations beyond the assumption of an isentropic process in an inviscid fluid. To the author's knowledge, the first appearance of this equation was in a text on aeroelasticity.[12] The derivation parallels that in a classic fluid mechanics treatise,[13] and an alternative derivation appears in an acoustics text.[14] After the field equation is derived, two general methods for solving it will be derived for the case of plane waves.

---

[12]R.L. Bisplinghoff, H. Ashley, and R.L. Halfman, *Aeroelasticity*, Addison Wesley (1955) Chap. 3

[13]S. Goldstein, *Lectures in Fluid Mechanics, Wiley-Interscience* (1960) Chap. 4

[14]D.T. Blackstock, *Physical Acoustics*, Wiley (2000) pp. 84–86.

13.3.1   A Nonlinear Wave Equation

The basic state variables are the acoustic pressure p, the particle velocity $\bar{v}$, and the current density ρ. The ambient state corresponds to $p = 0$, $\bar{v} = \bar{0}$, and $\rho = \rho_0$. The field equation we seek governs the velocity potential $\phi(\bar{x}, t)$, whose gradient is the particle velocity,

$$\bar{v} = \nabla\phi \tag{13.3.1}$$

In terms of this function, the continuity equation, Eq. (4.1.5), is

$$\frac{1}{\rho}\frac{\partial\rho}{\partial t} + \frac{\nabla\rho}{\rho}\cdot\nabla\phi + \nabla^2\phi = 0 \tag{13.3.2}$$

and the momentum equation, Eq. (4.1.7), is

$$\frac{\partial}{\partial t}\nabla\phi + \frac{1}{2}\nabla(\nabla\phi\cdot\nabla\phi) = -\frac{\nabla p}{\rho} \tag{13.3.3}$$

The overall task now is to replace terms that contain p or ρ with terms that solely contain ϕ. Our strategy entails manipulating the momentum equation and the equation of state to obtain suitable representations of the first two terms in the continuity equation. Removal of the second term is the more direct. The corollary of the assumption of constant entropy is that p is a known function of ρ. Taking the gradient of this function shows that ∇p is proportional to $\nabla\rho$, specifically,

$$\nabla p = \tilde{c}^2\nabla\rho, \quad \tilde{c} = \left(\frac{dp}{d\rho}\right)^{1/2} \tag{13.3.4}$$

The quantity $\tilde{c}$ is the tangential sound speed we first encountered in the derivation of the Riemann solution. The result of substituting this definition into the momentum equation is an expression we will use to replace the second term in the continuity equation, Eq. (13.3.2), specifically,

$$\frac{\nabla\rho}{\rho} = -\frac{1}{\tilde{c}^2}\left[\frac{\partial}{\partial t}\nabla\phi + \frac{1}{2}\nabla(\nabla\phi\cdot\nabla\phi)\right] \tag{13.3.5}$$

The task now is to find a suitable description, the first term in the continuity equation using relations derived from the momentum equation and/or equation of state. The means for doing so entails working with the pressure, rather than the density, using an ingenious application of Leibnitz rule for differentiating an integral. According to it

$$\frac{1}{\rho}\frac{\partial p}{\partial t} \equiv \frac{\partial}{\partial t}\int_0^p \frac{dp'}{\rho(p')}$$
$$\frac{\nabla p}{\rho} \equiv \nabla\int_0^p \frac{dp'}{\rho(p')} \tag{13.3.6}$$

Both identities follow from the fact that only the upper limit of the integral depends explicitly on position and time.

These relations merely replace one form of dependence on ρ with another. The reason they are useful is that the right sides are respectively the time derivative or gradient of a quantity. Substitution of the gradient form into the original momentum equation, Eq. (13.3.3), gives

$$\nabla \left(\frac{\partial \phi}{\partial t} + \frac{1}{2} \nabla \phi \cdot \nabla \phi + \int_0^p \frac{dp'}{\rho(p')} \right) = 0 \tag{13.3.7}$$

This equation requires that the term within the parentheses be independent of position, but it could be a function of time. However, it was stipulated that the fluid is initially at rest in the ambient state. The term in parentheses is zero at that initial state and therefore always. Thus, we have

$$\boxed{\frac{\partial \phi}{\partial t} + \frac{1}{2} \nabla \phi \cdot \nabla \phi + \int_0^p \frac{dp'}{\rho(p')} = 0} \tag{13.3.8}$$

In later developments, this equation will lead to a relation for the pressure corresponding to a known solution for the velocity potential. Here, we use it to eliminate the density from the continuity equation. Toward that objective, we differentiate the equation with respect to time and then recall the first of Eq. (13.3.6). The result is

$$\frac{\partial^2 \phi}{\partial t^2} + \frac{\partial}{\partial t} \left(\frac{1}{2} \nabla \phi \cdot \nabla \phi \right) + \frac{1}{\rho} \frac{\partial p}{\partial t} = 0 \tag{13.3.9}$$

The pressure is eliminated by applying the chain rule for differentiation to the equation of state, which gives $\partial p / \partial t = \tilde{c}^2 \partial \rho / \partial t$. Thus, these manipulations of the momentum equation have led us to a relation that matches the first density term in Eq. (13.3.2),

$$\frac{1}{\rho} \frac{\partial \rho}{\partial t} = -\frac{1}{\tilde{c}^2} \left[\frac{\partial^2 \phi}{\partial t^2} + \frac{\partial}{\partial t} \left(\frac{1}{2} \nabla \phi \cdot \nabla \phi \right) \right] \tag{13.3.10}$$

The result of substituting this relation and Eq. (13.3.5) into Eq. (13.3.2), followed by collection and rearrangement of some terms, is

$$\tilde{c}^2 \nabla^2 \phi - \frac{\partial^2 \phi}{\partial t^2} - \frac{\partial}{\partial t} (\nabla \phi \cdot \nabla \phi) - \frac{1}{2} \nabla (\nabla \phi \cdot \nabla \phi) = 0 \tag{13.3.11}$$

All product terms in the preceding are the result of kinematical nonlinearity. Constitutive nonlinearity is contained in the equation of state, but that effect is buried in $\tilde{c}^2$. Differentiation of the equation of state will give $\tilde{c}^2$ as a function of either p or ρ, depending on how the equation of state is written. Unfortunately, neither form is acceptable because we seek a field equation in which ϕ is the sole dependent variable.

Thus, we cannot proceed until we specify the equation of state. Let us begin by considering an ideal gas. For an isentropic process, the relation is

$$\frac{\rho}{\rho_0} = \left(\frac{p}{p_0} + 1\right)^{1/\gamma} \tag{13.3.12}$$

The definition of $\tilde{c}$ gives

$$\tilde{c}^2 \equiv \frac{1}{d\rho/dp} = c^2 \left(\frac{p}{p_0} + 1\right)^{(\gamma-1)/\gamma} \tag{13.3.13}$$

We wish to eliminate p in favor of ϕ in this equation. The means for doing so is to substitute the equation of state into Eq. (13.3.8). This operation makes it possible to evaluate the integral, specifically

$$\frac{\partial \phi}{\partial t} + \frac{1}{2}\nabla\phi \cdot \nabla\phi + \frac{1}{\rho_0}\int_0^p \left(\frac{p'}{p_0}+1\right)^{-1/\gamma} dp'$$
$$\equiv \frac{\partial \phi}{\partial t} + \frac{1}{2}\nabla\phi \cdot \nabla\phi + \left(\frac{c^2}{\gamma-1}\right)\left[\left(\frac{p}{p_0}+1\right)^{(\gamma-1)/\gamma} - 1\right] = 0 \tag{13.3.14}$$

where the relation for the linear speed of sound, $c^2 = \gamma p_0/\rho_0$, has been used. Solving this equation for p in terms of ϕ yields

$$\boxed{\frac{p}{p_0} = \left[1 - \frac{\gamma-1}{c^2}\left(\frac{\partial \phi}{\partial t} + \frac{1}{2}\nabla\phi \cdot \nabla\phi\right)\right]^{\gamma/(\gamma-1)} - 1} \tag{13.3.15}$$

We seek an expression for $\tilde{c}^2$ in terms of p, so we substitute the above expression for p into Eq. (13.3.13). The result is

$$\tilde{c}^2 = c^2 - (\gamma-1)\left(\frac{\partial \phi}{\partial t} + \frac{1}{2}\nabla\phi \cdot \nabla\phi\right) \tag{13.3.16}$$

Substitution of $\tilde{c}^2$ into Eq. (13.3.11) yields a nonlinear field equation,

$$\boxed{\begin{aligned} c^2\nabla^2\phi - \frac{\partial^2\phi}{\partial t^2} - &\left[(\gamma-1)\nabla^2\phi\frac{\partial \phi}{\partial t} + \frac{\partial}{\partial t}(\nabla\phi \cdot \nabla\phi)\right. \\ &\left. + \frac{(\gamma-1)}{2}(\nabla\phi \cdot \nabla\phi)\nabla^2\phi + \frac{1}{2}\nabla\phi \cdot \nabla(\nabla\phi \cdot \nabla\phi)\right] = 0 \end{aligned}} \tag{13.3.17}$$

This is the *nonlinear wave equation*. When the terms contained within the bracket are dropped, what remains is the linear wave equation. The full equation governs any wave in an ideal gas. Its appearance is quite daunting. We may simplify it somewhat by invoking the small-signal approximations. The implication of $|p|$ being much less than the bulk modulus is that ϕ also is small. If that is so, then the cubic terms are much less than the quadratic terms, which are much less than the linear terms. It is necessary to retain the quadratic products as the first approximation of nonlinear effects, but the cubic terms may be dropped.

In addition to simplification of the field equation, dropping the cubic terms has another beneficial feature. When the equation of state for an ideal gas is truncated at quadratic terms, it is the same as the equation for a liquid if we state both in terms of the coefficient of nonlinearity in Eq. (13.1.34). Hence, setting $\gamma + 1 = 2\beta$ in the nonlinear wave equation and deleting the cubic terms yield a field equation that is descriptive of weakly nonlinear waves in an ideal gas, as well as almost any wave in a liquid. This version of the nonlinear wave equation is

$$c^2 \nabla^2 \phi - \frac{\partial^2 \phi}{\partial t^2} - \left[2 (\beta - 1) \nabla^2 \phi \frac{\partial \phi}{\partial t} + \frac{\partial}{\partial t} (\nabla \phi \cdot \nabla \phi) \right] = 0 \qquad (13.3.18)$$

We will seek solutions of this equation. However, the state variable of interest is the pressure. Equation (13.3.15) describes p in terms of ϕ. This relation also is simplified by the small-signal approximations. The bracketed term is expanded in a binomial series. Truncation of the expansion at quadratic terms gives

$$\frac{p}{p_0} = -\frac{\gamma}{c^2} \left[\frac{\partial \phi}{\partial t} + \frac{1}{2} \nabla \phi \cdot \nabla \phi - \frac{1}{2c^2} \left(\frac{\partial \phi}{\partial t} \right)^2 \right] \qquad (13.3.19)$$

A minor simplification comes from the relation for the linear speed of sound in an ideal gas, $\gamma p_o / \rho_0 = c^2$. This converts the preceding to

$$p = -\rho_0 \left[\frac{\partial \phi}{\partial t} + \frac{1}{2} \nabla \phi \cdot \nabla \phi - \frac{1}{2c^2} \left(\frac{\partial \phi}{\partial t} \right)^2 \right] \qquad (13.3.20)$$

The absence of γ or β from this expression tells us that the nonlinear effect contained in this relation is kinematical in origin and may be used for both liquids and gases. We also see that the relation reduces to the linear one when the quadratic terms are dropped.

In summary, we have found that any acoustic signal in a liquid and all except the loudest acoustic signals in an ideal gas are governed by the quadratic version of the nonlinear wave equation, Eq. (13.3.18). Boundary conditions typically impose conditions on the particle velocity, which is $\bar{v} = \nabla \phi$. The pressure corresponding to a solution for ϕ is given by Eq. (13.3.20). In the exceptional case of a signal in an ideal gas whose amplitude is deemed to be too large to use a small-signal approximation, the appropriate nonlinear wave equation is Eq. (13.3.17) and Eq. (13.3.15) should be used to determine the pressure.

The derivation of the nonlinear wave equation is based on the assumptions that dissipation mechanisms are insignificant, and that the signal propagates isentropically. It is possible to modify this equation by adding terms that capture dissipative effects. Linear terms are adequate for this purpose, because dissipation, relaxation, and such are small effects that are comparable to the effect of nonlinearity. Suitable dissipation terms may be found in the modified versions of the linear wave equation that were developed in Sect. 3.3.

Even if we exclude nonideal effects, a direct attempt to solve the nonlinear wave equation by standard techniques, such as separation of variables, would not be successful. For this reason, several simplified field equations have been developed.[15] They differ in their assumptions regarding scales over which state variables change in various directions and the relative order of magnitude of terms representing various effects. Some model equations are particularly amenable to analysis by finite difference methods. Our concentration will be on systems that are amenable to analytical techniques. The quadratic nonlinear wave equation will be the framework for our investigations, but the same techniques may be applied to other model equations.

13.3.2 Frequency-Domain Formulation

Numerical techniques have been used to solve the nonlinear wave equation. The first thought might be to use finite difference formulas to approximate the derivatives in the nonlinear wave equation. However, that approach has inherent limitations, not the least of which is the accumulation of error with increasing distance and time. This is a crucial issue because the error might overwhelm the nonlinear effects, especially for a weakly nonlinear signal. Another issue is the large number of mesh points resulting from the requirements that the spatial mesh size be considerably smaller than a wavelength and the temporal time increment be a fraction of the interval over which the signal fluctuates.

The best computational approaches begin with an ansatz that captures as many aspects as possible of the fundamental physics of a system. If we are solely interested in situations where the excitation is periodic in time, a formulation that inherently recognizes the periodicity of the steady-state signal can reasonably be expected to be optimal. Periodicity allows the time signature to be represented by a Fourier series. In linear theory, each harmonic in that series would be a solution of the linear wave equation, and the series coefficients would be constants. We have seen with the Fubini-Ghiron solution that the amplitude of each harmonic in a plane wave is position dependent. This suggests that the periodic signal in any system may be described as a Fourier series whose harmonic functions are solutions of the linear wave equation for that system, and whose coefficients are position dependent. Making this ansatz satisfy the nonlinear wave equation will lead to a coupled set of first-order differential equations for the coefficients. Such equations are readily solved by standard numerical methods.

---

[15]M.F. Hamilton and C.L. Morfey, "Model Equations," M.F. Hamilton and D.T. Blackstock, *Nonlinear Acoustics*, eds., Acoustical Society of America (2008) Chap. 3.

This approach is said to work in the frequency domain because the signal is decomposed into its harmonics. However, the most useful property of linear equations, which is the principle of superposition, is not applicable to nonlinear equations. A corollary is that one cannot simplify matters by using real or imaginary parts and factoring out complex exponential factors. We shall develop the basic approach by applying it to the analysis of a plane wave.

The period is T, so the fundamental frequency is $\omega_1 = 2\pi/T$. The corresponding phase variable for a linear wave is

$$\boxed{\theta_1 = \omega_1 t - kx, \ \ k = \frac{\omega_1}{c}}$$

(13.3.21)

Thus, the Fourier series representation of a plane wave is taken to be

$$\phi = \frac{1}{2} \sum_{n=-\infty}^{\infty} \Phi_n(x) \, e^{in\theta_1}, \ \ \Phi_{(-n)} = \Phi_n^*$$

(13.3.22)

Equations for the Φ_n functions are obtained by requiring that this expression satisfies the quadratic nonlinear wave equation, Eq. (13.3.18). An important aspect of those operations arises when we consider $\nabla^2 \phi$, which is

$$\nabla^2 \phi = \frac{\partial^2 \phi}{\partial x^2} = \frac{1}{2} \sum_{n=-\infty}^{\infty} \left[\Phi_n'' - 2ink\Phi_n' - n^2 k^2 \Phi_n \right] e^{in\theta_1}$$

(13.3.23)

where a prime will denote an ordinary derivative with respect to x. Based on our experience thus far, as well as the expectation that nonlinear effects grow gradually, we consider $\left| \Phi_n'(x) \right| \ll k \left| \Phi_n(x) \right|$. In other words, we take each of the amplitude factors to be a *slowly varying function*. Correspondingly, second derivatives of Φ_n are ignored, so the linear terms in the wave equation become

$$c^2 \frac{\partial^2 \phi}{\partial x^2} - \frac{\partial^2 \phi}{\partial t^2} = -ic^2 k \sum_{n=-\infty}^{\infty} n\Phi_n' e^{in\theta_1}$$

(13.3.24)

The nonlinear terms in the wave equation, Eq. (13.3.18), require that derivatives of the Fourier series be multiplied. The manipulations are simplified by the slowness property, which allows derivatives of the Φ_n coefficients to be dropped when the product is formed. This means that only variation of the complex exponential is considered, so that

$$\frac{\partial \phi}{\partial t} \approx \frac{\partial \phi}{\partial \theta_1} \frac{\partial \theta_1}{\partial t} = \omega_1 \frac{\partial \phi}{\partial \theta_1}, \ \ \frac{\partial \phi}{\partial x} \approx \frac{\partial \phi}{\partial \theta_1} \frac{\partial \theta_1}{\partial x} = -k \frac{\partial \phi}{\partial \theta_1}$$

(13.3.25)

In turn, these properties lead to

$$\nabla^2\phi\frac{\partial\phi}{\partial t} = \frac{\partial^2\phi}{\partial x^2}\frac{\partial\phi}{\partial t} \approx k^2\omega_1\frac{\partial^2\phi}{\partial\theta_1^2}\frac{\partial\phi}{\partial\theta_1}$$

$$\frac{\partial}{\partial t}\left(\nabla\phi\cdot\nabla\phi\right) = 2\frac{\partial^2\phi}{\partial x\partial t}\frac{\partial\phi}{\partial x} \approx 2k^2\omega_1\frac{\partial^2\phi}{\partial\theta_1^2}\frac{\partial\phi}{\partial\theta_1}$$

(13.3.26)

The preceding relations allow us to combine the product terms in the wave equation. Each factor in the product must be described by a different summation index in order to assure that all individual products are described. Thus, the product terms become

$$2\left(\beta-1\right)\nabla^2\phi\frac{\partial\phi}{\partial t} + \frac{\partial}{\partial t}\left(\nabla\phi\cdot\nabla\phi\right) = 2\beta k^2\omega_1\frac{\partial^2\phi}{\partial\theta_1^2}\frac{\partial\phi}{\partial\theta_1}$$

$$= -\frac{i}{2}\beta\sum_{n=-\infty}^{\infty}\sum_{m=-\infty}^{\infty} mn^2k^2\omega_1\Phi_m\Phi_n e^{i(n+m)\theta_1}$$

(13.3.27)

Substitution of this expression and Eq. (13.3.24) into the nonlinear wave equation, Eq. (13.3.18), gives

$$-ic^2k\sum_{n=-\infty}^{\infty} n\Phi_n' e^{in\theta_1} = -\frac{i}{2}\beta k^2\omega_1\sum_{n=-\infty}^{\infty}\sum_{m=-\infty}^{\infty} mn^2\Phi_m\Phi_n e^{i(n+m)\theta_1}$$ (13.3.28)

The frequencies of the harmonic functions are multiples, so the functions are orthogonal over their common period. This property makes it possible to obtain a set of first-order differential equations for the amplitude functions. To that end, we multiply the equation by $\exp\left(-ij\theta\right)$, where j may be any integer. We then integrate over a period, $-\pi < \theta_1 < \pi$. All terms in the single summation, other than the one for which $n = j$, are orthogonal to this complex exponential. The only terms in the double summation that are not orthogonal to $\exp\left(-ij\theta_1\right)$ are those for which $m = j - n$. Thus, these operations filter a single term out of the single summation and reduce the double summation to a single one. The result is a set of coupled ordinary differential equations,

$$jk\Phi_j' = \frac{1}{2}\beta\frac{k^2\omega}{c^2}\sum_{n=-\infty}^{\infty} (j-n)\,n^2\Phi_{(j-n)}\Phi_n$$ (13.3.29)

It would be more meaningful if we worked with the Fourier pressure coefficients. We may describe $p\left(x, t\right)$ as a Fourier series,

$$p = \frac{1}{2} \sum_{n=-\infty}^{\infty} P_n(x) e^{in\theta_1}, \quad P_{(-n)} = P_n^* \qquad (13.3.30)$$

The pressure is related to the velocity potential by $p = -\rho_0 \partial \phi / \partial t$. (Nonlinearity is uniformly a second-order effect here.) The result of replacing p and ϕ by their respective Fourier series and matching like harmonics shows that, for any n,

$$P_n = -in\omega \rho_0 \Phi_n \Longrightarrow nk\Phi_n = \left(\frac{i}{\rho_0 c}\right) P_n \qquad (13.3.31)$$

When this relation is substituted into Eq. (13.3.29), the result is

$$\frac{P_j'}{\rho_0 c} = \frac{i}{2} \beta \frac{\omega}{c^2} \sum_{n=-\infty}^{\infty} n \frac{P_{(j-n)} P_n}{(\rho_0 c)^2} \qquad (13.3.32)$$

This form suggests nondimensionalization of the position and pressure coefficients according to

$$\tilde{x} = kx, \quad \tilde{P}_n = \frac{P_n}{\rho_0 c^2} \qquad (13.3.33)$$

This replacement converts the differential equations to

$$\frac{d\tilde{P}_j}{d\tilde{x}} = \frac{i}{2} \beta \sum_{n=-\infty}^{\infty} n \tilde{P}_{(j-n)} \tilde{P}_n \qquad (13.3.34)$$

These differential equations apply for all j, but it would be repetitive to evaluate the coefficients for $j < 0$ because $\tilde{P}_{(-j)}$ is the complex conjugate of $\tilde{P}_j$. If we are to make use of this property to reduce the number of variables, we must explicitly account for the complex conjugate property within the summation. The first step is to break up the summation into two parts, $n > 0$ and $n < 0$. (The contribution from $n = 0$ is zero.) For the second part, n is replaced by $-n$, which gives

$$\frac{d\tilde{P}_j}{d\tilde{x}} = \frac{i}{2} \beta \left[\sum_{n=1}^{\infty} n \tilde{P}_{(j-n)} \tilde{P}_n - \sum_{n=1}^{\infty} n \tilde{P}_{(j+n)} \tilde{P}_{(-n)} \right] \qquad (13.3.35)$$

In the equation for $j = 0$, the sums cancel, so that

$$\frac{d\tilde{P}_0}{d\tilde{x}} = 0 \qquad (13.3.36)$$

In other words, the mean pressure $\tilde{P}_0$ is independent of the position.

For $j > 0$, the second summation in Eq. (13.3.35) features a coefficient with a negative subscript. The complex conjugate property gives $\tilde{P}_{(-n)} = \tilde{P}_n^*$. The first summation also has a coefficient with a negative subscript when $n > j$. This is handled by splitting the summation into two ranges: $n \le j$ and $n > j$, with $\tilde{P}_{(j-n)} = \tilde{P}_{(n-j)}^*$ for the latter range. These operations lead to

$$\frac{d\tilde{P}_j}{d\tilde{x}} = \frac{i}{2}\beta \left[\sum_{n=1}^{j} n\tilde{P}_{(j-n)}\tilde{P}_n + \sum_{n=j+1}^{\infty} n\tilde{P}_{(n-j)}^*\tilde{P}_n - \sum_{n=1}^{\infty} n\tilde{P}_{(j+n)}\tilde{P}_n^* \right], \quad j > 0 \quad (13.3.37)$$

Replacing n with $j + n$ in the middle summation reduces this expression to

$$\frac{d\tilde{P}_j}{d\tilde{x}} = \frac{i}{2}\beta \left[\sum_{n=1}^{j} n\tilde{P}_{(j-n)}\tilde{P}_n + \sum_{n=1}^{\infty} j\tilde{P}_n^*\tilde{P}_{(j+n)} \right], \quad j > 0 \quad (13.3.38)$$

This is the form we seek because it only features P_n values for $n > 0$.

The first summation in this equation describes the growth or decay of harmonic j due to the interaction of harmonics below j. The second summation captures the influence on harmonic j of a higher harmonic interacting with all harmonics. For example, the first summation indicates that second harmonics ($j = 2$) are generated by the interaction of the first harmonic with itself, while the second summation indicates that second harmonics are also generated by the interaction of many harmonic pairs consisting of first and third, second and fourth, and so on.

The differential equations must be solved numerically, so it is necessary to truncate the pressure series. Let N be the highest harmonic that is retained. Thus, we set $\tilde{P}_n = 0$ for $n > N$. Neither subscript in the first summation will exceed N, so that summation requires no adjustment. The lowest value of n in the second summation that leads to a factor being zero is that for which $j + n = N + 1$. Thus, that summation is truncated at $n = N - j$, which means that the equations to be solved numerically are

$$\frac{d\tilde{P}_j}{d\tilde{x}} = \frac{i}{2}\beta \left[\sum_{n=1}^{j} n\tilde{P}_{j-n}\tilde{P}_n + \sum_{n=1}^{N-j} j\tilde{P}_n^*\tilde{P}_{(j+n)} \right], \quad j = 1, 2, ..., N \quad (13.3.39)$$

The second summation is not executed if $j = N$. The value of $\tilde{P}_0$ is set at whatever constant value it has at $x = 0$, so there are N pressure coefficients to determine and N differential equations to solve.

An attractive feature of the frequency-domain analysis is its generality. Monitoring the term on the right side of Eq. (13.3.39) enables one to delve into the interaction of harmonics at a fundamental level, rather than merely as a computed result. Another beneficial aspect of the formulation is that it is readily modified to incorporate the effects of dissipation. Section 3.3 discusses how dissipation is manifested in the frequency domain. It was shown that weak dissipation attenuates the complex amplitude of a plane wave. Dissipation is important for nonlinear propagation because the decay constant typically increases with increasing frequency. Thus, attenuation due to dissipation counters the nonlinear tendency for harmonics to grow as the wave propagates.

For propagation in the direction of increasing x, the attenuation factor is $e^{-\alpha x}$. The absorption coefficient α is a real quantity that depends on frequency. Different dissipation effects, such as bulk viscosity or wall friction, lead to different dependencies, so we will write it as $\alpha(\omega)$. If nonlinear effects were not present, this exponential decay would be obtained if the coefficients were governed by

$$\frac{d\tilde{P}_j}{d(\tilde{x}/k)} + \alpha(j\omega_1)\tilde{P}_j = 0 \qquad (13.3.40)$$

The assumption that dissipation is weak implies that it is not altered by nonlinear effects, and *vice versa*. Consequently, we may insert the dissipation term into the nonlinear amplitude equations, with the result that Eq. (13.3.39) becomes

$$\boxed{\frac{d\tilde{P}_j}{d\tilde{x}} = -\frac{\alpha(j\omega_1)}{k}\tilde{P}_j + \frac{i}{2}\beta\left[\sum_{n=1}^{j} n\tilde{P}_{j-n}\tilde{P}_n + \sum_{n=1}^{N-j} j\tilde{P}_n^*\tilde{P}_{(j+n)}\right], \quad j = 1, 2, ..., N}$$

$$(13.3.41)$$

Note that the presence of dissipation does not alter the fact that $\tilde{P}_0$ is independent of x.

Starting values for these differential equations are obtained from the boundary condition. In the weakly nonlinear regime, $\varepsilon \ll 1$, the particle velocity at $x = 0$ equals the motion of the boundary. Let $\varepsilon cV(t)$ denote the velocity there. The small-signal approximation gives the corresponding pressure $\rho_0 c(\varepsilon cV)$. The phase variable at $x = 0$ is $\theta = \omega_1 t$, so matching the Fourier series to the motion that is imposed gives

$$\frac{1}{2}\rho_0 c^2 \sum_{n=-\infty}^{\infty} \tilde{P}_n(\tilde{x} = 0)e^{in\omega_1 t} = \rho_0 c[\varepsilon cV(t)] \qquad (13.3.42)$$

Orthogonality of the harmonic functions in a Fourier series then leads to

$$\boxed{\tilde{P}_j(x = 0) = \frac{\varepsilon}{2\pi}\int_{-\pi/\omega_1}^{\pi/\omega_1} V(t)e^{-ij\omega_1 t}d(\omega_1 t)} \qquad (13.3.43)$$

EXAMPLE 13.5 Determine the Fourier series pressure coefficients for the steady-state plane wave generated by a harmonic source at $x = 0$. The particle velocity at that location is $v = \varepsilon c \sin(\omega t)$ with $\omega = 2\pi(5000)$ rad/s and

$\varepsilon = 2\left(10^{-5}\right)$. The fluid is air at standard conditions. Let the decay constant be $\alpha = 2.54\left(10^{-13}\right)\omega^2$ neper/m. Carry out the evaluation to the distance where a shock would form if there were no dissipations, which is $(x_{shock})_{sine} = 1/(\beta\varepsilon k)$. Compare the results to those obtained from the Fubini-Ghiron solution for the dissipationless case. Also evaluate the waveform at $(x_{shock})_{sine}$.

Significance

Issues such as implementation of a numerical procedure for solving the coupled differential equations and the number of harmonics that are required shall be covered. Equally important is the demonstration of the interplay of dissipation and nonlinearity effects.

Solution

The analysis entails solving Eq. (13.3.41) subject to the initial conditions described by Eq. (13.3.43). The velocity at $x = 0$ is specified to be $v = \varepsilon c\left(e^{i\omega t} - e^{-i\omega t}\right)/(2i)$. Matching $\rho_0 cv$ to the Fourier series for p at $x = 0$ leads to

$$\tilde{P}_1\left(x = 0\right) = -i\varepsilon, \text{ otherwise } P_n\left(x = 0\right) = 0 \tag{1}$$

The mean value, $\tilde{P}_0$, is zero. Nevertheless, we shall treat it as a variable because doing so leads to the implementation of a numerical procedure that is valid for any (periodic) boundary velocity.

The coupled differential equations may be solved with a Runge–Kutta numerical integrator. Several versions are available; we shall use a forth-fifth order scheme. The form that is required is

$$\frac{d\tilde{P}_j}{d\tilde{x}} = F_j\left(\tilde{P}_n, \tilde{x}\right), \quad j = 0, ..., N \tag{2}$$

Equation (13.3.41) matches this general form, with

$$F_0\left(\tilde{P}_n, x\right) = 0$$

$$F_j\left(\tilde{P}_n, x\right) = -\frac{\alpha(j\omega_1)}{k}\tilde{P}_j + \frac{i}{2}\beta\left[\sum_{n=1}^{j}n\tilde{P}_{j-n}\tilde{P}_n + \sum_{n=1}^{N-j}j\tilde{P}_n^*\tilde{P}_{(j+n)}\right], \quad j = 1, ..., N-1$$

$$F_j\left(\tilde{P}_n, x\right) = -\frac{\alpha(N\omega_1)}{k}\tilde{P}_j + \frac{i}{2}\beta\left[\sum_{n=1}^{N}n\tilde{P}_{j-n}\tilde{P}_n\right], \quad j = N \tag{3}$$

There is no simple way to vectorize computation of the F_j terms, so the operations are implemented term by term. Within a loop that increments j from 1 to N, the computation begins by equating F_j to the linear term. A loop over n from 1 to j adds the terms in the first summation. Then, if j is less than N, a loop over n from one to $N - j$ adds the terms in the summations. The only caution is that if one's software requires that subscripts begin at one, as is the case with MATLAB, then the subscript number must be one greater than the associated harmonic number.

The dissipation factor that appears in the differential equations is $\alpha (j\omega_1) k$. For the given α dependence, this is

$$\frac{\alpha (j\omega)}{k} = 2.54 \left(10^{-13}\right) \frac{j^2 \omega_1^2}{k} = 2.54 \left(10^{-13}\right) (\omega_1 c) j^2 \tag{4}$$

The units of α are nepers/m, so α/k is a dimensionless quantity. This means that the numerical coefficient has units of s^2/m.

The computation requires selection of N. We know from Fig. 13.8 that $\left|\tilde{P}_n\right| \ll \left|\tilde{P}_1\right|$ if $n > 5$, and dissipation will affect the higher harmonics more than $\tilde{P}_1$ because the dissipation factor increases monotonically with increasing harmonic number. We shall begin with $N = 10$, which would seem to be adequate to assure that all significant nonlinear interactions are properly described. As one should always do when making an a priori decision like this, we will verify the adequacy of this truncation by comparing the results to those obtained from a larger N.

It is requested that the series coefficients be compared to those of the Fubini-Ghiron solution of the dissipationless case. That solution is described in Eq. (13.2.13) in terms of the retarded time $t' = t - x/c$ and x nondimensionalized relative to the shock formation distance for harmonic excitation, $(x_{\mathrm{shock}})_{\mathrm{sine}} = 1/(\beta\varepsilon k)$. In terms of the present variables, the complex Fourier series for the signal in the case of an ideal fluid is

$$p = \frac{1}{2}\rho_0 c^2 \sum_{n=1}^{\infty} \tilde{P}_n e^{in\theta_1}, \quad \tilde{P}_n = \frac{2 J_n (n\beta\varepsilon\tilde{x})}{in\beta\tilde{x}} \frac{P_n}{\rho_0 c^2} = \frac{2\varepsilon}{n\upsilon} J_n (n\sigma) \tag{5}$$

One computational check is to compare the solution of the differential equations for $\alpha = 0$ to this expression. Evaluation of $\left|\left(\tilde{P}_n\right)_{\mathrm{comp}} - \left(\tilde{P}_n\right)_{\mathrm{F\text{-}G}}\right| / \varepsilon$ showed an error that rose from 0.02% for $n = 1$ to 0.14% for $n = 9$. The error for $n = N = 10$ was found to be 4.8%, which is typical behavior when a Fourier series is truncated. For reasons that will become apparent when we examine waveforms, the computation was repeated with $N = 50$. The error for the first ten harmonics in that case was found to be less than 0.005%.

The pressure coefficients when $\alpha(j\omega)$ is given by Eq. (4) are the solid curves in Fig. 1. The dashed curves are the values for the dissipationless case. Each amplitude is attenuated. For clarity, both linear and logarithmic scales are used. Viewed as a ratio, the loss is greater as the harmonic number increases. Whether this is a drastic effect depends on the appearance of the waveform at the shock formation distance.

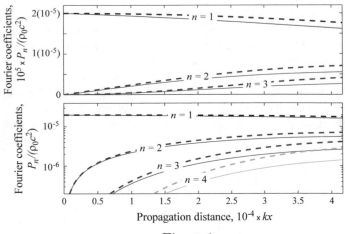

Figure 1.

A waveform is obtained by evaluating Eq. (13.3.30). The result at the shock formation distance, which is $\tilde{x} = k(x_{\text{shock}})_{\text{sine}} = 1/(\beta\varepsilon)$, appears in Fig. 2. The result for the ideal case could be obtained by evaluating the Riemann solution; instead, it also was obtained by computing the series. The cutoff number N was raised to obtain these waveforms because a much higher resolution is required to capture the steep slope of the $\alpha = 0$ curve. This is unlike the behavior of the curve for $\alpha \neq 0$, which was found to differ negligibly from the one obtained from $N = 10$.

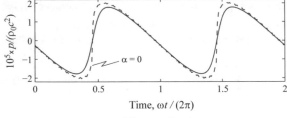

Figure 2.

The waveform in Fig. 2 shows that the presence of dissipation has two effects. The first is to attenuate the overall amplitude, which would be $|p/\rho_0 c^2| = \varepsilon$ if α were zero. The second effect is to reduce the slope in the phase where $\partial p/\partial t > 0$. Thus, it forestalls the formation of a shock. The differential equations lose validity when a shock forms because they do not account for the Rankine–Hugoniot conditions. Conversely, they can be used for any $\tilde{x}$ in the $\alpha \neq 0$ case, provided $\partial p/\partial t$ remains finite at lesser $\tilde{x}$. Figure 3 is the waveform at $\tilde{x} = 4/(\beta\varepsilon)$. It appears that the attenuation is sufficiently great that a shock will not form at any location. Because the attenuation increases with increasing harmonic number, at a sufficiently large distance the signal will be reduced to a greatly attenuated fundamental. This is one of the reasons that sounds that are generated in everyday life do not form shocks.

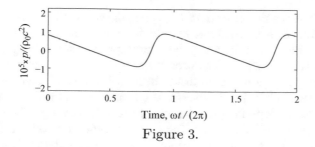

Figure 3.

13.3.3 Regular Perturbation Series Expansion

There are a number of fundamental systems and phenomena that can be solved analytical, rather than by recourse to numerical methods. One approach that has had success is a *perturbation series* representation of the pressure or velocity potential. The concept is limited by the requirement that the linearized version of the problem be solvable. In actuality, this is not much of a limitation, because it is unreasonable to expect to solve the nonlinear equations if the linearized version of those equations cannot be solved. The starting point is identification of a basic parameter that can be used as a metric of the relative importance of linear and nonlinear terms in the field equation. We have already encountered such a parameter, the acoustic Mach number ε.

In a linear problem, any constant that multiplies the excitation scales the response in the same way. This is not true for a nonlinear signal, so it follows that the potential function and therefore, the acoustic pressure and particle velocity are unknown functions of the Mach number ε, as well as position and time. This dependence on ε can be expected to be analytical, at least in regions where shocks do no occur. An analytical dependence on ε implies that the variables can be represented as a Taylor series relative to $\varepsilon - 0$. The response variable for the nonlinear wave equation is ϕ, so its Taylor series is

$$\phi(\bar{x}, t, \varepsilon) = \phi(\bar{x}, t, \varepsilon)|_{\varepsilon=0} + \varepsilon \left(\frac{\partial}{\partial \varepsilon} \phi(\bar{x}, t, \varepsilon) \right)\bigg|_{\varepsilon=0} + \frac{1}{2} \varepsilon^2 \left(\frac{\partial^2}{\partial \varepsilon^2} \phi(\bar{x}, t, \varepsilon) \right)\bigg|_{\varepsilon=0} + \cdots$$

$$(13.3.44)$$

The functional dependence of ϕ is not known when the analysis is initiated, so the derivatives may be replaced by functions to be determined. Furthermore, because ε is the scaling factor of the excitation, setting $\varepsilon = 0$ means there is no excitation, in which case ϕ will be zero. This means that the series should not contain a term that is independent of ε. Thus, the series reduces to

$$\phi(\bar{x}, t, \varepsilon) = \varepsilon \phi_1(\bar{x}, t) + \varepsilon^2 \phi_2(\bar{x}, t) + \varepsilon^3 \phi_3(\bar{x}, t) + \cdots \qquad (13.3.45)$$

When this series is used as a Taylor series, the concept is to add terms until it converges. Adding more terms will decrease the error of the truncated series relative to the true ϕ. This is not the viewpoint if the representation is considered to be a perturbation series. Rather, the number of terms is fixed. If $|\varepsilon| \ll 1$, we know that the difference between the series that is truncated beyond ε^N and the true ϕ will be proportional to ε^{N+1}. In other words, lengthening a perturbation solution would reduce the error at a specified ε, rather than increasing the range of ε for which the error is less than a specified amount. The change in viewpoint from a Taylor series to a perturbation series is fortunate, because the analysis of each function ϕ_n becomes progressively more difficult as n increases. It follows that there is no intent to increase the length of the series, even though doing so might lead to a convergent answer.

Frequent reference will be made to the order of magnitude of a term, so it is important to have a clear definition of that metric. A function $f(\varepsilon)$ is said to be $O(\varepsilon^n)$ if, for two very small ε values, it is found that $f(\varepsilon_1)/f(\varepsilon_2)$ is essentially $(\varepsilon_1/\varepsilon_2)^n$. We will use the weakly nonlinear version of the wave equation, Eq. (13.3.18). Cubic terms have been omitted from this equation. The velocity potential is $O(\varepsilon)$, so the omitted terms are $O(\varepsilon^3)$. It would be inconsistent to seek terms at that order in the response, so the perturbation series for the velocity potential pressure and particle velocity are truncated at $O(\varepsilon^2)$. The forms are

$$
\boxed{
\begin{aligned}
\phi &= \varepsilon \phi_1(x, t) + \varepsilon^2 \phi_2(x, t) + \cdots \\
p &= \varepsilon p_1(x, t) + \varepsilon^2 p_2(x, t) + \cdots \\
v &= \varepsilon v_1(x, t) + \varepsilon^2 v_2(x, t) + \cdots
\end{aligned}
}
\tag{13.3.46}
$$

We have no intrinsic interest in the density as a response variable, but if we did, its perturbation series would begin with ρ_0, which is the $O(1)$ term.

The next step is to assemble the equations to be satisfied. For a planar wave, they are

$$
v = \frac{\partial \phi}{\partial x}
$$

$$
p = -\rho_0 \left[\frac{\partial \phi}{\partial t} + \frac{1}{2} \left(\frac{\partial \phi}{\partial x} \right)^2 - \frac{1}{2c^2} \left(\frac{\partial \phi}{\partial t} \right)^2 \right]
\tag{13.3.47}
$$

$$
c^2 \frac{\partial^2 \phi}{\partial x^2} - \frac{\partial^2 \phi}{\partial t^2} - \left[2(\beta - 1) \frac{\partial^2 \phi}{\partial x^2} \frac{\partial \phi}{\partial t} + 2 \frac{\partial \phi}{\partial x} \frac{\partial^2 \phi}{\partial x \partial t} \right] = 0
$$

The early development of the method of characteristics showed that the effect of a moving boundary condition is always $O(\varepsilon^2)$, see Eq. (13.1.17). Nevertheless, it is useful to see how such effects may be incorporated in a perturbation analysis. We have defined ε to measure the magnitude of the boundary motion, so we take the displacement of the boundary to be $\varepsilon X(t)$. Thus, the boundary condition is

$$
\left. \frac{\partial \phi}{\partial x} \right|_{x = \varepsilon X(t)} = \varepsilon \dot{X}(t)
\tag{13.3.48}
$$

Statement of the problem is completed by imposing the radiation condition, which requires that p be an outgoing wave as $x \to \infty$, and that the fluid was at the ambient state for $t < 0$, so that $p = v = 0$ at $t = 0$.

The fundamental step in the analysis stems from the requirement that the perturbation series constitute a solution for any ε. When any such series is substituted into an equation, terms like order of ε may be collected. The result will have the form $\varepsilon \times$ (term 1) $+ \varepsilon^2 \times$ (term 2) $= 0$. It is necessary that this equation be satisfied for many different values of ε. This requirement can be met only if the coefficient of each power of ε is zero. Thus, a single equation becomes an equation for each power of ε: term 1 $= 0$, term 2 $= 0$, ... An equivalent statement is that the coefficients of like powers of ε in each equation must match.

In this way, we obtain a sequence of problems in which ε does not appear. Each set of equations is described as being the equations at the order of the associated power of ε. The first-order set consists of term 1 from each governing equation, the second-order set consists of term 2 for each equation, and so on. Thus, the first-order equations derived from Eq. (13.3.47) are

$$
\boxed{
\begin{aligned}
v_1 &= \frac{\partial \phi_1}{\partial x} \\[2mm]
p_1 &= -\rho_0 \frac{\partial \phi_1}{\partial t} \\[2mm]
c^2 \frac{\partial^2 \phi_1}{\partial x^2} &- \frac{\partial^2 \phi_1}{\partial t^2} = 0
\end{aligned}
}
\tag{13.3.49}
$$

The second-order equations are

$$
\boxed{
\begin{aligned}
v_2 &= \frac{\partial \phi_2}{\partial x} \\[2mm]
p_2 &= -\rho_0 \frac{\partial \psi_2}{\partial t} - \left[\frac{1}{2} \left(\frac{\partial \phi_1}{\partial x} \right)^2 - \frac{1}{2c^2} \left(\frac{\partial \phi_1}{\partial t} \right)^2 \right] \\[2mm]
c^2 \frac{\partial^2 \phi_2}{\partial x^2} - \frac{\partial^2 \phi_2}{\partial t^2} &= 2 \left(\beta - 1 \right) \frac{\partial^2 \phi_1}{\partial x^2} \frac{\partial \phi_1}{\partial t} + 2 \frac{\partial \phi_1}{\partial x} \frac{\partial^2 \phi_1}{\partial x \partial t}
\end{aligned}
}
\tag{13.3.50}
$$

A slightly simpler differential equation for ϕ_2 is obtained if the first-order equation is used to replace $\partial^2 \phi_1 / \partial x^2$ with $\left(1/c^2 \right) \partial^2 \phi_1 / \partial t^2$. Doing so leads to

$$
c^2 \frac{\partial^2 \phi_2}{\partial x^2} - \frac{\partial^2 \phi_2}{\partial t^2} = \frac{\partial}{\partial t} \left[\frac{(\beta - 1)}{c^2} \left(\frac{\partial \phi_1}{\partial t} \right)^2 + \left(\frac{\partial \phi_1}{\partial x} \right)^2 \right]
\tag{13.3.51}
$$

It will be noted that at each order the terms that are lower order have been placed to the right of the equality sign. Doing so anticipates the way in which the equations are solved. The first-order equations are linear in the dependent variables, which means we should be able to solve them. The left side of the first- and second-order equations is the same, but each second-order equation is inhomogeneous. Substitution of the first-order solutions will convert them into terms that represent a spatial distribution of sources.

The moving boundary condition also is transformed to a set of perturbation equations by means of a Taylor series relative to $x = 0$, specifically,

$$\frac{\partial \phi}{\partial x}\bigg|_{x=\varepsilon X(t)} = \frac{\partial \phi}{\partial x}\bigg|_{x=0} + \varepsilon X(t)\, \frac{\partial^2 \phi}{\partial x^2}\bigg|_{x=0} + \frac{1}{2}\, (\varepsilon X(t))^2\, \frac{\partial^2 \phi}{\partial x^2}\bigg|_{x=0} + \cdots = \varepsilon \dot{X}(t)$$

$$(13.3.52)$$

Only effects up to $O\left(\varepsilon^2\right)$ are of interest. Thus, we substitute the perturbation series for ϕ and match the $O\left(\varepsilon\right)$ and $O\left(\varepsilon^2\right)$ terms on either side of the equality. The result is a sequence of boundary conditions,

$$\frac{\partial \phi_1}{\partial x}\bigg|_{x=0} = \dot{X}(t)$$

$$\frac{\partial \phi_2}{\partial x}\bigg|_{x=0} = -X(t)\, \frac{\partial^2 \phi_1}{\partial x^2}\bigg|_{x=0}$$

$$(13.3.53)$$

In addition, the radiation and initial conditions must be satisfied for any ε, which leads to the requirements that each order be an outgoing wave, and that each order of p and v be zero for $t < 0$.

In most systems, solution of the linear problem requires specification of the excitation. However, the d'Alembert solution, which is general, is available for plane waves. An outgoing wave that satisfies the linear wave equation in Eq. (13.3.49) is $\phi_1 = f(t - x/c)$. The function f is determined by satisfying the first-order boundary condition, Eq. (13.3.53),

$$-\frac{1}{c} f'(t) = \dot{X} \qquad (13.3.54)$$

The notation f' prime denotes differentiation with respect to the argument of f, which is $t - x/c$. The chain rule for differentiation leads to $f'(t - x/c) \equiv (\partial/\partial t) f(t - x/c)$, so the boundary condition may be integrated with respect to time to find that

$$\phi_1 = -cX\left(t - \frac{x}{c}\right) \qquad (13.3.55)$$

The corresponding state variables are

$$v_1 = \dot{X}\left(t - \frac{x}{c}\right), \quad p_1 = \rho_0 c \dot{X}\left(t - \frac{x}{c}\right) \qquad (13.3.56)$$

It is implicit to this solution that $X(t) = 0$ for $t < 0$, so these terms are consistent with the initial condition.

Now that we have determined the first-order terms, we proceed to the next order. Substitution of ϕ_1 into the second-order wave equation, Eq. (13.3.51), gives

$$c^2 \frac{\partial^2 \phi_2}{\partial x^2} - \frac{\partial^2 \phi_2}{\partial t^2} = \beta \frac{\partial}{\partial t} \left[\dot{X} \left(t - \frac{x}{c} \right)^2 \right] \tag{13.3.57}$$

The corresponding second-order boundary condition is

$$\left. \frac{\partial \phi_2}{\partial x} \right|_{x=0} = \frac{1}{c} X(t) \ddot{X}(t) \tag{13.3.58}$$

The solution of the second-order wave equation consists of a complementary equation solution $(\phi_2)_c$ and a particular solution $(\phi_2)_p$. The definition of $(\phi_2)_p$ is that its substitution into the left side of the equation exactly matches the right side, whereas substitution of $(\phi_2)_c$ gives zero. The left side of the equation is the linear wave equation, so any term that gives zero on the right side must be a function of either characteristic variable. The requirement that such a term be a forward propagating wave removes dependence on the backward characteristic. Therefore, we know that

$$(\phi_2)_c = F_c \left(t - \frac{x}{c} \right) \tag{13.3.59}$$

We now are presented with a small dilemma, because the source term on the right side of Eq. (13.3.57) is a function of $t - x/c$. Any function with such dependence is a homogeneous solution, so we cannot use that combination of variables to form a particular solution. Let us try a particular solution that is modified by introducing a factor x, specifically,

$$(\phi_2)_p = x F_p \left(t - \frac{x}{c} \right) \tag{13.3.60}$$

Substitution of this ansatz into the second-order wave equation yields

$$- 2c F_p' \left(t - \frac{x}{c} \right) = \beta \frac{\partial}{\partial t} \left[\dot{X} \left(t - \frac{x}{c} \right)^2 \right] \tag{13.3.61}$$

According to the chain rule for differentiation, $(\partial/\partial t) F_p (t - x/c) = F_p' (t - x/c)$. Thus, the preceding may be integrated directly. The resulting particular solution is

$$(\phi_2)_p = - \frac{\beta}{2c} x \dot{X} \left(t - \frac{x}{c} \right)^2 \tag{13.3.62}$$

One might wonder at this juncture why the trial form of the particular solution was not taken to be $t F_p (t - x/c)$, which could also be made to match the source term. The term that was selected could be justified by noting that it matches the behavior observed in the Riemann solution, specifically that distortion of the pressure profile increases with increasing x. However, invocation of this argument would violate the intention of solving the problem without having prior knowledge of the Riemann solution. A justification fitting this criterion is that the form selected for the particular solution must be valid for any $X(t)$. If $X(t)$ is periodic for $t > 0$, then regardless of the fact that the system is nonlinear, the response must evolve to have that same temporal periodicity. Such a property cannot be obtained if t is a factor in the trial solution.

The function F_c for the complementary solution in Eq. (13.3.59) is arbitrary in regard to satisfying the second-order wave equation. It is set by satisfying the second-order boundary condition, Eq. (13.3.58). This condition must be satisfied by the sum of the complementary and particular solutions, which is why the particular solution for ϕ_2 was found first. Thus, we seek a function $F_c(t - x/c)$ for which

$$\frac{\partial}{\partial x}\left[(\phi_2)_c + (\phi_2)_p\right]\bigg|_{x=0} \equiv -\frac{1}{c}F_c'(t) - \frac{\beta}{2c}\dot{X}(t)^2 = \frac{1}{c}X(t)\ddot{X}(t) \qquad (13.3.63)$$

Because $F_c'(t) \equiv \dot{F}_c(t)$ at $x = 0$, this equation may be integrated, although the integral can be evaluated only upon specification of $X(t)$. Thus, we find that

$$F_c(t) = -\int_0^t \left[\frac{\beta}{2}\dot{X}(t')^2 dt' + X(t')\ddot{X}(t')\right]dt' \qquad (13.3.64)$$

The second-order potential at any x is obtained by replacing t with $t - x/c$ in this expression and then combining it with the particular solution. Doing so yields

$$\phi_2 = -\frac{\beta}{2c}x\dot{X}\left(t - \frac{x}{c}\right)^2 - \int_0^{t-x/c}\left[\frac{\beta}{2}\dot{X}(t')^2 dt' + X(t')\ddot{X}(t')\right]dt' \qquad (13.3.65)$$

The second-order pressure and particle velocity may be obtained from Eq. (13.3.50). The result shall not be listed here because doing so would not advance the dual objective of demonstrating how to implement a perturbation analysis, and how to handle a moving boundary condition. However, the expression for ϕ_2 that we have derived serves several purposes.

It is evident that the full solution, which is obtained by combining first- and second-order effects, is not at all like the Riemann solution. The combination is not consistent with the small-signal approximation that $p = \rho_0 cv$. However, the discrepancy is $O(\varepsilon^2)$ everywhere, so the difference is not significant. It is far more important that the perturbation solution has a limited range of validity. This is so because $(\phi_2)_p$ is proportional to x, so it grows in magnitude. In the vicinity of the shock formation distance, $x = O(1/\varepsilon)$. This makes $\varepsilon^2\phi_2$ have the same magnitude as $\varepsilon\phi_1$, but the basic concept of a perturbation series is that each succeeding term is small compared to the preceding terms. Therefore, the velocity potential we have derived is only valid in regions where $x = O(1)$. A perturbation series that is not equally accurate at all locations is said to be *nonuniformly valid*. In contrast, the Riemann solution is valid everywhere up to the shock formation distance. Some individuals say that ϕ_2 has a singularity at $x = \infty$ because it grows without bound as x increases.

Despite these faults, a nonuniformly valid series might be quite useful. Suppose we were interested in mapping the field around the boundary and we had no knowledge of the Riemann solution. The perturbation solution would explain certain nonlinear effects, such as the occurrence of second harmonics if $X(t)$ is sinusoidal, and the fact that the difference between the linear and nonlinear solutions grows with increasing propagation distance. The nonuniformly valid solution also illustrates a general fact that nonlinear effects arising from moving boundary conditions do not grow with increasing distance from the boundary. Ultimately, the most important justification for the analysis is that its faults can be corrected. Methods for doing so are the subject of the next section.

EXAMPLE 13.6 A plane wave is induced by a wall vibration that constitutes a beat. The steady state motion is $v = \varepsilon c \sin(\sigma t) \sin(\omega t)$, where $\sigma \ll \omega$. Identify the terms in the second-order velocity potential that are not uniformly valid.

Solution

An analysis based on the d'Alembert solution, like the one for the preceding development, will not be suitable for more general systems. Frequency-domain tools often are likely to be employed in such circumstances. We will see how to do so here.

Solution

Although the basic perturbation equations have already been stated, it is good practice to assemble them anew. We seek a solution for the velocity potential whose form is

$$\phi = \varepsilon \phi_1(x, t) + \varepsilon^2 \phi_2(x, t) + \cdots \tag{1}$$

The first-order term must satisfy the linear wave equation in Eq. (13.3.49), and the gradient of this term at $x = 0$ must match the wall vibration. Thus, the first-order problem is

$$c^2 \frac{\partial^2 \phi_1}{\partial x^2} - \frac{\partial^2 \phi_1}{\partial t^2} = 0$$

$$\left. \frac{\partial \phi_1}{\partial x} \right|_{x=0} = c \sin(\sigma t) \sin(\omega t)$$

The velocity of the wall is a beat. We describe it in terms of complex exponentials. Doing so expedites solving the first-order equations, and it will also allow us to avoid recalling trigonometric identities for products when we go on to the second order. Thus, the boundary condition is converted to

$$\left. \frac{\partial \phi_1}{\partial x} \right|_{x=0} = -\frac{1}{4} \varepsilon c \left(e^{i\sigma t} - e^{-i\sigma t} \right) \left(e^{i\omega t} - e^{-i\omega t} \right)$$

$$= \frac{1}{4} \varepsilon c \left(e^{i(\omega-\sigma)t} - e^{i(\omega+\sigma)t} \right) + \text{c.c.} \tag{2}$$

This is a familiar problem. Each harmonic in the excitation generates a forward propagating wave at that frequency. Because there are two uncorrelated frequencies, it is preferable to use the retarded time to represent each term,

$$\tau = t - \frac{x}{c}$$

Thus, a suitable trial solution is

$$\phi_1 = a_1 e^{i\Omega_1 \tau} + a_2 e^{i\Omega_2 \tau} + \text{c.c.} \tag{3}$$

where

$$\Omega_1 = \omega - \sigma, \quad \Omega_2 = \omega + \sigma$$

Because $\partial \tau / \partial x = -1/c$, substitution of this form into Eq. (2) yields

$$a_1 = \frac{ic^2}{4\Omega_1}, \quad a_2 = -\frac{ic^2}{4\Omega_2} \tag{4}$$

We have seen that the moving boundary condition leads to second-order terms in ϕ and p that are uniformly valid. Therefore, we may ignore that effect. Because the gradient of the first-order term matches the specified wall velocity, it must be that $\partial \phi_2 / \partial x = 0$ at $x = 0$. The resulting second-order problem is

$$c^2 \frac{\partial^2 \phi_2}{\partial x^2} - \frac{\partial^2 \phi_2}{\partial t^2} = 2(\beta - 1) \frac{\partial^2 \phi_1}{\partial x^2} \frac{\partial \phi_1}{\partial t} + 2 \frac{\partial \phi_1}{\partial x} \frac{\partial^2 \phi_1}{\partial x \partial t}$$
$$\left. \frac{\partial \phi_2}{\partial x} \right|_{x=0} = 0 \tag{5}$$

The differential equation is simplified by the properties that

$$\frac{\partial \phi_1}{\partial t} = \frac{d\phi_1}{d\tau}, \quad \frac{\partial \phi_1}{\partial x} = -\frac{1}{c} \frac{d\phi_1}{d\tau}$$

Consequently, the differential equation in Eq. (5) may be written as

$$c^2 \frac{\partial^2 \phi_2}{\partial x^2} - \frac{\partial^2 \phi_2}{\partial t^2} = 2 \frac{\beta}{c^2} \frac{d\phi_1}{d\tau} \frac{d^2 \phi_1}{d\tau^2} \equiv \frac{\beta}{c^2} \frac{d}{d\tau} \left[\left(\frac{d\phi_1}{d\tau} \right)^2 \right]$$

The term on the right side is formed by substituting for ϕ_1 from Eq. (3). The use of c.c. in that expression is a convenience; care must be taken to account for the unlisted terms.

$$c^2 \frac{\partial^2 \phi_2}{\partial x^2} - \frac{\partial^2 \phi_2}{\partial t^2} = \frac{\beta}{c^2} \frac{d}{d\tau} \left\{ \left[i\Omega_1 a_1 e^{i\Omega_1 \tau} + i\Omega_2 a_2 e^{i\Omega_2 \tau} + \text{c.c.} \right]^2 \right\}$$

$$= \frac{\beta}{c^2} \frac{d}{d\tau} \left[-\Omega_1^2 a_1^2 e^{2i\Omega_1 \tau} - \Omega_2^2 a_2^2 e^{2i\Omega_2 \tau} - 2\Omega_1 \Omega_2 a_1 a_2 e^{2i\omega\tau} \right.$$

$$\left. +2\Omega_1 \Omega_2 a_1^* a_2 e^{2i\sigma\tau} + \Omega_1^2 a_1 a_1^* + \Omega_2^2 a_2 a_2^* + \text{c.c.} \right]$$

$$= -\frac{2i\beta}{c^2} \left[\Omega_1^3 a_1^2 e^{2i\Omega_1 \tau} + \Omega_2^3 a_2^2 e^{2i\Omega_2 \tau} \right.$$

$$\left. +2\omega\Omega_1 \Omega_2 a_1 a_2 e^{2i\omega\tau} - 2\sigma\Omega_1 \Omega_2 a_1^* a_2 e^{2i\sigma\tau} \right] + \text{c.c.}$$

$$(6)$$

The task now is to determine the portion of the particular solution for ϕ_2 that grows as x increases. Each term on the right side is solely a function of τ, so each identically satisfies the homogeneous terms to the left of the equality side. Accordingly, we form a trial solution by multiplying each term by x. (The alternative of multiplying each term by t must be rejected because the excitation is a periodic function of time, so the response must share that property.) Consequently, the trial form for $(\phi_2)_p$ is

$$(\phi_2)_p = x \left[b_1 e^{2i\Omega_1 \tau} + b_2 e^{2i\Omega_2 \tau} + b_3 e^{2i\omega\tau} + b_4 e^{2i\sigma\tau} \right] + \text{c.c.} \qquad (7)$$

Substitution of this ansatz into the left side of Eq. (6) generates terms that match the right side if

$$b_1 = \frac{\beta}{2c^3} \Omega_1^2 a_1^2, \quad b_2 = \frac{\beta}{2c^3} \Omega_2^2 a_2^2$$

$$b_3 = \frac{\beta}{c^3} \Omega_1 \Omega_2 a_1 a_2, \quad b_4 = -\frac{\beta}{c^3} \Omega_1 \Omega_2 a_1^* a_2$$

The complementary solution of Eq. (6) is the d'Alembert solution, which does not grow. Hence, we have completed the analysis. The velocity potential is constructed by substituting Eqs. (3) and (7) into Eq. (1), which yields

$$\phi = \varepsilon(a_1 e^{i\Omega_1 \tau} + a_2 e^{i\Omega_2 \tau}) + \varepsilon^2 \left[b_1 \left(x + \frac{c}{2i\Omega_1} \right) e^{2i\Omega_1 \tau} + b_2 \left(x + \frac{c}{2i\Omega_2} \right) e^{2i\Omega_2 \tau} \right.$$

$$\left. + b_3 \left(x + \frac{c}{2i\omega} \right) e^{2i\omega\tau} + b_4 \left(x + \frac{c}{2i\sigma} \right) e^{2i\sigma\tau} + \text{c.c} + O\left(\varepsilon^2\right) \right.$$

$$(8)$$

This expression shows that the beating excitation at the boundary, which consists of two harmonic signals at close frequencies $\Omega_1 = \omega - \sigma$ and $\Omega_2 = \omega + \sigma$, generates double-frequency harmonics at $2(\omega + \sigma)$ and $2(\omega - \sigma)$, as well as a sum frequency harmonic at 2ω, and a difference frequency harmonic at 2σ. All frequencies except 2σ are close, so it seems that the waveform should display a beating pattern. These frequencies appear in both the particular and complementary solutions for ϕ_2. The second-order terms grow with increasing x, so the representation is valid only for locations at which x is small relative to some reference length, like the shock formation distance.

The expression we have derived does not provide much insight regarding the appearance of a waveform at substantial distances from the source. Although it was not requested, Fig. 1 shows the waveform obtained from the Riemann solution at $x_{shock} = c/(\beta \varepsilon \omega)$. The frequencies are $\omega/(2\pi) = 1000\,\text{Hz}$ and $\sigma/(2\pi) = 40\,\text{Hz}$. The envelope function that has been plotted is $\pm \varepsilon c \sin(\sigma t)$. This is the true envelope for linear theory, but a very close examination would show that the maxima and minima of each oscillation of p sometimes do not reach the envelope, and sometimes it exceeds the envelope. These discrepancies are minor. On the other hand, the oscillation within the envelope shows the advancement and retardation that causes the waveform to lean over.

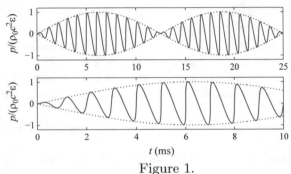

Figure 1.

Before we see how a nonuniformly valid perturbation series can be corrected, let us consider applying the Fourier series analysis in Sect. 13.3.2 to this problem. The condition satisfied by a period is that it contain an integer number of cycles at each frequency. In the case of the present excitation, it must be such that an interval T contains n oscillations at $\omega - \sigma$, and m oscillations at $\omega + \sigma$. In other words, $T = n2\pi/(\omega - \sigma) = m2\pi/(\omega + \sigma)$. The smallest integers meeting this condition for the frequencies in Fig. 1 are $m = 13$ and $n = 12$, which corresponds to a period of $T = 2\pi/80\,\text{s}$. In other words, in a Fourier series analysis the frequencies contained in the excitation are the 12th and 13th harmonics of the fundamental frequency. The highest harmonic for the frequency series must be many times the frequency of the excitation if the propagation distance is comparable to x_{shock}. Thus, a very long series would be required if that formulation were to be applied to analyze the pressure generated by a beating vibration. The formulation that follows provides an alternative.

13.3.4 Method of Strained Coordinates

The occurrence of a nonuniformly valid description of a variable is not unique to analyses of nonlinearity. Consider a harmonic waveform whose frequency is slightly different from a nominal value ω_0, for example, $y = \sin((\omega_0 + \varepsilon)t) \equiv$

$\sin(\omega_0 t)\cos(\varepsilon t) + \cos(\omega_0 t)\sin(\varepsilon t)$. This expression may be expanded in a Taylor series of increasing powers of ε. Truncation of that series at the first power gives $y \approx \sin(\omega_0 t) + \varepsilon t \cos(\omega_0 t)$. This is a nonuniformly valid perturbation series. Rather than considering the nonuniformity to stem from a small difference of the frequency from ω_0, we could consider the frequency to be ω_0 but the actual time variable to be slightly different from t. This is the essential idea underlying *singular perturbation methods*. Specifically, the nonuniformly valid perturbation series is regarded as the correct solution in the wrong place in space and/or time.

A widely employed method for correcting a nonuniformly valid perturbation series is Lighthill's method of strained coordinates.[16] It distorts the space-time grid in a manner that is gradual when viewed over a small distance, but accumulates to a significant level at long range. If the nonuniformly valid term grows with increasing x, then the coordinate straining transformation replaces x with a variable ξ that differs from x at a rate that matches the nonuniformly valid term(s), specifically,

$$x = \xi + \varepsilon x F(\xi, t) \tag{13.3.66}$$

The concept is to introduce this transformation into an expression that is not uniformly valid and then select the function F that regularizes the expression.

We will use the regular perturbation series for plane waves derived in the previous section as the framework for the development. The second-order velocity potential in Eq. (13.3.65) is not required to be uniformly valid because it is an abstract quantity that does not directly characterize the state of the fluid. Rather, the expressions for the pressure and particle velocity derived by differentiating ϕ must behave properly. An expression for ϕ is obtained by combining the first-order potential in Eq. (13.3.55) and the second-order potential in Eq. (13.3.65). It is permissible to ignore terms in ϕ that will lead to terms in p and v that are $O(\varepsilon^2)$ everywhere, because those terms will decrease in importance relative to the nonuniformly valid terms as the wave propagates. Therefore, we may begin with

$$\phi = -\varepsilon c X\left(t - \frac{x}{c}\right) - \varepsilon^2 \frac{\beta}{2c} x \dot{X}\left(t - \frac{x}{c}\right)^2 + O\left(\varepsilon^2\right) \tag{13.3.67}$$

A corollary of being allowed to drop terms that uniformly are $O(\varepsilon^2)$ or smaller is that we may use $p = -\rho_0 (\partial\phi/\partial t)$. No approximation is contained in $v = \partial\phi/\partial x$. The state variables that result are

$$p = \varepsilon \rho_0 c \dot{X}\left(t - \frac{x}{c}\right) + \varepsilon^2 \rho_0 \frac{\beta}{c} x \dot{X}\left(t - \frac{x}{c}\right)\ddot{X}\left(t - \frac{x}{c}\right) + O\left(\varepsilon^2\right)$$

$$v = \varepsilon \dot{X}\left(t - \frac{x}{c}\right) + \varepsilon^2 \frac{\beta}{2c}\left[\frac{2}{c} x \dot{X}\left(t - \frac{x}{c}\right)\ddot{X}\left(t - \frac{x}{c}\right) + \dot{X}\left(t - \frac{x}{c}\right)^2\right] + O\left(\varepsilon^2\right)$$

$$\tag{13.3.68}$$

[16]M.J. Lighthill, "A technique for rendering approximate solutions to physical problems uniformly valid," Philosophical Mag., Vol. 40 (1949) 1179–1201.

The last term within the bracket for v does not grow as x increases. It is uniformly valid, so it may be placed in the group of terms that are uniformly $O\left(\varepsilon^2\right)$. Doing so leads to $p = \rho_0 c v + O\left(\varepsilon^2\right)$, which is like the relation according to linear theory. This is not merely an interesting observation. Rather, it tells us that whatever we do to make p uniformly valid will do the same for v.

We are now ready to introduce the coordinate straining transformation. We substitute the expression for x in Eq. (13.3.66) into the preceding equation for p. The $O\left(\varepsilon\right)$ term is expanded in a Taylor series in powers of ε, according to

$$
\begin{aligned}
\dot{X}\left(t - \frac{x}{c}\right) &= \dot{X}\left(t - \frac{\xi}{c} - \varepsilon\frac{x}{c}F\left(\xi, t\right)\right) \\
&= \dot{X}\left(t - \frac{\xi}{c}\right) - \varepsilon\frac{x}{c}F\left(\xi, t\right)\ddot{X}\left(t - \frac{\xi}{c}\right) + \cdots
\end{aligned}
\tag{13.3.69}
$$

Because $O\left(\varepsilon^3\right)$ terms are not being considered, the Taylor series may be truncated at the listed terms. For the same reason, we can simply replace x with ξ in the phase of the $O(\varepsilon^2)$ part of ϕ. The result of these operations is

$$
\begin{aligned}
p = \varepsilon\rho_0 c&\left[\dot{X}\left(t - \frac{\xi}{c}\right) - \varepsilon\frac{x}{c}F\left(\xi, t\right)\ddot{X}\left(t - \frac{\xi}{c}\right)\right] \\
&+ \varepsilon^2\rho_0\frac{\beta}{c}x\dot{X}\left(t - \frac{\xi}{c}\right)\ddot{X}\left(t - \frac{\xi}{c}\right) + O\left(\varepsilon^2\right)
\end{aligned}
\tag{13.3.70}
$$

The role of the F function is to make the perturbation series regular. Thus, we select this function to cancel the growing term, that is

$$
\rho_0 c\varepsilon\left[-\varepsilon\frac{\xi}{c}F\left(\xi, t\right)\ddot{X}\left(t - \frac{\xi}{c}\right)\right] + \varepsilon^2\rho_0\left[\frac{\beta\xi}{c}\dot{X}\left(t - \frac{\xi}{c}\right)\ddot{X}\left(t - \frac{\xi}{c}\right)\right] = 0
\tag{13.3.71}
$$

In other words

$$
F\left(\xi, t\right) = \frac{\beta}{c}\dot{X}\left(t - \frac{\xi}{c}\right)
\tag{13.3.72}
$$

The assembly of the results is

$$
\begin{aligned}
p = \rho_0 c v + O\left(\varepsilon^2\right) &= \rho_0 c\varepsilon\dot{X}\left(\tau\right) + O\left(\varepsilon^2\right) \\
\tau = t - \frac{\xi}{c}, \quad x &= \xi + \varepsilon\beta\frac{x}{c}\dot{X}\left(\tau\right)
\end{aligned}
\tag{13.3.73}
$$

The second-order terms that remain in the expression for p are those in Eq. (13.3.68) that do not grow with increasing x. They represent effects that always are small relative to the overall signal and therefore seldom are of interest. Some individuals refer to the process of removing the growing terms as a *renormalization* of the perturbation series. ("Normal" here refers to regularity, as opposed to singularity.)

Is the solution consistent with the Riemann solution? To show this to be true, we write the last of Eq. (13.3.73) as

$$\xi = x \left(1 - \varepsilon\beta\frac{\dot{X}(\tau)}{c} \right) \tag{13.3.74}$$

Correspondingly, the first and second of Eq. (13.3.73) are

$$p = \rho_0 c v = \rho_0 c \varepsilon \dot{X}(\tau) + O\left(\varepsilon^2\right)$$
$$\tau = t - \frac{x}{c}\left(1 - \varepsilon\beta\frac{\dot{X}(\tau)}{c} \right) = t - \frac{x}{c}\left(1 - \beta\frac{p}{\rho_0 c^2} \right) \tag{13.3.75}$$

For comparison recall Eq. (13.1.37), which is the small-signal approximation of the Riemann solution. The term containing p in the denominator is small compared to one, so the denominator may be expanded in a binomial series. Doing so leads to

$$p = \rho_0 c \dot{X}(\tau), \quad \tau = t - \frac{x}{c\left(1 + \beta p/\left(\rho_0 c^2\right)\right)} = t - \frac{x}{c}\left(1 - \beta\frac{p}{\rho_0 c^2} + O\left(\varepsilon^2\right) \right) \tag{13.3.76}$$

For second order, this is the same as the expressions derived by the method of renormalization. It is useful to have this confirmation of the renormalization formulation, but for more general systems any confirmation must come from experiments.

Before we consider such generalizations, it is appropriate that we consider another singular perturbation technique that has been employed for nonlinear acoustics analyses. The *method of multiple scales* had been used in other areas for many years prior to Nayfeh and Kluwick's application[17] of the method to analyze a finite amplitude plane wave. It recognizes that the spatial scale over which nonlinear effects become important is much greater than the scale of a wavelength. Each scale is defined to be an independent spatial variable ξ_n that changes more slowly as n increases. For the nonlinear wave equation, these variables are

$$\xi_n = \varepsilon^n x, \quad n = 0, 1, \ldots \tag{13.3.77}$$

The velocity potential is considered to depend on each of these scaled variables, even though the value of each coordinate is set when x and ε are set. Thus, a derivative with respect to x becomes

$$\frac{\partial\phi}{\partial x} = \frac{\partial\phi}{\partial\xi_0} + \varepsilon^2\frac{\partial\phi}{\partial\xi_1} + \varepsilon^2\frac{\partial\phi}{\partial\xi_2} + \cdots \tag{13.3.78}$$

---

[17] A.H. Nayfeh and A. Kluwick, "A comparison of three perturbation methods for nonlinear waves," J. Sound and Vibration (1976) pp. 293–299.

This transformation is introduced to the nonlinear wave equation and the boundary condition, after which the velocity potential is expanded in the perturbation series in Eq. (13.3.46). The first-order equations are

$$c^2 \frac{\partial^2 \phi_1}{\partial \xi_0^2} - \frac{\partial^2 \phi_1}{\partial t^2} = 0$$

$$\left. \frac{\partial \phi_1}{\partial \xi_0} \right|_{x=0} = \dot{X}(t)$$

(13.3.79)

and the second-order equations are

$$c^2 \frac{\partial^2 \phi_2}{\partial \xi_0^2} - \frac{\partial^2 \phi_2}{\partial t^2} = -2c^2 \frac{\partial^2 \phi_1}{\partial \xi_0 \partial \xi_1} + 2(\beta - 1) \frac{\partial^2 \phi_1}{\partial \xi_0^2} \frac{\partial \phi_1}{\partial t} + 2 \frac{\partial \phi_1}{\partial \xi_0} \frac{\partial^2 \phi_1}{\partial \xi_0 \partial t}$$

$$\left. \frac{\partial \phi_2}{\partial \xi_0} \right|_{x=0} = - \left. \frac{\partial \phi_2}{\partial \xi_0} \right|_{x=0} - X(t) \left. \frac{\partial^2 \phi_1}{\partial \xi_0^2} \right|_{x=0}$$

(13.3.80)

Note that the second-order boundary condition accounts for the displaced location of the boundary, even though we know that effect does not lead to growing nonlinear effects.

Because we are not interested in ϕ terms that are $O(\varepsilon^3)$ or smaller, only the physical scale ξ_0 and the slow scale ξ_1 enter into the analysis. The first-order equations have the same form as those for a regular perturbation, but they allow ϕ_1 to have arbitrary dependence on ξ_1. It is easier to satisfy boundary conditions after the general solution has been identified. Hence, the first-order solution is written as

$$\phi_1 = f(t', \xi_1), \quad t' = t - \frac{\xi_0}{c}$$

(13.3.81)

This form is used to generate the second-order equations. Equation (13.3.57) may be used to describe the quadratic terms, so the second-order wave equation becomes

$$c^2 \frac{\partial^2 \phi_2}{\partial \xi_0^2} - \frac{\partial^2 \phi_2}{\partial t^2} = 2c \frac{\partial}{\partial \xi_1} \left[\frac{\partial}{\partial t'} f(t', \xi_1) \right] + \beta \frac{\partial}{\partial t'} \left[\frac{\partial}{\partial t'} f(t', \xi_1) \right]^2$$

(13.3.82)

We know from the regular perturbation series analysis that the term whose coefficient is β will lead to a particular solution for ϕ_2 that is not uniformly valid. The role of the first term on the right side is to prevent this condition from occurring. That is, we set

$$\frac{\partial}{\partial \xi_1} \left[\frac{\partial}{\partial t'} f(t', \xi_1) \right] = -\frac{\beta}{c} \frac{\partial}{\partial t'} \left[\frac{\partial}{\partial t'} f(t', \xi_1) \right]^2$$

(13.3.83)

Fulfillment of this condition represents the primary difficulty in using the method of multiple scales. If considered in isolation, without prior knowledge of the nonlinear behavior of a plane wave, solution of this equation would not be a simple task. The result, after it is made to satisfy the first-order boundary condition, is

$$\phi_1 = -cX\left(\tau\right) + \frac{\beta}{2c}\xi_1\dot{X}\left(\tau\right)^2$$
$$t' = \tau - \frac{\beta}{c^2}\xi_1\dot{X}\left(\tau\right) \tag{13.3.84}$$

From here, expressions for the pressure and particle velocity are obtained by differentiation of this function. We shall not pursue the details because the result will be equivalent to the Riemann solution. The primary motivation for this discussion of the method of multiple scales is to demonstrate that it will be substantially more difficult to implement than the method of strained coordinates. Such is especially the case when we consider situations where a signal depends on more than one spatial coordinate. Nevertheless, multiple scales method is a more general method than strained coordinates, so it is conceivable that only it would be suitable in some as yet unexplored situation.

13.4 Multidimensional Systems

Most linear acoustic systems of practical interest feature signals that are not truly planar waves. Higher-order modes in a rectangular waveguide are best characterized by Cartesian coordinates, whereas the signal within a cylindrical waveguide or radiated by a circular piston is best described in cylindrical coordinates. In the latter systems, the axial direction of the coordinate system would be the direction of propagation, whereas radiation from a cylinder features cylindrical coordinates with propagation in the transverse direction. Spherical coordinates are appropriate for radiation from a vibrating sphere. Nonlinear effects arise in each of these systems. Each has been successfully researched, some by analytical techniques, some by computational techniques, and some by both approaches. The scope of the investigations is too vast to survey here. An excellent starting point is the monograph edited by Hamilton and Blackstock,[18] but even that work is merely an entry point to an enormous body of research. Our objective here is not to delve deeply into specific topics. Rather, we will consider a few systems that highlight methods and phenomena not seen in plane waves.

13.4.1 Finite Amplitude Spherical Wave

The choice of coordinate system to use for the analysis of any wave usually is dictated by the shape of the boundary. Here, we shall study how nonlinearity affects the propagation of a radially symmetric spherical wave. The signal is generated by a spherical body whose radius oscillates relative to a mean value a, such that $r = a + \varepsilon Y\left(t\right)$. The

---

[18]M.F. Hamilton & D.T. Blackstock, *Nonlinear Acoustics*, eds., Acoustical Society of America (2008).

magnitude of the radial velocity $\varepsilon \dot{Y}$ is much less than the speed of sound c, but it is sufficiently large to warrant consideration of nonlinear effects. In a sense, the signal is not a multidimensional wave because it depends on a single spatial coordinate, the radial distance r. Indeed, the basic line of investigation we shall pursue parallels the perturbation analysis of plane waves. However, dependence on the spherical coordinate r rather than the Cartesian coordinate x considerably complicates each stage of the analysis, and these complications are typical of those encountered in formulations using more general curvilinear coordinates. It is important to note that a much simpler analysis of spherical waves is available.[19] The analysis here will show how the method of strained coordinates may be employed when simpler approaches are not available.

The radiated signal is radially symmetric, so ϕ is a function of radial distance r and time t. The perturbation series for the velocity potential is

$$\phi = \varepsilon \phi_1 (r, t) + \varepsilon^2 \phi_2 (r, t) \tag{13.4.1}$$

The first-order equation derived from the quadratic nonlinear wave, Eq. (13.3.18), is

$$\boxed{c^2 \nabla^2 \phi_1 - \frac{\partial^2 \phi_1}{\partial t^2} = 0} \tag{13.4.2}$$

We use this relation to replace $\nabla^2 \phi_1$ in the second-order equation, which thereby is simplified to

$$\boxed{c^2 \nabla^2 \phi_2 - \frac{\partial^2 \phi_2}{\partial t^2} = \frac{\partial}{\partial t} \left[\frac{(\beta - 1)}{c^2} \left(\frac{\partial \phi_1}{\partial t} \right)^2 + \nabla \phi_1 \cdot \nabla \phi_1 \right]} \tag{13.4.3}$$

[Equations (13.4.2) and (13.4.3) have been highlighted because they apply in any situation.] The operators in radially symmetric spherical coordinates are

$$\nabla^2 \phi = \frac{1}{r} \frac{\partial^2}{\partial r^2} (r\phi), \quad \nabla \phi = \bar{e}_r \frac{\partial \phi}{\partial r} \tag{13.4.4}$$

It is convenient to address satisfaction of the boundary condition at the surface, $r = a$, after the general solution has been obtained. The radiation condition requires that the signal be an outgoing wave, so we write the first-order solution as

$$\phi_1 = \frac{a}{r} f (t') \tag{13.4.5}$$

---

[19]D.T. Blackstock, "On plane, spherical, and cylindrical waves of finite amplitude in lossless fluids," J. Acoust. Soc. Am., Vol. 36 (1966) pp. 217–219.

where the nature of the function f is not specified at this juncture. The phase variable t' is the retarded time, which accounts for the time required for a signal to travel from the sphere's surface to the field point, that is,

$$t' = t - \frac{r - a}{c} \tag{13.4.6}$$

Because $\partial(\ )/\partial t \equiv \partial(\ )/\partial t'$, substitution of the expression for ϕ_1 into the second-order wave equation leads to

$$\frac{1}{r}\left[c^2 \frac{\partial^2}{\partial r^2}(r\phi_2) - \frac{\partial^2}{\partial(t')^2}(r\phi_2) \right] = \frac{\partial}{\partial t'}\left[\frac{\beta a^2}{c^2 r^2}\dot{f}(t')^2 \right.$$
$$\left. + \frac{2a^2}{cr^3}f(t')\dot{f}(t') + \frac{a^2}{r^4}f(t')^2 \right] \tag{13.4.7}$$

A consequence of the interchangeability of derivatives with respect to t and t' is that an overdot will be used to denote differentiation with respect to the argument of f.

The preceding is a linear partial differential equation, so we may obtain its particular solution by superposing the contribution of each term on the right side. Each of those terms has the general form $r^{-n}Q(t')$. Let us examine the possibility that the particular solution has a similar form by considering $r\phi_2 = z_n$, where

$$z_n = g(r)G(t') \tag{13.4.8}$$

Then, the left side of Eq. (13.4.7) would be

$$\frac{1}{r}\left[c^2 \frac{\partial^2}{\partial r^2}z_n - \frac{\partial^2}{\partial(t')^2}z_n \right] = -2\frac{g'(r)}{r}\dot{G}(t') + \frac{g''(r)}{r}G(t') \tag{13.4.9}$$

If we set $g'(r) = 1/r$, then $\dot{G}(t')$ can be selected to match the first term within the bracket of Eq. (13.4.7). However, doing so generates a term that is like the second term in the bracket because $g''(r) = -1/r^2$. Suppose we add to z_n another term for which $g'(r) = 1/r^2$ in order to match the second term in the bracket plus the residual from the first term. Doing so generates a new residual term that resembles the third term in the bracket because it gives $g''(r) = -2/r^3$. Addition of another term to z_n for which $g'(r) = 1/r^3$ would handle the $1/r^3$ terms to the right, but also add another residual at $1/r^4$. The process continues *ad infinitum*, with each added z_n term selected to match the residual from the previous term. Hence, we conclude that we may construct a particular solution as an infinite series,

$$(r\phi_2)_p = \sum_{n=1}^{\infty} g_n(r)G_n(t') \tag{13.4.10}$$

$$g'_n(r) = r^{-n}$$

If we were to carry out the full analysis, we would be led to a recurrence relation wherein $\dot{G}_n(t')$ for $n > 1$ is defined in terms of $G_{n-1}(t')$. Such an analysis would be required if it were necessary to identify all parts of ϕ_2. Fortunately, this seldom is the case because of the nature of the $g_n(r)$ functions. Integration of the second of Eq. (13.4.10) leads to

$$g_1(r) = \ln(r), \quad g_n(r) = -\frac{1}{n-1}r^{1-n} \text{ if } n > 1 \qquad (13.4.11)$$

Hence, each term in Eq. (13.4.10) other than $n = 1$ represents a contribution to $r\phi_2$ that decays at least as rapidly as $1/r$. Because $r\phi_1$ does not decay with increasing r, all of those terms will be such that $\varepsilon^2\phi_2$ is much less than $\varepsilon\phi_1$ everywhere.

The same is not true for the $n = 1$ term. Its contribution to $r\phi_2$ is proportional to $\ln(r)$. Although logarithmic growth is relatively slow, there is some large value of r that would lead to $\varepsilon^2\phi_2$ having the same order of magnitude as $\varepsilon\phi_1$. This term represents the part that is responsible for the perturbation series not being uniformly valid.

Our interest is identification of terms that describe cumulative growth of nonlinear effects, so we modify Eq. (13.4.10) to be

$$r\phi_2 = a\ln\left(\frac{r}{a}\right)G_1(t') + O\left(\frac{1}{r}\right) \qquad (13.4.10)$$

where the introduction of the factor a will help avoid confusion regarding the dimensionality of quantities. Substitution of this expression converts the left side of Eq. (13.4.7) to

$$\frac{1}{r}\left[c^2\frac{\partial^2}{\partial r^2}(r\phi_2)_\text{p} - \frac{\partial^2}{\partial(t')^2}(r\phi_2)_\text{p}\right] = -\frac{2ac}{r^2}\dot{G}_1(t') \qquad (13.4.12)$$

We equate this term to the first term on the right side of Eq. (13.4.7), which leads to

$$-\frac{2ac}{r^2}G_1(t') = \frac{\beta a^2}{c^2 r}\dot{f}(t')^2 \qquad (13.4.13)$$

The terms on the right side of Eq. (13.4.7) that have not been matched would lead to $O(1/r)$ and $O(1/r^2)$ contributions to $r\phi_2$. Neither grows relative to $r\phi_1$, so these terms may be ignored. Thus, we have found that

$$(r\phi_2)_\text{p} = -\frac{\beta a^2}{2c^3}\ln\left(\frac{r}{a}\right)\dot{f}(t')^2 + O\left(\frac{1}{r}\right) \qquad (13.4.14)$$

A corollary of this development is that there is no need to consider the complementary solution for ϕ_2 because it is a solution of the linear wave equation. Such a term cannot lead to $\varepsilon^2\phi_2$ being comparable to $\varepsilon\phi_1$. In turn, because the role of the complementary solution is solely to satisfy the second-order boundary condition, only the first-order condition need be satisfied. This is fortunate, because satisfying the second-order boundary condition, possibly accounting for the moving boundary that the surface is at $r = a + \varepsilon Y$, would require that we employ the full particular solution in Eq. (13.4.10).

The velocity potential obtained by substitution of the first- and second-order velocity potentials, Eqs. (13.4.5) and (13.4.14), into the perturbation series is

$$r\phi = \varepsilon a f\left(t'\right) - \varepsilon^2 \frac{\beta a^2}{2c^3} \ln\left(\frac{r}{a}\right) \dot{f}\left(t'\right)^2 + O\left(\varepsilon^2\right) \qquad (13.4.15)$$

As a check, note that the dimensionality of $r\phi$ must be L^3/T. The $O\left(\varepsilon\right)$ term has those dimensions if f is L^2/T, in which case the $O\left(\varepsilon^2\right)$ also is consistent. Identification of the coordinate straining transformation requires expressions for p and v_r. Terms that are uniformly $O\left(\varepsilon^2/r\right)$ or less may be ignored, so we use $p = -\rho_0 \partial \phi / \partial t'$. In combination with $v_r = \partial \phi / \partial r$, we obtain

$$rp = -\rho_0 \varepsilon a \dot{f}\left(t'\right) + \rho_0 \varepsilon^2 \frac{\beta a^2}{c^3} \ln\left(\frac{r}{a}\right) \dot{f}\left(t'\right) \ddot{f}\left(t'\right) + O\left(\varepsilon^2\right)$$

$$rv_r = -\varepsilon \left[\frac{a}{c} \dot{f}\left(t'\right)\left(\tilde{t}\right) + \frac{a}{r} f\left(t'\right)\right] + \varepsilon^2 \frac{\beta a^2}{c^4} \ln\left(\frac{r}{a}\right) \dot{f}\left(t'\right) \ddot{f}\left(t'\right) + O\left(\varepsilon^2\right)$$

$$\qquad (13.4.16)$$

The loss of uniform validity for rp and rv_r is caused by a term that grows as $\ln\left(r/a\right)$ relative to the $O\left(\varepsilon\right)$ terms, so we shall strain the r coordinate with a growth factor that is logarithmic. Thus, we try

$$r = \xi + \varepsilon a \ln\left(\frac{r}{a}\right) \mathcal{R} \qquad (13.4.17)$$

where $\mathcal{R}$ is an unknown function to be determined. The coordinate straining converts the retarded time to

$$t' \equiv t - \frac{r - a}{c} = \tau - \varepsilon \frac{a}{c} \ln\left(\frac{r}{a}\right) \mathcal{R} \qquad (13.4.18)$$

The variable τ is a nonlinear retarded time. Its definition is

$$\boxed{\tau = t - \frac{\xi - a}{c}} \qquad (13.4.19)$$

The task now is to determine the function $\mathcal{R}$ that makes the expressions for p and v_r uniformly valid. Toward that end, Eq. (13.4.18) is used to replace $\tilde{t}$ as the argument of the F and $\dot{F}$ functions. Taylor series expansions give

$$f\left(t'\right) = f\left(\tau\right) - \varepsilon \left(\frac{a}{c}\right) \ln\left(\frac{r}{a}\right) \mathcal{R} \dot{f}\left(\tau\right) + O\left(\varepsilon^2\right)$$

$$\dot{f}\left(t'\right) = \dot{f}\left(\tau\right) - \varepsilon \left(\frac{a}{c}\right) \ln\left(\frac{r}{a}\right) \mathcal{R} \ddot{f}\left(\tau\right) + O\left(\varepsilon^2\right)$$

$$\qquad (13.4.20)$$

We substitute the transformation in Eq. (13.4.18) into Eq. (13.4.16). The result for p is

$$
rp = -\rho_0 \varepsilon a \left[\dot{f}(\tau) - \varepsilon \left(\frac{a}{c}\right) \ln\left(\frac{r}{a}\right) \mathcal{R}\ddot{f}(\tau) \right]
$$

$$
+ \rho_0 \varepsilon^2 \frac{\beta a^2}{c^3} \ln\left(\frac{r}{a}\right) \dot{f}(\tau)\ddot{f}(\tau) + O\left(\frac{\varepsilon^2 a}{r}\right)
$$

$$(13.4.21)$$

Similar operations for v_r give

$$
rv_r = -\frac{\varepsilon}{c} a \left[\dot{f}(\tau) - \varepsilon \left(\frac{a}{c}\right) \ln\left(\frac{r}{a}\right) \mathcal{R}\ddot{f}(\tau) \right]
$$

$$
- \varepsilon \frac{a}{r} \left[f(\tau) - \varepsilon \left(\frac{a}{c}\right) \ln\left(\frac{r}{a}\right) \mathcal{R}\dot{f}(\tau) \right]
$$

$$
+ \varepsilon^2 \frac{\beta a^2}{c^4} \ln\left(\frac{r}{a}\right) \dot{f}(\tau)\ddot{f}(\tau) + O\left(\varepsilon^2\right)
$$

$$(13.4.22)$$

The role of $\mathcal{R}$ is to cancel all terms in rp that have the factor $\varepsilon^2 \ln(r/a)$. This condition occurs if

$$
\mathcal{R} = -\frac{\beta}{c^2} \dot{f}(\tau)
$$

$$(13.4.23)$$

The pressure term that remains is

$$
rp = -\rho_0 \varepsilon a \dot{f}(\tau)
$$

$$(13.4.24)$$

The expression for rv_r corresponding to the selection of $\mathcal{R}$ is

$$
rv_r = -\frac{\varepsilon a}{c} \left[\dot{f}(\tau) + \frac{c}{r} f(\tau) \right] - \varepsilon^2 \beta \frac{a^2}{c^3 r} \ln\left(\frac{r}{a}\right) \dot{f}(\tau)^2 + O\left(\varepsilon^2\right)
$$

$$(13.4.25)$$

The presence of $\ln(r/a)$ as a factor in the $O(\varepsilon^2)$ term might seem to mean that this expression is not uniformly valid. However, that factor is divided by r and the limit as $r \to \infty$ of $\ln(r/a)/r$ is zero.

At this juncture, we have not addressed the determination of the f function. It is somewhat less confusing to carry out that task if we work with a function that is dimensionless. Accordingly, we let

$$
f(t) = -c^2 F(t)
$$

$$(13.4.26)$$

This substitution converts the coordinate straining transformation to

$$
r = \xi + \varepsilon \beta a \ln\left(\frac{r}{a}\right) \dot{F}(\tau)
$$

$$(13.4.27)$$

The corresponding nonlinear retarded time is

$$t' \equiv t - \frac{r-a}{c} = \tau - \varepsilon\beta\frac{a}{c}\ln\left(\frac{r}{a}\right)\dot{F}(\tau)$$

(13.4.28)

The corresponding uniformly valid expressions for the state variables are

$$p = \rho_0 c^2 \varepsilon \frac{a}{r}\dot{F}(\tau) + O\left(\frac{\varepsilon^2 a}{r}\right)$$

$$v_r = \varepsilon c\frac{a}{r}\left[\dot{F}(\tau) + \frac{c}{r}F(\tau)\right] + \varepsilon^2 c\beta\ln\left(\frac{r}{a}\right)\left(\frac{a^2}{r^2}\right)\dot{F}(\tau)^2 + O\left(\frac{\varepsilon^2 a}{r}\right)$$

(13.4.29)

Aside from identification of τ dependence of F, the pressure has been determined. To see that this is so, suppose we specify r and t, which sets $t - r/c$. Then, Eq. (13.4.28) is an algebraic equation for τ. Substitution of the value of τ into the first of Eq. (13.4.29) gives the value of p at that r and t.

Determination of the F function is done by satisfying the boundary condition on the sphere. At $r = a$, the coordinate straining gives $\tau = t$. To be consistent with the notation for plane waves, we let the specified velocity on the boundary be $\varepsilon c V(\tau)$. Velocity continuity at the surface requires that

$$v_r|_{r=a} = \varepsilon c V(t)$$

(13.4.30)

Equation (13.4.29) describes the particle velocity at $r = a$. We only need to substitute the $O(\varepsilon)$ terms, because the terms that are $O(\varepsilon^2)$ will merely lead to small corrections of $F(\tau)$. Thus, the substitution leads to a first-order differential equation for $F(\tau)$,

$$\dot{F}(\tau) + \frac{c}{a}F(\tau) = V(\tau)$$

(13.4.31)

This is the same differential equation as that which must be solved in linear theory when the radial velocity is known at $r = a$. Indeed, the expressions for p is the same as the description of the pressure for an outgoing wave according to linear theory, see Eq. (6.2.5), except that the retarded time has been replaced by τ.

Before we proceed to an example, let us consider some qualitative aspects. Because nonlinear effects become progressively more significant at large r, it is reasonable to wonder whether they only occur in the farfield. A convenient reference distance marking where nonlinear effects are significant is that at which a shock forms. This is the closest location at which $\partial t/\partial \tau = 0$ occurs at any instant. Differentiation of Eq. (13.4.28) gives

$$\frac{\partial t}{\partial \tau} = 1 - \varepsilon\beta\frac{a}{c}\ln\left(\frac{r_{shock}}{a}\right)\ddot{F}(\tau) = 0$$

(13.4.32)

from which we find that

$$\ln\left(\frac{r_{\text{shock}}}{a}\right) = \frac{c/a}{\varepsilon\beta\max\left(\ddot{F}\right)} \qquad (13.4.33)$$

The farfield of a spherical wave is characterized by p being essentially $\rho_0 c v_r$. To identify where this condition occurs let T be a timescale representing the rate at which F changes, that is, $O\left(|\dot{F}|\right) = O\left(|F|\right)/T$. We can assert from Eq. (13.4.29) that $p \approx \rho_0 c v_r$ if $O\left(|F|/T\right) \gg (c/r) O\left(|F|\right)$, which is equivalent to $r \gg cT$. (If $a \gg cT$, then the farfield condition occurs everywhere.) Thus, the shock formation distance is inversely proportional to the maximum rate at which the pressure changes, whereas the farfield distance is independent of the pressure amplitude. This observation suggests that it might be possible to adjust the pressure to attain a shock at any range.

This hypothesis can be tested by considering the steady-state harmonic pressure corresponding to setting $\dot{f}(t) = \sin(\omega\tau)$. The steady-state nonlinear signal is described by

$$p = \varepsilon\rho_0 c^2 \left(\frac{a}{r}\right)\sin(\omega\tau) \qquad (13.4.34)$$

An appropriate reference time is $1/\omega$, so the farfield occurs at $kr \gg 1$. The maximum value of $\ddot{f}$ is ω, so the shock formation distance in this case is

$$r_{\text{shock}} = ae^{1/(\varepsilon\beta ka)} \qquad (13.4.35)$$

If we arbitrarily define the farfield to occur when $kr > 10$, we obtain

$$\frac{r_{\text{shock}}}{r_{\text{ff}}} = \frac{ka}{10}e^{1/(\varepsilon_p\beta ka)} \qquad (13.4.36)$$

The value of ε is much less than one and β is $O(1)$. Hence, it is evident that r_{shock} will always be much greater than r_{ff}. For example, a large value of ε in air is 0.001, in which case the above expression leads to $r_{\text{shock}} > 170 r_{\text{ff}}$ for $ka < 100$. A corollary of r_{shock} being much greater than r_{ff} is that wherever a spherical wave is significantly affected by nonlinearity, it will be true that $p \approx \rho_0 c v_r$.

Although the solution for a nonlinear spherical wave appears to be different from the Riemann solution, they actually are quite similar. To demonstrate the resemblance, we note that the first of Eq. (13.4.29) gives $\dot{F}(\tau) = (r/a)p/\left(\rho_0 c^2\right)$. Substitution of this expression into Eq. (13.4.28) shows that the nonlinear retarded time is

$$\tau = t - \frac{r-a}{c} + \beta\frac{r}{c}\ln\left(\frac{r}{a}\right)\left(\frac{p}{\rho_0 c^2}\right) \qquad (13.4.37)$$

For comparison, the characteristic variable for a weakly nonlinear plane wave was written in Eq. (13.3.76) as

$$\tau = t - \frac{x}{c} + \beta \frac{x}{c} \left(\frac{p}{\rho_0 c^2} \right) + O\left(\varepsilon^2\right) \qquad (13.4.38)$$

The quantity $(r - a)/c$ in Eq. (13.4.37) is the time required for a linear signal to propagate from the surface to the field point. This time is analogous to x/c, which is the time required for a linear plane wave to travel from the boundary at which it is generated. One beneficial aspect of this similarity of the descriptions is that it is possible to adapt the earlier Fourier series analysis to handle situations where the boundary excitation is time-periodic. The adaptation is based on the fact that time occurs in the same manner in Eqs. (13.4.37) and (13.4.38). The implementation of this notion is the topic of Exercise 13.24.

The difference between linear and nonlinear retarded times for a spherical wave is proportional to $\ln(r/a)(r/c)p/\left(\rho_0 c^2\right)$. In contrast, this difference for a planar wave is proportional to $(x/c)p/\left(\rho_0 c^2\right)$. To assess the relative strength of the two effects, it is essential to bear in mind that spherical spreading leads to $r|p|$ being constant. Accordingly, we write the ratio of nonlinear effects as

$$\frac{\text{Spherical nonlinear term}}{\text{Planar nonlinear term}} = \frac{\ln(r/a)(r/c)p_{\text{spher}}}{(x/c)p_{\text{planar}}} \equiv \frac{a\ln(r/a)\left(rp_{\text{spher}}\right)/a}{xp_{\text{planar}}}$$

$$(13.4.39)$$

If p_{spher} at $r = a$ has the same order of magnitude as p_{planar}, then $\left(rp_{\text{spher}}\right)/a$ will have that magnitude at all r. This reduces the ratio of terms to $\ln(r/a)/(x/a)$. If r and x both equal some value $\ell \gg a$, then this ratio is much less than one. This observation leads to the conclusion that the nonlinear effect for a spherical wave is much weaker than it is for a plane wave.

Either of the computational algorithms in Sect. 13.1.5 for evaluating profiles and waveforms of a planar signal may be adapted with small modification to evaluate spherical waves. For example, if we select a value of τ, we may find p from the first of Eq. (13.4.29). Then, Eq. (13.4.37) may be solved algebraically for t if r is specified. Solution for r at a specified t is more difficult because Eq. (13.4.37) is a transcendental function of r. For the same reason, determination of p at specified r and t requires more effort. Like the case of planar waves, a nonlinear equation solver would be needed.

EXAMPLE 13.7 It is desired to compare the effect of nonlinearity in planar and spherical waves. Toward that end, consider a boundary motion in which the velocity normal to the surface is $\varepsilon c \left[\sin(\omega t) + (1/3)\sin(3\omega t)\right]$. For the spherical wave, this is v_r at $r = a$, whereas it is v_x at $x = 0$ for a plane wave propagating in the x direction. Parameters are $\varepsilon = 0.002$, $f = 12$ kHz, and $a = 400$ mm. The fluid is water, for which $c = 1480$ m/s and $\beta = 3.5$. Determine the steady-state planar waveform at $x = x_{\text{shock}}$. Compare that result to the waveform of a spherical wave at $r = x_{\text{shock}}$ and at $r = r_{\text{shock}}$.

Significance

Evaluation of the equations describing a nonlinear spherical wave is the primary emphasis. The results will vividly demonstrate the much slower advent of nonlinear effects that results from spherical spreading, as well as some surprising similarities to nonlinear distortion of plane waves.

Solution

The function F is obtained by solving the linear differential equation that is the boundary condition, Eq. (13.4.31). The specified boundary velocity is

$$
v_r = \varepsilon c V = \varepsilon c \left[\sin(\omega t) + \frac{1}{3} \sin(3\omega t) \right] = \frac{\varepsilon c}{2i} \left(e^{i\omega t} + \frac{1}{3} e^{3i\omega t} \right) + \text{c.c.}
$$

The corresponding differential equation is

$$
\dot{F}(\tau) + \frac{c}{a} F(\tau) = \frac{1}{2i} \left(e^{i\omega \tau} + \frac{1}{3} e^{3i\omega \tau} \right) + \text{c.c.} \tag{1}
$$

The form of the steady-state solution of Eq. (1) is

$$
F(\tau) = \frac{1}{2i} \left(B_1 e^{i\omega \tau} + B_3 e^{3i\omega \tau} \right) + \text{c.c.} \tag{2}
$$

Substitution of this ansatz into Eq. (1) leads to expressions for the coefficients, which are found to be

$$
B_1 = \frac{a/c}{1 + ika}, \quad B_3 = \frac{a/c}{3 + 9ika}
$$

where $k = \omega/c$, as always. It is useful to work with a nondimensional version of $\dot{F}$, which is defined according to

$$
F(\tau) = \frac{a}{c} \operatorname{Re} \left[\frac{1/i}{1 + ika} e^{i\omega \tau} + \frac{1/i}{3 + 9ika} e^{3i\omega \tau} \right] \tag{3}
$$

Determination of F allows us to finalize the expressions to evaluate. Equation (13.4.28) becomes

$$
t - \frac{r - a}{c} = \tau - \varepsilon \beta \frac{a}{c} \ln\left(\frac{r}{a}\right) \operatorname{Re} \left[\frac{ka}{1 + ika} e^{i\omega \tau} + \frac{ka}{1 + 3ika} e^{3i\omega \tau} \right] \tag{4}
$$

The pressure resulting from substitution of Eq. (2) into the first of Eq. (13.4.29) is

$$
p = \rho_0 c^2 \varepsilon \frac{a}{r} \dot{F}(\tau) = \rho_0 c^2 \varepsilon \frac{a}{r} \operatorname{Re} \left[\frac{ka}{1 + ika} e^{i\omega \tau} + \frac{ka}{1 + 3ika} e^{3i\omega \tau} \right] \tag{5}
$$

The radial distance at which an arbitrary wave forms a shock is given by Eq. (13.4.33). In terms of the current functional representation, that expression is

$$\ln\left(\frac{r_{shock}}{a}\right) = \frac{c^3}{\varepsilon\beta a \max\left(-\ddot{F}(\tau)\right)} \equiv \frac{1}{\varepsilon\beta ka \max(D(\tau))} \tag{6}$$

where

$$D(\tau) = \text{Re}\left[i\left(\frac{i}{1+ika}e^{i\omega\tau} + \frac{3i}{1+3ika}e^{3i\omega\tau}\right)\right]$$

We could determine the maximum value of $D(\tau)$ by seeking the value of $\omega\tau$ at which $\dot{D} = 0$. Such an evaluation is complicated by the fact that there is more than one local maximum. A simpler approach is to evaluate $D(\tau)$ for closely spaced values of $\omega\tau$ ranging over a period. The specified parameters give $ka = 20.38$. A data scan gives $\max(D(\tau)) = 1.9978$ at $\omega\tau/(2\pi) = 0.9984$. The shock formation distance obtained from Eq. (6) is $r_{shock} = 13.37$ m.

We are now ready to evaluate waveforms. There is no requirement to compute the pressure values at a uniform time increment. Consequently, we may follow the simpler procedure in which we select $\omega\tau_n$ at equal increments in the range of a period 2π. The corresponding value of p_n is obtained from Eq. (5). The instant t_n at which p_n occurs at a specified r is found from Eq. (4), which gives

$$\omega t_n = kr - ka + \omega\tau - \varepsilon\beta ka \ln\left(\frac{r}{a}\right)\dot{F}(\tau) \tag{7}$$

The plane wave also must be evaluated. The velocity at $x = 0$ in the Riemann solution was written as $\varepsilon cV(t)$. The function $V(t)$ in the present situation is

$$V(\tau) = \left.\frac{v_x}{\varepsilon c}\right|_{x=0} = \frac{1}{2i}\left(e^{i\omega t} + \frac{1}{3}e^{3i\omega t}\right) + \text{c.c.}$$

The Riemann solution for the corresponding plane wave pressure is

$$p = \varepsilon\rho_0 c^2 \text{Re}\left(\frac{1}{i}e^{i\omega\tau} + \frac{1}{3i}e^{3i\omega\tau}\right), \quad \omega t = \omega\tau + \frac{kx}{1+\varepsilon\beta V(\tau)}$$

The shock formation distance for a plane wave is given by Eq. (13.2.26). For the present $V(\tau)$, the maximum value of $dV/d\tau$ is 2ω at $\omega\tau = 0$. This leads to shock formation of the plane wave at

$$x_{shock} = \frac{c}{2\beta\varepsilon\omega} = 1.402 \text{ m}$$

All computations may be carried out in a vectorial manner. Figure 1 compares the boundary velocity $V(\tau)$ to the pressure function $\dot{F}(\tau)$. Both are very close, and suggestive of a square wave. Both functions differ little because ka is quite large. This means that the spherical wave exhibits its farfield behavior everywhere, so that $p \approx \rho_0 c v_r$ at the surface of the sphere.

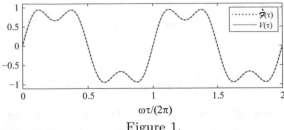

Figure 1.

The spherical waveforms at $r = x_{\text{shock}}$ and $r = r_{\text{shock}}$ in Fig. 2 have been delayed by the respective value of $(r - a)/c$, in order to place them in a common time base. It is difficult to compare them because of their vastly different magnitude. But it appears that the spherical wave is essentially undistorted at $r = x_{\text{shock}}$. This is not surprising because $r_{\text{shock}} \gg x_{\text{shock}}$.

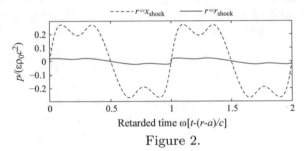

Figure 2.

Multiplying each pressure waveform in Fig. 2 by their respective value of r/a makes it possible to compare them and the plane wave at x_{shock}. This is the data in Fig. 3. Other than the spherical spreading factor, there is no perceptible difference between the spherical and plane waves at their respective shock formation distance. This behavior might seem to be surprising. It is attributable to the fact that $\ln(r/a)$ for a spherical wave and x/x_{shock} for a plane wave are merely factors that scale the difference between the nonlinear and linear retarded times.

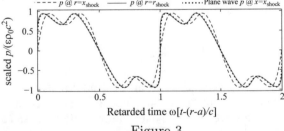

Figure 3.

To close this discussion, it is instructive to consider what would happen if ka were much smaller. Reducing the frequency by a factor of ten with all other parameters constant leads to $r_{shock} = 21.7 \left(10^{15}\right)$ m. On the other hand, lowering the frequency by a factor of ten and raising the acoustic Mach number by a factor of ten change the shock formation distance by a relatively small amount, to $r_{shock} = 18.86$ m. These trends show that nonlinear effects for a spherical wave are relatively small unless either the frequency, as measured by ka, or the acoustic Mach number is very large.

13.4.2 Waves in Cartesian Coordinates

There is no variation of pressure along the wavefronts of plane waves and radially symmetric spherical waves. Variation of pressure transverse to the direction of propagation gives rise to a phenomenon we have not yet encountered, specifically deviation of rays due to the signal associated with those rays. The system we shall investigate here exhibits this phenomenon, but it is somewhat academic. It concerns radiation from an infinite planar boundary, which is unrealizable. Nevertheless, with some modifications the solution can be used to describe the nonlinear propagation of signals in a hard-walled waveguide.[20]

Figure 13.21 depicts the system we will investigate. A flat plate extends infinitely in both directions. It is the boundary of the fluid in the infinite half-space above it. Knife-edge supports running perpendicularly to the plane of the diagram are spaced equidistantly at distance L. The plate vibrates in the direction normal to its surface, with the displacement w defined to be positive into the fluid, which is the sense of increasing z. The displacement must be zero at its supports. The longest possible half-wavelength is L, so the displacement is set to

$$w = \frac{\varepsilon c}{\omega} \sin\left(\omega t\right) \sin\left(\frac{\pi x}{L}\right) \tag{13.4.40}$$

(Shorter wavelengths along the plate may be studied by replacing L with L/n, $n = 1, 2, \ldots$ throughout the analysis.) We shall perform a steady-state analysis, so w is presumed to have existed for a long time prior to $t = 0$.

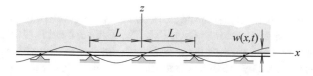

Fig. 13.21 A periodically supported vibrating plate, whose rest position is the plane $z = 0$, generates a two-dimensional acoustic field

---

[20]H.C. Miao and J.H. Ginsberg, "Finite amplitude distortion and dispersion of a nonplanar mode in a waveguide," J. Acoust. Soc. Amer., Vol. 80 (1986), pp. (911–920).

The task before us is to find a solution of the nonlinear wave equation such that the particle velocity and the velocity of the plate have equal components normal to the plate's surface. Another condition is that the acoustic signal must satisfy the Sommerfeld radiation condition, which requires that the pressure and particle velocity be outgoing or evanescent waves at large z. A further requirement follows from the fact that Fig. 13.21 does not change if we shift the coordinate system by any multiple of $2L$ in either sense of the x direction. Consequently, the solution must have period $2L$ in the x direction.

The condition of continuity of particle velocity at the fluid/plate interface constitutes a moving boundary condition. Eventually, the analysis will lead to a simplified boundary condition, as it did for plane and spherical waves. However, such an approximation might not be warranted in other systems, or in situations where nonlinear effects near the boundary are of interest. Hence, let us consider the "exact" condition for the plate. Figure 13.22 zooms in on a small portion of the surface at an arbitrary instant.

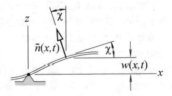

Fig. 13.22 Formulation of the continuity condition of particle velocity for fluid at the interface with a vibrating plate. The condition applies at the current position of the plate in the direction $\bar{n}$ normal to the current tangent to the plate

Figure 13.22 represents the displaced plate as the surface defined by $z = w(x, t)$. The normal to this surface is $\bar{n} = -\sin \chi \bar{e}_x + \cos \chi \bar{e}_z$, where χ is the angle between the normal to the plate and the z-axis. The particle velocity normal to the surface of the plate at its current position must equal the normal velocity of the plate at that position. In terms of the velocity potential, this condition is stated as

$$\bar{n} \cdot \nabla \phi|_{z=w} = \bar{n} \cdot \frac{\partial w}{\partial t} \bar{e}_z \qquad (13.4.41)$$

Figure 13.22 shows that χ also is the angle between the tangent to the surface in the xz-plane and the x-axis, so $\chi = \tan^{-1}(\partial w / \partial x)$. The displacement is $O(\varepsilon)$, so $|\chi| \ll 1$. Therefore, $\cos \chi \approx 1$ and $\sin \chi \approx \tan \chi \approx \partial w / \partial x$, so $\bar{n} = -(\partial w / \partial x) \bar{e}_x + \bar{e}_z$. When the gradient is replaced by the first two terms of a Taylor series relative to $z = 0$, the boundary condition becomes

$$\left(-\frac{\partial w}{\partial x} \bar{e}_x + \bar{e}_z \right) \cdot \left[\nabla \phi|_{z=0} + w \left(\frac{\partial}{\partial z} \nabla \phi \right)_{z=0} \right]_{z=0} = \frac{\partial w}{\partial t} \qquad (13.4.42)$$

The result of expanding this expression and dropping terms that are $O\left(\varepsilon^3\right)$ is

$$\frac{\partial\phi}{\partial z} - \frac{\partial w}{\partial x}\frac{\partial\phi}{\partial x} + w\frac{\partial^2\phi}{\partial z^2} = \frac{\partial w}{\partial t} \quad \text{at } z = 0 \tag{13.4.43}$$

The first term is the one that would be employed for a linear analysis, the second compensates for the fact that v_x has a normal component because $\bar{n}$ is rotated from the z-axis, and the third term compensates for the plate not being on the x-axis.

There is no variation of any feature of the system perpendicular to the plane of Fig. 13.21, so the velocity potential is a function of x, z, and t. The perturbation series is

$$\phi = \varepsilon\phi_1\left(x, z, t\right) + \varepsilon^2\phi_2\left(x, z, t\right) + \cdots \tag{13.4.44}$$

Equations (13.4.2) and (13.4.3) respectively govern the first and second-order potentials.

The boundary conditions for the first and second-order potentials are found by substituting Eqs. (13.4.40) and (13.4.44) into the moving boundary condition and then matching like powers of ε. It is useful to use a complex exponential representation to handle products of harmonic time functions. Thus, we write

$$\left.\frac{\partial\phi_1}{\partial z}\right|_{z=0} = \frac{\partial}{\partial t}\left(\frac{w}{\varepsilon}\right) = \frac{1}{2}ce^{i\omega t}\sin\left(\frac{\pi x}{L}\right) + \text{c.c.}$$

$$\left.\frac{\partial\phi_2}{\partial z}\right|_{z=0} = \frac{\partial}{\partial x}\left(\frac{w}{\varepsilon}\right)\left.\frac{\partial\phi_1}{\partial x}\right|_{z=0} - \left(\frac{w}{\varepsilon}\right)\left.\frac{\partial^2\phi_1}{\partial z^2}\right|_{z=0} \tag{13.4.45}$$

The sinusoidal dependence on x on the boundary is like a higher-order mode in a two-dimensional waveguide. This suggests that the general solution for ϕ_1 has the form

$$\phi_1 = \frac{1}{2}Be^{i(\omega t - \kappa z)}\sin\left(\frac{\pi x}{L}\right) + \text{c.c.} \tag{13.4.46}$$

This ansatz satisfies the first-order (linear) wave equation if the wavenumber is

$$\kappa = \left(k^2 - \frac{\pi^2}{L^2}\right)^{1/2} \tag{13.4.47}$$

If $kL < \pi$, the pressure will decay exponentially with increasing z. Such a situation is irrelevant to a study of nonlinearity at small ε because the pressure will decay to insignificance before ϕ_2 grows to a significant level. Thus, we shall only consider the case where the wavelength in the z direction is less than $2L$, that is, $kL > \pi$, which means that κ is real. The coefficient B is found by satisfying the first of Eq. (13.4.45). The resulting first-order potential is

$$\phi_1 = \frac{1}{2}\left(\frac{ic}{\kappa}e^{i(\omega t - \kappa z)} + \text{c.c.}\right)\sin\left(\frac{\pi x}{L}\right) + \text{c.c.} \tag{13.4.48}$$

To determine the second-order potential, we substitute ϕ_1 into the second-order inhomogeneous wave equation, Eq. (13.4.3), which becomes

$$c^2 \nabla^2 \phi_2 - \frac{\partial^2 \phi_2}{\partial t^2} = \frac{1}{4} \frac{\partial}{\partial t} \left[\frac{(\beta - 1)}{c^2} \left(-\frac{\omega c}{\kappa} e^{i(\omega t - \kappa z)} + \text{c.c.} \right)^2 \sin\left(\frac{\pi x}{L}\right)^2 \right.$$
$$+ \left(\frac{ic\pi}{\kappa L} e^{i(\omega t - \kappa z)} + \text{c.c.} \right)^2 \cos\left(\frac{\pi x}{L}\right)^2$$
$$\left. + \left(c e^{i(\omega t - \kappa z)} + \text{c.c.} \right)^2 \sin\left(\frac{\pi x}{L}\right)^2 \right] \tag{13.4.49}$$

The product of the complex exponential and its complex conjugate is a constant. There is no need to track such terms because each square is differentiated with respect to time. For the remaining terms we use the formulas for the cosine of a double angle to handle the x dependence. Doing so leads to

$$c^2 \nabla^2 \phi_2 - \frac{\partial^2 \phi_2}{\partial t^2} = \frac{1}{8} \frac{\partial}{\partial t} \left\{ e^{2i(\omega t - \kappa z)} \left[\frac{(\beta - 1)\omega^2}{\kappa^2} + c^2 \right] \left[1 - \cos\left(\frac{2\pi x}{L}\right) \right] \right.$$
$$\left. - \frac{\pi^2 c^2}{\kappa^2 L^2} \left[1 + \cos\left(\frac{2\pi x}{L}\right) \right] \right\} + \text{c.c.}$$
$$\tag{13.4.50}$$

The definition of κ in Eq. (13.4.47) reduces the second-order equation to

$$c^2 \nabla^2 \phi_2 - \frac{\partial^2 \phi_2}{\partial t^2} = \frac{i\omega}{4\kappa^2} e^{2i(\omega t - \kappa z)} \left\{ \left(\omega^2 \beta - 2 \frac{\pi^2 c^2}{L^2} \right) \right.$$
$$\left. - \omega^2 \beta \cos\left(\frac{2\pi x}{L}\right) \right\} + \text{c.c.} \tag{13.4.51}$$

The second term is a solution of the linear wave equation. We can generate a particular solution by multiplying its form by x, z, and/or t. Other considerations dictate which of these alternatives should be chosen. Specifically, we know that the pressure must be periodic in x and t because shifting t by $2\pi/\omega$ or x by $2\pi/L$ does not alter the excitation. Factors of t or x multiplying sinusoidal terms would violate these periodicity requirements. Although the presence of a z factor in ϕ_2 might seem to violate the Sommerfeld radiation condition, the intent is to identify terms that lead to nonuniform validity. Thus, we shall try as a particular solution

$$(\phi_2)_{\text{part}} = \left[B_0 + B_1 z \cos\left(\frac{2\pi x}{L}\right) \right] e^{2i(\omega t - \kappa z)} + \text{c.c.} \tag{13.4.52}$$

The coefficients are found by substituting this expression into Eq. (13.4.51) and then matching like terms. The result is

$$B_0 = \frac{ick}{16\kappa^2} \frac{\beta k^2 - 2\pi^2/L^2}{(k^2 - \kappa^2)}, \quad B_1 = \frac{c}{16} \beta \left(\frac{k^3}{\kappa^3} \right) \tag{13.4.53}$$

The B_1 term leads to growth of ϕ_2 relative to ϕ_1 as z increases, whereas the B_0 term remains bounded at all z. By definition, $(\phi_2)_{\text{comp}}$ is a solution of the (homogeneous) linear wave equation. As such, it too is bounded and therefore represents a contribution to the second-order potential that does not grow. For this reason, we need not determine it. (This is fortunate because substitution of w and ϕ_1 into the second of Eq. (13.4.45) would generate several terms that would need to be matched.) Consequently, the solution for the velocity potential reduces to

$$\phi = \varepsilon \frac{ic}{2\kappa} e^{i(\omega t - \kappa z)} \sin\left(\frac{\pi x}{L}\right) + \varepsilon^2 \frac{c\beta}{16}\left(\frac{k^3}{\kappa^3}\right) z e^{2i(\omega t - \kappa z)} \cos\left(\frac{2\pi x}{L}\right) + \text{c.c.} + O\left(\varepsilon^2\right)$$

$$(13.4.54)$$

Although we could work with this representation, the presence of complex conjugate terms would complicate the process of identifying the coordinate straining transformation. The equivalent real form is

$$\boxed{\begin{aligned} \phi = {}&-\varepsilon\frac{c}{\kappa}\sin\left(\omega t - \kappa z\right)\sin\left(\frac{\pi x}{L}\right) \\ &+ \varepsilon^2 \frac{c\beta}{8}\left(\frac{k^3}{\kappa^3}\right) z\cos\left(2\omega t - 2\kappa z\right)\cos\left(\frac{2\pi x}{L}\right) + O\left(\varepsilon^2\right) \end{aligned}}$$

$$(13.4.55)$$

As always, $O\left(\varepsilon^2\right)$ should be understood to denote terms that everywhere have that order of magnitude.

The nonuniform validity of this expression is not addressed directly, because the velocity potential is not a state variable for the state of the fluid. The fact that ϕ is not uniformly valid is passed on to the pressure and particle velocity, which are state variables. However, the growth factor will be different. This is a two-dimensional system, so there are two particle velocity components, given by

$$\begin{aligned} v_x = \frac{\partial\phi}{\partial x} = {}&\varepsilon c\left[-\left(\frac{\pi}{\kappa L}\right)\sin\left(\omega t - \kappa z\right)\cos\left(\frac{\pi x}{L}\right)\right. \\ &\left.- \varepsilon\frac{\beta}{4}\left(\frac{\pi}{L}\right)\left(\frac{k^3}{\kappa^3}\right) z\cos\left(2\omega t - 2\kappa z\right)\sin\left(\frac{2\pi x}{L}\right)\right] + O\left(\varepsilon^2\right) \\ v_z = \frac{\partial\phi}{\partial z} = {}&\varepsilon c\left[\cos\left(\omega t - \kappa z\right)\sin\left(\frac{\pi x}{L}\right)\right. \\ &\left.+ \varepsilon\frac{\beta}{4}\left(\frac{k^3}{\kappa^2}\right) z\sin\left(2\omega t - 2\kappa z\right)\cos\left(\frac{2\pi x}{L}\right)\right] + O\left(\varepsilon^2\right) \end{aligned}$$

$$(13.4.56)$$

The pressure may be evaluated according to $p = -\rho_0 \partial\phi/\partial t$ because the quadratic terms in Eq. (13.3.20), being formed from ϕ_1, would merely add $O\left(\varepsilon^2\right)$ terms that do not grow relative to the $O\left(\varepsilon\right)$ terms. Because time only occurs in ϕ as $\omega t - \kappa z$, we have $\left(\partial\phi/\partial t\right)/\omega = -\left(\partial\phi/\partial x\right)/\kappa + O\left(\varepsilon^2\right)$, from which it follows that

$$\boxed{p = \left(\rho_0\omega/\kappa\right) v_z + O\left(\varepsilon^2\right)}$$

$$(13.4.57)$$

The descriptions of p, v_x, and v_z that have been derived are not uniformly valid. However, because p is proportional to v_z, it is only necessary to find a coordinate straining transformation by considering v_x and either v_z or p. Both x and z should be transformed because two state variables require correction, and the second-order term for v_x and v_z is different. A different perspective is that the velocity components are comparable in magnitude, so we should anticipate that the spatial grid should be strained comparably in both directions. The growth factor is proportional to z, so let us try

$$
\boxed{
\begin{aligned}
x &= \eta + \varepsilon z F\left(\eta, \xi, t\right) \\
z &= \xi + \varepsilon z G\left(\eta, \xi, t\right)
\end{aligned}
}
\tag{13.4.58}
$$

We substitute the coordinate transformation into the real forms of the state variables, which gives

$$
\frac{v_x}{\varepsilon c} = -\left(\frac{\pi}{\kappa L}\right) \sin\left[\omega t - \kappa\left(\xi + \varepsilon z G\left(\eta, \xi, t\right)\right)\right] \cos\left[\frac{\pi}{L}\left(\eta + \varepsilon z F\left(\eta, \xi, t\right)\right)\right]
$$
$$
-\varepsilon \frac{\beta}{4}\left(\frac{\pi}{L}\right)\left(\frac{k^3}{\kappa^3}\right) z \cos\left(2\omega t - 2\kappa\xi\right) \sin\left(\frac{2\pi\eta}{L}\right) + O\left(\varepsilon\right)
$$
$$
\tag{13.4.59a}
$$

$$
\frac{v_z}{\varepsilon c} = \frac{\kappa}{k}\left(\frac{p}{\varepsilon\rho_0 c^2}\right) = \cos\left[\omega t - \kappa\left(\xi + \varepsilon z G\left(\eta, \xi, t\right)\right)\right] \sin\left[\frac{\pi}{L}\left(\eta + \varepsilon z F\left(\eta, \xi, t\right)\right)\right]
$$
$$
+\varepsilon \frac{\beta}{4} \frac{k^3}{\kappa^2} z \sin\left(2\omega t - 2\kappa\xi\right) \cos\left(\frac{2\pi\eta}{L}\right) + O\left(\varepsilon\right)
$$
$$
\tag{13.4.59b}
$$

The terms containing F or G are expanded in Taylor series that are truncated at $O\left(\varepsilon\right)$. In addition, the double angle formulas are applied to the $O\left(\varepsilon^2 z\right)$ terms,

$$
\frac{v_x}{\varepsilon c} = -\left(\frac{\pi}{\kappa L}\right)\left[\sin\left(\omega t - \kappa\xi\right)\cos\left(\frac{\pi\eta}{L}\right)\right.
$$
$$
-\left(\kappa\varepsilon z G\right)\cos\left(\omega t - \kappa\xi\right)\cos\left(\frac{\pi\eta}{L}\right)
$$
$$
\left.-\left(\frac{\pi}{L}\varepsilon z F\right)\sin\left(\omega t - \kappa\xi\right)\sin\left(\frac{\pi\eta}{L}\right)\right]
$$
$$
-\varepsilon\frac{\beta}{2}\left(\frac{\pi}{L}\right)\left(\frac{k^3}{\kappa^3}\right) z\left[\cos\left(\omega t - \kappa\xi\right)^2\right.
$$
$$
\left.-\sin\left(\omega t - \kappa\xi\right)^2\right]\sin\left(\frac{\pi\eta}{L}\right)\cos\left(\frac{\pi\eta}{L}\right) + O\left(\varepsilon^2\right)
$$

$$
\frac{v_z}{\varepsilon c} = \frac{\kappa}{k}\left(\frac{p}{\varepsilon\rho_0 c^2}\right) = \cos\left(\omega t - \kappa\xi\right)\sin\left(\frac{\pi\eta}{L}\right) + \left(\kappa\varepsilon z G\right)\sin\left(\omega t - \kappa\xi\right)\sin\left(\frac{\pi\eta}{L}\right)
$$
$$
+\left(\frac{\pi}{L}\varepsilon z F\right)\cos\left(\omega t - \kappa\xi\right)\cos\left(\frac{\pi\eta}{L}\right)
$$

$$+ \varepsilon \frac{\beta}{2} \frac{k^3}{\kappa^2} z \sin(\omega t - \kappa \xi) \cos(\omega t$$

$$- \kappa \xi) \left[\cos\left(\frac{\pi\eta}{L}\right)^2 - \sin\left(\frac{\pi\eta}{L}\right)^2 \right] + O\left(\varepsilon^2\right) \qquad (13.4.60)$$

The task now is to determine F and G functions that remove all terms in both v_x and v_z that have z as a factor. A clue suggesting the appropriate choice comes from recognition that in both expressions, the sinusoidal factors of G are shifted by 90° relative to those for F. This suggests that G also should be 90° out-of-phase relative to F in both its ξ and η dependencies. Each should match the nonuniformly valid term derived from ϕ. Thus, we try

$$F = C_F \sin(\omega t - \kappa \xi) \cos\left(\frac{\pi\eta}{L}\right)$$
$$G = C_G \cos(\omega t - \kappa \xi) \sin\left(\frac{\pi\eta}{L}\right) \qquad (13.4.61)$$

These forms are substituted into Eq. (13.4.60), and terms having like dependence on ξ and η are grouped. The values of C_F and C_G are those that result in annihilation of all terms that have z as a factor. This leads to

$$C_F = -\frac{\beta k L}{2\pi} \left(\frac{k}{\kappa}\right)^2, \quad C_G = \frac{\beta}{2} \left(\frac{k}{\kappa}\right)^3 \qquad (13.4.62)$$

The corresponding coordinate straining transformation is

$$\boxed{\begin{aligned} x &= \eta - \frac{1}{2} \varepsilon \beta z \frac{kL}{\pi} \left(\frac{k}{\kappa}\right)^2 \sin(\omega t - \kappa \xi) \cos\left(\frac{\pi\eta}{L}\right) \\ z &= \xi + \frac{1}{2} \varepsilon \beta z \left(\frac{k}{\kappa}\right)^3 \cos(\omega t - \kappa \xi) \sin\left(\frac{\pi\eta}{L}\right) \end{aligned}} \qquad (13.4.63)$$

The terms that remain in the state variables are

$$\boxed{\begin{aligned} \frac{v_x}{\varepsilon c} &= -\frac{\pi}{\kappa L} \sin(\omega t - \kappa \xi) \cos\left(\frac{\pi\eta}{L}\right) + O\left(\varepsilon^2\right) \\ \frac{v_z}{\varepsilon c} &= \frac{\kappa}{k} \left(\frac{p}{\varepsilon \rho_0 c^2}\right) = \cos(\omega t - \kappa \xi) \sin\left(\frac{\pi\eta}{L}\right) + O\left(\varepsilon^2\right) \end{aligned}} \qquad (13.4.64)$$

Evaluation of the pressure corresponding to specified x, z, and t requires that Eq. (13.4.63) be solved for η and ξ. Such a determination will require numerical methods because of the transcendental nature of the transformation. These matters will be addressed in the next example. Here, we shall focus on the meaning of the coordinate straining. The general concept of a coordinate straining transformation is that it shifts the location where the linear signal is observed. The linear solution is recovered by setting $\varepsilon = 0$ in Eq. (13.4.63), which leads to $\xi = z$ and $\eta = x$ in Eq. (13.4.64). Thus, one way to view the nonlinear effect is to view lines of constant ξ and constant η as a grid. Such a view is afforded by Fig. 13.23, which shows the transformation grid at two instants separated by a quarter-period. (The value of ε used there is quite large because doing so assists visualizations.)

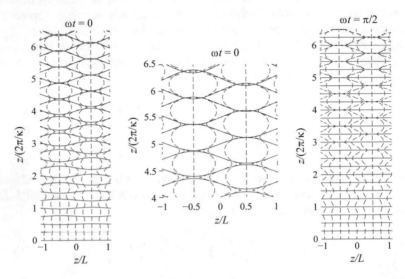

Fig. 13.23 Instantaneous visualizations of the coordinate straining transformation for the nonlinear signal radiated by a vibrating flat plate, $\varepsilon = 0.005$, $kL = 3.628$. Constant ξ lines are *solid* and *red*; constant η lines are *dashed* and *blue*

The phase variable for the signal is

$$\theta = \omega t - \kappa \xi \tag{13.4.65}$$

The coordinate transformation gives $\xi = 0$ at $z = 0$, so all signals on the wavefront at specific values of t and ξ left the plate at $t_0 = t - (\kappa/\omega)\,\xi$. Thus, lines of constant ξ are wavefronts. The transformation also states that $\eta = x$ at $z = 0$. The signal corresponding to specific ξ and η at time t left the plate from location $x = \eta$ at time $t - (\kappa/\omega)\,\xi$. Thus, a line of constant η is the locus of locations at instant t of signals that departed from $x = \eta$ prior to time t. (These lines were mistakenly referred to as rays in prior publications describing this system.[21] The rays are identified later in this discussion). Figure 13.23 shows that the constant ξ and η lines near the plate differ little from the linear grid, $z = \eta$ and $\xi = \eta$. The differences grow in an oscillatory manner with increasing z, but the lines $\xi = \pm L/2$ are straight.

---

[21] J.H. Ginsberg, "A new viewpoint for the two-dimensional nonlinear acoustic wave radiating from a harmonically vibrating flat plate," J. Sound and Vib., Vol. 63 (1978) pp. 151–154. The analysis of the plate problem by J.H. Ginsberg, "Perturbation Methods," M.F. Hamilton and D.T. Blackstock, eds., *Nonlinear Acoustics*, Acoustical Society of America (2008) Chap. 10 repeated the misinterpretation.

Each of these behaviors has a relatively simple physical explanation. Note that the combination of sinusoidal terms in each of Eq. (13.4.63) matches a velocity component in Eq. (13.4.64). When we use this fact to eliminate the sinusoidal terms in the coordinate straining, we see that

$$
\begin{aligned}
x &= \eta + \frac{1}{2}\beta z \left(\frac{kL}{\pi}\right)^2 \left(\frac{k}{\kappa}\right)\left(\frac{v_x}{c}\right) \\
z &= \xi + \frac{1}{2}\beta z \left(\frac{k}{\kappa}\right)^3 \left(\frac{v_z}{c}\right)
\end{aligned}
\tag{13.4.66}
$$

A signal corresponding to specific ξ and η would occur at $x = \xi$ and $x = \eta$ in the absence of nonlinearity. The nonlinear signal occurs at a location that is shifted in the x direction by an amount that is proportional to its v_x value. That location also is shifted in the z direction by an amount that is proportional to its v_z value. In other words, the manner in which a signal propagates is altered by the signal. The constant ξ lines in Fig. 13.23 are oscillatory because v_z depends sinusoidally on η. Similarly, the oscillatory nature of the constant η lines is due to the sinusoidal dependence of v_x on η. The lines $\eta = \pm L/2$ are straight because v_x is zero for signals that may be traced back to the plate at $= \pm L/2$.

Advancement and retardation of a wavefront is the *amplitude dispersion* process encountered in plane and spherical waves. Indeed, if we set $\kappa = k$, the difference between z and ξ is half the value for a plane wave. This process is weaker here because v_z fluctuates along the wavefront. Shifting of the signal in the direction transverse to the propagation due to the signal's particle velocity in that direction is the process of *self-refraction*. It does not occur in plane waves. We will see in the next example that the presence of both processes sometimes leads to waveform distortion that is unlike what we have observed thus far.

The zoomed view in Fig. 13.23 shows that at large z the grid lines for different ξ and η have regions where they intersect or are out of synchronization. For example, a wavefront for a lesser value of ξ might be at a larger value of z. This, of course, corresponds to multivaluedness of the signal. In other words, a shock is previously formed as some distance closer to the boundary. One way of identifying a shock is to extend the approach for plane waves in which the transformation is analyzed. Suppose we look at the signal at a point x, z that corresponds to strained coordinates η, ξ. If we increment the latter by differential amounts $d\eta$ and $d\xi$, we will obtain the signal at $x + dx, z + dz$. Let us write Eq. (13.4.63) generically as $x = f_x (\eta, \xi, t, z), x = f_x (\eta, \xi, t, z)$. Then the differential increments are related to

$$
\begin{aligned}
dx &= d\eta \frac{\partial f_x}{\partial \eta} + d\xi \frac{\partial f_x}{\partial \xi} \\
dz &= d\eta \frac{\partial f_z}{\partial \eta} + d\xi \frac{\partial f_z}{\partial \xi}
\end{aligned}
\tag{13.4.67}
$$

The matrix representation of these relations is

$$
\begin{Bmatrix} dx \\ dz \end{Bmatrix} = \begin{bmatrix} \dfrac{\partial f_x}{\partial \eta} & \dfrac{\partial f_x}{\partial \xi} \\[2mm] \dfrac{\partial f_z}{\partial \eta} & \dfrac{\partial f_z}{\partial \xi} \end{bmatrix} \begin{Bmatrix} d\eta \\ d\xi \end{Bmatrix} \tag{13.4.68}
$$

The coefficient matrix is the Jacobian of the transformation. If its determinant is zero, the transformation is not well-defined. This is the condition for the advent of a shock. Expansion of the determinant and the formula for the cosine of the sum of two angles show that either of the two conditions leads to the Jacobian being zero,

$$
\begin{aligned}
\frac{1}{2}\varepsilon\beta\,(kz)\left(\frac{k}{\kappa}\right)^2 \cos\left(\omega t - \kappa\xi + \frac{\pi}{L}\eta\right) &= 1 \\
\frac{1}{2}\varepsilon\beta\,(kz)\left(\frac{k}{\kappa}\right)^2 \cos\left(\omega t - \kappa\xi - \frac{\pi}{L}\eta\right) &= -1
\end{aligned} \tag{13.4.69}
$$

Either relation in combination with Eq. (13.4.63) constitutes three equations relating the values of η, ξ, x, at a specified t. Their simultaneous solution for fixed t defines the locus of points at which a shock forms at that instant.

The views in Fig. 13.23 provide an instantaneous picture of the location of signals that were radiated from the plate in the prior interval. In some respects, a more meaningful question is: What is the path followed by the signal that radiated from the plate at a specific instant? By definition, this path is a ray. All properties of the signal on a ray other than its amplitude are constant. Consequently, a ray in the present system is the locus of points at which a constant value of the phase variable $\theta = \omega t - \kappa\xi$ and strained coordinate η is observed as time elapses.

In linear theory, $\eta = x$ and $\xi = z$. Constant θ in that case is observed at $z = (\omega t - \theta)/\kappa$ on a line of constant x. Thus, the rays of a linearized description are perpendicular to the plate, and the signal propagates along the ray at phase speed ω/κ. The situation is different when nonlinear effects are included. Identification of rays and wavefronts begins by noting that the strained coordinate corresponding to designated values of θ and t is $\xi = (\omega t - \theta)/\kappa$. This relation is used to eliminate ξ from Eq. (13.4.63). Then the equation is solved for x and z as functions of θ, t, and η. The second equation gives

$$
\begin{aligned}
z &= \frac{\omega t - \theta}{\kappa\left[1 - \dfrac{1}{2}\varepsilon\beta\left(\dfrac{k^3}{\kappa^3}\right)\cos(\theta)\sin\left(\dfrac{\pi\eta}{L}\right)\right]} \\
&= \left(\frac{\omega t - \theta}{\kappa}\right)\left[1 + \frac{1}{2}\varepsilon\beta\left(\frac{k^3}{\kappa^3}\right)\cos(\theta)\sin\left(\frac{\pi\eta}{L}\right)\right] + O\left(\varepsilon^2\right)
\end{aligned} \tag{13.4.70}
$$

This expression is used to eliminate z in the first of Eq. (13.4.63), with the result that

$$x = \eta - \left(\frac{\omega t - \theta}{\kappa}\right) \varepsilon \beta \frac{kL}{2\pi} \left(\frac{k}{\kappa}\right)^2 \sin(\theta) \cos\left(\frac{\pi \eta}{L}\right) + O\left(\varepsilon^2\right) \qquad (13.4.71)$$

To obtain a ray, we fix θ and η, and consider t to be a variable parameter. Equations (13.4.70) and (13.4.71) give the x and z coordinates of the point at which this phase is situated at a specific t. Both equations are linear in t, so it follows that *a ray is a straight line*. The slope of this line relative to the z-axis is

$$\left(\frac{dx}{dz}\right)_{\text{ray}} = \frac{\partial x/\partial t}{\partial z/\partial t} = -\varepsilon \beta \frac{kL}{2\pi} \left(\frac{k^2}{\kappa^2}\right) \sin(\theta) \cos\left(\frac{\pi \eta}{L}\right) + O\left(\varepsilon^2\right) \qquad (13.4.72)$$

Self-refraction is evidenced here by the fact that the sinusoidal terms match v_x in Eq. (13.4.64), which leads to

$$\left(\frac{dx}{dz}\right)_{\text{ray}} = \frac{1}{2}\beta \left(\frac{kL}{\pi}\right)^2 \left(\frac{k}{\kappa}\right)^2 \left(\frac{v_x}{c}\right) \qquad (13.4.73)$$

The value of ε is small, so the slope essentially is the angle. From this we conclude that a ray is rotated from perpendicularity to the $z = 0$ plane by an amount that is proportional to the particle velocity v_x in the signal associated with that ray.

To determine the phase speed at which a signal propagates along its ray as t elapses, we observe that if θ and η are constant, then Eq. (13.4.70) is an explicit expression for z. Differentiating it gives

$$\left.\frac{dz}{dt}\right|_{\text{constant } \theta \text{ and } \eta} = \left(\frac{\omega}{\kappa}\right)\left[1 + \frac{1}{2}\varepsilon \beta \left(\frac{k^3}{\kappa^3}\right) \cos(\theta) \sin\left(\frac{\pi \eta}{L}\right)\right] \qquad (13.4.74)$$

The actual phase velocity is parallel to the ray; the preceding is the z component of that velocity. However the angle between the ray and the z-axis is $O(\varepsilon)$, so the z component is essentially the magnitude of the vector. Furthermore, the trigonometric terms in the preceding are the same as those for v_z in Eq. (13.4.64). This leads to the description of the phase speed as

$$c_{\text{phase}} = \left(\frac{\omega}{\kappa}\right)\left[1 + \frac{1}{2}\beta \left(\frac{k^3}{\kappa^3}\right) \left(\frac{v_z}{c}\right)\right] \qquad (13.4.75)$$

If κ were the same as k and the half factor were not present, this would be the same of the phase speed of a nonlinear planar wave.

An important corollary of this analysis is the observation that rays leaving a specific point on the boundary at different instants do not coincide. This behavior is unlike the properties of rays according to the linear version of geometrical acoustics in which the rays are time invariant. Let us consider the rays that depart from a specific location x on the plate's surface. Because the plate displacement is $O(\varepsilon)$, this surface is essentially $z = 0$. Thus, θ/ω is the instant when the signal departed from the plate. According to Eq. (13.4.64), $v_x = 0$ for $\theta = 0, \pi, 2\pi, \dots$. It follows from Eq. (13.4.72) that the rays for these values of θ are parallel to the z-axis, as are the rays obtained from linear theory. The maximum positive value of v_x corresponds to $\theta = \pi/2, 5\pi/2, \dots$. Equation (13.4.72) indicates that the rays that depart from location x on the plate at these instants undergo the greatest rotation to the right of the z-axis. Conversely, $\theta = 3\pi/2, 7\pi/2, \dots$ correspond to the largest negative v_x. The rays departing at these instants undergo the greatest rotation to the left of the z-axis. Thus, within a period the rays rotate from one side to the other as time elapses.

An analogy explains the difference between this description of a ray and the nature of lines of constant η in Fig. 13.23. Consider the stream of water that emerges from a hose. Suppose the nozzle is rotated about the vertical axis in an oscillatory manner. Any particle of water follows a straight line in the horizontal plane because the only force acting on that particle is gravity. (Obviously, this is not true if there is a substantial wind.) This straight line is the ray. Further suppose that we take a photograph looking down on the horizontal plane. The stream in such a photograph would be situated along a line that resembles the oscillatory constant η line in Fig. 13.23. Both views: the path followed by a particle and the instantaneous location of signals, are meaningful, but the ray picture might be more revealing.

Another corollary of the analysis is the observation that rays for a specific phase that depart from the boundary at the same time, but different x, are not parallel. The maximum plate displacement $|w|$ at any instant occurs at locations where $|\sin(\pi x/L)| = 1$. These are the midpoints between the locations where the plate is supported. Because $\eta = x$ at the boundary, these locations correspond to $\cos(\pi\eta/L) = 0$. The deviation angle for rays leaving these locations is $\psi = 0$ regardless of θ. In contrast, at the locations where the plate is supported, $w = 0$, which means that $\sin(\pi x/L) = \sin(\pi\eta/L) = 0$. Hence, $|\cos(\pi\eta/L)| = 1$ for these rays, so the fan of rays that radiate from these points is the widest.

A description of rays sometimes is accompanied by a depiction of wavefronts. Such a picture allows us to monitor the movement of the signal along its ray. We have already constructed a wavefront by fixing ξ and t and allowing η to be a parameter for the coordinate transformation. Here, we wish to construct wavefronts for a specified θ at many instants. A systematic method for doing so is to treat Eqs. (13.4.70) and (13.4.71) as a parametric description of a coordinate transformation with θ and t set, and η considered to be a free parameter. Figure 13.24 displays these properties at four instants separated by a quarter period. The response is periodic in the x direction, so only one interval needs to be described.)

If θ is a multiple of π, the particle velocity in the x direction is zero. Consequently, the rays departing from $z = 0$ at any x are perpendicular to the boundary. These values of θ also correspond to v_z having its maximum magnitude at a specified η, so the wavefronts for these phases undergo the most distortion as they propagate. The situation for θ being an odd multiple of $\pi/2$ is opposite. These phases correspond to the magnitude of v_x being its maximum, so the rays departing from $z = 0$ at any x is at the extreme angle relative to the z-axis. Concurrently, these phases give $v_z = 0$, so these wavefronts remain straight as the signal propagates. Intermediate phases feature rays that are rotated less than the extreme, and wavefronts that undergo less than maximum distortion.

A consequence of the rays being rotated by different amounts is that they will intersect. This is evidenced by the region of concentration along $x = -L/2$ for $\theta = \pi/2$ and along $x = L/2$ for $\theta = 3\pi/2$. The z value at which this condition occurs seems to be close to the value in Fig. 13.24 at which the coordinate straining transformation becomes multivalued. To interrogate the behavior of the rays, Fig. 13.25 zooms in on the region along $x = -L/2$ for $\theta = \pi/2$. This evaluation extends to a greater distance from the plate, and the density of rays close $x = 0$ has been increased. Aside from shifting the x values, the same picture would be obtained around the line $x = 0.5L$ for $\theta = 3\pi/2$.

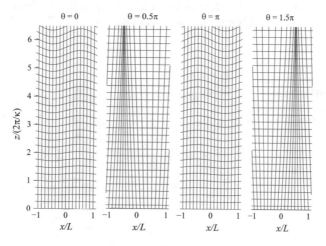

Fig. 13.24 Rays of the nonlinear signal radiated by a vibrating flat plate. The wavefronts are depicted at intervals of one-quarter period. $\varepsilon = 0.005$, $kL = 3.628$

Fig. 13.25 Zoomed view of a caustic formed by the rays of the nonlinear signal radiated by a vibrating flat plate, $\varepsilon = 0.005$, $kL = 3.628$

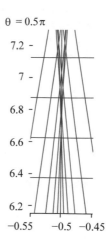

The enlarged figure shows that multiple rays at large angles do not have a common intersection, which means that the convergence of rays is not a focus. Rather, it is a caustic that results from the close intersection of rays at small angles relative to the line $x = -0.5L$. This feature marks shock formation. Let us consider a range of θ values. For $\theta = 0$, all rays are parallel, so the caustic is infinitely distant. As θ is increased, the caustic occurs closer to the plate somewhere along $x = -L/2$. The caustic is closest to the plate when $\theta = \pi/2$. Beyond that value, the caustic occurs at increasing distance, eventually ceasing to exist for $\theta = \pi$. Further increase of θ from π to 2π repeats this behavior along $x = 0.5L$

We can derive an expression for the distance to the caustic. The derivation is based on Fig. 13.26, which depicts a ray that intersects the line $x = -0.5L$. The ray departs

Fig. 13.26 Intersection of a ray departing from the plate close to the ray departing from $x = -L/2$; $\varepsilon = 0.005$, $kL = 3.628$, $\theta = \pi/2$

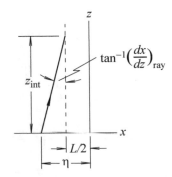

from the plate at $x = \eta$. A caustic on $x = -L/2$ is generated by rays for which $\eta < 0$, so the horizontal distance between the launch point and $x = -0.5L$ is $-\eta - L/2$. This is one side of a right triangle, with the opposite angle being $\tan^{-1}(dx/dz)_{\text{ray}}$, which is given by Eq. (13.4.72). The other side of the right triangle is the distance z_{int} to the intersection of the ray with $x = -L/2$. From trigonometry, we have

$$z_{\text{int}} = \frac{-\eta - L/2}{\tan\left(\tan^{-1}(dx/dz)_{\text{ray}}\right)} = \frac{\eta + L/2}{\varepsilon\beta\dfrac{kL}{2\pi}\left(\dfrac{k^2}{\kappa^2}\right)\sin(\theta)\cos\left(\dfrac{\pi\eta}{L}\right)} \tag{13.4.76}$$

The caustic occurs because the rays at small angles converge, and small angles correspond to η being close to $-L/2$. The arrete, which is the minimum distance to the caustic, is the limit as the angle goes to zero, that is, as $\eta \to -L/2$. Thus

$$z_{\text{arrete}} = \lim_{\eta \to -\pi/2}(z_{\text{int}}) = \frac{1}{\varepsilon\beta\dfrac{k}{2}\left(\dfrac{k^2}{\kappa^2}\right)\sin(\theta)} \tag{13.4.77}$$

The shock formation distance is the minimum distance to the caustic at any instant, which corresponds to $\sin(\theta) = 1$,

$$z_{\text{shock}} = \min\,(z_{\text{arrete}}) = \frac{2}{\varepsilon\beta k}\left(\frac{\kappa}{k}\right)^2 \tag{13.4.78}$$

For comparison, the plane wave shock formation distance for a harmonic excitation is $1/(\beta\varepsilon k)$. For the parameters used to construct Fig. 13.26, the above equation gives $z_{\text{sh}}/(2\pi/\kappa) = 6.631$, which is slightly less than the distance at which two rays in Fig. 13.25 intersect.

The motivation for the analyses of the coordinate straining grid, as well as of rays and wavefronts, is to understand the processes that affect how a signal propagates. The main concepts are the existence of the self-refraction phenomenon, as well as the altered behavior of rays as properties that depend on the phase of the signal that is transported along them. These occur concurrently with amplitude dispersion, which is the process that is responsible for distortion of plane waves. However, these observations are significant, and in some respects fascinating, we also are interested in the nature of waveforms. Evaluation of the signal at specified x, z, and t requires a quantitative solution of the coordinate straining transformation. The next example addresses the procedure for doing so, as well as the effects of the nonlinear processes.

EXAMPLE 13.8 A plate of large extent forms the surface of a large body of water. The plate is periodically supported at intervals of $L = 1.5\,\text{m}$. It is mechanically driven such that its displacement amplitude is $w = W\cos(\omega t)\sin(\pi x/L)$. The maximum displacement is $0.5\,\text{mm}$, the frequency is $2500\,\text{Hz}$, and it may be assumed that the water remains in contact with the plate throughout the vibration. Determine the steady-state waveforms of pressure at $x = 0$, $L/4$, and $L/2$ at a depth that is 95% of the shock formation distance. Also determine the particle velocity waveforms at those locations.

Significance

Thus far, we have only examined how the rays and waveforms are affected by self-refraction. Here, we will see its effects on waveforms. In doing so, we will see how to solve a coordinate transformation that features more than one position coordinate.

Solution

The displacement in Eq. (13.4.40) is $w = (\varepsilon c/\omega)\sin(\omega t)\sin(\pi x/L)$. A time shift of $\pi/(2\omega)$ will convert this expression to the given displacement. Accordingly, the derived formulas for the coordinate straining, as well as for the pressure and velocity components, are converted by setting $\varepsilon = Wk$ and replacing t with $t + \pi/(2\omega)$. It is convenient to work with a functional representation of the coordinate transformation. Upon introduction of the time shift, Eq. (13.4.63) becomes

$$x = f_x\left(\eta, \xi, t, z\right), \quad z = f_z\left(\eta, \xi, t, z\right) \tag{1}$$

where the functions are

$$f_x\left(\eta, \xi, z, t\right) = \eta - \frac{1}{2}\left(Wk\right)\beta z \frac{kL}{\pi}\left(\frac{k}{\kappa}\right)^2 \cos\left(\omega t - \kappa \xi\right) \cos\left(\frac{\pi\eta}{L}\right)$$

$$f_z\left(\eta, \xi, x, z, t\right) = \xi - \frac{1}{2}\left(Wk\right)\beta z \left(\frac{k}{\kappa}\right)^3 \sin\left(\omega t - \kappa \xi\right) \sin\left(\frac{\pi\eta}{L}\right) \tag{2}$$

Correspondingly, Eq. (13.4.64) becomes

$$\frac{v_x}{c} = -\left(Wk\right)\frac{\pi}{\kappa L}\cos\left(\omega t - \kappa \xi\right)\cos\left(\frac{\pi\eta}{L}\right) + O\left(\varepsilon^2\right)$$

$$\frac{v_z}{c} = \frac{\kappa}{k}\left(\frac{p}{\rho_0 c^2}\right) = -\left(Wk\right)\sin\left(\omega t - \kappa \xi\right)\sin\left(\frac{\pi\eta}{L}\right) + O\left(\varepsilon^2\right) \tag{3}$$

It is desired to determine waveforms at specified points. This requires that we solve the coordinate transformation, Eq. (13.4.63), for the values of η and ξ corresponding to specified values of x, z, and t. Repeating such a computation for a sequence of t values will yield one period of a waveform, which can be periodically replicated without requiring further solutions.

There are several numerical algorithms that may be used to solve two simultaneous nonlinear equations. We shall use the extended version of Newton's method. Suppose $\eta^{(n)}$ and $\xi^{(n)}$ are estimates for the values of η and ξ that satisfy Eq. (1) at specified x, z, and t. We seek to improve that estimate, such that $\eta^{(n+1)} = \eta^{(n)} + \Delta\eta$ and $\xi^{(n+1)} = \xi^{(n)} + \Delta\xi$ satisfy both transformations, that is,

$$f_x\left(\eta^{(n)} + \Delta\eta, \xi^{(n)} + \Delta\xi, t, z\right) = x, \quad f_z\left(\eta^{(n)} + \Delta\eta, \xi^{(n)} + \Delta\xi, t, z\right) = z$$

This constitutes two simultaneous nonlinear equations for $\Delta\eta$ and $\Delta\xi$. Let us expand each function in a first-order Taylor series relative to the values obtained with the prior estimates. Then the equations to solve are linear in the increments,

$$(f_x)^{(n)} + \left(\frac{\partial f_x}{\partial \eta}\right)^{(n)}\Delta\eta + \left(\frac{\partial f_x}{\partial \xi}\right)^{(n)}\Delta\xi = x$$

$$(f_z)^{(n)} + \left(\frac{\partial f_z}{\partial \eta}\right)^{(n)}\Delta\eta + \left(\frac{\partial f_z}{\partial \xi}\right)^{(n)}\Delta\xi = z$$

where the superscript for the functions and their derivatives means that the quantity is evaluated for the variables at the nth estimate. The matrix representation of these equations is

$$\left[J\left(f_x, f_z\right)^{(n)}\right]\begin{Bmatrix} \Delta\eta \\ \Delta\eta \end{Bmatrix} = \begin{Bmatrix} x - (f_x)^{(n)} \\ z - (f_z)^{(n)} \end{Bmatrix}$$

where $\left[J\left(F_x, F_z\right)^{(n)} \right]$ is the Jacobian of the transformation evaluated at the nth estimate,

$$\left[J\left(f_x, f_z\right)^{(n)} \right] = \begin{bmatrix} \left(\dfrac{\partial f_x}{\partial \eta}\right)^{(n)} & \left(\dfrac{\partial f_x}{\partial \xi}\right)^{(n)} \\ \left(\dfrac{\partial f_z}{\partial \eta}\right)^{(n)} & \left(\dfrac{\partial f_z}{\partial \xi}\right)^{(n)} \end{bmatrix}$$

Since $\Delta\eta = \eta^{(n+1)} - \eta^{(n)}$ and $\Delta\xi = \xi^{(n+1)} - \xi^{(n)}$, the improved estimates of the strained coordinates are given by

$$\left\{ \begin{matrix} \eta^{(n+1)} \\ \xi^{(n+1)} \end{matrix} \right\} = \left\{ \begin{matrix} \eta^{(n)} \\ \xi^{(n)} \end{matrix} \right\} + \left[J\left(f_x, f_z\right)^{(n)} \right]^{-1} \left\{ \begin{matrix} x - \left(f_x\right)^{(n)} \\ z - \left(f_z\right)^{(n)} \end{matrix} \right\}$$

The iterative procedure requires an initial estimate. All of the results that follow were obtained by setting $\eta^{(0)} = x$ and $\xi^{(0)} = z$. The procedure terminates when a convergence criterion has been met. The one used here is based on the observation that $|f_x - x|$ should be much less than εx and $|f_z - z|$ should be much less than εz. Both conditions will be met if the iteration is halted when $\left((f_x)^{(n)} - x\right)^2 + \left((f_z)^{(n)} - x\right)^2 < \varepsilon^4 \left(x^2 + z^2\right)$. The iterative algorithm is used to compute the waveforms of pressure and particle velocity at each of the specified locations. To do so the procedure is carried out in a loop of t values covering one period. After the values of η and ξ have been determined at each instant, the corresponding values of p, v_x, and v_z are determined from Eq. (13.4.64). All response variables have a period of $2\pi/\omega$, so the waveforms are replicated for another period in order to see the pattern more clearly.

The parameters for the evaluation are $c = 1480\,\text{m/s}$, $\beta = 3.5$ corresponding to $B/A = 5$, $\omega = 5000\pi\,\text{rad/s}$, $k = 10.613\,\text{m}^{-1}$, $\kappa = 10.405\,\text{m}^{-1}$, and $\varepsilon = Wk = 5.3067\left(10^{-3}\right)$. The wavelength in the z direction is $2\pi/\kappa = 0.6039\,\text{m}$. The shock formation distance is $x_{\text{shock}} = 9.705\,\text{m}$, so this is a fairly strong signal. The waveforms that are computed at $x = 0$, 0.375, and 0.75, with $z = 9.263\,\text{m}$ are shown in Fig. 1. They are best understood if they are compared to their linear analog. Linear theory recovered by setting $\varepsilon = 0$ in Eq. (2),

$$\begin{aligned} \frac{v_x}{c} &= -(Wk)\,\frac{\pi}{\kappa L}\,\cos\left(\omega t - \kappa z\right)\cos\left(\frac{\pi x}{L}\right) \\ \frac{v_z}{c} &= \frac{\kappa}{k}\left(\frac{p}{\rho_0 c^2}\right) = -(Wk)\sin\left(\omega t - \kappa z\right)\sin\left(\frac{\pi x}{L}\right) + O\left(\varepsilon^2\right) \end{aligned} \tag{5}$$

Thus, in linear theory p should be zero along $x = 0$ and v_x should exhibit its maximum excursion along that line, whereas the amplitude of p should be maximum and v_x should be zero along $x = L/2$. The situation at $x = L/4$ should be intermediate to these extremes.

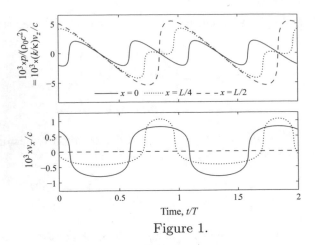

Figure 1.

Because $v_x = 0$ along $x = L/2$, the ray leaving the plate at that location does not rotate. The pressure signal evolves along a ray in a manner similar to planar wave. Hence, we see the same distortion effect along this line, with positive pressures arriving earlier, and negative pressures arriving later. Along $x = 0$, the pressure is nonzero because different rays pass through a point at $(0, z)$ at different instants as a result of the self refraction effect. Rays on which $v_x > 0$ arrive from $x < 0$ for half a period, and rays with $v_x < 0$ arrive from $x > 0$ for the other period. The variation of pressure along both sets of rays is similar, so the pressure undergoes a double oscillation in one period. Hence, the pressure at that location has half of the overall period.

The waveforms for v_x are interesting in that they are symmetric, rather than leaning over due to advancement and retardation. This symmetry occurs because v_x in a ray coming from the plate at $x > 0$ is merely 180° out-of-phase from v_x on a ray arriving from $x < 0$. If we were to examine v_x as a function of t at $x = L$, we would see the inverted version of v_x at $x = 0$ because all variables are shifted there by 180°.

13.5 Further Studies

Nonlinear phenomena have been the subject of a vast body of research. The topics covered here provide the primary tools and concepts required to go further. A historical perspective is available in a comprehensive treatise by Hamilton and Blackstock,[22] which is an excellent starting point for further exploration. The purpose of the discussion here is to make the reader aware of some interesting phenomena and applications without delving into the technical details.

---

[22]M.F. Hamilton and D.T. Blackstock, *ibid.*

• *Radiation pressure and acoustic levitation*

Our studies have concentrated on the growth of nonlinear effects as a wave prop-agates. The consequence was that a temporally periodic excitation with zero mean value was found to lead to a pressure field that has similar properties. This is not quite true. In actuality, there is a small mean value for the pressure at a point. This field may be traced back to the nonlinear terms that do not result in cumulative growth of distortion. For example, consider the nonlinear momentum equation, Eq. (13.3.3). It contains the term $\nabla \phi \cdot \nabla \phi$. If $\nabla \phi$ is harmonic, then $\nabla \phi \cdot \nabla \phi$ consists of a mean value and a second harmonic of equal magnitude.

The mean value of a pressure field is referred to as the *radiation pressure*. This leads to a *radiation force,* which is the mean value for the resultant force exerted on a body immersed in the pressure field. Although this concept might appear to be simple, the analysis of the concept raises some difficult issues. One concern is the obvious fact that placing a body in the field modifies the field due to scattering. That is, what is the radiation pressure distribution on the surface of a body that scatters the incident signal? Another question that arises pertains to the nature of this body. Is it fixed in space, or is it free to vibrate? Because the radiation pressure is small compared to the pressure amplitude, small differences are important, which is why the question of whether the body is restrained is an essential aspect.

One of the most interesting applications of radiation pressure is for *acoustic levitation.* As was noted, the radiation force is small. Using it to support a body at the earth's surface would require that the radiation force equals the body's weight, which would require an incredibly high acoustic pressure. However, for a body in orbit, for example, in the International Space Station, the weight is the centripetal force required to maintain the body in orbit. From the reference frame of the Space Station, the body is weightless and free to drift unless it is restrained. The radiation force has been used to restrain a body without contacting it. This is achieved by setting up a resonant field in a closed chamber. An unconstrained body then will drift to a pressure node. This concept is useful for any process in which contact with a container will lead to contamination or distortion of an object.

• *Sound beams*

Suppose you wish to send an acoustical message to listeners at a long range but in a limited direction. We select a loudspeaker or hydrophone for this purpose, embed it in a baffle, and set the frequency sufficiently high to obtain the desired directivity. To overcome amplitude reductions due to spherical spreading and dissipation, we drive the transducer harder. Several alternatives arise. As the input is increased, the amplifier that drives the transducer might begin to distort or cut out entirely. The speaker might be overdriven, which also might cause distortion or cause the speaker to fail. The first two possibilities can be avoided with proper selection of the apparatus, but the third possibility is not avoidable. Raising the amplitude of the signal radiated by the transducer leads to the growth of nonlinear effects as the signal propagates. If the drive amplitude is sufficiently large, shock waves might form. Raising the amplitude further will merely bring the location of shocks closer to the transducer, rather than further raising the amplitude that is received. The ultimate result might be saturation. Another possibility for an underwater signal is that the amplitude will be sufficiently large that the absolute pressure (ambient plus acoustic disturbance) is negative. The consequence of such an occurrence is creation of bubbles, which is called *cavitation*. This process also limits the power that can be transferred into the radiated signal. (Negative pressure is not an issue for atmospheric sound. To see why consider a plane wave for which the acoustic Mach number is $\varepsilon = |p| / (\rho_0 c^2)$. A negative absolute pressure corresponds to $|p| = 1$ atm, but $\rho_0 c^2$ is approximately 1 atm, and we know that $\varepsilon = 1$ is only attained with an explosion. In contrast $\rho_0 c^2 \approx 2$ GPa for water and the ambient pressure near the surface is approximately 1 atm. Setting $|p| = 1$ atm leads to $\varepsilon \approx 0.5 \left(10^{-4}\right)$, which is quite attainable.)

A primary application of sound beams is a method of long-range communication. They also are at the core of some imaging concepts, such as an acoustical microscope and some biomedical ultrasonic devices. In each case, it is imperative that nonlinear effects be understood. Our study of sound beams according to linear theory required numerical methods to evaluate the general field properties. Thus, it is obvious that studying nonlinear effects is even more challenging. Many analyses, particularly the early ones were based on a regular perturbation analysis, in which the linear solution for a sound beam is used to describe the second order, that is, quadratic, nonlinear terms. The Rayleigh integral is one way of describing the linear solution, but there are others. Furthermore, to reduce the level of difficulty various simplified field equations often provided the basis for the analysis. The simplifications were based on properties of sound beams at high frequency, such as that the spatial scale of diffraction is large compared to the axial wavelength, but small compared to the Rayleigh distance.

A regular perturbation analysis is limited in its range of validity. A more complete picture of nonlinear effects was obtained by using some of the methods developed here, notably the frequency domain Fourier series approach and the singular perturbation formulation.

- *Parametric arrays*

An application that relies on nonlinear effects is a concept that combats a basic aspect of the linear field of a sound beam. Specifically, if one wishes the beamwidth to be very small, then the frequency must be high, but high frequencies mean greater dissipation, thereby reducing the effective range. The concept entails imparting to the piston on the boundary a vibration that is the sum of two harmonic signals at closely spaced frequencies. The linear field, which is the first-order solution for a regular perturbation series, is the superposition of sound beams at each frequency. If both frequencies are high, both beams will be narrow, and their directivities will be quite similar because the frequencies are close. When the first-order solution is used to generate the quadratic terms in a nonlinear wave equation, approximate or exact, various combinations of harmonic signals are created. Second harmonics of each primary frequency are like the behavior of each beam if it were the only one present. Mixing of the two harmonics results in a sum frequency signal, which is close to the second harmonics of the individual beams, and a difference frequency signal. The latter part is the one that is of interest because proximity of the two drive frequencies means that the difference frequency is very small. Furthermore, because these terms are the product of two narrow directivities, the difference frequency terms are significant only near the axis of the sound beam. The homogeneous terms of the second-order wave equation are the same as those of the first-order equation, but the quadratic terms occur on the right side of the equation as virtual sources. These difference frequency sources are confined to the region close to the axis. In combination with the fact that the propagation is predominantly axial, the consequence is that the virtual sources act like a narrow end-fire array. The creation of this array is attributable to the parameters of the primary frequency sound beams and hence the name of the difference frequency sources as a *parametric array*. This concept has been used in some applications, notably as a narrowly confined megaphone. It has not been used widely because it relies on nonlinear interactions, which are relatively weak, with a magnitude that is the order of the square of the acoustic Mach number.

- *Biomedical applications*

The use of sound beams for ultrasonic diagnostics was mentioned earlier. Increasing the depth of penetration by raising the amplitude enhances nonlinear effects. One aspect of this is beneficial in that higher harmonics have smaller wavelengths. This serves to enhance the visual resolution. At the same time, larger amplitudes lead to greater tissue heating. This is a harmful effect as a diagnostic tool. However, it is useful as a method for destroying tissue that is harmful, such as cancer cells. One of the key requirements for understanding these phenomena is a description of how large amplitude sound waves propagate through living tissue, which is heterogeneous and dissipative through heat transport and viscosity. This is a subject of ongoing research.

Lithotripsy is a different way in that finite amplitude effects are exploited. Stones can form in several organs, such as kidneys. They can become extremely painful, in which case highly invasive surgery is the usual treatment. A lithotripter provides a nonsurgical alternative. The basic concept is to immerse the torso in a tub of water, with the walls of the tub driven as a pulse. This generates an inwardly propagating transient wave that converges on the stone. The consequence of convergence is that the pressure field applied to the stone is greatly enhanced and changes even more abruptly than the original pulse. (Recall the behavior in linear theory of an inwardly propagating spherical wave at the focus, see Sect. 6.2.3.) Hence, the pressure acting on the stone is large and the pressure gradient is very large. Furthermore, the stone typically is somewhat irregular in shape and has some internal faults. The combination of these attributes is that large stresses induced within the stone might cause it to fracture into smaller pieces that might cause it to pass through the duct. Transfer from a concept to a functioning apparatus required a considerable body of research concerning focusing of nonlinear waves and their interaction with solid objects.

13.6 Homework Exercises

Exercise 13.1 An explosion in air produces a pressure pulse in the form of a single lobe of a sine function. The amplitude is 20 kPa and the duration is 0.1 s. Draw the characteristics that track the peak pressure, and the zero pressure values at the beginning and end of the pulse. Identify the smallest distance at which any two of these lines intersect. Sketch the waveform predicted by the Riemann solution at this location. What does that sketch indicate regarding the location at which a shock forms?

Exercise 13.2 A transducer at the near end, $x = 0$, of a very long tube imparts a velocity wave form in the shape of a triangular pulse of duration T, given by

$$v = \begin{cases} \varepsilon c \dfrac{2t - T}{T}, & 0 < t < T \\ 0 \text{ otherwise} \end{cases}$$

Thus, $t = T$ is the last instant at which the transducer is moving. (a) Draw the spatial profile of this wave at $t = T$. (b) Evaluate the shock formation distance x_s for this wave. (c) Evaluate the waveform at $x = x_s$.

Exercise 13.3 At $t = 0$ the signal in a waveguide is observed to be a plane wave propagating in the direction of increasing x. Its spatial profile at that instant has the shape of an equilateral triangle,

$$\frac{p}{\rho_0 c^2} = \begin{cases} \varepsilon \dfrac{2x}{L}, & 0 < x < L/2 \\ \varepsilon \dfrac{2(L - x)}{L}, & L/2 < x < L \\ 0 \text{ otherwise} \end{cases}$$

Determine the shock formation distance x_{shock} according to weak shock theory. Then draw the waveform at $x = x_{shock}$ and $x = x_{shock}/2$.

Exercise 13.4 A projectile impacts the back of a wall, thereby inducing a displacement pulse $u = (4/3)\varepsilon c \left(t^2/T^2\right)(3T - 2t)$ that occurs over $0 \le t \le T$. The parameters are $\varepsilon = 0.0001$ and $T = 10$ ms, and the fluid is air at standard conditions. (a) Determine the minimum distance at which a shock forms according to the weakly nonlinear approximation. (b) Use the weakly nonlinear approximation of the Riemann solution to evaluate the waveform at $x = x_{shock}$. (c) Use Eq. (13.1.17), which accounts for movement of the wall, to evaluate the waveform at $x = x_{shock}$. Estimate the overall error in the result in Part (b) that results from using a stationary, rather than moving, boundary condition.

Exercise 13.5 The velocity of a vibrating wall consists of a pulse that occurs over an interval T. The waveform of that pulse is a single cycle of a sine function, $v = \varepsilon c \sin(2\pi t/T)$. The fluid is air at standard conditions. To test whether the limits of the weakly nonlinear approximation set $\varepsilon = 0.1$, $T = 4$ ms. (a) Use the weakly nonlinear approximation to evaluate the waveform at the shock formation distance. (b) Use the exact ideal gas relations to evaluate the waveform at the same location.

Exercise 13.6 If the particle velocity in a weakly nonlinear plane wave at $x = 0$ is $v = \varepsilon c \sin(\omega t)$, then the shock formation distance is $x_{shock} = 1/(\beta \varepsilon k)$. Consider the case of a 1 kHz signal in water with $\varepsilon = 0.0001$. (a) Use numerical methods to evaluate and graph the waveform at $x = x_{shock}$. (b) Evaluate and graph the waveform at $x = 1/(\beta \varepsilon k)$ when the input waveform (at $x = 0$) is the negative of the waveform at $x = x_{shock}$ found in Part (a). (c) Based on the properties of the waveforms at $x = 0$ and $x = 1/(\beta \varepsilon k)$, identify the minimum value of x at which the inverted signal will form a shock.

Exercise 13.7 An experiment in the Tunnel de Sainte-Marie-aux-Mines in eastern France measured the waveform of a pressure pulse at 3 Km from the western entrance. With $t = 0$ defined as the instant when the pulse arrived, a function fitting the measured data is $p = 42(t/T)^{1/2}(1 - t/T)[h(t) - h(t - 1.25T)]$ Pa, with $T = 12$ ms. Because $\partial p/\partial t$ is infinite at $t = 0$ and $t = 1.25T$, with p increasing at both instants, the location where this waveform was measured marks shock formation. Determine the waveform at the tunnel entrance, where the pulse was generated. It may be assumed that waves only propagate from the source to the measurement point, and that the propagation is planar with negligible dissipation.

Exercise 13.8 The pressure input at $x = 0$ consists of second and third harmonics, such that

$$p = \varepsilon \rho_0 c^2 [a_2 \sin(2\omega t) + a_3 \sin(3\omega t)]$$

When $a_3 = 0$, the shock forms at $x = 1/(2a_2 \varepsilon \beta k)$, whereas this location is $x = 1/(3a_3 \varepsilon \beta)$ when $a_2 = 0$, where $k = \omega/c$. Cases of interest are $a_2 = 1$, $a_3 = 0.1$, and $a_2 = 1$, $a_3 = 2/3$. For both cases $\varepsilon = 0.002$ and $\beta = 3.6$. Evaluate and graph $p/(\rho_0 c^2)$ as a function of $\omega t - kx$ at $kx = 0.25/(\beta \varepsilon a_2)$ for each case. Compare each waveform to its linear counterpart. Identify the instants at which $\partial p/\partial t$ is greatest in each case, and examine each waveform to determine whether a shock has formed.

Exercise 13.9 The pressure input at $x = 0$ consists of second and third harmonics, such that

$$p = \varepsilon \rho_0 c^2 [\sin(2\omega t) + (2/3)\sin(3\omega t)]$$

The period of this function is $\omega T = 2\pi$. The parameters are $\varepsilon = 0.002$ and $\beta = 3.5$. (a) Use numerical methods to analyze the waveform at a sequence of position $kx_n = 0.04n/(\beta \varepsilon)$, $n = 0, 1, ..., 6$. (b) Use FFT techniques to evaluate the amplitude of harmonics 1 to 5 at each position.

Exercise 13.10 Consider the situation in Exercise 13.9. Rather than using FFT analysis of computed waveforms, use Fourier series analysis of the Riemann solution to determine the amplitude of harmonics 1 to 5 at $kx_n = 0.04n/(\beta\varepsilon)$, $n = 0, 1, ..., 6$.

Exercise 13.11 The pressure input at $x = 0$ consists of second and third harmonics, such that

$$p = \varepsilon\rho_0 c^2 [\sin(2\omega t) + (2/3)\sin(3\omega t)]$$

The fluid is air with the attenuation described by Eq. (3.3.33). The parameters are $\omega = 800\,\text{rad/s}$, $\varepsilon = 0.002$, and $\beta = 1.2$. Solve the differential equations governing the waveform's Fourier series coefficients in order to evaluate the x dependence of the lowest four harmonic amplitudes. Halt the evaluation at $kx = 0.24/(\beta\varepsilon)$. Compare the waveform at the farthest location to the result that would be obtained if dissipation were negligible.

Exercise 13.12 The pressure generated by a periodic source at $x = 0$ is a combination of the ninth and tenth harmonics, specifically, $p = \rho_0 c^2 \varepsilon (\sin(9\omega t) + \sin(10\omega t))$. At the origin, this signal constitutes a beat, but an important aspect of the nonlinear process is harmonic and intermodulation distortion, in which higher and lower harmonics are generated. Derive an expression for the shock formation distance, x_{shock}. Then, solve the differential equations in Sect. 13.3.2 for $0 \le x \le 0.9x_{\text{shock}}$. Use Eq. (3.3.33) to describe the frequency dependence of the attenuation coefficient. The system parameters are $\omega/(2\pi) = 500\,\text{Hz}$ and $\varepsilon = 0.002$. The fluid is water, for which $\rho_0 = 1000\,\text{kg/m}^3$, $c = 1480\,\text{m/s}$, and $\beta = 3.6$. Plot the amplitudes of pressure harmonics 1 to 15 as a function of x. Use a logarithmic scale for the amplitude values.

Exercise 13.13 The spatial profile of a wave is described as p as a function of x at fixed t. Prove that the shock formation distance is manifested in a profile as an infinite value of $\partial p/\partial x$ with p decreasing.

Exercise 13.14 A piston at the end $x = 0$ of a semi-infinite waveguide vibrates periodically. Its steady-state velocity is the sawtooth waveform in the sketch. The waveform at any x preceding shock formation is a scalene triangle. (a) Use the graphical construction method to establish the shape of the triangle at an arbitrary x preceding shock formation. (b) Carry out a Fourier analysis of the waveform in Part (a), and thereby, derive an expression for the dependence of the harmonic amplitudes on x. (c) Prove that the shock formation distance for this signal is $x_{\text{shock}} = cT/(2\varepsilon\beta)$. (d) Evaluate the waveform and harmonic content at $x_n = nx_{\text{shock}}/4$, $n = 0, 1, ..., 4$. Which of these harmonics are enhanced and which are depleted as the signal propagates?

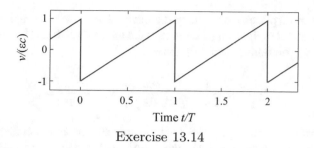

Exercise 13.14

Exercise 13.15 The periodic boundary velocity in Exercise 13.14 generates a steady-state plane wave in an infinitely long waveguide. (a) Determine the distance x_{old} at which the wave enters the old-age stage. (b) Determine the waveform at $x > x_{old}$. From that result, derive an expression for the maximum pressure as a function of x.

Exercise 13.16 A pressure surge occurs at $x = 0$ in a waveguide filled with water. The duration of the pulse is $0 \le t \le T$, and its signature in that interval is $p(x = 0, t) = 0.50 \left(10^6\right) (t/T) \exp\left(3\left(1 - t/T\right)\right)$ Pa where $T = 50$ ms. (a) Determine the shock formation distance x_{shock}. (b) Evaluate the Riemann solution at evenly spaced locations $x = x_{old}, 2x_{old}, 3x_{old}, \ldots$ Sequentially examine each waveform to estimate the smallest distance at which old age begins. (c) Evaluate the waveform at the location estimated in Part (b) to be the beginning of old age. Does the shock formation distance obtained from the analysis indicate that the estimated distance was too large or too small?

Exercise 13.17 When the spring in a one-degree-of-freedom oscillator is weakly nonlinear, the force it exerts deviates from proportionality to its elongation. A model that is descriptive of some systems adds a force term that is proportional to the square of the displacement. The resulting equation of motion is

$$M\ddot{y} + \mu\dot{y} + K_1 y + K_2 y^2 = f(t)$$

where $f(t)$ is the excitation force and μ is a viscosity coefficient that accounts for dissipation. In the case where $f(t)$ is a periodic function, the steady-state response must share that period. This means that y, as well as f, may be represented by Fourier series. Thus, let

$$y = \frac{1}{2}\sum_{n=-N}^{N} Y_n e^{in\omega t}, \quad Y_{(-n)} = Y_n^* \quad f = \frac{1}{2}\sum_{n=-N}^{N} F_n e^{in\omega t}, \quad F_{(-n)} = F_n^*$$

where the harmonic N for truncation is based on convergence of the series. Derive a set of algebraic equations whose solution would give the Y_n values when the F_n values are specified.

Exercise 13.18 If an elastic beam is fastened at its ends to support that are immobile, any displacement increases the length of the beam. This elongation is an accumulation of the effect of displacement w along the beams length, so it is represented by an integral. The effect of the elongation is to make the beam stiffer. The field equations and boundary conditions for the beam are

$$EI\frac{\partial^4 w}{\partial x^4} + \frac{1}{2}\frac{EA}{L^2}\int_0^L \left(\frac{\partial w}{\partial x}\right)^2 dx + \sigma\frac{\partial^2 w}{\partial t^2} = q$$

where E is Young's modulus, A and I are the area and the area moment of inertia of the cross section, σ is the mass per unit length, and q is the force loading per unit length of the beam. If the beam is supported at its ends by stationary pins that allow the beam to rotate freely, they are said to be *simple supports*. The associated boundary conditions are $w = \partial^2 w/\partial x^2 = 0$ at $x = 0$ and $x = L$. If the displacement is small relative to the span L, then it may be represented nondimensionally as $w = \varepsilon L \xi(x, t)$, where $|\varepsilon| \ll 1$. Write the dimensionless displacement as a regular perturbation series, $\xi = \xi_1 + \varepsilon\xi_2 + \cdots$. Derive the differential equation and boundary conditions governing ξ_1 and ξ_2.

Exercise 13.19 When the spring in a one-degree-of-freedom oscillator is weakly nonlinear, the force it exerts deviates from proportionality to its elongation. This effect often is modeled by adding a force term that is proportional to the cube of the displacement. The resulting equation of motion is

$$M\ddot{x} + K_1 x + K_3 x^3 = F(t)$$

where $F(t)$ is the excitation force. The nonlinear coefficient K_3 is much less than K_1, so it may be written as $K_3 = \varepsilon K_1$, $|\varepsilon| \ll 1$. A regular perturbation series for the oscillator displacement is $x = x_0 + \varepsilon x_1 + \cdots$. (a) Derive the differential equations governing x_1, x_2, and x_3. (b) When the force varies harmonically, $F(t) = \hat{F}\cos(\omega t)$, the steady-state response also must be periodic. Find the steady-state response of the differential equations obtained in Part (a). Is there any value of ω at which the solution fails to be uniformly valid?

Exercise 13.20 Free vibration of the oscillator in Exercise 13.19 corresponds to $f(t) = 0$. Consider an initial state in which $x = X_0$ and $\dot{x} = 0$ when $t = 0$. Division of the equation of motion by M leads to

$$\ddot{x} + \omega_{nat}^2 \left(x + \varepsilon x^3\right) = 0$$

where $\omega_{nat} = (K_1/M)^{1/2}$ is the natural frequency of the linear system. Use the regular perturbation series $x = x_0 + \varepsilon x_1 + \cdots$ to obtain a solution of the equation of motion that is consistent with the initial conditions. Then, apply the method of strained coordinates to the time variable in order to obtain a uniformly valid description of the response.

Exercise 13.21 The diagram is the pressure pulse measured on the surface of a sphere, $r = a$. This radius is much greater than the spatial extent of the pulse, that is, $a \gg cT$. (a) Use the graphical construction method to sketch the pressure waveform at an arbitrary radial distance. (b) Use the waveform obtained in Part (a) to derive an expression for the radial distance r_{shock} at which a shock forms. (c) Determine the waveforms of the radial particle velocity at $r = a$, $r = r_{\text{shock}}/2$, and $r = r_{\text{shock}}$.

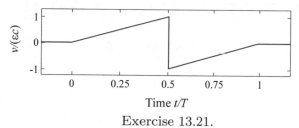

Exercise 13.21.

Exercise 13.22 The pressure resulting from a huge explosion in the atmosphere is measured at $r_1 = 2$ km from the ignition point. A function that correlates well with the measured data is $p = p_1 (t/T) (1 - 5t/T) \exp(-\mu t)$, with $p_1 = 20$ kPa and $\mu = 4.54$ s^{-1}. This function applies for $t < 2$ seconds, after which it may be set to zero. It is believed that the propagation is well described by the weakly nonlinear approximation beyond $r_0 = 5$ m. (Inside this distance, the signal is highly nonlinear and the water is in a nonideal state.) Plot p at r_0 as a function of the retarded time $t' = t - (r_0 - r_1)/c$. Without performing calculations, conjecture what the waveform would look like at a radial distance r that is significantly greater than r_1?

Exercise 13.23 A 10 m diameter sphere will be used as the living chamber of a deepwater submersible. One of its tests will apply an internal pressure fluctuation when it is submerged. This pressure induces a radially symmetric vibration of the outer surface in which the surface velocity is $v_r = 0$ for $t < 0$ and $t > T$, and $v_r = \varepsilon c(1 - 4/T)^2(5t/T - 1)$ for $0 < t < T$. Parameters are $\varepsilon = 0.0032$ and $T = 0.94$ ms (a) Determine the pressure waveform at the radial distance at which a shock develops. (b) Determine the radial velocity at $r = a$ and $r = r_{\text{shock}}$. Plot v_r/c at each location as a function of the retarded time $t - (r - a)/c$. Compare each waveform to $(r/a) p/(\rho_0 c^2)$ as a function of the retarded time. What can be deduced about the distance at which the farfield begins?

Exercise 13.24 If $-\dot{F}(\tau)$ in Eq. (13.4.29) is replaced by $c^2 V(\tau)$, the resulting expression for $(r/a) p$ in a spherical wave has the same appearance as $p = \rho_0 c^2 \varepsilon V(\tau)$ for a plane wave. Use this similarity to modify the general Fourier series analysis in Sect. 13.2.1 in order to determine the Fourier coefficients of a nonlinear spherical wave when the function $\dot{F}(t)$ is periodic. Specialize the general result to the case where the pressure at $x = 0$ is a single harmonic.

Exercise 13.25 At a reference radial distance r_0, the waveform of a radially symmetric spherical wave is observed to be the square wave depicted below. (a) Derive an expression for the radial distance r_{old} at which this wave enters old age. (b) Derive an expression for the maximum positive pressure in this wave if the radial distance exceeds r_{old}.

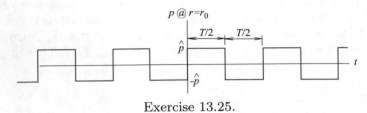

Exercise 13.25.

Correction to: Acoustics—A Textbook for Engineers and Physicists

Correction to:
J. H. Ginsberg, *Acoustics—A Textbook*
for Engineers and Physicists,
https://doi.org/10.1007/978-3-319-56847-8

The original version of the book was inadvertently published without including the Electronic Supplementary Material files in Chaps. 7–13.

The erratum book has been updated with the changes.

The updated version of the book can be found at
https://doi.org/10.1007/978-3-319-56847-8

Appendix A
Curvilinear Coordinates

A.1 Spherical Coordinates

When p is known in terms of the spherical coordinates of the location at which is observed, the most expedient evaluation of ∇p uses a form in terms of the radial, polar, and azimuthal components. This form is more complicated in appearance than merely the derivative of a component with respect to a coordinate because incrementing the angular variables alters the direction of the unit vectors. There are several approaches by which these effects may be described in a derivation of the gradient ∇p and the Laplacian $\nabla^2 p$ in spherical coordinates. We shall employ a direct approach based on the coordinate transformation to derive ∇p, and then switch to a slightly more subtle approach based on the properties of the spherical coordinate unit vectors to derive $\nabla^2 p$.

A.1.1 Transformations

The derivation of a gradient in spherical coordinates begins by considering the pressure to be a function of those coordinates. In turn, the spherical coordinates are treated as functions of the Cartesian coordinates. Doing so calls for the chain rule for partial differentiation. It is helpful to have the transformations at hand. The coordinates are defined in Fig. A.1.

The Cartesian coordinates corresponding to a given a set of spherical coordinates are

$$x = r \sin \psi \cos \theta$$
$$y = r \sin \psi \sin \theta \qquad (\text{A.1.1})$$
$$z = r \cos \psi$$

The inverse transformation gives the spherical values for a given set of Cartesian coordinates. It is

© Springer International Publishing AG 2018
J.H. Ginsberg, *Acoustics—A Textbook for Engineers and Physicists, Volume II*,
DOI 10.1007/978-3-319-56847-8

Fig. A.1 Spherical
coordinates and associated
unit vectors

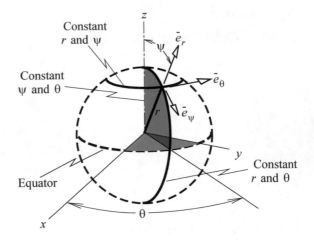

$$r = \left(x^2 + y^2 + z^2\right)^{1/2}$$

$$\psi = \cos^{-1}\left(\frac{z}{\left(x^2 + y^2 + z^2\right)^{1/2}}\right) \tag{A.1.2}$$

$$\theta = \tan^{-1}\left(\frac{y}{x}\right) = \sin^{-1}\left(\frac{y}{\left(x^2 + y^2\right)^{1/2}}\right) = \cos^{-1}\left(\frac{x}{\left(x^2 + y^2\right)^{1/2}}\right)$$

We also will have need for the definition of spherical unit vectors. Each vector is the
change of the position vector from the origin when the corresponding coordinate is
increased by a unit value, with the others held constant. The result of doing so is

$$\bar{e}_r = \sin\psi\cos\theta\bar{e}_x + \sin\psi\sin\theta\bar{e}_y + \cos\psi\bar{e}_z$$
$$\bar{e}_\psi = \cos\psi\cos\theta\bar{e}_x + \cos\psi\sin\theta\bar{e}_y - \sin\psi\bar{e}_z \tag{A.1.3}$$
$$\bar{e}_\theta = -\sin\theta\bar{e}_x + \cos\theta\bar{e}_y$$

A.1.2 Gradient

The Cartesian components of a gradient are the derivative with respect to the corresponding coordinate. The chain rule is needed to evaluate these derivatives when p
is known in terms of r, ψ, and θ, so we have

$$\frac{\partial p}{\partial x} = \frac{\partial p}{\partial r}\frac{\partial r}{\partial x} + \frac{\partial p}{\partial \psi}\frac{\partial \psi}{\partial x} + \frac{\partial p}{\partial \theta}\frac{\partial \theta}{\partial x}$$
$$\frac{\partial p}{\partial y} = \frac{\partial p}{\partial r}\frac{\partial r}{\partial y} + \frac{\partial p}{\partial \psi}\frac{\partial \psi}{\partial y} + \frac{\partial p}{\partial \theta}\frac{\partial \theta}{\partial y} \tag{A.1.4}$$
$$\frac{\partial p}{\partial z} = \frac{\partial p}{\partial r}\frac{\partial r}{\partial z} + \frac{\partial p}{\partial \psi}\frac{\partial \psi}{\partial z} + \frac{\partial p}{\partial \theta}\frac{\partial \theta}{\partial z}$$

The transformation from Cartesian to spherical coordinates is used to fill in the derivatives appearing in these equations. Implicit differentiation is useful for the angles. The derivatives of Eq. (A.1.2) with respect to x proceed as follows:

$$\frac{\partial r}{\partial x} = \frac{\partial}{\partial x}\left(x^2 + y^2 + z^2\right)^{1/2} = \frac{x}{\left(x^2 + y^2 + z^2\right)^{1/2}}$$

$$\frac{\partial}{\partial x}(\cos \psi) \equiv -\sin \psi \frac{\partial \psi}{\partial x} = \frac{\partial}{\partial x}\left(\frac{z}{r}\right) = \frac{\partial}{\partial x}\left(\frac{z}{\left(x^2 + y^2 + z^2\right)^{1/2}}\right)$$

$$= -\frac{xz}{\left(x^2 + y^2 + z^2\right)^{3/2}}$$

$$\frac{\partial}{\partial x}(\tan \theta) \equiv \frac{1}{(\cos \theta)^2}\frac{\partial \theta}{\partial x} = \frac{\partial}{\partial x}\left(\frac{y}{x}\right) = -\frac{y}{x^2} \tag{A.1.5}$$

Derivatives with respect to y and z are obtained by similar operations.

We seek a description of ∇p in which only variables associated with spherical coordinates appear. The transformations from Cartesian to spherical coordinates, along with $x^2 + y^2 + z^2 = r^2$, are used to eliminate x, y, and z from the preceding. Doing so leads to

$$\frac{\partial r}{\partial x} = \sin \psi \cos \theta, \quad \frac{\partial r}{\partial y} = \sin \psi \sin \theta, \quad \frac{\partial r}{\partial z} = \cos \psi$$

$$\frac{\partial \psi}{\partial x} = \frac{\cos \psi \cos \theta}{r}, \quad \frac{\partial \psi}{\partial y} = \frac{\cos \psi \sin \theta}{r}, \quad \frac{\partial \psi}{\partial z} = -\frac{\sin \psi}{r} \tag{A.1.6}$$

$$\frac{\partial \theta}{\partial x} = -\frac{\sin \theta}{r \sin \psi}, \quad \frac{\partial \theta}{\partial y} = \frac{\cos \theta}{r \sin \psi}, \quad \frac{\partial \theta}{\partial z} = 0$$

These expressions are substituted into Eq. (A.1.4), which then are used to form the gradient in Cartesian components. This step yields

$$\nabla p = \left(\frac{\partial p}{\partial r}\sin \psi \cos \theta + \frac{1}{r}\frac{\partial p}{\partial \psi}\cos \psi \cos \theta - \frac{1}{r}\frac{\partial p}{\partial \theta}\frac{\sin \theta}{\sin \psi}\right)\bar{e}_x$$

$$+ \left(\frac{\partial p}{\partial r}\sin \psi \sin \theta + \frac{1}{r}\frac{\partial p}{\partial \psi}\cos \psi \sin \theta + \frac{1}{r}\frac{\partial p}{\partial \theta}\frac{\cos \theta}{\sin \psi}\right)\bar{e}_y \tag{A.1.7}$$

$$+ \left(\frac{\partial p}{\partial r}\cos \psi - \frac{1}{r}\frac{\partial p}{\partial \psi}\sin \psi\right)\bar{e}_z$$

The last step is to convert this expression into spherical coordinate components. By definition, a component is the projection of a vector onto a coordinate axis, so we may write

$$\nabla p \equiv (\nabla p \cdot \bar{e}_r)\bar{e}_r + (\nabla p \cdot \bar{e}_\psi)\bar{e}_\psi + (\nabla p \cdot \bar{e}_\theta)\bar{e}_\theta \tag{A.1.8}$$

Substitution of the unit vectors in Eq. (A.1.3) and the gradient on Eq. (A.1.7) ultimately reduces to

$$\nabla p = \frac{\partial p}{\partial r}\bar{e}_r + \frac{1}{r}\frac{\partial p}{\partial \psi}\bar{e}_\psi + \frac{1}{r \sin \psi}\frac{\partial p}{\partial \theta}\bar{e}_\theta \qquad (A.1.9)$$

One device for remembering this expression is that dr is the displacement when r is incremented infinitesimally, $rd\psi$ is the meridional displacement for an infinitesimal increment of ψ, and $(r \sin \psi)\, d\theta$ is the azimuthal displacement when θ is incremented.

A.1.3 Laplacian

The final form of the gradient in spherical coordinates is understandable as changes in the field when the position is shifted infinitesimally in each coordinate direction. The Laplacian is less intuitive and somewhat more difficult to derive. One approach for deriving $\nabla^2 p$ parallels that used to describe ∇p. Specifically, it applies the chain rule to convert $(\partial/\partial x)(\partial p/\partial x)$, $(\partial/\partial y)(\partial p/\partial y)$, and $(\partial/\partial z)(\partial p/\partial z)$ to derivatives with respect to r, ψ, and θ. The derivation that follows proceeds differently. It uses Eq. (A.1.9) to describe both terms in $\nabla^2 p \equiv \nabla \cdot \nabla p$, so that

$$\nabla^2 p = \left(\frac{\partial}{\partial r}\bar{e}_r + \frac{1}{r}\frac{\partial}{\partial \psi}\bar{e}_\psi + \frac{1}{r \sin \psi}\frac{\partial}{\partial \theta}\bar{e}_\theta \right) \cdot \left(\frac{\partial p}{\partial r}\bar{e}_r \right.$$
$$\left. + \frac{1}{r}\frac{\partial p}{\partial \psi}\bar{e}_\psi + \frac{1}{r \sin \psi}\frac{\partial p}{\partial \theta}\bar{e}_\theta \right) \qquad (A.1.10)$$

Neither the components of ∇p nor the unit vectors are constant. For this reason, the derivatives must be evaluated prior to taking dot products. In other words,

$$\nabla^2 p = \bar{e}_r \cdot \frac{\partial}{\partial r}\left(\frac{\partial p}{\partial r}\bar{e}_r + \frac{1}{r}\frac{\partial p}{\partial \psi}\bar{e}_\psi + \frac{1}{r \sin \psi}\frac{\partial p}{\partial \theta}\bar{e}_\theta \right)$$
$$+ \frac{1}{r}\bar{e}_\psi \cdot \frac{\partial}{\partial \psi}\left(\frac{\partial p}{\partial r}\bar{e}_r + \frac{1}{r}\frac{\partial p}{\partial \psi}\bar{e}_\psi + \frac{1}{r \sin \psi}\frac{\partial p}{\partial \theta}\bar{e}_\theta \right) \qquad (A.1.11)$$
$$+ \frac{1}{r \sin \psi}\bar{e}_\theta \cdot \frac{\partial}{\partial \theta}\left(\frac{\partial p}{\partial r}\bar{e}_r + \frac{1}{r}\frac{\partial p}{\partial \psi}\bar{e}_\psi + \frac{1}{r \sin \psi}\frac{\partial p}{\partial \theta}\bar{e}_\theta \right)$$

Further progress entails evaluation of the derivatives of the unit vectors with respect to each spherical coordinate. These operations are applied to the unit vectors in Eq. (A.1.3). For example,

$$\frac{\partial \bar{e}_\psi}{\partial \psi} = -\sin \psi \cos \theta \bar{e}_x - \sin \psi \sin \theta \bar{e}_y - \cos \psi \bar{e}_z \qquad (A.1.12)$$

It will be easier to evaluate the dot products in Eq. (A.1.11) if the unit vector derivatives are expressed in terms of $\bar{e}_r$, $\bar{e}_\psi$, and $\bar{e}_\theta$. Toward that end we form the dot product of each unit vector with the derivative of each unit vector. For the preceding description of $\partial \bar{e}_\psi / \partial \psi$, these operations yield

$$\bar{e}_r \cdot \frac{\partial \bar{e}_\psi}{\partial \psi} = \left(\sin \psi \cos \theta \bar{e}_x + \sin \psi \sin \theta \bar{e}_y + \cos \psi \bar{e}_z \right) \cdot \frac{\partial \bar{e}_\theta}{\partial \psi} = -1$$

$$\bar{e}_\psi \cdot \frac{\partial \bar{e}_\psi}{\partial \psi} = \left(-\cos \psi \cos \theta \bar{e}_x - \cos \psi \sin \theta \bar{e}_y - \sin \psi \bar{e}_z \right) \cdot \frac{\partial \bar{e}_\psi}{\partial \psi} = 0 \quad \text{(A.1.13)}$$

$$\bar{e}_\theta \cdot \frac{\partial \bar{e}_\psi}{\partial \psi} = \left(-\sin \theta \bar{e}_x + \cos \theta \bar{e}_y \right) \cdot \frac{\partial \bar{e}_\psi}{\partial \psi} = 0$$

These components are used to construct the vector representation of $\partial \bar{e}_\theta / \partial \theta$ according to

$$\frac{\partial \bar{e}_\psi}{\partial \psi} = \left(\bar{e}_r \cdot \frac{\partial \bar{e}_\psi}{\partial \psi} \right) \bar{e}_r + \left(\bar{e}_\psi \cdot \frac{\partial \bar{e}_\psi}{\partial \psi} \right) \bar{e}_\psi + \left(\bar{e}_\theta \cdot \frac{\partial \bar{e}_\psi}{\partial \psi} \right) \bar{e}_\theta = -\bar{e}_r \quad \text{(A.1.14)}$$

The other derivatives are found by a similar process. The full set of results is

$$\frac{\partial \bar{e}_r}{\partial r} = \bar{0}, \quad \frac{\partial \bar{e}_\psi}{\partial r} = \bar{0}, \quad \frac{\partial \bar{e}_\theta}{\partial r} = \bar{0}$$

$$\frac{\partial \bar{e}_r}{\partial \psi} = \bar{e}_\psi, \quad \frac{\partial \bar{e}_\psi}{\partial \psi} = -\bar{e}_r, \quad \frac{\partial \bar{e}_\theta}{\partial \psi} = \bar{0} \quad \text{(A.1.15)}$$

$$\frac{\partial \bar{e}_r}{\partial \theta} = \sin \psi \, \bar{e}_\theta, \quad \frac{\partial \bar{e}_\psi}{\partial \theta} = \cos \psi \, \bar{e}_\theta, \quad \frac{\partial \bar{e}_\theta}{\partial \theta} = -\sin \psi \bar{e}_r - \cos \psi \bar{e}_\psi$$

Now that derivatives of the unit vectors have been characterized, we may proceed to carry out the derivatives in Eq. (A.1.11). Some terms are zero because the unit vectors form an orthonormal set, and others vanish because some partial derivatives of a unit vector are zero. What remains leads to

$$\boxed{\nabla^2 p = \frac{\partial^2 p}{\partial r^2} + \frac{2}{r} \frac{\partial p}{\partial r} + \frac{1}{r^2} \frac{\partial^2 p}{\partial \psi^2} + \frac{\cot \psi}{r^2} \frac{\partial p}{\partial \psi} + \frac{1}{r^2 (\sin \psi)^2} \frac{\partial^2 p}{\partial \theta^2}} \quad \text{(A.1.16)}$$

A.1.4 Velocity and Acceleration

To describe the velocity and acceleration of a particle, it is necessary to account for the time dependence of the spherical coordinates. The position of a particle depends explicitly on the value of r because $\bar{x} = r \bar{e}_r$. However, the position also depends on ψ and θ because the $\bar{e}_r$ depends on ψ and θ. Accordingly, differentiating position requires application of the chain rule,

$$\bar{v} \equiv \frac{d\bar{x}}{dt} = \dot{r} \bar{e}_r + r \frac{\partial \bar{e}_r}{\partial r} \dot{r} + r \frac{\partial \bar{e}_r}{\partial \psi} \dot{\psi} + r \frac{\partial \bar{e}_r}{\partial \theta} \dot{\theta} \quad \text{(A.1.17)}$$

Equation (A.1.15) describe the derivatives of the unit vectors, so the velocity is

$$\boxed{\bar{v} = \dot{r}\bar{e}_r + r\dot{\psi}\bar{e}_\psi + r\dot{\theta}\sin\psi\bar{e}_\theta} \tag{A.1.18}$$

An expression for the acceleration is derived similarly. However, the motion variable of interest in fluid motion is the particle velocity, so we consider the velocity components to be functions of time that are known. For this reason, the preceding is rewritten as

$$\bar{v} = v_r\bar{e} + v_\psi\bar{e}_\psi + v_\theta\bar{e}_\theta$$
$$\dot{r} = v_r, \quad \dot{\psi} = \frac{v_\psi}{r}, \quad \dot{\theta} = \frac{v_\theta}{r\sin\psi} \tag{A.1.19}$$

Differentiation of this description of velocity leads to

$$\bar{a} = \dot{v}_r\bar{e}_r + \dot{v}_\psi\bar{e}_\psi + \dot{v}_z\bar{e}_z + v_r\left(\frac{\partial\bar{e}_r}{\partial r}\dot{r} + \frac{\partial\bar{e}_r}{\partial\psi}\dot{\psi} + \frac{\partial\bar{e}_r}{\partial\theta}\dot{\theta}\right)$$
$$+ v_\psi\left(\frac{\partial\bar{e}_\psi}{\partial r}\dot{r} + \frac{\partial\bar{e}_\psi}{\partial\psi}\dot{\psi} + \frac{\partial\bar{e}_\psi}{\partial\theta}\dot{\theta}\right) + v_\theta\left(\frac{\partial\bar{e}_\theta}{\partial r}\dot{r} + \frac{\partial\bar{e}_\theta}{\partial\psi}\dot{\psi} + \frac{\partial\bar{e}_\theta}{\partial\theta}\dot{\theta}\right) \tag{A.1.20}$$

The rates of change of the spherical coordinates are replaced by the velocity components in Eq. (A.1.19), and Eq. (A.1.15) are used to describe how the unit vectors change. Collection of like components yields

$$\bar{a} = \dot{v}_r\bar{e}_r + \dot{v}_\psi\bar{e}_\psi + \dot{v}_\theta\bar{e}_\theta + v_r\left(\frac{v_\psi}{r}\bar{e}_\psi + \frac{v_\theta}{r}\bar{e}_\theta\right)$$
$$+ v_\psi\left(-\frac{v_\psi}{r}\bar{e}_r + \frac{v_\theta}{r\sin\psi}\cos\psi\,\bar{e}_\theta\right) + v_\theta\frac{v_\theta}{r\sin\psi}\left(-\sin\psi\bar{e}_r - \cos\psi\bar{e}_\psi\right) \tag{A.1.21}$$

$$\boxed{\bar{a} = \left(\dot{v}_r - \frac{v_\psi^2}{r} - \frac{v_\theta^2}{r}\right)\bar{e}_r + \left(\dot{v}_\psi - \frac{v_r^2}{r} - \frac{v_\theta^2}{r}\cot\psi\right)\bar{e}_\psi + \left(\dot{v}_\theta + \frac{v_r v_\theta}{r}\right)\bar{e}_\theta} \tag{A.1.22}$$

The primary use of Eq. (A.1.18) for acoustics is to formulate a velocity continuity condition at an interface, and to evaluate the intensity. There are few contexts in which the acceleration arises, so Eq. (A.1.22) seldom is needed.

A.2 Cylindrical Coordinates

Cylindrical coordinates are an extension of polar coordinates in which the distance from the plane, which is the axial distance z, provides the third coordinate. Thus, if a variable only occurs in the xy plane, or if a three-dimensional field variable is constant in the axial direction, setting all derivatives with respect to z will convert the formulas derived here to polar coordinates.

A.2.1   Transformations

The definition of cylindrical coordinates begins with the selection of a reference direction to be the axis of the coordinate system. In most cases, we are free to label directions as we wish, so we shall label this reference axis as z. The plane for the polar coordinates is defined such that z is perpendicular to it. The polar coordinates are R and θ, but here too, other labels might be used. In Fig. A.2, θ is defined relative to the x-axis in the sense of the right hand rule. (Point the thumb of the right hand parallel to the z-axis. The curled fingers of that hand define the sense in which θ increases.) This rule is somewhat arbitrary, for example, one might wish to define θ relative to the y-axis, but the definition in the figure will suffice for our purposes.

The coordinate transformations between (x, y, z) Cartesian coordinates and $(R.\theta, z)$ cylindrical coordinates may be derived by projecting a point onto the coordinate planes. Figure A.2 shows that the *axial distance* z locates a point above or below the xy plane; it is the same as the Cartesian coordinate. In a three-dimensional situation, we should think of R as the perpendicular distance from the point to the z-axis. Thus, we say that it is the *transverse distance*. In addition to being descriptive, using the adjective transverse aids us to avoid confusing it with the radial distance r in spherical coordinates, which is distance to a reference point. The angle θ measures how much the transverse line is rotated about the z-axis. It is the *circumferential angle* because it locates a point along the circumference of the circle whose radius is R.

Trigonometry shows that the forward transformation is

$$\begin{aligned} x &= R \cos \theta \\ y &= R \sin \theta \\ z &= z \end{aligned} \qquad (A.2.23)$$

We may determine the inverse transformation either by solving the above, or else through another trigonometric analysis. The result is

Fig. A.2 Definition of cylindrical coordinates when z defines the axial direction

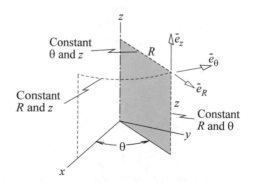

$$R = \left(x^2 + y^2\right)^{1/2}$$

$$\theta = \cos^{-1}\left(\frac{x}{\left(x^2 + y^2\right)^{1/2}}\right) = \sin^{-1}\left(\frac{y}{\left(x^2 + y^2\right)^{1/2}}\right) \qquad (A.2.24)$$

$$z = z$$

It should be noted that the circumferential angle has not been described as $\theta = \tan^{-1}(y/x)$ because doing so leads to ambiguity in the assignment of quadrant.

Every curvilinear coordinate system has an associated set of unit vectors. They describe the direction in which a point would displace if one coordinate were increased and the others held constant. The cylindrical coordinate directions are the transverse direction $\bar{e}_R$, the circumferential direction $\bar{e}_\theta$, and the axial direction $\bar{e}_z$. Increasing R displaces the point outward perpendicularly to the z-axis, and increasing z displaces the point parallel to the z-axis. When θ is increased with R and z held constant, the point moves along a circle of radius R parallel to the xy plane. Thus, $\bar{e}_R$, $\bar{e}_\theta$, and $\bar{e}_z$ form a mutually orthogonal set of unit vectors. A cylindrical coordinate description of a vector entails finding the projections of the vector along each of these unit vectors. In turn, these unit vectors may be defined as components relative to the Cartesian coordinate directions. Reference to Fig. A.2 shows that

$$\bar{e}_R = \bar{e}_x \cos\theta + \bar{e}_y \sin\theta$$
$$\bar{e}_\theta = -e_x \sin\theta + \bar{e}_y \cos\theta \qquad (A.2.25)$$
$$\bar{e}_z = \bar{e}_z$$

The orthogonal nature of the cylindrical coordinate directions may be verified by using the preceding to evaluate $\bar{e}_R \cdot \bar{e}_\theta$, $\bar{e}_R \cdot \bar{e}_z$, and $\bar{e}_\theta \cdot \bar{e}_z$.

A.2.2 Gradient

The derivation of ∇p in terms of cylindrical coordinates follows the general procedure for the spherical coordinate derivation. Because z is the same for cylindrical and Cartesian coordinates and z is independent of x and y, some terms in the chain rule for differentiation with respect to x, y, and z are identically zero. The nonzero terms that result when p is a function of R, θ, and z are

$$\frac{\partial p}{\partial x} = \frac{\partial p}{\partial R}\frac{\partial R}{\partial x} + \frac{\partial p}{\partial \theta}\frac{\partial \theta}{\partial x}$$
$$\frac{\partial p}{\partial y} = \frac{\partial p}{\partial R}\frac{\partial R}{\partial y} + \frac{\partial p}{\partial \theta}\frac{\partial \theta}{\partial y} \qquad (A.2.26)$$
$$\frac{\partial p}{\partial z} = \frac{\partial p}{\partial z}$$

We could evaluate the derivatives of the spherical coordinates by differentiating Eq. (A.2.24). However, an indirect approach that differentiates Eq. (A.2.23) is more efficient, because the resulting derivatives only depend on the spherical coordinate. The procedure forms

$$1 = \frac{\partial R}{\partial x}\cos\theta - R\frac{\partial\theta}{\partial x}\sin\theta, \quad 0 = \frac{\partial R}{\partial y}\cos\theta - R\frac{\partial\theta}{\partial y}\sin\theta$$
$$0 = \frac{\partial R}{\partial x}\sin\theta + R\frac{\partial\theta}{\partial x}\cos\theta, \quad 1 = \frac{\partial R}{\partial y}\sin\theta + R\frac{\partial\theta}{\partial y}\cos\theta$$

(A.2.27)

The matrix form of these equations is

$$\begin{bmatrix} \cos\theta & -R\sin\theta \\ \sin\theta & R\cos\theta \end{bmatrix} \begin{bmatrix} \dfrac{\partial R}{\partial x} & \dfrac{\partial R}{\partial y} \\ \dfrac{\partial\theta}{\partial x} & \dfrac{\partial\theta}{\partial y} \end{bmatrix} = \begin{bmatrix} 1 & 0 \\ 0 & 1 \end{bmatrix}$$

(A.2.28)

Inversion of the coefficient matrix gives

$$\begin{bmatrix} \dfrac{\partial R}{\partial x} & \dfrac{\partial R}{\partial y} \\ \dfrac{\partial\theta}{\partial x} & \dfrac{\partial\theta}{\partial y} \end{bmatrix} = \begin{bmatrix} \cos\theta & \sin\theta \\ -\dfrac{1}{R}\sin\theta & \dfrac{1}{R}\cos\theta \end{bmatrix}$$

(A.2.29)

Substitution of these derivatives into Eq. (A.2.26), followed by assembly of the individual terms to form ∇p leads to

$$\nabla p = \left(\frac{\partial p}{\partial R}\cos\theta - \frac{\partial p}{\partial\theta}\frac{1}{R}\sin\theta \right)\bar{e}_x$$
$$+ \left(\frac{\partial p}{\partial R}\sin\theta + \frac{\partial p}{\partial\theta}\frac{1}{R}\cos\theta \right)\bar{e}_y$$
$$+ \frac{\partial p}{\partial z}\bar{e}_z$$

(A.2.30)

When we collect the coefficients of each derivative of p, we obtain

$$\nabla p = \frac{\partial p}{\partial R}\left(\bar{e}_x\cos\theta + \bar{e}_y\sin\theta\right) + \frac{\partial p}{\partial\theta}\left(-\frac{\bar{e}_x}{R}\sin\theta + \frac{\bar{e}_y}{R}\cos\theta\right) + \frac{\partial p}{\partial z}\bar{e}_z \quad \text{(A.2.31)}$$

A comparison of these coefficients to Eq. (A.2.25) reveals that

$$\boxed{\nabla p = \frac{\partial p}{\partial R}\bar{e}_R + \frac{1}{R}\frac{\partial p}{\partial\theta}\bar{e}_\theta + \frac{\partial p}{\partial z}\bar{e}_z}$$

(A.2.32)

A.2.3 Laplacian

The procedure by which we will describe $\nabla^2 P \equiv \nabla \cdot \nabla P$ is to use Eq. (A.2.32). This will entail describing the derivatives of the cylindrical coordinate unit vectors. The axial direction $\bar{e}_z$ is constant, and Eq. (A.2.25) indicates that $\bar{e}_R$ and $\bar{e}_\theta$ depend solely on θ. Therefore, the derivatives are

$$\frac{\partial \bar{e}_R}{\partial R} = 0, \quad \frac{\partial \bar{e}_R}{\partial \theta} = -\bar{e}_x \sin \theta + \bar{e}_y \cos \theta = \bar{e}_\theta, \quad \frac{\partial \bar{e}_R}{\partial z} = 0$$

$$\frac{\partial \bar{e}_\theta}{\partial R} = 0, \quad \frac{\partial \bar{e}_\theta}{\partial \theta} = \bar{e}_x \cos \theta + \bar{e}_y \sin \theta = -\bar{e}_R, \quad \frac{\partial \bar{e}_\theta}{\partial z} = 0 \qquad (A.2.33)$$

$$\frac{\partial \bar{e}_z}{\partial R} = \frac{\partial \bar{e}_z}{\partial \theta} = \frac{\partial \bar{e}_z}{\partial z} = 0$$

The next step is to use Eq. (A.2.32) to describe both the divergence operator and ∇P. The operation requires that the derivatives be applied to each component of ∇p before the dot product is applied. Thus, the operations leading to the Laplacian are

$$\nabla^2 p \equiv \nabla \cdot \nabla p = \bar{e}_R \cdot \frac{\partial}{\partial R} \left(\frac{\partial p}{\partial R} \bar{e}_R + \frac{1}{R} \frac{\partial p}{\partial \theta} \bar{e}_\theta + \frac{\partial p}{\partial z} \bar{e}_z \right)$$

$$+ \frac{1}{R} \bar{e}_\theta \cdot \frac{\partial}{\partial \theta} \left(\frac{\partial p}{\partial R} \bar{e}_R + \frac{1}{R} \frac{\partial p}{\partial \theta} \bar{e}_\theta + \frac{\partial p}{\partial z} \bar{e}_z \right) \qquad (A.2.34)$$

$$+ \bar{e}_z \cdot \frac{\partial}{\partial z} \left(\frac{\partial p}{\partial R} \bar{e}_R + \frac{1}{R} \frac{\partial p}{\partial \theta} \bar{e}_\theta + \frac{\partial p}{\partial z} \bar{e}_z \right)$$

The unit vectors are mutually orthogonal, and they only depend on θ. Thus, the nonzero terms arising from the preceding are

$$\nabla^2 p = \frac{\partial}{\partial R} \left(\frac{\partial p}{\partial R} \right) \bar{e}_R \cdot \bar{e}_R + \frac{1}{R} \frac{\partial p}{\partial R} \bar{e}_\theta \cdot \frac{\partial \bar{e}_R}{\partial \theta} + \frac{1}{R^2} \frac{\partial}{\partial \theta} \left(\frac{\partial p}{\partial \theta} \right) \bar{e}_\theta \cdot \bar{e}_\theta + \frac{\partial^2 p}{\partial z^2} \bar{e}_z \cdot \bar{e}_z$$

$$(A.2.35)$$

The result of substituting Eq. (A.2.33) is

$$\boxed{\nabla^2 p = \frac{\partial^2 p}{\partial^2 R} + \frac{1}{R} \frac{\partial p}{\partial R} + \frac{1}{R^2} \frac{\partial^2 p}{\partial \theta^2} + \frac{\partial^2 p}{\partial z^2}} \qquad (A.2.36)$$

Some individuals prefer to write the Laplacian in a manner that groups the derivatives with respect to R,

$$\nabla^2 p = \frac{1}{R} \frac{\partial}{\partial R} \left(R \frac{\partial p}{\partial R} \right) + \frac{1}{R^2} \frac{\partial^2 p}{\partial \theta^2} + \frac{\partial^2 p}{\partial z^2} \qquad (A.2.37)$$

Other than brevity, there is little advantage to this form. For example, there is no power n for which $R^n f (t - R/c)$ satisfies the wave equation in a situation where p is independent of θ.

A.2.4 Velocity and Acceleration

The derivation of velocity and acceleration in terms of cylindrical coordinates follows the analysis for spherical coordinate. The first step is to construct the position by examining Fig. A.2. The field point is reached from the origin by traveling R units in the transverse direction and z units in the axial direction. The angle θ implicitly affects this construction because it defines the orientation of $\bar{e}_R$. Thus

$$\boxed{\bar{x} = R\bar{e}_R + z\bar{e}_z} \tag{A.2.38}$$

The axial unit vector points in a constant direction and $\bar{e}_R$ depends on θ, so differentiation of the position gives

$$\bar{v} = \dot{R}\bar{e}_R + \dot{z}\bar{e}_z + R\frac{\partial \bar{e}_R}{\partial \theta}\dot{\theta} \tag{A.2.39}$$

The derivative of $\bar{e}_R$ is described in Eq. (A.2.33). The result is that the velocity is

$$\boxed{\bar{v} = \dot{R}\bar{e}_R + R\dot{\theta}\bar{e}_\theta + \dot{z}\bar{e}_z} \tag{A.2.40}$$

In terms of the velocity components, the preceding is

$$\boxed{\begin{array}{l} \bar{v} = v_R\bar{e}_R + v_\theta\bar{e}_\theta + v_z\bar{e}_z \\ v_R = \dot{R}, \quad v_\theta = R\dot{\theta}, \quad v_z = \dot{z} \end{array}} \tag{A.2.41}$$

To describe the acceleration, only the velocity components are considered to depend explicitly on t, whereas the unit vectors depend on time implicitly through their dependence on θ. Hence, differentiation of the velocity expression leads to

$$\bar{a} = \dot{v}_R\bar{e}_R + \dot{v}_\theta\bar{e}_\theta + \dot{v}_z\bar{e}_z + v_R\frac{\partial \bar{e}_R}{\partial \theta}\dot{\theta} + v_\theta\frac{\partial \bar{e}_\theta}{\partial \theta}\dot{\theta} \tag{A.2.42}$$

Setting $\dot{\theta} = v_\theta/R$ and collecting like components lead to the acceleration being described by

$$\boxed{\bar{a} = \left(\dot{v}_R - \frac{v_\theta^2}{R}\right)\bar{e}_R + \left(\dot{v}_\theta + \frac{v_R v_\theta}{R}\right)\bar{e}_\theta + \dot{v}_z\bar{e}_z} \tag{A.2.43}$$

Appendix B
Fourier Transforms

A Fourier transform may be used to represent a function whose independent variable
extends from minus to plus infinity. A corollary is that a Fourier transform is suitable
for acoustic models of an infinite plane or an infinitely long cylinder. In both cases, the
Fourier transform may be used to describe the dependence of the field on coordinates
that measure position parallel to the boundary. In the time domain, the range for use
of a Fourier transform is $-\infty < t < \infty$. However, we usually take $t = 0$ to be the
initial time. We may bypass the range requirement in this case by defining the function
to be zero for $t < 0$, but doing so requires that we not consider initial conditions.
Moreover, response in the time domain is readily handled by standard methods, as
well the Laplace transforms, which is derived from the Fourier transform. For this
reason, the focus here is application of Fourier transforms to functions of position.

B.1 Derivation

We begin with a function $p(z)$ that is periodic in a distance ℓ. The Fourier transform
will emerge as $\ell \to \infty$. The complex amplitude of a plane wave propagating in the
direction of increasing z has been represented as e^{-ikz}, so we use the same convention
for the complex exponential of the Fourier series,

$$p(z) = \frac{1}{2} \sum_{m=-\infty}^{\infty} P_m e^{-im2\pi z/\ell} \tag{B.1.1}$$

Because we are dealing with a spatial function, we say that ℓ is the wavelength and
$2\pi/\ell$ is the fundamental wavenumber. The Fourier coefficients corresponding to a
specified function $p(z)$ are found by evaluating

$$P_m = \frac{2}{\ell} \int_{-\ell/2}^{\ell/2} p(z)\, e^{im2\pi z/\ell} dz \tag{B.1.2}$$

© Springer International Publishing AG 2018
J.H. Ginsberg, *Acoustics—A Textbook for Engineers and Physicists, Volume II*,
DOI 10.1007/978-3-319-56847-8

To obtain the limiting behavior as $\ell \to \infty$ we denote the wavenumber of the mth harmonic as $\mu_m = m2\pi/\ell$ and note that the separation between adjacent wavenumbers is $\Delta\mu = 2\pi/\ell$. In terms of these parameters, the coefficients are given by

$$P_m = \frac{\Delta\mu}{\pi} \int_{-\ell/2}^{\ell/2} p(z) e^{i\mu_m z} dz \tag{B.1.3}$$

For a given $p(z)$ and ℓ, the quantity obtained from the integral is solely a function of μ_m. Let us designate this integral as $\tilde{P}(\mu_m)$. Accordingly, we shall replace P_m with $(\Delta\mu/\pi) \tilde{P}(\mu_m)$. This changes the Fourier series to

$$p(z) = \frac{1}{2\pi} \sum_{m=-\infty}^{\infty} \tilde{P}(\mu_m) e^{-im2\pi z/\ell} \Delta\mu \tag{B.1.4}$$

Increasing ℓ decreases $\Delta\mu$, but $\Delta\mu$ is the wavenumber spacing between adjacent values of $\tilde{P}(\mu_m)$. Hence, a plot of $\tilde{P}(\mu_m)$ against its associated value of μ_m approaches a curve describing $\tilde{P}(\mu)$ as a function of μ. This function is the *Fourier transform* of $p(z)$. A variety of notations have been used to denote the operation of Fourier transforming a function. Our choice is $\mathcal{F}(p(z), \mu)$, which explicitly lists the function that is transformed and the transform variable. Hence, we find that

$$\boxed{\mathcal{F}(p(z), \mu) \equiv \tilde{P}(\mu) = \int_{-\infty}^{\infty} p(z) e^{i\mu z} dz} \tag{B.1.5}$$

To determine $p(z)$ when $\tilde{P}(\mu)$ is known we return to Eq. (B.1.4). Because $\Delta\mu$ becomes $d\mu$ as $\ell \to \infty$, the summation becomes an integral. The result is the *inverse Fourier transform*. This operation is designated as $\mathcal{F}^{-1}\left(\tilde{P}(\mu), z\right)$, so we have

$$\boxed{\mathcal{F}^{-1}\left(\tilde{P}(\mu), z\right) \equiv p(z) = \frac{1}{2\pi} \int_{-\infty}^{\infty} \tilde{P}(\mu) e^{-i\mu z} d\mu} \tag{B.1.6}$$

This is a heuristic derivation, but there is a formal mathematical foundation.[1] Not all functions have a Fourier transform. A sufficient condition for its evaluation is that the integral of $|p(z)|$ be finite. No periodic function, such as a sine or cosine, meets this requirement, but we will see in the next section that it is possible to use Fourier transforms to deal with such functions. The inverse Fourier transform requires values of $\tilde{P}(\mu)$ covering the full wavenumber spectrum $-\infty < \mu < \infty$. However, if the original function $p(z)$ is real, then $\tilde{P}(-\mu) = \tilde{P}(\mu)^*$, so only the positive spectrum needs to be evaluated.

---

[1] I.N. Sneddon, *Fourier Transforms*, Dover reprint (1951).
H.F. Davis, *Fourier Series and Orthogonal Functions*, Dover reprint (1963).

At the outset, we restricted our attention to the Fourier transformation of a spatial function based on the usage of $\exp(-i\mu z)$ for a wave propagating in the positive z direction. However, it is quite likely that you will eventually encounter a work in which the analysis is based on using $\exp(i\mu z)$. Furthermore, it might be appropriate to employ the Fourier transform to a describe function of time based, for which we have adopted $\exp(i\omega t)$ for harmonic functions. Both situations require a modification of the definition of a Fourier transform and its inverse. Let us denote this alternative definition as $\Phi(p(\xi), \mu)$ where ξ is the independent space or time variable and μ is the transform parameter. This alternative is

$$\Phi(p(\xi), \mu) \equiv \hat{P}(\mu) = \int_{-\infty}^{\infty} p(\xi) e^{-i\mu\xi} d\xi$$

$$\Phi^{-1}\left(\hat{P}(\mu), \xi\right) \equiv p(\xi) = \frac{1}{2\pi} \int_{-\infty}^{\infty} \hat{P}(\mu) e^{i\mu\xi} d\mu \tag{B.1.7}$$

When it is necessary to distinguish between the alternatives, we will refer to Eq. (B.1.5) as the *spatial Fourier transform,* and Eq. (B.1.7) as the *temporal Fourier transform.*

There is much similarity of this alternative definition to the one with which we began. Therefore, it should not be surprising that given one, the other is readily recovered. The following relations are readily verified

$$\Phi(p(\xi), -\mu) = \mathcal{F}(p(\xi), \mu)$$
$$\Phi(p(\xi), \mu)^* = \mathcal{F}\left(p(\xi)^*, \mu\right), \quad \mathcal{F}(p(\xi), \mu)^* = \Phi\left(p(\xi)^*, \mu\right) \tag{B.1.8}$$

Variants on the present Fourier transform definition multiply the integral by different coefficients. One that is often used leads to a transform and its inverse that have similar forms. This quality is obtained if the integral in Eq. (B.1.7) or (B.1.5) is multiplied by $1/(2\pi)^{1/2}$, which also becomes the factor multiplying the integral for the inverse transform. Another version, which is not often used, has a $1/(2\pi)$ factor for the transform and none for the inverse. Knowledge of the relationships between the various definitions is needed when we wish to use a work that has a definition that differs from our selection. Equation (B.1.8) can be quite useful in this regard because it allows us to employ any set of tables of functions and their Fourier transforms.

B.2 Evaluation Techniques

Our concern here is evaluation of the Fourier transform of a spatial function $p(z)$, as well as evaluation of the spatial function when we know a transform $\tilde{P}(\lambda)$. If $p(z)$ is not too complicated, we can evaluate the integrals directly, possibly with the aid of a table of integrals. However, there are alternative methods of evaluation that are likely to require less effort than that entailed for a formal integration. One is based on using tabulations of transform pairs rather than of integrals. The other makes use of FFTs.

B.2.1 Transform Pairs

If a function $p(z)$ is so familiar that its Fourier integral can be obtained with the aid of a table of integrals, it is likely that the result appears in a table of Fourier transform pairs. An entry in such a tabulation An entry in such a tabulation will list a function $p(z)$ and the adjacent entry will list the corresponding Fourier transform $\tilde{P}(\mu)$. Such a tabulation may be used in either direction: from the physical variable that is a function of z to the transformed function of μ, or vice versa. (Here we are using z as the independent variable and μ as the transform parameter. Obviously, other symbols may be used interchangeably.) The first consideration in using a table of Fourier transforms is to ascertain that it is based on a definition of a Fourier transform that is consistent with one's work. If not, then the tabulation will require conversion as described in the previous section.

Extensive tables of Fourier transforms are available in several books,[2] (1954), Ch. III. as well as some web sites. A condensed table, whose entries tend to be relevant to acoustic systems, are listed here. Table B.1 gives some basic properties that tell us how to adapt a known transform pair $p(z)$ and $\tilde{P}(\mu)$ to handle related functions. The last entry in Table B.1 is known as the convolution integral. It seldom is used to find a transform, but it is quite relevant for performing the inverse. One reason for this is that it fits a situation in which $\tilde{P}(\mu)$ is the transform of an excitation and $\tilde{Q}(\mu)$ is a transfer function. However, we will see in the next section that an FFT provides a much simpler alternative for evaluating a convolution integral.

Table B.2 lists some specific functions of z and their transform. Several entries contain the Dirac delta function $\delta(\mu)$. In each case, the associated function $p(z)$ does not meet the specification that the integral of $|p(z)|$ be finite, which is why the transform is singular. One way to verify these transforms is to evaluate the corresponding $p(z)$ function as an inverse transform. For example, to verify the entry for a complex exponential we observe that the integrand for the inverse transform is $2\pi\delta(\mu - \mu_0)e^{-i\mu z}$ over an infinite range of μ yields $2\pi e^{-i\mu z}$ evaluated at $\mu = \mu_0$, that is, $2\pi e^{-i\mu_0 z}$. The integral for the inverse transform is multiplied by $1/(2\pi)$, so we have shown that $\mathcal{F}^{-1}(2\pi\delta(\mu - \mu_0), z) = \exp(-i\mu_0 z)$.

The entries in Table B.2 may be extended by using the properties in Table B.1. For example, suppose we wish to transform $\sin(kz)$. The Euler identity and linearity leads to

$$\mathcal{F}(\sin(kz), \mu) \equiv \left(\frac{1}{2i}\right)\mathcal{F}(e^{ikz} - e^{-ikz}, \mu) = \frac{\pi}{i}[\delta(\mu + k) - \delta(\mu - k)] \quad \text{(B.2.9)}$$

---

[2]A. Erdelyi *Bateman Project Manuscript*, Vol. I - Table of Integral Transforms (1954) Ch. III.

Table B.1 Basic properties relating an arbitrary function and its Fourier transform

Name	Function	Fourier transform
Definition	$p(z) = \dfrac{1}{2\pi} \displaystyle\int_{-\infty}^{\infty} \tilde{P}(\mu) e^{-i\mu z} d\mu$	$\tilde{P}(\mu) = \displaystyle\int_{-\infty}^{\infty} p(z) e^{i\mu z} dz$
Linearity	$ap(z) + bq(z)$	$a\tilde{P}(z) + b\tilde{Q}(z)$
Coordinate scaling	$p(\alpha z)$	$\dfrac{1}{|\alpha|} \tilde{P}\left(\dfrac{\mu}{\alpha}\right)$
Coordinate shifting	$p(z - z_0)$	$\tilde{P}(\mu) e^{i\mu z_0}$
Modulation	$p(z) e^{-i\mu_0 z}$	$\tilde{P}(\mu - \mu_0)$
Complex conjugate	$p(z)^*$	$\tilde{P}(-\mu)^*$
nth derivative	$\dfrac{d^n}{dz^n} p(z)$	$(-i\mu)^n \tilde{P}(\mu)$
Definite integration	$\displaystyle\int_{-\infty}^{z} p(\eta) d\eta$	$-\dfrac{1}{i\mu} \tilde{P}(\mu) + \pi \tilde{P}(0) \delta(\mu)$
Differentiate the transform	$(iz)^n p(z)$	$\dfrac{d^n}{d\mu^n} \tilde{P}(\mu)$
Convolution	$\displaystyle\int_{-\infty}^{\infty} p(\xi) q(z - \xi) d\xi =$ $\displaystyle\int_{-\infty}^{\infty} p(\xi) q(z - \xi) d\xi$	$\tilde{P}(\mu) \tilde{Q}(\mu)$

A particularly useful function is sometimes referred to as the "box" function. It is defined as $\text{box}(z) = h(z - a) - h(z - b)$ with $b > a$. This function is one for $a < z < b$, zero outside that band, and one half at the discontinuities. We could find the corresponding transform by starting with the rect function, but application of the linearity and coordinate shifting properties to the definition is more direct. Thus,

$$\mathcal{F}(\text{box}(z), \mu) = \mathcal{F}(h(z - a), \mu) - \mathcal{F}(h(z - b), \mu) = e^{i\mu a} \mathcal{F}(h(z), \mu) - e^{i\mu b} \mathcal{F}(h(z), \mu)$$
$$= \left(e^{i\mu a} - e^{i\mu b}\right) \left[\pi \delta(\mu) - \frac{1}{i\mu}\right] = \frac{1}{i\mu}\left(e^{i\mu b} - e^{i\mu a}\right)$$

$$(B.2.10)$$

where the last form results from $\delta(\mu)$ being zero except at $\mu = 0$.

Table B.2 Common functions and their Fourier transform

Descriptive name	Function	Fourier transform
Constant	1	$2\pi\delta(\mu)$
Step	$h(z)$	$\pi\delta(\mu) - \dfrac{1}{i\mu}$
Impulse	$\delta(z - z_0)$	$e^{i\mu z_0}$
Complex exponential	$e^{-i\mu_0 z}$	$2\pi\delta(\mu - \mu_0)$
One-sided exponential decay	$e^{-\alpha z}h(z)$, $\mathrm{Re}(\alpha) > 0$	
Attenuated ramp	$ze^{-\alpha z}h(z)$	$\dfrac{1}{(\alpha - i\mu)^2}$
Gaussian	$e^{-(z/z_0)^2/2}$	$(2\pi)^{1/2} z_0 e^{-(\mu z_0)^2/2}$
Normalized sinc: $\mathrm{sinc}(z) \equiv \dfrac{\sin(\pi z)}{\pi z}$	$\mathrm{sinc}(\alpha z)$, $\alpha > 0$	$\dfrac{1}{\alpha}\mathrm{rect}\left(\dfrac{\mu}{2\pi\alpha}\right)$
Rectangle: $\mathrm{rect}(z) \equiv$ $h\left(z + \dfrac{1}{2}\right) - h\left(z - \dfrac{1}{2}\right)$	$\mathrm{rect}\left(\dfrac{z}{z_0}\right)$	$z_0\mathrm{sinc}\left(\dfrac{\mu z_0}{2\pi}\right)$
Symmetric exponential	$e^{-\alpha|t|}$, $\mathrm{Re}(\alpha) > 0$	$\dfrac{2\alpha}{\alpha^2 + \mu^2}$

B.2.2   Fast Fourier Transforms

For nonstandard functions analytical evaluation of a Fourier transform or inversion of a transform will require significant analytical effort, possibly involving contour integration in the complex space. If numerical results are sufficient, the Fast Fourier Transform (FFT) technology is ideal for that purpose. The presumption here is that the usage of Fast Fourier Transforms in conjunction with periodic time data, which was addressed in Sect. 1.4, is a familiar concept. The first objective here is to ascertain how an FFT computer routine may be used to evaluate the Fourier transform of an excitation whose spatial dependence is arbitrary. Then, we will examine how FFTs may be used in conjunction with transfer functions to evaluate an acoustical response.

Whether the Fourier transform of an arbitrary function $p(z)$ exists is a complicated issue. It is assured whether the function is zero outside some region centered around the origin. This is the situation for our studies, because the objects we are interested in have finite extent. More generally, the transform will be finite if the function is absolutely integrable, that is,

$$\int_{-\infty}^{\infty} |f(x)| \, dx < \infty \tag{B.2.11}$$

If we avoid processing periodic functions, this property will be satisfied for finite acoustical systems.

Usage of an FFT to treat nonperiodic data requires that we define a window whose width L serves as the effective period. There are two considerations to selecting this window. The selection begins by identifying the interval $-\ell/2 < x < \ell/2$ outside of which the data is negligible. For example, if the data is a Gaussian function, $p(z) = a \exp\left(-\beta x^2\right)$, then the criterion $|p(z)| < \max(|p|)/100$ if $|x| > \ell/2$ would be met by setting $\ell/2 \geq (\ln(100)/\beta)^{1/2}$. Data that suddenly drops off requires special care in this regard because the interval should be sufficient to capture the nature of such data. For example, if $p(z)$ is a rectangle function, then ℓ should be substantially larger than the width of the rectangle.

The value of ℓ that is identified should not be the period L. This prohibition stems from our ultimate objective of using transfer functions to determine the response to an excitation that has been processed with an FFT. This operation introduces *wraparound error* that narrows the interval in which the data is valid. It is possible to avoid contaminating the response with wraparound error. To do so, the length of the window must be at least twice ℓ. The data is set to zero in the extended range. This is a process of *zero padding*. In other words, if the data is such that $|p(z)| < \varepsilon \max(|p|)/100$ if $|x| > \ell/2$, then the data window for the FFT would be $-L/2 \leq x \leq L/2$, with $L \geq 2\ell$. In the intervals $-L/2 \leq x < -\ell/2$ and $\ell/2 < x \leq L/2$, we would set $p(z) = 0$. If we follow this protocol, then FFT results will be valid for the original data interval, $-\ell/2 < x < \ell/2$.

Let us first consider the formal definition of a DFT. After we have done so, we will examine how the definition is adjusted to a form that is suitable for an FFT algorithm. The period is L, and the data set consists of N samples. Therefore, the discrete wavenumbers for the Fourier series are

$$\mu_n = n\left(\frac{2\pi}{L}\right), \quad n = -N/2 + 1, \dots, N/2 \tag{B.2.12}$$

Suppose we knew the function $p(z)$ underlying the sampled data. The Fourier transform of that function would integrate over $-L/2 < z < L/2$, because we are considering $p(z)$ to be zero outside that interval. Thus, the Fourier transform at the series wavenumbers is

$$\mathcal{F}(p(z), \mu_n) \equiv \tilde{P}(\mu_n) = \int_{-L/2}^{L/2} p(z) e^{i\mu_n z} dz \equiv \int_{-L/2}^{L/2} p(z) e^{i(2n\pi/L)z} dz \tag{B.2.13}$$

Rather than attempting to carry out this integration analytically, let us consider a numerical approximation using a strip rule whose width is $\Delta z = L/N$. There are N strips and the sample locations are $z_j = j\Delta z, j = -N/2 + 1, ..., N/2$. The resulting approximation is

$$\mathcal{F}(p(z), \mu_n) \approx \frac{L}{N} \sum_{j=-N/2+1}^{N/2} p_j e^{i2\pi jn/N}, \quad n = -\frac{N}{2} + 1, ..., \frac{N}{2} \qquad \text{(B.2.14)}$$

The significant aspect is that although the definition of a DFT is based on this approximation, the operation itself exists independently of the concept of a Fourier transform. Rather, it is a formal prescription of a way in which data may be processed. This distinction is indicated by using a slightly different notation. A "caret" will denote DFT data, whereas a "tilda" has been used for a Fourier transform. The definition of the DFT is

$$\hat{P}_n \equiv \sum_{j=-N/2+1}^{N/2} p_j e^{i2\pi jn/N}, \quad n = -\frac{N}{2} + 1, ..., \frac{N}{2} \qquad \text{(B.2.15)}$$

The inverse spatial DFT evaluates the $p(z_j)$ data given a set of DFT values. It is

$$p_j = \frac{1}{N} \sum_{n=-N/2+1}^{N/2} \hat{P}_n e^{-i2\pi nj/N}, \quad j = -\frac{N}{2} + 1, ..., \frac{N}{2} \qquad \text{(B.2.16)}$$

One manifestation of the fact that a DFT is an independent operation, separate from a Fourier transform, is that the inverse operation is exact. That is, if we process a set of $p(z_j)$ data to obtain a DFT set $\hat{P}_n$, then take the inverse DFT of this data, the result will exactly be the $p(z_j)$ values (aside from numerical errors). No approximation is involved.

In order to implement an FFT algorithm and its inverse, the spatial data set and the DFT values are manipulated slightly. One alteration is that the spatial data is considered to be situated in the window $0 \leq z \leq L$, which constitutes a forward shift by half the fundamental wavelength. Shifting the spatial data leads to an algorithm that evaluates

$$P'_n = \sum_{j=0}^{N-1} p_j e^{i2\pi jn/N}, \quad n = 0, 1, ..., N - 1 \qquad \text{(B.2.17)}$$

The notation P'_n is intended to distinguish the FFT output from the DFT values, which are denoted as $\hat{P}_n$. The shift of the data window is unimportant for our current purpose. In the event that it is necessary to compare the FFT output to the DFT values obtained from the formal definition in Eq. (B.2.15), the coordinate shifting property in Table B.1 for a shift of $L/2$ multiplies the transform by $\exp(i\mu_n L/2) = (-1)^n$.

The other adjustment arises because the range of n appearing in the FFT definition above differs from the range of n in Eq. (B.2.15). The FFT exploits the periodicity properties of a DFT to place the data for negative wavenumbers after the data for zero and positive wavenumbers. This is the complex conjugate mirror image property first described in Eq. (1.4.40). The rearrangement is described by

$$
\begin{aligned}
P_n' &= \hat{P}_n, \quad n = 0, ..., N/2 \\
P_n' &= \hat{P}_{n-N}, \quad n = N/2 + 1, ..., N - 1
\end{aligned}
\tag{B.2.18}
$$

The rearrangement of wavenumbers follows the same rule

$$
\begin{aligned}
\mu_n' &= \mu_n = \frac{2\pi n}{L}, \quad n = 0, ..., N/2 \\
\mu_n' &= \mu_{n-N} = \frac{2\pi(n-N)}{L}, \quad n = N/2 + 1, ..., N - 1
\end{aligned}
\tag{B.2.19}
$$

This shift in the wavenumber spectrum is a crucial issue for the objective of using transfer functions to find a response.

Any routine that performs a Fast Fourier Transform will be accompanied by one that performs the inverse transform. FFT definition in Eq. (B.2.15) is

$$
P_n = \frac{1}{N} \sum_{j=0}^{N-1} \hat{P}_j e^{-i2\pi nj/N}, \quad n = 0, 1, ... N - 1
\tag{B.2.20}
$$

We will refer to a computer routine that implements the definition in Eq. (B.2.15) as a spatial FFT, whereas the FFT definition in Sect. 1.4.4 will be said to be a a temporal FFT. (The only difference is the sign of the argument of the complex exponential.) If the available software package only has a routine that implements a temporal FFT, it will be necessary to adjust its output. One way to do so is to use the first of Eq. (B.1.8), which calls for swapping the negative and positive index of the DFT data. This can be somewhat problematic because of the arrangement of data for FFT algorithms, as described by Eq. (B.2.18). An alternative is to follow the last of Eq. (B.1.8) which calls for using the available FFT routine to transform the complex conjugate of the spatial data, then taking the complex conjugate of that output.

The preceding addresses the fundamentals of FFT processing. Let us now turn our attention to the reason that an FFT might be useful for acoustical system analysis. Suppose we have used an FFT algorithm to transform the spatial dependence of an excitation. Examples of this would be transformation of an arbitrary vibrational pattern along one direction of an infinite plane, or transformation of the axial pattern of vibration on the surface of a circular cylinder. The transfer function in the first case would be the acoustic wave radiated into the fluid by a sinusoidal surface wave, which was determined in Chap. 5. In the case of a cylinder, an analysis in Chap. 7 found the wave radiated into the fluid by a helical wave surface wave at a fixed circumferential harmonic number. For both systems, the transfer function is the amplitude of a wave that is radiated due to an excitation that propagates along the surface in a certain direction, which we will designate as x. The wavenumber in that direction is designated as μ. Let us denote the amplitude of the wave that propagates in the x direction as $G(\mu)$. To be a transfer function, the excitation causing it must be a wave in the same direction with unit amplitude. Our task is to determine the signal when the excitation is an arbitrary function.

Let us analyze radiation from a plane to illustrate the process. Suppose the surface vibration is harmonic at frequency ω. The distribution in the x direction is arbitrary, such that the surface velocity normal to the plane is $\mathrm{Re}\,(v(x)\exp(i\omega t))$, where $v(x)$ may be complex. It is necessary that there is a DFT window L that accommodates the significant portion of $v(x)$. The transfer function is the y dependence of the wave radiated from the surface when $v(x) = \exp(-i\mu x)$. This quantity is described by Eq. (5.1.16), with the velocity amplitude $i\omega W$ set to one. Thus, the transfer function is such that

$$
P = G(\mu)\,e^{-i\mu x}, \quad G(\mu) = \rho_0 c \frac{k}{\kappa} e^{-i\kappa y}
$$

$$
\kappa = \begin{cases} \left(k^2 - \mu^2\right)^{1/2} & \text{if } k > \mu \\ -i\left(\mu^2 - k^2\right)^{1/2} & \text{if } k < \mu \end{cases}
\tag{B.2.21}
$$

If the amplitude of the surface vibration is $V(\mu) \neq 1$, then the wave radiated from the surface would be $V(\mu)\,G(\mu)\exp(-i\mu x)$. Addition of the contribution of the full spectrum of $V(\mu)$ leads to

$$
p(x, y, t) = \mathrm{Re}\left(P(x, y)\,e^{i\omega t}\right)
\tag{B.2.22}
$$

where the complex pressure amplitude is given by

$$
P(x, y) = \frac{1}{2\pi}\int_{-\infty}^{\infty} V(\mu)\,G(\mu)\,e^{-i\mu x}d\mu = \int_{-\infty}^{\infty}\left(\rho_0 c \frac{k(\mu)}{\kappa(\mu)} V(\mu)\,e^{-i\kappa(\mu)y}\right)e^{-i\mu x}d\mu
\tag{B.2.23}
$$

This expression is an inverse Fourier transform. It is readily adapted to fit the definition of an inverse FFT in Eq. (B.2.20). First we evaluate the FFT of the excitation distribution $V(\mu)$, which produces velocity coefficients V'_n. The FFT of the pressure is analogous to the integrand, specifically,

$$
P'_n = \rho_0 c \frac{k}{\kappa_n} V'_n e^{-i\kappa_n y}, \quad \kappa_n = \kappa(\mu_n)
\tag{B.2.24}
$$

The last step is to take the inverse FFT of the P'_n data. The output of the inverse FFT will be values of $P(x, y)$ at fixed y at the points x_n used to sample $v(x)$. The only precaution for implementing this procedure is associated with the manner in which the DFT data is arranged in an FFT algorithm. Specifically, the first $N + 1$ values are the associated with $\lambda_n \geq 0$. These values are followed by the data for $\lambda_n < 0$. The wavenumbers corresponding to the data that is input to the inverse FFT are

$$
\mu_n = \begin{cases} \dfrac{2n\pi}{L}, & n = 0, 1, ..., \dfrac{N}{2} \\ \dfrac{2(n - N)\pi}{L}, & n = \dfrac{N}{2} + 1, ..., N - 1 \end{cases} \tag{B.2.25}
$$

If the wavenumbers are arranged in this sequence as a vector array $\{\mu\}$ in a computer program, all operations may carried out vectorially, for example, kappa=conj(sqrt(k^2-mu.^2)) in MATLAB.

Although this FFT procedure is fairly straightforward, there is a profound difficulty that can inhibit its application for acoustics. The system we have set up consists of an interval along the boundary within which there is a vibration. This is the two-dimensional analog of a piston in an infinite baffle. In both cases, increasing distance from the boundary leads to ever increasing spreading of the acoustic field parallel to the surface. The inverse FFT that generates this distribution must have a window width that accommodates the spreading effect. In other words, the window length must be selected on the basis of the extent of the signal at the selected distance y to the boundary, rather than the nature of the surface vibration. This is where the difficulty arises, because the extent of the field is not known until the system has been analyzed.

Index

A

Absorbing boundary conditon, 123
Absorption coefficient, 607
Absorptive liner, 224
Acceleration
    in cylindrical coordinates, 675
    in spherical coordinates, 670
Acoustical imaging, 656
Acoustically compact, 99
Acoustic cavity, 85
Acoustic levitation, 655
Acoustic Mach number, 546, 553, 611
Acoustic perturbation, 409
Acoustic-structure interaction, 238
Added mass, 103, 501, *see also* Virtual mass
Adiabatic assumption, 541
Algorithm
    for a modal series, 252, 264, 282, 308
    for a nonlinear wave, 559, 563, 633
Ambient conditions, 409, 412
Amplitude
    dispersion, 545, 645
    slowly varying, 407, 428, 443
Angular spectrum, 176
Annular waveguide, 275
Antinode, 154
Anti-piston transducer, 156
Area function
    for a multivalued waveform, 585
    for the Webster horn equation, 191
Arrête, 417, 434
Asymptotic approximation
    of a spherical harmonic series, 21
    of Bessel functions, 58, 270, *see also*
        Farfield approximation
Atmosphere, propagation in, 406
Attenuation, 230, 521, 596, 607

Axial wavenumber, 57, 217, 267
    complex, 229
Axisymmetric body, scattering by, 497
Axisymmetric field, 19, 61, 71
Azimuthal angle, 94, 140, 353
Azimuthal harmonic, 16, 268, 351

B

Backscatter, 486, 502, 509, 511, 514
Backscatter cross section, 510, 512, 514, 517
Backward propagating wave, 199, 292, 298, 548
B/A coefficient, 551
Baffle, 29, 133, 160
Beam aspect, 97
Beamwidth, 148, 162
Beating signal
    for group velocity, 202
    from a supersonic source, 242
    nonlinear, 607
Bending of rays, 420
Bessel function
    cylindrical, 12, 57, 145
    derivative of, 14, 270
    for modes, 268, 271
    integral for, 570
    modified, 64, 273
    of negative order, 62
    spherical, 11, 15, 57, 312, 319
Bessel's equation, 57
    modified, 64, 273
Biomedical applications, 657
Blocked area, 491, 513, 520
Blocked pressure, 522, 527
Blue sky, 491
Body force, 409

Born approximation, 485, 490
Boundary condition
    for a cavity, 243, 300, 313
    for a perturbation series, 614, 639
    for a waveguide, 188, 196, 214, 218
    for farfield radiation, 125
    mixed, 244
    nonlinear, 638
Boundary elements, 111, 363
Box function, 681
Branch cut, 218, 226, 246
Breathing mode, 20
Bulk modulus, 40, 485, 491, 551
Burton-Miller formulation, 115

C
Calculus of variations, 365, 463
Cauchy principal part, 109, 113
Causality, principle of, 451
Caustic, 417, 432
    effect on amplitude, 450
    example, 454
    formed by nonlinear rays, 651
Cavitation, 577, 656
Cavity, 91, 93, 113, 291
Cavity mode, 293, 295, 297, 336, 365
Center of curvature, of a ray path, 423
Center wave, 153, 179
Chaladni lines, 346
Characteristic curve, 542, 544
Characteristic equation, 219, 225
    for a cylinder, 325
    for a spherical shell, 41
    for a waveguide, 219, 224, 238, 259
CHIEF computer code, 114
Circular disk, scattered field, 508, 513
Circular ray path, 424, 443
Circumferential angle, 671
Circumferential harmonic, 69, 77, 271, 323
Coefficient of nonlinearity, 552
Coincidence frequency, 236
Compact set of sources, 18
Complex eigenvalues, 225
Complex wavenumber, 226
Compliant boundary, 273, 298, 326, 379, 386
Concave reflector, 512
Concentric cylinders, 352
Concentric spheres, 319, 360
Conical horn, 188, 191
Connectivity matrix, 117

Conservation
    of energy, 572, 579
    of mass, 189, 541, 579
Constitutive nonlinearity, see Equation of state
Continuity condition
    at a cavity wall, 361, 380
    at a vibrating plate, 638
    on the surface of an elastic shell, 39
Continuity equation, 409, 540, 598
Control volume, 190, 578
Convective nonlinearity, 544
Convergence
    of a Legendre series, 8, 26
    of a modal series, 42, 47, 250, 262
    of a spherical harmonic series, 22, 526
    of Rayleigh-Ritz method, 374
Coordinate straining, 621, 629, 642
Corner, for Kirchhoff–Helmholtz integral theorem, 110
Coupled equations
    for a spherical shell, 40
    for cavity-structure interaction, 385
    for waveguide-structure interaction, 238
Creeping waves, 520
Critical incidence, 471
Cross-sectional area
    of a horn, 189
    of a ray tube, 445
Cumulative growth, 628
Curvature of rays, 423
Cutoff, 259
    for an exponential horn, 194, 204
    for radiation from a cylinder, 68
    of waveguide modes, 218, 260
Cylindrical Bessel function, see Bessel function
Cylindrical coordinates, 56, 144, 267
    definition of, 671
    unit vectors, 672, 674
Cylindrical spreading, 63

D
D'Alembert solution, 544, 548, 614
Decay constant, 220, 246
Decomposition of a propagating mode, 222
Degree of a Legendre function, 4, 16
Derivative of a unit vector
    for cylindrical coordinates, 675
    for spherical coordinates, 668
Differential scattering cross-section, 509, 517

Diffraction, 152, 473, 520
Dipole, 495
  for scattering, 491
  two-dimensional, 78
Dipole moment, 78, 100, 101
Dirac delta function, 295
Directivity
  for a transducer in a baffle, 137, 144, 166
  for radiation, 53, 97, 147
  for scattering, 484, 517
  from a spherical harmonic series, 21
  of a horn, 187, 196
  of a ring, 169
Disk
  radiation from, 103
  scattering from, 497, 501
Dispersion, 195, 202
Displacement
  of a plate, 637
  of a spherical shell, 37, 522
  structural, 383
Dissipation, 229, 567, 607
Divergence theorem, 86, 338, 340
Dome tweeter, 29
Dowell's approximation, 380

E
Earnshaw solution, 540, 546
Edge wave, 154, 179
Eigenray, 458
Eigenvalue
  complex, 225
  for a rigid-walled cylinder, 218, 271
  for a waveguide, 218, 225
  purely imaginary, 228, 274
Eigenvalue problem, 337
Eikonal equation, 439, 440
Elastic plate, 230, 637
Elastic spherical shell, 36, 523
Elastic structure, 380
Elliptic cavity, 374
End-fire array, 657
Energy absorption, 300
Energy and work, 579
Energy functionals, 366
Energy, in the old-age stage, 594
Entropy, 581
Equal area rule, 582, 592
Equal eigenvalues, 261
Equal natural frequencies, 339, 345, 354
Equation of state
  for a heterogeneous fluid, 410

  for a liquid, 550
  for an ideal gas, 549
  relation to sound speed, 593
Equations of motion
  for a spherical shell, 36
  for a structure, 383
Error residual, 106
Euler's equation, 190, 381, 541, 578
Euler-Lagrange equation, 468
Evanescent mode, 222, 248, 260, 262
Evanescent wave, 195, 202
Exponential horn, 194
  group velocity for, 204
Extensional elastic wave, 525

F
Farfield
  for a transducer, 127, 133, 144, 149, 154, 163
  for radiation from a cylinder, 69
  for radiation from a spherical shell, 41, 52
  of a nonlinear spherical wave, 632
  of a spherical harmonic series, 21
  of Kirchhoff–Helmholtz integral theorem, 94
  of the Rayleigh integral, 144
Fast Fourier Transform, 682
  for radiation, 84, 139
Fermat's principle, 463
FFT, see Fast Fourier Transform
Finite amplitude wave, see Nonlinear wave
Finite elements, 115, 363
Fitting shocks, 582, 587
Flexible wall, 42, 224, 236, 379
Fluid impedance, 44, 525
Fluid-loaded resonance, 51, 526
Fluid-loading effect of, 43
Fluid-structure interaction, 36, 243, 379
Focus, 431, 435
Focusing factor, 431
Force
  acting on a sphere, 38
  exerted on a piston, 166
  for low frequency radiation, 102
Forward and backward waves
  in a cavity, 292, 298
  in a horn, 194, 199
Forward characteristic, 544
Forward scattering, 509, 520
Fourier coefficient, 677
Fourier series, 677

for a cylindrical waveguide, 267
for a nonlinear wave, 569, 603
for radiation from a cylinder, 56, 69
Fourier transform, 82, 138, 177, 677
    alternative definitions, 679
    conditions for, 678
    of a surface velocity, 82, 139, 176
    table of, 679, 680
Fubini-Ghiron, 571

G

Galerkin method, 116, 118
Gamma function, 169
General eigenvalue problem, 373
Generalized coordinates, 383
Generalized force, 383
Generalized Fourier series, 6
Geometric boundary condition, 366
Geometrical acoustics, 405, 511
Gibb's phenomenon, 174
Gradient
    for scattering, 492
    in cylindrical coordinates, 672
    in spherical coordinates, 665
Graphical construction
    of a characteristic equation, 228
    of a spatial profile, 553
    of a waveform, 554
    of wavefronts and rays, 408
Grazing angle, 418
Green's function, 85, 481
    for a cavity, 295, 296
    for a rigid plane, 135
    free space, 87
    two-dimensional, 72, 87
Green's theorem, 118, 381
Group velocity, 202, 222, 233

H

Half-power point, 148
Hamilton's principle, 383
Hankel function
    cylindrical, 58, 268
    spherical, 11, 19
Hankel transform, 146
Harmonic amplitudes, nonlinear wave, 567,
        568, 603
Heavy fluid loading, 394
Helical surface wave, 61, 66, 69
Helmholtz equation, 10, 85, 217
    for scattering, 480, 492
    in cylindrical coordinates, 56, 267

in spherical coordinates, 2
inhomogeneous, 135
Hemisphere, in a baffle, 29, 165, 198
Hermitian, 106
Heterogeneity, 409
    arbitrary dependence, 438
    cylindrical region, 451
    scattering from, 484, 491
    vertically stratified, 418
High frequency approximation
    for a spherical wave, 24
    for cylindrical waves, 71
    for geometrical acoustics, 407
Horizontal tangency of a ray path, 421
Horn, 187
    conical, 188, 191
    exponential, 193
    WKB method, 207
Hyperbolic function, 226

I

Ideal gas law, 549
Images, method of, 30, 135
Impedance boundary condition
    in a cavity, 330
    in a waveguide, 254, 259, 273
    orthogonality condition with, 254
Impedance tube, 326
Imposed boundary condition, 366
Impulsive source, 347
Incident wave, 485
Index of refraction, 418, 465
Inertia matrix, 120, 383
Infinite elements, 123
Inhomogeneous Helmholtz equation, 294
Inhomogeneous wave equation, 340, 381
Initial conditions, for a nonlinear wave, 548
Integral equation, 111
Intensity,
    for a spherical harmonic series, 22
    for radiation of a hemisphere, 31
    from a baffled piston, 148, 166
    from a spherical harmonic series, 22
    from a vibrating cylinder, 70
    in a ray tube, 429, 437, 445
    in a scattered wave, 512, 514
    in a waveguide, 196, 247, 256
    in the old-age stage, 595
    radiated by a source, 429
    using a farfield approximation, 171
Interaction of harmonics, 175, 606
Interference

in a sound beam, 154, 161, 173
in a waveguide, 260, 266
in scattering, 521
Interpolating function, 111, 116
Inverse Fourier transform, 82, 678

J
Jacobian, 448, 646
Jump conditions, 580

K
KHIT, *see* Kirchhoff–Helmholtz integral theorem
Kinematical nonlinearity, 599
Kinetic energy, 364
Kirchhoff approximation, 510
Kirchhoff–Helmholtz integral theorem
  evaluation at a surface point, 90
  farfield approximation, 93, 121
  for a cavity, 85
  for a cylinder, 94
  for a half-space, 134
  for a spherical cavity, 88
  for radiation, 91
  for scattering, 481
  multipole expansion of, 99

L
Lagrange's equations, 383
Laplace equation, 492
Laplacian
  in cylindrical coordinates, 674
  in spherical coordinates, 668
Legendre function, 4, 5, 8, 16, 22
Legendre polynomial, 4
Legendre series, 6, 8
Legendre's equation, 3
Leibnitz' rule, 212, 580, 583, 585
Light fluid loading, 394
Lighthill's method, *see* Strained coordinates
Line source, 72
  finite length, 73
Linear least squares, 107, 114
Lithotripter, 657
Locally planar approximation, 408
Locally reacting
  cylindrical wall, 330
  effect on cavity resonance, 305
  end, 323
  for a waveguide, 253
  for geometrical acoustics, 511

Long range communication, 656
Loudspeaker, 133, 144, 332

M
Mass acceleration, 72, 74, 86, 105
Material derivative, 410
Material nonlinearity, 544
Matrix equation
  for a cavity, 309, 350
  for a waveguide, 250, 264, 266
  for finite elements, 120
  for radiation, 106, 113, 120
  for ray tracing, 440
  for Rayleigh-Ritz method, 373
Membrane shell model, 36
Modal amplitudes, *see* Modal series
Modal density, 346
Modal series
  for a cavity, 293, 298, 300, 313, 324, 331, 340
  for a circular waveguide, 243, 246, 255, 279
Modes, 336
  of a cylindrical cavity, 352
  of a cylindrical waveguide, 272
  of a rectangular cavity, 342
  of a spherical cavity, 358
  of a spherical shell, 41
  of a spherical waveguide, 272
  two-dimensional, 217, 224, 235
Modified Bessel function, 64, 273
Momentum equation
  for a heterogeneous fluid, 410
  for a horn, 190
  for a shock, 578
  in terms of a velocity potential, 597
  nonlinear wave, 544, 598
Monopole, *see* Green's function, Point source
Monopole amplitude, 73, 78
  for scattering from a rigid body, 450, 501
Monostatic scattering measurement, 505
Morse–Penrose inverse, 107
Moving boundary condition, 612, 638
  importance of, 547
Mulitvalued waveform, 565, 577
Multipath propagation, 458
Multiple scales, method of, 623
Multipole expansion, 99, 130, 496
Mutual scattering, 480

N

Natural frequency
  of a cylindrical cavity, 330, 353
  of a rectangular cavity, 291, 301, 342
  of a spherical cavity, 311, 358
  of a spherical shell, 41, 523
Neumann function
  cylindrical, 58, 268, 353
  spherical, 12, 15, 319
Newton's method, 563
Nondispersive wave, 202
Nonlinear propagation
  in two-dimensions, 637
  of a plane wave, 540
  of a spherical wave, 625
Nonlinear wave equation, 598, 604, 611, 623
Nonlinearity, effects of, 568, 574, 577
Nonplanar transducer in a baffle, 163
Nonuniform plane wave, 221
Nonuniform validity, 621, 641
Normal direction
  to a surface, 86, 91, 413
  to a vibrating plate, 638
  to a wavefront, 408
Normalization
  of cavity modes, 339
  of transverse mode functions, 245, 255,
    260, 278
Normal velocity, 133, 638
Numerical method
  for a horn, 213
  for a nonlinear wave, 559, 563, 649, 652
  for characteristic equation, 227, 231
  for nearfield of a sound beam, 158
  for ray tracing equations, 440, 452
  for ray tube area, 448
  for transport equation, 452
Nyquist criterion, 174
  for a nonlinear waveform, 567

O

Oblique plane waves, 123, 222
Ocean, propagation in, 405
Old-age stage, 594
Omnidirectional source, 430
Order
  of a Bessel function, 58
  of an associated Legendre function, 16
  of Neumann and Hankel functions, 11,
    59
Order of magnitude, definition, 612
Orthogonality

  of axial mode functions, 330
  of cavity modes, 338, 365
  of harmonic functions, 604
  of Legendre functions, 31
  of transverse mode functions, 245, 255,
    260, 278
Output from a horn, 188

P

Parametric arrays, 657
Parseval's theorem, 574
Particle velocity
  according to geometrical acoustics, 445
  for a rigid body, 101
  for a set of sources, 106
  in a cylindrical wave, 62
  in a nonlinear wave, 544, 548, 551
  in a ray tube, 429
Perfectly transmitting boundary conditon,
    123
Perturbation series, 612, 626, 639
Phase delay, due to a caustic, 450
Phase speed
  for a plate, 38
  for a waveguide, 223, 228, 243
  in a horn, 195, 202, 204, 207, 208
  in ray tracing equations, 438
  nonlinear, 543, 647
Phase variable, 209, 603, 644
Phase-reversal, 556
Piecewise linear waveform, 553, 574
Piston
  at the end of a cylinder, 279
  in the time domain, 176
  nearfield, 158
  on-axis pressure, 153, 159
  radiated power, 171
  radiation impedance, 166
Plane wave
  for group velocity, 202
  in a horn, 193
  in a waveguide, 215, 221, 268
  nonlinear, 541
  refraction by a cylindrical heterogeneity,
    451
Plane wave approximation, 62, 123, 149
Plane wave mode, 220, 225, 261
Point source, set of, 104
Poisson solution, 540
Position
  in cylindrical coordinates, 671
  in spherical coordinates, 665

Potential energy, 36, 120, 364
Potential function, *see* Velocity potential
Power flow, 256
    in a ray tube, 430, 446
    in a waveguide, 248, 257
Power input, 171, 197, 248, 303, 430
Power, Radiated
    at an interface, 256
    by a hemisphere, 32
    for a spherical harmonic series, 24
    from a baffled piston, 166, 171
    from a vibrating cylinder, 72
    in a scattered wave, 506, 512
    using a farfield approximation, 172
Pressure
    amplitude along a ray, 446
    from the ray tracing equations, 459, 497
    in a cylindrical wave, 62
    in a spherical cavity, 15
    in terms of strained coordinates, 621,
        630, 643
    in terms of velocity potential, 598, 599
    relation to particle velocity, *see* Euler's
        equation
    scattered, *see* Scattered pressure
Pressure loading, of a sphere, 38, 497, 524
Pressure-release
    baffle, 29
    end correction, 133
    wall, 226, 353
Principal curvature, 511
Product, of Fourier series, 604
Pulse, 596

R
Radial symmetry, 23, 625
Radiation
    from a piston, 157
    from a spherical shell, 522
    from a vibrating plate, 218, 637
    from a vibrating sphere, 15, 19
    from an infinite cylinder, 55
Radiation damping, 45, 122, 525
Radiation impedance
    of a piston, 166
    of an infinite cylinder, 63
    specific, 170
Radiation pressure, 655
Range determination, 506
Rankine–Hugoniot conditions, 580
Ray
    definition, 408

    for a nonlinear wave, 644, 647
    graphical construction of, 408
    in vertically stratified fluid, 417
    reflected, 223, 413, 421
    transmitted, 414
Ray tracing equations, 466
    derivation, 438
    numerical solution of, 440, 452
    relation to Fermat's principle, 467
Ray tube
    definition, 428, 445
    Jacobian for, 448, 459
Rayleigh distance, 146, 155, 162
Rayleigh integral
    application for nonplanar surface, 146,
        163
    derivation, 134
    farfield approximation of, 138, 144
    for a nonplanar transducer, 163
    for a surface point, 167
    for an axial field point, 153
    in Cartesian coordinates, 137
    in cylindrical coordinates, 144
    numerical approximation of, 158
    time domain version, 177
Rayleigh ratio, 364, 370
Rayleigh-Ritz method, 370
Rayleigh scattering limit, 490, 508, 517, 519
Reactive boundary, 227, 247
Reciprocity, 167, 388
Rectangular cavity, 291, 301, 342
Rectangular waveguide, 257, 298, 625
Recurrence relation
    for a gamma function, 169
    for Bessel functions, 270
    for Legendre polynomials, 6, 24
    for modified Bessel functions, 65
    for second order velocity potential, 629
    for spherical Bessel functions, 14, 65
Reflecting telescope, 415
Reflection
    effect on shadow zones, 435
    from a curved surface, 413, 511
    from a free surface, 421, 459
    in a hard-walled waveguide, 223
    of waveguide modes, 298
Reflection coefficient
    from a curved surface, 414, 450, 511
    modal, 300
Refraction
    according to geometrical acoustics, 413
Renormalization, *see* Strained coordinate
Resonance, 296, 300, 325

effect of fluid-loading, 44, 522, 526
in a cavity, 296, 300, 322, 325, 393
of a mechanical oscillator, 44, 526
of a spherical shell, 43, 46
Retarded time, 569
Riemann solution, 541, 623
compared to nonlinear spherical wave,
632
energy conservation, 572
evaluation of, 553, 559, 564
fitting shocks to, 588
for initial conditions, 548
interpretation, 543
Rigid body, scattering from, 494, 499, 507
Rigid surface, 337, 353
Rigid walled waveguide, 218, 271
Rigid-walled cavity, 336, 353
Ring, vibrating, 145
Rodrigue's formula, 6

S
Salinity, 412
Saturation, 595
Sawtooth waveform, 588, 592, 595
Scattering
Born approximation, 485
by a disk, 497, 501, 502
by a movable body, 499
by a sphere, 497, 515
due to strong heterogeneity, 497
from a rigid body, 494
governing equations, 492
Rayleigh limit, 490
Scattering cross section, 507, 512, 517
Scattering directivity function, 519
Scattering volume, 491
Second orthogonality condition, 339
Selection of trial functions, 367
Self-refraction, 645
Separation of variables, 2, 9, 16, 53, 56, 217
Separation theorem, 374
Series
for structural displacement, 383
of forced cavity modes, 297, 312, 323
of natural cavity modes, 342
of rigid cavity modes, 381
of spherical harmonics, 17, 19
of transverse mode functions, 245, 298
of trial functions, 370
Shadow zone, 433, 435, 521
Shedding, of creeping waves, 521
Shell theory, 37

SHIE, 111, see Surface Helmholtz integral
equation
Shock
condition for, 575, 646
for a sinusoidal input, 577
formation, 539, 555, 562, 564, 574
in a spherical wave, 632
in a two-dimensional wave, 651
location, 550, 577, 582, 631, 646
old age, 594
Shock formation, 545
Short wavelength behavior, 22
Side lobe, 98, 142, 147, 519
Sinc function, 141
Singular perturbation, see Strained coordinate
Slowly varying function, 603
Slowness, see Wave slowness
Small signal approximations
for linearization, 409
for nonlinear propagation, 554
for shock formation, 575
Snell's law
at an interface, 414
for vertical heterogeneity, 418, 424
from Fermat's principle, 463
Solid angle, 110, 508
Sommerfeld radiation condition
for a spherical wave, 15
for radiation, 92
for scattering, 480, 517, 522
for the Kirchhoff–Helmholtz integral, 91
from a plate, 638
in a waveguide, 218, 254
Sonar, 505
Sound beam, 142
axial pressure, 156
farfield pressure, 142, 146
intensity, 148, 171, 175
nearfield, 160
nonlinear, 655
qualitative description, 162
Sound speed in a heterogenous fluid, 412,
418, 439, 492
Source
in a cavity, 291, 381, see also Point
source
Source distribution, 105, 341
for surface vibration, 178
Source strength, see Monopole amplitude
Source superposition method, 104
Spatial Fourier transform, 679
Speaker cone, 148

Specular reflection, 521
Sphere
    radially vibrating, 19, 103, 625
    scattered field, 492, 496, 513, 516, 522
Spherical aberration, 417
Spherical cap, 188
Spherical cavity, 311, 358
Spherical coordinates, 137, 665
    for an elastic shell, 37, 515
    for scattering, 483
Spherical Hankel function, 11, 19, 319, 516
Spherical harmonics, 36, 493
Spherical spreading, 10, 62, 92
Square wave
    distortion of, 592
    radiation from a piston, 172, 180
Standing wave, 223, 268, 296
State vector, 386
Stationarity
    of a variational path, 463
    of the Rayleigh ratio, 371
Stiffness matrix, 120, 383
Strained coordinates, 621, 642
Stringed musical instrument, 79
Structural acoustics, 1, 379
Struve functions, 169
Sturm–Liouville analysis, 244
Subsonic surface wave, 63
Supercritical incidence, 414
Superposition
    of plane waves, 222, 260
    of rays, 459
    of standing waves, 293
Supersonic surface wave, 63, 242
Surface Helmholtz integral equation, 108,
    111
    derivation, 108
    discretization of, 113
Symmetry
    in a rectangular waveguide, 236, 266, 306
    of a ray path, 422
    of cavity-structure equations, 387
    radial, 22, 625

T
Table
    of Fourier transform pairs, 681, 682
Tangential stiffness, 543
Target identification, 505
Target strength, 510
Taylor series, 170, 410, 465, 543, 551, 611
Temporal Fourier transform, 679

Tensor, 102
Throat impedance, 197
Time delay, 177, 443
Time domain Rayleigh integral, 177
Time, along a ray, 420, 440, 442
Trace velocity, 222
Trace wavenumber, 63, 139
Transducer impedance, 248
Transfer function, 683
Transient response, 296
    in a cavity, 296
    of a piston, 176
Transmission coefficient, 414
Transport equation, 444
Transverse distance, 671
Transverse mode function, 219
    coupling with an elastic plate, 235
    relation to forced cavity modes, 254, 258,
        259, see also Mode function
Transverse wavenumber, 82, 353
Trial function, 365, 367
Truncation
    for Dowell's approximation, 394
    of a Fourier series, 602
    of a modal series, 280, 350
    of a perturbation series, 612
    of a spherical harmonic series, 23, 517,
        525
    of equation of state, 551
Turning point, 421

U
Ultrasonic imaging, 505, 656
Underwater acoustics, 418
Unit vector
    in cylindrical coordinates, 672
    in spherical coordinates, 666
Upper bound, for natural frequency, 366

V
Variational path, 463
Velocity
    in cylindrical coordinates, 675
    in spherical coordinates, 669
    of a shock, 581
Velocity potential, 598, 605, 623, 626
Vertical stratification, 417
Vertical tangency, of a waveform, 562, 576,
    578
Vibrating cylinder, 56
Vibrating pipe, 71
Vibrating plate, 637

Virtual displacement, 384, 464
Virtual mass, 103, 502
Virtual sources, 657
Virtual work, 383
Volume velocity, 340
    for multiple sources, 100, 105
    of a piston, 151, 166
    of a source, 23, 72, 74, 105

W
Water, standard properties of, 412, 551
Watson transformation, 521
Wave equation
    for heterogeneous fluid, 412
    nonlinear, 600
Wave packet, 204
Wave slowness, 439
Waveform, nonlinear, 539, 545, 553, 559,
        633, 651
Wavefront
    according to geometrical acoustics, 407,
        438, 448
    definition, 406
    for a nonlinear wave, 644
Waveguide
    circular, 267
    end correction, 199
    three-dimensional, 253
    two-dimensional, 217

with compliant walls, 224
Waveguide mode, *see* Transverse mode
        function
Wavenumber
    axial, 217
    complex, 195
    dispersion equation for, 202
    for a cylindrical wave, 62
    for a spherical shell, 37
    for a surface vibration, 137
    for an arbitrary surface vibration, 137
    vector, 61, 137, 221
Webster horn equation, 191, 209
    criteria for using, 193
    derivation, 189
    numerical solution of, 212
Window
    for Fast Fourier Transform, 683
WKB method, 207
Woofer, 332
Work-energy principle, 364
Wraparound error, 683
Wronskian, 524, 527

Z
Zero frequency mode, 367
Zero padding, 683
Zeros of a Bessel function, 151

Printed in the United States
by Baker & Taylor Publisher Services